CISM COURSES AND LECTURES

The series presents lecture notes, monographs, edited works and proceedings in the field of Mechanics, Engineering, Computer Science and Applied Mathematics.
Purpose of the series in to make known in the international scientific and technical community results obtained in some of the activities organized by CISM, the International Centre for Mechanical Sciences.

INTERNATIONAL CENTRE FOR MECHANICAL SCIENCES

COURSES AND LECTURES - No. 337

CONTINUUM MECHANICS IN ENVIRONMENTAL SCIENCES AND GEOPHYSICS

EDITED BY

K. HUTTER
TECHNISCHE HOCHSCHULE DARMSTADT

SPRINGER - VERLAG WIEN - NEW YORK

Le spese di stampa di questo volume sono in parte coperte da
contributi del Consiglio Nazionale delle Ricerche.

This volume contains 154 illustrations.

Printed in Italy

In order to make this volume available as economically and as rapidly as possible the authors' typescripts have been reproduced in their original forms. This method unfortunately has its typographical limitations but it is hoped that they in no way distract the reader.

ISBN 3-211-82449-9 Springer Verlag Wien-New York
ISBN 0-387-82449-9 Springer Verlag New York-Wien

PREFACE

The effects of human's activities on the environment, both locally and globally have in recent years led to new and challenging professional activities in the applied physical and engineering sciences. A large number of physical processes can be described by classical continuum solid and fluid mechanics, however often complicated small scale behaviour is involved, requiring knowledge in nonlinear material behaviour, and various microstructural effects.

This book is an attempt to describe processes that arise in the geophysical environment, are either caused by human's activities or affect mankind, and can be described physically by the methods of continuum physics. Obviously, the topics that are treated are limited and the selection is subjective. Many other, important topics could equally be collected under, or introduced by the same subject heading. The book is an outgrowth of the lecture notes on "Continuum Mechanics in Environmental Sciences and Geophysics" at the "IUTAM International Summer School" held in Udine from 22 - 26 June 1992 by five internationally well known specialists. The purpose is to (hopefully successfully) demonstrate that methods of continuum mechanics can advantagously be used to describe rather diverse and complex fields of geophysical and environmental mechanics.

Chapter 1 deals with the "Foundation of Continuum Mechanics": Kinematics, balance laws and constitutive relations of a single constituent body as well as mixtures are treated. This chapter will set the tone and lay the foundations from which subsequent chapters will draw.

Chapter 2 on "Waves and Oscillations in the Ocean and in Lakes" demonstrates how these concepts are used in physical oceanography or limnology. In particular, it will be studied how the effects of the rotation of the Earth manifest themselves in linear and nonlinear processes of a homogeneous and inhomogeneous water body and how these wave phenomena manifest themselves in observations.

"Continuum Description of Natural Slopes in Slow Movement" is dealt with in Chapter 3. The description of the mechanical behaviour of soils under quasistatic loadings requires non-classical mixture concepts and encounters generally complicated elastoplastic type

constitutive behaviour, while effects of body acceleration are negligible, so that conditions of Stokes flow prevail. Nonlinearities are introduced here through the geometry and the constitutive behaviour. However, considerable simplifications are possible when the geometries of the moving and deforming slides are shallow.

In Chapter 4 on "Constitutive Equations for Granular Materials in Geomechanical Context" we focus attention on the elasto-(visco)plastic response of a granular material whose pores are, or are not filled with a fluid. Elastic and elasto-plastic constitutive relations for the viscous pore fluid and the respective interaction mechanisms between the constituents are described. The theory is based on the concept of second-grade materials, and minimal forms of the constitutive relations are deduced by invoking the second law of thermodynamics.

Chapter 5 presents a theoretical model for "The Flow of Ice Sheets and Ice Shelves" thermomechanical coupling paired with nonlinear rheology affects the flow of such ice masses as e.g. Antarctica and Greenland. Its response to the climate forcing in an environment of greenhouse warming may significantly affect the lives of future generations.

The final Chapter 6 deals with the "Mechanics of Granular Flows", the flows of bulk solids comprised of discrete particles. Particle-particle and particle-fluid interactions are studied both qualitatively and quantitatively. Kinetic theory and molecular dynamics modelling are presented, and deduced models are applied to simple flow situations. Rockfalls and ice avalanches are studied as are phenomena of particle segregation as evidenced in shear flows of particles of different sizes.

It is my pleasant duty to thank (i) the IUTAM and CISM authorities to provide me with the opportunity to organise the course, (ii) to the lecturers for their cooperation in preparing the lecture notes and the respective chapters of this book, (iii) to the audience of the course for the many interesting discussions and (iv) to the many secretaries and students who were involved in the production of the final manuscript. Let us hope that the book will find a large and interested readership.

K. Hutter

Contents

Continuum Description of Natural Slopes in Slow Movement

Foundation of Continuum Mechanics[1]

Peter Haupt
Institute of Mechanics , University of Kassel
Mönchebergstr. 7, 3500 Kassel, Germany

The following lectures give a short indroduction into modern continuum mechanics. The presentation omits many details and should motivate the reader to study the references.

First of all, the kinematics of deformable bodies is elaborated. Basic concepts are the material body and its various configurations, material and spatial representations, the deformation gradient and velocity gradient, both containing all informations of the local properties of a continuous motion as well as the definition of strain and strain rate tensors. Then the general structure of a balance relation is explained. Local forms of a general balance relation are given including jump conditions, which must hold on singular surfaces, where the field may be discontinuous. The balance relations of mechanics concern the mass, the linear momentum and the moment of momentum. An immediate implication is the balance of mechanical energy. The balance relations of thermodynamics are related to the total energy of a body (First Law) and the entropy, where a positive entropy production is postulated (Second Law). The spatial representation of the local balance laws can be interpreted in terms of an open system and thus motivate their generalization to multi-component continua (mixtures).

Whereas the balance relations express universally valid natural laws, the individual material properties of a body are modelled by means of constitutive equations. Systematic methods to formulate constitutive models can be obtained from the general theory of material behavior. As one basic result of this theory the constitutive equation of a linear viscous fluid can be understood to be an asymptotic approximation of a quite general constitutive functional, which is continuous in the sense of a fading memory norm. This approximation is arbitrarily close for sufficiently slow motions.

1. Introduction

The main subject of mechanics is the investigation of the motion of material bodies. As a part of natural science, mechanics tries to understand these motions, which can be observed in the real world. The principal goal in this endeavor of scientific understanding is thereby focussed on the identification of the basic causes and the description of the related natural laws. Mechanics, as a part of physics, regards the force to be the principal cause for any motion of material bodies. As far as the physical processes on the planet Earth are

[1] A precursor to this manuscript was the basis of a part of a graduate course held in August 1990 at the Institute of Applied Mechanics, National Taiwan University, Taipei. The Author is thankful to Professor YIH-HSING PAO for the invitation and the opportunity to teach the course at his institute.

concerned, incidentally a rather small part of the universe, we encounter flows, deformations and fracture processes of fluids (air, water), solids (lithosphere, soils, ice sheets and glaciers) and mixtures of these. Events in these media of the geophysical environment take place on extremely different time scales, but they are all interpretable as motions of certain continua in the sense of mechanics. In fact, they are not only related to forces as driving agents, but equally so to changes in temperature and energy, diffusion phenomena and chemical reaction processes. However, since the forces seem to play a central role in a great many of relevant processes, mechanics can justly be conjectured to embrace a large variety of concepts and methods appropriate to analyze the behavior of our globe as an environmental system. This point of view is motivation enough for us to begin a monograph on the dynamics of the environment with an introduction into continuum mechanics. Continuum mechanics is based on the presumption that the spatial distribution of matter is *continuous*. As a consequence this basic idea leads to the construction of field theories. The deduction of such field theories, applicable to environmental processes, is the main topic of this text.

In view of its general methodology the entire subject separates into three different global topics, namely:

- Kinematics
- Balance relations
- Material behavior

Among these the **kinematics** of continuous media analyses the *geometric aspects of the motion* of a continuous body. In particular, it deals with the local properties of a motion or deformation. Basic notions are the idea of the *configuration*, the *deformation gradient*, the *velocity gradient* and the concepts of *strains* and *strain rates*.

The **balance relations** express the basic principles or the *universal laws of nature* by which all mechanical processes are governed. These fundamental laws – when restricted to the principles of mechanics and thermodynamics – are the balance relations of mass, momentum, moment of momentum, energy and entropy.

Whereas the statements of the kinematics as well as the balance principles are accepted to be *universally valid for all material systems*, the particular property of a peculiar material body must be singled out and characterized individually. This is achieved by means of the **theory of material behavior**. Within this theory *general principles* and *systematic methods* have been developed to represent a great variety of different material behaviors. Major results are the principal distinction of **solids** and **fluids** on the basis of the material symmetry properties and the systematic elaboration of the theories of elasticity, plasticity, viscoelasticity and viscoplasticity.

In each special case, a particular field theory of continuum mechanics is formulated and an initial boundary value problem emerges, which combines the general kinematics and balance principles with the modelization of the individual material properties.

In the following sections some basic concepts of continuum mechanics are discussed. All the presented material can be found in the references given at the end of this text. Among these references the books of ERINGEN [1], JAUNZEMIS [3], LEIGH [4], MALVERN [5] and SPENCER [6] are useful introductions into the subject, which was summarized in the comprising monograph of TRUESDELL and NOLL [7]. A more axiomatic approach was developed by GURTIN [2] and TRUEDELL [8]. The theory of multi–component systems (mixtures) is presented according to HUTTER [10] and LAVENDA [11], where the reader is referred to for more details. The fading memory theory, briefly sketched in the last section, rests on the fundamental paper of COLEMAN and NOLL [9]. Other topics and some more resent results in the theory of material behavior are summarized by HAUPT [12].

2. Kinematics

2.1 Bodies and motions

All statements of continuum mechanics refer to material bodies. Therefore, an introduction should start with an outline of this basic concept. For simplicity we shall not go into a mathematically rigorous definition of a material body; this would necessitate the application of the theory of differentiable manifolds. However, it is most important that a material point of a body be not confused with its (material or spatial) coordinates or with its position in the physical space, which in the context of classical mechanics is a Euclidean space. The object of all considerations in continuum mechanics and the domain of all physical quantities is the material body.

A **material body** $\mathcal{B} = \{\mathcal{X}\}$ is a set of elements $\mathcal{X}$, which are called *material elements* or *material points*. The motion of a material body is represented by vector functions of the form

$$\mathbf{x} = \chi_R(\mathbf{X}, t). \tag{2.1}$$

Here, $\mathbf{x}$ is the position vector associated to the material point $\mathcal{X}$ at the time t. $\mathbf{x}$ denotes the present place of the material element $\mathcal{X}$. The notion of a position vector is based on the choice of a reference frame, consisting of an origin and three linearly independent directions. Without loss of generality, these directions can be identified with a base system of orthogonal unit vectors. Essentially, a reference frame facilitates a one–to–one correspondence between the points P of the threedimensional Euclidean space and the position vectors $\mathbf{x}$,

$$P \quad \longleftrightarrow \quad \mathbf{x}.$$

Usually, an inertial system is utilized as a reference frame. The set of all present position vectors $\mathbf{x}$, the image of the material body $\mathcal{B} = \{\mathcal{X}\}$, is called the current or **present configuration**. The present configuration represents the actual picture of the material body, realized by an observer.

The material point $\mathcal{X}$ itself, moving throughout the deformation process (2.1), is denoted by the argument vector $\mathbf{X}$. Thus a one–to–one corespondence

$$\mathcal{X} \quad \longleftrightarrow \quad \mathbf{X}$$

between material points and vectors is assumed. This correspondence is called the **reference configuration**.

It should be noted that **X** is not necessarily a position vector denoting a point in the space of physical observation. In the most general material description of the motion, **X** is only a name or an identifier for the material element $\mathcal{X}$. Quite contrary, **x** is always a position vector, since the particle $\mathcal{X}$ occupies a place at the present instant t. The situation is vizualized in Figure 2.1. As a reference frame relating the position vectors **x** to the points P in the Euclidean space of physical observation, an **inertial frame** is used.

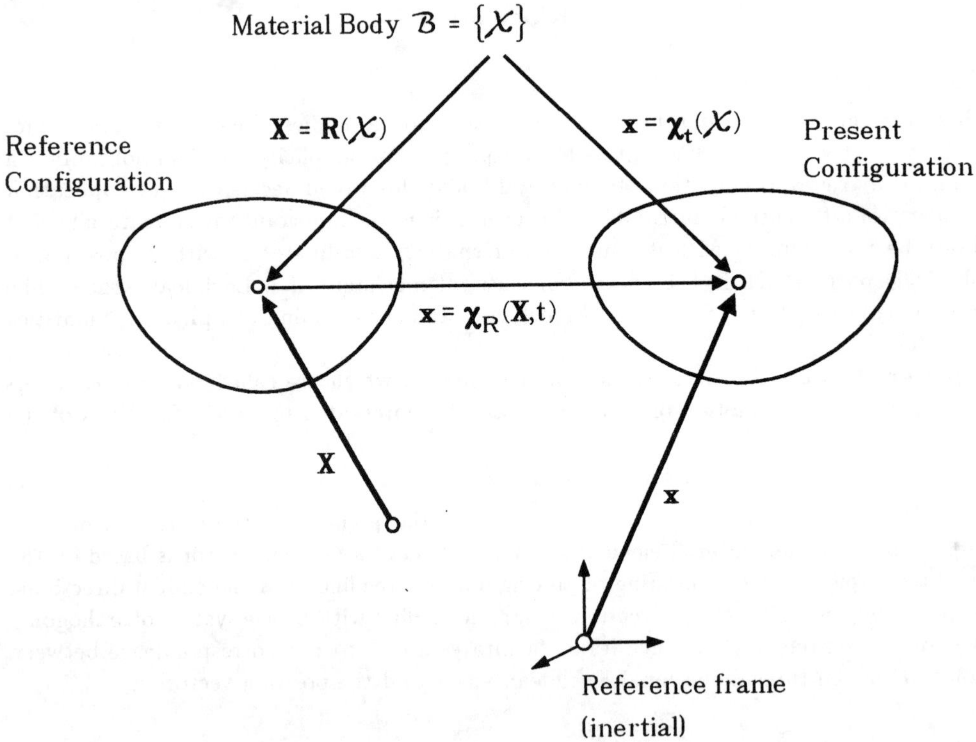

Fig.2.1: Motion of a material body

2.2 Material and spatial coordinates

To facilitate the mathematical representation of the motion, the names **X** of the particles as well as their position vectors **x** are identified with triples of real numbers. This is done with the choice of **coordinate systems**. In the most general case two independent systems of curvilinear coordinates are defined, which each imply corresponding **fields** of **tangent** and **gradient vectors**, dual to each other. These coordinate systems are called **material** and **spatial** coordinates, respectively.

The identification of the names $\mathbf{X}$ of material points with triples of real numbers is called a system of **material coordinates**:

$$\mathbf{X} \quad \longleftrightarrow \quad (X^1, X^2, X^3)$$

The material coordinate system implies a one–to–one correspondence,

$$\mathbf{X} = \hat{\mathbf{X}}(X^1, X^2, X^3) \quad \Longleftrightarrow \quad X^\alpha = \hat{X}^\alpha(\mathbf{X}) \qquad (\alpha = 1, 2, 3), \tag{2.2}$$

with tangent vectors,

$$\mathbf{G}_\alpha = \frac{\partial}{\partial X^\alpha} \hat{\mathbf{X}}(X^1, X^2, X^3) \tag{2.3}$$

and gradient vectors,

$$\mathbf{G}^\alpha = \operatorname{Grad} \hat{X}^\alpha(\mathbf{X}). \tag{2.4}$$

The tangent and gradient vectors are dual to each other:

$$\mathbf{G}_\alpha \cdot \mathbf{G}^\beta = \delta_\alpha^\beta \tag{2.5}$$

The identification of the position vectors $\mathbf{x}$ with triples of real numbers is called a system of **spatial coordinates**:

$$\mathbf{x} \quad \longleftrightarrow \quad (x^1, x^2, x^3)$$

The spatial coordinate system implies a one–to–one correspondence,

$$\mathbf{x} = \hat{\mathbf{x}}(x^1, x^2, x^3) \quad \Longleftrightarrow \quad x^k = \hat{x}^k(\mathbf{x}) \qquad (k = 1, 2, 3), \tag{2.6}$$

with tangent vectors,

$$\mathbf{g}_k = \frac{\partial}{\partial x^k} \hat{\mathbf{x}}(x^1, x^2, x^3) \tag{2.7}$$

and gradient vectors,

$$\mathbf{g}^k = \operatorname{grad} \hat{x}^k(\mathbf{x}), \tag{2.8}$$

dual to each other:

$$\mathbf{g}_i \cdot \mathbf{g}^k = \delta_i^k \tag{2.9}$$

With the choice of these two coordinate systems, the representation of the motion is a time–dependent transformation of the material coordinates into the spatial coordinates:

$$x^k = \chi_R^k(X^1, X^2, X^3, t) \qquad (k = 1, 2, 3) \tag{2.10}$$

As already mentioned, the material and spatial coordinate systems are by definition independent. For instance, the material coordinates could be Cartesian and the spatial ones cylindrical, or vice versa. Of course, the two systems can be related: So we may e.g. choose one and the same system for both the material and the spatial coordinates. In such a case, the most simple special choice is a common Cartesian coordinate system, i.e.

$$\mathbf{X}, \mathbf{x} \quad \longleftrightarrow \quad (X, Y, Z), (x, y, z)\,,$$

for which the motion is represented as

$$x = f(X,Y,Z,t), \quad y = g(X,Y,Z,t), \quad z = h(X,Y,Z,t). \tag{2.11}$$

A particular coordinate representation, very useful for theoretical considerations, arises from the choice of **convected coordinates**. In this special case the coordinate lines of the spatial coordinate system are defined as material lines, i.e., they are the deformed lines of the material coordinates. In connection with such different possibilities it turns out that there are many ways to establish special equations in various coordinate–dependent forms, which might express the same facts while being quite different in appearance. For this reason we prefer the symbolic coordinate–free notation according to e.g. (2.1), viz.

$$\mathbf{x} = \chi_R(\mathbf{X}, t).$$

Independent from any choice of particular coordinate systems, $\mathbf{x}$ is the present position vector of the material point with the name $\mathbf{X}$.

It is quite important that for any fixed time t the vector function χ_R is assumed to be invertible with respect to $\mathbf{X}$:

$$\mathbf{X} = \chi_R^{-1}(\mathbf{x}, t)\Big|_{(t \text{ fixed})} \tag{2.12}$$

Utilizing the coordinate free symbolic notation, all statements and their consequences are invariant with respect to any change of coordinate systems and base vectors.

A special choice of the reference configuration arises, when it is identified with the place, occupied by the body at some fixed instant t_0. In this case we have

$$\mathbf{X}_0 = \chi_R(\mathbf{X}, t_0) \quad \longleftrightarrow \quad \mathbf{X} = \chi_R^{-1}(\mathbf{X}_0, t_0)$$

and

$$\begin{aligned} \mathbf{x} &= \chi_R\left(\chi_R^{-1}(\mathbf{X}_0, t_0), t\right), \quad \text{i.e.,} \\ \mathbf{x} &= \phi_{t_0}(\mathbf{X}_0, t). \end{aligned} \tag{2.13}$$

By definition, the function ϕ_{t_0} possesses the property that

$$\phi_{t_0}(\mathbf{X}_0, t_0) = \mathbf{X}_0 \tag{2.14}$$

for all $\mathbf{X}_0$, and the representation $\mathbf{x} = \phi_{t_0}(\mathbf{X}_0, t)$ yields the present place of the particle $\mathcal{X}$, which at time t_0 has been found at the place $\mathbf{X}_0$. In comparison to $\mathbf{x} = \chi_R(\mathbf{X}, t)$, where the reference configuration never needs to be occupied by the body, the above representation is more special. This special choice of reference configuration motivates the introduction of the **displacement vector**

$$\mathbf{u}(\mathbf{X}_0, t) = \phi_{t_0}(\mathbf{X}_0, t) = \mathbf{x} - \mathbf{X}_0 \tag{2.15}$$

(see Figure 2.2), which is quite common in solid mechanics.

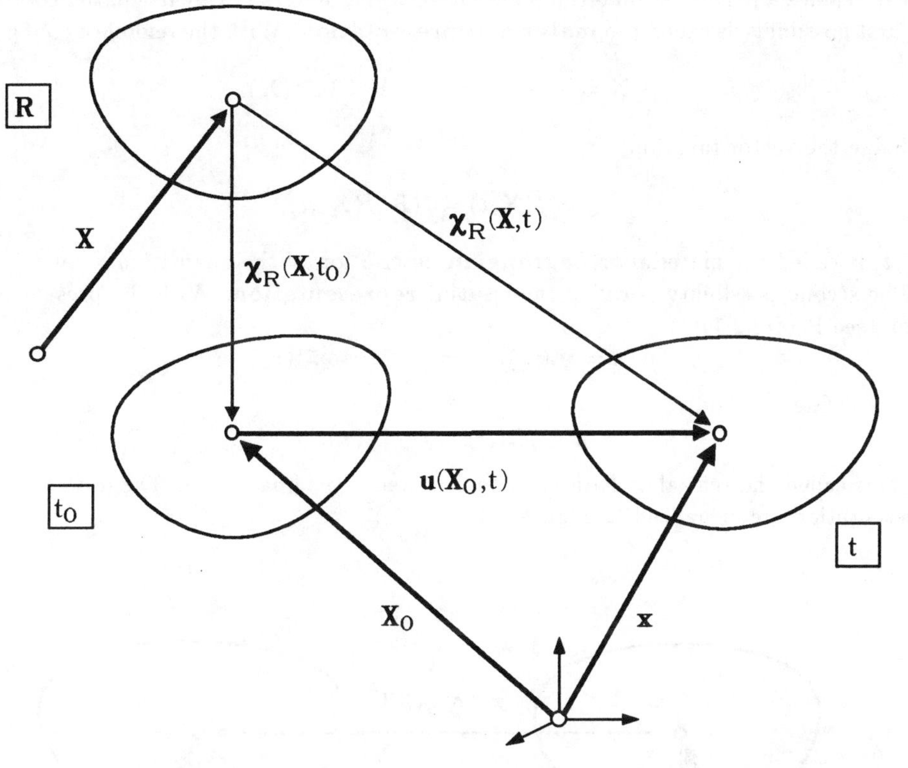

Fig.2.2: Displacement vector

2.3 Material and spatial representations

In principle, all physical quantities are defined on material bodies and processes relating such quantities take place within them. Therefore, the material body is the basic domain, where any function for the representation of a physical process has to be defined. In this sense a physical quantity g is a function f, defined as follows:

$$\begin{aligned} f: \quad \mathcal{B} \times \mathcal{R} &\longrightarrow \mathcal{M} \\ (\mathcal{X}, t) &\longrightarrow g = f(\mathcal{X}, t) \end{aligned} \tag{2.16}$$

g is called the value of the physical quantity at $\mathcal{X}$ and t. For any fixed value of time t the function f maps the set of material points $\mathcal{X}$ into a certain range $\mathcal{M}$, which can be identified with the space of real numbers, vectors, tensors or with any other mathematical objects.

In order to arrive at a field theory, the material elements $\mathcal{X}$ must be replaced by the vectors or coordinates of their configuration. To this end there are two basic possibilities:

We can replace a particular material point either by its material or by its spatial coordinates. The first possibility is called the **material representation**: With the reference configuration

$$\mathbf{X} = \mathbf{R}(\mathcal{X}) \quad \Longleftrightarrow \quad \mathcal{X} = \mathbf{R}^{-1}(\mathbf{X}) \tag{2.17}$$

we define the vector function

$$g = \hat{f}(\mathbf{X}, t) = f(\mathbf{R}^{-1}(\mathbf{X}), t). \tag{2.18}$$

$\hat{f}(\mathbf{X}, t)$ is called the material or **Lagrangian** description of the quantity g.

The second possibility is called the **spatial representation**: With the present configuration (see Figure 2.1),

$$\mathbf{x} = \chi_t(\mathcal{X}) \quad \Longleftrightarrow \quad \mathcal{X} = \chi_t^{-1}(\mathbf{x}), \tag{2.19}$$

we may define the field

$$g = \bar{f}(\mathbf{x}, t) = f(\chi_t^{-1}(\mathbf{x}), t). \tag{2.20}$$

$\bar{f}(\mathbf{x}, t)$ is called the spatial or **Eulerian** description of the quantity g. The different kinds of representation are vizualized in Figure 2.3.

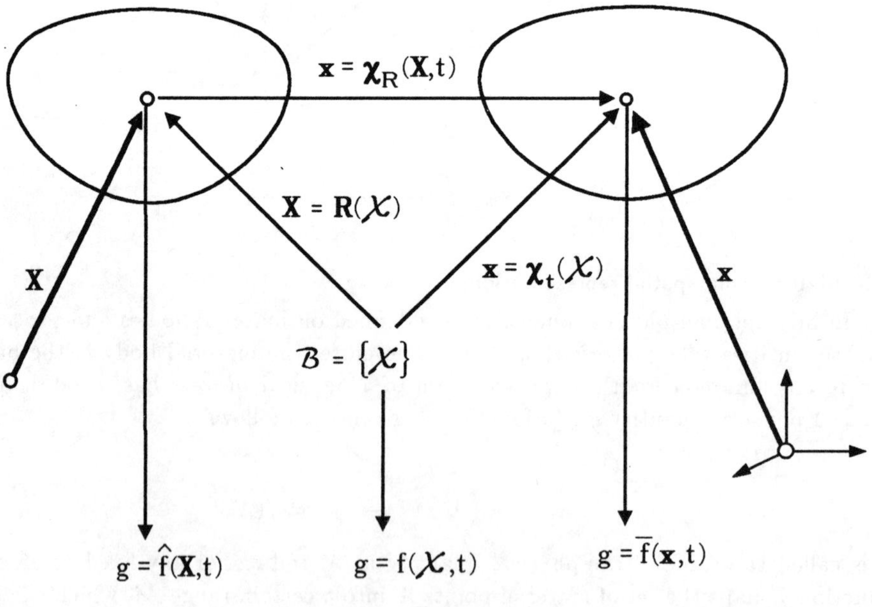

Fig.2.3: Material and spatial repesentation

For a fixed material point $\mathcal{X}$, the derivative with respect to time is called **material derivative:**

$$\dot{g} = \frac{\mathrm{d}}{\mathrm{d}t} f(\mathcal{X}, t)\Big|_{(\mathcal{X}\text{ fixed})} \tag{2.21}$$

In the context of the material representation the material derivative agrees with the partial derivative:

$$\frac{\mathrm{d}}{\mathrm{d}t} f(\mathcal{X}, t) = \frac{\partial}{\partial t} \hat{f}(\mathbf{X}, t) \tag{2.22}$$

In the spatial representation, the time–dependence of the position vector of a particle must be taken into account, and thus the chain rule of differentation must be observed to calculate

$$\frac{\mathrm{d}}{\mathrm{d}t} f(\mathcal{X}, t) = \frac{\mathrm{d}}{\mathrm{d}t} \bar{f}(\mathbf{x}(t), t). \tag{2.23}$$

Thus we have

$$\frac{\mathrm{d}}{\mathrm{d}t} f(\mathcal{X}, t) = \operatorname{grad} \bar{f}(\mathbf{x}, t)\, \dot{\mathbf{x}} + \frac{\partial \bar{f}}{\partial t}(\mathbf{x}, t), \tag{2.24}$$

where $\dot{\mathbf{x}}$ is the velocity of a material point, i.e. $\dot{\mathbf{x}} = \partial \chi_R / \partial t$. The special representation of the gradient operator 'grad' depends on the mathematical structure of the range of the function $\bar{f}$ (it may be the set of scalars, vectors or tensors etc.).

2.4 Deformation gradient

For any motion $\mathbf{x} = \chi_R(\mathbf{X}, t)$ of a material body, the **deformation gradient** $\mathbf{F}(\mathbf{X}, t)$ is defined through the linear approximation of $\chi_R(\mathbf{X}, t)$ at a fixed time t:

$$\chi_R(\mathbf{X} + \Delta\mathbf{X}, t) = \chi_R(\mathbf{X}, t) + \mathbf{F}(\mathbf{X}, t)\Delta\mathbf{X} + |\Delta\mathbf{X}|\, \mathbf{r}(\mathbf{X}, \Delta\mathbf{X}, t) \tag{2.25}$$

Linear approximation is tantamount to the property

$$\lim_{|\Delta\mathbf{X}| \to 0} \mathbf{r}(\mathbf{X}, \Delta\mathbf{X}, t) = \mathbf{0},$$

and the deformation gradient can be calculated via the GATEAUX–derivative:

$$\begin{aligned} \mathbf{F}(\mathbf{X}, t)\mathbf{H} &= \{\operatorname{Grad} \chi_R(\mathbf{X}, t)\}\mathbf{H} = \\ &= \frac{\partial}{\partial \nu} \chi_R(\mathbf{X} + \nu\mathbf{H}, t)\Big|_{\nu=0} \end{aligned} \tag{2.26}$$

A matrix representation of $\mathbf{F}$, valid for **two** systems of general material and spatial coordinates, is easily obtained: To this end, we identify $\mathbf{H}$ with the tangent vectors $\mathbf{G}_\alpha$ of the material coordinate lines to get

$$\mathbf{F}\mathbf{G}_\alpha = \frac{\partial}{\partial \nu} \chi_R(\mathbf{X} + \nu\mathbf{G}_\alpha, t)\Big|_{\nu=0} = \frac{\partial}{\partial X^\alpha} \chi_R(\mathbf{X}, t) \qquad (\alpha = 1, 2, 3). \tag{2.27}$$

Each vector $\mathbf{F}\mathbf{G}_\alpha$ can thus be represented in terms of the base system of the tangent vectors of the spatial coordinate lines. This yields

$$\mathbf{F}\mathbf{G}_\alpha = \left(\frac{\partial \chi_R}{\partial X^\alpha} \cdot \mathbf{g}^k\right) \mathbf{g}_k = \frac{\partial \chi_R^k}{\partial X^\alpha} \mathbf{g}_k. \tag{2.28}$$

Introducing the material gradient vectors $\mathbf{G}^\alpha$ into this formula, we have

$$\mathbf{F}\mathbf{G}_\alpha = \left[\frac{\partial \chi_R^k}{\partial X^\beta} \mathbf{g}_k \otimes \mathbf{G}^\beta\right] \mathbf{G}_\alpha.$$

Thus we arrive at the general representation of the deformation gradient,

$$\mathbf{F}(\mathbf{X}, t) = F^k{}_\alpha \, \mathbf{g}_k \otimes \mathbf{G}^\alpha, \tag{2.29}$$

where the matrix $(F^k{}_\alpha)$ is exactly the functional matrix of the mixed coordinate representation of the motion: With eq. (2.10),

$$x^k = \chi_R^k(X^1, X^2, X^3, t)$$

the components of $\mathbf{F}$ read

$$F^k{}_\alpha = \frac{\partial \chi_R^k}{\partial X^\alpha} \qquad (\alpha,\, k = 1,\, 2,\, 3). \tag{2.30}$$

Note that the elements of this matrix, when interpreted as components of a second order tensor, are related to a mixed system of base tensors, namely $\mathbf{g}_k \otimes \mathbf{G}^\alpha$, $(k,\, \alpha = 1,\, 2,\, 3)$.

In the simplest case, when the material and spatial coordinates are chosen to be one and the same Cartesian coordinate system, the motion is represented by the three functions

$$x = f(X, Y, Z, t), \quad y = g(X, Y, Z, t), \quad z = h(X, Y, Z, t),$$

and the representation of the deformation gradient is simply given by

$$\mathbf{F}(\mathbf{X}, t) \mathrel{\widehat{=}} \begin{pmatrix} \dfrac{\partial f}{\partial X} & \dfrac{\partial f}{\partial Y} & \dfrac{\partial f}{\partial Z} \\ \dfrac{\partial g}{\partial X} & \dfrac{\partial g}{\partial Y} & \dfrac{\partial g}{\partial Z} \\ \dfrac{\partial h}{\partial X} & \dfrac{\partial h}{\partial Y} & \dfrac{\partial h}{\partial Z} \end{pmatrix}. \tag{2.31}$$

Because the motion $\mathbf{x} = \chi_R(\mathbf{X}, t)$ for any fixed t is invertible, the deformation gradient is nonsingular, i.e., $\det \mathbf{F} \neq 0$. Equivalently, the determinant of the functional matrix, the Jacobian, does not vanish during any motion:

$$J = \det\left(\frac{\partial \chi_R^k}{\partial X^\alpha}\right) \neq 0$$

Note that the Jacobian J is, in general, not identical with the determinant of $\mathbf{F}$: Indeed, from the identity

$$(\det \mathbf{F})[(\mathbf{a} \times \mathbf{b}) \cdot \mathbf{c}] = (\mathbf{Fa} \times \mathbf{Fb}) \cdot \mathbf{Fc} \quad \text{(for all vectors } \mathbf{a}, \mathbf{b}, \mathbf{c}) \tag{2.32}$$

we conclude by referring to the mixed component representation (2.29),

$$\mathbf{F} = F^k{}_\alpha \, \mathbf{g}_k \otimes \mathbf{G}^\alpha,$$

and by introducing the tangent vectors $\mathbf{G}_1$, $\mathbf{G}_2$, $\mathbf{G}_3$, that

$$\begin{aligned}(\det \mathbf{F})[(\mathbf{G}_1 \times \mathbf{G}_2) \cdot \mathbf{G}_3] &= [(F^k{}_1 \mathbf{g}_k) \times (F^l{}_2 \mathbf{g}_l)] \cdot (F^m{}_3 \mathbf{g}_m) \quad = \\ &= \det(F^k{}_\alpha)[(\mathbf{g}_1 \times \mathbf{g}_2) \cdot \mathbf{g}_3].\end{aligned}$$

Thus we have

$$\det \mathbf{F} = \frac{(\mathbf{g}_1 \times \mathbf{g}_2) \cdot \mathbf{g}_3}{(\mathbf{G}_1 \times \mathbf{G}_2) \cdot \mathbf{G}_3} \det \left(\frac{\partial \chi_R^k}{\partial X^\alpha} \right). \tag{2.33}$$

The deformation gradient is a fundamental quantity in the kinematics of continuous media. In particular, it represents all **local properties of the deformation**. To see this consider a **material line**, i.e. a curve relating one and the same set of material points. Let its representation be given by $\alpha \rightarrow \mathbf{C}(\alpha)$, where the parameter α runs through some real interval while the values of $\mathbf{C}(\alpha)$ trace the material elements in the reference configuration. By definition, in the present configuration the material line has the representation

$$\alpha \quad \longrightarrow \quad \mathbf{c}(\alpha) = \chi_R(\mathbf{C}(\alpha), t). \tag{2.34}$$

A **material line element** at the point $\mathbf{X} = \mathbf{C}(\alpha)$ in the reference configuration is understood to be the tangent vector of an arbitrary material line crossing this point, i.e.,

$$d\mathbf{X} = \mathbf{C}'(\alpha) \, d\alpha. \tag{2.35}$$

To identify the same material line element in the present configuration,

$$d\mathbf{x} = \mathbf{c}'(\alpha) \, d\alpha, \tag{2.36}$$

the chain rule of differentiation must be applied to give

$$\mathbf{c}'(\alpha) = \frac{\mathrm{d}}{\mathrm{d}\alpha} \chi_R(\mathbf{C}(\alpha), t) = \mathbf{F}(\mathbf{C}(\alpha), t) \, \mathbf{C}'(\alpha)$$

or

$$d\mathbf{x} = \mathbf{F}(\mathbf{X}, t) \, d\mathbf{X}, \tag{2.37}$$

where $\mathbf{X} = \mathbf{C}(\alpha)$. This relation sets in evidence the fundamental property of the deformation gradient: It transforms a material line element between the reference and the spatial configurations (see Figure 2.4). The transformation (2.37) implies corresponding transformations of surface elements and volume elements: For an **oriented material surface element** one has

$$d\mathbf{A}_R = d\mathbf{X}_1 \times d\mathbf{X}_2. \tag{2.38}$$

In the present configuration, this becomes

$$d\mathbf{A} = d\mathbf{x}_1 \times d\mathbf{x}_2 = \mathbf{F}d\mathbf{X}_1 \times \mathbf{F}d\mathbf{X}_2 = (\det \mathbf{F})\mathbf{F}^{T-1}(d\mathbf{X}_1 \times d\mathbf{X}_2),$$

or

$$d\mathbf{A} = (\det \mathbf{F})\mathbf{F}^{T-1}\, d\mathbf{A}_R. \tag{2.39}$$

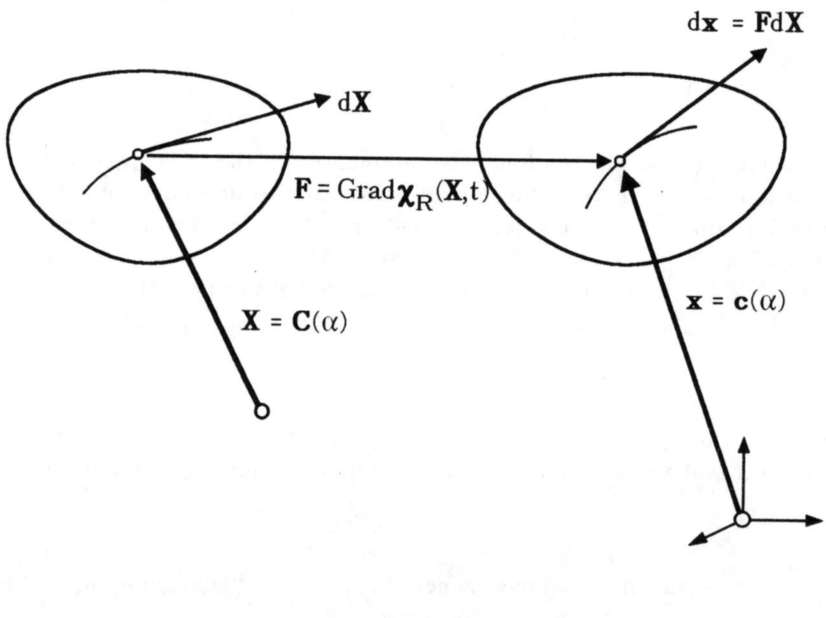

Fig.2.4: Deformation gradient and line elements

A **material volume element** in the reference configuration,

$$dV_R = |(d\mathbf{X}_1 \times d\mathbf{X}_2) \cdot d\mathbf{X}_3|, \tag{2.40}$$

transforms according to eq. (2.32) into

$$\begin{aligned} dV &= |(\mathbf{F}d\mathbf{X}_1 \times \mathbf{F}d\mathbf{X}_2) \cdot \mathbf{F}d\mathbf{X}_3| = |(\det \mathbf{F})|\,|(d\mathbf{X}_1 \times d\mathbf{X}_2) \cdot d\mathbf{X}_3|, \quad \text{i.e.,} \\ dV &= |(\det \mathbf{F})|\, dV_R = (\det \mathbf{F})\, dV_R. \end{aligned} \tag{2.41}$$

Here we have used the fact that $\mathbf{F}$ is continuously connected to the unit tensor $\mathbf{1}$ and thus $\det \mathbf{F}$ is always positive.

To understand the above derivations, identity (2.32) has to be applied, which leads directly to eq. (2.41). Moreover, (2.32) implies

$$(\det \mathbf{F})[(\mathbf{a} \times \mathbf{b}) \cdot \mathbf{c}] = \mathbf{F}^T(\mathbf{Fa} \times \mathbf{Fb}) \cdot \mathbf{c} \quad \text{identically in } \mathbf{a}, \mathbf{b}, \mathbf{c} \text{ or}$$

$$(\det \mathbf{F})(\mathbf{a} \times \mathbf{b}) = \mathbf{F}^T(\mathbf{Fa} \times \mathbf{Fb}) \quad \text{or}$$

$$\mathbf{Fa} \times \mathbf{Fb} = (\det \mathbf{F})\, \mathbf{F}^{T-1}(\mathbf{a} \times \mathbf{b}) \quad \text{identically in } \mathbf{a}, \mathbf{b}, \text{ which leads to eq. (2.39).}$$

2.5 Velocity fields

The velocity vector (or the velocity) of a material element $\mathcal{X}$ is the derivative of its position vector with respect to time,

$$\mathbf{v} = \dot{\mathbf{x}}(t) = \frac{\mathrm{d}}{\mathrm{d}t}\chi_t(\mathcal{X}). \tag{2.42}$$

It is a fundamental physical quantity, and its distribution over the body is of main interest. The material representation of the velocity is

$$\mathbf{v} = \hat{\mathbf{v}}(\mathbf{X}, t) = \frac{\partial}{\partial t}\chi_R(\mathbf{X}, t), \tag{2.43}$$

and its spatial representation becomes

$$\mathbf{v} = \bar{\mathbf{v}}(\mathbf{x}, t) = \left.\frac{\mathrm{d}}{\mathrm{d}t}\chi_t(\mathcal{X})\right|_{\mathcal{X}=\chi_t^{-1}(\mathbf{x})}.$$

In this section the spatial representation will be investigated in more detail. Thus we regard $\mathbf{v} = \bar{\mathbf{v}}(\mathbf{x}, t)$ as the velocity of the material element which occupies the position $\mathbf{x}$ at time t.

To obtain the motion of a body from the supposedly known field $\bar{\mathbf{v}}(\mathbf{x}, t)$, an integration process is required; the motion $\mathbf{x}(t)$ of the material point follows from the solution of the nonlinear system of ordinary differential equations

$$\frac{\mathrm{d}}{\mathrm{d}t}\mathbf{x}(t) = \bar{\mathbf{v}}(\mathbf{x}(t), t). \tag{2.44}$$

These differential equations define the **trajectories** or **path lines** of the particles, i.e., the motion of the material points. The differential equations (2.44) must be completed by the initial conditions

$$\mathbf{x}(t_0) = \mathbf{X}_0. \tag{2.45}$$

Thus the general integral of the differential equations (2.44) can be expressed as

$$\mathbf{x}(t) = \phi_{t_0}(\mathbf{X}_0, t), \tag{2.46}$$

where the initial conditions are incorporated as the identity

$$\mathbf{X}_0 = \phi_{t_0}(\mathbf{X}_0, t_0) \tag{2.47}$$

(see eq. (2.14)). Geometrically, the path line can be interpreted as the set of all vectors $\mathbf{x}(t)$ representing the present position of a particle which occupied position $\mathbf{X}_0$ at the time t_0.

There are two other sets of curves, being of kinematical importance and determined by the velocity field $\bar{\mathbf{v}}(\mathbf{x}, t)$: The first are the **streamlines**; these are defined by the system of differential equations,

$$\frac{\mathrm{d}}{\mathrm{d}\sigma}\mathbf{x}(\sigma) = \bar{\mathbf{v}}(\mathbf{x}(\sigma), t), \tag{2.48}$$

subjected to the initial conditions

$$\mathbf{x}(\sigma_0) = \mathbf{x}_0. \tag{2.49}$$

Geometrically, the streamlines are the integral curves of the velocity field at a *fixed* time t. Their definition differs from that of the trajectories, which are the integral curves of the velocity field *followed* through the time.

For a special case of velocity fields trajectories and streamlines are identical. Indeed, the reader may show that this class is defined by

$$\bar{\mathbf{v}}(\mathbf{x}, t) = \alpha(\mathbf{x}, t)\mathbf{v}_1(\mathbf{x}), \tag{2.50}$$

where α is a scalar–valued field and $\mathbf{v}_1(\mathbf{x})$ a time–independent vector field.

The general solution of the differential equations for the streamlines takes the form

$$\mathbf{x}(\sigma) = \psi_t(\mathbf{x}_0, \sigma), \tag{2.51}$$

where the initial condition corresponds to the identity

$$\psi_t(\mathbf{x}_0, \sigma_0) = \mathbf{x}_0. \tag{2.52}$$

As already mentioned, the present time t is a fixed parameter in the differential equation for the streamlines. Thus the defining differential equations for the streamlines form an *autonomous* system, whereas those for the trajectories or path lines are *non–autonomous.*

A third set of curves arising from the velocity field are the so–called **streaklines**: These are derived from the trajectories

$$\mathbf{x}(t) = \phi_{t_0}(\mathbf{X}_0, t)$$

through the following definition: For a given $\mathbf{X}_0 = \boldsymbol{\xi}$ we fix the present instant t, vary the reference time t_0 and replace it by a new variable τ. This yields the representation of a streakline:

$$\mathbf{x}(\tau) = \phi_\tau(\boldsymbol{\xi}, t) \tag{2.53}$$

This definition can be interpreted as follows: A streakline through the point $\boldsymbol{\xi}$ is the set of the current positions $\mathbf{x}(\tau)$ of all those material elements, which at time τ have been at position $\boldsymbol{\xi}$.

Summarizing the principal content of the three definitions we recognize that the trajectories or path lines provide a material representation of the motion of a material body. The streamlines are the integral curves of the velocity field at a fixed instant t, and the streaklines represent some kind of long range influence.

2.6 Velocity gradient

A very important quantity in continuum mechnanics is the acceleration of a material element. In the spatial description, it is given as a material derivative of the velocity field, i.e.,

$$\dot{\mathbf{v}} = \frac{\mathrm{d}}{\mathrm{d}t}\bar{\mathbf{v}}(\mathbf{x},t) = \frac{\partial v}{\partial t} + (\,\mathrm{grad}\,\bar{\mathbf{v}})\bar{\mathbf{v}}. \tag{2.54}$$

The second term on the right hand side, the so–called **convective part of the material derivative,** is governed by the velocity and the **spatial velocity gradient**. It is common to use a separate symbol $\mathbf{L}$ for this second order tensor field:

$$\mathbf{L}(\mathbf{x},t) = \mathrm{grad}\,\bar{\mathbf{v}}(\mathbf{x},t) = \left\{\frac{\mathrm{d}}{\mathrm{d}x^k}\bar{\mathbf{v}}(\mathbf{x},t)\right\} \otimes \mathbf{g}^k \tag{2.55}$$

The component representation of $\mathbf{L}$ is a 3×3–matrix; with respect to a Cartesian coordinate system we have $\mathbf{v} = v_x\mathbf{e}_x + v_y\mathbf{e}_y + v_z\mathbf{e}_z$ and thus $\mathbf{L}$ takes the form

$$\mathbf{L} = \mathrm{grad}\,\mathbf{v} \mathrel{\widehat{=}} \begin{pmatrix} \dfrac{\partial v_x}{\partial x} & \dfrac{\partial v_x}{\partial y} & \dfrac{\partial v_x}{\partial z} \\ \dfrac{\partial v_y}{\partial x} & \dfrac{\partial v_y}{\partial y} & \dfrac{\partial v_y}{\partial z} \\ \dfrac{\partial v_z}{\partial x} & \dfrac{\partial v_z}{\partial y} & \dfrac{\partial v_z}{\partial z} \end{pmatrix}. \tag{2.56}$$

Apart from its approximating property as a gradient field, the spatial velocity gradient enjoys distinct kinematic qualities. These are similar to the deformation gradient, however, related to rates and to the spatial description:
The velocity gradient $\mathbf{L} = \mathrm{grad}\,\mathbf{v}(\mathbf{x},t)$ determines the time rates of material line elements $d\mathbf{x}$, surface elements $d\mathbf{A}$ and volume elements dV according to the following formulae:

Line elements: $$(d\mathbf{x})^{\cdot} = \mathbf{L}d\mathbf{x} \tag{2.57}$$

Surface elements: $$(d\mathbf{A})^{\cdot} = [(\,\mathrm{div}\,\mathbf{v})\mathbf{1} - \mathbf{L}^T]d\mathbf{A} \tag{2.58}$$

Volume elements: $$(d\mathbf{V})^{\cdot} = (\,\mathrm{div}\,\mathbf{v})d\mathbf{V} \tag{2.59}$$

To prove these statements, consider the material representation of the velocity field,

$$\hat{\mathbf{v}}(\mathbf{X},t) = \bar{\mathbf{v}}(\chi_R(\mathbf{X},t),t). \tag{2.60}$$

This identity can be differentiated with respect to $\mathbf{X}$; by applying the chain rule of differentiation, this yields

$$\begin{aligned}\mathrm{Grad}\,\hat{\mathbf{v}}(\mathbf{X},t) &= \mathrm{grad}\,\bar{\mathbf{v}}(\mathbf{x},t)\,\mathrm{Grad}\,\chi_R(\mathbf{X},t) \\ &= \mathbf{L}(\mathbf{x},t)\,\mathbf{F}(\mathbf{X},t).\end{aligned} \tag{2.61}$$

On the other hand, the material velocity gradient is the time rate of the deformation gradient,

$$\mathrm{Grad}\,\hat{\mathbf{v}}(\mathbf{X},t) = \mathrm{Grad}\left[\frac{\partial}{\partial t}\chi_R(\mathbf{X},t)\right] = \frac{\partial}{\partial t}\mathrm{Grad}\,\chi_R(\mathbf{X},t) = \dot{\mathbf{F}}(\mathbf{X},t).$$

Thus we have $\dot{\mathbf{F}} = \mathbf{LF}$, or

$$\mathbf{L} = \operatorname{grad}\mathbf{v}(\mathbf{x},t) = \dot{\mathbf{F}}\,\mathbf{F}^{-1}. \tag{2.62}$$

In this relation $\mathbf{L}$ is (as before) the gradient of the velocity field with respect to the spatial coordinates; however, since the most natural representation of $\mathbf{F}$ is material, $\mathbf{F} = \mathbf{F}(\mathbf{X},t)$, the spatial gradient $\mathbf{L}$ is delivered in its material representation, i.e.,

$$\mathbf{L} = \hat{\mathbf{L}}(\mathbf{X},t) = \dot{\mathbf{F}}(\mathbf{X},t)\,\mathbf{F}^{-1}(\mathbf{X},t).$$

With these preliminaries we now return to the above statements. We commence with the first statement and start with relation (2.37), $d\mathbf{x} = \mathbf{F}(\mathbf{X},t)\,d\mathbf{X}$. It can be differentiated with respect to time t to yield

$$(d\mathbf{x})^{\cdot} = \frac{\partial}{\partial t}\mathbf{F}(\mathbf{X},t)\,d\mathbf{X}.$$

Replacing $d\mathbf{X}$ by $d\mathbf{x}$, this becomes

$$(d\mathbf{x})^{\cdot} = \dot{\mathbf{F}}\,\mathbf{F}^{-1}\,d\mathbf{x} = \mathbf{L}\,d\mathbf{x},$$

which proves the first statement.

The third statement will now be verified by differentiating the relation (2.41), i.e., $dV = [\det\mathbf{F}(\mathbf{X},t)]\,dV_R$, with respect to time. To this end we need the identity

$$\frac{\mathrm{d}}{\mathrm{d}\mathbf{T}}(\det\mathbf{T}) = (\det\mathbf{T})\mathbf{T}^{T-1}, \tag{2.63}$$

valid for all invertible tensors $\mathbf{T}$ of second order. Application of the chain rule yields

$$\begin{aligned}(dV)^{\cdot} &= \left[\frac{\mathrm{d}}{\mathrm{d}\mathbf{F}}(\det\mathbf{F})\right]\cdot(\dot{\mathbf{F}}\,dV_R) = (\det\mathbf{F})\mathbf{F}^{T-1}\cdot(\dot{\mathbf{F}}\,dV_R) = \\ &= \operatorname{tr}(\dot{\mathbf{F}}\,\mathbf{F}^{-1})(\det\mathbf{F})\,dV_R = \operatorname{tr}(\operatorname{grad}\mathbf{v})\,dV = (\operatorname{div}\mathbf{v})\,dV.\end{aligned}$$

This result can be utilized in a simple proof of the remaining second statement by differentiating relation (2.39), $d\mathbf{A} = (\det\mathbf{F})\mathbf{F}^{T-1}\,d\mathbf{A}_R$, according to the product rule:

$$\begin{aligned}(d\mathbf{A})^{\cdot} &= \left[(\det\mathbf{F})^{\cdot}\mathbf{F}^{T-1} + (\det\mathbf{F})\,(\mathbf{F}^{T-1})^{\cdot}\right]\,d\mathbf{A}_R \\ &= (\det\mathbf{F})\,\left[(\operatorname{div}\mathbf{v})\,\mathbf{F}^{T-1} - \mathbf{F}^{T-1}\,\dot{\mathbf{F}}^{T}\,\mathbf{F}^{T-1}\right]\,d\mathbf{A}_R\end{aligned}$$

The last step follows from the differentiation of $\mathbf{F}^{T-1}\,\mathbf{F}^{T} = \mathbf{1}$, implying

$$(\mathbf{F}^{T-1})^{\cdot} = -\mathbf{F}^{T-1}\,\dot{\mathbf{F}}^{T}\,\mathbf{F}^{T-1}. \tag{2.64}$$

Replacing $d\mathbf{A}_R$ by $d\mathbf{A}$ we finally arrive at

$$(d\mathbf{A})^{\cdot} = \left[(\operatorname{div}\mathbf{v})\,\mathbf{1} - \left(\dot{\mathbf{F}}\,\mathbf{F}^{-1}\right)^{T}\right]\,d\mathbf{A} = \left[(\operatorname{div}\mathbf{v})\,\mathbf{1} - \mathbf{L}^{T}\right]\,d\mathbf{A},$$

which completes the proof of the statements.

As a result we summarize the properties of the (material) deformation gradient and the (spatial) velocity gradient as follows: The **deformation gradient F** operates on the set of material line elements in the reference configuration. It determines the present configuration of such a line element, which in the reference configuration corresponds to the vector $d\mathbf{X}$. The **velocity gradient L** operates on the set of material line elements in the present configuration. It determines the rate of range of a line element, which in the present configuration corresponds to the vector $d\mathbf{x}$. Corresponding interpretations apply to the material surface and volume elements in their various configurations. Thus it is now clear that the two gradients $\mathbf{F}$ and $\mathbf{L}$ include all informations about the local properties of a continuous motion or deformation. In the next section, these properties will be investigated in more detail.

2.7 Strain tensors

A further analysis of the local behavior of the deformation leads to the **concept of strains**. In this context, the basic properties emerge from the possibility to decompose the deformation into a rotation and a stretch which, roughly speaking, is a change of the shape of the volume element. This decomposition is called **polar decomposition of the deformation gradient**, and it is summarized in the following theorem:

A non-singular tensor $\mathbf{F}$ ($\det \mathbf{F} \neq 0$) admits the **polar decompositions**

$$\mathbf{F} = \mathbf{R}\mathbf{U} = \mathbf{V}\mathbf{R}, \tag{2.65}$$

where the particular factors have the following properties:

1. The tensor $\mathbf{R}$ is orthogonal: $\mathbf{R}\mathbf{R}^T = \mathbf{R}^T\mathbf{R} = \mathbf{1}$
2. The tensors $\mathbf{U}$ and $\mathbf{V}$ are symmetric and positive definite.
3. $\mathbf{U}$,$\mathbf{V}$ and $\mathbf{R}$ are uniquely determined.
4. The eigenvalues of $\mathbf{U}$ and $\mathbf{V}$ are identical; if $\mathbf{e}$ is an eigenvector of $\mathbf{U}$, then $\mathbf{R}\,\mathbf{e}$ is an eigenvector of $\mathbf{V}$.

To prove these statements, we first define the tensors $\mathbf{C} = \mathbf{F}^T\mathbf{F}$ and $\mathbf{B} = \mathbf{F}\mathbf{F}^T$. These are positive definite: For all vectors $\mathbf{w}$ we have

$$\mathbf{w} \cdot (\mathbf{F}^T\mathbf{F}\mathbf{w}) = (\mathbf{F}\mathbf{w}) \cdot (\mathbf{F}\mathbf{w}) \geq 0,$$

where the equality holds only for $\mathbf{w} = \mathbf{0}$, since $\mathbf{F}$ is assumed to be nonsingular. It follows that there are tensors $\mathbf{U}$ and $\mathbf{V}$, which are symmetric, positive definite and uniquely determined, defined by $\mathbf{U}^2 = \mathbf{C}$ and $\mathbf{V}^2 = \mathbf{B}$. These tensors are called the positive roots of $\mathbf{C}$ and $\mathbf{B}$:

$$\mathbf{U} = \sqrt{\mathbf{C}} = \sqrt{\mathbf{F}^T\mathbf{F}}\ ; \quad \mathbf{V} = \sqrt{\mathbf{B}} = \sqrt{\mathbf{F}\mathbf{F}^T}$$

Next, two tensors $\mathbf{R}$ and $\tilde{\mathbf{R}}$ are defined by

$$\mathbf{R} = \mathbf{F}\mathbf{U}^{-1} \quad \text{and} \quad \tilde{\mathbf{R}} = \mathbf{V}^{-1}\mathbf{F}.$$

We recognize that both are orthogonal, since by definition

$$\begin{aligned} \mathbf{R}\mathbf{R}^T &= (\mathbf{F}\mathbf{U}^{-1})(\mathbf{F}\mathbf{U}^{-1})^T = \mathbf{F}\mathbf{U}^{-1}\,\mathbf{U}^{-1}\,\mathbf{F}^T = \mathbf{F}(\mathbf{U}^2)^{-1}\,\mathbf{F}^T \\ &= \mathbf{F}(\mathbf{F}^T\,\mathbf{F})^{-1}\,\mathbf{F}^T = \mathbf{F}(\mathbf{F}^{-1}\,\mathbf{F}^{T-1})\mathbf{F}^T = \mathbf{1}. \end{aligned}$$

A similar proof holds for $\tilde{\mathbf{R}}$. So far, we have the decompositions $\mathbf{F} = \mathbf{R}\mathbf{U} = \mathbf{V}\,\tilde{\mathbf{R}}$, where $\mathbf{U}$ and $\mathbf{V}$ are symmetric, positive definite and $\mathbf{R}$, $\tilde{\mathbf{R}}$ are orthogonal. From

$$\mathbf{F} = \mathbf{V}\tilde{\mathbf{R}} = (\tilde{\mathbf{R}}\tilde{\mathbf{R}}^T)\mathbf{V}\tilde{\mathbf{R}} = \tilde{\mathbf{R}}(\tilde{\mathbf{R}}^T\,\mathbf{V}\tilde{\mathbf{R}}) = \tilde{\mathbf{R}}\tilde{\mathbf{U}}$$

it may be concluded that there might be two decompositions of $\mathbf{F}$, namely $\mathbf{F} = \tilde{\mathbf{R}}\tilde{\mathbf{U}}$ and $\mathbf{F} = \mathbf{R}\mathbf{U}$. However, if this were true we were forced to conclude that $\mathbf{C} = \mathbf{F}^T\,\mathbf{F} = \tilde{\mathbf{U}}^2 = \mathbf{U}^2$, whence follows that $\tilde{\mathbf{U}} = \mathbf{U}$, because of the uniquencess of the positive root. This implies $\tilde{\mathbf{R}} = \mathbf{R}$; consequently, $\mathbf{U}$,$\mathbf{V}$ and $\mathbf{R}$ are unique.

Finally, we assume $\mathbf{e}$ and λ to be an eigenvector and eigenvalue of $\mathbf{U}$. Then, we have $\lambda\mathbf{e} = \mathbf{U}\,\mathbf{e}$, as well as $\lambda\mathbf{R}\,\mathbf{e} = (\mathbf{R}\mathbf{U})\mathbf{e} = (\mathbf{V}\mathbf{R})\mathbf{e} = \mathbf{V}(\mathbf{R}\,\mathbf{e})$. Thus λ is also eigenvalue of $\mathbf{V}$ and $\mathbf{R}\mathbf{e}$ eigenvector. This completes the proof of the theorem.

The theorem of polar decomposition makes the idea precise that the deformation of a small part of a material body is the result of a stretch and a rotation, where the sequence of these two steps may be interchanged (see Figure 2.5). The combination of rotation and stretch corresponds to the multiplication of two tensors, namely $\mathbf{V}$ and $\mathbf{R}$ or $\mathbf{R}$ and $\mathbf{U}$. However, $\mathbf{R}$ should not be understood as a rigid body rotation, since in the general case it varies from point to point. Thus the polar decomposition theorem reflects only a local property of the motion.

The tensors $\mathbf{U}$ and $\mathbf{V}$ are called right and left stretch tensors. The squares

$$\mathbf{C} = \mathbf{F}^T\,\mathbf{F} = \mathbf{U}^2 \tag{2.66}$$

and

$$\mathbf{B} = \mathbf{F}\mathbf{F}^T = \mathbf{V}^2 \tag{2.67}$$

are called right and left CAUCHY–GREEN tensors, respectively.

The polar decomposition of the deformation gradient is the starting point for the definition of various **strain tensors**. The notion of a strain is based on the idea that the deviation of a deformation from a rigid–body motion should be measured. Since the motion of a rigid body preserves all lengths and angles, its manifestation is nothing else than a superposition of translation and rotation, namely

$$\mathbf{x} = \chi_R(\mathbf{X}, t) = \mathbf{Q}(t)(\mathbf{X} - \mathbf{X}_m) + \mathbf{c}(t). \tag{2.68}$$

Here, $\mathbf{X}_m$ denotes the center of mass in the reference configuration, $\mathbf{c}(t)$ the time–dependent translation and $\mathbf{Q}(t)$ an orthogonal tensor, which depends only on time t, because $\mathbf{Q}(t)$ represents the global rotation. Clearly, for a rigid–body motion the deformation gradient reads

$$\mathbf{F}(\mathbf{X}, t) = \mathbf{Q}(t), \tag{2.69}$$

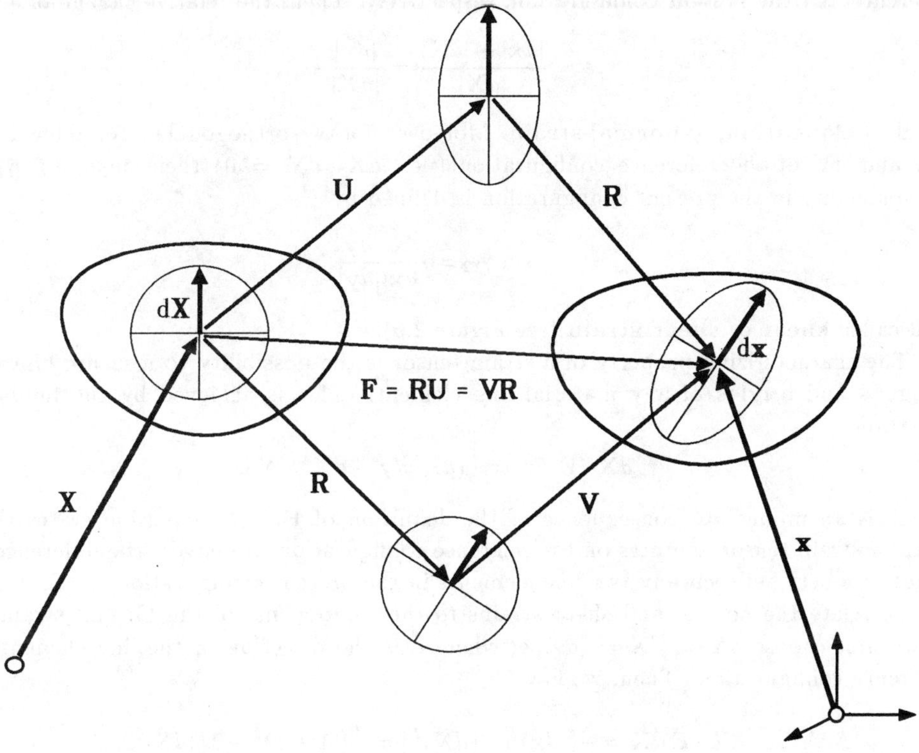

Fig.2.5: Polar decomposition

i.e., it is orthogonal and spatially homogeneous. In case of a rigid body motion the CAUCHY–GREEN tensors and stretch tensors are equal to the unit tensor, viz.,

$$\mathbf{C} = \mathbf{B} = \mathbf{U} = \mathbf{V} = \mathbf{1}. \tag{2.70}$$

If a particular deformation of a body differs from a rigid body motion, strains will occur; these are measured by means of a strain tensor. Among other possibilities of definition, the strain tensor

$$\mathbf{E} = \frac{1}{2}(\mathbf{C} - \mathbf{1}) = \frac{1}{2}\left(\mathbf{F}^T\,\mathbf{F} - \mathbf{1}\right) \tag{2.71}$$

is called the **GREEN strain tensor.**

For a justification and interpretation of this definition the properties of $\mathbf{E}$ should be identified. To this end, we consider material line elements $d\mathbf{X}$ and $d\mathbf{x}$, occuring in the reference and the present configuration, respectively. Then, the relative change of length,

$$\varepsilon = \frac{|d\mathbf{x}| - |d\mathbf{X}|}{|d\mathbf{X}|} = \frac{|d\mathbf{x}|}{|d\mathbf{X}|} - 1, \tag{2.72}$$

is called **elongation**, or **normal strain**. Moreover, for two orthogonal material line elements $d\mathbf{X}$ and $d\mathbf{Y}$ of the reference configuration (i.e., $d\mathbf{X} \cdot d\mathbf{Y} = 0$) their deviation γ_{12} from orthogonality in the present configuration is defined by

$$\sin \gamma_{12} = \frac{d\mathbf{x} \cdot d\mathbf{y}}{|d\mathbf{x}||d\mathbf{y}|} \tag{2.73}$$

and called **shear or shear strain** (see Figure 2.6).

The characterizing property of a strain tensor is the possibility to measure **changes in lengths** and **angles** of any material line element. This is achieved by the fundamental equation

$$d\mathbf{X} \cdot \mathbf{E}\, d\mathbf{Y} = \frac{1}{2}(d\mathbf{x} \cdot d\mathbf{y} - d\mathbf{X} \cdot d\mathbf{Y}), \tag{2.74}$$

which is an immediate consequence of the definition of $\mathbf{E}$. The equation states that the GREEN strain tensor operates on the reference configuration; it delivers the difference of the inner products between any two line elements in the present configuration.

To relate the normal and shear strains to the components of the GREEN strain tensor $\mathbf{E}$ we first choose $d\mathbf{Y} = d\mathbf{X} = |d\mathbf{X}|\,\mathbf{e}$, where $\mathbf{e}$ is the direction of the line element in the reference configuration. Then, we have

$$d\mathbf{X} \cdot \mathbf{E}\, d\mathbf{X} = \frac{1}{2}[(d\mathbf{x})^2 - (d\mathbf{X})^2] = \frac{1}{2}[(1+\varepsilon)^2 - 1]\,|d\mathbf{X}|^2,$$

which implies

$$\mathbf{e} \cdot \mathbf{E}\mathbf{e} = \varepsilon + \frac{\varepsilon^2}{2}, \tag{2.75}$$

or

$$\varepsilon = \sqrt{1 + 2\mathbf{e} \cdot \mathbf{E}\mathbf{e}} - 1. \tag{2.76}$$

The choice of $d\mathbf{X} = d\mathbf{X}_1 = |d\mathbf{X}_1|\mathbf{e}_1$ and $d\mathbf{Y} = d\mathbf{X}_2 = |d\mathbf{X}_2|\mathbf{e}_2$ with $\mathbf{e}_1 \cdot \mathbf{e}_2 = 0$ leads to

$$\begin{aligned} d\mathbf{X}_1 \cdot \mathbf{E}\, d\mathbf{X}_2 &= \frac{1}{2} d\mathbf{x}_1 \cdot d\mathbf{x}_2 = \\ &= \frac{1}{2}[(1+\varepsilon_1)(1+\varepsilon_2)] \sin\gamma_{12}\, |d\mathbf{X}_1||d\mathbf{X}_2|. \end{aligned} \tag{2.77}$$

Here, $\varepsilon_1 = \frac{|d\mathbf{x}_1|}{|d\mathbf{X}_1|} - 1$ and $\varepsilon_2 = \frac{|d\mathbf{x}_2|}{|d\mathbf{X}_2|} - 1$ are the elongations of the two line elements $d\mathbf{X}_{1,2}$, i.e., $1 + \varepsilon_{1,2} = \sqrt{1 + 2\mathbf{e}_{1,2} \cdot \mathbf{E}\,\mathbf{e}_{1,2}}$.

As a result, the shear strain is given by

$$\sin\gamma_{12} = \frac{2\mathbf{e}_1 \cdot \mathbf{E}\,\mathbf{e}_2}{\sqrt{1 + 2\mathbf{e}_1 \cdot \mathbf{E}\,\mathbf{e}_1}\,\sqrt{1 + 2\mathbf{e}_2 \cdot \mathbf{E}\,\mathbf{e}_2}}. \tag{2.78}$$

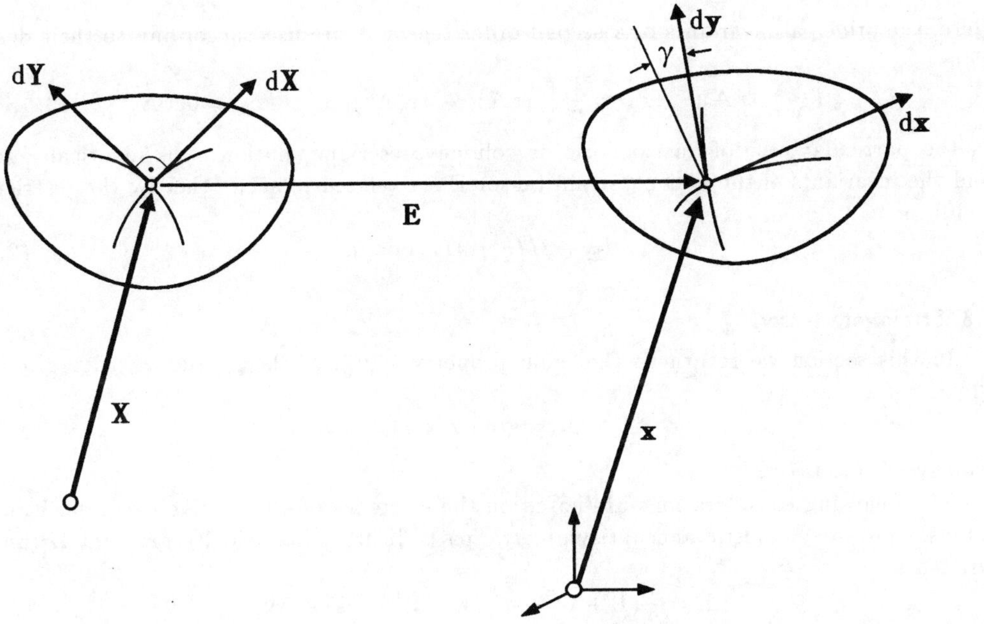

Fig.2.6: Strain tensor

The above relations demonstrate that **E** is a strain tensor in the intrinsic sense of the word. In particular, **E** vanishes identically for a rigid body motion, since the deformation gradient is orthogonal in this case. Accordingly, there is no elongation and shear in the motion of a rigid body: All lengths and angles are preserved.

Another important quantity is the volume strain, i.e., the change of a material volume element. This is given by the determinant of the deformation gradient,

$$\varepsilon_v = \frac{dV - dV_R}{dV_R} = \det \mathbf{F} - 1. \tag{2.79}$$

In terms of the GREEN strain tensor we evaluate

$$\begin{aligned} \varepsilon_v + 1 &= \det \mathbf{F} = \sqrt{\det \mathbf{C}} = \sqrt{\det(\mathbf{1} + 2\mathbf{E})} \\ &= \sqrt{1 + 2I_{\mathbf{E}} + 4II_{\mathbf{E}} + 8III_{\mathbf{E}}}, \end{aligned}$$

or

$$\varepsilon_v = \sqrt{1 + 2I_{\mathbf{E}} + 4II_{\mathbf{E}} + 8III_{\mathbf{E}}} - 1. \tag{2.80}$$

Here, the principal invariants of a second order tensor **A** are used according to their definitions

$$I_{\mathbf{A}} = \operatorname{tr}\mathbf{A}\,, \quad II_{\mathbf{A}} = \frac{1}{2}\left[(\operatorname{tr}\mathbf{A})^2 - \operatorname{tr}(\mathbf{A}^2)\right]\,, \quad III_{\mathbf{A}} = \det\mathbf{A}.$$

In the particular case of an isochoric or volume–preserving motion ε_v is identically zero, and the invariants of the GREEN strain tensor **E** are related to each other by the restrictive condition

$$I_{\mathbf{E}} + 2II_{\mathbf{E}} + 4III_{\mathbf{E}} = 0. \tag{2.81}$$

2.8 Strain rate tensors

In this section we return to the main property (2.57) of the spatial velocity gradient (2.55)

$$\mathbf{L} = \operatorname{grad}\mathbf{v}(\mathbf{x},t),$$

namely $(d\mathbf{x})^{\cdot} = \mathbf{L}\,d\mathbf{x}$.

The following considerations are based on the existence of an additive decomposition of a tensor into a symmetric and antisymmetric part. In fact, the velocity gradient **L** can be written as

$$\mathbf{L} = \frac{1}{2}(\mathbf{L}+\mathbf{L}^T) + \frac{1}{2}(\mathbf{L}-\mathbf{L}^T) = \mathbf{D}+\mathbf{W}, \tag{2.82}$$

where both **D** and **W** are uniquely determined:

$$\mathbf{D} = \frac{1}{2}(\mathbf{L}+\mathbf{L}^T) \qquad \text{is symmetric} \qquad (\mathbf{D}=\mathbf{D}^T) \tag{2.83}$$

and

$$\mathbf{W} = \frac{1}{2}(\mathbf{L}-\mathbf{L}^T) \qquad \text{is antisymmetric} \qquad (\mathbf{W}=-\mathbf{W}^T). \tag{2.84}$$

The physical significance of the additive decomposition of the velocity gradient into symmetric and antisymmetric parts is quite similar to that of the multiplicative decomposition of the deformation gradient into orthogonal and symmetric positive definite factors. Essentially, the additive decomposition of **L** implies the decomposition of the rate of deformation into a **rate of strain** and a **rate of rotation.** The symmetric part **D** of the spatial velocity gradient is called **strain rate tensor** or **stretching tensor** and the antisymmetric part **W** is called **rate of rotation** or **vorticity tensor.** The rates of change in lengths and angles of material line elements are related to the components of **D**: If a line element

$$d\mathbf{x} = |d\mathbf{x}|\,\mathbf{e} \tag{2.85}$$

is given, where **e** is its direction in the present configuration, differentiation of the identity $|d\mathbf{x}|^2 = d\mathbf{x}\cdot d\mathbf{x}$ with respect to time yields

$$|d\mathbf{x}||d\mathbf{x}|^{\cdot} = d\mathbf{x}\cdot(d\mathbf{x})^{\cdot} = d\mathbf{x}\cdot(\mathbf{L}\,d\mathbf{x}) = d\mathbf{x}\cdot(\mathbf{D}\,d\mathbf{x})\ , \tag{2.86}$$

or

$$\frac{|d\mathbf{x}|^{\cdot}}{|d\mathbf{x}|} = \mathbf{e}\cdot(\mathbf{D}\,\mathbf{e}). \tag{2.87}$$

For two line elements

$$d\mathbf{x}_{1,2} = |d\mathbf{x}_{1,2}|\,\mathbf{e}_{1,2}, \tag{2.88}$$

which, in the present configuration, span the relative angle $\frac{\pi}{2} + \gamma_{12}$, we have

$$|d\mathbf{x}_1||d\mathbf{x}_2| \sin\gamma_{12} = d\mathbf{x}_1 \cdot d\mathbf{x}_2.$$

Differentiation with respect to time implies

$$\begin{aligned} \dot{\gamma}_{12} \cos\gamma_{12}|d\mathbf{x}_1||d\mathbf{x}_2| + (|d\mathbf{x}_1||d\mathbf{x}_2|)^{\cdot} \sin\gamma_{12} &= (d\mathbf{x}_1)^{\cdot} \cdot d\mathbf{x}_2 + d\mathbf{x}_1 \cdot (d\mathbf{x}_2)^{\cdot} \\ &= d\mathbf{x}_1 \cdot [(\mathbf{L}^T + \mathbf{L})\, d\mathbf{x}_2] = 2 d\mathbf{x}_1 \cdot \mathbf{D}\, d\mathbf{x}_2. \end{aligned}$$

If we assume orthogonality of the two line elements in the present configuration, i.e., $d\mathbf{x}_1 \cdot d\mathbf{x}_2 = 0$ ($\gamma_{12} = 0$), we arrive at

$$\dot{\gamma}_{12}(t) = 2\mathbf{e}_1 \cdot \mathbf{D}\,\mathbf{e}_2. \tag{2.89}$$

Thus for any pairs of line elements of given directions, the strain rate tensor $\mathbf{D}$ determines the rate of change in length and relative angle.

The antisymmetric part $\mathbf{W}$ of the velocity gradient corresponds to the *mean rotation* rate of material elements. To make this statement precise, the time rate of change of a material direction

$$\mathbf{e} = \frac{d\mathbf{x}}{|d\mathbf{x}|} \tag{2.90}$$

is calculated; this yields

$$\dot{\mathbf{e}} = \frac{\mathrm{d}}{dt}\left(\frac{d\mathbf{x}}{|d\mathbf{x}|}\right) = \frac{1}{|d\mathbf{x}|}(d\mathbf{x})^{\cdot} - \frac{|d\mathbf{x}|^{\cdot}}{|d\mathbf{x}|^2}\, d\mathbf{x} = \mathbf{L}\,\mathbf{e} - (\mathbf{e}\cdot\mathbf{D}\,\mathbf{e})\,\mathbf{e}.$$

Invoking once more the decomposition $\mathbf{L} = \mathbf{D} + \mathbf{W}$, this implies

$$\dot{\mathbf{e}} = \mathbf{W}\,\mathbf{e} + [\mathbf{D}\,\mathbf{e} - (\mathbf{e}\cdot\mathbf{D}\,\mathbf{e})\,\mathbf{e}]. \tag{2.91}$$

This formula admits the following interpretation: If $\mathbf{e}$ is an eigenvector of $\mathbf{D}$, i.e.,

$$\mathbf{D}\,\mathbf{e} = \lambda\mathbf{e}, \tag{2.92}$$

the bracketted term on the right hand side of (2.91) is zero, so that we have

$$\dot{\mathbf{e}} = \mathbf{W}\,\mathbf{e} \tag{2.93}$$

for all material line elements, whose directions in the present configuration concide with the eigenvectors of the strain rate tensor $\mathbf{D}$. In this sense, the antisymmetric part

$$\mathbf{W} = \frac{1}{2}(\mathbf{L} - \mathbf{L}^T) \tag{2.94}$$

of the velocity gradient represents a mean rotation rate of the material line elements. The term vorticity tensor is motivated by the fact that $\mathbf{W}$ corresponds to the vorticity vector

$$\mathbf{w} = \operatorname{rot} \mathbf{v}(\mathbf{x}, t), \tag{2.95}$$

which is the axial vector of $2\mathbf{W}$. Indeed, for all vectors $\mathbf{a}$, $\mathbf{w}$ and $\mathbf{W}$ are related by the identity

$$\begin{aligned} \mathbf{W}\mathbf{a} &= \frac{1}{2}\left[\operatorname{grad} \mathbf{v} - (\operatorname{grad} \mathbf{v})^T\right] \mathbf{a} = \\ &= \frac{1}{2}(\operatorname{rot} \mathbf{v}) \times \mathbf{a} = \frac{1}{2}\mathbf{w} \times \mathbf{a}. \end{aligned} \tag{2.96}$$

For rigid body motions we infer $\mathbf{D} = \mathbf{0}$ and $\mathbf{W} = \dot{\mathbf{Q}}(t)\mathbf{Q}^T(t)$. Isochoric motions – defined by zero volume changes – imply

$$\operatorname{tr} \mathbf{D} = \operatorname{div} \mathbf{v} = 0. \tag{2.97}$$

The additive decomposition $\mathbf{L} = \mathbf{D} + \mathbf{W}$ is another expression for the idea that the deformation of a small piece of matter consists locally of a stretch and a rotation. Mathematically, the combination of these two processes corrresponds to the addition of two tensors, namely the strain rate tensor and the vorticity tensor.

The strain rate tensor $\mathbf{D}$ operates on the **present configuration** in the sense that

$$d\mathbf{x} \cdot \mathbf{D}\, d\mathbf{x} = |d\mathbf{x}||d\mathbf{x}|^{\cdot}. \tag{2.98}$$

Of course, it is quite simple to construct a strain rate tensor with the same property, acting on the **reference configuration**. In fact, we must only replace the line element $d\mathbf{x}$ by $d\mathbf{X}$ via $d\mathbf{x} = \mathbf{F}\, d\mathbf{X}$. This yields

$$|d\mathbf{x}||d\mathbf{x}|^{\cdot} = (\mathbf{F}\, d\mathbf{X}) \cdot (\mathbf{D}\mathbf{F}\, d\mathbf{X}) = d\mathbf{X} \cdot \left[(\mathbf{F}^T \mathbf{D}\mathbf{F})\, d\mathbf{X}\right].$$

This strain rate tensor can be identified as the material time derivative of the GREEN strain tensor:

$$\begin{aligned} \mathbf{F}^T \mathbf{D}\mathbf{F} &= \mathbf{F}^T \left[\frac{1}{2}(\mathbf{L} + \mathbf{L}^T)\right] \mathbf{F} = \mathbf{F}^T \left[\frac{1}{2}(\dot{\mathbf{F}}\mathbf{F}^{-1} + \mathbf{F}^{T-1}\dot{\mathbf{F}}^T)\right] \mathbf{F} \\ &= \frac{1}{2}\left[(\mathbf{F}^T\dot{\mathbf{F}} + \dot{\mathbf{F}}^T\mathbf{F})\right] = \frac{1}{2}(\mathbf{F}^T\mathbf{F})^{\cdot} = \frac{1}{2}(\mathbf{F}^T\mathbf{F} - 1)^{\cdot} = \dot{\mathbf{E}} \end{aligned}$$

Thus we end up with

$$\dot{\mathbf{E}} = \mathbf{F}^T \mathbf{D}\mathbf{F}. \tag{2.99}$$

The tensor

$$\dot{\mathbf{E}} = \frac{\partial}{\partial t}\mathbf{E}(\mathbf{X}, t) \tag{2.100}$$

with the property

$$d\mathbf{X} \cdot (\dot{\mathbf{E}}\, d\mathbf{X}) = |d\mathbf{x}||d\mathbf{x}|^{\cdot} \tag{2.101}$$

is called **GREEN strain rate** or **material strain rate tensor** in contrast to the **spatial strain rate tensor D**.

2.9 Displacement vector and gradient

The kinematics of deformable bodies can also be formulated by utilizing the displacement vector. This is a quite useful concept in special problems of solid mechanics. The definition of the displacement vector is

$$\mathbf{u}(\mathbf{X},t) = \mathbf{x} - \mathbf{X} = \chi_R(\mathbf{X},t) - \mathbf{X}, \tag{2.102}$$

where the reference configuration should be occupied by the body at some instant t_0 in the sense

$$\chi_R(\mathbf{X},t_0) = \mathbf{X}. \tag{2.103}$$

As an immediate consequence of the definition of $\mathbf{u}$ the material gradient of the displacement vector, or the **displacement gradient**

$$\mathbf{H}(\mathbf{X},t) = \text{Grad}\,\mathbf{u}(\mathbf{X},t) \tag{2.104}$$

is related to the deformation gradient by

$$\mathbf{H} = \mathbf{F} - \mathbf{1}. \tag{2.105}$$

In principle, all physical quantities, representing the kinematics of deformable bodies (such as **C**, **B**, **U**, **V**, **E**, **L**, **D**, **W**), can be expressed in terms of the displacement vector **u** and its gradient **H**, respectively:

For the right and left CAUCHY GREEN tensors **C** and **B** we have

$$\begin{aligned}\mathbf{C} &= \mathbf{F}^T\mathbf{F} = (\mathbf{1}+\mathbf{H}^T)(\mathbf{1}+\mathbf{H}) = \mathbf{1}+\mathbf{H}+\mathbf{H}^T+\mathbf{H}^T\mathbf{H}\\ \mathbf{B} &= \mathbf{F}\mathbf{F}^T = (\mathbf{1}+\mathbf{H})(\mathbf{1}+\mathbf{H}^T) = \mathbf{1}+\mathbf{H}+\mathbf{H}^T+\mathbf{H}\mathbf{H}^T;\end{aligned}$$

the GREEN strain tensor **E** can be expressed as

$$\mathbf{E} = \frac{1}{2}(\mathbf{H}+\mathbf{H}^T+\mathbf{H}^T\mathbf{H}). \tag{2.106}$$

It is straightforward, but quite lengthy, to express the other kinematic quantities (**U**, **V**, **R**, **L** etc.) in terms of the displacement gradient and, eventually, their material time derivatives. Since these relations are of little practical use, they will be not be given here.

2.10 Geometric linearization

The whole kinematics of deformable bodies is considerably simplified, if deformations are assumed to be small. In this context, a convenient measure of smallness is the norm of the displacement gradient, i.e.,

$$\delta = |\mathbf{H}| = \sqrt{\text{tr}\,(\mathbf{H}\mathbf{H}^T)}. \tag{2.107}$$

With respect to Cartesian coordinates, this yields

$$\delta = \sqrt{\sum_{i,k}\left(\frac{\partial u_i}{\partial X^k}\right)^2}. \tag{2.108}$$

In the following, the term **small deformations** will correspond to small displacement gradients. This means that all first derivatives of the displacements with respect to the coordinates are sufficiently small such that lineariziation is justified, i.e., all terms of higher order are ignored. In its geometrical interpretation a small value of δ implies small strains as well as small rotations.

In the context of small strains and small rotations **geometric linerarization** is the process of developing all kinematics quantities with respect to δ and omitting all terms of order higher than $\mathbf{O}(\delta)$. This definition implies the following asymptotic relations:

$$\mathbf{F} = \mathbf{1} + \mathbf{H} \quad \text{(exact)} \tag{2.109}$$

$$\mathbf{F}^{-1} = \mathbf{1} - \mathbf{H} + \mathbf{O}(\delta^2) \tag{2.110}$$

$$\det \mathbf{F} = 1 + \operatorname{tr}\mathbf{H} + \mathbf{O}(\delta^2) \tag{2.111}$$

$$\mathbf{C} = \mathbf{1} + \mathbf{H} + \mathbf{H}^T + \mathbf{O}(\delta^2) \tag{2.112}$$

$$\mathbf{B} = \mathbf{1} + \mathbf{H} + \mathbf{H}^T + \mathbf{O}(\delta^2) \tag{2.113}$$

$$\mathbf{U} = \mathbf{1} + \frac{1}{2}(\mathbf{H} + \mathbf{H}^T) + \mathbf{O}(\delta^2) \tag{2.114}$$

$$\mathbf{V} = \mathbf{1} + \frac{1}{2}(\mathbf{H} + \mathbf{H}^T) + \mathbf{O}(\delta^2) \tag{2.115}$$

$$\mathbf{R} = \mathbf{1} + \frac{1}{2}(\mathbf{H} - \mathbf{H}^T) + \mathbf{O}(\delta^2) \tag{2.116}$$

From the last three relations we infer

$$\mathbf{RU} = \left[\mathbf{1} + \frac{1}{2}(\mathbf{H} - \mathbf{H}^T) + \mathbf{O}(\delta^2)\right]\left[\mathbf{1} + \frac{1}{2}(\mathbf{H} + \mathbf{H}^T) + \mathbf{O}(\delta^2)\right]$$

$$\mathbf{RU} = \mathbf{1} + \mathbf{H} + \mathbf{O}(\delta^2) \tag{2.117}$$

and

$$\mathbf{VR} = \left[\mathbf{1} + \frac{1}{2}(\mathbf{H} + \mathbf{H}^T) + \mathbf{O}(\delta^2)\right]\left[\mathbf{1} + \frac{1}{2}(\mathbf{H} - \mathbf{H}^T) + \mathbf{O}(\delta^2)\right]$$

$$\mathbf{VR} = \mathbf{1} + \mathbf{H} + \mathbf{O}(\delta^2). \tag{2.118}$$

Thus we see that for small deformations the multiplicative decomposition of the deformation gradient into orthogonal and positive definite symmetric factors is approximated by the additive decomposition of the displacement gradient into symmetric and antisymmetric parts:

$$\mathbf{F} = \mathbf{RU} = \mathbf{VR} = \mathbf{1} + \mathbf{H}$$

$$\mathbf{F} = \mathbf{1} + \frac{1}{2}(\mathbf{H} + \mathbf{H}^T) + \frac{1}{2}(\mathbf{H} - \mathbf{H}^T) + \mathbf{O}(\delta^2) \tag{2.119}$$

The symmetric part of the displacement gradient,

$$\tilde{\mathbf{E}} = \frac{1}{2}(\mathbf{H} + \mathbf{H}^T) = \frac{1}{2}\left[\operatorname{Grad}\mathbf{u} + (\operatorname{Grad}\mathbf{u})^T\right], \tag{2.120}$$

is called the **linearized strain tensor**, and the antisymmetric part plus unit tensor,

$$\tilde{\mathbf{R}} = \mathbf{1} + \frac{1}{2}(\mathbf{H} - \mathbf{H}^T), \tag{2.121}$$

is called the **linearized rotation tensor.** Quite the same procedure of geometric linearization leads to the asymptotic approximation of the spatial velocity gradient,

$$\mathbf{L} = \dot{\mathbf{F}}\mathbf{F}^{-1} = \dot{\mathbf{H}} + \mathrm{O}(\delta^2), \tag{2.122}$$

the spatial strain rate tensor,

$$\mathbf{D} = \frac{1}{2}(\mathbf{L} + \mathbf{L}^T) = \frac{1}{2}(\dot{\mathbf{H}} + \dot{\mathbf{H}}^T) + \mathrm{O}(\delta^2) \tag{2.123}$$

and the vorticity tensor,

$$\mathbf{W} = \frac{1}{2}(\mathbf{L} - \mathbf{L}^T) = \frac{1}{2}(\dot{\mathbf{H}} - \dot{\mathbf{H}}^T) + \mathrm{O}(\delta^2). \tag{2.124}$$

It is seen that in the context of geometric linearization the GREEN strain rate is asymptotically identical with the spatial strain rate tensor:

$$\begin{aligned} \dot{\mathbf{E}} &= \frac{1}{2}(\dot{\mathbf{F}}^T\mathbf{F} + \mathbf{F}^T\dot{\mathbf{F}}) = \frac{1}{2}(\dot{\mathbf{H}}^T\mathbf{F} + \mathbf{F}^T\dot{\mathbf{H}}) = \frac{1}{2}(\dot{\mathbf{H}} + \dot{\mathbf{H}}^T) + \mathrm{O}(\delta^2) \\ \dot{\mathbf{E}} &= \mathbf{D} + \mathrm{O}(\delta^2) \end{aligned} \tag{2.125}$$

3. Balance relations: General structure

3.1 General structure of a balance relation

The general concept of a balance relation is based on the idea that a material body is thought to be separated from its exterior surroundings; in this imagination the *influence of the exterior world* on the body is replaced by a set of physical quantities, which give rise to a rate of change of the *present status of the material system.* Of course, this separation is by no means a real process, but rather an abstract idea that facilitates the mathematical formulation of natural laws. In the context of continuum mechanics and thermodynamics the *present status* of a body is characterized by its *mass, momentum, moment of momentum, internal energy* and *entropy.* Changes of these quantities are caused by actions like *force, torque, energy flux, entropy flux* and *entropy production.* In all particular cases the present status of a body is represented by volume integrals of corresponding density distributions. The action of the exterior world on the body is generally exhibited in volume and surface integrals of convenient volume and surface densities. All volume and surface integrals, respectively, extend over the volume and the surface of the material body in its current or reference configuration.

To explain the general structure of a balance relation, let z be a real valued quantity characterizing the present status of a material body such that

$$z(t) = \int_V \bar{\psi}(\mathbf{x}, t)\, dV, \tag{3.1}$$

or

$$z(t) = \int_{V_R} \hat{\psi}_R(\mathbf{X}, t)\, dV_R, \tag{3.2}$$

where ψ and ψ_R are corresponding densities with respect to the volume element in the current and reference configurations, respectively. The representations (3.1) and (3.2) assume $z(t)$ to be additive with respect to volume measure. In view of eq. (2.41) ψ and ψ_R are related by the equation

$$\hat{\psi}_R = \bar{\psi} \det \mathbf{F}. \tag{3.3}$$

The most general formulation of a scalar valued balance relation postulates that the material time rate of $z(t)$,

$$\dot{z}(t) = \frac{\mathrm{d}}{\mathrm{d}t} \int_V \bar{\psi}(\mathbf{x}, t)\, dV = \frac{\partial}{\partial t} \int_{V_R} \hat{\psi}_R(\mathbf{X}, t)\, dV_R,$$

equals the action of the exterior world:

$$\frac{\mathrm{d}}{\mathrm{d}t} \int_V \bar{\psi}(\mathbf{x}, t)\, dV = \int_A \phi\, dA + \int_V \sigma\, dV \tag{3.4}$$

$$\frac{\partial}{\partial t} \int_{V_R} \hat{\psi}_R(\mathbf{X}, t)\, dV_R = \int_{A_R} \phi_R\, dA_R + \int_{V_R} \sigma_R\, dV_R \tag{3.5}$$

On the right–hand sides, ϕ or ϕ_R is the surface density of the action on the body, referred to the surface element in the present and reference configuration, respectively; σ and σ_R are the corresponding volume densities of the action, referred to the volume element in the current and reference configuration. According to eq. (3.4) all actions are assumed to be distributed over the volume as well as over the surface of the material body in its present configuration; transformation of the surface and volume integrals to the domain of the reference configuration leads to the equivalent formulation (3.5). Generally, the surface density of the action is related to a flow or flux of the corresponding quantity, whereas the volume density may incorporate supplies and productions. If the balance law is vector valued, it has essentially the same general structure, namely

$$\frac{\mathrm{d}}{\mathrm{d}t} \int_V \bar{\boldsymbol{\psi}}(\mathbf{x}, t)\, dV = \int_A \boldsymbol{\phi}\, dA + \int_V \boldsymbol{\sigma}\, dV \tag{3.6}$$

or

$$\frac{\partial}{\partial t} \int_{V_R} \hat{\boldsymbol{\psi}}_R(\mathbf{X}, t)\, dV_R = \int_{A_R} \boldsymbol{\phi}_R\, dA_R + \int_{V_R} \boldsymbol{\sigma}_R\, dV_R. \tag{3.7}$$

Note that the surface densities ϕ, ϕ_R, $\boldsymbol{\phi}$, $\boldsymbol{\phi}_R$ are neither scalar nor vector fields, because apart from position and time they must also depend on the orientation of the surface element, i.e.,

$$\phi_R = \hat{\phi}_R(\mathbf{X}, t, \mathbf{n}_R)\,, \qquad \phi = \bar{\phi}(\mathbf{x}, t, \mathbf{n})\,, \quad \text{etc.,} \tag{3.8}$$

where $\mathbf{X}$ and $\mathbf{x}$ denote a point of the surface and $\mathbf{n}_R$, $\mathbf{n}$ the outer unit normals at that material surface point (see Figure 3.1). As a first implication of the general balance law we

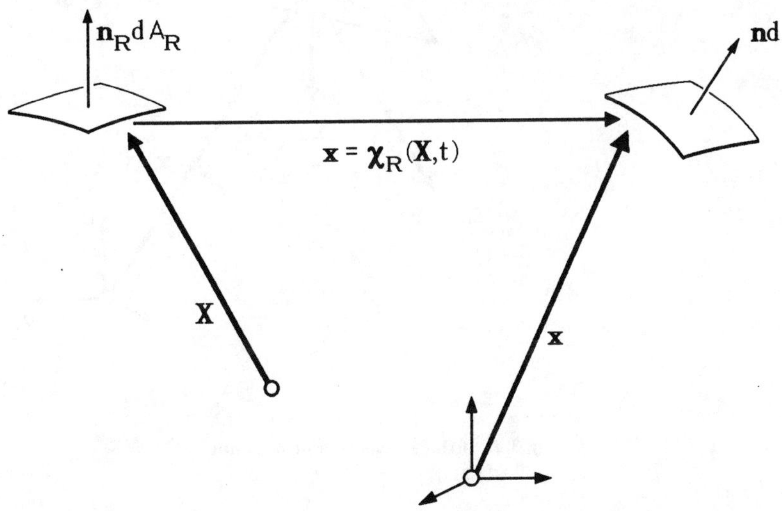

Fig.3.1: Material surface

note the following theorem, which is called CAUCHY's Lemma:

If the surface density ϕ is assumed to depend on the unit normal $\mathbf{n}$, it must be a linear function of $\mathbf{n}$.

The proof of this statement makes use of the fact that a general balance law is postulated to hold for all material bodies: In particular, eq. (3.4) or eq. (3.6) must hold for all parts of the body, which can be thought to be separated. In this sense we apply eq. (3.6) to a sequence of tetrahedra, which are shrunk down to a single point $\mathbf{x}$. As visualized in Figure 3.2, the surfaces ΔA_x, ΔA_y, ΔA_z are taken parallel to the coordinate planes x, y, $z =$ const., the fourth surface has an arbitrary (but fixed) orientation $\mathbf{n}$. The balance relation (3.6) is now applied to this particular material body; in doing so, the integrals are calculated (in the sense of the mean-value-theorem), (i) as products of the volume times the integrand at an appropriate inner point or, (ii) as products of the plane surfaces times the values of the

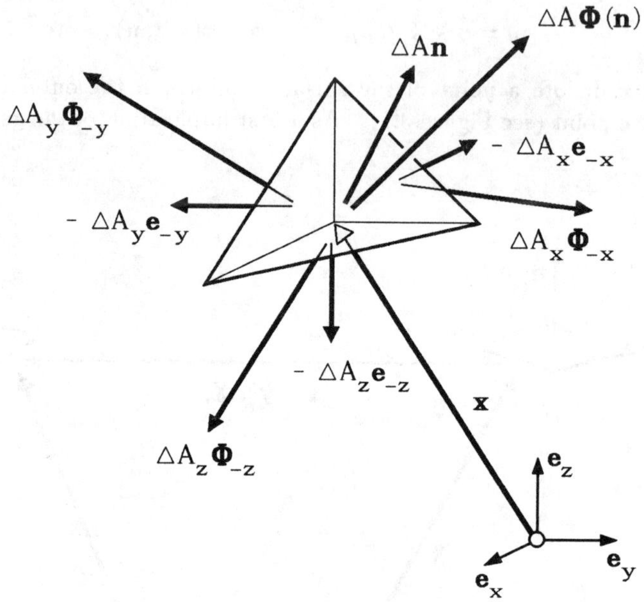

Fig.3.2: Surface density and unit normal

surface density at appropriate surface points. This leads to

$$\Delta A\boldsymbol{\phi}(\mathbf{n}) + \Delta A_x\boldsymbol{\phi}_{-x} + \Delta A_y\boldsymbol{\phi}_{-y} + \Delta A_z\boldsymbol{\phi}_{-z} = \int_{\Delta V} \boldsymbol{\sigma}\, dV - \frac{\mathrm{d}}{\mathrm{d}t}\int_{\Delta V} \boldsymbol{\psi}\, dV = \Delta V\boldsymbol{\xi}, \quad (3.9)$$

where $\boldsymbol{\xi}$ is an appropriate vector in the sense that $\Delta V\boldsymbol{\xi}$ is equal to the sum of both volume integrals. The notation $\boldsymbol{\phi}_{-x}$, $\boldsymbol{\phi}_{-y}$, $\boldsymbol{\phi}_{-z}$ indicates that these densities are acting on the surfaces with the outer normals $-\mathbf{e}_x$, $-\mathbf{e}_y$ and $-\mathbf{e}_z$, respectively. Since the tetrahedron is a closed surface, we have

$$\Delta A\mathbf{n} - (\Delta A_x\mathbf{e}_x + \Delta A_y\mathbf{e}_y + \Delta A_z\mathbf{e}_z) = \mathbf{0},$$

or

$$\mathbf{n} = \frac{\Delta A_x}{\Delta A}\,\mathbf{e}_x + \frac{\Delta A_y}{\Delta A}\,\mathbf{e}_y + \frac{\Delta A_z}{\Delta A}\,\mathbf{e}_z,$$

i.e.,

$$n_x = \frac{\Delta A_x}{\Delta A}\,, \quad n_y = \frac{\Delta A_y}{\Delta A}\,, \quad n_z = \frac{\Delta A_y}{\Delta A}\,. \quad (3.10)$$

With this property, eq. (3.9) reads

$$\phi(\mathbf{n}) = -n_x\phi_{-x} - n_y\phi_{-y} - n_z\phi_{-z} + \frac{\Delta V}{\Delta A}\boldsymbol{\xi}.$$

If we let ΔV tend to zero with the normal $\mathbf{n}$ held constant, the last term vanishes, and we have the intermediate result

$$\phi(\mathbf{n}) = -n_x\phi_{-x} - n_y\phi_{-y} - n_z\phi_{-z}. \tag{3.11}$$

The special cases $\mathbf{n} = \mathbf{e}_x$, $\mathbf{n} = \mathbf{e}_y$, $\mathbf{n} = \mathbf{e}_z$ are included in eq. (3.11); introducing the notation $\phi_x = \phi(\mathbf{e}_x)$, $\phi_y = \phi(\mathbf{e}_y)$, $\phi_z = \phi(\mathbf{e}_z)$ for the densities acting on surfaces with the outer normals $\mathbf{e}_x$, $\mathbf{e}_y$, $\mathbf{e}_z$, we deduce from (3.11)

$$\phi_x = -\phi_{-x}\,, \quad \phi_y = -\phi_{-y}\,, \quad \phi_z = -\phi_{-z}\,. \tag{3.12}$$

Putting this back into (3.11) leads to

$$\phi(\mathbf{n}) = n_x\phi_x + n_y\phi_y + n_z\phi_z. \tag{3.13}$$

Thus $\phi(\mathbf{n})$ is *linear* in $\mathbf{n}$ and we are able to define a tensor of second order, namely

$$\boldsymbol{\Phi} = \phi_x \otimes \mathbf{e}_x + \phi_y \otimes \mathbf{e}_y + \phi_z \otimes \mathbf{e}_z, \tag{3.14}$$

which is composed of three vector valued fields and maps $\mathbf{n}$ into $\phi(\mathbf{n})$ according to $\phi(\mathbf{n}) = \boldsymbol{\Phi}\,\mathbf{n}$ or

$$\phi(\mathbf{x}, t, \mathbf{n}) = \boldsymbol{\Phi}(\mathbf{x}, t)\,\mathbf{n}. \tag{3.15}$$

This completes the proof of the theorem in the case of a vector valued surface density. If the surface density is scalar valued (as in eq. (3.4)) the same argumentation leads to a real valued linear function, which corresponds to three scalar valued fields or to one vector field $\boldsymbol{\Phi}(\mathbf{x}, t)$ with the property that

$$\phi(\mathbf{x}, t, \mathbf{n}) = \boldsymbol{\Phi}(\mathbf{x}, t) \cdot \mathbf{n}. \tag{3.16}$$

Moreover in view of the formulations (3.5) and (3.7) exactly the same results hold with respect to the reference configuration:

$$\phi_R(\mathbf{x}, t, \mathbf{n}_R) = \boldsymbol{\Phi}_R(\mathbf{x}, t) \cdot \mathbf{n}_R, \tag{3.17}$$
$$\phi_R(\mathbf{x}, t, \mathbf{n}_R) = \boldsymbol{\Phi}_R(\mathbf{x}, t)\,\mathbf{n}_R. \tag{3.18}$$

We call the fields $\boldsymbol{\Phi}$ and $\boldsymbol{\Phi}_R$ (vector or tensor valued) *fluxes*, referred to the current and reference configuration, respectively. The fluxes $\boldsymbol{\Phi}$ and $\boldsymbol{\Phi}_R$ are related to each other; indeed, these relations can be inferred from eq. (2.39): In case of a scalar valued surface density we have the identity $\phi_R\,dA_R = \phi\,dA$ or, equivalently, since $d\mathbf{A} = \mathbf{n}\,dA$, $d\mathbf{A}_R = \mathbf{n}_R\,dA_R$,

$$\boldsymbol{\Phi}_R \cdot d\mathbf{A}_R = \boldsymbol{\Phi} \cdot d\mathbf{A} = \boldsymbol{\Phi} \cdot \left[(\det \mathbf{F})\mathbf{F}^{T-1}\,d\mathbf{A}_R\right] = (\det \mathbf{F})\mathbf{F}^{-1}\boldsymbol{\Phi}\,d\mathbf{A}_R.$$

Thus for vector valued fluxes we arrive at the relation

$$\mathbf{\Phi}_R = (\det \mathbf{F})\mathbf{F}^{-1}\mathbf{\Phi}. \tag{3.19}$$

The same reasoning applies to tensor valued fluxes with the result

$$\mathbf{\Phi}_R = (\det \mathbf{F})\mathbf{\Phi}\mathbf{F}^{T-1}. \tag{3.20}$$

We note that eqs. (3.12) corroborate the **reaction principle for surface densities**: Two surface densities acting on surfaces through the same material point, however with opposite normals of the surface elements, have the same absolute value but opposite signs.

As a result the general structure of a scalar or vector valued balance relation, referred to the spatial or material representation corresponds to one of the following equations:

$$\frac{\mathrm{d}}{\mathrm{d}t}\int_V \psi \, dV = \int_A \mathbf{\Phi}\cdot\mathbf{n}\, dA + \int_V \sigma \, dV \tag{3.21}$$

$$\frac{\mathrm{d}}{\mathrm{d}t}\int_V \boldsymbol{\psi} \, dV = \int_A \mathbf{\Phi}\mathbf{n}\, dA + \int_V \boldsymbol{\sigma} \, dV \tag{3.22}$$

$$\frac{\partial}{\partial t}\int_{V_R} \psi_R \, dV_R = \int_{A_R} \mathbf{\Phi}_R\cdot\mathbf{n}_R\, dA_R + \int_{V_R} \sigma_R \, dV_R \tag{3.23}$$

$$\frac{\partial}{\partial t}\int_{V_R} \boldsymbol{\psi}_R \, dV_R = \int_{A_R} \mathbf{\Phi}_R\mathbf{n}_R\, dA_R + \int_{V_R} \boldsymbol{\sigma}_R \, dV_R \tag{3.24}$$

To clarify the procedure is should be mentioned that σ, σ_R, $\boldsymbol{\sigma}$ and $\boldsymbol{\sigma}_R$ comprise contributions from two different physical origins: Source and supply terms from outside agents and production terms from inside the body. Equivalent local formulations correspond to the global balance relations, *if sufficient smoothness properties of the related fields are assumed.* For the derivations of local forms three steps must be carried out, namely

- differentation of volume integrals,
- transformation of surface integrals to volume integrals and
- evaluation of continuity assumptions.

3.2 Transport theorems

Differentiation of volume integrals

The goal of this subchapter is the evaluation of the material time derivative of the quantities $\psi(t)$ and $\boldsymbol{\psi}(t)$, defined by eqs. (3.1) and (3.2). Referring first to the material description (3.2), we have

$$\dot{z}(t) = \frac{\partial}{\partial t}\int_{V_R} \hat{\psi}_R(\mathbf{X},t)\, dV_R = \int_{V_R} \frac{\partial \hat{\psi}_R(\mathbf{X},t)}{\partial t}\, dV_R. \tag{3.25}$$

To differentiate the spatial representation (3.1) the volume integral is transformed to the reference configuration,

$$z(t) = \int_V \bar{\psi}(\mathbf{x},t)\,dV = \int_{V_R} \hat{\psi}(\mathbf{X},t)\det\mathbf{F}(\mathbf{X},t)\,dV_R, \tag{3.26}$$

where eq. (2.41) has been applied. Differentiation with respect to time yields

$$\begin{aligned}\dot{z}(t) &= \int_{V_R}\left\{\frac{\partial\hat{\psi}(\mathbf{X},t)}{\partial t}\det\mathbf{F}(\mathbf{X},t) + \hat{\psi}(\mathbf{X},t)\frac{\partial}{\partial t}\det\mathbf{F}(\mathbf{X},t)\right\}dV_R =\\ &= \int_{V_R}\left\{\frac{\partial\hat{\psi}(\mathbf{X},t)}{\partial t} + \hat{\psi}(\mathbf{X},t)\operatorname{div}\bar{\mathbf{v}}\right\}(\det\mathbf{F})\,dV_R,\end{aligned}$$

where use has been made of (2.59). Observing the relation

$$\frac{\partial\hat{\psi}(\mathbf{X},t)}{\partial t} = \frac{\mathrm{d}}{\mathrm{d}t}\bar{\psi}(\mathbf{x},t) = \dot{\bar{\psi}}\,,$$

and transforming the above integral back to the current configuration leads to

$$\dot{z}(t) = \frac{\mathrm{d}}{\mathrm{d}t}\int_V \bar{\psi}\,dV = \int_V(\dot{\bar{\psi}} + \bar{\psi}\operatorname{div}\bar{\mathbf{v}})\,dV. \tag{3.27}$$

Here, $\dot{\bar{\psi}}$ denotes the material derivative of the density $\bar{\psi}$, calculated according to eq. (2.24):

$$\dot{\psi} = \dot{\bar{\psi}}\,(\mathbf{x},t) = \frac{\partial\bar{\psi}(\mathbf{x},t)}{\partial t} + [\operatorname{grad}\bar{\psi}(\mathbf{x},t)]\cdot\bar{\mathbf{v}}(\mathbf{x},t) \tag{3.28}$$

Similar formulae hold for the material derivative of vector valued volume integrals:

$$\dot{\mathbf{z}} = \int_{V_R}\frac{\partial}{\partial t}\hat{\boldsymbol{\psi}}\,(\mathbf{X},t)\,dV = \int_V\left(\dot{\bar{\boldsymbol{\psi}}} + \bar{\boldsymbol{\psi}}\,\operatorname{div}\bar{\mathbf{v}}\right)dV \tag{3.29}$$

Here,

$$\dot{\boldsymbol{\psi}} = \dot{\bar{\boldsymbol{\psi}}}\,(\mathbf{x},t) = \frac{\partial}{\partial t}\bar{\boldsymbol{\psi}}\,(\mathbf{x},t) + (\operatorname{grad}\bar{\boldsymbol{\psi}}\,)\,\bar{\mathbf{v}} \tag{3.30}$$

is the material derivative of the vector field $\bar{\boldsymbol{\psi}}$. A brief summary of the above derivation can be noted as follows:

$$\begin{aligned}\frac{\mathrm{d}}{\mathrm{d}t}\int_V \boldsymbol{\psi}\,dV &= \int_V(\boldsymbol{\psi}\,dV)^{\cdot} = \int_V\left[\dot{\boldsymbol{\psi}}\,dV + \boldsymbol{\psi}(dV)^{\cdot}\right] =\\ &= \int_V\left[\dot{\boldsymbol{\psi}} + \boldsymbol{\psi}(\,\operatorname{div}\mathbf{v})\right]dV\end{aligned} \tag{3.31}$$

The bars, indicating the spatial representation, will be omitted in the sequel.

Reynolds' transport theorem

Eqs. (3.27) and (3.29) can be interpreted in terms of a transport equation. To see this interpretation we calculate the material derivative of the density ψ,

$$\dot{\psi}(\mathbf{x}, t) = \frac{\partial \psi}{\partial t} + (\operatorname{grad} \psi) \cdot \mathbf{v},$$

and substitute this back into (3.27). With the product rule

$$\psi \operatorname{div} \mathbf{v} + (\operatorname{grad} \psi)\mathbf{v} = \operatorname{div}(\psi \mathbf{v}) \tag{3.32}$$

we then arrive at

$$\frac{\mathrm{d}}{\mathrm{d}t} \int_V \psi \, dV = \int_V \left[\frac{\partial \psi}{\partial t} + \operatorname{div}(\psi \mathbf{v}) \right] dV.$$

Applying the divergence theorem to the second term, we get

$$\frac{\mathrm{d}}{\mathrm{d}t} \int_V \psi \, dV = \int_V \frac{\partial \psi}{\partial t} \, dV + \int_A \psi(\mathbf{v} \cdot \mathbf{n}) \, dA. \tag{3.33}$$

This relation is called **Reynolds' transport theorem** and is interpreted as follows: The material time rate of change of a volume integral equals the sum of two terms: The first term is the integral of the local derivative of the density field ψ. The second term can be understood to be the rate of the transport or the flux of the quantity ψ through the surface A of the volume V, which in this interpretation is imagined to be fixed in space. In this way the time rate of change of a physical quantity in a material body can be interpreted in terms of an **open system**.

The same interpretation holds for the material derivative of a vector valued quantity:

$$\frac{\mathrm{d}}{\mathrm{d}t} \int_V \boldsymbol{\psi} \, dV = \int_V \left[\dot{\boldsymbol{\psi}} + \boldsymbol{\psi}(\operatorname{div} \mathbf{v}) \right] dV = \int_V \left[\frac{\partial \boldsymbol{\psi}}{\partial t} + (\operatorname{grad} \boldsymbol{\psi})\mathbf{v} + \boldsymbol{\psi} \operatorname{div} \mathbf{v} \right] dV,$$

the only difference being that the product rule

$$\operatorname{div}(\boldsymbol{\psi} \otimes \mathbf{v}) = (\operatorname{grad} \boldsymbol{\psi})\mathbf{v} + \boldsymbol{\psi}(\operatorname{div} \mathbf{v}) \tag{3.34}$$

must be used in this case, which leads to the analogous formula

$$\frac{\mathrm{d}}{\mathrm{d}t} \int_V \boldsymbol{\psi} \, dV = \int_V \frac{\partial \boldsymbol{\psi}}{\partial t} \, dV + \int_A \boldsymbol{\psi}(\mathbf{v} \cdot \mathbf{n}) \, dA. \tag{3.35}$$

Modified Reynolds' transport theorem

In some situations a modification of the transport theorem (3.33) or (3.35) is needed, which allows for an arbitrary motion of the domain of integration. To obtain this modification we recall that, according to (3.35),

$$\frac{\mathrm{d}}{\mathrm{d}t} \int_V \boldsymbol{\psi} \, dV = \int_V \frac{\partial \boldsymbol{\psi}}{\partial t} \, dV + \int_A \boldsymbol{\psi}(\mathbf{v} \cdot \mathbf{n}) \, dA,$$

the material derivative calculates the rate of change of the integral as it is recognized by an observer moving exactly with the material; this is because the velocity $\mathbf{v}$ in the surface integral is the material velocity field $\mathbf{v}(\mathbf{x},t)$. Another observer moving with a different (non–material) velocity $\mathbf{w}(\mathbf{x},t)$ would recognize a different time rate, namely

$$\left(\frac{\mathrm{d}}{\mathrm{d}t}\right)_{\mathbf{w}} \int_V \psi \, dV = \int_V \frac{\partial \psi}{\partial t} \, dV + \int_A \psi (\mathbf{w} \cdot \mathbf{n}) \, dA. \tag{3.36}$$

Eliminating the local derivative from the above two equations we arrive at the **modified transport theorem**

$$\frac{\mathrm{d}}{\mathrm{d}t} \int_V \psi \, dV = \left(\frac{\mathrm{d}}{\mathrm{d}t}\right)_{\mathbf{w}} \int_{V(t)} \psi \, dV + \int_{A(t)} \psi [(\mathbf{v} - \mathbf{w}) \cdot \mathbf{n}] \, dA. \tag{3.37}$$

According to this formula the material derivative of a volume integral is calculated with respect to a volume which, at time t, coincides with the current configuration of the body and moves with an arbitrary velocity $\mathbf{w}(\mathbf{x},t)$. The first term is the time rate of change with respect to the moving volume, and the second term is the flux of the quantity ψ across the moving surface. We see that the derivative $\left(\frac{\mathrm{d}}{\mathrm{d}t}\right)_{\mathbf{w}}$ equals the local derivative for $\mathbf{w} = \mathbf{0}$ and the material derivative for $\mathbf{w} = \mathbf{v}$. For both special cases $\mathbf{w} = \mathbf{0}$ and $\mathbf{w} = \mathbf{v}$ the modified transport theorem coincides with the original formulation.

3.3 Continuous fields: Local formulations

We now suppose all fields to be sufficiently smooth, such that all differentiations are permissible.

Transformation of surface integrals

The surface integrals on the right–hand side of the balance equations (3.21) – (3.24) must be transformed to volume integrals. To this end, the **divergence theorem** is applied to four special cases:

$$\begin{aligned}
\int_A \phi \, dA &= \int_A \mathbf{\Phi} \cdot \mathbf{n} \, dA = \int_V \operatorname{div} \mathbf{\Phi} \, dV, \\
\int_A \boldsymbol{\phi} \, dA &= \int_A \mathbf{\Phi} \mathbf{n} \, dA = \int_V \operatorname{div} \mathbf{\Phi} \, dV, \\
\int_{A_R} \phi_R \, dA_R &= \int_{A_R} \mathbf{\Phi}_R \cdot \mathbf{n}_R \, dA_R = \int_{V_R} \operatorname{Div} \mathbf{\Phi}_R \, dV_R, \\
\int_{A_R} \boldsymbol{\phi}_R \, dA_R &= \int_{A_R} \mathbf{\Phi}_R \mathbf{n}_R \, dA_R = \int_{V_R} \operatorname{Div} \mathbf{\Phi}_R \, dV_R.
\end{aligned} \tag{3.38}$$

In these equations the various divergence–operators are defined as follows:

$$\operatorname{div} \mathbf{\Phi}(\mathbf{x},t) = \begin{cases} \dfrac{\partial \mathbf{\Phi}}{\partial x^k} \cdot \mathbf{g}^k & \text{if } \mathbf{\Phi} \text{ is vector valued} \\[2ex] \dfrac{\partial \mathbf{\Phi}}{\partial x^k} \mathbf{g}^k & \text{if } \mathbf{\Phi} \text{ is tensor valued} \end{cases} \tag{3.39}$$

$$\operatorname{Div} \mathbf{\Phi}_R(\mathbf{X},t) = \begin{cases} \dfrac{\partial \mathbf{\Phi}_R}{\partial X^\alpha} \cdot \mathbf{G}^\alpha & \text{if } \mathbf{\Phi}_R \text{ is vector valued} \\ \dfrac{\partial \mathbf{\Phi}_R}{\partial X^\alpha} \mathbf{G}^\alpha & \text{if } \mathbf{\Phi}_R \text{ is tensor valued} \end{cases} \tag{3.40}$$

In these definitions, EINSTEIN's summation convention is employed.

Evaluation of the continuity assumptions

Combination of (3.6) with (3.33), (3.16) and $(3.38)_1$ leads to the identity

$$\int_V \left[\dot{\psi} + \psi \operatorname{div} \mathbf{v} - \operatorname{div} \mathbf{\Phi} - \sigma\right] dV = 0, \tag{3.41}$$

which must hold for all material bodies or for all spatial volumina V. If it is assumed that all terms of the integrand are continuous functions of $\mathbf{x}$, a standard argument leads to the conclusion that the integrand itself must vanish identically; what results is the local form of the balance relation (3.6):

$$\frac{\partial \boldsymbol{\psi}}{\partial t} + (\operatorname{grad} \boldsymbol{\psi})\, \mathbf{v} + \boldsymbol{\psi}(\operatorname{div} \mathbf{v}) = \operatorname{div} \mathbf{\Phi} + \boldsymbol{\sigma} \tag{3.42}$$

This local formulation corresponds to the *spatial description.*

In the *material description* we arrive in the same way at the local form of eq. (3.7):

$$\frac{\partial \boldsymbol{\psi}_R}{\partial t} = \operatorname{Div} \mathbf{\Phi}_R + \boldsymbol{\sigma}_R \tag{3.43}$$

For the scalar valued balance relations (3.4) and (3.5) the same local equivalents must hold, namely

$$\frac{\partial \psi}{\partial t} + (\operatorname{grad} \psi) \cdot \mathbf{v} + \psi(\operatorname{div} \mathbf{v}) = \operatorname{div} \Phi + \sigma \tag{3.44}$$

and

$$\frac{\partial \psi_R}{\partial t} = \operatorname{Div} \mathbf{\Phi}_R + \sigma_R. \tag{3.45}$$

The structure of the global balance relations and their corresponding local forms is the basis for the formulation of all particular balance relations of continuum mechanics and thermodynamics.

3.4 Discontinuous fields: Jump conditions

Discontinuity surfaces

The local balance relations, derived in the previous paragraphs, were based on the assumption that the involved fields be smooth functions, such that the vanishing of an integral over an arbitrary volume implies that the integrand must be zero. In the case that the integrand fails to be continuous, this deduction is not permissible. Therefore, we must investigate the situation where discontinuous fields occur.

In the following it will be assumed that the discontinuities of an arbitrary field $\psi = \psi(\mathbf{x}, t)$ are distributed along a surface $\mathcal{S}(t)$ that is moving through the material body. The discontinuity surface is assumed to be smooth and orientable such that each surface element has a unit normal $\mathbf{n}_s$. The surface $\mathcal{S}(t)$ is illustrated in Figure 3.3. Each point $\mathbf{x}$ of the surface is supposed to have a velocity $\mathbf{w}(\mathbf{x}, t)$. Generally, $\mathbf{w}$ differs from the velocity $\mathbf{v}$ of the material points.

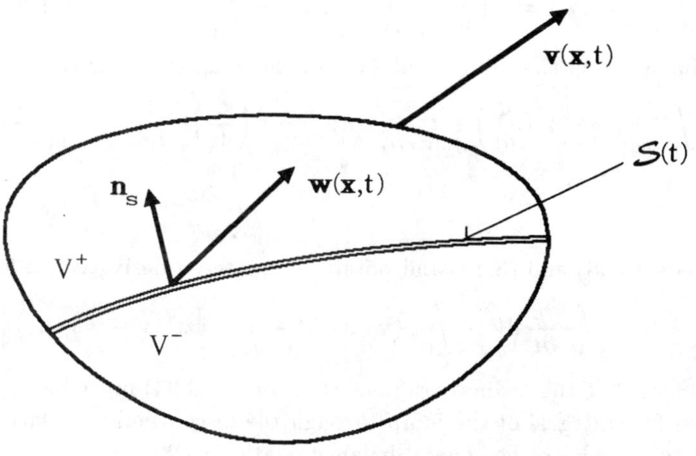

Fig.3.3: Singular surface

The surface $\mathcal{S}(t)$ divides the domain of the current configuration into two parts, namely V^+ on the side of the normal $\mathbf{n}_s$, in the direction of which the surface is travelling, and V^- on the other side. The limit values of the function ψ at one point of the surface $\mathcal{S}$ are denoted by $\psi^+(\mathbf{x}, t)$ and $\psi^-(\mathbf{x}, t)$. ψ^+ and ψ^- are the limit values if the point $\mathbf{x}$ is approached from the $(+)$ and the $(-)$ side, respectively. The difference of these two values is the **jump** of the field ψ and denoted by

$$[\![\psi]\!] = \psi^+ - \psi^-. \tag{3.46}$$

Since ψ is assumed to be discontinuous along the surface $\mathcal{S}$ the jump $[\![\psi]\!]$ differs from zero: Usually, it is assumed that this jump is bounded away from infinity and that on the surface $\mathcal{S}(t)$ the jump $[\![\psi]\!](\mathbf{x}, t)$ is a continuous function of $\mathbf{x}$ and t. Finally, it is assumed that outside the surface $\mathcal{S}(t)$ the field $\psi(\mathbf{x}, t)$ is continuously differentiable as in the previous considerations.

Starting from these assumptions, we consider the integral of the field $\psi(\mathbf{x},t)$ over the total volume of the body. Clearly, this integral is the sum of two parts:

$$\int_V \psi\, dV = \int_{V+} \psi\, dV + \int_{V-} \psi\, dV \tag{3.47}$$

Discontinuity relations (jump conditions)

We now proceed with the calculation of the material derivative of eq. (3.47) and apply the modified transport theorem (3.37) to the two terms on the right–hand side:

$$\frac{\mathrm{d}}{\mathrm{d}t}\int_{V-} \psi\, dV = \left(\frac{\mathrm{d}}{\mathrm{d}t}\right)_{\mathbf{w}} \int_{V-(t)} \psi\, dV + \int_{\mathcal{S}(t)} \psi^-[(\mathbf{v}^- - \mathbf{w})\cdot \mathbf{n}_s]\, dA \tag{3.48}$$

$$\frac{\mathrm{d}}{\mathrm{d}t}\int_{V+} \psi\, dV = \left(\frac{\mathrm{d}}{\mathrm{d}t}\right)_{\mathbf{w}} \int_{V+(t)} \psi\, dV - \int_{\mathcal{S}(t)} \psi^+[(\mathbf{v}^+ - \mathbf{w})\cdot \mathbf{n}_s]\, dA \tag{3.49}$$

The volume integrals on the right–hand sides of these equations are added to give

$$\left(\frac{\mathrm{d}}{\mathrm{d}t}\right)_{\mathbf{w}} \int_{V-(t)} \psi\, dV + \left(\frac{\mathrm{d}}{\mathrm{d}t}\right)_{\mathbf{w}} \int_{V+(t)} \psi\, dV = \left(\frac{\mathrm{d}}{\mathrm{d}t}\right)_{\mathbf{v}} \int_{V(t)=V^+ + V^-} \psi\, dV =$$

$$= \int_V \frac{\partial \psi}{\partial t}\, dV + \int_A \psi(\mathbf{v}\cdot\mathbf{n})\, dA. \tag{3.50}$$

Now we add eqs. (3.48) and (3.49) and obtain the material derivative as follows:

$$\frac{\mathrm{d}}{\mathrm{d}t}\int_V \psi\, dV = \int_V \frac{\partial \psi}{\partial t}\, dV + \int_A \psi(\mathbf{v}\cdot\mathbf{n})\, dA - \int_{\mathcal{S}(t)} \Big[\!\Big[\psi\big((\mathbf{v}-\mathbf{w})\cdot\mathbf{n}_s\big)\Big]\!\Big]\, dA \tag{3.51}$$

This result shows that the ordinary transport theorem (3.33) must be supplemented by an appropriate surface integral of the jump through the discontinuity surface $\mathcal{S}(t)$.

We apply this result to the general balance relation (3.22),

$$\frac{\mathrm{d}}{\mathrm{d}t}\int_V \psi\, dV = \int_A \mathbf{\Phi}\, \mathbf{n}\, dA + \int_V \sigma\, dV, \tag{3.52}$$

where we assume that the fields $\psi(\mathbf{x},t)$ and $\mathbf{\Phi}(\mathbf{x},t)$ are discontinuous across the surface $\mathcal{S}(t)$, while the jumps are continuous on $\mathcal{S}$. Application of the balance relation (3.52) to V^- and V^+ yields

$$\frac{\mathrm{d}}{\mathrm{d}t}\int_{V-} \psi\, dV = \int_{A-} \mathbf{\Phi}\, \mathbf{n}\, dA + \int_{\mathcal{S}(t)} \mathbf{\Phi}^-\, \mathbf{n}_s\, dA + \int_{V-} \sigma\, dV$$

and

$$\frac{\mathrm{d}}{\mathrm{d}t}\int_{V+} \psi\, dV = \int_{A+} \mathbf{\Phi}\, \mathbf{n}\, dA - \int_{\mathcal{S}(t)} \mathbf{\Phi}^+\, \mathbf{n}_s\, dA + \int_{V+} \sigma\, dV.$$

Addition and application of (3.49), (3.50) and (3.51) leads to

$$\int_V \frac{\partial \psi}{\partial t}\, dV + \int_A \psi(\mathbf{v}\cdot\mathbf{n})\, dA - \int_{\mathcal{S}(t)} \Big[\!\Big[\psi\big((\mathbf{v}-\mathbf{w})\cdot\mathbf{n}\big)\Big]\!\Big]\, dA =$$

$$= \int_V (\operatorname{div}\mathbf{\Phi} + \sigma)\, dV - \int_{\mathcal{S}(t)} [\![\mathbf{\Phi}\, \mathbf{n}_s]\!]\, dA \tag{3.53}$$

or, equivalently, to

$$\int_V \left[\dot{\psi} + \psi \operatorname{div} \mathbf{v} - \operatorname{div} \mathbf{\Phi} - \sigma\right] dV = \int_{\mathcal{S}(t)} \left[\!\left[\psi\Big((\mathbf{v}-\mathbf{w})\cdot\mathbf{n}_s\Big) - \mathbf{\Phi}\,\mathbf{n}_s \right]\!\right] dA. \tag{3.54}$$

The integrand on the left–hand side is continuous outside the surface $\mathcal{S}(t)$. According to the standard argument this integrand must identically vanish; this leads to (3.42) for all $\mathbf{x}$ outside the surface and, in addition, to

$$\int_{\mathcal{S}(t)} \left[\!\left[\psi\Big((\mathbf{v}-\mathbf{w})\cdot\mathbf{n}_s\Big) - \mathbf{\Phi}\,\mathbf{n}_s \right]\!\right] dA = \mathbf{0}. \tag{3.55}$$

Again, this surface integral must vanish for all subsurfaces, which, in view of the continuity of the jumps along the surface, implies that the integrand must vanish identically. Therefore we conclude

$$\left[\!\left[\psi\Big((\mathbf{v}-\mathbf{w})\cdot\mathbf{n}_s\Big) - \mathbf{\Phi}\,\mathbf{n}_s \right]\!\right] = \mathbf{0} \tag{3.56}$$

or

$$[\![\psi \otimes (\mathbf{v}-\mathbf{w}) - \mathbf{\Phi}]\!]\, \mathbf{n}_s = \mathbf{0} \tag{3.57}$$

for all $\mathbf{x}$ on $\mathcal{S}(t)$. In case of a scalar valued balance relation we have

$$[\![\psi(\mathbf{v}-\mathbf{w}) - \mathbf{\Phi}]\!] \cdot \mathbf{n}_s = 0. \tag{3.58}$$

Eqs. (3.57) and (3.58) are the general **discontinuity relations** or **jump conditions** which complete the local formulation of the balance relations in case of singular surfaces.

4. Balance relations of mechanics

4.1 Mass

The **mass** is a characteristic quantity of each material body. Physically, it is responsible for the *gravitational force* as well as for the *inertia*, i.e., the resistence against any change of the state of motion. The total mass of a material body is represented as the volume integral of its density:

$$m = \int_V \rho\, dV = \int_{V_R} \rho_R\, dV_R \tag{4.1}$$

Here, ρ denotes the mass density of a material point $\mathcal{X}$ in the current configuration, while ρ_R is the corresponding density of the same point $\mathcal{X}$ in the reference configuration. To formulate the balance relation for the mass, we refer to the previous section and set $\psi = \rho$, as well as

$$\mathbf{\Phi} = \mathbf{0}\,, \quad \sigma = 0 \qquad \text{(no flux, supply and production).} \tag{4.2}$$

Thus we arrive at the global formulation

$$\frac{\mathrm{d}}{\mathrm{d}t} m(t) = 0, \tag{4.3}$$

i.e.,

$$\frac{\mathrm{d}}{\mathrm{d}t}\int_V \bar{\rho}(\mathbf{x},t)\,dV = \frac{\partial}{\partial t}\int_{V_R} \hat{\rho}_R(\mathbf{X},t)\,dV_R = 0. \tag{4.4}$$

In view of eq. (3.44) this implies

$$\dot{\rho} + \rho\,\mathrm{div}\,\mathbf{v} = 0 \tag{4.5}$$

or

$$\frac{\partial \rho}{\partial t} + \mathrm{div}\,(\rho\mathbf{v}) = 0; \tag{4.6}$$

eq. (3.45) yields

$$\frac{\partial \hat{\rho}_R}{\partial t} = 0 \quad \Longleftrightarrow \quad \rho_R = \hat{\rho}_R(\mathbf{X}), \tag{4.7}$$

i.e., the mass density of the reference configuration is time–independent. In case of a discontinuity surface, these local forms must be supplemented by the jump condition

$$[\![\rho(\mathbf{v}-\mathbf{w})\cdot\mathbf{n}_s]\!] = 0, \tag{4.8}$$

according to (3.58). Because mass production and supply is zero, the global balance relation (4.3) is called the **conservation of mass**. The local forms (4.5), (4.6) and (4.7) are called **continuity equation** in the spatial and material description, respectively. According to its conservation property the mass of a material is the same for all configurations, such that the identity

$$\int_{V_R} \rho_R\,dV_R = \int_V \rho\,dV = \int_{V_R} \rho(\det\mathbf{F})\,dV_R$$

or

$$\int_{V_R} [\rho_R - \rho(\det\mathbf{F})]\,dV_R = 0$$

holds for all volumina V_R. With the usual argument of continuity we may deduce

$$\rho_R = \rho\det\mathbf{F}. \tag{4.9}$$

This equation is another formulation of the conservation of mass; it can be expressed as the differential form

$$dm = \rho\,dV = \rho_R\,dV_R. \tag{4.10}$$

Material differentiation of (4.9) leads to (4.5), if we take into account that $\rho_R(\mathbf{X})$ is independent of t.

4.2 Linear momentum

The **linear** or **translational momentum**

$$\mathbf{J} = \int_V \mathbf{v}\rho\,dV = \int_{V_R} \mathbf{v}\rho_R\,dV_R = \int \mathbf{v}\,dm \tag{4.11}$$

combines the velocity and mass distribution of a moving material body. It is one global characterization of the dynamic state of motion. The balance relation for linear momentum postulates that this quantity is influenced by the resultant force, which acts on the body in the sense of its separation from the exterior world, i.e.,

$$\dot{\mathbf{J}} = \mathbf{F}, \tag{4.12}$$

or

$$\frac{\mathrm{d}}{\mathrm{d}t}\int_V \mathbf{v}\rho\, dV = \int_A \mathbf{t}\, dA + \int_V \mathbf{k}\rho\, dV, \tag{4.13}$$

or

$$\frac{\partial}{\partial t}\int_{V_R} \mathbf{v}\rho_R\, dV_R = \int_{A_R} \mathbf{t}_R\, dA_R + \int_{V_R} \mathbf{k}\rho_R\, dV_R. \tag{4.14}$$

In particle mechanics this statement corresponds to NEWTON's first law of motion. In the context of continuum mechanics it is due to CAUCHY, and in an earlier special case to EULER. On the right–hand sides the total force is the sum of the *surface force* and the *volume force.* The surface densities $\mathbf{t}$ and $\mathbf{t}_R$ are called stress vectors related to the surface element in the spatial and reference configuration, respectively. $\mathbf{k}$ is the specific volume force per unit mass. The tensor valued fluxes corresponding to the stress vectors are called stress tensors: The tensor $\mathbf{T}$ with the property

$$\mathbf{t} = \mathbf{T}\mathbf{n} \tag{4.15}$$

is called the **CAUCHY stress tensor**, and the tensor $\mathbf{T}_R$ with the property

$$\mathbf{t}_R = \mathbf{T}_R\mathbf{n}_R \tag{4.16}$$

is called the **First PIOLA–KIRCHOFF stress tensor**. According to eq. (3.20) $\mathbf{T}$ and $\mathbf{T}_R$ are related by

$$\mathbf{T}_R = (\det \mathbf{F})\, \mathbf{T}\mathbf{F}^{T-1}. \tag{4.17}$$

In the context of the general balance laws, $\mathbf{T}$ and $\mathbf{T}_R$ are the *momentum fluxes*, $\rho\mathbf{k}$ and $\rho_R\mathbf{k}$ the *momentum supplies*, while the *momentum production* terms are equal to *zero*, all in the current and reference configurations respectively: In this sense, the *translational momentum is conserved.* In view of the general structure of a balance relation we have the identifications

$$\psi = \rho\mathbf{v}, \qquad \psi_R = \rho_R\mathbf{v}, \tag{4.18}$$

$$\mathbf{\Phi} = \mathbf{T}, \qquad \mathbf{\Phi}_R = \mathbf{T}_R, \tag{4.19}$$

$$\sigma = \rho\mathbf{k}, \qquad \sigma_R = \rho_R\mathbf{k}, \tag{4.20}$$

and the local balance relations for linear momentum read as follows:

$$\rho\left[\frac{\partial \mathbf{v}(\mathbf{x},t)}{\partial t} + (\operatorname{grad}\mathbf{v})\mathbf{v}\right] = \operatorname{div}\mathbf{T} + \rho\mathbf{k} \tag{4.21}$$

$$\rho_R\frac{\partial \mathbf{v}(\mathbf{X},t)}{\partial t} = \operatorname{Div}\mathbf{T}_R + \rho_R\mathbf{k} \tag{4.22}$$

(In the derivation of (4.21) from (3.42) the continuity equation (4.5) has been used.)

In case of a discontinuity surface the jump condition

$$[\![\rho \mathbf{v}(\mathbf{v}-\mathbf{w})\cdot \mathbf{n}_S - \mathbf{T}\mathbf{n}_s]\!] = \mathbf{0} \tag{4.23}$$

holds in addition to the differential equation (4.21) for all $\mathbf{x}$ on $\mathcal{S}(t)$. If the discontinuity surface is material ($\mathbf{v}=\mathbf{w}$), the stress vector is continuous: $[\![\mathbf{T}\mathbf{n}_s]\!]=\mathbf{0}$.

4.3 Moment of momentum

Beside linear momentum the **moment of momentum** or **rotational** or **angular momentum**

$$\mathbf{D}_{\mathbf{x}_0} = \int_V (\mathbf{x}-\mathbf{x}_0)\times(\rho\mathbf{v})\,dV \tag{4.24}$$

is a second quantity, which gives a global characterization of the dynamic state of motion. The balance equation for moment of momentum postulates that this quantity is governed by the resultant torque exerted on the body by surface and volume forces. Both the moment of momentum and the total torque of the exterior forces are calculated with repect to the fixed point $\mathbf{x}_0$. Then the balance equation of angular momentum is

$$\dot{\mathbf{D}}_{\mathbf{x}_0} = \mathbf{M}_{\mathbf{x}_0}, \tag{4.25}$$

or

$$\frac{\mathrm{d}}{\mathrm{d}t}\int_V (\mathbf{x}-\mathbf{x}_0)\times(\rho\mathbf{v})\,dV = \int_A (\mathbf{x}-\mathbf{x}_0)\times(\mathbf{T}\mathbf{n})\,dA + \int_V (\mathbf{x}-\mathbf{x}_0)\times(\rho\mathbf{k})\,dV. \tag{4.26}$$

The balance relations of linear and angular momentum are completely independent from one another. Therefore there is no possibility to derive eq. (4.26) from (4.12) or (4.21) without further assumptions. The independence of these two balance laws is the generalization of the simple and well–known fact that a necessary condition for the mechanical equilibrium of a body is a vanishing resultant force as well as a vanishing resultant torque.

To derive the local form, we identify in this case the following densities:

$$\begin{aligned} \boldsymbol{\psi} &= (\mathbf{x}-\mathbf{x}_0)\times(\rho\mathbf{v}) \\ \boldsymbol{\Phi} &= (\mathbf{x}-\mathbf{x}_0)\times\mathbf{T} \\ \boldsymbol{\sigma} &= (\mathbf{x}-\mathbf{x}_0)\times(\rho\mathbf{k}) \end{aligned} \tag{4.27}$$

In the second equation the cross product of a vector and a tensor of second order has been introduced in the sense

$$\mathbf{a}\times\mathbf{A} = \mathbf{a}\times\left(A^{ik}\mathbf{g}_i\otimes\mathbf{g}_k\right) = A^{ik}\,(\mathbf{a}\times\mathbf{g}_i)\otimes\mathbf{g}_k \tag{4.28}$$

(Summation convention). All these are moment of momentum quantities; there is no production term: Angular momentum is conserved, and eq. (3.42) implies

$$(\mathbf{x}-\mathbf{x}_0)\times(\rho\dot{\mathbf{v}}) = \operatorname{div}\left[(\mathbf{x}-\mathbf{x}_0)\times\mathbf{T}\right] + (\mathbf{x}-\mathbf{x}_0)\times(\rho\mathbf{k}), \tag{4.29}$$

where the continuity equation (4.5) was also applied. The first term on the right–hand side needs careful attention and must be evaluated according to the product rule of differentiation:

$$\begin{aligned}\operatorname{div}[(\mathbf{x}-\mathbf{x}_0)\times\mathbf{T}] &= \left\{\frac{\partial}{\partial x^k}[(\mathbf{x}-\mathbf{x}_0)\times\mathbf{T}]\right\}\mathbf{g}^k = \\ &= \left\{\left[\frac{\partial}{\partial x^k}(\mathbf{x}-\mathbf{x}_0)\right]\times\mathbf{T}+(\mathbf{x}-\mathbf{x}_0)\times\frac{\partial\mathbf{T}}{\partial x^k}\right\}\mathbf{g}^k = \\ &= (\mathbf{g}_k\times\mathbf{T})\,\mathbf{g}^k+(\mathbf{x}-\mathbf{x}_0)\times\operatorname{div}\mathbf{T}\end{aligned}$$

Inserting this into equation (4.29) yields

$$(\mathbf{x}-\mathbf{x}_0)\times[\rho\dot{\mathbf{v}}-\operatorname{div}\mathbf{T}-\rho\mathbf{k}]+\mathbf{g}_k\times\mathbf{T}\mathbf{g}^k=\mathbf{0},$$

or, with the local balance (4.21) of linear momentum,

$$\mathbf{g}_k\times\mathbf{T}\mathbf{g}^k=\mathbf{0}.$$

With $\mathbf{T}\mathbf{g}^k=[\mathbf{g}^i\cdot(\mathbf{T}\mathbf{g}^k)]\mathbf{g}_i=T^{ik}\mathbf{g}_i$ this is equivalent to

$$T^{ik}\mathbf{g}_i\times\mathbf{g}_k=\mathbf{0}. \tag{4.30}$$

In other words, the axial vector of the antisymmetric part of $\mathbf{T}$ vanishes, i.e., the CAUCHY stress tensor must by symmetric:

$$\mathbf{T}=\mathbf{T}^T \tag{4.31}$$

We note that in view of relation (4.17) the First PIOLA–KIRCHHOFF stress tensor is not symmetric; however, we have the identity

$$\mathbf{T}_R\mathbf{F}^T=\mathbf{F}\mathbf{T}_R^T. \tag{4.32}$$

In the context of singular surfaces the general jump condition (3.57) yields

$$\left[\!\left[\Big((\mathbf{x}-\mathbf{x}_0)\times(\rho\mathbf{v})\Big)\otimes(\mathbf{v}-\mathbf{w})-(\mathbf{x}-\mathbf{x}_0)\times\mathbf{T}\right]\!\right]\mathbf{n}_s=\mathbf{0},$$

or

$$\left[\!\left[(\mathbf{x}-\mathbf{x}_0)\times\Big(\rho\mathbf{v}\otimes(\mathbf{v}-\mathbf{w})-\mathbf{T}\Big)\right]\!\right]\mathbf{n}_s=\mathbf{0}.$$

If the motion itself, $\mathbf{x}=\chi_R(\mathbf{X},t)$, is continuous, this implies

$$(\mathbf{x}-\mathbf{x}_0)\times\Big\{[\![\rho\mathbf{v}\otimes(\mathbf{v}-\mathbf{w})-\mathbf{T}]\!]\,\mathbf{n}_s\Big\}=\mathbf{0}, \tag{4.33}$$

and we see that this jump condition is statisfied if the jump condition (4.23) for the linear momentum holds.

4.4 Balance of mechanical energy

As a *consequence* of the balance relations of mass, linear momentum and moment of momentum a further balance relation is derived, namely the **balance of mechanical energy**. In order to derive this balance law we start from the local balance (4.21) of linear momentum, viz.

$$\rho\dot{\mathbf{v}}=\operatorname{div}\mathbf{T}+\rho\mathbf{k}.$$

Scalar multiplication with the velocity vector $\mathbf{v}$ yields

$$\rho\mathbf{v}\cdot\dot{\mathbf{v}} = \mathbf{v}\cdot\operatorname{div}\mathbf{T} + \rho\mathbf{v}\cdot\mathbf{k}. \tag{4.34}$$

With the product rule

$$\operatorname{div}(\mathbf{T}^T\mathbf{v}) = \mathbf{v}\cdot\operatorname{div}\mathbf{T} + \mathbf{T}\cdot\operatorname{grad}\mathbf{v} \tag{4.35}$$

and $\mathbf{v}\cdot\dot{\mathbf{v}} = \left(\frac{\mathbf{v}^2}{2}\right)^{\cdot}$, eq. (4.34) is equivalent to

$$\rho\left(\frac{\mathbf{v}^2}{2}\right)^{\cdot} = \operatorname{div}(\mathbf{T}^T\mathbf{v}) - \mathbf{T}\cdot\operatorname{grad}\mathbf{v} + \rho\mathbf{k}\cdot\mathbf{v}. \tag{4.36}$$

We note that this equation is an identity, valid if the local balance of linear momentum is satisfied.

The following is restricted to differentiable fields: Multiplication of (4.36) by the volume element dV and integration over the whole volume of the body in the current configuration yields

$$\int_V \left[\operatorname{div}(\mathbf{T}^T\mathbf{v}) + \rho\mathbf{k}\cdot\mathbf{v} - \mathbf{T}\cdot\operatorname{grad}\mathbf{v}\right] dV = \int_V \rho\left(\frac{\mathbf{v}^2}{2}\right)^{\cdot} dV.$$

Application of the divergence theorem in the first term on the left–hand side and invoking the continuity equation $(\rho\, dV)^{\cdot} = 0$ on the right–hand side implies

$$\int_A (\mathbf{T}\mathbf{n})\cdot\mathbf{v}\, dA + \int_V \mathbf{k}\cdot\mathbf{v}\rho\, dV = \frac{\mathrm{d}}{\mathrm{d}t}\int_V \frac{1}{2}\mathbf{v}^2\rho\, dV + \int_V \mathbf{T}\cdot\operatorname{grad}\mathbf{v}\, dV. \tag{4.37}$$

In this equation, we identify

(i) the **power of working** of the surface and body forces,

$$P_e = \int_A (\mathbf{T}\mathbf{n})\cdot\mathbf{v}\, dA + \int_V \mathbf{k}\cdot\mathbf{v}\rho\, dV, \tag{4.38}$$

(ii) the **kinetic energy**

$$K = \frac{1}{2}\int_V \mathbf{v}^2\rho\, dV \tag{4.39}$$

and

(iii) the **stress power**

$$P_i = \int_V \mathbf{T}\cdot\operatorname{grad}\mathbf{v}\, dV = \int_V p_i\rho\, dV. \tag{4.40}$$

With these quantities the balance of mechanical energy states that the power of the exterior forces transforms into a change of the kinetic energy plus the stress power or the power of the internal forces:

$$P_e = \dot{K} + P_i \tag{4.41}$$

The *specific stress power per unit mass*,

$$p_i = \frac{1}{\rho}\mathbf{T}\cdot \operatorname{grad}\mathbf{v} = \frac{1}{\rho}\mathbf{T}\cdot\mathbf{L} \tag{4.42}$$

can be written as the inner product of the CAUCHY stress and the strain rate tensor $\mathbf{D}$ which, according to eq. (2.83), is the symmetric part of the spatial velocity gradient $\mathbf{L}$; in fact, this is a consequence of the local balance of moment of momentum or the symmetry of the CAUCHY stress $\mathbf{T}$: The additive decomposition (2.82), $\mathbf{L} = \mathbf{D} + \mathbf{W}$, into a (symmetric) strain rate and a (skew) vorticity tensor implies $\mathbf{T}\cdot\mathbf{L} = \mathbf{T}\cdot(\mathbf{D}+\mathbf{W}) = \mathbf{T}\cdot\mathbf{D}$, since $\mathbf{T}\cdot\mathbf{W} = 0$. Thus the specific stress power is calculated according to

$$p_i = \frac{1}{\rho}\mathbf{T}\cdot\mathbf{D}. \tag{4.43}$$

We note that the stress power can also be expressed in terms of the GREEN strain rate tensor

$$\dot{\mathbf{E}} = \frac{\mathrm{d}}{\mathrm{d}t}\left\{\frac{1}{2}\left(\mathbf{F}^T\mathbf{F} - 1\right)\right\},$$

which is related to $\mathbf{D}$ by eq. (2.99): $\dot{\mathbf{E}} = \mathbf{F}^T\mathbf{D}\mathbf{F}$. Together with the continuity equation (4.9), $\rho \det \mathbf{F} = \rho_R$, we derive

$$p_i = \frac{1}{\rho}\mathbf{T}\cdot\mathbf{D} = \frac{1}{\rho_R}(\det \mathbf{F})\mathbf{T}\cdot\left(\mathbf{F}^{T-1}\dot{\mathbf{E}}\mathbf{F}^{-1}\right) = \frac{1}{\rho_R}(\det \mathbf{F})\left(\mathbf{F}^{-1}\mathbf{T}\mathbf{F}^{T-1}\right)\cdot\dot{\mathbf{E}},$$

or

$$p_i = \frac{1}{\rho}\mathbf{T}\cdot\mathbf{D} = \frac{1}{\rho_R}\tilde{\mathbf{T}}\cdot\dot{\mathbf{E}}. \tag{4.44}$$

Here, a further stress tensor has been introduced, which is most useful in continuum mechanics, namely the **Second PIOLA–KIRCHHOFF stress tensor**

$$\tilde{\mathbf{T}} = (\det \mathbf{F})\mathbf{F}^{-1}\mathbf{T}\mathbf{F}^{T-1}. \tag{4.45}$$

Clearly, $\tilde{\mathbf{T}}$ is symmetric, since $\mathbf{T}$ is symmetric.

The balance of mechanical energy (4.41) fits into the general structure of a balance relation: To recognize this we write it as

$$\dot{K}(t) = P_e - P_i \quad \text{, i.e.,}$$

$$\frac{\mathrm{d}}{\mathrm{d}t}\int_V \frac{\mathbf{v}^2}{2}\rho\, dV = \int_A (\mathbf{T}\mathbf{v})\cdot\mathbf{n}\, dA + \int_V \left(\mathbf{k}\cdot\mathbf{v} - \frac{1}{\rho}\mathbf{T}\cdot\mathbf{D}\right)\rho\, dV \tag{4.46}$$

or

$$\frac{\partial}{\partial t}\int_{V_R} \frac{\mathbf{v}^2}{2}\rho_R\, dV_R = \int_{A_R}\left(\mathbf{T}_R^T\mathbf{v}\right)\cdot\mathbf{n}_R\, dA_R + \int_{V_R}\left(\mathbf{k}\cdot\mathbf{v} - \frac{1}{\rho_R}\tilde{\mathbf{T}}\cdot\dot{\mathbf{E}}\right)\rho_R\, dV_R. \tag{4.47}$$

Then we have the identifications

$$\psi = \frac{\rho}{2}\mathbf{v}^2, \quad \mathbf{\Phi} = \mathbf{T}\mathbf{v}, \quad \sigma = \rho\mathbf{k}\cdot\mathbf{v} - \mathbf{T}\cdot\mathbf{D} \tag{4.48}$$

and

$$\psi_R = \frac{\rho_R}{2}\mathbf{v}^2, \quad \boldsymbol{\Phi}_R = \mathbf{T}_R^T\mathbf{v}, \quad \sigma_R = \rho_R\mathbf{k}\cdot\mathbf{v} - \tilde{\mathbf{T}}\cdot\dot{\mathbf{E}}. \tag{4.49}$$

In particular, σ (or σ_R) is the sum of a supply $\rho_R\mathbf{k}\cdot\mathbf{v}$ ($\rho_R\mathbf{k}\cdot\mathbf{v}$) and a production term $-\mathbf{T}\cdot\mathbf{D}$ $\left(-\tilde{\mathbf{T}}\cdot\dot{\mathbf{E}}\right)$. Thus in contrast to mass, translational momentum and angular momentum, the **mechanical energy is not conserved** in general.

4.5 Conservative systems

A mechanical system is said to be a **conservative system,** if the power of the exterior forces and the stress power can be expressed as the total time derivative of scalar valued potential functions.

In this case we have

$$P_e = -\dot{U}(t), \tag{4.50}$$
$$P_i = \dot{W}(t); \tag{4.51}$$

U is called the **potential energy** of the external forces and W the **strain energy** or the work of the internal forces or stresses. Generally, the potential energy is uniquely determined by the current position of a material body, whereas the strain energy depends on the strain field. Insertion of the special assumption (4.50) and (4.51) into the general balance equation (4.41) yields

$$\begin{aligned} \dot{K}(t) + \dot{U}(t) + \dot{W}(t) &= 0, \\ K(t) + U(t) + W(t) &= \text{konst.}\,. \end{aligned} \tag{4.52}$$

We recognize that the total mechanical energy, i.e. the kinetic plus the potential plus the strain energy is conserved during the motion of the body. This motivates the terminology conservative system.

5. Balance relations of thermodynamics

5.1 Preliminary remarks

It is an elementary experience that the mechanical energy is not conserved in real processes: Most external forces, in particular all kind of friction phenomena, are not conservative in the sense that they cannot be associated with a potential energy. Because internal friction or dissipation occurs, it can equally not be expected that the stress power is fully stored in the material body as strain energy. In general, a conversion of mechanical energy into other non–mechanical forms must take place: The external and internal friction or *dissipation* gives rise to an amount of non–mechanical energy, which may be stored within the material system or be transported outside of it. In the context of *thermomechanics* attention is restricted to *thermal energy* as the only form of non–mechanical energy. In particular, the union of thermal and strain energy, stored in a material body, is called *internal energy*, and the thermal energy, supplied to the body from the exterior world or flowing out of the body, is called *heat supply* and *heat flux*, depending upon whether it is a volume supply or surface flow. Generally, the amount of internal energy is connected with a change of *temperature* of a body, and we know that mechanical and thermal entities, such as forces, motions, temperature changes and energy fluxes are coupled in real processes. As a well-known fact, mechanical energy can be converted into thermal energy without any limitation, but the converse is not true: For the transformation of heat into mechanical energy definite restrictions exist.

The goal of thermodynamics is to make these ideas precise and to develop an appropriate theory which completes the balance relations of continuum mechanics such that a rigorous investigation of coupled mechanical and thermal phenomena becomes possible. To this end the basic dynamic structure by which a material body is endowed is now analysed by two further balance relations for *energy* and *entropy.*

5.2 Conservation of energy – First Law of Thermodynamics

The total energy of a material body is the sum of its thermal and its kinetic part. The **internal energy** is

$$E = \int_V e\rho \, dV, \tag{5.1}$$

where the density e is called the specific internal energy per unit mass. The kinetic energy is

$$K = \frac{1}{2}\int_V \mathbf{v}^2 \rho \, dV. \tag{5.2}$$

The balance relation for the total energy of a material body is called the **First Law of thermodynamics**. It postulates that the rate of change of the total energy equals the total energy supply from the exterior world; in this context the mechanical energy supply is the power of the external forces and the thermal energy supply is the heat flux plus the radiation. Thus the First Law takes the form

$$\dot{E} + \dot{K} = P_e + Q, \tag{5.3}$$

i.e.,

$$\frac{\mathrm{d}}{\mathrm{d}t}\int_V \left(e + \frac{\mathbf{v}^2}{2}\right)\rho \, dV = \int_A \mathbf{t}\cdot\mathbf{v}\, dA + \int_V \mathbf{k}\cdot\mathbf{v}\rho \, dV + \int_A q \, dA + \int_V r\rho \, dV, \tag{5.4}$$

or

$$\frac{\partial}{\partial t}\int_{V_R} \left(e + \frac{\mathbf{v}^2}{2}\right)\rho_R \, dV_R = \int_{A_R} \mathbf{t}_R\cdot\mathbf{v}\, dA_R + \int_{V_R} \mathbf{k}_R\cdot\mathbf{v}\rho_R \, dV_R + \int_{A_R} q_R \, dA_R + \int_{V_R} r\rho_R \, dV_R. \tag{5.5}$$

The surface densities q and q_R of the thermal energy supply are called **heat flow**, and the volume density r is called **radiation**. In view of the structure of this balance relation we identify

$$\psi = \rho\left(e + \frac{\mathbf{v}^2}{2}\right) \quad , \quad \psi_R = \rho_R\left(e + \frac{\mathbf{v}^2}{2}\right) , \tag{5.6}$$

$$\phi = \mathbf{t}\cdot\mathbf{v} + q \quad , \quad \phi_R = \mathbf{t}_R\cdot\mathbf{v} + q_R , \tag{5.7}$$

$$\mathbf{\Phi} = \mathbf{T}\mathbf{v} - \mathbf{q} \quad , \quad \mathbf{\Phi}_R = \mathbf{T}_R^T\mathbf{v} - \mathbf{q}_R , \tag{5.8}$$

$$\sigma = \rho(\mathbf{k}\cdot\mathbf{v} + r) \quad , \quad \sigma_R = \rho_R(\mathbf{k}\cdot\mathbf{v} + r) . \tag{5.9}$$

Because there are no production terms, the First Law of Thermodynamics states the conservation of (mechanical and thermal) energy. The vector field $\mathbf{q}$ is called the heat flux vector; it is related to the heat flow by

$$q = -\mathbf{q}\cdot\mathbf{n}. \tag{5.10}$$

Obviously, $\mathbf{q}$ is the heat flux vector with respect to the surface element in the current configuration. The heat flux $\mathbf{q}_R$ with respect to the surface in the reference configuration is

$$q_R = -\mathbf{q}_R\cdot\mathbf{n}_R. \tag{5.11}$$

In the sense of eq. (3.19) the two heat flux vectors are related by

$$\mathbf{q}_R = (\det\mathbf{F})\mathbf{F}^{-1}\mathbf{q}. \tag{5.12}$$

Application of the general local form (3.42) together with the continuity equation (4.3) leads to the following local form of the First Law:

$$\rho\left(e + \frac{\mathbf{v}^2}{2}\right)^{\cdot} = \operatorname{div}(\mathbf{T}\mathbf{v} - \mathbf{q}) + \rho(\mathbf{k}\cdot\mathbf{v} + r). \tag{5.13}$$

In the context of the material description we have the equivalent formulation

$$\rho_R\frac{\partial}{\partial t}\left(e + \frac{\mathbf{v}^2}{2}\right) = \operatorname{Div}(\mathbf{T}_R^T\mathbf{v} - \mathbf{q}_R) + \rho_R(\mathbf{k}\cdot\mathbf{v} + r). \tag{5.14}$$

The kinetic energy can be eliminated from these relations, if the balance of mechanical power is applied, which holds for all thermomechanical processes. Indeed, combination of (4.41), $P_e = \dot{K} + P_i$, with (5.3) yields

$$\dot{E} = P_i + Q, \tag{5.15}$$

i.e.,

$$\frac{\mathrm{d}}{\mathrm{d}t}\int_V e\rho\,dV = -\int_A \mathbf{q}\cdot\mathbf{n}\,dA + \int_V\left(\frac{1}{\rho}\mathbf{T}\cdot\mathbf{D} + r\right)\rho\,dV, \tag{5.16}$$

or

$$\frac{\partial}{\partial t}\int_{V_R} e\rho_R\,dV_R = -\int_{A_R}\mathbf{q}_R\cdot\mathbf{n}_R\,dA_R + \int_{V_R}\left(\frac{1}{\rho_R}\tilde{\mathbf{T}}\cdot\dot{\mathbf{E}} + r\right)\rho_R\,dV_R. \tag{5.17}$$

These relations imply the local forms

$$\dot{e} = \frac{1}{\rho}\mathbf{T}\cdot\mathbf{D} - \frac{1}{\rho}\operatorname{div}\mathbf{q} + r \tag{5.18}$$

and

$$\dot{e} = \frac{1}{\rho_R}\tilde{\mathbf{T}}\cdot\dot{\mathbf{E}} - \frac{1}{\rho_R}\operatorname{Div}\mathbf{q}_R + r. \tag{5.19}$$

Moreover, inserting eqs. (5.6) to (5.9) into eq. (3.58) we arrive at the energy jump condition

$$\left[\!\left[\rho\left(e + \frac{\mathbf{v}^2}{2}\right)(\mathbf{v} - \mathbf{w}) - \mathbf{T}\mathbf{v} + \mathbf{q}\right]\!\right] \cdot \mathbf{n}_s = 0. \tag{5.20}$$

For material surfaces ($\mathbf{v} = \mathbf{w}$) the jump condition reduces to $[\![\mathbf{q} - \mathbf{T}\mathbf{v}]\!] \cdot \mathbf{n}_s = 0$. Since the stress vector is continuous in this case (see Section 4.2), this implies $[\![\mathbf{q}\cdot\mathbf{n}_s]\!] = 0$ if $\mathbf{v}$ is also continuous.

5.3 Balance of Entropy – Second Law of Thermodynamics

With the **Second Law of thermodynamics** a general principle of dissipation or irreversibilty is postulated. In contrast to the other balance relations there is up to now no formulation of such a principle, which is generally accepted. The particular version of the dissipation principle, adopted in this monograph, employs the **concepts of entropy** and **temperature** as primitive. In fact, these concepts can be motivated from equilibirium thermodynamics, however, they cannot be proved. This dissipation principle is known as the **CLAUSIUS–DUHEM–inequality**. The balance relation of entropy postulates that the rate of change of the total entropy of a material body equals the entropy supply plus the entropy flux plus the entropy production, viz.,

$$\frac{\mathrm{d}}{\mathrm{d}t}\int_V s\rho\, dV = -\int_A \mathbf{\Sigma}\cdot\mathbf{n}\, dA + \int_V (\delta + \gamma)\rho\, dV. \tag{5.21}$$

Here, s is the specific entropy per unit mass, $\mathbf{\Sigma}$ the entropy flux vector, δ the entropy supply per unit mass and γ the entropy production. As the **Second Law of Thermodynamics** we assume

$$\gamma \geq 0 \tag{5.22}$$

for all admissible thermomechanical processes. In this context, an admissible thermomechanical process is any solution of the balance equations of linear momentum and energy, which is compatible with the constitutive equations.

In rational thermodynamics as well as in classical thermodynamics the entropy supply is taken to be the heat supply divided by the **absolute** (non–negative) **temperature** θ, i.e.,

$$\mathbf{\Sigma} = \frac{\mathbf{q}}{\theta} \tag{5.23}$$

and

$$\delta = \frac{r}{\theta}. \tag{5.24}$$

In equilibrium thermodynamics, it can be proven that the entropy supply has this structure, but within a more general thermodynamic theory eqs. (5.23) and (5.24) are assumptions, which may fail to hold in particular cases. With (5.23), (5.24) and (5.22) the dissipation principle is given by the CLAUSIUS–DUHEM–inequality,

$$\frac{\mathrm{d}}{\mathrm{d}t}\int_V s\rho\, dV \geq -\int_A \frac{\mathbf{q}\cdot\mathbf{n}}{\theta}\, dA + \int_V \rho\frac{r}{\theta}\, dV, \tag{5.25}$$

or

$$\frac{\partial}{\partial t}\int_{V_R} s\rho_R\, dV_R \geq -\int_{A_R} \frac{\mathbf{q}_R\cdot\mathbf{n}_R}{\theta}\, dA_R + \int_{V_R} \rho_R\frac{r}{\theta}\, dV_R. \tag{5.26}$$

A corresponding local form is

$$\rho\dot{s} \geq -\operatorname{div}\left(\frac{\mathbf{q}}{\theta}\right) + \rho\frac{r}{\theta}. \tag{5.27}$$

Equivalent formulations are

$$\theta\dot{s} \geq -\frac{1}{\rho}\operatorname{div}\mathbf{q} + r + \frac{1}{\rho\theta}\mathbf{q}\cdot\operatorname{grad}\theta \tag{5.28}$$

and

$$\theta\dot{s} \geq -\frac{1}{\rho_R}\operatorname{Div}\mathbf{q}_R + r + \frac{1}{\rho_R\theta}\mathbf{q}_R\cdot\operatorname{Grad}\theta. \tag{5.29}$$

The local heat supply on the right–hand side of these inequalities can be eliminated by means of the the First Law as formulated in eqs. (5.18) and (5.19). This yields two other formulations of the dissipation inequality, namely

$$\theta\dot{s} - \dot{e} + \frac{1}{\rho}\mathbf{T}\cdot\mathbf{D} - \frac{1}{\rho\theta}\mathbf{q}\cdot\operatorname{grad}\theta \geq 0 \tag{5.30}$$

and

$$\theta\dot{s} - \dot{e} + \frac{1}{\rho_R}\tilde{\mathbf{T}}\cdot\dot{\mathbf{E}} - \frac{1}{\rho_R\theta}\mathbf{q}_R\cdot\operatorname{Grad}\theta \geq 0. \tag{5.31}$$

In case of discontinuous fields we have the jump condition

$$\left[\!\left[\rho s(\mathbf{v}-\mathbf{w}) + \frac{\mathbf{q}}{\theta}\right]\!\right]\cdot\mathbf{n}_s + \gamma_s = 0, \tag{5.32}$$

with

$$\gamma_s \geq 0 \tag{5.33}$$

according to the dissipation principle: A non–negative entropy production must be incorporated on discontinuity surfaces.

6. Summary: Balance relations for one–component systems

In this section, the local forms of thermomechanical balance laws derived so far are briefly summarized under two aspects. For shortness, only the spatial representation is considered below.

Mass:
$$\dot{\rho} + \rho \operatorname{div} \mathbf{v} = 0 \tag{6.1}$$

Linear momentum:
$$\rho \dot{\mathbf{v}} = \operatorname{div} \mathbf{T} + \rho \mathbf{k} \tag{6.2}$$

Moment of momentum:
$$\mathbf{T} = \mathbf{T}^T \tag{6.3}$$

Energy:
$$\dot{e} = \frac{1}{\rho} \mathbf{T} \cdot \mathbf{D} - \frac{1}{\rho} \operatorname{div} \mathbf{q} + r \tag{6.4}$$

Entropy inequality:
$$\dot{s} \geq -\frac{1}{\rho} \operatorname{div} \left(\frac{\mathbf{q}}{\theta} \right) + \frac{r}{\theta} \tag{6.5}$$

or
$$\theta \dot{s} - \dot{e} + \frac{1}{\rho} \mathbf{T} \cdot \mathbf{D} - \frac{1}{\rho\theta} \mathbf{q} \cdot \operatorname{grad} \theta \geq 0 \tag{6.6}$$

A more general form of the entropy inequality (6.5) is given by
$$\dot{s} \geq -\frac{1}{\rho} \operatorname{div} \mathbf{\Sigma} + \delta. \tag{6.7}$$

The balance relations (6.1) to (6.5) concern the material body as a **closed system**, where no matter is transported across the boundary surface. They were derived under the basic assumption that each material point consists of only a single material species; in other words, a one–component system was presumed. In case of a mixture more than one component may occur: Each point of the material body consists of different subsystems which can exchange mass, momentum, energy and entropy. Therefore general balance relation for multicomponent systems (mixtures) are needed.

In order to motivate balance relations for mixtures, which will be formulated in the next section, we return to the general form of the balance relation (3.21) in connection with the REYNOLDS' transport theorem (3.33):
$$\int_V \frac{\partial \psi}{\partial t} \, dV + \int_A \psi (\mathbf{v} \cdot \mathbf{n}) \, dA = \int_A \mathbf{\Phi} \cdot \mathbf{n} \, dA + \int_V \sigma dV \tag{6.8}$$

This balance relation can be interpreted in terms of an **open system**: We identify the present configuration of a material system with a fixed volume; then, the action of the exterior world on the system is not only given by the fields $\mathbf{\Phi}$ and σ, but also by a flow of matter through the boundary surface; this flow contributes significantly to the flux $\mathbf{\Phi}$,
$$\int_V \frac{\partial \psi}{\partial t} \, dV = \int_A (\mathbf{\Phi} - \psi \mathbf{v}) \cdot \mathbf{n} \, dA + \int_V \sigma \, dV, \tag{6.9}$$

which corresponds to the local form

$$\frac{\partial \psi}{\partial t} + \operatorname{div}(\psi \mathbf{v} - \mathbf{\Phi}) - \sigma = 0. \tag{6.10}$$

In case of a vector valued balance relation similar results are obtained, with the only difference being that the product rule

$$\operatorname{div}(\boldsymbol{\psi} \otimes \mathbf{v}) = (\operatorname{grad} \boldsymbol{\psi})\mathbf{v} + \boldsymbol{\psi} \operatorname{div} \mathbf{v}$$

must be used, which leads to

$$\int_V \frac{\partial \boldsymbol{\psi}}{\partial t} dV = \int_A (\mathbf{\Phi} - \boldsymbol{\psi} \otimes \mathbf{v})\mathbf{n}\, dA + \int_V \boldsymbol{\sigma}\, dV, \tag{6.11}$$

corresponding to the local form

$$\frac{\partial \boldsymbol{\psi}}{\partial t} + \operatorname{div}(\boldsymbol{\psi} \otimes \mathbf{v} - \mathbf{\Phi}) - \boldsymbol{\sigma} = \mathbf{0}. \tag{6.12}$$

The main purpose to consider eqs. (6.10), (6.12) once more is the interpretation of these equations in the context of the transport theorem: An arbitrary material body or even a single material point may be regarded as an open system that is capable to exchange with neighboring systems the physical quantities, represented by the density fields $\psi(\mathbf{x}, t)$ or $\boldsymbol{\psi}(\mathbf{x}, t)$. These considerations are now applied to the reformulation of the local forms of the balance relations of thermomechanics. The basic structure of the balance laws is now characterized by a *local derivative* of the density and a *divergence term* to which the exchange with neighboring systems is added. In this sense we have the following list of local forms, which of course are equivalent expressions for the relations eqs. (6.1) to (6.5).

Mass:
$$\frac{\partial \rho}{\partial t} + \operatorname{div}(\rho \mathbf{v}) = 0, \tag{6.13}$$

Linear momentum:
$$\frac{\partial}{\partial t}(\rho \mathbf{v}) + \operatorname{div}[\rho \mathbf{v} \otimes \mathbf{v} - \mathbf{T}] - \rho \mathbf{k} = \mathbf{0}, \tag{6.14}$$

Moment of momentum:
$$\mathbf{T} - \mathbf{T}^T = \mathbf{0}, \tag{6.15}$$

Energy:
$$\frac{\partial}{\partial t}\left(\rho e + \frac{\rho}{2}\mathbf{v}^2\right) + \operatorname{div}\left[\left(\rho e + \frac{\rho}{2}\mathbf{v}^2\right)\mathbf{v} - \mathbf{T}\mathbf{v} + \mathbf{q}\right] - \rho(\mathbf{k} \cdot \mathbf{v} + r) = 0, \tag{6.16}$$

Entropy inequality:
$$\frac{\partial}{\partial t}(\rho s) + \operatorname{div}\left(\rho s \mathbf{v} + \frac{\mathbf{q}}{\theta}\right) - \rho \frac{r}{\theta} = \gamma, \tag{6.17}$$

$$\gamma \geq 0. \tag{6.18}$$

In ensuing developments these interpretations will be applied to the balance relations of multi–component systems.

7. Multi–component systems (mixtures)

7.1 Definition of a mixture

A **multi–component material body** or a so–called **mixture** consists of a finite number of components. Each particular **component** or **constituent** is characterized by a **partial density** field $\rho_\alpha(\mathbf{x}, t)$ ($\alpha = 1, ..., n$). ρ_α is not the material density of the constituent α: It is the density of the constituent α in the mixture. (Of course, when no other components are present, it is the true density.) If ρ denotes the mass density of the mixture as a whole, the **concentrations** $c_\alpha(\mathbf{x}, t)$ by mass are defined as

$$c_\alpha = \frac{\rho_\alpha}{\rho} \qquad (\alpha = 1, ...n). \tag{7.1}$$

The conservation of mass, which is quite a general principle, implies that the sum of the partial densities equals the total density, i.e.,

$$\sum_\alpha \rho_\alpha = \rho, \tag{7.2}$$

or

$$\sum_\alpha c_\alpha = 1. \tag{7.3}$$

The partial densities ρ_α themselves or the concentrations c_α may change according to possible interactions between the components, such as chemical reactions, phase transitions or diffusion processes by molecular migration. Further interactions can exist in terms of local interaction forces (exchange of linear momentum) and local interaction torques (exchange of moment of momentum). Finally, local energy and entropy productions are possible.

A representation of these ideas by means of a convenient field theory requires a certain modification of the basic concepts of classical continuum mechanics: Principally, a generic material point of a mixture contains all n components simultaneously; the kinematics of each particular component is characterized by its own constituent velocity field $\mathbf{v}_\alpha(\mathbf{x}, t)$ ($\alpha = 1, ...n$), and by its partial CAUCHY stress $\mathbf{T}_\alpha(\mathbf{x}, t)$ ($\alpha = 1, ...n$). Moreover, the existence of separate energy– and entropy–densities $e_\alpha(\mathbf{x}, t)$ and $s_\alpha(\mathbf{x}, t)$ are postulated for each component ($\alpha = 1, ...n$), as well as the heat flux vectors $\mathbf{q}_\alpha$, volume force densities $\mathbf{k}_\alpha$ and partial radiations r_α. In view of the general structure of a balance relation these additional fields should be accompanied by associated production terms, which represent the material, the dynamic, or the energetic interaction of the individual components; these components are understood as material elements, which are open systems on a local scale.

7.2 Balance relations

Because each component of a mixture is regarded as an open system, the local balance relations (6.13) to (6.17) are postulated for each individual constituent; however, in this case their production densities must now differ from zero, since interactions between the individual constituents must be incorporated. Thus for $\alpha = 1, \ldots n$ we may postulate the following set of equations:

$$\frac{\partial \rho_\alpha}{\partial t} + \operatorname{div}(\rho_\alpha \mathbf{v}_\alpha) = C_\alpha \tag{7.4}$$

$$\frac{\partial}{\partial t}(\rho_\alpha \mathbf{v}_\alpha) + \operatorname{div}\left[\rho_\alpha \mathbf{v}_\alpha \otimes \mathbf{v}_\alpha - \mathbf{T}_\alpha\right] - \rho_\alpha \mathbf{k}_\alpha = \mathbf{m}_\alpha \tag{7.5}$$

$$\mathbf{T}_\alpha - \mathbf{T}_\alpha^T = \mathbf{M}_\alpha \tag{7.6}$$

$$\frac{\partial}{\partial t}\left(\rho_\alpha e_\alpha + \frac{\rho_\alpha}{2}\mathbf{v}_\alpha^2\right) + \operatorname{div}\left[\left(\rho_\alpha e_\alpha + \frac{\rho_\alpha}{2}\mathbf{v}_\alpha^2\right)\mathbf{v}_\alpha - \mathbf{T}_\alpha^T \mathbf{v}_\alpha + \mathbf{q}_\alpha\right] - \\ - \rho_\alpha(\mathbf{k}_\alpha \cdot \mathbf{v}_\alpha + r_\alpha) = p_\alpha \tag{7.7}$$

$$\frac{\partial}{\partial t}(\rho_\alpha s_\alpha) + \operatorname{div}\left(\rho_\alpha s_\alpha \mathbf{v}_\alpha + \frac{\mathbf{q}_\alpha}{\theta}\right) - \frac{\rho_\alpha r_\alpha}{\theta} = \gamma_\alpha \tag{7.8}$$

It is natural to expect that the mixture as a whole acts like a one–component single body. Therefore the following restrictive conditions are postulated as **conservation principles**:

1. *Conservation of mass*

$$\sum_\alpha C_\alpha = 0 \tag{7.9}$$

2. *Conservation of linear momentum*

$$\sum_\alpha \mathbf{m}_\alpha = \mathbf{0} \tag{7.10}$$

3. *Conservation of moment of momentum*

$$\sum_\alpha \mathbf{M}_\alpha = \mathbf{0} \tag{7.11}$$

4. *Conservation of energy*

$$\sum_\alpha p_\alpha = 0 \tag{7.12}$$

5. *Dissipation principle*

$$\sum_\alpha \gamma_\alpha \geq 0 \tag{7.13}$$

7.3 Inferences from the balance relations

It is obvious that in the balance relations (7.4) to (7.8) and in the conservation (and production) principles (7.9) to (7.13) the definitions of the partial fields ρ_α, $\mathbf{v}_\alpha$, $\mathbf{T}_\alpha$, e_α and $\mathbf{q}_\alpha$ must be related to the macroscopic fields ρ, $\mathbf{v}$, $\mathbf{T}$, e, $\mathbf{q}$, such that no contradiction with the balance relations (6.13) to (6.18) arises: In particular, it must be possible to identify the mixture as a whole with a one–component system and thus to recover the corresponding balance relations (6.13) to (6.18), which must hold for all material systems. In the following paragraphs we introduce these additional relations such that the local balance laws are satisfied.

Balance of mass

We start from the definition (7.2) of the total density $\rho = \sum_\alpha \rho_\alpha$ and define the **barycentric velocity**

$$\mathbf{v} = \frac{1}{\rho}\sum_\alpha \rho_\alpha \mathbf{v}_\alpha = \sum_\alpha c_\alpha \mathbf{v}_\alpha \tag{7.14}$$

and the **diffusion velocities**

$$\mathbf{u}_\alpha = \mathbf{v}_\alpha - \mathbf{v} \quad (\alpha = 1, ...n). \tag{7.15}$$

By definition of $\mathbf{u}_\alpha$ we have

$$\sum_\alpha \rho_\alpha \mathbf{u}_\alpha = \mathbf{0}. \tag{7.16}$$

Addition of the individual continuity equations (7.4) together with the conservation assumption (7.9) yields

$$\sum_\alpha \left\{\frac{\partial \rho_\alpha}{\partial t} + \text{div}\,(\rho_\alpha \mathbf{v}_\alpha)\right\} = \sum_\alpha C_\alpha = 0,$$

from which we infer

$$\frac{\partial \rho}{\partial t} + \text{div}\,(\rho \mathbf{v}) = 0, \tag{7.17}$$

applying eqs. (7.2) and (7.14). Thus the definitions of the mixture density ρ and the barycentric velocity $\mathbf{v}$ are compatible with the continuity equation (6.13).

It is convenient to modify the balance relations (7.4) for the partial densities as follows: With eq. (7.1) we may deduce

$$\frac{\partial \rho_\alpha}{\partial t} = \frac{\partial}{\partial t}(\rho c_\alpha) = \rho\frac{\partial c_\alpha}{\partial t} + c_\alpha \frac{\partial \rho}{\partial t}. \tag{7.18}$$

Introducing the **barycentric derivative,**

$$\dot{c}_\alpha = \frac{\partial c_\alpha}{\partial t} + (\,\text{grad}\, c_\alpha) \cdot \mathbf{v}, \tag{7.19}$$

and applying the continuity equation (7.17), we obtain

$$\frac{\partial \rho_\alpha}{\partial t} = \rho\,[\dot{c}_\alpha - (\,\text{grad}\, c_\alpha)\cdot \mathbf{v}] - c_\alpha\,\text{div}\,(\rho \mathbf{v}) = \rho \dot{c}_\alpha - \,\text{div}\,(\rho c_\alpha \mathbf{v}).$$

Therefore, eqs. (7.4), $\frac{\partial \rho_\alpha}{\partial t} + \text{div}\,(\rho c_\alpha \mathbf{v}_\alpha) = C_\alpha \quad (\alpha = 1, ..., n)$ are equivalent to

$$\rho \dot{c}_\alpha + \,\text{div}\,[\rho c_\alpha(\mathbf{v}_\alpha - \mathbf{v})] = C_\alpha. \tag{7.20}$$

With the definition of the **diffusive mass flux** of component α,

$$\mathbf{j}_\alpha = \rho c_\alpha(\mathbf{v}_\alpha - \mathbf{v}) = \rho_\alpha \mathbf{u}_\alpha, \tag{7.21}$$

the balance relations for the partial densities ρ_α are transformed into evolution equation for the concentrations $c_\alpha(\mathbf{x}, t)$:

$$\rho \dot{c}_\alpha + \operatorname{div} \mathbf{j}_\alpha = C_\alpha \tag{7.22}$$

These will be used in later chapters.

Balance of linear momentum

Given the diffusion velocities (7.15), we define the total or the mixture stress tensor as the sum of partial stress tensors minus the diffusion stresses:

$$\mathbf{T} = \sum_\alpha (\mathbf{T}_\alpha - \rho_\alpha \mathbf{u}_\alpha \otimes \mathbf{u}_\alpha) \tag{7.23}$$

Moreover, the total volume force is defined as the sum of the partial volume forces:

$$\rho \mathbf{k} = \sum_\alpha \rho_\alpha \mathbf{k}_\alpha \tag{7.24}$$

These two definitions suffice to guarantee the validity of the momentum balance (6.14) for the mixture as whole; to verify this, we sum the balance relations (7.5) over all components. Then, on using eqs. (7.14), (7.24) and (7.23) we obtain

$$\begin{aligned}
\frac{\partial}{\partial t}(\rho \mathbf{v}) - \rho \mathbf{k} &= \operatorname{div} \left\{ \sum_\alpha [\mathbf{T}_\alpha - \rho_\alpha \mathbf{v}_\alpha \otimes \mathbf{v}_\alpha] \right\} = \\
&= \operatorname{div} \left\{ \mathbf{T} + \sum_\alpha \rho_\alpha (\mathbf{u}_\alpha \otimes \mathbf{u}_\alpha - \mathbf{v}_\alpha \otimes \mathbf{v}_\alpha) \right\}.
\end{aligned}$$

Here, the last sum reduces to

$$\begin{aligned}
\sum_\alpha \rho_\alpha (\mathbf{u}_\alpha \otimes \mathbf{u}_\alpha - \mathbf{v}_\alpha \otimes \mathbf{v}_\alpha) &= \sum_\alpha \rho_\alpha [(\mathbf{v}_\alpha - \mathbf{v}) \otimes (\mathbf{v}_\alpha - \mathbf{v}) - \mathbf{v}_\alpha \otimes \mathbf{v}_\alpha] \\
&= \sum_\alpha \rho_\alpha [\mathbf{v} \otimes \mathbf{v} - \mathbf{v}_\alpha \otimes \mathbf{v} - \mathbf{v} \otimes \mathbf{v}_\alpha] = -\rho \mathbf{v} \otimes \mathbf{v},
\end{aligned}$$

where eqs. (7.2) and (7.14) have been applied. Thus (6.14) is verified with the barycentric velocity $\mathbf{v}$ and the definitions (7.23) and (7.24).

Balance of moment of momentum

According to eqs. (7.6) the partial stress tensors $\mathbf{T}_\alpha$ can be non–symmetric, however the total stress $\mathbf{T}$ as defined by (7.23) is symmetric if (7.11) is observed.

Balance of energy

In order to be compatible with the total energy balance (6.16) we define from the partial quantities ρ_α, e_α, $\mathbf{v}_\alpha$, $\mathbf{u}_\alpha$, $\mathbf{T}_\alpha$ and $\mathbf{q}_\alpha$ the **total internal energy**

$$\rho e = \sum_\alpha \rho_\alpha \left(e_\alpha + \frac{1}{2}\mathbf{u}_\alpha^2\right), \tag{7.25}$$

the **total** or **mixture heat flux vector**

$$\mathbf{q} = \sum_\alpha \left\{\mathbf{q}_\alpha + \rho_\alpha \left(e_\alpha + \frac{1}{2}\mathbf{u}_\alpha^2\right)\mathbf{u}_\alpha - \mathbf{T}_\alpha^T \mathbf{u}_\alpha\right\}, \tag{7.26}$$

and the **overall heat supply**

$$\rho r = \sum_\alpha \rho_\alpha (r_\alpha + \mathbf{k}_\alpha \cdot \mathbf{u}_\alpha). \tag{7.27}$$

We then must verify that the total energy balance (6.16) is satisfied. This is done in three steps:

1. With (7.2), (7.15) and (7.16) the definition (7.25) yields

$$\begin{aligned}
\sum_\alpha \rho_\alpha e_\alpha &= \rho e - \sum_\alpha \frac{\rho_\alpha}{2}\mathbf{u}_\alpha^2 = \\
&= \rho e - \sum_\alpha \frac{\rho_\alpha}{2}\left(\mathbf{v}_\alpha^2 - 2\mathbf{v}_\alpha \cdot \mathbf{v} + \mathbf{v}^2\right) = \\
&= \rho\left(e + \frac{\mathbf{v}^2}{2}\right) - \sum_\alpha \frac{\rho_\alpha}{2}\mathbf{v}_\alpha^2,
\end{aligned}$$

i.e.,

$$\sum_\alpha \rho_\alpha \left(e_\alpha + \frac{\mathbf{v}_\alpha^2}{2}\right) = \rho\left(e + \frac{\mathbf{v}^2}{2}\right). \tag{7.28}$$

2. Next, we evaluate the vector

$$\tilde{\mathbf{q}} = \sum_\alpha \left\{\rho_\alpha \left(e_\alpha + \frac{1}{2}\mathbf{v}_\alpha^2\right)\mathbf{v}_\alpha - \mathbf{T}_\alpha^T \mathbf{v}_\alpha + \mathbf{q}_\alpha\right\}. \tag{7.29}$$

Application of (7.15) and (7.28) yields

$$\begin{aligned}
\tilde{\mathbf{q}} &= \sum_\alpha \left\{\rho_\alpha \left(e_\alpha + \tfrac{1}{2}\mathbf{v}_\alpha^2\right)(\mathbf{v} + \mathbf{u}_\alpha) - \mathbf{T}_\alpha^T(\mathbf{v} + \mathbf{u}_\alpha) + \mathbf{q}_\alpha\right\} \\
&= \rho\left(e + \frac{1}{2}\mathbf{v}^2\right)\mathbf{v} + \sum_\alpha \left\{\rho_\alpha \left[e_\alpha + \tfrac{1}{2}(\mathbf{v} + \mathbf{u}_\alpha)^2\right]\mathbf{u}_\alpha - \mathbf{T}_\alpha^T \mathbf{v}\right\} + \sum_\alpha \left\{\mathbf{q}_\alpha - \mathbf{T}_\alpha^T \mathbf{u}_\alpha\right\}
\end{aligned}$$

$$
\begin{aligned}
&= \rho\left(e+\frac{1}{2}\mathbf{v}^2\right)\mathbf{v} + \sum_\alpha\left\{\rho_\alpha\left[\tfrac{1}{2}\mathbf{v}^2+(\mathbf{v}\cdot\mathbf{u}_\alpha)\right]\mathbf{u}_\alpha - \mathbf{T}_\alpha^T\mathbf{v}\right\}+\\
&\quad + \sum_\alpha\left\{\mathbf{q}_\alpha+\rho_\alpha\left(e_\alpha+\tfrac{1}{2}\mathbf{u}_\alpha^2\right)\mathbf{u}_\alpha - \mathbf{T}_\alpha^T\mathbf{u}_\alpha\right\},\\
\tilde{\mathbf{q}} &= \rho\left(e+\frac{1}{2}\mathbf{v}^2\right)\mathbf{v} + \left\{\sum_\alpha\left(\rho_\alpha\,\mathbf{u}_\alpha\otimes\mathbf{u}_\alpha - \mathbf{T}_\alpha^T\right)\right\}\mathbf{v}+\\
&\quad + \sum_\alpha\left\{\mathbf{q}_\alpha+\rho_\alpha\left(e_\alpha+\tfrac{1}{2}\mathbf{u}_\alpha^2\right)\mathbf{u}_\alpha - \mathbf{T}_\alpha^T\mathbf{u}_\alpha\right\}.
\end{aligned}
$$

By comparison with the definitions (7.23) and (7.26) we obtain

$$
\tilde{\mathbf{q}} = \rho\left(e+\frac{1}{2}\mathbf{v}^2\right)\mathbf{v} - \mathbf{T}^T\mathbf{v} + \mathbf{q}. \tag{7.30}
$$

3. Finally, we deduce from (7.15) and (7.24)

$$
\sum_\alpha \rho_\alpha(\mathbf{k}_\alpha\cdot\mathbf{v}_\alpha + r_\alpha) = \sum_\alpha \rho_\alpha[\mathbf{k}_\alpha\cdot(\mathbf{v}+\mathbf{u}_\alpha)+r_\alpha] = \rho\mathbf{k}\cdot\mathbf{v} + \sum_\alpha \rho_\alpha(\mathbf{k}_\alpha\cdot\mathbf{u}_\alpha + r_\alpha),
$$

which, with the definition (7.27), implies

$$
\sum_\alpha \rho_\alpha(\mathbf{k}_\alpha\cdot\mathbf{v}_\alpha + r_\alpha) = \rho\mathbf{k}\cdot\mathbf{v} + r. \tag{7.31}
$$

Now, adding all relations (7.7) and taking in account the balance equation (7.12) and eqs. (7.28) – (7.31) leads to (6.16) for the total internal energy.

Balance of entropy

From the partial entropies s_α we define the total entropy

$$
\rho s = \sum_\alpha \rho_\alpha s_\alpha. \tag{7.32}
$$

Again, we might expect a balance relation for the total entropy of the form (6.17). Addition of the partial entropy relations (7.8) and application of (7.15) and (7.32) yields the entropy balance of the mixture as follows:

$$
\begin{aligned}
\gamma &= \sum_\alpha \gamma_\alpha = \sum_\alpha\left\{\frac{\partial}{\partial t}(\rho_\alpha s_\alpha) + \operatorname{div}\left[\rho_\alpha s_\alpha(\mathbf{v}+\mathbf{u}_\alpha)+\frac{\mathbf{q}_\alpha}{\theta}\right] - \frac{\rho_\alpha r_\alpha}{\theta}\right\}\\
&= \frac{\partial}{\partial t}(\rho s) + \operatorname{div}(\rho s\mathbf{v}) + \sum_\alpha\left\{\operatorname{div}\left[\rho_\alpha s_\alpha\mathbf{u}_\alpha+\frac{\mathbf{q}_\alpha}{\theta}\right] - \frac{\rho_\alpha r_\alpha}{\theta}\right\}
\end{aligned}
$$

Here, we can introduce the total heat flux and heat supply by means of definitions (7.26) and (7.27), and then obtain

$$
\gamma = \frac{\partial}{\partial t}(\rho s) + \operatorname{div}(\rho s\mathbf{v}) +
$$

$$+\operatorname{div}\left\{\frac{\mathbf{q}}{\theta}+\frac{1}{\theta}\sum_\alpha\left[(\rho_\alpha s_\alpha\theta)\mathbf{u}_\alpha-\rho_\alpha\left(e_\alpha+\frac{1}{2}\mathbf{u}_\alpha^2\right)\mathbf{u}_\alpha+\mathbf{T}_\alpha^T\mathbf{u}_\alpha\right]\right\}-$$

$$-\frac{1}{\theta}\left(\rho r-\sum_\alpha\rho_\alpha\mathbf{k}_\alpha\cdot\mathbf{u}_\alpha\right). \tag{7.33}$$

This result shows that for a multi–component body the entropy supply cannot be equal to the heat supply divided by the absolute temperature: The entropy production is given by the more general equation

$$\gamma=\rho\dot{s}+\operatorname{div}\mathbf{\Sigma}-\rho\delta \tag{7.34}$$

with the entropy flux

$$\mathbf{\Sigma}=\frac{1}{\theta}\left\{\mathbf{q}+\sum_\alpha\left[\rho_\alpha\left(\theta s_\alpha-e_\alpha-\frac{1}{2}\mathbf{u}_\alpha^2\right)\mathbf{1}+\mathbf{T}_\alpha^T\right]\mathbf{u}_\alpha\right\} \tag{7.35}$$

and the entropy supply

$$\delta=\frac{1}{\theta}\left\{r-\sum_\alpha\frac{\rho_\alpha}{\rho}\mathbf{k}_\alpha\cdot\mathbf{u}_\alpha\right\}. \tag{7.36}$$

In case of a mixture, it is therefore impossible to express the entropy inequality as eq. (6.5) or eq. (6.17): The entropy supply must be different from its classical form in equilibrium thermodynamics. Obviously, this deviation is due to the diffusion process, i.e., $\mathbf{u}_\alpha\neq\mathbf{0}$, which is a non–equilibrium phenomenon.

8. Material Behavior

8.1 Preliminary remarks

In the preceding sections general principles of continuum mechanics were elaborated. These principles referred to the kinematics of a material system and the balance relations. It was shown how the concepts of continuum mechanics led to a systematic formulation of field theories representing the behavior of the real world. In such an approach different levels of generality can be chosen: For example, from the very beginning we may decide between a purely mechanical theory and a thermomechanical formulation, which is more general but at the same time far more complicated. Another basic decision is to dwell on a mixture instead of a one–component system. In this sense, continuum mechanics is acting with a hierarchy of general field theories; the appropriate choice of a particular theory will depend on the main characteristics of the special system under consideration.

In spite of these basic differences arising at a quite early stage of the modelization, all field theories formulated so far have in common that they do not refer to individual properties of the material. As pointed out in the introduction, kinematics and balance laws are quite general statements, universally valid for all material bodies and for all different kinds of materials.

This point of view corresponds to the observation that the local forms of the balance equations are not sufficient to determine uniquely the involved fields: The local balance relations for mass, momentum, energy and entropy, written down for a one–component system,

$$
\begin{aligned}
\dot\rho \; + \; \rho \operatorname{div} \mathbf{v} \; &= \; 0, && (8.1)\\
\rho\dot{\mathbf{v}} &= \operatorname{div}\mathbf{T} + \rho\mathbf{k}\,, \quad \mathbf{T} = \mathbf{T}^T, && (8.2)\\
\dot e &= -\operatorname{div}\mathbf{q} + r + \frac{1}{\rho}\mathbf{T}\cdot\mathbf{D} && (8.3)\\
\gamma &= -\dot e + \frac{1}{\rho}\mathbf{T}\cdot\mathbf{D} + \theta\dot s - \frac{1}{\rho\theta}\mathbf{q}\cdot\mathbf{g} \;\geq\; 0 && (8.4)\\
(\mathbf{g} &= \operatorname{grad}\theta(\mathbf{x},t))
\end{aligned}
$$

imply five scalar valued equations for 16 unknown fields, namely the mass density ρ, the velocity $\mathbf{v}$, the CAUCHY stress tensor $\mathbf{T}$, the internal energy e, the heat flux vector $\mathbf{q}$, the temperature θ and the entropy s. The last relationship (8.4) does not help to determine these unknowns: The entropy inequality is merely a restrictive condition for all solutions of initial–boundary value problems. In this general case eleven further equations are felt to be needed in order to close the system in the sense that the number of equations agrees with the number of unknown functions. Usually, these equations are called **constitutive equations**. The particular mathematical form of the constitutive equations including a set of material parameters represents the individual properties of a material body. The situation is similar in a mechanical theory or even in the case of a mixture with the only difference that a different number of independent and dependent variables are involved: In each particular case constitutive relations are required to express the intrinsic material properties and to close the balance relations such that they may determine unique solutions.

8.2 Thermoelastic fluids

The most simple example how to construct a consistent set of constitutive equations is the theory of a one–component thermoelastic fluids. In an elastic fluid the state of stress is always hydrostatic:

$$\mathbf{T} = -p\mathbf{1} \tag{8.5}$$

In the context of thermodynamics, an elastic fluid may be characterized through the following constitutive equations:

$$
\begin{aligned}
e &= \hat e(v,s) && (8.6)\\
p &= \hat p(v,s) && (8.7)\\
\theta &= \hat\theta(v,s) && (8.8)\\
\mathbf{q} &= \mathbf{q}(v,s,\mathbf{g}) && (8.9)
\end{aligned}
$$

In these equations the specific volume

$$v = \frac{1}{\rho} \tag{8.10}$$

has been introduced instead of the density. The total time derivative of the internal energy,

$$\dot{e} = \frac{\partial \hat{e}}{\partial v}\dot{v} + \frac{\partial \hat{e}}{\partial s}\dot{s}, \tag{8.11}$$

together with the stress power,

$$\frac{1}{\rho}\mathbf{T}\cdot\mathbf{D} = -p\dot{v}, \tag{8.12}$$

must be inserted into the entropy inequality. This yields the inequality

$$-\left(\frac{\partial \hat{e}}{\partial v} + p\right)\dot{v} + \left(-\frac{\partial \hat{e}}{\partial s} + \theta\right)\dot{s} - \frac{1}{\rho\theta}\mathbf{q}\cdot\mathbf{g} \geq 0, \tag{8.13}$$

which must be satisfied for all solutions of the field equations. Standard arguments suggest the potential relations

$$p = \hat{p}(v,s) = -\frac{\partial}{\partial v}\hat{e}(v,s), \tag{8.14}$$

$$\theta = \hat{\theta}(v,s) = \frac{\partial}{\partial s}\hat{e}(v,s), \tag{8.15}$$

as well as the remaining heat conduction inequality:

$$\mathbf{q}\cdot\mathbf{g} = \hat{\mathbf{q}}(v,s,\mathbf{g})\cdot\mathbf{g} \leq 0 \tag{8.16}$$

Consequently, a thermoelastic fluid is completely defined, if two constitutive functions are given, namely the internal energy $e = \hat{e}(v,s)$ and the heat flux vector $\mathbf{q} = \hat{\mathbf{q}}(v,s,\mathbf{g})$. The constitutive function $\hat{\mathbf{q}}(v,s,\mathbf{g})$ for the heat flux vector is restricted by the heat conduction inequality (8.16). A simple realization is given by FOURIER's law

$$\mathbf{q} = -\lambda\mathbf{g}, \tag{8.17}$$

where the (non-negative) number of heat conduction λ may depend on v, s and $|\mathbf{g}|$.

The chain rule (8.11) together with the potential relations (8.14), (8.15) imply the GIBBS relation,

$$\dot{e} = \theta\dot{s} - p\dot{v}. \tag{8.18}$$

Because of the properties (8.14), (8.15) the specific *internal energy* is called a *thermodynamic potential.* There are three other thermodynamic potentials, related to e by means of LEGENDRE transforms. These are the *free energy*

$$\psi = e - \theta s = \hat{\psi}(v,\theta), \tag{8.19}$$

$$\dot{\psi} = -p\dot{v} - s\dot{\theta}, \tag{8.20}$$

the *enthalpy*

$$h = e + pv = \hat{h}(p,s), \tag{8.21}$$

$$\dot{h} = v\dot{p} + \theta\dot{s}, \tag{8.22}$$

and the *free enthalpy*

$$g = h - \theta s = \hat{g}(p, \theta), \tag{8.23}$$
$$\dot{g} = v\dot{p} - s\dot{\theta}. \tag{8.24}$$

Inserting the GIBBS relation (8.18) into the energy balance yields the equation of heat conduction:

$$\theta \dot{s} = -\frac{1}{\rho}\operatorname{div}\mathbf{q} + r \tag{8.25}$$

This equation can be put into a more convenient form, if we observe the temperature relation $\theta = \frac{\partial \hat{e}}{\partial s}(v, s) = \hat{\theta}(v, s)$ to express the entropy as a function of the specific volume and the temperature:

$$s = \hat{s}(v, \theta) \tag{8.26}$$

Then we have

$$\theta \dot{s} = c_v(v, \theta)\dot{\theta} + \theta \frac{\partial \hat{s}(v, \theta)}{\partial v}\dot{v},$$

where the specific heat at constant volume,

$$c_v(v, \theta) = \theta \frac{\partial}{\partial \theta}\hat{s}(v, \theta), \tag{8.27}$$

has been indroduced.

The balance of mass $\dot{\rho} = \rho \operatorname{div}\mathbf{v}$ implies

$$\dot{v} = v \operatorname{div}\mathbf{v}. \tag{8.28}$$

Now, the equation of heat conduction takes the form

$$c_v(v, \theta)\dot{\theta} + \theta \frac{\partial \hat{s}(v, \theta)}{\partial v} v \operatorname{div}\mathbf{v} = -\frac{1}{\rho}\operatorname{div}\mathbf{q} + r. \tag{8.29}$$

In many cases of application an elastic fluid is assumed to be incompressible ($\operatorname{div}\mathbf{v} = \mathbf{0}$), and the specific heat is assumed to be a constant ($c_v = c$). Then, the equation of heat conduction reduces to

$$\dot{\theta}(\mathbf{x}, t) = -\frac{1}{\rho c}\operatorname{div}\mathbf{q} + \frac{r}{c}. \tag{8.30}$$

With FOURIER's law (8.17), $\mathbf{q} = -\lambda \operatorname{grad}\theta$, the classical equation of heat conduction is derived:

$$\frac{\partial \theta}{\partial t} + (\operatorname{grad}\theta)\cdot\mathbf{v} = \frac{\lambda}{\rho c}\operatorname{div}\operatorname{grad}\theta + \frac{r}{c} \tag{8.31}$$

Together with convenient constitutive equations the balance relations for mass (8.1), momentum (8.2) and energy (8.3) (or its specializations (8.3), (8.29)) form a system of five partial differential equations to determine five scalar valued fields, namely $\rho(\mathbf{x}, t)$, $\mathbf{v}(\mathbf{x}, t)$ and $\theta(\mathbf{x}, t)$; the solutions are unique if appropriate initial and boundary conditions are prescribed.

The basic idea of the foregoing analysis applies also to thermoelastic solids. Observing some additional points of view a similar analysis can be carried out even for dissipative materials such as thermoviscoelastic, thermoplastic, and thermoviscoplastic bodies (see [12] and the literature cited there).

In the following sections purely mechanical material properties will be considered. These considerations will give only a sketch of some basic points of view. The reader will find more informations in the references (see, e.g., [1], [5], [7] and [12]).

8.3 Theory of material behavior

The development of modern continuum mechanics has stimulated the formulation of a general theory of material behavior. This theory provides systematic and rational methods to model material properties, such that this modelling can be combined with the general balance relations. In the tradition of classical continuum mechanics only a few special classes of materials have been analysed, namely linear–elastic, viscoelastic and ideally plastic solids, elastic and linear–viscous fluids and ideal gases. In the light of the general theory of material behavior these special theories turn out to be first approximations of a very general constitutive model under certain restrictive conditions, which can be precisely defined. The theory of material behavior delivers specific guidelines, how to generalize the classical theories of continuum mechanics. In this context it is possible to choose between different levels of generality and idealization in accordance with experimental observations.

In the context of a mathematical representation of material properties three principal methods are utilized, namely the specification of

- Constitutive equations,
- Material symmetry and
- Kinematical constraints.

We will briefly outline these three topics for the purely mechanical case of a one–component system.

In *continuum mechanics* a **constitutive equation** is a relation between strain and stress tensors. For elastic bodies the stress is a function of strain or vice versa. The inelastic material behavior may be viscoelastic, plastic or viscoplastic. In these cases functional relations are needed to represent the stress as a functional of the strain process or to determine the current strain from the loading history.

The concept of **material symmetry** or **isotropy** expresses the fact that the material behavior can more or less depend on the direction of a material element. To represent the material symmetry, invariance properties of the constitutive relation with respect to changes of the reference configuration are postulated. The consideration of material symmetry properties leads to the definition of *fluids* and *solids* and to a division of the non-isotropic behavior of solids into different classes of *anisotropy*.

A condition of **kinematical constraint** is a geometric restriction of the class of motions which are possible for a material body. A well–known example is the constraint of *incompressibility*. If incompressibility is assumed, all changes of volume are a priori excluded, i.e., only isochoric motions are allowed for the material body. Another example of a kinematical constraint is the definition of a *rigid body*: It postulates that the distances between any two material points are preserved throughout the motion. In the general theory of kinematical constraints it is assumed that there is one part of the stress tensor which is not determined by the deformation process. The undetermined part of the stress is a reaction force corresponding directly to the kinematical constraint. It is postulated that this part of the stress performs no work under any motion compatible with the constraint condition.

The combination of a particular constitutive equation with convenient assumptions about symmetry properties and kinematical constraints leads to a set of equations which complete the local form of the balance laws to initial– and boundary–value problems. Of course, the number of constitutive equations as well as the special choice of independent and dependent variables depends on the particular field theory, which formulates the problem: In a thermomechanical approach, further constitutive equations are required in addition to that for the stress tensor, and in the general case of a mixture even more constitutive equations are needed. However, the main ingredients — *constitutive equations, material symmetry, kinematical constraints* — may occur in all different situations. In each particular case the assumption of a special *constitutive equation* corresponds to the definition of an *ideal material.*

In order to have a physical meaning, a constitutive equation of any kind should have some basic properties, which usually are postulated a priori. Among those postulates the assumptions of *determinism, material objectivity* and *local action* are most important (see [7]). Moreover, a constitutive equation should be compatible with the general balance relations. In particular, it is not permitted to contradict the Second Law of thermodynamics; this argument applies to the thermomechanical as well as to the purely mechanical case.

8.4 Constitutive equations

A *constitutive equation* is a relation between a set of thermomechanical variables. In the purely mechanical theory of a one–component system, which is taken as an example, a constitutive equation relates stress and strain tensors. More specifically, the process histories of stresses and strains are related in form of a general functional:

$$\underset{\substack{\tau \leq t \\ \sigma \leq t}}{\mathcal{R}} \; [\mathbf{T}(\tau), \mathbf{F}(\sigma)] = \mathbf{0}. \tag{8.32}$$

As a starting assumption this relation states an interaction between the past histories of the internal forces and the deformation. The internal forces are represented by the CAUCHY stress tensor $\mathbf{T}$; this may be justified by the local form of the balance of linear momentum. The choice of the deformation gradient $\mathbf{F}$ expresses the idea that only the motion of an infinitesimal neighborhood of a material point is related to the state of stress. The main idea behind this assumption is called the **principle of local action**.

The above equation is a *functional relation*, i.e., the past *histories* of the deformation gradient and the CAUCHY stress are involved. The particular definition of the tensor–valued functional $\mathcal{R}$ is not yet specified: It can be a system of differential equations or integral equations or equations of a mixed type. However, it is generally assumed that for any given deformation history $\mathbf{F}(\cdot)$ the implicit functional equation can be solved uniquely for the present CAUCHY stress $\mathbf{T}(t)$; in this sense we replace it by the explicit functional equation

$$\mathbf{T}(t) = \underset{\tau \leq t}{\mathcal{F}} [\mathbf{F}(\tau)]. \tag{8.33}$$

On physical reasons which are inferred from the idea of causality, only past histories enter the functional $\mathcal{R}$ or $\mathcal{F}$. The underlying ideas are summarized by the **principle of determinism:** *The present state of the stress is uniquely determined by the past history of the deformation.*

The general functional $\mathcal{F}[\cdot]$ is restricted by a further basic postulate, the assumption of **frame indifference** or **material objectivity**. According to this postulate it is assumed that *a constitutive equation must be invariant under a change of frame.*

To evaluate this latter assumption we note that a (time–dependent) change of frame corresponds to a superimposed rigid body motion, i.e., if the motion of a material body is given,

$$\mathbf{x} = \chi_R(\mathbf{X}, t), \tag{8.34}$$

then, after a change of frame, it transforms into

$$\mathbf{x}^* = \mathbf{Q}(t)\,\mathbf{x} + \mathbf{c}(t) = \mathbf{Q}(t)\,\chi_R(\mathbf{X}, t) + \mathbf{c}(t). \tag{8.35}$$

Here, $\mathbf{Q}(t)$ is a time–dependent orthogonal tensor ($\mathbf{Q}(t)\,\mathbf{Q}^T(t) = \mathbf{1}$) and $\mathbf{c}(t)$ a time–dependent vector. In fact, $\mathbf{c}(t)$ corresponds to the translation and Q(t) to the rotation of the transformed frame of reference.

Differentiating eq. (8.35) with respect to $\mathbf{X}$ we conclude that, in the context of a change of frame, the deformation gradient $\mathbf{F}$ transforms according to

$$\mathbf{F}^*(\mathbf{X}, t) = \mathbf{Q}(t)\,\mathbf{F}(\mathbf{X}, t). \tag{8.36}$$

Furthermore, a force is rotated with $\mathbf{Q}$; in particular, the stress vector $\mathbf{t}$ transforms according to

$$\mathbf{t}^* = \mathbf{Q}\mathbf{t}. \tag{8.37}$$

Because the normal $\mathbf{n}$ of a surface element is also rotated, i.e.,

$$\mathbf{n}^* = \mathbf{Q}\mathbf{n}, \tag{8.38}$$

and since we have

$$\mathbf{t} = \mathbf{T}\mathbf{n}, \tag{8.39}$$

we infer from the assumption

$$\mathbf{t}^* = \mathbf{T}^*\,\mathbf{n}^* \tag{8.40}$$

and $\mathbf{t}^* = \mathbf{Q}\mathbf{t} = \mathbf{Q}\mathbf{T}\mathbf{n} = \mathbf{T}^*\,\mathbf{n}^* = \mathbf{T}^*\,\mathbf{Q}\mathbf{n}$ for the CAUCHY stress tensor the transformation rule

$$\mathbf{T}^* = \mathbf{Q}\mathbf{T}\mathbf{Q}^T. \tag{8.41}$$

In view of the transformation of physical quantities in the context of a change of frame we call scalar-, vector- and tensorvalued physical quantities ϕ, $\mathbf{v}$ and $\mathbf{A}$ **objective**, if under a change of frame they transform according to

$$\phi^* = \phi, \tag{8.42}$$
$$\mathbf{v}^* = \mathbf{Q}\mathbf{v}, \tag{8.43}$$
$$\mathbf{A}^* = \mathbf{Q}\mathbf{A}\mathbf{Q}^T. \tag{8.44}$$

In the sense of this definition the mass of a material body is an objective scalar, the force is an objective vector and the CAUCHY stress tensor is an objective tensor. Quantities which are not objective are among others the position vector, the velocity vector and the deformation gradient. The property of a particular quantity to be objective or not is either a priori postulated or derived from other definitions.

We apply the transformation rules for $\mathbf{F}$ and $\mathbf{T}$ to derive the inferences implied by the assumption of objectivity. Formally, the **assumption of objectivity** or **frame indifference** reads as follows: If the constitutive equation (8.33) is given relative to one frame, then the same constitutive equation relative to another transformed frame should read

$$\mathbf{T}^*(t) = \underset{\tau \le t}{\mathcal{F}} [\mathbf{F}^*(\tau)]. \tag{8.45}$$

In this formulation it is essential that no star is attached to the functional $\mathcal{F}$: The assumption of frame indifference states that an arbitrary change of frame has no influence on the functional $\mathcal{F}$, which maps the deformation history into the present state of stress. From equations (8.33) and (8.45) we conclude that the functional $\mathcal{F}$ must satisfy a restrictive condition, namely

$$\mathbf{Q}(t) \cdot \underset{\tau \le t}{\mathcal{F}} [\mathbf{F}(\tau)]\mathbf{Q}^T(t) = \underset{\tau \le t}{\mathcal{F}} [\mathbf{Q}(\tau)\,\mathbf{F}(\tau)] \tag{8.46}$$

for all orthogonal tensor valued functions $\mathbf{Q}(\cdot)$ and all deformation processes $\mathbf{F}(\cdot)$. It is not difficult to solve this functional equation (see [7]). To this end, we recall the polar decomposition (2.65) of the deformation gradient

$$\mathbf{F}(\tau) = \mathbf{R}(\tau)\,\mathbf{U}(\tau) \tag{8.47}$$

into a rotation tensor $\mathbf{R}$ and the right stretch tensor $\mathbf{U} = \sqrt{\mathbf{C}} = \sqrt{\mathbf{F}^T\,\mathbf{F}}$. Since the functional equation (8.46) is postulated to hold for all rotation histories $\mathbf{Q}(\cdot)$, we set $\mathbf{Q}(\tau) = \mathbf{R}^T(\tau)$ and arrive at

$$\underset{\tau \le t}{\mathcal{F}} [\mathbf{F}(\tau)] = \mathbf{R}(t) \underset{\tau \le t}{\mathcal{F}} [\mathbf{U}(\tau)]\,\mathbf{R}^T(t). \tag{8.48}$$

Here, the rotation $\mathbf{R}(t)$ can be expressed through the deformation gradient via $\mathbf{R} = \mathbf{F}\mathbf{U}^{-1}$; this yields the following representation for the CAUCHY stress:

$$\mathbf{T}(t) = \mathbf{F}(t)\,\mathbf{U}^{-1}(t) \underset{\tau \le t}{\mathcal{F}} [\mathbf{U}(\tau)]\,\mathbf{U}^{-1}(t)\mathbf{F}^T(t) \tag{8.49}$$

The above equation can be formulated in terms of the second PIOLA–KIRCHHOFF stress tensor $\tilde{\mathbf{T}}$, defined by

$$\tilde{\mathbf{T}} = (\det \mathbf{F})\mathbf{F}^{-1}\mathbf{T}\mathbf{F}^{T-1}. \tag{8.50}$$

With $\det \mathbf{F} = \det(\mathbf{RU}) = \det \mathbf{U}$ we have

$$\tilde{\mathbf{T}}(t) = (\det \mathbf{U}(t))\, \mathbf{U}^{-1}(t) \underset{\tau \le t}{\mathcal{F}} [\mathbf{U}(\tau)]\, \mathbf{U}^{-1}(t), \tag{8.51}$$

and we see that the right–hand side is a functional of the stretch history $\mathbf{U}(\cdot)$ alone. Now, the stretch $\mathbf{U}$ is replaced by the right CAUCHY–GREEN tensor

$$\begin{aligned} \mathbf{C}(\tau) &= \mathbf{F}^T(\tau)\,\mathbf{F}(\tau) = \mathbf{U}^2(\tau) \\ \text{i.e.,}\quad \mathbf{U}(\tau) &= \sqrt{\mathbf{C}(\tau)} \qquad (\tau \le t). \end{aligned} \tag{8.52}$$

We put this into eq. (8.51) and define a new functional $\mathcal{G}$ of the history of $\mathbf{C}$. Thus we obtain as a final result the following **reduced form** of the general constitutive equation (8.33):

$$\tilde{\mathbf{T}}(t) = \underset{\tau \le t}{\mathcal{G}} [\mathbf{C}(\tau)] \tag{8.53}$$

According to this result the second PIOLA–KIRCHHOFF stress tensor is a functional of the past history of the right CAUCHY–GREEN tensor. Moreover, in view of $\mathbf{F}^* = \mathbf{QF}$ and $\mathbf{T}^* = \mathbf{QTQ}^T$ it is easily verified that the identities

$$\tilde{\mathbf{T}}^* = \tilde{\mathbf{T}} \tag{8.54}$$

$$\mathbf{C}^* = \mathbf{C} \tag{8.55}$$

hold. Therefore the reduced form (8.53) is necessary and sufficient to satisfy the assumption of frame indifference.

Another useful reduced form, equivalent to (8.53), arises if we express the past strain history $\mathbf{C}(\tau) = \mathbf{C}(t-s) \quad (s \ge 0)$ in terms of the relative difference history,

$$\mathbf{C}_d^t(s) = \mathbf{C}(t-s) - \mathbf{C}(t). \tag{8.56}$$

We insert $\mathbf{C}(\tau) = \mathbf{C}_d^t(s) + \mathbf{C}(t)$ in (8.53) and obtain

$$\tilde{\mathbf{T}}(t) = \underset{\tau \le t}{\mathcal{G}} [\mathbf{C}(t)] = \underset{s \ge 0}{\mathcal{H}} [\mathbf{C}_d^t(s); \mathbf{C}(t)], \tag{8.57}$$

where $\mathcal{H}$ is a new functional of the difference history $\mathbf{C}_d^t(\cdot)$, depending on the present strain $\mathbf{C}(t)$ as a parameter. If we introduce the *static part*

$$\mathbf{f}(\mathbf{C}) = \underset{s \ge 0}{\mathcal{H}} [\mathbf{0}(s); \mathbf{C}(t)] \tag{8.58}$$

and the *memory part*

$$\underset{s\geq 0}{\mathcal{F}}\,[\mathbf{C}_d^t(s);\mathbf{C}(t)] = \underset{s\geq 0}{\mathcal{H}}\,[\mathbf{C}_d^t(s);\mathbf{C}(t)] - \mathbf{f}(\mathbf{C}), \tag{8.59}$$

we may deduce a second version of the reduced form (8.57), namely

$$\tilde{\mathbf{T}}(t) = \mathbf{f}(\mathbf{C}(t)) + \underset{s\geq 0}{\mathcal{F}}\,[\mathbf{C}_d^t(s);\mathbf{C}(t)]. \tag{8.60}$$

According to this equation the second PIOLA–KIRCHHOFF stress is decomposed into a static part and a memory part. The *static part* is a *function* of the present strain $\mathbf{C}$ and the *memory part* is a *functional* of the past history of the difference strain $\mathbf{C}_d^t(\cdot)$, depending on the present strain $\mathbf{C}(t)$ as a parameter. In particular, the memory part vanishes for strain histories which are identically constant ("static histories"), i.e.,

$$\underset{s\geq 0}{\mathcal{F}}\,[\mathbf{0}(s);\mathbf{C}(t)] = \mathbf{0}. \tag{8.61}$$

We close this section with a brief interpretation of the underlying ideas. The functional formulations of the most general mechanical constitutive equation (8.53) or (8.60) corresponds to the asumption that a material body is able to memorize past events of its deformation. If there is no memory of the material, the memory functional $\mathcal{F}$ in the reduced form (8.60) degenerates to zero, i.e.,

$$\tilde{\mathbf{T}}(t) = \mathbf{f}(\mathbf{C}(t)). \tag{8.62}$$

In this special case, a material body is called **elastic**.

Conversely, any kind of **inelastic** behavior corresponds to a **memory property** of the material. The specific memory behavior of a particular inelastic material can be very different in its intensity and character: It may be a long–range memory, as it is usually the case in **plasticity**. The material memory can fade away for those events which are long ago; this characteristics is associated to the term **viscoelasticity**. In some cases it may happen that only an infinitesimal part of the strain history influences the present stress. Then the stress tensor depends on strain rates and the term **viscosity** is the appropriate characterization. The general theory of material behavior provides a large variety of possibilities to obtain special representations of the functionals $\mathcal{G}$ or $\mathcal{H}$ in the reduced forms (8.53) or (8.60). Functional relations can be defined **explicitely** by integrals or by tensor–valued functions of strain and strain rate tensors. They can also be defined **implicitly** by means of ordinary differential equations. In each case, the special realization of a functional relation between $\tilde{\mathbf{T}}$ and $\mathbf{C}$ corresponds to the definition of an *ideal material body*.

8.5 Material symmetry

The idea of material symmetry or isotropy, as explained by TRUESDELL and NOLL [7], is expressed by invariance properties of the constitutive equation relative to a change of the reference configuration. A change of the reference configuration is a one–to–one transformation

$$\hat{\mathbf{X}} = \mathbf{\Lambda}(\mathbf{X}) \quad \Longleftrightarrow \quad \mathbf{X} = \mathbf{\Lambda}^{-1}(\hat{\mathbf{X}}), \tag{8.63}$$

which associates to the material point $\mathcal{X}$ the "name" $\hat{\mathbf{X}}$ instead of $\mathbf{X}$. Then, the motion $\mathbf{x} = \chi_R(\mathbf{X}, t)$ takes the form

$$\mathbf{x} = \chi_{\hat{R}}(\hat{\mathbf{X}}, t), \tag{8.64}$$

and we have for all $\mathbf{X}$ and t the identity

$$\chi_R(\mathbf{X}, t) = \chi_{\hat{R}}(\Lambda(\mathbf{X}), t). \tag{8.65}$$

Differentiating this with respect to $\mathbf{X}$ leads to the following transformation rule for the deformation gradient:

$$\mathbf{F} = \hat{\mathbf{F}}\,\mathbf{P} \tag{8.66}$$

Here,

$$\hat{\mathbf{F}}(\hat{\mathbf{X}}, t) = \operatorname{Grad} \chi_{\hat{R}}(\hat{\mathbf{X}}, t) \tag{8.67}$$

denotes the transformed deformation gradient;

$$\mathbf{P}(\mathbf{X}, t) = \operatorname{Grad} \mathbf{\Lambda}(\mathbf{X}) \tag{8.68}$$

is the gradient of the the transformation $\mathbf{\Lambda}(\cdot)$, which maps the "old" configuration $(\mathbf{X})$ onto the "new" configuration $(\hat{\mathbf{X}})$.

Because the basic constitutive equation (8.33) was assumed as a relation between the CAUCHY stress and the deformation gradient, the functional $\mathcal{F}$ depends on the choice of the reference configuration. We have

$$\mathbf{T}(t) = \underset{\tau \le t}{\mathcal{F}_R}[\mathbf{F}(\tau)] = \underset{\tau \le t}{\mathcal{F}_{\hat{R}}}\left[\hat{\mathbf{F}}(\tau)\right], \tag{8.69}$$

where the indices R and $\hat{R}$ refer to the underlying reference configurations. The relation between the different functionals $\mathcal{F}_R$ and $\mathcal{F}_{\hat{R}}$ is given by the identity

$$\underset{\tau \le t}{\mathcal{F}_R}[\mathbf{F}(\tau)] = \underset{\tau \le t}{\mathcal{F}_{\hat{R}}}\left[\mathbf{F}(\tau)\mathbf{P}^{-1}\right]. \tag{8.70}$$

If for two different configurations R or $\hat{R}$ the constitutive functional happens to be unchanged; then $\mathcal{F}_R = \mathcal{F}_{\hat{R}}$ or

$$\underset{\tau \le t}{\mathcal{F}_R}[\mathbf{F}(\tau)] = \underset{\tau \le t}{\mathcal{F}_R}\left[\mathbf{F}(\tau)\mathbf{P}^{-1}\right], \tag{8.71}$$

which is an identy for all deformation histories $\mathbf{F}(\cdot)$. In this case the two configurations are indistinguishable and express a certain symmetry of the body. On physical reasons only volume preserving transformations of the reference configurations are taken into account, i.e. transformations with

$$|\det \mathbf{P}| = |\det(\operatorname{Grad} \mathbf{\Lambda}(\mathbf{X}))| = 1. \tag{8.72}$$

A transformation with this property is called **unimodular**. The properties of material symmetry are represented in terms of the symmetry group: The **symmetry group** g_R, referred to the reference configuration R, is the set of all unimodular transformations, which leave the constitutive equation relative to R invariant:

$$g_R = \left\{ \mathbf{H} \,\middle|\, |\det \mathbf{H}| = 1 \quad \text{and} \quad \underset{\tau \le t}{\mathcal{F}_R}[\mathbf{F}(\tau)] = \underset{\tau \le t}{\mathcal{F}_R}[\mathbf{F}(\tau)\,\mathbf{H}] \ \text{ for all deformation processes } \mathbf{F}(\tau) \right\} \tag{8.73}$$

It can be shown that g_R is an algebraic group, which motivates the denotation. Moreover, it can be proved that a transformation of the reference configuration from R to $\hat{R}$ changes the symmetry group g_R into

$$g_{\hat{R}} = \mathbf{P}\, g_R\, \mathbf{P}^{-1}, \tag{8.74}$$

where $\mathbf{P} = \operatorname{Grad} \mathbf{\Lambda}(\mathbf{X})$ is not required to be unimodular.

From the definition of the symmetry group we infer the following criterion for the reduced form (8.53): A unimodular transformation $\mathbf{H}$ ($\det \mathbf{H} = \pm 1$) belongs to the symmetry group g_R if and only if the functional $\mathcal{G}_R$ satisfies

$$(\det \mathbf{H})\, \mathbf{H}^{-1}\, \underset{\tau \le t}{\mathcal{G}_R}[\mathbf{C}(\tau)]\, \mathbf{H}^{T-1} = \underset{\tau \le t}{\mathcal{G}_R}\left[\mathbf{H}^T \mathbf{C}(\tau)\, \mathbf{H}\right] \tag{8.75}$$

for all strain processes $\mathbf{C}(\tau)$.

Accordingly, the reduced form (8.60) implies the criterion

$$\mathbf{H} \in g_R \quad \Leftrightarrow$$

$$(\det \mathbf{H})\, \mathbf{H}^{-1}\, \tilde{\mathbf{T}}\, \mathbf{H}^{T-1} = \mathbf{f}\left[\mathbf{H}^T \mathbf{C}\, \mathbf{H}\right] + \underset{s \ge 0}{\mathcal{F}}\left[\mathbf{H}^T \mathbf{C}_d^t(s)\, \mathbf{H}; \mathbf{H}^T \mathbf{C}\, \mathbf{H}\right] \tag{8.76}$$

By definition, the symmetry group is a subgroup of the group of all unimodular tensors. The "size" of a specific symmetry group is measure for the amount of material symmetry.

The concept of the symmetry group can be utilized to distinguish between solids and fluids: in case of a **solid** there exists a reference configuration R, such that the symmetry group is a subgroup of the orthogonal group. The smallest symmetry group

$$g_R = \{\mathbf{1}\ ,\ -\mathbf{1}\}$$

defines the *triclinic solid*; if (for an appropriate reference configuration) the symmetry group coincides with the full set of orthogonal tensors, we have the case of an **isotropic solid**.

A **fluid** is characterized through the largest symmetry group, being identical with the set of all unimodular tensors. This formal characterization can be applied to derive the most general constitutive equation of a fluid. To this end, we insert the special unimodular tensor

$$\mathbf{H} = (\det \mathbf{F})^{\frac{1}{3}} \mathbf{F}^{-1} = \left(\frac{\rho_R}{\rho}\right)^{\frac{1}{3}} \mathbf{F}^{-1}. \tag{8.77}$$

into (8.76), which in view of equations (8.50) and (8.52) implies

$$\left(\frac{\rho_R}{\rho}\right)^{\frac{1}{3}} \mathbf{T} = \mathbf{f}\left(\left(\frac{\rho_R}{\rho}\right)^{\frac{2}{3}} \mathbf{1}\right) + \underset{s\geq 0}{\mathcal{F}} \left[\left(\frac{\rho_R}{\rho}\right)^{\frac{2}{3}} \mathbf{F}^{T-1}\mathbf{C}_d^t(s)\mathbf{F}^{-1};\ \left(\frac{\rho_R}{\rho}\right)^{\frac{2}{3}} \mathbf{1}\right].$$

Next, we introduce the **relative strain history**

$$\mathbf{G}^t(s) = \mathbf{F}^{T-1}\mathbf{C}_d^t(s)\,\mathbf{F}^{-1} = \mathbf{F}^{T-1}(t)\,\mathbf{F}^T(t-s)\,\mathbf{F}(t-s)\,\mathbf{F}^{-1}(t) - \mathbf{1} \tag{8.78}$$

and rewrite the above equation with new denotations:

$$\mathbf{T}(t) = \mathbf{g}(\rho(t)) + \underset{s\geq 0}{\mathcal{Q}} \left[\mathbf{G}^t(s);\ \rho(t)\right] \tag{8.79}$$

This shows that for a fluid the memory part of the CAUCHY stress depends on the relative strain history and on the present mass density as a parameter. The static part is a function of the present mass density alone. This function can be further reduced: According to the assumption of material frame indifference we have for the static part

$$\mathbf{T}_s = \mathbf{g}(\rho), \tag{8.80}$$

as well as

$$\mathbf{T}_s^* = \mathbf{g}(\rho^*) \tag{8.81}$$

under a change of frame. Because of $\rho = \rho^*$ and $\mathbf{T}_s^* = \mathbf{Q}\mathbf{T}_s\mathbf{Q}^T$ this implies

$$\mathbf{Q}\mathbf{g}(\rho)\mathbf{Q}^T = \mathbf{g}(\rho) \tag{8.82}$$

for all orthogonal tensors $\mathbf{Q}$. This identity forces the static part of the stress to be a spherical tensor, i.e.,

$$\mathbf{g}(\rho) = -p(\rho)\,\mathbf{1}. \tag{8.83}$$

The pressure p depends on the density; the material function $p(\rho)$ represents the elastic or the static properties of a fluid. As a result, the symmetry properties of a fluid imply the following most general form of a constitutive equation:

$$\mathbf{T} = -p(\rho)\,\mathbf{1} + \underset{s\geq 0}{\mathcal{Q}} \left[\mathbf{G}^t(s);\ \rho\right] \tag{8.84}$$

According to this equation the static part of the state of stress in a fluid is always a hydrostatic pressure depending on the present mass density. The memory part is a functional of the relative strain history depending on the mass density as a parameter. The functional $\mathcal{Q}$ must be isotropic in the sense that

$$\underset{s\geq 0}{\mathcal{Q}} \left[\mathbf{Q}\mathbf{G}^t(s)\mathbf{Q}^T;\ \rho\right] = \mathbf{Q}\ \underset{s\geq 0}{\mathcal{Q}} \left[\mathbf{G}^t(s);\ \rho\right] \mathbf{Q}^T \tag{8.85}$$

holds for all orthogonal tensors $\mathbf{Q}$. We get this equation if we apply once more the assumption of objectivity and observe the transformation rule for the relative strain history,

$$\mathbf{G}^{t*}(s) = \mathbf{Q}(t)\,\mathbf{G}^t(s)\mathbf{Q}^T(t), \tag{8.86}$$

which can easily be verified from its definition (8.78) and the transformation rule (8.36) for the deformation gradient.

The special case of an elastic fluid is characterized by $\mathcal{Q} = \mathbf{0}$, which leads to eq. (8.5).

If there exists a reference configuration such that the isotropy group is a subgroup of the orthogonal group, the material body is called a **solid**. From eq. (8.75) we infer that an orthogonal tensor $\mathbf{Q}$ is contained in the isotropy group g_R of a solid, if and only if the functional $\mathcal{G}_R$ has the property

$$\mathbf{Q}\ \underset{\tau\le t}{\mathcal{G}_R}[\mathbf{C}(\tau)]\ \mathbf{Q}^T = \underset{\tau\le t}{\mathcal{G}_R}\left[\mathbf{Q}\mathbf{C}(\tau)\,\mathbf{Q}^T\right] \tag{8.87}$$

identically in all strain histories $\mathbf{C}(\cdot)$. If the symmetry group of a solid equals the full orthogonal group, the material is called an **isotropic solid**. If the symmetry group of a solid is smaller than the full orthogonal group the solid is called **anisotropic**.

In case of an **isotropic elastic solid** the stress–strain–relation is represented by an isotropic tensor function: According to eqs. (8.62) the most general constitutive equation for an elastic material is given by

$$\tilde{\mathbf{T}} = \mathbf{f}(\mathbf{C}).$$

For an isotropic elastic solid $\mathbf{f}$ has the property

$$\mathbf{f}\left(\mathbf{Q}\mathbf{C}\mathbf{Q}^T\right) = \mathbf{Q}\,\mathbf{f}(\mathbf{C})\,\mathbf{Q}^T \tag{8.88}$$

for all orthogonal tensors $\mathbf{Q}$. It can be shown that it is represented by

$$\mathbf{f}(\mathbf{C}) = \psi_0\mathbf{1} + \psi_1\mathbf{C} + \psi_2\mathbf{C}^2, \tag{8.89}$$

where the material functions ψ_i depend on the invariants of the tensor $\mathbf{C}$:

$$\psi_i = \psi_i(I_{\mathbf{C}}, II_{\mathbf{C}}, III_{\mathbf{C}}) \qquad (i = 1, 2, 3) \tag{8.90}$$

8.6 Kinematic constraints

A condition of kinematic constraint is a geometric restriction on the set of all motions which are possible for a material body. A condition of constraint can be defined as a restriction on the deformation gradients:

$$\lambda(\mathbf{F}) = 0 \tag{8.91}$$

Since the definition of a constraint is a constitutive equation, the assumption of frame indifference should be applied, and eq. (8.91) be replaced by

$$\lambda(\mathbf{C}) = 0. \tag{8.92}$$

In the context of kinematic constraints the principle of determinism is modified; it is postulated that only a part of the stress tensor is related to the history of the deformation. Thus instead of (8.33), we write

$$\mathbf{T}(t) = \mathbf{N}(t) + \underset{\tau \leq t}{\mathcal{F}}\,[\mathbf{F}(\tau)]. \tag{8.93}$$

On the right–hand side $\mathbf{N}$ denotes a stress tensor, which is not determined from the deformation process by means of a constitutive equation. Physically, $\mathbf{N}$ is a reaction stress, related to the constraint (8.91). In analogy to analytical mechanics it is assumed that the reaction stress $\mathbf{N}$ performs no work, i.e., the stress power vanishes for all motions compatible with the constraint condition:

$$\mathbf{N} \cdot \mathbf{D} = 0 \tag{8.94}$$

To derive an alternative form of this we start from eq. (8.92) and modify the reduced form (8.53) to obtain

$$\tilde{\mathbf{T}} = \tilde{\mathbf{N}} + \underset{\tau \leq t}{\mathcal{G}}\,[\mathbf{C}(\tau)], \tag{8.95}$$

where the reaction stress $\tilde{\mathbf{N}}$ is now left unspecified, but its power of working required to vanish, i.e., we have

$$\tilde{\mathbf{N}} \cdot \dot{\mathbf{C}} = 0 \tag{8.96}$$

for all motions, compatible with the constraint (8.92). This compatibility can be evaluated to put the undetermined part of the stress in a more explicit form: Differentiating (8.92) with respect to time, we obtain

$$\left[\frac{\mathrm{d}}{\mathrm{d}\mathbf{C}}\lambda(\mathbf{C}(t))\right] \cdot \dot{\mathbf{C}}(t) = 0. \tag{8.97}$$

We conclude that the normal $\mathrm{d}\lambda/\mathrm{d}\mathbf{C}$ to the surface $\lambda(\mathbf{C}) =$ const. is orthogonal to all strain rates which are allowed by the constraint condition. Eq. (8.96) suggests that just the same orthogonality holds for the reaction stress $\tilde{\mathbf{N}}$, whence follows that $\tilde{\mathbf{N}}$ must be parallel to $\mathrm{d}\lambda/\mathrm{d}\mathbf{C}$:

$$\tilde{\mathbf{N}} = \alpha \frac{\mathrm{d}}{\mathrm{d}\mathbf{C}}\lambda(\mathbf{C}) \tag{8.98}$$

Here, the factor α is left undetermined and must be regarded as an independent field in the balance law of linear momentum.

The same arguments apply, if simultaneously more constraints are specified: For

$$\lambda_i(\mathbf{C}) = 0 \qquad (i = 1, 2, 3, \ldots) \tag{8.99}$$

we have

$$\tilde{\mathbf{T}} = \sum_i \tilde{\mathbf{N}}_i + \underset{\tau \leq t}{\mathcal{G}}\,[\mathbf{C}(\tau)], \tag{8.100}$$

with

$$\tilde{\mathbf{N}}_i = \alpha_i \frac{\mathrm{d}\lambda_i(\mathbf{C})}{\mathrm{d}\mathbf{C}} \qquad (i = 1, 2, 3, \ldots) \tag{8.101}$$

We note that at most six constraint conditions can be defined. The maximum of six constraints corresponds to a rigid body; in this case the state of stress is completely undetermined.

As an example we investigate the constraint of incompressibility. If a body is assumed to be incompressible, only volume–preserving motions are possible, and we have the condition

$$\det \mathbf{C} = 1, \tag{8.102}$$

i.e.,

$$\lambda(\mathbf{C}) = \det \mathbf{C} - 1. \tag{8.103}$$

Then, we derive $\frac{\mathrm{d}}{\mathrm{d}\mathbf{C}}(\det \mathbf{C} - 1) = (\det \mathbf{C})\mathbf{C}^{-1} = \mathbf{C}^{-1}$, and eq. (8.98) implies

$$\tilde{\mathbf{N}} = \alpha \mathbf{C}^{-1}. \tag{8.104}$$

Instead of α we write $-p$ in order to indicate that p is a pressure. Then, eq. (8.95) reads

$$\tilde{\mathbf{T}} = -p\mathbf{C}^{-1} + \underset{\tau \leq t}{\mathcal{G}}\,[\mathbf{C}(\tau)], \tag{8.105}$$

and the CAUCHY stress tensor is given by

$$\mathbf{T} = -p\mathbf{1} + \mathbf{F}\,\underset{\tau \leq t}{\mathcal{G}}\,[\mathbf{C}(\tau)]\mathbf{F}^T. \tag{8.106}$$

This equation shows that in case of incompressibility the general constitutive equation determines the stress tensor up to an hydrostatic pressure.

In particular, we conclude in view of eq. (8.84) that the most general constitutive equation for an incompressible fluid is given by

$$\mathbf{T} = -p\mathbf{1} + \underset{s \geq 0}{\mathcal{Q}}\,[\mathbf{G}^t(s)]. \tag{8.107}$$

8.7 Fading memory: Approximation of memory functionals

The most general constitutive equation of a solid can be expressed as eq. (8.60), i.e.,

$$\tilde{\mathbf{T}} = \mathbf{f}(\mathbf{C}) + \underset{s \geq 0}{\mathcal{F}}\,[\mathbf{C}_d^t(s); \mathbf{C}].$$

For a fluid the general constitutive equation is given by eq. (8.84),

$$\mathbf{T} = -p(\rho)\,\mathbf{1} + \underset{s \geq 0}{\mathcal{Q}}\,[\mathbf{G}^t(s); \rho],$$

and for an incompressible fluid by eq. (8.107). According to these equations the present stress decomposes into an elastic part and a memory part: The elastic part depends on the present strain and the memory part on the strain history. The memory parts vanish for static histories: $\mathcal{F}$ and $\mathcal{Q}$ are normalized to

$$\underset{s \geq 0}{\mathcal{F}}\,[\mathbf{0}(s)\,;\,\mathbf{C}] \;=\; \underset{s \geq 0}{\mathcal{Q}}\,[\mathbf{0}(s)\,;\,\rho] = \mathbf{0}.$$

The reduced forms (8.60), (8.84) and (8.107) are most convenient for materials with fading memory. For an inelastic material with a fading memory property deformation events have a very small influence on the present stress if they are sufficiently long ago. The mathematical theory of fading memory was formulated be COLEMAN and NOLL [9]. They introduced the fading memory norm of a strain history,

$$\left\|\mathbf{C}_d^t(s)\right\|_h = \left\{\int_0^\infty \left|\mathbf{C}_d^t(s)\right|^2 h^2(s)\,ds\right\}^{\frac{1}{2}}, \tag{8.108}$$

where $h(s)$ is an influence function, which goes "fast enough" to zero as s tends to infinity (The difference history $\mathbf{C}_d^t(s)$ may be replaced by the relative strain history $\mathbf{G}^t(s)$.). The norm (8.108) measures a strain history, where the norm $|\mathbf{C}_d^t(s)| = \sqrt{\mathrm{tr}\,(\mathbf{C}_d^t(s)^2)}$ of each past deformation is multiplied with a weighting factor $h(s)$ which decreases monotonically to zero. Thus the idea of fading memory is included in the definition (8.108). In [9] the mathematical expression of a fading memory property was introduced as a continuity assumption for the functionals $\mathcal{F}$ or $\mathcal{Q}$: These functionals may be continuous, which implies the relaxation property, or they may be differentiable; in this case they can be approximated by linear functionals for which explicit representations are known.

The most important result of the fading memory theory of COLEMAN and NOLL is the possibility to approximate a retarded (slow) strain history by its TAYLOR–expansion; for sufficiently slow motions this approximation is arbitrarily close in the sense of the fading memory norm (8.108).

The TAYLOR–expansion of the relative strain history (8.78),

$$\mathbf{G}^t(s) = \mathbf{F}^{T-1}(t)\,\mathbf{C}(t-s)\,\mathbf{F}^{-1}(t) - \mathbf{1},$$

is calculated to

$$\begin{aligned}
\mathbf{G}^t(s) &= \mathbf{G}^t(0) + s\,\dot{\mathbf{G}}^t(0) + \frac{s^2}{2!}\,\ddot{\mathbf{G}}^t(0) + \;\dots\;, \\
\mathbf{G}^t(s) &= s\,\mathbf{A}_1 + \frac{s^2}{2}\,\mathbf{A}_2 - \;\dots\; + (-1)^n \frac{s^n}{n!}\,\mathbf{A}_n + \;\dots
\end{aligned} \tag{8.109}$$

$\left(\mathbf{G}^t(0) = \mathbf{0},\ \mathbf{A}_k = \overset{(k)}{\mathbf{G}^t}(0)\right)$, where

$$\mathbf{A}_1 = \mathbf{L} + \mathbf{L}^T = 2\mathbf{D} \tag{8.110}$$

$$\left(\mathbf{L} = \dot{\mathbf{F}}\mathbf{F}^{-1} = \mathrm{grad}\,\mathbf{v}(\mathbf{x},t)\right)$$

$$\mathbf{A}_2 = \mathbf{L}_2 + \mathbf{L}_2^T + 2\mathbf{L}^T\mathbf{L} \tag{8.111}$$

$$\left(\mathbf{L}_2 = \ddot{\mathbf{F}}\mathbf{F}^{-1} = \mathrm{grad}\,\dot{\mathbf{v}}(\mathbf{x},t)\right)$$

are the first and second RIVLIN–ERICKSON tensors, respectively.

As a fundmental result of COLEMAN and NOLL, the TAYLOR–expansion (8.109) can be inserted into the functional $\mathcal{Q}$, such that, for sufficiently slow motions, an asymptotic approximation of $\mathcal{Q}$ is derived, which is an isotropic function of the RIVLIN–ERICKSON tensors and the mass density:

$$\underset{s \geq 0}{\mathcal{Q}} [\mathbf{G}^t(s)] \approx \mathbf{g}(\mathbf{A}_1, \mathbf{A}_2, ..., \mathbf{A}_n, \rho) \tag{8.112}$$

This leads to the definition of the REINER–RIVLIN fluid of order n,

$$\mathbf{T} = -p(\rho)\,\mathbf{1} + \mathbf{g}_n(\mathbf{A}_1, \mathbf{A}_2, ..., \mathbf{A}_n, \rho). \tag{8.113}$$

More details are found in [7] and [9]. For $n = 1$ we have ($\mathbf{A}_1 = 2\mathbf{D}$)

$$\mathbf{T} = -p(\rho)\,\mathbf{1} + \mathbf{g}_1(\mathbf{D}), \tag{8.114}$$

where $\mathbf{g}_1$ is an isotropic function with the representation

$$\begin{aligned} \mathbf{g}_1(\mathbf{D}, \rho) = \; & \phi_0(I_{\mathbf{D}}, II_{\mathbf{D}}, III_{\mathbf{D}}, \rho)\mathbf{1} + \phi_1(I_{\mathbf{D}}, II_{\mathbf{D}}, III_{\mathbf{D}}, \rho)\mathbf{D} + \\ & + \phi_2(I_{\mathbf{D}}, II_{\mathbf{D}}, III_{\mathbf{D}}, \rho)\mathbf{D}^2 \end{aligned} \tag{8.115}$$

(compare (8.89)). Through linearization of $\mathbf{g}_1$ the definition of the **NAVIER STOKES fluid** (linear viscous fluid) is derived,

$$\mathbf{T} = -p(\rho)\,\mathbf{1} + \kappa(\mathrm{tr}\,\mathbf{D})\mathbf{1} + 2\eta\mathbf{D}, \tag{8.116}$$

where η and κ correspond to the shear and bulk viscosities. Of course, no bulk viscosity exists for an **incompressible NAVIER STOKES fluid**. In case of incompressibility the constitutive equation of a linear viscous fluid reads

$$\mathbf{T} = -p(\rho)\,\mathbf{1} + 2\eta\mathbf{D}, \tag{8.117}$$

where no constitutive equation exists for the hydrostatic pressure p. As a consequence of the theory of COLEMAN and NOLL, the constitutive equations (8.116) and (8.117) can be understood to be asymptotic approximations of the most general constitutive equations (8.84) and (8.107) for a fluid with fading memory. This approximations are valid for deformation processes, which are sufficiently slow.

References

[1] A. C. ERINGEN: Nonlinear Theory of Continuous Media, McGraw–Hill Book Company, New York, Toronto, London, 1962

[2] M. E. GURTIN: An Introduction to Continuum Mechanics, Academic Press, New York, London, Toronto, Sydney, San Francisco, 1981

[3] W. JAUNZEMIS: Continuum Mechanics, The Macmillan Company, New York, Collier–Macmillan Limited, London, 1967

[4] D. C. LEIGH: Nonlinear Continuum Mechanics, McGraw–Hill Book Company, New York, St. Louis, San Francisco, Toronto, London, Sydney, 1968

[5] L. E. MALVERN: Introduction of the Mechanics of a Continuous Medium, Prentice–Hall, Englewood Cliffs, N. Y., 1969

[6] A. I. M. SPENCER: Continuum Mechanics, Longman, London and New York, 1980

[7] C. TRUESDELL, W. NOLL: The Non–Linear Field Theories of Mechanics, Encyclopedia of Physics III/3, Springer 1965

[8] C. TRUESDELL: A First Course in Rational Continuum Mechanics, Volume 1, Academic Press, New York, San Francisco, London, 1977

[9] D. COLEMAN, W. NOLL: An Approximation Theorem for Functionals with Applications in Continuum Mechanics, Arch. Rat. Mech. Anal. 6 (1960), 355–370

[10] K. HUTTER: The Foundation of Thermodynamics, its Basic Postulates and Implications. A Review of Modern Thermodynamics, Acta Mech. 27 (1977), 1–54

[11] B. H. LAVENDA: Thermodynamics of Irreversible Processes, The Macmillan Company, New York, Collier–Macmillan Limited, London, 1978

[12] P. HAUPT: Thermodynamics of Solids, CISM–Course No. 59: Non–Equilibrium Thermodynamics with Applications to Solids (Ed.: W. MUSCHIK), to appear

Waves and Oscillations in the Ocean and in Lakes[1]

Kolumban Hutter
Institut für Mechanik, Technische Hochschule Darmstadt
Hochschulstraße 1, D - 6100 Darmstadt, Germany

We study linear wave motions in lakes and ocean bays whose characteristic length scales extend over most parts or all of the water masses. We present the governing equations and motivate, by means of a scale analysis, the various simplified versions of model equations that are in use in linearized lake dynamics. This scale analysis permits rational deduction of the reduced equations - the most important of these are the shallow water equations - and delimits their applicability.
We first introduce the reader into the analysis of linear gravity and Rossby waves. Surface gravity waves without rotation of the Earth are studied and a classification of Rossby waves is given. Second, plane linear waves in a rotating stratified fluid are looked at, and the motions of barotropic and baroclinic responses are defined by separating the full three-dimensional equations in the shallow water approximation into a vertical-stratification-dependent eigenvalue problem for the (barotropic and) baroclinic mode structure and the horizontal wave propagation problem. This justifies the introduction of layered models, in particular the use of the two layer approximation of the dynamics of a lake under summer stratification. Third, we study gravity waves in domains that are influenced by boundaries and, in particular, discuss Kelvin and Poincaré waves in infinite channels and how these combine if they reflect at the wall of a semi-infinite channel.

---

[1]This paper is an extended version of a 1/2-semester course on *"Physical Limnology and Oceanography"* which I hold at ETH Zürich to Environmental Physicists in their 6th semester of education. As far as referencing current literature is concerned I am neither exhaustive nor complete and only refer to literature that is directly related to the material presented. My focus is to explain phenomena and to verify them on field observations; it is then only natural that I draw from my own work. I apologize to all those whose important contributions are not mentioned and would have to be included in a more embracing text. For a list with more than 300 relevant references, the reader may consult e.g. HUTTER (1984b, 1987). Obviously also, most aspects, including a great many others are for instance treated in DEFANT (1961), GREENSPAN (1968), KRAUSS (1966), LEBLOND and MYSAK (1980), LIGHTHILL (1978), MORTIMER (1974), PHILLIPS (1966), TURNER (1973), WITHAM (1974) and YIH (1965).
An earlier version of the text was prepared for a graduate course held in August 1991 at the Institute of Applied Mechanics, National Taiwan University, Taipei. The author is grateful to Prof. Y. H. Pao for the opportunity to draft these lecture notes.
This article is dedicated to C. H. MORTIMER, from whom I have learned much of what is reported here on limnology of lakes in general and the Swiss Lakes in particular.

Topographic Rossby waves are treated separately and restricted to the situation without barotropic-baroclinic coupling. After a brief presentation of observations in Lake Michigan and the Lake of Lugano we slowly build-up a physical understanding by constructing exact solutions to free topographic waves in circular and elliptical basins with parabolic bottom, channels, shelves and trenches. Most significant is the study of topographic waves in rectangular closed basins with smoothly varying bathymetry: Three types of such waves exist; (1) large scale basin-wide modes with a few gyres filling the entire basin, (2) small scale modes, also filling the entire basin, but consisting of a large number of gyres, and (3) bay modes with wave activity restricted to the bay ends of the basin. These permit interpretation of many observations.

In the last paragraph we treat arbitrary enclosed basins under the effect of gravitational oscillations. Special features of external and internal wave motions are discussed. For barotropic oscillations of a lake system (Lake of Lugano) it is shown that substantial water masses are exchanged between the individual basins at the resonating periods. External and internal seiches are computed and results compared with detailed observations in various lakes and the effects of the rotation of the Earth are discussed. Long periodic oscillating features may be attributed to topographic Rossby waves or higher baroclinic internal gravity waves.

1. Introduction[2]

The water motion in the ocean and in lakes is driven by a variety of external forces, and depending upon their forms of appearance these motions are termed *currents, circulations, flows* or *waves.* The driving forces may be meteorological activity through *winds, pressure variations* and *radiative heat*; they may be the effects of the variation of the gravitational force due to the relative positions of the Sun, the Earth, the Moon (and to a much lesser extent also the other planets) and thus give rise to the *tidal motion*; however, they are also the result of the *rotation of the Earth.* This non-inertiality has a non-generic component insofar as oceanic, limnologic (and atmospheric) motions are modified - sometimes substantially, sometimes only peripherically - by the additional forces due to the motion of the frame of reference. However, it also gives rise to a generic component, because motions do exist in such systems that are absent when the frame of reference is inertial. Accordingly, one may classify fluid motions in the geophysical environment respectively as *first class* and as *second class* motions; among the first are *gravity waves*, with manifestation of embodied rotational effects in the so-called KELVIN and POINCARÉ WAVES, among the second are the so-called ROSSBY WAVES, manifested as *planetary waves* in the atmosphere and as *shelf waves, topographic, vorticity waves* in the ocean and in lakes.

Water motions in the ocean and in lakes are either *localized* i.e. restricted in their activity to a finite region, or *global* i.e. embracing the entire water domain. The processes may be *episodic*, i e., of short duration and then die out very quickly, or *persistent* and then last long. Waves are episodic in nature and often localized and then unaffected by boundaries. By contrast, circulation flows are generally global with water motions whose characteristic lengths extend over most parts or all of the water masses. The topography and the boundaries of the domains, in which the dynamic processes take place affect their evolution. Freely evolving

[2]The subject treated in this and the following section, i.e., a rigorous rational deduction of the equations governing the motion of water in lakes and the ocean (and also in the atmosphere) is similarly treated, for instance also by GREENSPAN (1968), GILL (1982), HUTTER (1984, 1991) and MARCHUK and SARKISYAN (1989).

dynamic processes that are unaffected by boundaries are significant only as transients in establishing the overall response; the ocean or lake dynamics reflect as much the properties of the domain geometry as they are governed by their driving forces. Similarly, *time scales* that typify the governing processes are determined by both the underlying physics and the length scales of the basin dimensions in which they take place. Given this tremendeous span of length scales it is understandable that the variety of phenomena that can be classified as large scale over-all circulation is immense. Generally all water motions contain local and global, episodic and persistent components. A full analysis which would allow, at least in principle, the study of the dynamic processes over all these length and time scales is not only out of reach; it would also be of little use because the complexity of the equations could hardly uncover the physics in mind. Therefore, simplifications are introduced by eliminating small scale-short time features, local phenomena such as *steep fronts*, *strong upwelling and downwelling*, *local billow formation*, *thermocline erosion* etc. This filter - incidentally introduced on all levels of computational ocean and lake dynamics - is accomplished by the introduction of two basic assumptions, the *shallow water assumption* and the *Boussinesq assumption.* The former effectively states that the dynamics are primarily horizontal so that the vertical momentum equation reduces to a balance between pressure gradients and gravity forces. The second states that density variations are unimportant except in the buoyancy forces; its other most important manifestation is the statement that the water may effectively be treated as incompressible which eliminates under-water accoustics.
This leads us, naturally, to the characterization of the two most important classes of motions that are distinguished in lake and ocean dynamics: the *barotropic* and *baroclinic* processes. In the former the density variations are irrelevant for the description of the processes involved, in the latter they are crucial. Except for *spin-up* and *quasi-steady circulation processes* time scales of the barotropic and baroclinic circulation processes lie sufficiently apart (minutes and hours, versus days, weeks and months); this separation explains why often free surface deformations are ignored when studying baroclinic (or even barotropic) circulation patterns. Such a *rigid-lid* suppresses *surface gravity waves* and restricts barotropic motions to *topographic vorticity-dominated circulation patterns.* In most global ocean and lake problems, the rigid-lid assumption is justified, but surprisingly, in the water exchange mechanisms of basin *systems* the rigid-lid assumption may lead to erroneous predictions. The reason is that barotropic gravity waves are responsible for these processes.
It is the aim of this overview article to identify these separate forms of water motions and to relate them to the external driving forces. We shall in particular give an account of linear waves in the ocean and in lakes.

2. Governing Equations

The governing equations describing the water motions and the dispersion of substances in lakes and the ocean comprise of field equations (mass, momentum and energy balances) and boundary conditions at the free surface and the bottom. Here we list them, introduce through scalings, various simplifications and discuss the properties of the simplified system of equations.

In what follows we shall use symbolic and Cartesian notation interchangingly. The coordinate system will be rotating with the Earth as shown in *Fig. 1.* All fields will be regarded as turbulent averages and flux terms will, as usual, refer to the sums of molecular and turbulent correlation quantities. Symbols are collectively defined in a Notations-list and will, with exceptions not individually be defined.
With reference to *Fig. 1* the acceleration of a fluid particle will be given by

$$\mathbf{a} = \dot{\mathbf{v}} + 2\boldsymbol{\Omega} \times \mathbf{v} + \boldsymbol{\Omega} \times (\boldsymbol{\Omega} \times (\mathbf{r} + \mathbf{x})) + \dot{\boldsymbol{\Omega}} \times \mathbf{x} \simeq \dot{\mathbf{v}} + 2\boldsymbol{\Omega} \times \mathbf{v}, \tag{2.1}$$

thus ignoring[3] the centrifugal and the Euler accelerations but maintaining the Coriolis term. In component form,

$$\mathbf{a} = \left\{ \left(\dot{u} + \tilde{f}w - fv\right), \left(\dot{v} + fu\right), \left(\dot{w} - \tilde{f}u\right) \right\} \tag{2.2}$$

where

$$f = 2\Omega \sin\phi, \qquad \tilde{f} = 2\Omega \cos\phi \tag{2.3}$$

and

$$\dot{u} = \frac{\partial u}{\partial t} + \frac{\partial u}{\partial x}u + \frac{\partial u}{\partial y}v + \frac{\partial u}{\partial z}w, \ldots . \tag{2.4}$$

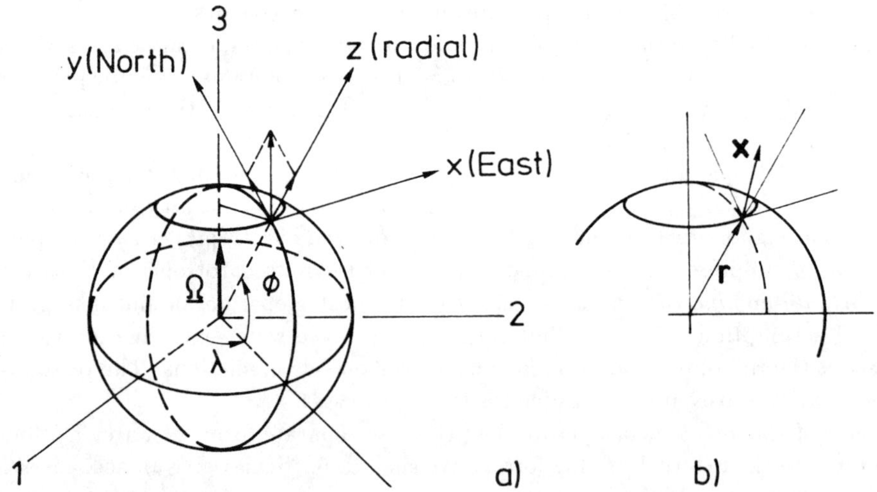

Figure 2.1: Sketch of the rotating Earth indicating two co-rotating coordinate systems (1, 2, 3) and (x, y, z) with axes pointing towards East, North and radially. λ and ϕ are the angles of longitude and latitude.

[3]An alternative way to argue is a follows: The centrifugal acceleration may, in the momentum equation, be combined with the external force to form the effective external force, here effective gravity force. On the earth this effectively is done, as gravity is always measured as a combination of true gravity and gravity due to centrifugal forces. This makes the globe a geoid, different from a sphere.

The two parameters, f and $\tilde{f}$ are called Coriolis parameters, however in view of the shallow water approximation only f will enter later equations. For most applications f may be assumed to be constant, however when f is small or the North-South extent of the basin large then $\beta = df/dy$ must not be ignored[4].

2.1 The basic boundary value problem

The balance laws of mass, momentum and energy comprise the relevant *field equations*;

$$\begin{aligned}
&\text{mass:} && \frac{d\rho}{dt} + \rho \,\mathrm{div}\mathbf{v} = 0, \\
&\text{tracer mass:} && \rho\, \dot{c}_\alpha = -\mathrm{div}\mathbf{j}_\alpha + \pi_\alpha, \qquad (\alpha = 1, \ldots, N), \\
&\text{momentum:} && \rho\{(\dot{\mathbf{v}} + 2\mathbf{\Omega} \times \mathbf{v})\} = -\mathrm{grad}p + \mathrm{div}\mathbf{t} + \rho\mathbf{f}, \\
&\text{energy:} && \rho c_v \dot{T} = -\mathrm{div}\mathbf{q}.
\end{aligned} \tag{2.5}$$

Here ρ, $\mathbf{v}$, c_α, $\mathbf{j}_\alpha$, π_α, $\mathbf{\Omega}$, p, $\mathbf{t}$, $\mathbf{q}$ are the density (ρ^* = reference value), the (relative) fluid velocity, concentration (ratio of masses) of tracer α, the mass flux of tracer α [5], the production density of tracer α, the vector of angular velocity of the Earth's rotation, the pressure, the stress deviator and heat flux vector, respectively. Furthermore, internal and external heat sources have been ignored, and the tracer mass balance has been written for N different conservative ($\pi_\alpha = 0$) or nonconservative ($\pi_\alpha \neq 0$) tracers; c_1 could, for instance, be salinity, c_2 particulate phosphate concentration and c_3 a certain concentration of a plankton with $-\pi_3$ being the rate of annihilation of the phosphate nutrient and π_3 the rate of production of planktons. Their functional relations must be provided by the biologist.

It is assumed that equations (2.5) are valid for ensemble or time averages of corresponding formally identical equations that are describing the fine scale processes of otherwise turbulent motions. These fine scale processes are filtered out in equations (2.5) by applying appropriate closure conditions.

We shall choose the simplest possible *closure conditions*, viz[6].

$$\mathbf{j}_\alpha = -\rho_* \left(D^H \nabla c_\alpha, D^V \frac{\partial c_\alpha}{\partial z} \right), \qquad \mathbf{q} = -\rho_* c_v \left(D^H \nabla T, D^V \frac{\partial T}{\partial z} \right),$$

[4] In the above a Cartesian coordinate system is used. In such a system f is constant. The β-plane approximation stems from a first order approximation of the equations of motion in spherical coordinates (in which the Coriolis parameter varies with latitude). The linear correction to constant f then corresponds to the β-plane approximation.

[5] We employ the simplest possible mixture concept with $N + 1$ balance laws of mass for each constituent and one momentum and one energy balance law for the mixture as a whole. Thus, $\mathbf{v}$ is the barycentric velocity, ρ the mixture density, $()^{\cdot}$ the time derivative following the barycentric velocity and $\mathbf{j}_\alpha = \rho(\mathbf{v}_\alpha - \mathbf{v})$ the diffusive mass flux of the constituent α. For small concentrations of the tracer substances the barycentric velocity will only slightly differ from the water velocity. In subsequent calculations we shall often refer to the barycentric velocity as the (water) velocity; confusion is not likely to arise.

[6] The form of this stress tensor is awkward - in order not say wrong - in two respects. First it is not symmetric even though it should be and second it is inobjective through its postulate. Oceanographers never seem to have been bothered by this. We shall later only use $\mathbf{t} = \mathbf{0}$ and so will not dwell upon this inconsistency.

$$\mathbf{t} = \rho_* \begin{pmatrix} A\dfrac{\partial u}{\partial x} & A\dfrac{\partial u}{\partial y} & \nu\dfrac{\partial u}{\partial z} \\ A\dfrac{\partial v}{\partial x} & A\dfrac{\partial v}{\partial y} & \nu\dfrac{\partial v}{\partial z} \\ A\dfrac{\partial w}{\partial x} & A\dfrac{\partial w}{\partial y} & \nu\dfrac{\partial w}{\partial z} \end{pmatrix} \tag{2.6}$$

Here and henceforth $\nabla = (\partial/\partial x, \partial/\partial y)$ is the two - dimensional Nabla operator. Later, we shall also employ the notation $A := \nu_H$, $\nu := \nu_V$. D_H, D_V $[m^2 s^{-1}]$ are (turbulent) *mass* and *thermal diffusivities* in the horizontal and vertical directions, respectively, and A, ν $[m^2 s^{-1}]$ are horizontal and vertical momentum diffusivities. As is evident we have assumed that *turbulent Fickian* and *Fourier diffusivities* are the same, but they differ in the horizontal and vertical directions. Moreover, the Reynolds-stress tensor $\mathbf{t}$ in (2.6) is a physically inobjective quantity and the horizontal momentum exchange term, A, differs from the corresponding vertical one, ν.

Finally we must add an *equation of state*, e.g.

$$\rho = \rho(T, p, c_1, \ldots, c_N), \quad \text{or} \quad \rho = \rho(T, S), \qquad S = c_1 .$$

Usually one can ignore a pressure dependence and also may suppress all tracers other than salinity. Often the equation of state is written in terms of *density anomaly*

$$\sigma = \frac{\rho - \rho_*}{\rho_*} = \sigma(T, S). \tag{2.7}$$

The following formulas have, among others, been proposed:

- For fresh water calculations HUTTER, (1984a)

$$\sigma = -\varepsilon\Theta^2, \qquad \Theta = T - T^*, \qquad \varepsilon = 6.8 \times 10^{-6}[C^{-2}] \tag{2.8}$$

  or according to BÜHRER and AMBÜHL (1975):

$$\rho = \left\{0.059385T^3 - 8.56272T^2 + 65.4891T\right\} \times 10^{-6} + 0.99984298 \tag{2.9}$$

- For ocean waters according to the UNESCO recommendations

$$\begin{aligned} \rho \;=\; & 800.969062 \times 10^{-4} + 588.194023 \times 10^{-4} T + 797.018644 \times 10^{-3} S \\ & -811.465413 \times 10^{-3} T^2 S - 325.310441 \times 10^{-5} TS + 131.710842 \times 10^{-6} S^2 \\ & +476.600414 \times 10^{-7} ST^3 + 389.187483 \times 10^{-7} ST^2 + 287.971530 \times 10^{-8} S^2 T \\ & -611.831499 \times 10^{-10} S^3 . \end{aligned} \tag{2.10}$$

T in (2.9) and (2.10) is in centigrades and S in mass fraction. It is easily checked that the above set forms $(5 + N)$ equations for $(5 + N)$ unknowns, namely p, $\mathbf{v}$, T, c_α, if $\mathbf{f}$ and π_α $(\alpha = 1, \ldots, N)$ are prescribed quantities.

Boundary conditions must be formulated at the free surface and the bottom and comprise *kinematic* and *dynamic* statements. At the *free surface*, $z = \zeta(x,y,t)$ they are

$$\left.\begin{array}{rcl} \dfrac{\partial\zeta}{\partial t} + \dfrac{\partial\zeta}{\partial x}u + \dfrac{\partial\zeta}{\partial y}v - w & = & 0, \\ (-p\mathbf{1} + \mathbf{t})\,\mathbf{n} & = & \mathbf{t}^{atm}_{(\mathbf{n})}, \\ T = T_f(x,t), \qquad \text{or} & & \mathbf{q}\cdot\mathbf{n} = Q^{atm}(x,t), \\ c_\alpha = c^f_\alpha(x,t), \qquad \text{or} & & \mathbf{j}_\alpha\cdot\mathbf{n} = \mathcal{M}^f_\alpha(x,t), \end{array}\right\} \text{at } z = \zeta(x,y,t) \tag{2.11}$$

in which $\mathbf{t}^{atm}_{(\mathbf{n})}$, T_f, c^f_α, Q^{atm} and $\mathcal{M}^{atm}_\alpha$ are presribed functions of position, $\mathbf{x}$, and time, t. The first of equations (2.11) is a kinematic statement expressing the fact that the free surface is material. Strictly, this is only correct if there is no mass added at the free surface and requires $\mathcal{M}^{atm}_\alpha = 0, \alpha = 1,2,\ldots,N$. The vector $\mathbf{t}^{atm}_{(\mathbf{n})}$ describes the surface traction consisting of atmospheric pressure p^{atm} and wind stress components τ^w_x and τ^w_y. The thermal boundary condition either requires the surface temperature, T_f, or the heat flow, Q^{atm}, to be prescribed, and the boundary conditions on the tracer mass consists either of a prescription of c^f_α or its flow $\mathcal{M}^f_\alpha$.

At the *bottom surface*, $z = -H(x,y)$, the kinematic and dynamic boundary conditions are

$$\left.\begin{array}{ll} \dfrac{\partial H}{\partial x}u + \dfrac{\partial H}{\partial y}v + w = 0, & \\ \mathbf{t}^*_{(\mathbf{n})} = -\rho_* C(|\mathbf{v}|)\,\mathbf{v}, & \mathbf{t}^*_{(\mathbf{n})} := \mathbf{tn} - (\mathbf{n}\cdot\mathbf{tn})\,\mathbf{n}, \\ T = T_h(x,t), & \\ c_\alpha = c^H_\alpha(x,t), \qquad \text{or} & \mathbf{j}_\alpha\cdot\mathbf{n} = \mathcal{M}^H_\alpha(x,t). \end{array}\right\} \text{at} \quad z = -H(x) \tag{2.12}$$

The first of these equations is the kinematic equation of motion and expresses the tangency of the barycentric velocity to the ground; as before it is exact only provided there is no mass flow (no drainage, $\mathcal{M}^H_\alpha = 0, \alpha = 1,2,\ldots,N$) through the basal boundary. Relation $(2.12)_2$ is a viscous sliding law, and states that the shear traction $\mathbf{t}^*_{(\mathbf{n})}$ is tangential and opposite in direction to the basal velocity, $\mathbf{v}$, with a drag coefficient C that may depend upon the modulus of the velocity; because $\mathbf{n}\cdot\mathbf{t}^*_{(\mathbf{n})} \equiv 0$ this sliding law automatically satisfies the tangency condition, $\mathbf{v}\cdot\mathbf{n} = 0$. For the temperature we have assumed a Dirichlet condition, whereas the tracer boundary conditions are of either Dirichlet or Neumann-type. In general the salinity is set constant, $c_1 = c^H_1$, while for tracers the mass flow into the sediment is prescribed ($\mathcal{M}_\alpha = \mathcal{M}^H_\alpha, \alpha = 2,3,\ldots,N$). The kinematic boundary conditions are approximate, but in general sufficiently accurate for the situation at hand.

2.2 Scaling analysis - Reduced equations

To compare the magnitudes of the various terms arising in the equations for the processes of interest, we transform them into dimensionless form by writing any physical quantity as $\psi = \psi_0 \bar{\psi}$. ψ_0 is a typical value of the field ψ for the processes under consideration, and $\bar{\psi}$ its dimensionless counterpart, of necessity of order unity, if ψ_0 is appropriately chosen. In ensuing developments the overbar will be omitted for simplicity.

f_0	$\cong$	$10^{-4}\ s^{-1}$	Coriolis parameter
ρ_*	$=$	$10^3\ kgm^{-3}$	Density
σ_0	$\cong$	10^{-3}	Density anomaly
ν_V	$\cong$	$10^{-2}\ m^2s^{-1}$	
A	$\cong$	$10^0 - 10^1\ m^2s^{-1}$	Turbulent
D_V	$\cong$	$10^{-2}\ m^2s^{-1}$	Diffusivities
D_H	$\cong$	$10^0 - 10^1\ m^2s^{-1}$	
T_0	$\cong$	$10°C$	Reference temperature
ΔT_0	$\cong$	$10°C$	Temperature range
L_0	$\cong$	$10^4 - 10^6\ m$	Horizontal length scale
H_0	$\cong$	$10 - 10^3\ m$	Vertical length scale
V_0	$\cong$	$10^{-2} - 10\ ms^{-1}$	Horizontal velocity scale
c_α^0			Scale for constituent α

Table 1) Physical parameters and typical orders of magnitude for the scales (2.13)[7]

We shall select the scales according to the transformations (HUTTER et al., 1991)

$$\begin{aligned}
(x,y,z) &\rightarrow (L_0x, L_0y, H_0z), & p &\rightarrow -\rho_* gz + \rho_* f_0 V_0 L_0 p, \\
(u,v,w) &\rightarrow \left(V_0u, V_0v, V_0\frac{H_0}{L_0}w\right), & \rho &\rightarrow \rho_*(1+\sigma_0\sigma), \\
(t,f,\tilde{f}) &\rightarrow \left(\frac{1}{f_0}t, f_0f, f_0\tilde{f}\right), & T &\rightarrow T_0 + \Delta T_0\Theta, \quad c_\alpha \rightarrow c_\alpha^0.
\end{aligned} \tag{2.13}$$

It is seen, horizontal and vertical lengths and velocities are scaled differently and time is scaled with the Coriolis parameter. The pressure is decomposed into a hydrostatic part and a dynamic component using a reference density ρ_*, the remaining scales being obvious. Typical values for these scales in ocean and lake applications are as indicated in *Table 1.* If the governing equations are being non-dimensionalized with the scales (2.13) they take the forms

Balance of mass

$$\frac{\sigma_0 \partial\sigma}{\mathbb{R}_0 \partial t} + \mathrm{div}\mathbf{v} + \sigma_0 div(\sigma\mathbf{v}) = 0, \tag{2.14}$$

[7]This value of the Coriolis parameter corresponds to mid - latitude conditions

Balance of momentum

$$
\begin{aligned}
(1+\sigma_o\sigma)\left\{\frac{\partial u}{\partial t}+\mathbb{R}_0\,(\mathrm{grad}u)\cdot\mathbf{v}+\mathbf{A}\tilde{f}w-fv\right\} &= -\frac{\partial p}{\partial x}+\nabla\cdot(\mathbb{E}_H\nabla u)+\frac{\partial}{\partial z}\left(\mathbb{E}_V\frac{\partial u}{\partial z}\right),\\
(1+\sigma_0\sigma)\left\{\frac{\partial v}{\partial t}+\mathbb{R}_0\,(\mathrm{grad}v)\cdot\mathbf{v}+fu\right\} &= -\frac{\partial p}{\partial y}+\nabla\cdot(\mathbb{E}_H\nabla v)+\frac{\partial}{\partial z}\left(\mathbb{E}_V\frac{\partial v}{\partial z}\right),\\
(1+\sigma_0\sigma)\left\{\mathbf{A}^2\left[\frac{\partial w}{\partial t}+\mathbb{R}_0\,(\mathrm{grad}w)\cdot\mathbf{v}\right]-\mathbf{A}\tilde{f}u\right\} &= -\frac{\partial p}{\partial z}+\mathbf{A}^2\Bigg[\nabla\cdot(\mathbb{E}_H\nabla w)\\
&\quad+\frac{\partial}{\partial z}\left(\mathbb{E}_V\frac{\partial w}{\partial z}\right)\Bigg]-\mathbb{B}\sigma, \qquad (2.15)
\end{aligned}
$$

Balance of energy or tracer mass

$$
(1+\sigma_0\sigma)\left\{\frac{\partial\Theta_\alpha}{\partial t}+\mathbb{R}_0\,(\mathrm{grad}\Theta_\alpha)\cdot\mathbf{v}\right\}=\nabla\cdot\left(\frac{\mathbb{E}_H}{\mathbb{P}_H}\nabla\Theta_\alpha\right)+\frac{\partial}{\partial z}\left(\frac{\mathbb{E}_V}{\mathbb{P}_V}\frac{\partial\Theta_\alpha}{\partial z}\right)+\Pi_\alpha\pi_\alpha \qquad (2.16)
$$

in which $\Theta_\alpha=(\Theta,c_\beta),\beta=1,2,\ldots,N,\alpha=1,2,\ldots,N+1$. Moreover, grad, div are three dimensional operators, while ∇ only operates on the horizontal coordinates and all of these operate on the dimensionless coordinates. The dimensionless hallow parameters are listed in *Table 2* with their common nomenclature and orders of magnitude as obtained with the scales of *Table 1*. Depending upon the orders of magnitude of the dimensionless parameters, various simplified versions of field equations emerge. Later we shall also nondimensionalize the boundary conditions.

The BOUSSINESQ-*assumption* corresponds to the limit

$$
\sigma_0\to 0 \qquad \text{and} \qquad \frac{\sigma_0}{\mathbb{R}_0}\to 0. \qquad (2.17)
$$

It is tantamount to the incompressiblity assumption (div$\mathbf{v}=0$, see (2.14)) and eliminates the density variations from all rate terms in equations (2.14) - (2.16). In other words, the BOUSSINESQ assumption assumes density variations to be negligible except in the buoyancy term. In ocean and lake applications this assumption is always made even though *Table 2* suggests that simultaneous satisfaction of both limits (2.17) may not always be justified.
In the *shallow water approximation* one supposes the asymptotic limits

$$
\mathbf{A}\to 0, \qquad \mathbf{A}^2\mathbb{E}_H\to 0, \qquad \mathbf{A}^2\mathbb{E}_V\to 0. \qquad (2.18)
$$

Inspection of (2.15) indicates that this assumption eliminates $\tilde{f}$ from the governing equations and reduces $(2.15)_3$, to a *hydrostatic force balance.*
With the BOUSSINESQ and the shallow water assumptions the most general field equations are obtained that are in use in global ocean and lake dynamic modelling. They read
Balance of mass

$$
\mathrm{div}\mathbf{v}=0, \qquad (2.19)
$$

$\mathrm{A} = \dfrac{H_0}{L_0}$		10^{-5} – 10^{-2}	Aspect ratio
$\mathrm{R_0} = \dfrac{V_0}{f_0 L_0}$		10^{-4} – 10^{0}	ROSSBY number
σ_0		10^{-3}	Density anomaly
$\mathrm{E}_H = \dfrac{A}{f_0 L_0^2}$		$\underline{10^{-4}}$ – 10^{-2}	Horizontal EKMAN number
$\mathrm{E}_V = \dfrac{\nu}{f_0 H_0^2}$		10^{-4} – $\underline{10^{-2}}$	Vertical EKMAN number
$\mathrm{B} = \dfrac{g\sigma_0 H_0}{f_0 V_0 L_0}$	$\sim$	10^{-1}	Buoyancy (BOUSSINESQ) parameter
$\mathrm{P}_H = \dfrac{v}{D_H}$	$\sim$	10^{-1} – 10^{1}	Horizontal PRANDTL number
$\mathrm{P}_V = \dfrac{v}{D_V}$	$\sim$	2	Vertical PRANDTL number
$\Pi_\alpha = \dfrac{\pi_\alpha^0}{c_\alpha^0 f_0}$	$\leq$	1	Production number
$\mathrm{L} = \dfrac{\zeta_0}{H_0}$	$\sim$	10^{-2}	Rigid Lid parameter

Table 2) Dimensionless parameters

Balance of momentum

$$\begin{aligned}
\frac{\partial u}{\partial t} + \mathbb{R}_0\,(\mathrm{grad}u)\cdot\mathbf{v} - fv &= -\frac{\partial p}{\partial x} + \nabla\cdot(\mathbb{E}_H\nabla u) + \frac{\partial}{\partial z}\left(\mathbb{E}_V\frac{\partial u}{\partial z}\right), \\
\frac{\partial v}{\partial t} + \mathbb{R}_0\,(\mathrm{grad}v)\cdot\mathbf{v} + fu &= -\frac{\partial p}{\partial y} + \nabla\cdot(\mathbb{E}_H\nabla v) + \frac{\partial}{\partial z}\left(\mathbb{E}_V\frac{\partial v}{\partial z}\right), \\
0 &= -\frac{\partial p}{\partial z} - \mathbb{B}\sigma.
\end{aligned} \tag{2.20}$$

Balance of energy (tracer mass)

$$\frac{\partial\Theta_\alpha}{\partial t} + \mathbb{R}_0\,(\mathrm{grad}\Theta_\alpha)\cdot\mathbf{v} = \nabla\cdot\left(\frac{\mathbb{E}_H}{\mathbb{P}_H}\nabla\Theta_\alpha\right) + \frac{\partial}{\partial z}\left(\frac{\mathbb{E}_V}{\mathbb{P}_V}\frac{\partial\Theta_\alpha}{\partial z}\right) + \Pi_\alpha\pi_\alpha, \qquad \alpha = 1,2,\ldots,N+1 \tag{2.21}$$

and, obviously, *do not apply for problems of strong upwelling and downwelling* as mentioned in the Introduction. Further reductions of these equations have been proposed and, in fact, been in use prior to the application of the above equations. For instance, one may ignore horizontal in comparison to vertical diffusion,

$$\mathbb{E}_H \to 0, \qquad \mathbb{E}_V \not\to 0, \tag{2.22}$$

as done in all oceanographic works prior to large scale computer applications. Such a simplification has been proved useful and adequate for storm surge predictions, and, in conjunction with the *small* ROSSBY *number* limit,

$$\mathbb{R}_0 \to 0, \tag{2.23}$$

forms the basis for the famous EKMAN *problem* in particular, and the determination of the vertical current distribution in general. However, this limit is not useful for dispersion of tracer substances, but may again be applied for computations of the development of the seasonal variation of the thermocline.
Another popular limit of equations (2.19) - (2.21) applies when time scales are too short *that turbulent diffusive mechanisms and tracer mass production could effectively develop.* For the former processes, such a time scale is

$$T^H_{diffusion} = \frac{H_0^2}{D_V} \simeq \mathcal{O}(\text{hours, days}).$$

That of the latter depends on the biochemical (decay) rate constants and is generally larger. Thus, one may in equation (2.21) ignore the terms on the right hand side and thus has approximately

$$\frac{d\Theta_\alpha}{dt} = 0, \qquad (\alpha = 1, 2, \ldots, N+1). \tag{2.24}$$

The temperature and the tracer concentrations are constant along particle paths. The statement (2.24) is often referred to as the *adiabaticity assumption*, even though this is strictly not correct. In view of the equation of state $\rho = \rho(\Theta_\alpha)$ it implies

$$\frac{d\rho}{dt} = 0, \quad (\alpha = 1, 2, \ldots, N+1), \tag{2.25}$$

an equation that is often and erroneously claimed to be a consequence of the Boussinesq assumption. We shall refer to this as the *generalized adiabatic assumption.*
The most simple field equations *in dimensional form* that ignore dissipation and advective

non-linearities are thus

$$\begin{aligned}
&\mathrm{div}\mathbf{v} = 0, && \text{mass balance}\\
&\left.\begin{aligned}\rho^*\left(\frac{\partial u}{\partial t} - fv\right) &= -\frac{\partial p}{\partial x},\\ \rho^*\left(\frac{\partial v}{\partial t} + fu\right) &= -\frac{\partial p}{\partial y},\end{aligned}\right\} && \text{horizontal momentum balances}\\
&\frac{\partial p}{\partial z} + \rho g = 0, && \text{vertical momentum balance}\\
&\frac{d\rho}{dt} = 0. && \text{generalized adiabaticity}
\end{aligned} \tag{2.26}$$

These incorporate the *shallow water (hydrostatic) hypothesis*, the BOUSSINESQ and the *adiabaticity assumptions* and are the basis for most wave models in geophysical fluid mechanics. The last of equations (2.26), incidentally is only spuriously non-linear because the common lake and ocean situation is

$$\rho(\mathbf{x},t) = \rho_0(z) + \rho'(\mathbf{x},t), \tag{2.27a}$$

where $\rho_0(z)$ is a basic long term non-uniform density field so that the linearized version of the fifth of equations (2.26) is

$$\frac{\partial \rho'}{\partial t} + \frac{d\rho_0}{dz} w = 0. \tag{2.27b}$$

For general analyses the shallow water or hydrostatic hypothesis is not invoked, and for non-linear large amplitude waves the advective transport is significant. Under these circumstances the governing equations take the form

$$\begin{aligned}
div\mathbf{v} &= 0,\\
\rho^*\left(\frac{du}{dt} - fv\right) &= -\frac{\partial p}{\partial x},\\
\rho^*\left(\frac{dv}{dt} + fu\right) &= -\frac{\partial p}{\partial y},\\
\rho^*\frac{dw}{dt} &= -\frac{\partial p}{\partial z} - \rho g,\\
\frac{d\rho}{dt} &= 0.
\end{aligned} \tag{2.28}$$

When density variations play no significant role (which is the case for all barotropic processes), then the last of equations (2.28) is dropped.

As has been demonstrated for the field equations, the boundary conditions can also be put into dimensionless form by using the scales (2.13). As an example, consider the boundary condition of stress $(2.11)_2$ at the free surface, that takes the form

$$(p - t_{xx})\frac{\partial \varsigma}{\partial x} - t_{xy}\frac{\partial \varsigma}{\partial y} + t_{xz} = N t_x^{atm},$$

$$
\begin{aligned}
-t_{xy}\frac{\partial\zeta}{\partial x}+(p-t_{yy})\frac{\partial\zeta}{\partial y}+t_{yz} &= Nt_y^{atm},\\
-t_{xz}\frac{\partial\zeta}{\partial x}-t_{yz}\frac{\partial\zeta}{\partial y}+(p-t_{zz}) &= Nt_z^{atm},
\end{aligned}
\tag{2.29}
$$

with

$$
N=\left\{1+\left(\frac{\partial\zeta}{\partial x}\right)^2+\left(\frac{\partial\zeta}{\partial y}\right)^2\right\}^{\frac{1}{2}}.
\tag{2.30}
$$

Using the scales $(2.13)_1$, it may then be shown that $N=1+\mathcal{O}(\mathbf{A}^2)$. Alternatively, (2.29) may be written as

$$
\begin{aligned}
\mathbf{A}[\,]+t_{xz} &= \tau_x^w,\\
\mathbf{A}[\,]+t_{yz} &= \tau_y^w,\\
\mathbf{A}[\,]-p+t_{zz} &= -p^{atm},
\end{aligned}
\tag{2.31}
$$

in which the content of the bracketed terms is of order unity, and for which, by using the representation (2.6), it is easily shown that $t_{zz}=\mathcal{O}(\mathbf{A})$. The shallow water approximation therefore states that the horizontal shear stresses t_{xz}, t_{yz} equal the corresponding wind stresses τ_x^w, τ_y^w and the surface pressure equals the atmospheric pressure p^{atm}. A similar approach can also be used for the boundary conditions of energy flux and mass flux for constituent α, yielding

$$
\mathbf{A}[\,]+q_z=Q^{atm},\qquad \mathbf{A}[\,]_\alpha+j_z^\alpha=\mathcal{M}_\alpha^f.
\tag{2.32}
$$

These results make it plausible that under the prerequisites of the shallow water approximation, $\mathbf{A}\to 0$, the boundary conditions may be written in the following form: At the *free surface*

$$
\left.
\begin{aligned}
&\frac{\partial\zeta}{\partial t}+\frac{\partial\zeta}{\partial x}u+\frac{\partial\zeta}{\partial y}v-w=0,\\
&p=p_{atm},\qquad \rho_*\nu\left(\frac{\partial u}{\partial z},\frac{\partial v}{\partial z}\right)=\left(\tau_x^w,\tau_y^w\right),\\
&T=T_f,\qquad S=S_f,\\
&\frac{\partial c_\alpha}{\partial z}=\frac{\partial c_\alpha^f}{\partial z},\qquad (\alpha=2,3,\ldots,N),
\end{aligned}
\right\}\quad \text{at}\quad z=\zeta(x,y,t)
\tag{2.33}
$$

and *at the base*

$$
\left.
\begin{aligned}
&\frac{\partial H}{\partial x}u+\frac{\partial H}{\partial y}v+w=0,\\
&(\tau_x,\tau_y)=-\rho_*C\sqrt{u^2+v^2}\,(u,v),\\
&T=T_H,\qquad S=S_H,\\
&\frac{\partial c_\alpha}{\partial z}=\frac{\partial c_\alpha^H}{\partial z},\qquad (\alpha=2,3,\ldots,N),
\end{aligned}
\right\}\quad \text{at}\quad z=-H(x,y).
\tag{2.34}
$$

Here, T_f, S_f, $\partial c_\alpha^f/\partial z$, T_H, S_H, $\partial c_\alpha^H/\partial z$ are prescribed functions of position and time. The first in (2.33) and (2.34) are kinematic statements and express the fact that the free and

the bottom surfaces are material. Eqs. $(2.33)_{2,3}$ relate atmospheric pressure and windstress to the surface traction. Correspondingly, relation $(2.34)_2$ is a sliding law for the bottom boundary layer. For temperature and salinity we have assumed Dirichlet conditions, whereas the remaining tracer boundary conditions are of Neumann type.
New insight can be brought by, if the scale for the surface elevation is differently chosen from the vertical length scale H_0. Indeed if

$$\zeta \rightarrow \zeta_0 \bar{\zeta} \tag{2.35}$$

in which ζ_0 is a typical amplitude of the free surface elevation, the dimensionless form of the kinematic surface equation reads

$$\mathbb{L}\left\{\frac{\partial \zeta}{\partial t} + \mathbb{R}_0 \nabla \zeta v\right\} - \mathbb{R}_0 w = 0, \qquad z = \zeta(x, y, t). \tag{2.36}$$

$\mathbb{L} = \zeta_0 / H_0$ is the *rigid lid parameter*, and, in the limit as $\mathbb{L} \rightarrow 0$ equation (2.36) gives rise to the *rigid lid assumption* with the kinematic boundary condition

$$w = 0, \qquad z = 0. \tag{2.37}$$

It can further be shown that with this new free surface scaling the remaining, dynamic boundary conditions (2.33) in the shallow water approximation are not altered.
The rigid lid assumption eliminates barotropic gravity waves. This becomes obvious when the vertically integrated version of the balance equations of mass and momentum are looked at. The only vertical length scale in those equations arises in the kinematic equation; a balance between w and ζ must then necessarily yield $\zeta_0 = H_0$ or $\mathbb{L} = 1$.

2.3 Remarks on Models

Fundamental to the judgement whether a mathematical description of a physical process is adequate is the selection of variables for which the comparison is performed. For instance in storm surge analyses surface-elevation-time series must be accurately predicted, but it is less important that velocity profiles are well reproduced. Adequacy of mathematical models depends in general on the target variables.
The accepted reduced models range from the field equations (2.14) - (2.21) to those listed in (2.28). Generally, simple models are useful for diagnostic purposes; they help to understand the physics. In attempts to develop prognostic tools models need be more complete. However, *even the most complex accepted version* (eqs. (2.19) - (2.21)) *is inappropriate or insufficient to properly model some important features in ocean and lake dynamics, in particular when turbulent vertical and horizontal mixing mechanisms are significant.* Steep internal fronts, up-welling and down-welling, billowing in the thermocline region are among these features and could be incorporated, at least in principle, by dispensing with the shallow water assumption. Most likely more limiting is the form of the turbulent closure conditions (2.6). These gradient-type relationships are known to be inappropriate e.g. when sedimentation is significant. They could in those circumstances be replaced by higher order closure conditions. The difficulties, however lie already on a lower level. Very little is known about numerical values of the

diffusivities. D_H and A can probably be taken as constants; the vertical diffusivities D_V and ν vary with depth and seem to have a pronounced minimum in the thermocline. They are usually postulated to depend on *fetch* and the RICHARDSON *number*, e.g.

$$\mathbb{R}_i = \frac{\frac{g}{\rho}\frac{d\rho_0}{dz}}{\left|\frac{\partial v}{\partial z}\right|^2}. \tag{2.38a}$$

A widely adopted form for ν is

$$\nu = \frac{F\left(\nu_0, \frac{\nu_1}{\rho_0}|\tau_{wind}|\right)}{\left(1+\frac{10}{3}\mathbb{R}_i\right)^4}, \qquad \nu_0 = 0.0025\ m^2 s^{-1}, \qquad \nu_1 = 25 - 100\ s. \tag{2.38b}$$

with $F(x,y) = Max(x,y)$. Such relations are still inappropriate and insufficient in modelling episodic thermocline erosions that are so important in the vertical tracer transport. I know of no paper in which this process would be adequately modelled.

Another complexity directly related to the turbulent closure conditions (2.6) is the additional introduction of numerical diffusion when the differential equations are replaced by finite difference or finite element approximations or other numerical schemes. Obviously the mesh size defines the scales of the turbulent eddies that can be resolved; large 'activity' in the thermocline region requires a dense layering in its vicinity. On the other hand, numerical stability also demands a certain amount of diffusion, particularly in the horizontal direction. In present numerical models this numerical diffusion is comparable in size to the physical diffusion or exceeds it; for instance, OMAN (1982) and HUTTER (1991) use a horizontal turbulent diffusivity $A = f\Delta x \Delta y/4$ which is far bigger than corresponding physical values. This, of course, raises some questions about the physical appropriateness of results. However, when advection is very strong, i.e. during storms and their secession, turbulent diffusion may be hidden behind the strong advective processes. In addition in those circumstances the turbulent scales tend to be large so that subgrid scales are of lesser importance. Both effects work in favor of the model equations. It is therefore understandable that such events are better reproduced in computational attempts than the oscillatory behavior under usual wind forcing.

The following recommendations may serve as rules of thumb for the application of the different equation sets:

- Advection and diffusion may be ignored in oscillation problems. All *seiche* analyses are performed on this level. They have primarily diagnostic value and provide information about the time scales of the barotropic and baroclinic motions that persistently exist in enclosed basins.

- When strong wind induced motions are considered and the amplitude of the surface elevation response is important, but details in the structure of the horizontal water transport are relatively insignificant then advection and *horizontal* diffusion may be ignored. Clearly, the kinematic surface equation is essential in this case and the rigid lid assumption should not be made. Such a model is appropriate neither for diagnostic nor prognostic tracer transport studies.

- When transport of heat or particulate matter is considered then the full equations (2.18) - (2.20) must be taken into account. Often the rigid lid assumption may be justifed; it simplifies numerical computations considerably. Dispersion of a tracer through a narrow channel in a coupled lake system may, however, be strongly influenced by the barotropic co-oscillation of the connected basins. In such a case the rigid lid assumption eliminates the gravity oscillation and thus filters out exactly the mentioned resonating motion.

We conclude this section with a few remarks on the coupling of the fields. Clearly the velocity distribution has an effect on the temperature and vice versa, and such a coupling also exists between the velocity field and the tracer mass distribution. The degree of this coupling may, however, largely simplify the solution procedure of the governing equations.

- The mechanical and thermal equations have a two-sided coupling through the buoyancy term. They must be solved simultaneously; however, a full baroclinic problem with radiative balance at the surface through the seasons cannot yet be solved with sufficient satisfaction. On time scales that are small in comparison to the seasonal evolution of the thermocline the thermal state, i.e. the stratification may be regarded as frozen-in, i.e., a ground stratification is assumed to be prescribed. Such models are best termed *semi-baroclinic.* Of course, equations are still dynamically coupled.

- On time scales of the seasonal development of the thermocline horizontal exchange processes may be regarded as infinitely fast in comparison to the vertical processes. Here vertically one-dimensional problems are adequate.

- The mechanical and tracer-mass-transport equations do, in general, have a weaker coupling, since the buoyancy term is not influenced by the concentration. Mathematically, the mechanical (and the thermal) equations separate in this case from the tracer diffusion equations; the two sets can be solved in succession. This rule is only occasionally violated e.g. when a river carries a heavy suspended sediment load or when very high density plankton develops. Note that the presented equations are not appropriate in this latter case, since the plankton has its own motion relative to the fluid. An additional plankton momentum evolution equation would have to be added in this case.

2.4 Vertically integrated transport equations

The linear adiabatic equations lie at one end of the range of approximations; much of the theory of wave motions in lakes and the ocean has, however been developed on the basis of models in which the vertical current structure is ignored except for finite jumps from one layer to another. In a *one-layer model* the procedure is to average the equations by integrating them over water depth. The basic unknown in this case is not the local velocity field, but the *integrated volume transport*

$$\mathbf{V} = (U, V) = \int_{-H}^{\zeta} \mathbf{v}\, dz = \int_{-H}^{\zeta} (u, v)\, dz. \tag{2.39}$$

Note that $\mathbf{v}$ possesses here only two components. By integrating the continuity equation and the *horizontal* momentum equations over depth and thereby using the boundary conditions, (2.33) and (2.34), one can prove that these equations reduce to

$$\begin{aligned} &\frac{\partial \zeta}{\partial t} + \nabla \cdot \mathbf{V} = 0, \\ &\frac{\partial \mathbf{V}}{\partial t} + f\hat{\mathbf{k}} \times \mathbf{V} = -\frac{(H+\zeta)}{\rho^*}\nabla p_e + \mathbf{F}, \end{aligned} \tag{2.40}$$

in which $\hat{\mathbf{k}}$ is a unit vector in the z-direction; the cross product $\hat{\mathbf{k}} \times \mathbf{V}$ has to be understood to be three dimensional with $\hat{\mathbf{k}} = (0, 0, 1)$ and $\mathbf{V} = (U, V, 0)$, so that its result is $(-V, U, 0)$, however in (2.40) only the two nontrivial components $(-V, U)$ are assumed to be reproduced. Furthermore,

$$\begin{aligned} p_e &= p^{atm} + \rho_* g\,(\zeta - z)\,, \\ p_i &= \rho_* g \int_z^{\zeta} \frac{\rho - \rho_*}{\rho_*} d\bar{z} = \rho_* g \int_z^{\zeta} \sigma(\bar{z})\, d\bar{z}, \\ \mathbf{F} &= -\frac{1}{\rho^*} \int_{-H}^{\zeta} \nabla p_i dz - \nabla \cdot \int_{-H}^{\zeta} (\mathbf{v} \otimes \mathbf{v} - \mathbf{\Gamma})\, dz + \frac{\mathbf{t}^{wind} - \mathbf{t}^{bottom}}{\rho^*}. \end{aligned} \tag{2.41}$$

The pressures p_e and p_i are called the *barotropic (external)* and the *baroclinic (internal) pressures*, and the force $\mathbf{F}$ consists of an integrated baroclinic pressure, an integrated advective term $(\mathbf{v} \otimes \mathbf{v})$, the integrated horizontal turbulent REYNOLDS stresses[8] $(\mathbf{\Gamma})$ and the contributions from wind stress and bottom friction.
We prove the first of relations (2.40) and leave the second, more complicated one to the reader as an exercise. To this end we start from the mass balance $\nabla \cdot \mathbf{v} + \partial w / \partial z = 0$ and integrate this equation over water depth

$$\begin{aligned} &\int_{-H}^{\zeta} \left(\nabla \cdot \mathbf{v} + \frac{\partial w}{\partial z} \right) dz = 0, \\ &\nabla \cdot \int_{-H}^{\zeta} \mathbf{v}\, dz - \{(\nabla \zeta) \cdot \mathbf{v} - w\}_{|\zeta} - \{(\nabla H) \cdot \mathbf{v} + w\}_{|-H} = 0, \end{aligned} \tag{2.42}$$

[8] $\mathbf{\Gamma}$ consists of $\frac{1}{\rho_*}\begin{bmatrix} t_{xx} & t_{xy} \\ t_{xy} & t_{yy} \end{bmatrix}$, which are the horizontal components of the turbulent stress $\mathbf{t}$.

in the second of which the orders of integration and differention have been intercharged and LEIBNIZ' rule has been used. With the kinematic boundary conditions (2.33), and (2.34), and the definition (2.39), equation (2.40), is obtained, qed.
The following additional assumptions are needed to transform (2.40) to an integrable system of equations for the surface elevation ζ and the volume transport $\mathbf{V}$:

- Set
$$\int_{-H}^{\zeta} \mathbf{v} \otimes \mathbf{v}\, dz = \frac{\alpha}{H+\zeta} \mathbf{V} \otimes \mathbf{V},$$
where α is a constant. Usually one chooses $\alpha = 1$, this is the so-called BOUSSINESQ *coefficient* and describes the non-uniformity of the velocity profile with depth. For $\alpha = 1$ uniformity is assumed.

- Ignore the internal pressure ($p_i \simeq 0$) and the horizontal Reynold-stresses ($\mathbf{\Gamma} \simeq \mathbf{0}$).

- Assume the bottom friction $\mathbf{t}^{bottom}$ in the form
$$\mathbf{t}^{bottom} = \rho^* C \frac{\|\mathbf{V}\|\mathbf{V}}{(H+\zeta)^2} \qquad \left(C \equiv 2 \times 10^{-3}\right).$$

With these assumptions the following boundary value problem emerges from (2.40):

$$\begin{aligned} &\frac{\partial \zeta}{\partial t} + \nabla \cdot \mathbf{V} = 0, \\ &\frac{\partial \mathbf{V}}{\partial t} + f\hat{\mathbf{k}} \times \mathbf{V} = -\nabla \cdot \left(\alpha \frac{\mathbf{V} \otimes \mathbf{V}}{H+\zeta}\right) - C \frac{\|\mathbf{V}\|\mathbf{V}}{(H+\zeta)^2} - \frac{H+\zeta}{\rho^*} \nabla p_e + \frac{\mathbf{t}^{wind}}{\rho^*}, \end{aligned} \tag{2.43}$$

to be solved in the lake or ocean domain $\mathcal{D}$ and subject to the boundary conditions

$$\mathbf{V} \cdot \mathbf{N} \text{ is prescribed along the shore } \partial\mathcal{D}. \tag{2.44}$$

Equations (2.43), (2.44) can only be meaningful if the vertical profiles of the horizontal velocities do not deviate too much from uniformity. In other words, *times for the vertical diffusion*

$$T^{V}_{diff} = \frac{H_0^2}{4\nu_V},$$

where H_0 is a typical water depth and ν_V a vertical diffusivity, must be small in comparison to *times of horizontal advection*,

$$T^{H}_{adv} = \frac{L}{U},$$

in which L is a typical horizontal length and U a characteristic horizontal velocity. Consequently,

$$\frac{H_0^2 U}{4\nu_V L} \ll 1. \tag{2.45}$$

With this inequality it can be demonstrated that wind induced currents can be studied with equations (2.43), (2.44) e.g. in Neusiedlersee[9], but probably not in the homogeneous Zürichsee. For the latter a one-layer model is apparently too simple.
Nevertheless, the linearized version of (2.43)

$$\frac{\partial \zeta}{\partial t} + \nabla \cdot \mathbf{V} = 0$$

$$\frac{\partial \mathbf{V}}{\partial t} + f\hat{\mathbf{k}} \times \mathbf{V} = -gH\nabla\zeta - \frac{H\nabla p_{atm}}{\rho^*} + \frac{1}{\rho^*}\mathbf{t}^{wind} \tag{2.46}$$

is very popular, in particular when eigensolutions ($\nabla p_{atm} = \mathbf{0}$, $\mathbf{t}^{wind} = \mathbf{0}$) are sought. This set is the basic set of equations for *surface seiches.* Under normal circumstances the atmospheric pressure gradients are small enough that over the extents of common lakes atmospheric pressure variations cannot become large. Thus, ∇p_{atm} is generally ignored in equations like (2.46). This may not be appropriate when weather fronts cross a lake or ocean basin.

2.5 Vorticity and potential vorticity

2.5.1 Vorticity

By taking the curl of the momentum equation, $(2.5)_3$,

$$\frac{d\mathbf{v}}{dt} + 2\Omega \times \mathbf{v} = -\frac{1}{\rho}\left(\text{grad}p + \text{div}\mathbf{t} + \rho\mathbf{f}\right), \tag{2.47}$$

and assuming that the body forces are derivable from a potential, the following extension of the HELMHOLTZ *vorticity equation* can be derived:

$$\frac{d}{dt}\left(\frac{\omega_a}{\rho}\right) = (\text{grad}\mathbf{v})\frac{\omega_a}{\rho} + \frac{\text{grad}\rho \times \text{grad}p}{\rho^3} + \frac{1}{\rho}\text{curl}\left(\frac{\text{div}\mathbf{t}}{\rho}\right) \tag{2.48}$$

in which ω_a is the absolute vorticity, defined by

$$\omega_a = 2\boldsymbol{\Omega} + \omega, \qquad \omega = \text{curl}\mathbf{v}. \tag{2.49}$$

As is evident from (2.48), the time rate of change of the specific absolute vorticity is due to three terms: the production due to viscous and turbulent stresses (last term on the right of (2.48)), the baroclinicity i.e. the fact that the *isobar* and *isopycnal* surfaces are not parallel ($\text{grad}\rho \times \text{grad}p \neq 0$) and a production that is due to *vortex stretching* and *vortex tilting* (the first term on the right). This last property shall now be proved.
A fluid is called barotropic if its pressure depends at most on density $p = p(\rho)$. For such a case and when viscous and turbulent stresses are insignificant (i.e. when the adiabaticity assumption is justified), equation (2.47) reduces to

$$\frac{d}{dt}\left(\frac{\omega_a}{\rho}\right) = (\text{grad}\mathbf{v})\left(\frac{\omega_a}{\rho}\right), \tag{2.50a}$$

[9]For Neusiedelersee $H_0 \sim 2m, L \simeq 10^5 m$, for Zürichsee $H_0 \simeq 70m, L \simeq 10^4 m$.

or for an *incompressible fluid* (or when the BOUSSINESQ assumption is justified)

$$\frac{d}{dt}\omega_a = (\text{grad}\mathbf{v})\,\omega_a. \tag{2.50b}$$

Equations (2.50a,b) comprise the content of the original HELMHOLTZ *vorticity theorem*. Let $\mathbf{x} = \chi(\mathbf{X}, t)$ be the *trajectory* of a fluid particle that was initially at position $\mathbf{X}$. Then

$$\mathbf{F} = \frac{\partial \chi(\mathbf{X}, t)}{\partial \mathbf{X}} \tag{2.51}$$

is called its *deformation gradient* at time t. With (2.51) it is easy to prove that (2.50a) possesses the solution

$$\frac{\omega_a}{\rho} = \mathbf{F}(X,t)\left(\frac{\omega_a}{\rho}\right)_0, \tag{2.52}$$

where $(\omega_a/\rho)_0$ is the value of the specific absolute vorticity for the particle at its initial position. Two important consequences are immediate from (2.52):

- With the *polar decomposition* of $\mathbf{F} = \mathbf{RU}$, where $\mathbf{R}$ is orthogonal and $\mathbf{U}$ symmetric and positive definite, one immediately recognizes that vorticity is produced by vortex stretching (through $\mathbf{U}$) and tilting (rotation through $\mathbf{R}$)[10].

- Consider a fluid referred to an inertial frame of reference. Then $\omega_a = \omega$. If such a fluid is barotropic or incompressible and inviscid, and if at any particular time the fluid motion at a particle is *irrotational*, then it remains irrotational for all time, viz.,

  $$\text{curl}\mathbf{v} = \mathbf{0}, \qquad \text{at } t = 0 \qquad \Rightarrow \text{curl}\mathbf{v} = \mathbf{0} \qquad \text{for all time.}$$

  If this irrotationality prevails through the entire flow region, it persists so for all time. Such a fluid possesses a velocity potential ϕ such that

  $$\mathbf{v} = \text{grad}\phi. \tag{2.53}$$

  Incompressible flows ($\text{div}\mathbf{v} = 0$) are then governed by the *Laplace equation*

  $$\text{div grad}\phi = \Delta\phi = 0 \tag{2.54}$$

  for the velocity potential.

[10]It is proved in texts on continuum-mechanics, see lecture by Prof. HAUPT, that $\mathbf{U}$ is typical for a strain, while $\mathbf{R}$ represents the rotation of the material point. If a vortex line element is then in the direction of the principal axes of the symmetric tensor $\mathbf{U}$ then its motion consists of a rotation (tilting) and a pure extension or contraction (stretching). Incidentally, this property can already be seen in equation (2.50), if we write $\text{grad}\mathbf{v} = \mathbf{L} = \mathbf{D} + \mathbf{W}$, where $\mathbf{D}$ is the symmetric stretching tensor and $\mathbf{W}$ the strew-symmetric vorticity tensor. Then the rate of change of the specific vorticity is the sum of contributions due to stretching and vorticity that corresponds to the tilt.

2.5.2 Potential vorticity

Consider a barotropic inviscid fluid (or a fluid in which the BOUSSINESQ- and adiabaticity assumptions are justified). Then the HELMHOLTZ theorem (2.50) must hold. Consider, moreover, a scalar λ for which the balance law

$$\frac{d\lambda}{dt} = \psi_\lambda, \qquad \text{with} \qquad \text{grad}\psi_\lambda = 0 \tag{2.55}$$

holds; ψ_λ is a prescribed function. Taking the inner product of (2.50a) with gradλ yields

$$\frac{d}{dt}\left(\frac{\omega_a}{\rho}\right) \cdot \text{grad}\lambda = (\text{grad}\mathbf{v})\frac{\omega\ \omega_a}{\rho} \cdot \text{grad}\lambda.$$

Adding to this the identity

$$\frac{\omega_a}{\rho} \cdot \frac{d}{dt}(\text{grad}\lambda) \equiv \frac{\omega_a}{\rho} \cdot \text{grad}\left(\frac{d\lambda}{dt}\right) - (\text{grad}\mathbf{v})\frac{\omega_a}{\rho} \cdot \text{grad}\lambda$$

shows that

$$\frac{d\Pi_\lambda}{dt} = 0, \qquad \Pi_\lambda := \frac{\omega_a}{\rho} \cdot \text{grad}\lambda. \tag{2.56}$$

In the derivation of this equation use has been made of the fact that *grad*$\psi_\lambda = 0$. The quantity Π_λ is called the *potential vorticity belonging to the scalar* λ, and equation $(2.56)_1$ is the *conservation law of potential vorticity* or ERTEL'S (1942) *vorticity theorem*. Accordingly, the potential vorticity of any scalar λ obeying a balance law (2.55) is conserved along particle trajectories.

As an example, consider an ocean or lake whose free and bottom surfaces are described geometrically by the equations

$$\begin{aligned} F_S(x,t) \equiv \zeta(x,y,t) - z &= c_1, \\ F_B(x) \equiv H(x,y) + z &= c_2, \end{aligned} \tag{2.57}$$

in which c_1 and c_2 are constants, see *Fig. 2.1.* Then λ defined by

$$\lambda := \frac{F_B}{F_B + F_S} = \frac{c_2}{c_1 + c_2} = \frac{H + z}{h + \zeta} = const \tag{2.58}$$

satisfies the balance law (2.55) with $\psi_\lambda = 0$. In an incompressible fluid we may thus construct the so called *barotropic potential vorticity*

$$\Pi_{bt} = (\omega + 2\mathbf{\Omega}) \cdot \text{grad}\left(\frac{H + z}{H + \zeta}\right). \tag{2.59}$$

Non-dimensionalizing this expression with the scales (2.13) and imposing the shallow water approximation ($\mathbf{A} \to 0$) will show that the dominant term of (2.59) is

$$\Pi_{bt} \equiv \frac{\omega_z + f}{\zeta + H} \simeq \frac{\omega_z + f}{H}, \tag{2.60}$$

in which ω_z is the z-component of the vorticity vector ω. It follows that in the approximation (2.60)

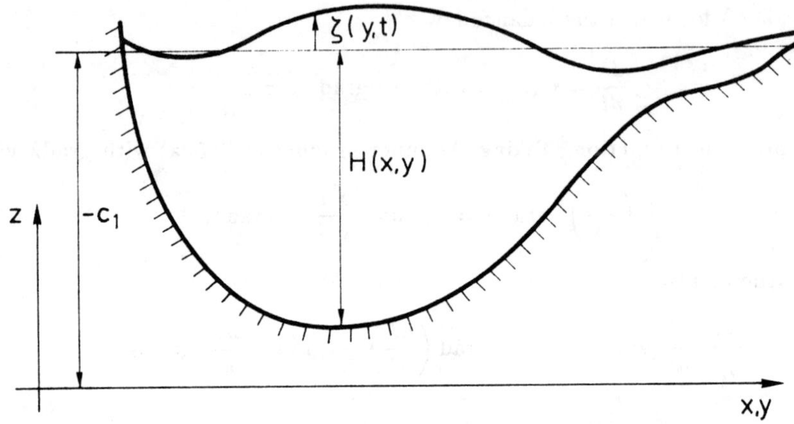

Figure 2.2: Vertical cut through a lake or ocean indicating the free and the bottom surfaces.

$$\begin{aligned}\frac{d\Pi_{bt}}{dt} &= \frac{d}{dt}\left(\frac{\omega_z+f}{H+\zeta}\right)\\ &\simeq \frac{d}{dt}\left(\frac{\omega_z+f}{H}\right) = \frac{1}{H}\frac{\partial\omega_z}{\partial t}+\frac{\partial}{\partial x}\left(\frac{\omega_z+f}{H}\right)u+\frac{\partial}{\partial y}\left(\frac{\omega_z+f}{H}\right)v=0,\end{aligned} \tag{2.61}$$

in which the terms involving the z-derivative are again omitted in the shallow water assumption.

The expressions on the far right in (2.60) and (2.61) are valid when the rigid lid assumption is made. Equation (2.61) is the basis for the derivation of various versions of the so-called ROSSBY-*wave equation.* We shall return to this equation in Chapters 3 and 4.

3. Linear Gravity and Rossby Waves[11]

3.1 Preliminary Remarks and Review

When effects of compressibility, surface tension, and electrical conductivity are ignored, lake water as a fluid on the rotating Earth with free surface permits propagation of *gravity waves* and *rotational waves.* Gravity waves arise through the restoring action of gravity on water particles displaced from equilibrium levels, such as a free surface or an internal isopycnal surface. The rotation of the Earth introduces the Coriolis force, which leads to the existence of *inertial* or *gyroscopic waves* and also gives rise to *planetary* and/or *topographic* ROSSBY

[11]This paragraph heavily draws from HUTTER (1984b, 1987, 1991), LAMB (1932), LIGHTHILL (1978), LEBLOND and MYSAK (1980) and GILL (1982).

waves. None of these exists without the rotation of the Earth justifying the term "rotational wave". More common, however, is to refer to gravity waves as ***first class waves*** and to rotational waves as *second class waves.*

Short surface waves represent the most conspicuous oscillatory response of a lake to wind action, and they are detectable directly by the eye. Their periods are seconds, and their wavelengths small as compared to the water depth. Those waves contribute much to the upper layer turbulent fluctuating motion and thus influence the various turbulent diffusivities. Gravity waves, which affect the global overall response of a lake have often periods of longer duration and wavelengths which are large as compared to the water depth. These waves essentially arise in two forms depending on whether the restoring force is primarily the cause of surface elevations from an undeformed equilibrium level or whether water particles in a (stably) stratified fluid are displaced from their neutral positions. The former are the long *surface gravity waves*; density inhomogeneities are irrelevant for their existence and modes are referred to as external or *barotropic.* The latter are *internal gravity waves,* inhomogeneities in the density distribution are crucial for their existence and modes are called *baroclinic.* A characteristic speed of barotropic gravity waves is $c = \sqrt{gh}$, where g is the gravity acceleration and h the water depth; thus it is typically from 10 to 50 ms^{-1}. Corresponding phase speeds of internal waves are about one to two orders of magnitude smaller. Periods of oscillations in an elongated lake of 10 to 100 km length, say, are therefore between minutes and hours for barotropic oscillations, but tens of hours or days for baroclinic processes. This separation of time scales is also manifest in the significance of the Coriolis effects which modify the corresponding gravity waves and now become KELVIN and POINCARÉ *waves,* respectively. The parameter, which measures this significance is the *external* or *internal Rossby radius* defined by $R = c/f$, in which c is the phase speed and f the Coriolis parameter. Typically, $R = (500\ km, 5\ km)$ for external and internal waves, respectively. A lake must have a width of one or serveral Rossby radii in order that rotational effects have an influence on gravity waves. Thus, the rotation of the Earth can often be ignored in the study of surface waves but may have to be taken into account in internal gravity waves. Another consequence of the spectral gap between external and internal oscillations is that equations for the description of one class of waves can be simplified such that processes of the other class are filtered out. This is done either by using the vertically integrated transport equations and yields the barotropic response, or by employing a "rigid lid" assumption by which surface gravity waves are eliminated. If this is done the long-period *topographic waves* (also called *second class, vortex* or *quasi-geostrophic* waves) are isolated from the surface gravity waves. Stratification affects the topographic waves (TW), in general, so that barotropic and baroclinic TW must be differentiated, but in deep lakes the period and the structure of the mean transport-stream function are only slightly affected by the internal motion (of the thermocline), even though this associated motion may be quite substantial.

Because of the boundedness of the lake domain, free waves combine and form *standing waves.* These are often called *seiches* and represent the eigenmodes of the particular basin under consideration. The only difference between these and free waves is that the frequencies are quantized according to the size and geometry of the basin. Mathematically, the quantization manifests itself as an eigenvalue problem and the eigenvalues are the (seiche) periods. Observationally, a substantial amount of the total energy is concentrated at these periods. The exact distribution depends on the wind forcing but the few lowest modes are generally the most likely excited ones. As is the case for free waves, seiches exist as barotropic surface seiches - involving motions of the whole water masses and attaining their maximum amplitude at or near the free surface - or baroclinic internal seiches associated with the density stratification and attaining their maximum amplitude at or near the thermocline. Closed basin solutions of TW are not called seiches, but these long-period oscillations are also quantized.

3.2 Linear Waves in the Ocean and in Lakes

Here in this section we are concerned with a simple introductory account of linear surface waves, first by illustrating how the short-wave and long-wave approximations of linear surface waves are derived from the general surface wave equations and, second, by discussing two classes of waves that do arise in a rotating fluid with free surface and are subject to the shallow ice approximation.

3.2.1 Surface Gravity Waves

Consider a layer of an ideal incompressible fluid referred to an inertial frame in two dimensional plane motion with free surface $z = \zeta(x,t)$, *Fig. 3.1.*
In the absence of the rotation of the frame of reference the linearized Euler equations take the form

$$\mathrm{div}\mathbf{v} = 0, \qquad \text{and} \qquad \rho\frac{\partial \mathbf{v}}{\partial t} = -\mathrm{grad}p', \tag{3.1}$$

in which p' is the perturbation pressure such that the total pressure p equals

$$p = -\rho g z + p'. \tag{3.2}$$

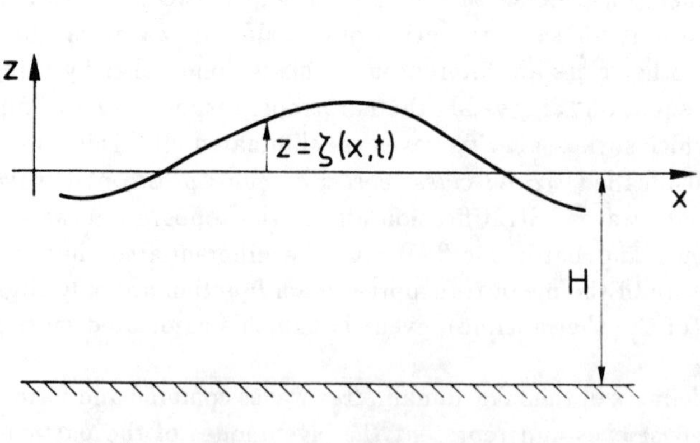

Figure 3.1: Plane surface wave in a layer of ideal incompressible fluid

By taking the divergence of the momentum equation $(3.1)_2$ we see that $\operatorname{div}\operatorname{grad} p' = \Delta p' = 0$. Thus, any continuous plane motion within the layer will be a solution of the boundary value problem

$$\begin{aligned}
&\Delta p' = \frac{\partial^2 p'}{\partial x^2} + \frac{\partial^2 p'}{\partial z^2} = 0, && \text{in } \mathcal{D}, \\
&\left.\begin{aligned} &p = -\rho g \zeta(x,t) + p' = 0, \\ &\frac{\partial \zeta}{\partial t} + \frac{\partial \zeta}{\partial x} u - w = 0, \end{aligned}\right\} && \text{at} \quad z = \zeta(x,t), \\
&w = 0, && \text{at} \quad z = -H.
\end{aligned} \tag{3.3}$$

Thus, the perturbation pressure satisfies the *Laplace equation.* At the free surface the pressure must vanish (or the perturbation pressure equals $\rho g \zeta(x,t)$), and the kinematic surface condition must hold, and at the base there cannot be any water flow through the basal surface. Except for the convective acceleration terms which are ignored, the above equations are exact.

Assuming surface elevation amplitudes to be small in comparison to water depth and wavelength equations (3.3) can be linearized and written as follows:

$$\begin{aligned}
&\Delta p' = \frac{\partial^2 p'}{\partial x^2} + \frac{\partial^2 p'}{\partial z^2} = 0, \qquad x \in (-\infty, \infty), && z \in [0, -H], \\
&\left.\begin{aligned} &p' = \rho g \zeta(x,t), \\ &\frac{\partial w}{\partial t} = \frac{\partial^2 \zeta}{\partial t^2} = -\frac{1}{\rho}\frac{\partial p'}{\partial z}, \end{aligned}\right\} && z = 0, \\
&\frac{\partial p'}{\partial z} = 0, && z = -H.
\end{aligned} \tag{3.4}$$

The boundary conditions at the upper free surface now hold at $z = 0$, the undeformed equilibrium surface.

Consider a *plane wave* solution

$$(p', \zeta) = (p_0(z), \zeta_0)\, e^{i(kx - \omega t)}, \tag{3.5}$$

in which $p_o(z)$ and ζ_0 are amplitudes the former being obviously z-dependent; k is the wavenumber and ω the frequency. Substituting the expressions (3.5) into (3.4) shows that $p_0(z)$ must satisfy the differential equation

$$\frac{d^2 p_0}{dz^2} - k^2 p_0 = 0$$

with the general solution

$$p_0(z) = A \cosh\left[k(z + H)\right] + B \sinh\left[k(z + H)\right].$$

Imposing the boundary conditions $(3.4)_{2,4}$ yields

$$A = \frac{\rho g \zeta_0}{\cosh(kH)}, \qquad B = 0$$

and $(3.4)_3$ determines a relation between the frequency ω, and the wavenumber k, viz.,

$$\omega^2 = gk \tanh(kH). \tag{3.6}$$

This is the *dispersion relation* of the linearized surface gravity waves on a liquid layer of constant depth. It follows that surface gravity waves are generally dispersive, since the *phase velocity* $\mathbf{c}_{ph} = (\omega/k^2)\,\mathbf{k}$ and the *group velocity* $\mathbf{c}_{gr} = d\omega/d\mathbf{k}$,

$$\mathbf{c}_{ph} = \sqrt{\frac{g}{k}\tanh(kH)}\frac{\mathbf{k}}{k}, \tag{3.7a}$$

$$\mathbf{c}_{gr} = \frac{1}{2}\left\{\sqrt{\frac{g\tanh(kH)}{k}} + \sqrt{\frac{gk}{\tanh(kH)}}\frac{H}{\cosh^2(kH)}\right\}\frac{\mathbf{k}}{k} \tag{3.7b}$$

differ from one another. Dimensionless representations of these speeds, valid for all H, are

$$\frac{c_{ph}}{\sqrt{gH}} = \sqrt{\frac{\tanh(kH)}{kH}} \quad \text{or} \quad \frac{\omega}{\sqrt{g/H}} = \sqrt{kH\tanh(kH)}, \tag{3.7c}$$

$$\frac{c_{gr}}{\sqrt{gH}} = \frac{1}{2}\left\{\sqrt{\frac{\tanh(kH)}{kH}} + \frac{\sqrt{kH}}{\cosh^2(kH)\sqrt{\tanh(kH)}}\right\} \tag{3.7d}$$

and these are graphically displayed in *Fig. 3.2.* It is seen that the short-wave limits for the phase and group velocities are sufficiently accurate approximations if $kH > 2$, while the long-wave limit is appropriate for $kH < 0.25$. Similar inferences also hold for the frequencies. Pressure, velocity components and surface elevation are now given by

$$\begin{aligned} p' &= \rho g\zeta_0\frac{\cosh(k(z+H))}{\cosh(kH)}e^{i(kx-\omega t)}, \\ \zeta &= \zeta_0 e^{i(kx-\omega t)}, \\ u &= \frac{kg}{\omega}\zeta_0\frac{\cosh(k(z+H))}{\cosh(kH)}e^{i(kx-\omega t)}, \\ w &= \frac{kg}{\omega}\zeta_0\frac{\sinh(k(z+H))}{\cosh(kH)}e^{(i(kx-\omega t)-i\pi/2)}. \end{aligned} \tag{3.8}$$

It is seen that the vertical velocity component lags behind the other three variables by a quarter period.

Interesting physical insight is gained from evaluating the particles' trajectories from a linearized analysis. To this end, we suppose that the particles do not move far away from their position at rest so that on the right handsides of $(3.8)_{3,4}$ x may be replaced by x_0, the particle's rest position. With this the differential equations of the particle trajectories are given by

$$\frac{dx}{d\tau} = U_0(z)e^{-i\omega\tau}, \qquad \frac{dz}{d\tau} = -iU_0(z)\tanh(k(z+H))\,e^{-i\omega\tau}, \tag{3.9}$$

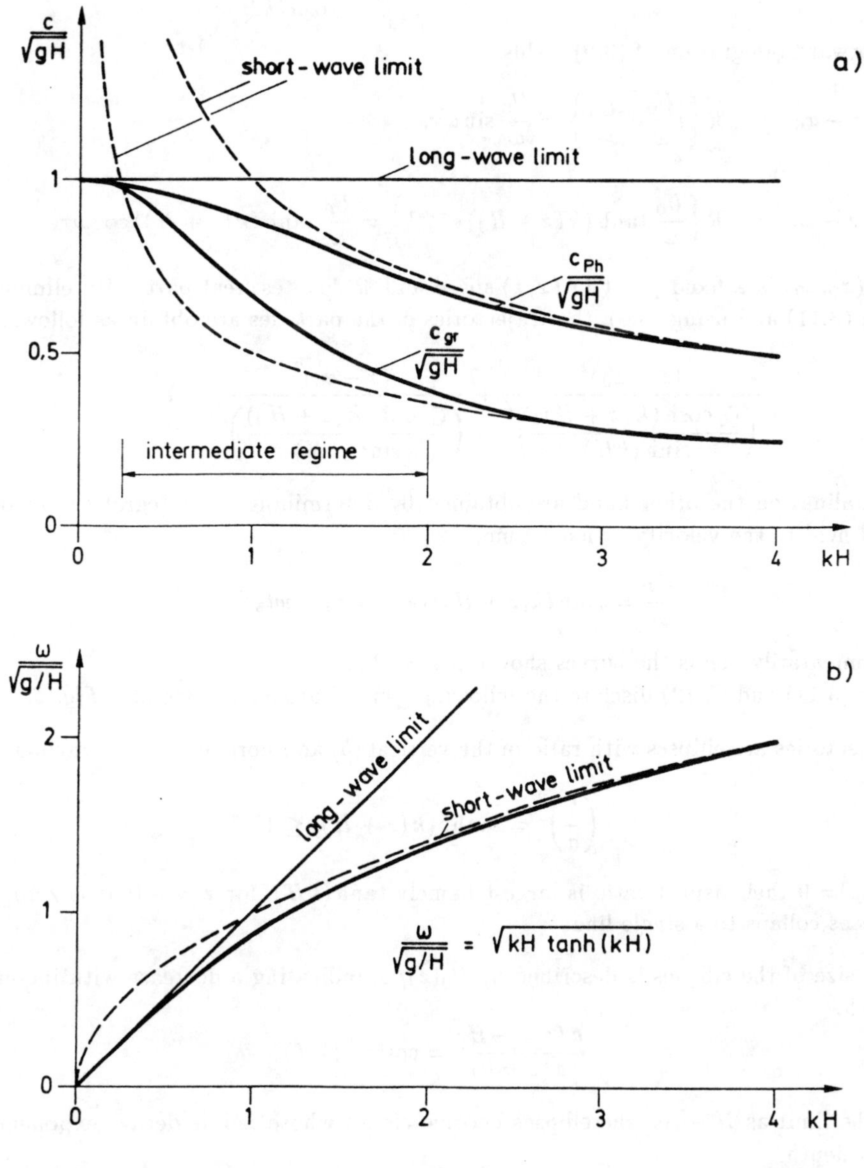

Figure 3.2: (a) Dimensionless phase- and group speeds plotted against dimensionless wavenumber kH.
(b) Dimensionless frequency $\omega/\sqrt{g/H}$ plotted against dimensionless wavenumber kH. In both graphs long-wave and short-wave limits are also indicated.

in which

$$\tau = -\frac{x_0}{c_{ph}} + t, \qquad U_0(z) = \frac{kg}{\omega}\zeta_0 \frac{\cosh(k(z+H))}{\cosh(kH)}. \tag{3.10}$$

Straightforward integration of (3.9) yields

$$\begin{aligned} x - x_0 &= \Re\left\{ i\frac{U_0}{\omega} e^{-i\omega\tau} \right\} = \frac{U_0}{\omega}\sin\omega\tau, \\ z - z_0 &= \Re\left\{ \frac{U_0}{\omega}\tanh(k(z+H))\, e^{-i\omega\tau} \right\} = \frac{U_0}{\omega}\tanh(k(z+H))\cos\omega\tau, \end{aligned} \tag{3.11}$$

in which (x_0, z_0) is a fixed point in (x,z)-space and $\Re$ denotes 'real part'. By eliminating time from (3.11) and using (3.6), the trajectories of the particles are obtain as follows:

$$\frac{(x-x_0)^2}{\left(\dfrac{\zeta_0\cosh(k(z+H))}{\sinh(kH)}\right)^2} + \frac{(z-z_0)^2}{\left(\dfrac{\zeta_0\sinh(k(z+H))}{\sinh(kH)}\right)^2} = 1. \tag{3.12}$$

The streamlines on the other hand are obtained by determining the integral curves of the tangential field to the velocity at fixed time,

$$\frac{dz}{dx} = \tanh(k(z+H))\tanh(kx_0 - wt_0)$$

which, momentarily yields the curves shown in *Fig. 3.3.*
Equations (3.11) and (3.12) disclose the following general properties, see also *Fig. 3.3*

- Trajectories are ellipses with ratio of the vertical (b) and horizontal (a) principal semi-axes

$$\left(\frac{b}{a}\right)^2 = \tanh^2(k(z+H)) \le 1.$$

- For $z = 0$ their aspect ratio is largest namely $\tanh(kH)$, for $z = -H$ it is zero, i. e., ellipses collaps to a single line.
- The size of the ellipses is described by $U_0(z)/\omega$, indicating a decrease with increasing depth:

$$\frac{a(z=-H)}{a(z=0)} = \cosh^{-1}(kH).$$

- In the limit as $H \to \infty$ the ellipses become circles whose radius decays exponentially with depth.

Sketches of the trajectories are shown in *Fig. 3.3.* Photographs of trajectories are shown in *Fig. 3.4.*

Problem *Standing waves:*

Consider two surface water waves of equal amplitude that propagate in opposite directions.

Using (3.8) with exponential factors $e^{i(\pm kx-\omega t)}$ it can be shown that the perturbation pressure, surface elevation and velocity components, when written in real notation, are given by

$$
\begin{aligned}
p' &= 2\rho g\zeta_0 \frac{\cosh(k(z+H))}{\cosh(kH)} \cos kx \cos \omega t, \\
\zeta &= 2\zeta_0 \cos kx \cos \omega t, \\
u &= 2\frac{kg}{\omega}\zeta_0 \frac{\cosh(k(z+H))}{\cosh(kH)} \sin kx \sin \omega t, \\
w &= -\ 2\frac{kg}{\omega}\zeta_0 \frac{\sinh(k(z+H))}{\cosh(kH)} \cos kx \sin \omega t.
\end{aligned}
\tag{3.13}
$$

This wave is indeed standing, because for $kx = (2n+1)\pi/2$ the surface elevation is zero (nodal points) and the velocity is strictly horizontal. Alternatively, for $kx = n\pi(n = 1, 2, \ldots)$ the surface elevation is a maximum and the velocity strictly vertical. To determine the particle trajectories, the equations

$$
\frac{dx}{dt} = u(x, z, t) \cong u(x_0, z_0, t), \qquad \frac{dz}{dt} = w(x, z, t) \cong w(x_0, z_0, t)
$$

must be integrated in time, where (x_0, z_0) is the initial position of the particle, the result being

$$
\begin{aligned}
x - x_0 &= -\frac{2gk\zeta_0}{\omega^2} \sin(kx_0) \cos \omega t \frac{\cosh(k(z_0+H))}{\cosh(kH)}, \\
z - z_0 &= \frac{2gk\zeta_0}{\omega^2} \cos(kx_0) \cos \omega t \frac{\sinh(k(z_0+H))}{\cosh(kH)},
\end{aligned}
\tag{3.14}
$$

It is easily seen from these relations that particle trajectories are straight lines with inclination

$$
\tan(\alpha) = \frac{z - z_0}{x - x_0} = -\cot(kx_0) \tanh(k(z_0+H)).
$$

The reader may easiliy convince himself that the streamlines are the intgral curves of these lines; they are graphically displayed in Fig. 3.3b.

At nodal points, $\alpha = 0$ and particles move horizontally, at crests/troughs $\alpha = 90°$, and the motion is vertical. *Fig. 3.3b* and *Fig. 3.4* illustrate this behavior.

Short-wave and long-wave approximation. The length scale that appears in the dispersion relation (3.6) and hence determines the character of the waves is the fluid depth H. Different approximations apply depending on how the wavenumber k relates to H. For the case of *short waves*, i.e., for $k^{-1} \ll H$, (3.6) is approximated by

$$
\omega^2 = gk, \qquad \tanh(kH) \to 1 \tag{3.15}
$$

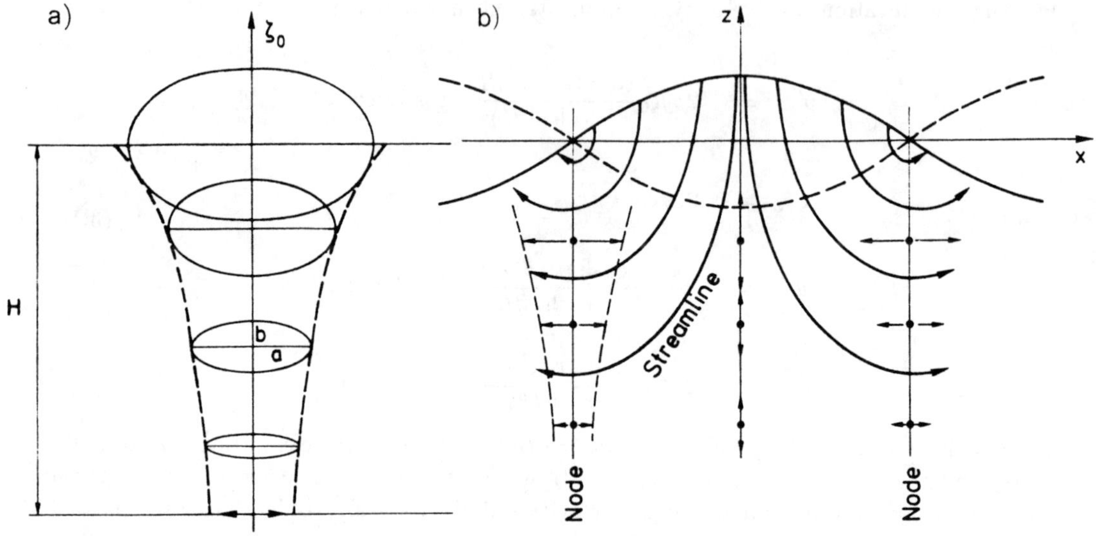

Figure 3.3: (a) Particle trajectories for fluid particles at various depths as determined by eq. (3.12)
(b) Streamlines in a standing gravity wave. Particle trajectories are straight lines tangential to the streamlines

and (3.8) reduce to

$$\begin{aligned} p' &= \rho g \zeta_0 e^{kz} e^{i(kx-\omega t)}, \\ \zeta &= \zeta_0 e^{i(kx-\omega t)}, \\ u &= \frac{kg}{\omega}\zeta_0 e^{kz} e^{i(kx-\omega t)}, \\ w &= \frac{kg}{\omega}\zeta_0 e^{kz} e^{(i(kx-\omega t)-i\pi/2)}. \end{aligned} \tag{3.16}$$

These are also called *deep-water waves* because $H \gg k^{-1}$. From the exponential z-dependence of the pressure and the velocity field it is seen that these are confined to a distance (in z) on the order k^{-1} from the surface, so propagation is unaffected by the bottom. It follows from the dispersion relation (3.7) that

$$c_{ph} = \sqrt{\frac{g}{k}}, \qquad c_{gr} = \frac{1}{2}\sqrt{\frac{g}{k}}. \tag{3.17}$$

The group velocity is half as large as the phase velocity.

Figure 3.4: (see next page) Particle trajectories in plane periodic water waves. Two wave trains of the same frequency travelling in opposite directions are produced by a progressive wave coming from the left that is reflected by a partially absorbent barrier. The top photograph on the left shows the pure progressive wave with no reflection. Its amplitude is four per cent of the wavelength, and the water depth is 22 per cent. White particles suspended in the water are photographed during one period. Their trajectories are practically ellipses traversed clockwise, circular at the free surface and flattened toward the bottom. Some open loops indicate a slow drift to the right near the surface and left near the bottom. As the reflection is increased, the orbits become increasingly flattened and inclined. Complete reflection gives a pure standing wave in the last photograph, where the trajectories are streamlines. There the upper and lower envelopes of the water surface show that the vertical motion does not vanish at the nodes, where the horizontal velocity is nil, from VAN DYKE, 1976.

Example:

The dominant waves that one sees on the ocean surface have periods $2\pi\omega^{-1}$ of order 10 s. By (3.15) a deep water wave has a wave-length of $2\pi k^{-1} = 2\pi g/\omega^2 \sim 150\ m$, its e-folding depth is $k^{-1} \simeq 25\ m$, and its phase speed is 15 ms^{-1}! The deep water approximation is therefore reasonable for such waves when the depth is greater than 25 m. When such a wave approaches a shallow shore region its frequency remains constant, but owing to the finite-depth-dispersion relation the wave will become shorter and the phase speed decreases. It follows that deductions about wave-length and phase speed made by observing waves on a beach can lead to erroneous conclusions about their properties in deep water.

The approximation to (3.6) for *long waves*, i.e., for $k^{-1} \gg H$ is obtained by a Taylor series expansion of $\tanh(kH)$ for small arguments, leading to

$$\omega = \sqrt{gH}k, \qquad (k \ll H), \tag{3.18}$$

from which one easily deduces

$$\mathbf{c}_{ph} = \mathbf{c}_{gr} = \sqrt{gH}\frac{\mathbf{k}}{k}. \tag{3.19}$$

These are the *shallow-water waves* because wavelengths are much larger than H; they are *non-dispersive* because the phase speed is independent of wavenumber. This speed depends on g and water depth and is given in the table below:

depth [m]	:	100	500	1000	5000
speed [ms^{-1}]	:	30	70	100	200

Such waves travel forth and back Lake Zürich (12 km long) in 15 min and could cross the Atlantic ocean in 7 hr. The corresponding approximation to (3.8) is

$$p' = \rho g \zeta_0 e^{i(kx-\omega t)},$$

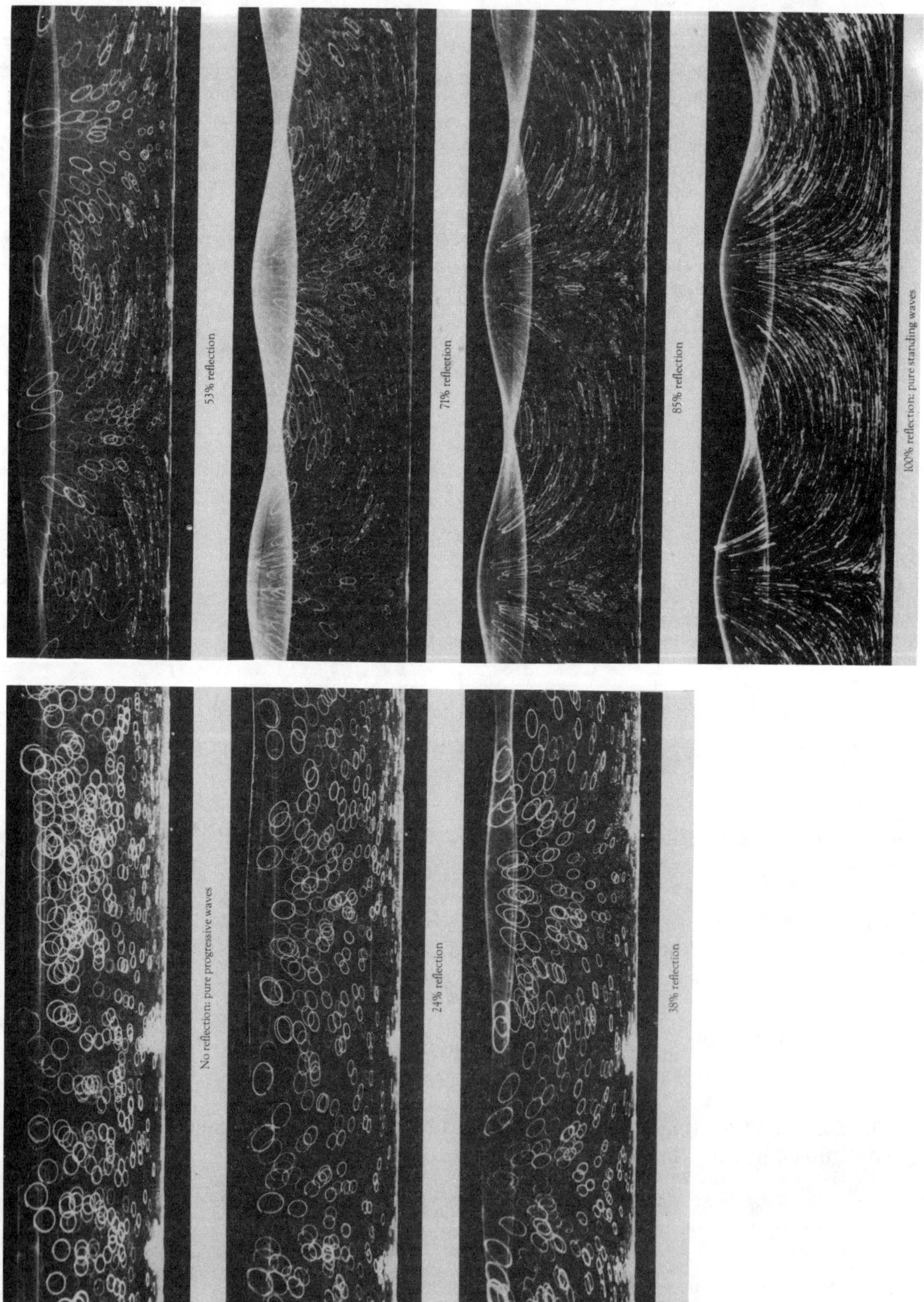
53% reflection
71% reflection
85% reflection
100% reflection: pure standing waves
No reflection: pure progressive waves
24% reflection
38% reflection

$$
\begin{aligned}
\zeta &= \zeta_0 e^{i(kx-\omega t)}, \\
u &= \sqrt{\frac{g}{H}}\zeta_0 e^{i(kx-\omega t)}, \\
w &= \sqrt{\frac{g}{H}}\zeta_0 k\,(z+H)\,e^{(i(kx-\omega t)-i\pi/2)},
\end{aligned}
\tag{3.20}
$$

i.e., the pressure perturbation and the horizontal velocity are independent of depth. Since the density perturbation is zero, this is precisely the result that would be obtained if the pressure were calculated from the hydrostatic equation. Thus we have explicitly demonstrated what has been proved in section 2 by a scaling analysis, that the long-wave and shallow-water approximation are equivalent. We also see from $(3.20)_4$ that the vertical velocity increases linearly with z from zero at the bottom to $\partial\zeta/\partial t$ at the surface.
Comparing the pressure formulas for general, deep and shallow water waves, $(3.8)_1$, $(3.16)_1$ and $(3.20)_1$, respectively, shows that pressures are attenuated with depth except in the shallow water approximation. It follows that a pressure gauge positioned at the bottom can be used to record long-wave surface fluctuations. It will filter out all short surface waves and, if not properly corrected, give erroneous results for intermediate wavelengths.

3.2.2 Shallow water approximation

The emphasis in this chapter on waves in lakes and the ocean is on motions with horizontal scale large enough compared with the vertical scale for the hydrostatic approximation to be valid. In section 2 a variety of further assumptions has been discussed, and here we wish to study small amplitude linear surface waves in a layer of fluid on the rotating Earth. The simplifying assumptions we shall impose are:

- barotropicity (no density stratification)
- adiabaticity (no turbulent Reynolds stresses and no turbulent heat flux)
- shallow water approximation (equivalent to the hydrostatic pressure assumption)
- linearity (no advective terms, small surface elevations)

They permit the governing equations of motion for the surface elevation ζ and the volume transport $\mathbf{V}$ to be written in the form (2.46). When atmospheric pressure and wind stress are absent these equations read

$$
\begin{gathered}
\frac{\partial u}{\partial t} - fv = -g\frac{\partial \zeta}{\partial x}, \qquad \frac{\partial v}{\partial t} + fu = -g\frac{\partial \zeta}{\partial y}, \\
\frac{\partial \zeta}{\partial t} + \frac{\partial}{\partial x}(Hu) + \frac{\partial}{\partial y}(Hu) = 0,
\end{gathered}
\tag{3.21}
$$

where $\mathbf{v}H = \mathbf{V}$, so (u, v) are the vertically averaged horizontal velocity components[12]. Equations (3.21) comprise the governing equations for free long waves in a basin of finite depth, $H(x, y)$. It is, in general, more convenient to deduce inferences from their transformed variants. For instance, if the velocity components u, v are eliminated from (3.21), then these equations take the form

$$\begin{gathered}\nabla \cdot \left\{ H \nabla \frac{\partial \zeta}{\partial t} \right\} + \mathcal{J}\left[fH, \zeta \right] - \frac{1}{g}\left(\frac{\partial^2}{\partial t^2} + f^2 \right) \frac{\partial \zeta}{\partial t} = 0, \\ \mathcal{J}\left[a, b \right] = \frac{\partial a}{\partial x}\frac{\partial b}{\partial y} - \frac{\partial a}{\partial y}\frac{\partial b}{\partial x}. \end{gathered} \tag{3.22}$$

This partial differential equation is third order in time and second order in space. The operator $\mathcal{J}$ is called the Jacobian operator, and the Coriolis-parameter need not be constant; it follows that (3.22) is valid for both the f-plane and the β-plane approximations. The differential equation (3.22), because it is third order in time, possesses in general three wave solutions; however, these belong only to two classes, as we shall now see.

- *Waves of the first class*: When $\mathcal{J} = 0$ then (3.22) reduces to an equation of the second order in $\partial \zeta / \partial t$. Since these waves obviously depend on gravity (the constant g arises explicitly in the resulting equation), and because the waves do also exist when $f = 0$ (non-rotating frame of reference), they are called *surface gravity waves.* It is easy to show that the dispersion relation agrees with (3.18) in this case. In the limit of no rotation these waves agree with the long wave-length shallow water waves.
  When the rotation of the Earth is present these surface gravity waves are *modified* - perhaps slightly, perhaps substantially - by the presence of the Coriolis parameter f.

- *Second class waves*, ROSSBY *waves*, *quasi-geostrophic waves* or *vorticity waves*[13]. These waves are governed by the third solution of the wave equation (3.22) and only exist when the Jacobian operator does not identically vanish, and they are approximately obtained from (3.22) by balancing the first two terms (and ignoring the third, gravity dependent term):

$$\nabla \cdot \left\{ H \nabla \frac{\partial \zeta}{\partial t} \right\} + \mathcal{J}\left[fH, \zeta \right] = 0. \tag{3.23}$$

  This is the famous ROSSBY *wave equation.* The reader may prove by himself that it may be obtained from equations $(3.21)_{1-3}$ by ignoring in $(3.21)_3$ the time derivative $\partial \zeta / \partial t$,

[12]If equations (3.21) are not viewed as approximations, then they imply that the horizontal velocity components cannot vary with depth. Indeed, since the right hand sides of (3.21) are functions of x, y and t alone, so must be the left hand sides, implying that $u = u(x, y, t)$ and $v = v(x, y, t)$.

[13]The question of how a fluid, not initially in equilibrium, adjusts in a uniformly rotating system was not completely discussed until the time of Rossby. In a series of papers, ROSSBY (1936, 1937, 1938a, b, 1940, ROSSBY et al., 1939) was concerned with how the mass and resulting pressure distributions in the ocean and atmosphere are established. In particular, the problem in which momentum was put into the ocean to result in a non equilibrium velocity distribution was discussed. ROSSBY (1938a) then considered the process of adjusting to equilibrium. It is in this paper where his famous equation first arises.

which is tantamount to the *rigid lid assumption*. This explanation makes equally clear that the rigid lid assumption eliminates surface gravity waves; indeed in (3.23) the gravity term is missing. Furthermore, ROSSBY waves do only exist when $f \neq 0$; this demonstrates that they only exist in a fluid on a rotating non-inertial frame of reference; thus, they necessarily possess non-vanishing absolute vorticity for which reason they are sometimes also called *vorticity waves*. Depending upon whether $\beta = 0$ or $\beta \neq 0$ and H = constant, $H \neq$ constant there are several special cases which we shall discuss below.

In the above analysis the surface gravity waves and the ROSSBY waves were obtained from (3.22) by balancing the appropriate two of the three terms of the differential equation. These balances can be obtained more formally by non-dimensionalizing equation (3.22). To this end, let L_0 and H_0 be typical horizontal and vertical length scales such that

$$H = H_0 h, \qquad x = L_0 \bar{x}, \qquad t = \frac{1}{f_0}\bar{t}, \qquad f = f_0 \bar{f}.$$

Time and the Coriolis parameter have been scaled with a reference value f_0. With these, (3.22) assumes the form

$$\bar{\nabla} \cdot \left\{ h \bar{\nabla} \frac{\partial \bar{\zeta}}{\partial \bar{t}} \right\} + \mathcal{J}\left[\bar{f} h, \bar{\zeta}\right] - \frac{L_0^2}{R^2}\left(\frac{\partial^2}{\partial \bar{t}^2} + \bar{f}^2\right)\frac{\partial \bar{\zeta}}{\partial \bar{t}} = 0 \tag{3.24}$$

in which

$$R := \frac{\sqrt{g H_0}}{f_0} \tag{3.25}$$

is the ROSSBY *radius of deformation*[14]. For $f_0 = 10^{-4}\ s^{-1}$, $H_0 = 5000\ m$ it is of the order of 2500 km; when $H_0 = 100\ m$ one obtains 330 km. Rotational and gravity effects uncouple when corresponding typical wave lengths are sufficiently separate. In particular, the effects of the rotation of the Earth upon surface gravity waves are significant if the horizontal length scale H_0 is comparable to the Rossby radius of deformation. For barotropic waves this is the case for lake and ocean basins of the order of the Great Lakes, Lake Baikal, the Baltic Sea, but certanly neither for Lake Constance nor Lake Geneva nor smaller basins.

Equation (3.24) also provides the conditions where second class waves uncouple from first class waves as stipulated by equation (3.23). This is so when $R \gg L_0$, which is certainly the case in many ocean basins and all lakes under barotropic conditions. Barotropic Rossby waves can accurately be described almost by the reduced equation (3.23). Baroclinic Rossby

[14]Later, we shall also introduce an *internal* ROSSBY radius of deformation. For a two-layer system (upper layer depth H_1, density ρ_1; lower layer depth H_2, density ρ_2) it is given by

$$R_i = \frac{\sqrt{g' H_E}}{f_0}, \quad g' = g\frac{\rho_2 - \rho_1}{\rho_2}, \quad H_E = \frac{H_1 H_2}{H_1 + H_2}.$$

These internal ROSSBY radii apply for a stratified fluid in two layers and are much smaller. So, whereas in certain lake basins the effects of the rotation of the Earth may be ignored for *surface gravity waves* this is not so for *internal gravity waves*. We shall come back to this property later on.

- In the f-plane approximation ($\beta = 0$) at mid to high latitudes ($f \simeq 10^{-4}$), the Jacobian operator differs from zero when $\nabla H \neq 0$. In this case ROSSBY waves can only exist because of the variable topography. These waves are called *topographic* (ROSSBY) *waves.* Because the variation of the topography is much richer than that of the Coriolis parameter, topographic waves are more complex in structure than planetary ROSSBY waves. Signals are particularly pronounced where topographic slopes are large, in the atmosphere near mountains, in the ocean in the vicinity of the shelf- and shore regions and near sea mountains. When the waves travelling along the continental shelves are addressed, then they are referred to as *shelf-waves.*

The planetary and the topographic ROSSBY waves are the two main ROSSBY wave types; there are special cases of these which, frequently, bear their own names:

- When topographic waves cover large NS-distances they may be called *planetary topographic waves.* Shelf waves travelling through the entire Atlantic or Pacific Ocean are of this type.

- Close to the Equator (to within $\pm 5°$ latitude) f is so small that it may be ignored in comparison to βL_0. The corresponding waves enjoy special features and are called *equatorial planetary* (ROSSBY) *waves* or *equatorial topographic* (ROSSBY) *waves.*

A first physical grasp of the ROSSBY wave equation is gained if time-independent *steady solutions* to $\mathcal{T}[\Psi] = 0$ are sought. These are given by

$$\mathcal{J}\left[\Psi, \frac{f}{H}\right] = 0,$$

or equivalently

$$\nabla\Psi \times \nabla\left(\frac{f}{H}\right) = 0. \tag{3.35}$$

Lines of constant f/H-values are called *isotrophs.* They can immediately be constructed if the bathymetry to an ocean basin is given together with the location of the basin on the globe, see *Fig. 3.5.* Equation (3.35) now tells us that the streamlines of a steady current must be parallel to the isotrophs: *the isotrophs are necessarily the streamlines of the steady vorticity motions.* Alternatively, whenever the streamlines are not parallel to the isotrophs then a ROSSBY wave must develop. In fact, if we envisage the topographic wave equation $T[\Psi] = 0$ as a simple oscillator model, then the Jacobian $\mathcal{J}[\Psi, f/H]$, whose value is a measure of the angle between streamlines and isotrophs, acts as a "restoring force" that drives the system back to this "equilibrium flow configuration". ROSSBY waves arise because of the fluid particles' tendency to follow the isotrophs as much as possible.

Another very informative example is obtained, when we consider planetary waves in a constant depth ocean or atmosphere, in which case equation (3.30) takes the form

$$\nabla^2\frac{\partial\Psi}{\partial t} + \beta\frac{\partial\Psi}{\partial x} = 0 \tag{3.36}$$

waves may well be influenced by gravity waves unless $L_0 \ll R_i$.
Equation (3.24) also discloses a *scale invariance* of the original Rossby wave equation (3.23). Indeed, when the last term in (3.24) is omitted and when $\bar{f} = 1$ (f-plane approximation), then its remainder is independent of any horizontal length scale. As a consequence, the solutions for the scaled variables of the reduced equation (3.24) with $\bar{f} = 1$ will be the same for all geometrically similar basin bathymetries (irrespective of their size). This is not so when the last term in (3.24) is kept because it contains the factor L_0^2/R^2 which introduces a dependence on size.

3.2.3 A brief classification of Rossby waves

Equation (3.23) is written for the free surface elevation ζ; more convenient and better suited for practical applications is an equation for the mass transport stream function. To introduce it, consider $(3.21)_3$ subject to the *rigid lid assumption* so that

$$\frac{\partial}{\partial x}(Hu) + \frac{\partial}{\partial y}(Hv) = 0. \tag{3.26}$$

It follows that the volume transport $H\mathbf{v}$ is derivable from the relations

$$Hu = -\frac{\partial \Psi}{\partial y}, \quad Hv = \frac{\partial \Psi}{\partial x}. \tag{3.27}$$

Ψ is called the *mass transport streamfunction.* If (3.27) is substituted into $(3.21)_{1,2}$ and the surface elevation ζ is eliminated from the emerging equations, what results reads as follows:

$$\nabla \cdot \left\{ \frac{1}{H} \nabla \frac{\partial \Psi}{\partial t} \right\} + \mathcal{J}\left[\Psi, \frac{f}{H}\right] = 0. \tag{3.28}$$

In form (3.28) is identical to the ROSSBY wave equation (3.23). Introducing the (elliptic) operator

$$E[\Psi] = \nabla \cdot \left\{ \frac{1}{H} \nabla \Psi \right\}, \tag{3.29}$$

we now write (3.28) as

$$\mathcal{T}[\Psi] := \frac{\partial}{\partial t} E[\Psi] + \mathcal{J}\left[\Psi, \frac{f}{H}\right] = 0. \tag{3.30}$$

Before discussing this equation, let us re-derive it in an alternative way from the conservation law of *barotropic potential vorticity*, given in (2.61):

$$\frac{d}{dt}\left(\frac{\omega_z + f}{H}\right) = 0. \tag{3.31}$$

Recall that ω_z is the z-component of the vorticity vector,

$$\omega_z = \frac{\partial v}{\partial x} - \frac{\partial u}{\partial y} = E[\Psi]. \tag{3.32}$$

Substituting this into (3.31) and linearizing yields

$$\frac{\partial}{\partial t} E[\Psi] + \frac{\partial \Psi}{\partial x}\frac{\partial}{\partial y}\left(\frac{f}{H}\right) - \frac{\partial \Psi}{\partial y}\frac{\partial}{\partial x}\left(\frac{f}{H}\right) = 0$$

or

$$\frac{\partial}{\partial t} E[\Psi] + \mathcal{J}\left[\Psi, \frac{f}{H}\right] = 0,$$

in which terms that are quadratic in Ψ have been ignored.

It is obvious that in the above derivation, all simplifying assumptions already imposed in (3.21) and stated immediately above that equation are equally imposed. In particular, (3.30) is a free wave equation, in which external driving forces have been omitted. In an ocean or lake domain $\mathcal{D}$ it is generally solved subject to the boundary condition of vanishing flow through the basin boundary $\partial\mathcal{D}$. In other words, the shore line is a mass transport streamline . Now, since mass transport streamlines are the lines along which Ψ = constant, and absolute values of Ψ are insignificant, a solution of the ROSSBY wave equation is determined by solving the boundary value problem

$$\mathcal{T}[\Psi] = 0, \quad \text{in } \mathcal{D}, \text{ with } \Psi = 0, \quad \text{in } \partial\mathcal{D} \tag{3.33}$$

We close this section with a few remarks and a *classification of* ROSSBY *waves*:

- The ROSSBY wave equation in the f-plane approximation is scale invariant, as already mentioned before. This scale invariance is physically valid when there is a sufficient separation of the length scales of the surface gravity and the ROSSBY waves.
- It can be shown that the ROSSBY wave equation is invariant under *conformal transformations.* This property can constructively be used to determine a great variety of solutions from any other one. However, because the depth distribution is transformed along with the domain, this invariance property is less useful from a practical viewpoint than a first sight might surmise.
- If the Jacobian $\mathcal{J}[\Psi, f/H] = 0$, then (3.33) reduces to

  $$E[\Psi] = 0, \quad \text{in } \mathcal{D}, \text{ with } \Psi = 0, \quad \text{on } \partial\mathcal{D}. \tag{3.34}$$

  It follows from this, because $E[\cdot]$ is an elliptic operator that Ψ must assume its maximum and its minimum at the boundary where $\Psi = 0$. So, (3.34) possesses only the zero solution; there are no ROSSBY waves when $\mathcal{J}[\Psi, f/H] = 0$.

Evidently, the condition that $\mathcal{J}[\cdot] \neq 0$ can only be fulfilled for $f \neq 0$. Depending upon the variations of f and/or H the following *classification of the (general)* ROSSBY *waves* is made:

- If $\nabla H = 0$, i.e., if the layer depth is constant, then necessarily $\beta \neq 0$ in order that (3.33) possesses a non-trivial solution. In a constant depth atmosphere, ocean or lake ROSSBY waves exist in the β-plane, but not in the f-plane approximation. Such waves can develop provided the NS-extent of a basin with constant depth is large such that $\beta L_0 \simeq \mathcal{O}(f)$, implying $L_0 \geq 500 - 1000\ km$. These waves cover a large portion of the globe and are thus called *planetary* ROSSBY *waves.* They are primarily seen in the atmosphere and the deep ocean.

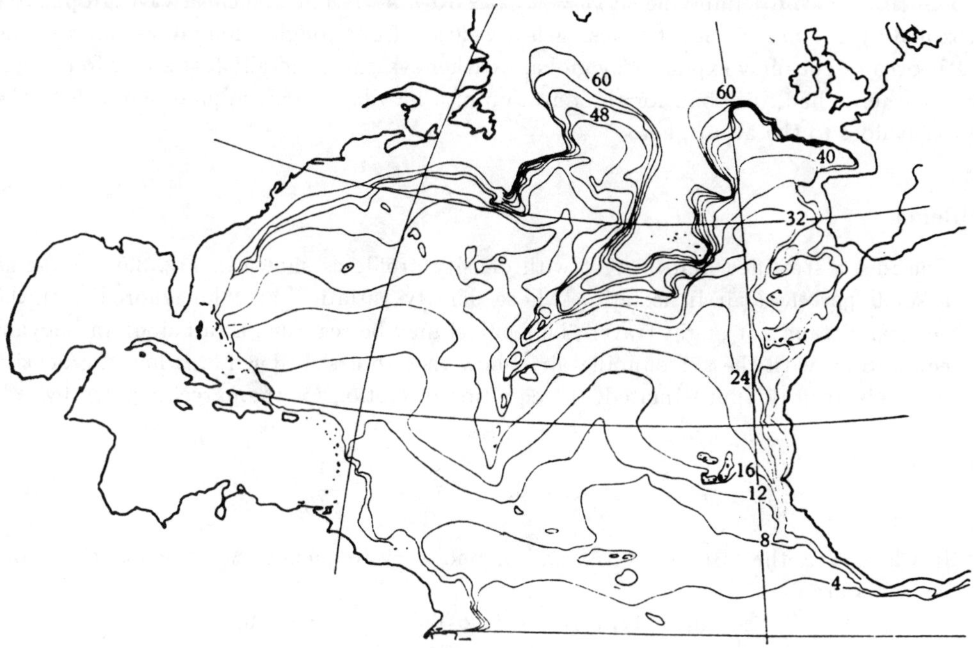

Figure 3.5: Contours of f/H ("isotrophes") for the North Atlantic Ocean.
The contours are separated by 4° of latitude at $H = 4000\ m$ (the numbers refer to this base). Many closed contours, and all contours north of 60°N, $H = 4000\ m$ have been omitted (from RHINES, 1969).

(H drops out from the equation). Assuming harmonic waves propagation in the x-direction, $\Psi = \Psi_0 e^{i(kx-\omega t)}$, yields the dispersion relation

$$\omega = -\frac{\beta}{k} \tag{3.37}$$

from which we may deduce

$$c_{ph} = -c_{gr} = -\frac{\beta}{k^2}. \tag{3.38}$$

These equations state that the phase necessarily travels from East to West whereas the group does so from West to East[15]. Thus, on the Northern or Southern hemisphere the energy of planetary ROSSBY waves travels across the mid to northern Atlantic from the American coast towards Europe with a group speed which equals in magnitude the phase speed that travels in the opposite direction.
At mid-latitudes typical diameters of a gyre of a storm-low are $10^6\ m$ (so $k \sim 10^{-6}\ m^{-1}$)

[15]With $e^{i(kx-\omega t)}$ a positive ω means propagation into the positive x-direction.

and $\beta \simeq 10^{-11}\ s^{-1}m^{-1}$ implying $|c_{ph}| = |c_{gr}| \simeq 10\ ms^{-1}$. Thus, such a wave propagates approximately $10^6\ md^{-1}$. So, it takes such a weather front roughly four days to cross the Atlantic ocean. This may explain why global weather systems generally last about four days: If the weather is bad, it is bad for the duration of a few days. This applies as much to the ocean as it does to the atmosphere.

Problems

1. Consider a stratified fluid at rest with density profile as shown in Fig. 3.6a. Imagine a small parcel of air in isolation whose density variation may be ignored over the vertical distances it might travel. This parcel may be regarded as a baloon in buoyant equilibrium with the surrounding air. Show that to first order, i.e., when the density profile is linearly approximated, the equation of motion for the parcel is given by

$$\ddot{\zeta}_s + N^2\zeta_s = 0, \qquad N^2 = -\frac{\frac{d\rho}{dz}(0)\,g}{\rho(0)} \tag{P.1.1}$$

in which N is the "BRUNT-VÄISÄLÄ" or buoyancy frequency. Solutions to the above equation are

$$\begin{aligned} \zeta_s &= A\sin(Nt) + B\cos(Nt), && N^2 > 0, \\ \zeta_s &= ae^{|N|t} + b^{-|N|t}, && N^2 < 0, \\ \zeta_s &= A + Bt, && N = 0. \end{aligned} \tag{P.1.2}$$

with constants of integration A and B. This proves that the unperturbed position $\xi_0 = 0$ is stable when $N^2 > 0$, neutrally stable when $N = 0$ and unstable when $N^2 < 0$.

2. The equations of motion for an ideal fluid referred to a frame of reference that is steadily rotating with constant angular frequency Ω are given by

$$\rho\{\dot{\mathbf{v}} + 2\mathbf{\Omega}\times\mathbf{v} + \mathbf{\Omega}\times(\mathbf{\Omega}\times\mathbf{x})\} = -\mathrm{grad}p + \rho\mathbf{g}. \tag{P.2.1}$$

and the fluid is at rest provided

$$\mathrm{grad}p = -\rho g\hat{\mathbf{e}}_z + \rho r\Omega^2\hat{\mathbf{e}}_r. \tag{P.2.2}$$

Here we have set $\dot{\mathbf{v}} = \mathbf{0}$, $2\mathbf{\Omega}\times\mathbf{v} = \mathbf{0}$ and have replaced

$$\begin{aligned} \mathbf{\Omega}\times(\mathbf{\Omega}\times\mathbf{x}) \quad &\text{by} \quad -\Omega^2 r\hat{\mathbf{e}}_r, \\ \rho\mathbf{g} \quad &by \quad -\rho g\hat{\mathbf{e}}_z \end{aligned}$$

$\hat{\mathbf{e}}_r$ and $\hat{\mathbf{e}}_z$ are unit vectors in the r- and z-directions of the cylindrical coordinate system of *Fig. 3.6b.*

Prove equation (P.2.2) and show that curl (grad$p$) = 0 implies the statement

$$\frac{\partial \rho}{\partial r} = -\frac{\Omega^2}{g} r \frac{\partial \rho}{\partial z}. \tag{P.2.3}$$

This shows that when centrifugal forces are taken into account (the terms involving Ω^2) then an axially (vertically) stratified fluid cannot be in equilibrium, because (P.2.3) implies a radial density gradient.

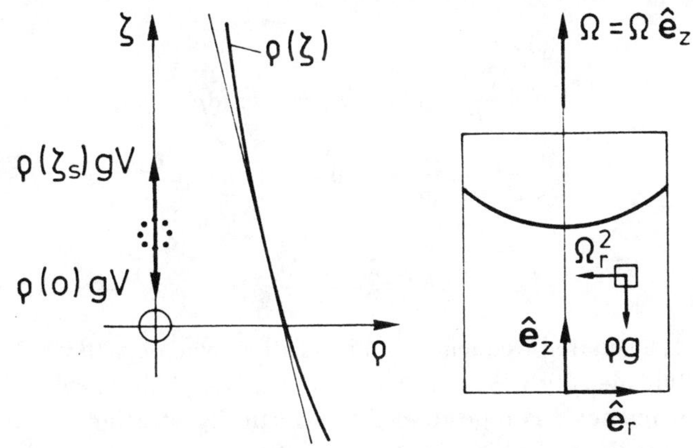

Figure 3.6: a) Vertical density profile in a stratified fluid. If a parcel of fluid at its rest position $\zeta = 0$ is displaced by the amount ζ_s it will experience a change in buoyancy force.
b) Hydrostatics in a fluid referred to a rotating system of coordinates. The free surfaces in a cylinder rotating about its axis of symmetry are axisymmetric paraboloids.

If the fluid obeys the thermal equation of state $\rho = \rho(T)$, then (P.2.3) implies

$$\frac{\partial T}{\partial r} = -\frac{\Omega^2}{g} r \frac{\partial T}{\partial z}. \tag{P.2.4}$$

Such a fluid, when being at rest must be radially and axially thermally stratified. Besides the above equation the temperature field also obeys the heat conduction equation,

$$\nabla 2T = 0 \Rightarrow \frac{1}{r}\frac{\partial}{\partial r}\left(r\frac{\partial T}{\partial r}\right) + \frac{1}{r^2}\frac{\partial^2 T}{\partial \phi^2} + \frac{\partial^2 T}{\partial z^2} = 0. \tag{P.2.5}$$

(This is Laplace's equation in cylindrical coordinates) subject to certain thermal boundary conditions.
Prove that equations (P.2.4) and (P.2.5) do not, in general have, a common non-trivial solution. It follows that in a rotating stratified fluid *with* centrifugal force there can in general be no state of rest. However, when there is no thermal stratification, (P.2.4) and (P.2.5) possess the common trivial solution $T = 0$. In a homogeneous rotating

fluid a state of rest can exist.
When centrifugal forces are ignored ($\Omega^2 \cong 0$), then (P.2.4) and (P.2.5) do permit a common non-trivial solution.

3.3 Plane linear waves in a rotating stratified fluid

3.3.1 General Boussinesq-fluid

We start our investigation here by restricting attention to the linearized equations in which the adiabaticity assumption and the BOUSSINESQ assumption are invoked, but no shallowness assumption is made. Thus the governing equations take the form

$$\mathrm{div}\mathbf{v} = 0, \tag{3.39}$$

$$\rho^* \left(\frac{\partial \mathbf{v}}{\partial t} + 2\boldsymbol{\Omega} \times \mathbf{v} \right) = -\mathrm{grad}p' + \rho' \mathbf{g}, \tag{3.40}$$

$$\frac{\partial \rho'}{\partial t} - \rho^* \frac{N^2}{g} w = 0, \tag{3.41}$$

in which

$$N^2 = -\frac{g}{\rho^*} \frac{d\rho_0}{dz} \tag{3.42}$$

has the dimension of a squared frequency and is positive, when $d\rho_0/dz < 0$; this corresponds to an increase in fluid density in the direction of gravity. N is known as the *buoyancy* or BRUNT-VÄISÄLÄ *frequency*. For positive N^2 a vertically stratified fluid at rest is *stably stratified*, for negative N^2 it is *unstably stratified*. This means that a small perturbation of the stratified fluid from its state at rest, will be restored to this "equilibrium state" when $N^2 > 0$; when $N^2 < 0$, however, the perturbed system will not return to its rest position, from which it was perturbed. The reader is asked to prove this by himself in the Problem set.
Equations (3.39) - (3.41) are the continuity equation, the momentum equation and energy equation (the latter in its adiabatic form) as explained in equations (2.26). Primed quantities denote small deviations from the state at rest, ρ^* is a constant density (as usual) and $\rho_0(z)$ is the density distribution in this state at rest. It is easy to show that with gravity being the only external force ρ_0 cannot have a dependence on the horizontal coordinates x, and y[16].
To study plane wave solutions of (3.39) - (3.42), let

$$\mathbf{v} = \mathbf{V} e^{i(\mathbf{k}\cdot\mathbf{x} - \omega t)}. \tag{3.43}$$

$\mathbf{V}$ is the amplitude vector, $\mathbf{k}$ the wavenumber vector, indicating the direction of propagation, and ω the frequency. If (3.43) is substituted into (3.39), one obtains

$$\mathbf{k} \cdot \mathbf{V} = 0, \tag{3.44}$$

[16]In general, i.e., when the rotation of the frame of reference is fast such that centrifugal accelerations cannot be ignored, it can be proved that there is no state of relative rest when the fluid is stratified, for a proof see the problems.

i.e., *in a* BOUSSINESQ *(or an incompressible) fluid in an infinite domain all motion is transverse to the direction of propagation; there can be no longitudinal waves in such a fluid.* This is exactly why acoustic waves are eliminated in a BOUSSINESQ fluid. Alternatively, differentiating equation (3.40) with respect to time and taking the curl of the resulting equation (to eliminate the pressure) yields

$$\rho^* \left[\operatorname{curl}\frac{\partial^2 \mathbf{v}}{\partial t^2} - 2\left(\operatorname{grad}\frac{\partial \mathbf{v}}{\partial t}\right)\Omega\right] = \operatorname{curl}\left[\frac{\partial \rho'}{\partial t}\mathbf{g}\right] = \operatorname{curl}\left[\frac{\rho^* N^2}{g}\mathbf{g}w\right]$$

in which (3.41) has also been used. With (3.43) and for *constant* N or *linear stratification* this equation can be put into the form

$$\omega^2 \mathbf{k} \times \mathbf{V} - i(2\mathbf{\Omega}\cdot\mathbf{k})\omega\mathbf{V} + \frac{N^2 W}{g}\mathbf{k}\times\mathbf{g} = \mathbf{0}. \qquad (3.45)$$

This equation may be regarded as a homogeneous, linear system for the components of the amplitude vector $\mathbf{V}$, which possesses only a nontrivial solution if its determinant vanishes. The deduction of this characteristic equation - the *dispersion relation* - is simplified, if it is recognized that owing to (3.44) $\mathbf{V}$ has only two independent components which are orthogonal to $\mathbf{k}$. With reference to *Fig. 3.7* one may thus choose

$$\begin{aligned} V &= U_i\hat{\mathbf{i}} + U_j\hat{\mathbf{j}}, \\ W &= -U_i \sin\Theta', \\ \mathbf{g} &= -g\left(\sin\Theta\hat{\mathbf{k}} + \sin\Theta'\hat{\mathbf{i}}\right), \end{aligned} \qquad (3.46)$$

in which $\left\{\hat{\mathbf{i}},\hat{\mathbf{j}},\hat{\mathbf{k}}\right\}$ is a unit orthogonal triad whose $\hat{\mathbf{k}}$ - vector is parallel to $\mathbf{k}$, and $\hat{\mathbf{i}}$ lies in the plane formed by $\mathbf{k}$ and $\mathbf{g}$, is orthogonal to $\mathbf{k}$, and $\hat{\mathbf{j}}$ is horizontal. Substituting (3.46) into (3.45) reduces this equation to the homogeneous linear system

$$\begin{bmatrix} 2i(\Omega\cdot\omega) & \omega^2 k \\ (\omega^2 - N^2\cos^2\Theta)\,k & -2i(\Omega\cdot\omega) \end{bmatrix}\begin{bmatrix} U_i \\ U_j \end{bmatrix} = 0 \qquad (3.47)$$

for the components U_i and U_j of the velocity vector in the directions $\hat{\mathbf{i}}$ and $\hat{\mathbf{j}}$ which possesses a nontrivial solution provided that

$$\omega^2\left\{k^2\left(\omega^2 - N^2\cos^2\Theta\right) - 4(\mathbf{\Omega}\cdot\mathbf{k})^2\right\} = 0. \qquad (3.48)$$

This is the *dispersion relation* for a linearly stratified fluid that is referred to a frame of reference rotating with constant angular velocity. A double solution to (3.48) is $\omega = 0$, corresponding to no wave; the remaining solutions can be written as

$$\omega^2 = N^2\cos^2\Theta + 4\Omega^2\cos^2\alpha, \qquad (3.49)$$

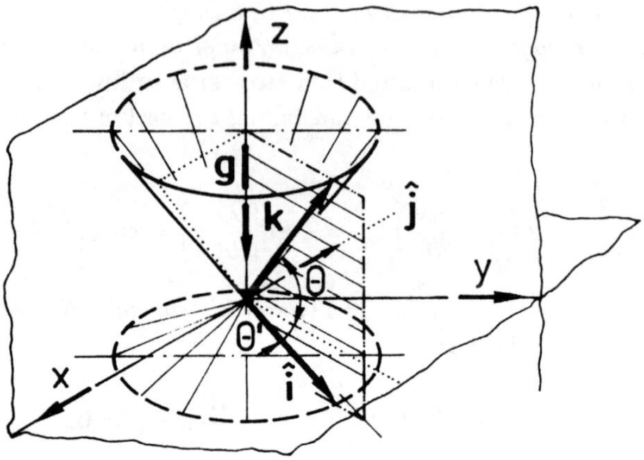

Figure 3.7: For a wave with wavenumber vector **k** (lying on the mantle of the upper cone), the displacement vector in the $\hat{\mathbf{i}}$-direction (which is perpendicular to **k** and lies on the mantle of the lower cone) and the $\hat{\mathbf{j}}$-direction. g, **k**, $\hat{\mathbf{i}}$ are all in the same vertical plane to which $\hat{\mathbf{j}}$ is orthogonal (and therefore in the (x,y)-plane). The angle between **k** and the horizontal plane is Θ , its complement to $\pi/2$ is Θ'.

in which α is the angle between $\boldsymbol{\Omega}$ and $\hat{\mathbf{k}}$. Evidently, for the squared frequency, stratification and rotation effects are additive. Given the wavenumber vector $\mathbf{k}$, the frequency ω, phase and group velocities follow from (3.49), while the velocity amplitudes U_i, U_j can be computed from (3.47). Qualitatively, the behaviour of the waves is as follows:

- When there is *no rotation*, $\boldsymbol{\Omega} = \mathbf{0}$, then $\omega = N\cos\Theta$; this does not mean that the frequency is independent of $\mathbf{k}$, because $\cos\Theta = \{(k_1^2 + k_2^2)\,/\,(k_1^2 + k_2^2 + k_3^2)\}^{\frac{1}{2}}$. Here the indices 1, 2, 3 refer to the x-, y- and z-directions, respectively. With $\boldsymbol{\Omega} = \mathbf{0}$ and (3.49) satisfied, equations (3.47) can be solved for the velocity amplitudes with the solution $U_j = 0$, $U_i =$ arbitrary. The fluid particles move persistently in the same direction: *Internal waves in a stratified fluid on an inertial frame are linearly polarized.*

- In a *nonstratified homogeneous* ($N = 0$) *rotating fluid* one has

$$\omega = 2\boldsymbol{\Omega}\cdot\hat{\mathbf{k}}, \qquad \mathbf{c}_{ph} = 2\frac{\boldsymbol{\Omega}\cdot\hat{\mathbf{k}}}{k}, \qquad \mathbf{c}_{gr} = 2\left[\frac{\boldsymbol{\Omega}}{\mathbf{k}} - \left(\frac{\boldsymbol{\Omega}\cdot\hat{\mathbf{k}}}{\mathbf{k}}\right)\frac{\hat{\mathbf{k}}}{k}\right]. \tag{3.50}$$

  This implies that true rotational waves propagate only when the direction of propagation is not perpendicular to $\boldsymbol{\Omega}$. If $\mathbf{k} \perp \boldsymbol{\Omega}$ then $\omega = 0$ and $c_{ph} = 0$.
  Moreover, substituting $(3.50)_1$ into (3.47), (in which $N = 0$), yields $U_j = -iU_i$, so that

$$\mathbf{v} = U_i\left(\hat{\mathbf{i}} - i\hat{\mathbf{j}}\right)e^{i(\mathbf{k}\cdot\mathbf{x}-\omega t)}. \tag{3.51}$$

After separating real and imaginary parts this becomes

$$\begin{aligned} \mathbf{v} &= U_i \left\{ \cos(\mathbf{k}\cdot\mathbf{x} - \omega t)\,\hat{\mathbf{i}} + \sin(\mathbf{k}\cdot\mathbf{x} - \omega t)\,\hat{\mathbf{j}} \right\} \quad \text{(real part)}, \\ \mathbf{v} &= U_i \left\{ \sin(\mathbf{k}\cdot\mathbf{x} - \omega t)\,\hat{\mathbf{i}} - \cos(\mathbf{k}\cdot\mathbf{x} - \omega t)\,\hat{\mathbf{j}} \right\} \quad \text{(imaginary part)}. \end{aligned} \tag{3.52}$$

Both are acceptable solutions to the wave equation, and it is easy to prove that for $\boldsymbol{\Omega}\cdot\mathbf{k} > 0$, corresponding to $\omega > 0$ the velocity vector at fixed x traces, in the course of the motion, a circle *in the clockwise direction*, performing a complete revolution in the period $2\pi/\omega$; see *Fig. 3.8. The waves are circularly polarized.*

- In the general case when the fluid is stratified ($N \neq 0$) and the frame of reference is rotating ($\boldsymbol{\Omega} \neq \mathbf{0}$), then the polarisation is a combination of the linear and circular polarisation. The waves are now *elliptically polarized* and the sense of revolution is - as above - against that of the circular frequency.

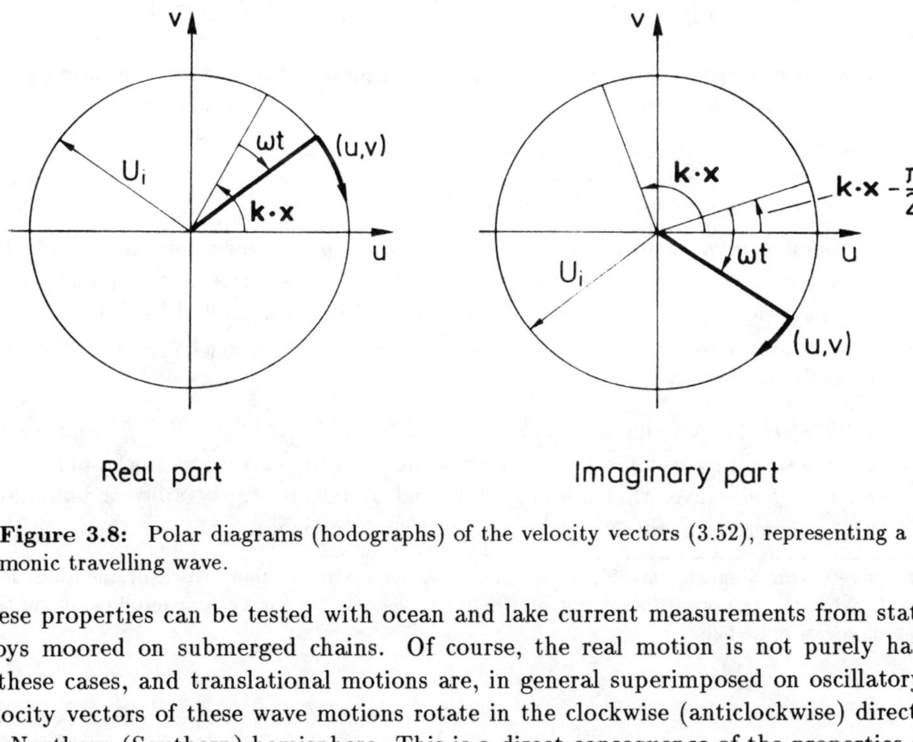

Figure 3.8: Polar diagrams (hodographs) of the velocity vectors (3.52), representing a harmonic travelling wave.

These properties can be tested with ocean and lake current measurements from stationary buoys moored on submerged chains. Of course, the real motion is not purely harmonic in these cases, and translational motions are, in general superimposed on oscillatory ones. Velocity vectors of these wave motions rotate in the clockwise (anticlockwise) direction on the Northern (Southern) hemisphere. This is a direct consequence of the properties of such rotational waves. If stratification is ignored, then gravity plays no role and the direction of the triad $(\hat{\mathbf{i}}, \hat{\mathbf{j}}, \hat{\mathbf{k}})$ relative to that of $\boldsymbol{\Omega}$ is alone significant. For $\omega > 0$, $\mathbf{k}$ points in the upper halfspace on the Northern hemisphere and the rotation of the velocity vector seen from above where $\mathbf{k}$ points at is clockwise. For $\omega < 0$, $\mathbf{k}$ points in the Southern halfspace and thus seen from below the velocity vector rotates counterclockwise (corresponding to clockwise rotation seen from above). A similar argument holds for the Southern hemisphere where the rotation is in the opposite direction. Progressive vector diagrams[17]constructed from time series of horizontal velocity vectors of such current measurements will thus form loops and be traversed in the clockwise (anticlockwise) direction. In small basins, such as mountainous lakes the relative narrowness of the shore will inhibit the circular component of this "garlandes", and loops may not close and be pointed, see *Fig. 3.9* showing examples from various lake and ocean basins.

A further physically very useful result emerges from (3.48), if $\mathbf{\Omega}$ and $\mathbf{k}$ are explicitly expressed in their component forms referred to the (x, y, z)-system (x is East- and y is Northward)

$$\mathbf{\Omega} = \frac{1}{2}\left(0, \tilde{f}, f\right) , \qquad \mathbf{k} = (k_1, k_2, k_3) .$$

With these, (3.48) becomes[18]

$$\left(k_1^2 + k_2^2\right)\left(\omega^2 - N^2\right) + k_3^2\left(\omega^2 - f^2\right) - k_2^2 \tilde{f}^2 - 2k_2 k_3 f \tilde{f} = 0 \tag{3.53}$$

In the shallow water approximation where $\tilde{f}$ may be ignored, the last two terms in (3.53) may be dropped, and (3.53) reduces to

$$\frac{k_1^2 + k_2^2}{k_3^2} = \frac{\omega^2 - f^2}{N^2 - \omega^2} . \tag{3.54}$$

This is the *dispersion relation in the shallow water approximation* for a fluid on the rotating Earth which is linearly stratified in the z-direction. It discloses significant physical results: k_1 and k_2 are the x- and y-components of the wavenumber vector, and the left hand side of equation (3.54) is necessarily positive; thus, the right hand side must also be positive. Two cases can be distinguished:

- When $N^2 > \omega^2$ then one must also have $\omega^2 > f^2$ implying $f < \omega < N$. f forms a lower bound for internal gyroscopic waves to exist and N is the corresponding upper bound. For constant f all waves that are not influenced by bottom topography or boundaries are thus *superinertial.* Should ω be larger than N or smaller than f then the right

[17]Progressive vector diagrams are 'displacement' plots from current meter measurements at fixed positions. If Δt is the temporal increment at which the velocity vector $\mathbf{v}_i$ (i = number of the time increment) is measured, then

$$\mathbf{s}_n = \sum_{i=0}^{n} \mathbf{v}_i \,\Delta t \qquad (n = 0, 1, 2, \ldots)$$

determines the trajectory of a 'virtual particle' that passed the current meter at n = 0. The particle is not real, because the current meter does not follow it (Eulerian description). An alternative is to follow a marker through time (Lagrangian description)

[18]This equation is most easily deduced from (3.48) if it is recognized that $k^2 \cos^2 \Theta = (k_1^2 + k_2^2)$.

hand side of (3.54) is negative, and this requires the wavenumbers to be complex-valued. Their imaginary parts will give rise to exponential behaviour in the x-direction (attenuation). The existence of an upper bound N and a lower bound f for ω thus explains why the metalimnion in a stably stratified lake may serve as a wave guide to internal waves. *Fig. 3.10* shows an N^2-profile obtained from density measurements in the Lake of Zürich. Often N^2-distributions are as shown schematically in the inset map. For the selected value of ω, plane unbounded internal waves can only propagate in the indicated band of the metalimnion.
The transition from the wave guide regime to the regimes outside of it must be special. It is characterized by $|\omega| \sim N$. It will be shown below that whenever $|\omega| \sim N$ then the conditions of the shallow water approximation are not satisfied. Hence, equation (3.54) should be replaced in these instances by (3.53). These transition regions are known in the literature as *critical layers*. They are particulary prone to instabilities and are the locations where localized overturning, circulating motions may arise. In these notes we shall not specifically look at these.

- There is another, and less important, frequency range for which (3.54) admits real solutions for frequency and wavenumbers: $N < \omega < f$. Here, waves must be *subinertial* and stratification is extremely weak. This is the reason why such subinertial plane free waves are less frequent. They can equally be ruled out in the shallow water approximation.

There are, however, also other waves with subinertial frequencies $\omega < f$. These incorporate β-effects and topographic variations or are bounded by the shore. The former are of ROSSBY-type, the latter are the KELVIN waves and are of first class, as we shall shortly see.

3.3.2 Characteristic wave solutions - shallow water approximation

The preceeding discussion pertains to waves with arbitrary wavelengths and arbitrary orientation. In this section we confine attention to *long waves* and thus make use of the shallow water approximation, for which the governing equations (3.39) - (3.41) reduce to

$$\nabla \cdot \mathrm{v} + \frac{\partial w}{\partial z} = 0 ,$$

Figure 3.9: (see next page) Examples of looping current tracks (plotted on a common scale and with distance from shore and depths indicated in brackets) from : Central Baltic (31 *km*, 14 *m*, GUSTAFSON and KULLENBERG, 1936); Lake Baikal (6 *km*, 10 *m*, GALAZII, 1969); Lake Michigan (12 *km*, 12 *m*, SATO and MORTIMER, 1975); Mediterranean (100 *km*, 10 *m*, GONELLA, 1974) (from MORTIMER, 1980)

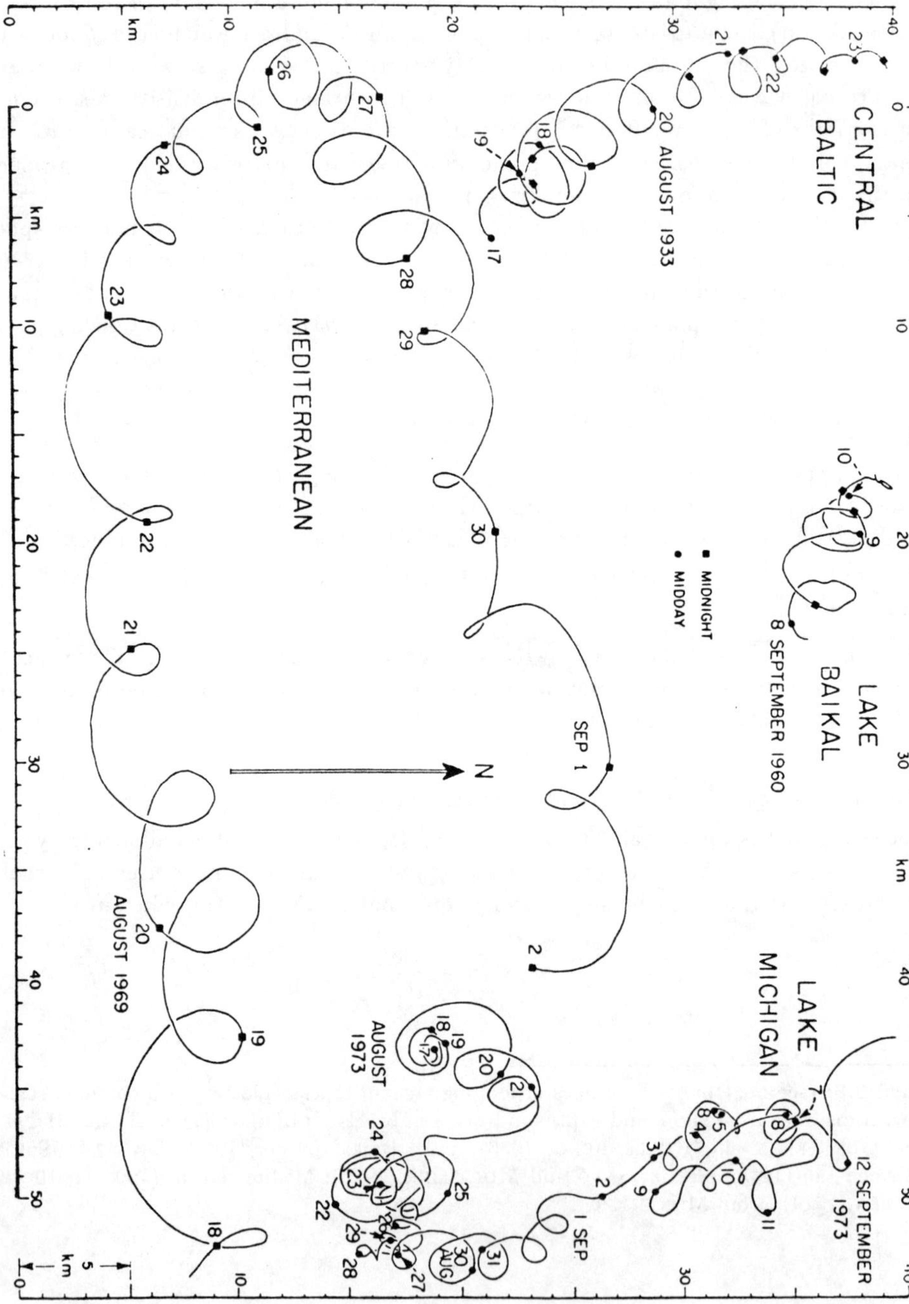
CENTRAL
BALTIC
20 AUGUST 1933
LAKE
BAIKAL
8 SEPTEMBER 1960
LAKE
MICHIGAN
12 SEPTEMBER
1973
AUGUST
1973
MEDITERRANEAN
20
AUGUST 1969
SEP 1
1 SEP
■ MIDNIGHT
● MIDDAY
N
km

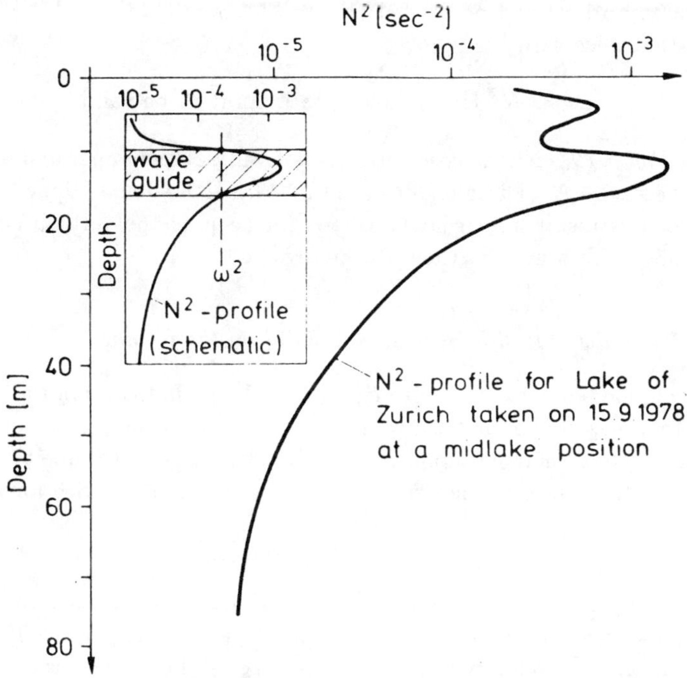

Figure 3.10: Square BRUNT-VÄISÄLÄ frequency N^2 as determined from measurements at a midlake position in Lake of Zürich, taken on September 15, 1978. The inset map shows its schematic distribution. For the selected ω waves can only exist in the indicated band (from HUTTER, 1984b).

$$\begin{aligned} \rho^* \frac{\partial \mathbf{v}}{\partial t} + f\rho^* \hat{\mathbf{k}} \times \mathbf{v} + \nabla p' &= 0 , \\ \underline{\rho^* \frac{\partial w}{\partial t}} + \frac{\partial p'}{\partial z} - \rho' g &= 0 , \\ \frac{\partial \rho'}{\partial t} - \rho^* \frac{N^2}{g} w &= 0. \end{aligned} \tag{3.55}$$

The first is the continuity equation, the second are the horizontal components of the momentum equation, the third is the vertical component of the momentum balance in which the underlined acceleration term is absent when the shallow water approximation is made, and the last equation is the adiabatic form of the energy equation.
From a combination of the last two of equations (3.55), on eliminating the perturbation density ρ', one may deduce the relation

$$\left(\underline{\frac{\partial^2}{\partial t^2}} + N^2 \right) w = -\frac{1}{\rho^*} \frac{\partial^2 p'}{\partial t \partial z} \tag{3.56}$$

of which the term $\partial^2 w / \partial t^2$ would be missing if the hydrostatic pressure assumption (the full shallow water assumption) had been used. From a harmonic ansatz $p' = p_0 e^{i\omega t}, w = w_0 e^{i\omega t}$ it then follows that necessarily

$$\omega^2 \ll N^2 \text{ for the hydrostatic limit to be valid.}$$

As is seen from *Fig. 3.10*, such an assumption is not valid at the upper and lower edges of the metalimnion where $\omega \sim N$. These regions are the *critical layers* mentioned above. However, the derivation of this result has been the reason for keeping the vertical acceleration terms in equation $(3.55)_3$. From now on they will be dropped.

Free waves in a shallow stratified BOUSSINESQ *fluid on the f-plane:*

It is somewhat cumbersome to work with equations (3.55) that contain the variables u, v, w, p' and ρ'. For this reason one may ask whether a single equation for one of these unknowns would be advantageous. Furthermore, for a vertically stratified fluid it would be natural to separate the vertical from the horizontal dependencies and to write for any of the above variables

$$\psi(x,y,z,t) = \Psi_n(z)\,\psi_n(x,y,t)$$

and to uncouple the equations for $\Psi_n(z)$ and $\psi_n(x,y,t)$. For a constant depth fluid and when the shallow water approximation is made this separation is possible if the perturbation pressure p' or a horizontal velocity component are taken to be the governing field variables[19]. Combination of the hydrostatic version of equations (3.55) permits this elimination of all but one variable. This elimination is rather cumbersome even though it is straightforward. Here we merely quote the result and ask the reader to perform the computations himself. For the shallow water equations on the f-plane the following alternatives are obtained:

$$\begin{aligned}
\nabla^2 v + \mathcal{L}\tfrac{\partial}{\partial z}\left(N^{-2}\tfrac{\partial v}{\partial z}\right) &= 0\,, \quad (v = \text{horizontal velocity component}),\\
\nabla^2 p' + \mathcal{L}\tfrac{\partial}{\partial z}\left(N^{-2}\tfrac{\partial p'}{\partial z}\right) &= 0\,, \quad (p' = \text{pressure}),\\
N^2\nabla^2 w + \mathcal{L}\tfrac{\partial^2 w}{\partial z^2} &= 0\,, \quad (w = \text{vertical velocity component}),
\end{aligned} \tag{3.57}$$

in which

$$\mathcal{L} = \frac{\partial^2}{\partial t^2} + f^2. \tag{3.58}$$

Had we kept the underlined term in $(3.55)_3$ then the equation for the vertical velocity component would be instead

$$\left(N^2 + \underline{\frac{\partial^2}{\partial t^2}}\right)\nabla^2 w + \mathcal{L}\frac{\partial^2 w}{\partial z^2} = 0. \tag{3.59}$$

[19]This separation also works in the β-plane approximation, if the perturbation pressure or the y-component of the velocity field are taken as the independent fields.

In a constant depth basin solutions of equations (3.57) are sought by *separation of variables* in the form

$$\begin{aligned} v(x,y,z,t) &= Y_n(z)v_n(x,y,t)\,, \\ w(x,y,z,t) &= Z_n(z)w_n(x,y,t)\,, \\ p'(x,y,z,t) &= \Pi_n(z)p_n(x,y,z,t)\,. \end{aligned} \tag{3.60}$$

The equations for the vertical and the horizontal problems then become

$$\begin{aligned} &\frac{d^2 Z_n}{dz^2}+\frac{N^2}{gh_n}Z_n=0, && \mathcal{L}w_n-gh_n\nabla^2 w_n=0,\\ &\frac{d}{dz}\left(N^{-2}\frac{dY}{dz}\right)+\frac{Y}{gh_n}=0, && \mathcal{L}Y-gh_n\nabla^2 Y=0, \end{aligned} \tag{3.61}$$

in which Y stands for Y_n and Π_n, respectively, and gh_n is a constant of separation; h_n has the dimension of a length and is called *equivalent depth.* By the separation of variables the original three-dimensional equations (3.57), which must be solved in the lake domain, have been split into two separate problems, one horizontal and two-dimensional, the other vertical and one-dimensional. It is not difficult to see that (3.59) cannot be separated by equation $(3.60)_2$.

The separation constant gh_n can be determined by solving for the vertical distributions Y_n, Z_n, or Π_n subject to the appropriate boundary conditions at the lake bottom, $z=-H$, where $w=0$ and at the free surface, $z=\zeta$, where equations $(3.3)_{3,4}$ must hold:

$$\begin{aligned} &p'-\rho_0 g\zeta(x,y,t)=0, && \text{at} \quad z=0,\\ &w-\frac{\partial\zeta}{\partial t}(x,y,t)=0, && \text{at} \quad z=0,\\ &w=0, && \text{at} \quad z=-H. \end{aligned} \tag{3.62}$$

These conditions must then be expressed in terms of v, w or p' alone, if a separation of a full boundary value problem should be possible. The corresponding expressions for v and w are awkward, but when ζ is eliminated from $(3.62)_{1,2}$ and the resulting equation is combined with (3.56), the pressure boundary conditions read

$$\left(\underline{\frac{\partial^2}{\partial t^2}}+N^2\right)p'+g\frac{\partial p'}{\partial z}=0, \quad z=0, \qquad \frac{\partial p'}{\partial z}=0, \quad z=-H. \tag{3.63}$$

The underlined term must be dropped when the shallow water approximation is made. Only in this case will the separation solution $(3.60)_3$ yield the separable boundary conditions

$$\frac{d\Pi_n}{dz}+\frac{N^2}{g}\Pi_n=0, \quad z=0 \qquad \frac{d\Pi_n}{dz}=0, \quad \text{at} \quad z=-H. \tag{3.64}$$

These conditions involve only functions of the z-variable. We leave it as an exercise to the reader to demonstrate that such a separation of the surface boundary condition cannot be achieved when w is chosen as the basic field variable.

This demonstrates why the pressure equation is advantageous: the differential equation *and* the boundary conditions for Π_n depend on the stratification alone and may be solved once and for all, and then in a second step the horizontal problem may be attacked. If this is done by a plane-wave ansatz

$$Y = Y_0 e^{i(\mathbf{k}_h \cdot \mathbf{x} - \omega t)} \tag{3.65}$$

in which $\mathbf{k}_h$ is a horizontal wavenumber vector, equations $(3.61)_{3,4}$ imply the *dispersion relation*

$$\omega_n^2 = f^2 + g h_n k_h^2 \tag{3.66}$$

from which the phase and group velocities can be determined

$$\mathbf{c} = \frac{\sqrt{g h_n k_h^2 + f^2}}{k_h^2} \mathbf{k}_h , \qquad \mathbf{c}_g = \frac{g h_n}{\omega_n} \mathbf{k}_h \, . \tag{3.67}$$

They are parallel to each other and obey the inequality $|c_g| < \sqrt{g h_n} < |c|$.These waves were first analyzed by POINCARÉ and are, therefore called POINCARÉ *waves*.
A complete separation of the differential equations and boundary conditions is also possible when the other variables (v or w) are used, but then a plane-wave ansatz must be used *ab initio*. For instance, equation (3.59) and $w = Z_n(z) e^{(i(\mathbf{k}_h \cdot \mathbf{x} - \omega t))}$ together with a rigid-lid boundary condition at the free surface ($w = 0$, at $z = 0$) yield the eigenvalue problem

$$\begin{aligned} \frac{d^2 Z_n}{dz^2} + \frac{N^2 - \omega^2}{\omega^2 - f^2} k_h^2 Z_n = 0 \quad &, \quad -H \le z \le 0 \ \text{(non-hydrostatic fluid)} \\ Z_n = 0 \quad &, \quad \text{at } z = (0, -H) \, . \end{aligned} \tag{3.68}$$

It contains N, k_h, and ω as parameters; thus apart from stratification the eigenvalue and eigenfunctions also depend on the frequency.

Vertical Mode Structure:

Even though N = constant is a poor approximation of realistic N-distributions, the solution of equation $(3.61)_2$ and equation (3.64) with this simplification gives good qualitative insight into the mode structure. With $\varepsilon = N^2 H/g$, $\lambda_n^2 = N^2 H^2/(g h_n)$, the vertical distribution of the perturbation pressure reads

$$\Pi_n = A_n \left[\cos \lambda_n z' - \varepsilon \lambda_n^{-1} \sin \lambda_n z' \right] , \qquad z' = \frac{z}{H} , \tag{3.69}$$

where λ_n satisfies the eigenvalue equation

$$\tan \lambda_n = \varepsilon \lambda_n^{-1}, \qquad n = 1, 2, 3, \ldots , \tag{3.70}$$

which has a countably infinite number of solutions. Because $\varepsilon = \mathcal{O}(10^{-3})$, approximate satisfaction of the characteristic equation yields

$$\lambda_0^2 = \varepsilon + \mathcal{O}\left(\varepsilon^2\right), \qquad \lambda_n^2 = n^2 \pi^2 + \mathcal{O}(\varepsilon), \qquad n = 1, 2, 3, \ldots , \tag{3.71a}$$

corresponding to

$$h_0 = H\left(1+\mathcal{O}\left(\varepsilon^2\right)\right), \qquad h_n = \frac{\varepsilon H}{n^2\pi^2}\left(1+\mathcal{O}(\varepsilon)\right), \qquad n = 1,2,3,\ldots, \tag{3.71b}$$

$$\Pi_0 = A_0\left(1+\mathcal{O}(\varepsilon)\right), \qquad \Pi_n = A_n\left[\cos\frac{n\pi z}{H} + \mathcal{O}(\varepsilon)\right], \qquad n = 1,2,3,\ldots.$$

For the lowest *barotropic* eigenfunction the equivalent depth nearly equals H, and the perturbation pressure and horizontal velocities, if they are calculated, are nearly independent of z. For all other, *baroclinic*, modes the equivalent depth is much smaller, $h_0 \gg h_1 > h_2 \ldots$, and the pressure distribution $\Pi_n(z)$ in $-H \le z \le 0$ changes its sign n-times. When $f = 0$ phase and group velocities are

$$c = c_g = \sqrt{gh_n}\ , \ n = 1,2,3,\ldots,$$

and the waves are nondispersive. For barotropic waves the speeds are from 10 to 200 ms^{-1}, for baroclinic waves they are $\mathcal{O}(\sqrt{\varepsilon})$ smaller and typically from 0.3 to 10 ms^{-1}; the first of these numbers is representative for shallow lakes, the second for the deep ocean.
Qualitatively, similar inferences can be drawn from equations (3.68). When $f < \omega < N$ then the differential equation $(3.68)_1$ is essentially of the form of the harmonic oscillator equation. The solution has oscillatory behaviour. This is the case within the wave guide shown in *Fig. 3.10.* Outside of it, equation (3.68) has exponential character.
These solutions must be patched together such that Z_n and dZ_n/dz are continuous at the positions z where $N = \omega$. The combined solution must finally satisfy the boundary conditions $(3.68)_2$, from which the spectrum of the eigenvalues can be calculated. The reader will profit if he makes this clear to himself by assuming the following z-distribution of the buoyancy frequency

$$\begin{array}{lll} N = 0, & 0 \ge z \ge -z_u, & \text{epilimnion,} \\ N = N_0, & -z_u \ge z \ge -z_l, & \text{metalimnion,} \\ N = 0, & -z_l \ge z \ge -H, & \text{hypolimnion.} \end{array} \tag{3.72}$$

Here N_0 is a constant, representative for the buoyancy frequency in the metalimnion and z_u and z_l are the upper and lower boundary of the metalimnion, respectively. This qualitative behaviour can be proved for a general vertical profile of the BRUNT-VÄISÄLÄ frequency. In the oscillatory portion of the solution Z_n has $(n-1)$ modes. When the rotation of the Earth is ignored the vertical distribution of an unidirectional horizontal velocity component is given by dZ_n/dz and has n zeros. This behaviour is shown in *Fig. 3.11.* For the first baroclinic mode the horizontal current is in one direction in the epilimnion and in the opposite direction in the hypolimnion. At the higher modes the horizontal current is to and fro according to the number of nodal points. When $\omega > N$, solutions are exponentially decaying. This is outside the indicated wave guide (shaded in *Fig. 3.11*).
For smooth distributions of the buoyancy frequency N these properties can mathematically be proven to be a consequence of the Sturm-Liouville property of the eigenvalue problem for Π_n (or Z_n): There is a computably infinte number of positive eigenvalues and corresponding

eigenfunctions that are complete in the sense that any twice differentiable function over $-H \leq z \leq 0$ can be expressed as a linear combination of these. Of course, when N is not smooth then these properties cannot be expected to hold.

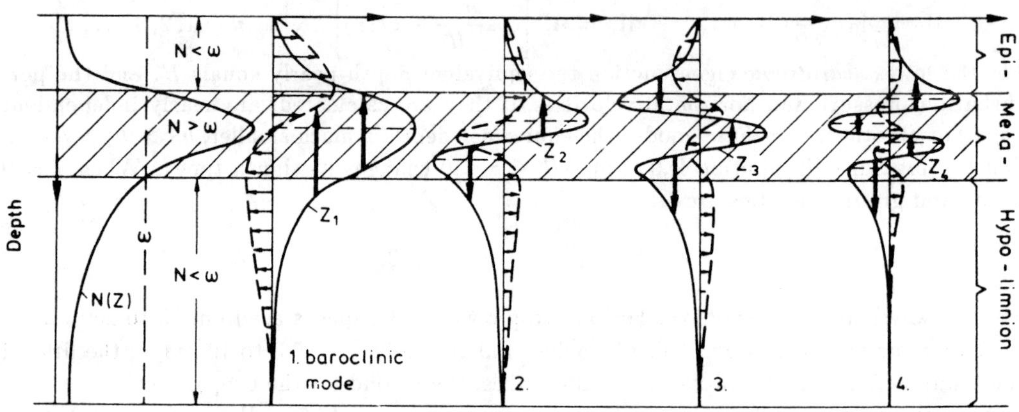

Figure 3.11: Typical vertical distribution of the BRUNT-VÄISÄLÄ frequency N (left) and the four lowest baroclinic modes (qualitative). Solid curves show the distribution of the vertical velocity component, dashed curves indicate the distribution of the longitudinal velocity component (when $f = 0$). Most energy is usually concentrated in the first baroclinic mode, the exact distribution must, however, be determined by continuous profils of horizontal velocities.

The Two-Layer Model:

In strongly stratified lakes or oceans with a well mixed upper layer the BRUNT-VÄISÄLÄ frequency is practically zero in the epilimnion and small in the hypolimnion but large in a narrow metalimnion. As an approximation one may choose the distribution of the buoyancy frequency as given in (3.72). An even farther reaching assumption would be to choose $N = 0$ everywhere except at the thermocline in which N is a DIRAC pulse. Such an approximation corresponds to a two-layer model with an upper layer of constant depth H_1 and density ρ_1 and a second layer with density $\rho_2 > \rho_1$ and (variable) depth H_2, see *Fig. 3.12.* It is also physically obvious that with a metalimnion of vanishing extent (it is restricted to the discontinuity surface where the density changes from ρ_1 to ρ_2) there can be no oscillatory part of the solution within the metalimnetic wave band: there is only *one* barotropic and *one* baroclinic mode.

To substantiate these claims, consider the linearized equations of motion

$$\begin{aligned} \nabla \cdot \mathbf{v} + \frac{\partial w}{\partial z} &= 0\,, \\ \rho^* \left(\frac{\partial \mathbf{v}}{\partial t} + f \hat{\mathbf{k}} \times \mathbf{v} \right) &= -\nabla p\,. \end{aligned} \tag{3.73}$$

$$\frac{\partial p}{\partial z} = -\rho g ,$$

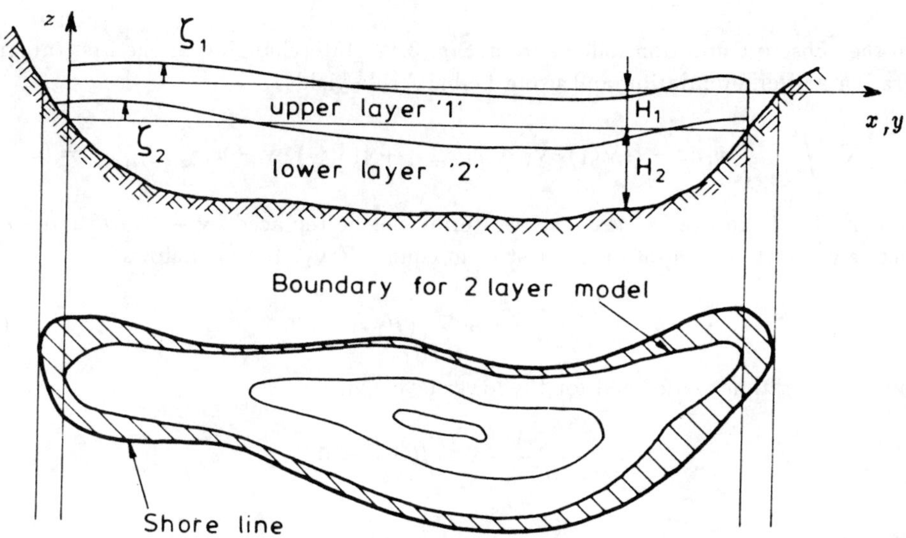

Figure 3.12: Two-layer model with upper layer depth H_1 and lower depth H_2. The mathematical models often use a vertical wall as boundary at the depth contour of H_1. The dashed portion is then ignored in the model calculations.

which must hold in each layer and solved subject to the boundary conditions

$$\left.\begin{aligned} &\frac{\partial \zeta_1}{\partial t} + (\nabla \zeta_1) \cdot \mathbf{v}_1 - w_1 = 0, \\ &p = 0, \end{aligned}\right\} \quad \text{at} \quad z = \zeta_1 , \tag{3.74}$$

$$\left.\begin{aligned} &\frac{\partial \zeta_2}{\partial t} + (\nabla \zeta_2) \cdot \mathbf{v}_1 - w_1 = 0, \text{ or} \\ &\frac{\partial \zeta_2}{\partial t} + (\nabla \zeta_2) \cdot \mathbf{v}_2 - w_2 = 0, \\ &p_1 = p_2, \end{aligned}\right\} \quad \text{at} \quad z = -H_1 + \zeta_2, \tag{3.75}$$

$$(\nabla H_2) \cdot \mathbf{v}_2 - w_2 = 0, \qquad \text{at} \quad z = -(H_1 + H_2) \tag{3.76}$$

These equations shall now be reduced to a set of two-layer equations. To this end, we first integrate the continuity equation over the upper and lower layer, respectively: In the upper layer this yields

$$\int_{-H_1+\zeta_2}^{\zeta_1} (\nabla \cdot \mathbf{v}_1)\, dz + w_1(\zeta_1) - w_1(-H_1+\zeta_2) = 0 \ ,$$

where the subscript notation follows from *Fig.* 3.12. Interchanging in the first integral the integration and differentiation and using Leibniz' rule yields

$$\nabla \cdot \int_{-H_1+\zeta_2}^{\zeta_1} v_1 dz - \{(\nabla\zeta_1)\cdot \mathbf{v}_1 - w_1\}_{z=\zeta_1} + \{(\nabla\zeta_2)\cdot \mathbf{v} - w_1\}_{z=-H_1+\zeta_2} = 0.$$

In view of $(3.74)_1$ and (3.75) the last two terms can be replaced by $-\partial\zeta_1/\partial t$ and $-\partial\zeta_2/\partial t$, respectively, and the integral in the first term equals $H_1\mathbf{v}_1$. It thus follows

$$\frac{\partial(\zeta_1 - \zeta_2)}{\partial t} + \nabla \cdot (H_1 \mathbf{v}_1) = 0. \tag{3.77a}$$

A similar integration, performed for the lower layer, yields

$$\frac{\partial(\zeta_2)}{\partial t} + \nabla \cdot (H_2 \mathbf{v}_2) = 0. \tag{3.77b}$$

These two equations express the mass balance in the upper and lower layers, respectively. Next, we integrate the hydrostatic pressure equation from the top surface to an arbitrary depth z. This yields

$$\begin{aligned} p &= -\rho_1 g(z-\zeta_1), && \text{in layer 1,} \\ p &= \rho_1 g(H_1+\zeta_1-\zeta_2) + \rho_2 g(z - H_1 - \zeta_1)\ , && \text{in layer 2.} \end{aligned}$$

Taking the horizontal gradient of each of these and substituting the resulting expressions into $(3.73)_2$ leads to the equations[20]

$$\frac{\partial \mathbf{v}_1}{\partial t} + f\hat{\mathbf{k}} \times \mathbf{v}_1 = -g\nabla\zeta_1 \ , \tag{3.77c}$$

$$\frac{\partial \mathbf{v}_2}{\partial t} + f\hat{\mathbf{k}} \times \mathbf{v}_2 = -g'\nabla(\zeta_2 - \zeta_1) + g\nabla\zeta_1 \ , \tag{3.77d}$$

in which

$$g' = g\frac{\rho_2 - \rho_1}{\rho_2} = g\varepsilon \tag{3.78}$$

is the *reduced gravity constant* and ε is the *relative density difference*. In usual summer stratifications ε is of the order of 10^{-3}, or slightly larger. Equations (3.77a, b, c, d) comprise the governing equations for the two layer model.

For constant depth of the basin the two-layer equations will now be reduced to two separate

[20]In this evaluation we set $\rho^* = \rho_2$ and approximate ρ_1/ρ_2 by 1.

one-layer models. To this end, we add ν_α times the lower layer continuity and momentum equations to the corresponding upper layer equations, define auxiliary variables according to

$$\mathbf{V}_\alpha = H_1 \mathbf{v}_1 + \nu_\alpha H_2 \mathbf{v}_2, \qquad \zeta_\alpha = \zeta_1 + (\nu_\alpha - 1)\zeta_2, \tag{3.79}$$

and require that the emerging equations have the same form as the one-layer equations. The mass balance equation of this one-layer model then has the form $(3.82)_1$ below, whereas the momentum balance takes the form

$$\frac{\partial \mathbf{V}_\alpha}{\partial t} + f\hat{\mathbf{k}} \times \mathbf{V}_\alpha = -g\nabla \left\{ [H_1 + \nu_\alpha (1-\varepsilon) H_2]\zeta_1 + \nu_\alpha \varepsilon H_2 \zeta_2 \right\} .$$

This equations has the form of a one-layer equation if

$$[H_1 + \nu_\alpha (1-\varepsilon) H_2]\zeta_1 + \nu_\alpha \varepsilon H_2 \zeta \equiv h_\alpha \zeta_\alpha$$

for some h_α. Using the definition $(3.79)_2$ of ζ_α this equation implies the identities

$$\begin{aligned} H_1 + \nu_\alpha (1-\varepsilon) H_2 &= h_\alpha, \\ H_2 \varepsilon \nu_\alpha &= h_\alpha (\nu_\alpha - 1), \end{aligned} \tag{3.80}$$

which form two equations for ν_α and h_α. On eliminating h_α, the following quadratic equation for ν_α is obtained

$$\nu_\alpha^2 + \frac{H_1 - H_2}{(1-\varepsilon) H_2}\nu_\alpha - \frac{1}{1-\varepsilon}\frac{H_1}{H_2} = 0 . \tag{3.81}$$

In summary the *one-layer equivalent depth model* for the two-layer system (3.77a-d) is described by the equations

$$\frac{\partial \zeta_\alpha}{\partial t} + \nabla \cdot \mathbf{V}_\alpha = 0, \qquad \frac{\partial \mathbf{V}_\alpha}{\partial t} + f\hat{\mathbf{k}} \times \mathbf{V}_\alpha = -g h_\alpha \nabla \zeta_\alpha, \tag{3.82}$$

in which the one-layer variables are defined by (3.79), ν_α follows from (3.81) and h_α from one of the equations (3.80).
Since $\varepsilon = \mathcal{O}(10^{-3})$, equations (3.81), (3.80) and (3.79) may be solved by perturbation methods. The solutions are

- *Barotropic, external mode* $\nu_\alpha := \nu_e$, $H = H_1 + H_2$

$$\begin{aligned} \nu_e &= 1 + \varepsilon \frac{H_2}{H} + \mathcal{O}\left(\varepsilon^2\right), \\ h_e &= H - \varepsilon \frac{H_1}{H} + \mathcal{O}\left(\varepsilon^2\right), \\ \mathbf{V}_e &= H_1 \mathbf{v}_1 + H_2 \mathbf{v}_2 + \varepsilon \frac{H_2^2}{H}\mathbf{v}_2 + \mathcal{O}\left(\varepsilon^2\right) , \\ \zeta_e &= \zeta_1 + \varepsilon \frac{H_2}{H}\zeta_2 . \end{aligned} \tag{3.83}$$

- *Baroclinic, internal mode* $\nu_\alpha := \nu_i$

$$\begin{aligned}
\nu_i &= -\frac{H_1}{H_2}\left\{1 - \varepsilon\frac{H_1 - H_2}{H}\right\} + \mathcal{O}\left(\varepsilon^2\right), \\
h_i &= \varepsilon\frac{H_1 H_2}{H} + \mathcal{O}\left(\varepsilon^2\right), \\
\mathbf{V}_i &= H_1\left(\mathbf{v}_1 - \mathbf{v}_2\right) + \varepsilon\frac{H_1}{H}\left(H_1 - H_2\right)\mathbf{v}_2 + \mathcal{O}\left(\varepsilon^2\right), \\
\zeta_i &= -\frac{H}{H_2}\zeta_2 + \zeta_1 + \varepsilon\frac{H_1}{H_2}\frac{H_1 - H_2}{H}\zeta_2 + \mathcal{O}\left(\varepsilon^2\right).
\end{aligned} \tag{3.84}$$

From these expressions we may deduce the inversions

$$\begin{aligned}
\mathbf{v}_1 &= \frac{\mathbf{V}_e + \frac{H_2}{H_1}\mathbf{V}_i}{H}, \qquad \zeta_1 = \zeta_e + \varepsilon\frac{H_2^2}{H^2}\zeta_i + \mathcal{O}\left(\varepsilon^2\right), \\
\mathbf{v}_2 &= \frac{\mathbf{V}_e - \mathbf{V}_i}{H}, \qquad \zeta_2 = \frac{H_2}{H}\left(\zeta_e - \zeta_i\right) + \mathcal{O}\left(\varepsilon\right).
\end{aligned} \tag{3.85}$$

In these expressions $\varepsilon = (\rho_2 - \rho_1)/\rho_2$, and the indices $\alpha = e$ and $\alpha = i$ refer to the external and internal modes, respectively. It is also seen that for barotropic modes the equivalent depth h_e equals $H = H_1 + H_2$, and that the equivalent baroclinic depth h_i corresponds to the first baroclinic equivalent depth.
Suppose now that $\mathbf{V}_i = \mathbf{0}$, $\zeta_i = 0$ (no baroclinic motion); then up to order ε-terms

$$\begin{aligned}
\mathbf{v}_1 &= \mathbf{v}_2 = \frac{1}{H_1 + H_2}\mathbf{V}_e, \\
\zeta_1 &= \zeta_e, \\
\zeta_2 &= \frac{H_2}{H_1 + H_2}\zeta_e.
\end{aligned} \tag{3.86}$$

On the other hand, when $\mathbf{V}_e = \mathbf{0}$, $\zeta_e = 0$ (no barotropic motion), then

$$\begin{aligned}
H_1 \cdot \mathbf{v}_1 &= -H_2 \cdot \mathbf{v}_2 = \frac{H_2}{H_1 + H_2}\mathbf{V}_i, \\
\zeta_1 &= \varepsilon\left(\frac{H_2}{H_1 + H_2}\right)^2 \zeta_i, \\
\zeta_2 &= -\frac{H_2}{H_1 + H_2}\zeta_i.
\end{aligned} \tag{3.87}$$

For pure barotropic processes the horizontal velocity is uniform with depth and for baroclinic processes the upper layer volume flux equals the lower layer volume flux, but the two have opposite direction, reminiscent of the first baroclinic mode in *Fig. 3.11*.
The two-layer equivalent depth model is an attractive one; however it is strictly only valid for constant depth and eliminates all higher-order baroclinic modes. Replacing the variable

depth by a mean constant depth and using the constant equivalent depth model, equations (3.82) seem to be reasonable but higher order baroclinicity may require multilayer models. LONGUET-HIGGINS in MORTIMER (1952) and HEAPS (1961) therefore introduced a three-layer model and LIGHTHILL (1969) showed, how an N-layer model can be reduced to the equivalent depth form.
Often in ensuing developments we shall discuss barotropic or baroclinic behaviours simultaneously. In these instances we always have an equivalent depth model such as (3.82) in mind and will distinguish between barotrópic and baroclinic processes according to whether h_α is the water depth or an equivalent depth belonging to a certain stratification.

Problems

1. Consider plane waves in a homogeneous rotating fluid. Verify with the aid of (3.52) that the velocity vector in the hodograph plane traverses a circle in the clockwise direction when $\mathbf{\Omega} \cdot \mathbf{k} > 0$ and in the counter-clockwise direction when $\mathbf{\Omega} \cdot \mathbf{k} < 0$.

2. Show that the hodographs of a plane wave in a linearly stratified rotating fluid are ellipses.

3. In this problem we demonstrate, how the linearized adiabatic shallow water equations in a stratified fluid on the β-plane can be reduced to a single equation for the NS-component of the velocity field. We shall not impose the BOUSSINESQ assumption[21]. The governing equations are

$$\frac{\partial u}{\partial t} - fv + \frac{1}{\rho_0}\frac{\partial p}{\partial x} = 0, \qquad \frac{\partial u}{\partial x} + \frac{\partial v}{\partial y} + \frac{\partial w}{\partial z} = 0,$$

$$\frac{\partial v}{\partial t} + fv + \frac{1}{\rho_0}\frac{\partial p}{\partial y} = 0, \qquad \frac{\partial \rho}{\partial t} + \frac{d\rho_0}{dz}w = 0, \tag{P1}$$

$$\frac{\partial p}{\partial z} = -\rho g,$$

in which $\rho_0 = \rho_0(z)$.
We first combine these equations to obtain an equation for v. To this end we differentiate $(\mathrm{P1})_1$ with respect to y and $(\mathrm{P1})_2$ with respect to x, subtract the two equations from one another to obtain

$$\frac{\partial}{\partial t}\left(\frac{\partial v}{\partial x} - \frac{\partial u}{\partial y}\right) + \beta v - f\frac{\partial w}{\partial z} = 0. \tag{P2}$$

This is the vertical component of the vorticity balance. Differentiating this equation with respect to x and eliminating from the emerging equation the velocity component

[21] In these equations we have dispensed with the BOUSSINESQ assumption in the momentum equations, but still assume $\mathrm{div}\mathbf{v} = 0$. This is not entirely consistent with our scaling in section 2.

u (by means of the continuity equation) yields

$$\nabla_h^2 \frac{\partial v}{\partial t} + \beta \frac{\partial v}{\partial x} + \frac{\partial}{\partial z}\left\{ \frac{\partial^2 w}{\partial y \partial t} - f \frac{\partial w}{\partial x} \right\} = 0. \tag{P3}$$

∇_h^2 is the two dimensional Laplacian. Combining $(\mathrm{P1})_{3,5}$, on the other hand, yields

$$\rho_0(z) N^2 w = -\frac{\partial^2 p}{\partial z \partial t}, \quad N^2 = -\frac{g d\rho_0/dz}{\rho_0(z)}, \tag{P4}$$

so that (P3) may also be written as

$$\nabla_h^2 \frac{\partial v}{\partial t} + \beta \frac{\partial v}{\partial x} + \frac{\partial}{\partial z}\left\{ \frac{1}{N^2 \rho_0}\left[f \frac{\partial^3 p}{\partial x \partial z \partial t} - \frac{\partial^4 p}{\partial y \partial z \partial t^2} \right] \right\} = 0. \tag{P5}$$

To eliminate the pressure, we notice that $(\mathrm{P1})_{1,2}$ can be combined to yield

$$\mathcal{L} \frac{\partial v}{\partial t} = \frac{1}{\rho_0}\left\{ f \frac{\partial^2 p}{\partial x \partial t} - \frac{\partial^3 p}{\partial y \partial t^2} \right\}, \qquad \mathcal{L} := \left(\frac{\partial^2}{\partial t^2} + f^2 \right). \tag{P6}$$

It is now easily seen how (P5) and (P6) must be combined to obtain an equation for v alone:

$$\frac{\partial}{\partial t} \nabla_h^2 v + \beta \frac{\partial v}{\partial x} + \left(\frac{\partial^3}{\partial t^3} + f^2 \frac{\partial}{\partial t} \right) \frac{\partial}{\partial z}\left\{ \frac{1}{N^2}\left(\frac{\partial v}{\partial z} - \underline{\frac{N^2}{g} v} \right) \right\} = 0. \tag{P7}$$

Here the underlined term is due to the fact that the BOUSSINESQ assumption has not been invoked.

Equation (P7) is simply the x-derivative of the vorticity equation. Therefore it represents a vorticity balance equation, describing how temporal changes in the relative vorticity may be compensated by its latitudinal repositioning within the planetary vorticity field (the second term), and by vertical stretching of interior vortex lines (the last term). Clearly these latter two terms are due respectively to the variability of f and to the presence of the mean density field.

When the BOUSSINESQ assumption is made the last term in (P7) is absent and $N^2 = g d\rho_0/dz/\rho^*$, where ρ^* is a constant references density. This implies that the BOUSSINESQ assumption can only hold as long as $|\partial v/\partial z| \gg N^2 v/g$ implying

$$H \ll \frac{g}{N^2}, \tag{P8}$$

where H is a representative depth scale. Moreover, (P7) is a partial differential equation third order in time, implying that it incorporates both gravity and Rossby waves. Indeed, when $f = 0$, $\beta = 0$ it is becoming second order in time ($\partial/\partial t$ becomes a common factor and the second class waves become steady currents).

Equation (P7) has the advantage that the horizontal and vertical processes can be separated. Indeed, with the ansatz

$$v(x, y, z, t) = Z_n(z) v_n(x, y, t) \tag{P9}$$

it is straightforward to show that $Z_n(z)$ and $v_n(x,y,t)$ satisfy the equations

$$\frac{d}{dz}\left\{\frac{1}{N^2}\left(\frac{dZ_n}{dz}-\frac{N^2}{g}Z_n\right)\right\}+\frac{1}{gh_n}Z_n=0, \tag{P10}$$

$$\nabla_h^2\frac{\partial v_n}{\partial t}+\beta\frac{\partial v_n}{\partial x}-\frac{1}{gh_n}\mathcal{L}\frac{\partial v_n}{\partial t}=0,$$

where gh_n is the separation parameter and h_n the equivalent depth. Notice that in the f-plane approximation ($\beta = 0$) $(\text{P10})_2$ is isotropic in the horizontal coordinates; i.e., v_n need no longer be the NS-component of the velocity field but can be any horizontal velocity component that one pleases.
By taking a wave-ansatz

$$v_n = V_n e^{i(\mathbf{k}_h\cdot\mathbf{x}-\omega t)} \tag{P11}$$

one may also show that (P11) leads to the dispersion relation

$$\omega^3-\left(gh_n\mathbf{k}_h^2+f^2\right)\omega-\beta gh_n k_1=0. \tag{P12}$$

Here, $\mathbf{x}=(x,y)$ and $\mathbf{k}_h=(k_1,k_2)$. It is readily shown that it contains first and second class wave solutions.

4. Consider the boundary value problem (3.68) for the vertical mode structure for w and assume a stratification as given in (3.72), in which N_0 is a constant, such that $N > \omega$. Construct the solution of the resulting eigenvalue problem.

5. Consider the eigenvalue problem

$$\frac{d}{dz}\left(N^{-2}\frac{d\Pi_n}{dz}\right)+\frac{\Pi_n}{gh_n}=0,\quad 0>z>-H,$$

$$\frac{d\Pi_n}{dz}+\frac{N^2}{g}\Pi_n=0,\quad z=0,\qquad \frac{d\Pi_n}{dz}=0,\qquad z=-H$$

for the vertical distribution of the perturbation pressure. Write a program that permits determination of the eigenvalues gh_n and mode structure.
Hint: It may be advantageous to nondimensionalize the equation and to transform it to a first order system.

3.4 Kelvin Waves

For small to medium size lakes rotational effects in gravity waves are usually manifested as modifications of the solutions in which the rotation of the Earth is ignored. To demonstrate this consider equations (3.82)

$$\begin{aligned}&\frac{\partial U}{\partial t}-fV=-gh\frac{\partial\zeta}{\partial x},\\&\frac{\partial\zeta}{\partial t}+\frac{\partial U}{\partial x}+\frac{\partial V}{\partial y}=0,\\&\frac{\partial V}{\partial t}+fU=-gh\frac{\partial\zeta}{\partial y},\end{aligned} \tag{3.88}$$

in which the index α has been dropped for convenience. In a long, narrow channel the dominant motion is likely to be longitudinal, so $|V| \ll |U|$ and V-terms can be ignored. The first two of equations (3.88) then read

$$\frac{\partial U_0}{\partial t} + gh\frac{\partial \zeta_0}{\partial x} = 0\ , \quad \frac{\partial \zeta_0}{\partial t} + \frac{\partial U_0}{\partial x} = 0 \tag{3.89}$$

and determine a zeroth order approximation $U_0(x,t)$, $\zeta_0(x,t)$. By eliminating either U_0 or ζ_0 from the two equations (3.89) the shallow water wave equation

$$\frac{\partial^2 f}{\partial t^2} = gh\frac{\partial^2 f}{\partial x^2}\ , \quad f = (U_0, \zeta_0) \tag{3.90}$$

is obtained. In a constant depth basin of length L, the standing wave solution satisfying the boundary conditions $U_0 = 0$, at $x = 0,\ L$, reads[22]

$$U_0 = A_n \sin\left(\frac{n\pi x}{L}\right)\sin\omega t, \qquad \zeta_0 = \frac{A_n}{\sqrt{gh}}\cos\left(\frac{n\pi x}{L}\right)\cos\omega t, \qquad n = 1,2,3,\ldots, \tag{3.91}$$

with

$$\omega_n = \sqrt{gh}\frac{n\pi}{L}, \qquad T_n = \frac{2L}{n\sqrt{gh}}, \qquad n = 1,2,3,\ldots, \tag{3.92}$$

which are the MERIAN frequencies and periods, respectively.
Equations (3.91), (3.92) describe the standing wave solutions in a rectangular basin with constant depth h and length L. Neither the surface elevation ζ_0 nor the velocity U_0 enjoy a transverse dependence nor is an explicit dependence on the rotation of the Earth discernible (f does not arise in the formulas). However in the horizontal direction these waves have an oscillatory structure whose wavelength $\lambda = 2L/n$ becomes smaller and smaller the larger the counting index n becomes. *Fig. 3.13* illustrates how the surface elevation and the longitudinal velocity are distributed over a wavelength of the channel.
The last of equations (3.88) is left untouched at this level of approximation, however it can be used to determine a first order correction to the surface elevation (by still assuming that $V = 0$). This yields

$$\frac{\partial \zeta_1}{\partial y} = -\frac{f}{gh}U_0 \Rightarrow \zeta_1(x,y,z) = -\frac{f}{gh}U_0(x,t)\,y, \tag{3.93}$$

and demonstrates that for longitudinal motions the surface *must be transversely inclined* in order that the transverse pressure gradient can balance the Coriolis force. This approximate procedure is known as KELVIN *wave dynamics* and was introduced by F. DEFANT (1953) and later used by PLATZMAN and RAO (1964 a, b) in an analysis of surface gravity oscillations of Lake Erie.

[22] The propagating wave solution $U_0 = A_n e^{i(\mathbf{k}\cdot\mathbf{x}-\omega t)}$ would simply yield the dispersion relation of shallow waters, $\omega = \sqrt{gh}\,k$.

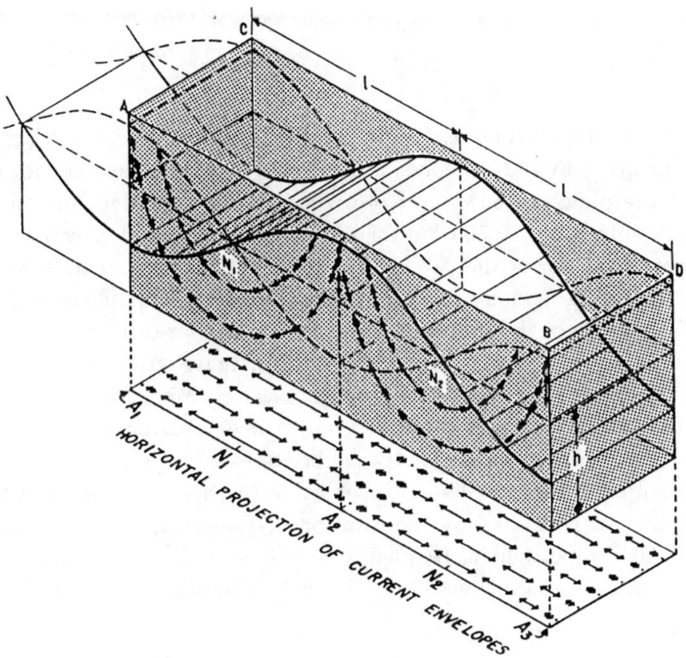

Figure 3.13: A long standing wave without rotation in a uniform depth model.(from Mortimer, 1974)

This iterative procedure can be continued. For instance by substituting the result (3.93) into $(3.88)_1$ or $(3.88)_2$, U_1 can be computed:

$$U_1 = \frac{f}{\sqrt{gh}} A_n \cos\left(\frac{n\pi x}{L}\right) \cos(\omega t)\, y.$$

Collecting all the results yields

$$U = U_0 + U_1 + \ldots = A_n \left\{ \sin\left(\frac{n\pi x}{L}\right) \sin\omega t + \frac{fy}{\sqrt{gh}} \cos\left(\frac{n\pi x}{L}\right) \cos\omega t + \mathcal{O}\left(y^2\right) \right\},$$

$$\zeta = \zeta_0 + \zeta_1 + \ldots = \frac{A_n}{\sqrt{gh}} \left\{ \cos\left(\frac{n\pi x}{L}\right) \cos\omega t - \frac{fy}{\sqrt{gh}} \sin\left(\frac{n\pi x}{L}\right) \sin\omega t + \mathcal{O}\left(y^2\right) \right\}.$$

Continuing in this iterative fashion the entire power series for U and ζ can be constructed. It is our next intention to construct exact solutions of equations (3.88) that enjoy a transverse variatiation of ζ in conformity with the Coriolis terms. To this end, consider a half-infinite basin with constant depth and bounded at $y = 0$; we seek wave solutions of the form

$$U = u_0 h \Phi(y) e^{i(kx-\omega t)}, \qquad V \equiv 0, \qquad \zeta = \zeta_0 \Phi(y) e^{i(kx-\omega t)}. \tag{3.94}$$

This ansatz satisfies the no-flux condition through the side boundary; a forteriori, it requires V to vanish everywhere in the halfspace. Substituting (3.94) into (3.88) yields

$$c = c_{ph} = \frac{\omega}{k} = \pm\sqrt{gh},$$

$$\Phi' + \frac{f}{c}\Phi = 0 \quad \Rightarrow \quad \Phi = e^{-fy/c}, \tag{3.95}$$
$$\zeta_0 = \frac{c}{g}u_0.$$

The amplitudes of U and ζ thus decay exponentially as one moves away from the boundary, the decay rate being given by the ROSSBY *radius* $R = c_{ph}/f$; its value depends on the Coriolis parameter and on the phase speed $c_{ph} = \sqrt{gh}$ whose value itself depends on the size of the equivalent depth. Typically, $R \approx 500\ km$ for barotropic waves and $R \approx 5\ km$ for baroclinic waves (of the lowest order, or of the two layer system). A lake or ocean basin needs to have the width of one to several ROSSBY radii in order that rotational effects come to bear. *Fig. 3.14* depicts the behaviour of the solution (3.95); in the honour of Lord KELVIN it is called a KELVIN *wave*. These waves are non-dispersive and propagate with the same phase speed as do shallow water waves without rotation. They are *coastal trapped* and often have boundary layer structure; the boundary layer width lessens with decreasing f and increasing c_{ph}; in particular, then $\Phi(y) \simeq 1$, and the KELVIN wave becomes an ordinary long gravity wave, see *Fig. 3.13*. On the other hand, the dispersion relation $(3.95)_1$ indicates that the frequencies of KELVIN waves are not restricted by the rotation or the stratification; so they can propagate at subinertial and superinertial frequencies. This should be no surprise, because they constitute the longitudinal shallow water gravity waves that are *modified* by the rotation of the Earth.

KELVIN waves can also exist in channels with finite width, since $V = 0$ everywhere; depending upon how close this second vertical wall is to the first one the effects of the rotation of the Earth are observable in the surface elevation or velocity profiles, see *Fig. 3.14*. Moreover, on the Northern (Southern) hemisphere, their crests lie to the right (left) when looking in the direction of propagation of the wave.
In a non-rotating fluid layer of constant depth standing wave solutions do exist. In fact such a standing wave solution has been constructed in (3.91). Its distinctive feature is that spatially fixed *nodal lines* $x =$ const do exist, across which no water will move at any one time. For (3.91) these positions are at $x = 0$ and $x = L/n$ (for fixed n). It is also readily seen that such a standing wave solution can be obtained by adding forward and backward moving longitudinal waves of the form

$$U^+ = A\sin(kx - \omega t), \qquad U^- = A\sin(kx + \omega t)\ . \tag{3.96}$$

Adding a forward and a backward moving KELVIN wave in an infinite channel results in what is called a *standing double* KELVIN *wave*; however these waves are not "standing" in the strict sense of the meaning of the word because there are no lines $x =$ const. across which no water will move at all times. To see this consider the two KELVIN waves

$$\begin{aligned}(\zeta_1, u_1) &= \left(\frac{c}{g}u_0, u_0\right) e^{-fy/c} e^{i(kx-\omega t)}, \\ (\zeta_2, u_2) &= \left(\frac{c}{g}u_0, -u_0\right) e^{-f(B-y)/c} e^{-i(kx+\omega t)}\ .\end{aligned} \tag{3.97}$$

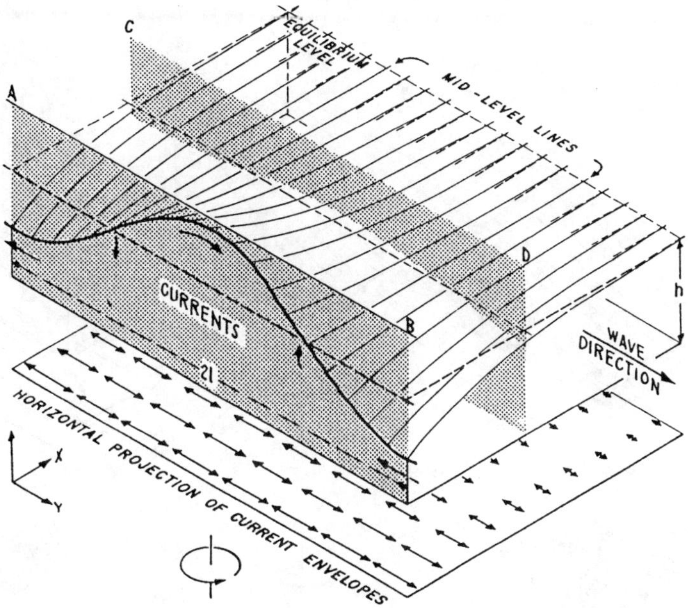

Figure 3.14: Long KELVIN wave progressing in the positive x-direction in a uniform depth model. Velocities are strictly longitudinal. The vertical planes AB and CD may be the vertical shores of a channel. (from Mortimer, 1974)

The first of these moves in the positive x-direction with the wave crest at the shore $y = 0$, the second moves in the opposite direction and has its wave crest at the shore $y = B$ and wave activity decays as $(B - y)$ increases. Adding the two waves and taking real parts yields

$$\begin{aligned} \zeta = \Re\left(\zeta_1 + \zeta_2\right) = \frac{2c}{g} u_0 e^{-\frac{fB}{2c}} & \left\{\cosh\left[\tfrac{f}{c}\left(\tfrac{B}{2} - y\right)\right] \cos kx \cos \omega t + \right. \\ & \left. \sinh\left[\tfrac{f}{c}\left(\tfrac{B}{2} - y\right)\right] \sin kx \sin \omega t\right\} , \\ u = \Re\left(u_1 + u_2\right) = 2u_0 e^{-\frac{fB}{2c}} & \left\{\sinh\left[\tfrac{f}{c}\left(\tfrac{B}{2} - y\right)\right] \cos kx \cos \omega t + \right. \\ & \left. \cosh\left[\tfrac{f}{c}\left(\tfrac{B}{2} - y\right)\right] \sin kx \sin \omega t\right\} . \end{aligned} \tag{3.98}$$

This solution enjoys the following properties, compare also *Figs. 3.15* and *3.16*:

- For $y = B/2$ and $x = (2n + 1)\pi/(2k)$, $\zeta = 0$ for all time. This point, called *amphidromic point* or *amphidromy* is characterized by the fact that its surface never experiences any eleviation from its equilibrium level.
- For $y = B/2$ and $x = n\pi/k$ (these are the points on the center line between two amphidromic points) the velocity vanishes at all times.
- At times $t = n\pi/\omega$ the surface elevation ζ (velocity u) is symmetrically (antisymmetrically) distributed across the channel width with maximum velocities arising at the

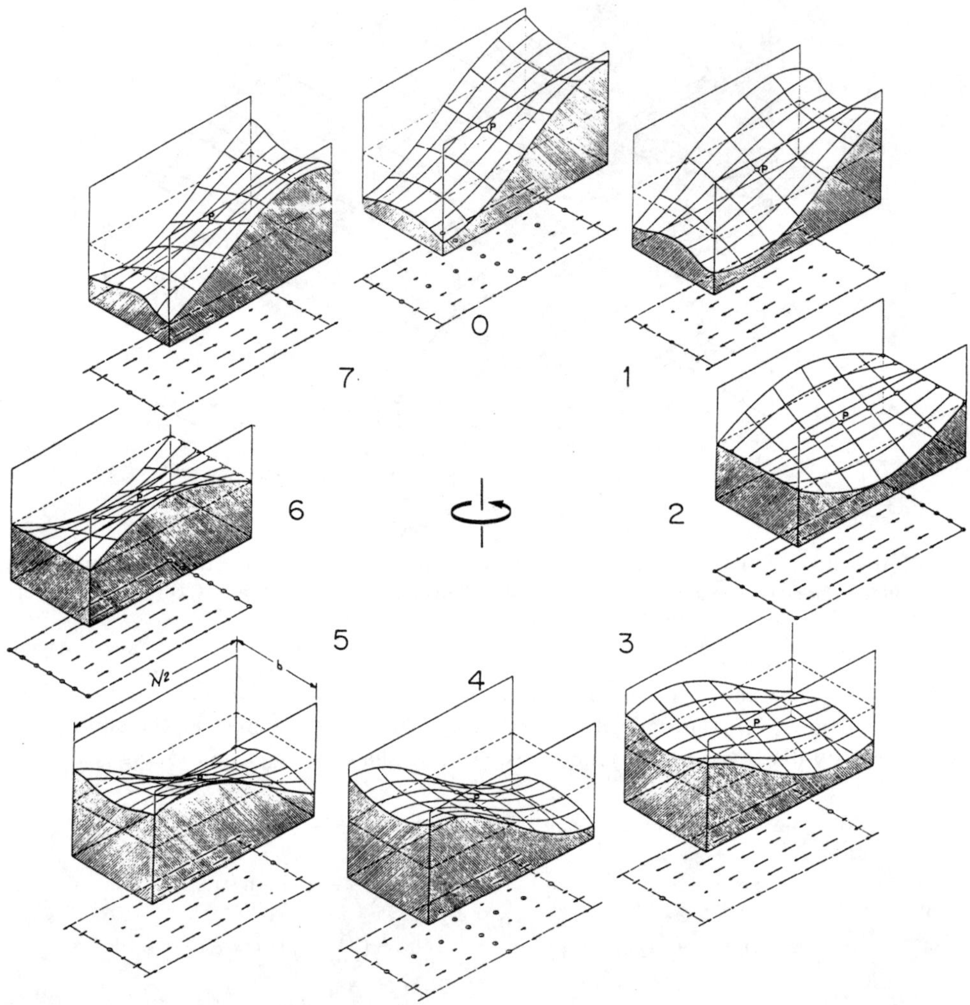

Figure 3.15: Successive phases (1/8 cycle) of a long, standing, amphidromic wave in a straight, rotating channel of rectangular cross-section. Horizontal components of the wave current are projected on the plane below the channel. The amphidromic point of zero elevation change, around which the wave rotates counter-clockwise, is in the center (from MORTIMER, 1975).

shore lines at positions x midway between the amphidromic points.
Similarly, for $t = ((2n+1)\pi/(2\omega))\,\zeta$ is antisymmetrically and u is symmetrically distributed across the channel.

- Lines of *constant elevation amplitudes* - the so-called *co-range lines* - are given by[23]

$$\cosh^2\left[\frac{f}{c}\left(\frac{B}{2}-y\right)\right]\cos^2(kx)+\sinh^2\left[\frac{f}{c}\left(\frac{B}{2}-y\right)\right]\sin^2(kx)=\text{const.} \tag{3.99}$$

  These lines are dashed in *Fig. 3.16*. In the vicinity of the amphidromy they encircle the amphidromic point, farther away they are oscillatory.

- Lines of *constant phase* - the so-called *co-tidal lines* - are described instead by

$$\tan\phi=\tan(kx)\tanh\left[\frac{f}{c}\left(\frac{B}{2}-y\right)\right]=\text{const.} \tag{3.100}$$

  It is easily seen that these are lines through the amphidromic points. Co-tidal lines differing in phase by the angle π from on another are centro-symmetrically arranged with respect to the amphidromic point.

All these properties can with little effort be deduced from the respective formulas, and they are easily corroborated in *Figs. 3.15* and *3.16*. The former of these displays the distribution of the surface elevation and longitudinal velocity in successive phases of (1/8)-cycle of such an amphidromic wave. The figure illustrates the cyclonic (anticlockwise) propagation of the wave: In one period of its motion the wave crest moves around the amphidromic point and the velocity moves forth and back along the channel. The velocities are weakest along lines transverse to the channel and midway between the amphidromic points; however *there is no line across the channel at which the longitudinal velocity would vanish for all time.* It follows that gravity waves in a gulf or a closed basin cannot simply be obtained by superimposing Kelvin waves. We shall return to this point.

3.5 Inertial waves

KELVIN waves are long shallow water waves on the rotating Earth whose particle motion is purely longitudinal. Because of the Coriolis forces such an unidirectional motion can only be maintained when the surface elevation experiences an inclination in the direction perpendicular to the direction of wave propagation. In an infinite basin there are no boundaries that can sustain a deviation of ζ from its position at rest. It is thus reasonable to construct solutions to (3.88) with the property that $\zeta=0$. They are easily shown to enjoy the following properties.

[23] To derive (3.99) and (3.100) it is advantageous to write $(3.98)_1$ as

$$\zeta=\alpha(x,y)\cos(\omega t-\phi)\,,\qquad u=\beta(x,y)\cos(\omega t-\phi)$$

and to identify α, β and ϕ with the help of (3.98).

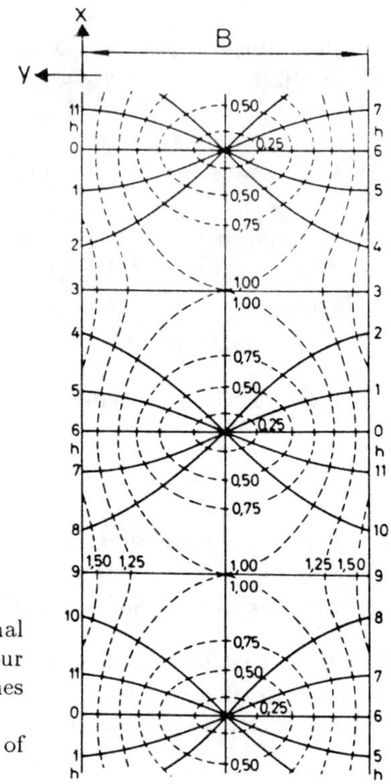

Figure 3.16: Amphidromic system in an infinitely long canal resulting from superposition of two KELVIN waves of a 12-hour period travelling in opposite directions. Full lines: co-tidal lines in hours; broken lines: co-range lines.
($Z_0 = 0.5$, $k = 1$, $f/c = 0.7$ corresponding to a canal width of 400 km, $H = 40$ m), (from KRAUSS, 1966).

- The angular frequency equals the inertial frequency: $\omega = f$. This fact explains why these waves are called *inertial waves.*

- The wave speed is not restricted, i.e., these waves may have an arbitrary wave-number.

- Particle trajectories are horizontal circles with diameter $R = c/f$ where c is the phase speed; this radius is akin to the ROSSBY radius of deformation, its difference to the latter being that here c is not explicitly specified, but may have any value.

3.6 Poincaré waves

Kelvin waves have the property that water particle displacements are strictly longitudinal. There are other wave solutions of equations (3.88) which satisfy the boundary conditions $V = 0$ at the channel sides, however, may have a non-trivial transverse velocity inside the channel. Eliminating U and ζ from equations (3.88) and seeking plane wave solutions for V,

$$V = V_o(y)\, e^{i(kx - \omega t)}\ , \tag{3.101}$$

yields the eigenvalue problem[24]

$$
\begin{aligned}
&V_0'' + \left(\frac{\omega^2 - f^2}{gh} - k^2\right) V_0 = 0, \qquad 0 \le y \le B \, , \\
&V_0 = 0, \qquad \text{at} \qquad y = 0 \; , \; y = B \; .
\end{aligned}
\tag{3.102}
$$

for the distribution of V across the channel. Depending on the value of h ($= H$ for the barotropic mode, $= h_n$ for the baroclinic mode) equations (3.101) and (3.102) describe barotropic or baroclinic progressive POINCARÉ *waves* (POINCARÉ, 1910):
The solution of (3.102) is

$$
V_0 = \overline{V} \sin\left(\frac{m\pi y}{B}\right) \; , \; (m = 1, 2, 3, \ldots) \tag{3.103}
$$

with the dispersion relation

$$
k^2 = \frac{\omega^2 - f^2}{gh_n} - \frac{m^2\pi^2}{B^2} \; , \; (m = 1, 2, 3, \ldots), \; (n = 0, 1, 2, \ldots), \tag{3.104}
$$

or alternatively

$$
\omega^2 = f^2 + gh_n \left(k^2 + \frac{m^2\pi^2}{B^2}\right) > f^2 + gh_n \frac{m^2\pi^2}{B^2} =: \omega_c^2 \; . \tag{3.105}
$$

This shows that, even though ω may continuously vary, there are quantized branches of k depending upon the horizontal (m) and vertical (n) mode structure. Moreover, the reader may show that associated with the solution (3.101) - (3.104) are the following progressing waves of the free surface elevation ζ and the longitudinal velocity component n:

$$
\begin{aligned}
\zeta &= \left\{\alpha_m \sin\left(\frac{m\pi y}{B}\right) - \beta_m \cos\left(\frac{m\pi y}{B}\right)\right\} e^{i(kx - \omega t + \pi/2)}, \\
\alpha_m &= \phi_{km} \frac{k}{f} \overline{V}, \qquad \beta_m = \phi_{km} \frac{m\pi}{Bf} \frac{\omega}{f} \overline{V} \; ,
\end{aligned}
\tag{3.106}
$$

$$
\begin{aligned}
U &= \left\{\bar{\alpha}_m \sin\left(\frac{m\pi y}{B}\right) - \bar{\beta}_m \cos\left(\frac{m\pi y}{B}\right)\right\} e^{i(kx - \omega t + \pi/2)}, \\
\bar{\alpha}_m &= \phi_{km} \frac{\omega}{f} \overline{V} \; , \qquad \bar{\beta}_m = -\frac{gh}{f} \phi_{km} \frac{m\pi}{Bf} \frac{\omega}{f} \overline{V},
\end{aligned}
\tag{3.107}
$$

[24]In a non-BOUSSINESQ fluid on the β-plane, this elimination has been performed in Problem 3 of section 3. The relevant equations are (P7) and (P10). For $\beta = 0$ the equation for the transverse component of the transport V ($= v_n h_n$ in equation $(\text{P10})_2$) reads

$$
\nabla_{h.}^2 V - \frac{1}{gh} \mathcal{L} V = 0 \; ,
$$

where we have also set $h_n = h$ and have performed one integration in time. Substituting (3.101) in the above equation then yields $(3.102)_1$.
This derivation also makes clear that, depending upon the value of h_n, there are barotropic and baroclinic POINCARÉ waves in a channel of constant water depth.

where

$$\phi_{km} = \frac{k^2 + \dfrac{m^2\pi^2}{B^2}}{k^2 + \dfrac{m^2\pi^2}{B^2}\dfrac{\omega^2}{f^2}} \tag{3.108}$$

Let us discuss inferences that can readily be deduced from the above formulae:

- For given frequency the squared wavenumber, k^2, is only positive provided the transverse modenumber, m, is small enough. Thus, for

$$m < m_0 := \frac{B}{\pi}\sqrt{\frac{\omega^2 - f^2}{gh}} \rightarrow k^2 > 0 \ ,$$

the POINCARÉ waves are oscillatory in the x-direction. On the other hand, modes with mode number

$$m > m_0 \ \rightarrow \ k^2 < 0 \ ,$$

are exponentially evanescing as $x \rightarrow \infty$ or $x \rightarrow -\infty$. Such waves can only exist in a semi-infinite channel or a channel of finite length. These are important in the solution of the reflection problem. This point will be taken up later again.

- POINCARÉ waves are necessarily *superinertial*, i.e., their frequency ω is larger than the inertial frequency f. A forteriori, in an infinite channel (for which $k^2 > 0$) ω must be larger than the *cut-off frequency*, ω_c, whose square is the sum of f^2 and the square of the frequency of the transverse oscillation, w_c^2, in a non-rotating channel.

- In the direction of wave progression the variable V lags behind ζ and U by $\pi/2$; so the motion is purely transverse where U and ζ vanish and purely longitudinal where ζ experiences the largest transverse variations. This can also be inferred from *Fig. 3.17a*.

Figure 3.17 : (See next Page) a) Progressing POINCARÉ wave with a uninodal ($m = 1$) cross channel distribution of the elevation ζ. Shown is a semi-period (wave length $\lambda/2$) with the velocity distribution in the lower rectangle. The dotted line (CCL) indicates the middle line of the channel, NL indicates the along channel nodal line shifted towards the wall to the right of the wave progression; MCL marks the maximum current line. The current is along shore and sinusoidally distributed (with asymmetry) where the surface elevation shows the strongest transverse variations; it is transverse where $\zeta = 0$ across the entire channel.
b) Same for a trinodal progressive POINCARÉ wave. The nodal line of the elevation is shifted $\Delta\zeta$ to the right of the channel side, while MCL is shiftet Δ_v to the left of the nodal line (N). (from MORTIMER, 1980)

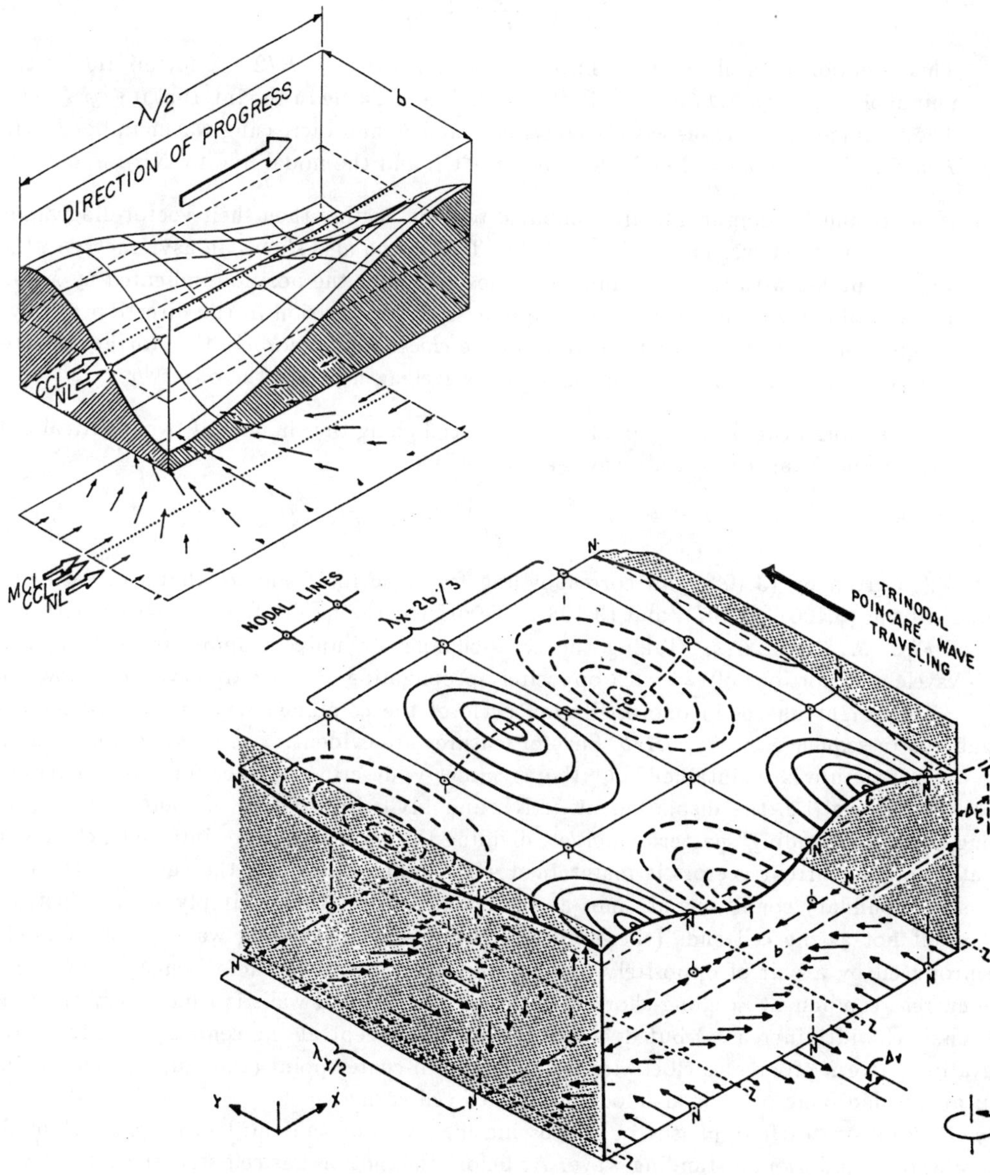
λ/2
b
DIRECTION OF PROGRESS
CCL
NL
MCL
CCL
NL
NODAL LINES
$\lambda_x = 2b/3$
TRINODAL POINCARÉ WAVE TRAVELING
$\lambda_y/2$
Y
X

- The progressive POINCARÉ wave is asymmetric across the channel width. It possesses nodal lines of ζ where

$$\tan\left(\frac{m\pi y}{B}\right) = \frac{m\pi}{B}\frac{\omega}{k} \, . \tag{3.109}$$

  This equation is easily seen to have no solution for $y = B/2$. A forteriori, for the uninodal case (*Fig. 3.17 a*, $m = 1$) the nodal line must lie in the internal $0 \le y \le B/2$. The asymmetry increases with decreasing ω and m and increasing wavenumber k. (In *Fig. 3.17* the symmetry line is denoted by CCL and the nodal line by NL).

- If the U and V components are combined to form vectors then their vector-diagram is as shown in the lower panel of *Fig. 3.17a*. This graph also shows an asymmetry in the velocity profile with the maximum velocities arising at the position indicated by MCL. For a stationary observer, as the wave is passing his position in the direction of wave progression, the velocity vector rotates in the *clockwise direction* in the Northern hemisphere, performing a full revolution as one wavelength passes the observer.

A pseudo-standing wave pattern can also be produced along the channel, if two identical but oppositely-propagating POINCARÉ waves are combined:

$$V = V_0\,(y)\left\{e^{i(kx-\omega t)} + e^{-i(kx+\omega t)}\right\} = 2V_0\,(y)\cos\,(kx)\,e^{-i\omega t}, \tag{3.110}$$

with $V_0(y)$ given by (3.103) and corresponding formulae for ζ and U that can easily be deduced from (3.106), (3.107), and (3.108). We discuss the qualitative behaviour with the aid of *Figs. 3.18* and *3.19*. Taking the cross-channel uninodal example from *Fig. 3.17*, half-wavelength portions of two such oppositely-propagating POINCARÉ waves are shown in *Fig. 3.18a*. Eight phases in one oscillation cycle of the combined wave they produce are numbered in sequence in *Fig. 3.18b*. Several features are evident: The cross channel standing wave structure is maintained, but the asymmetry, described earlier for the progressive POINCARÉ wave (i.e. the displacement of NL and MCL from CCL), is cancelled out; an along channel "standing" pattern emerges, dividing the wave structures into cells; clockwise rotation of the current vectors is maintained with highest speeds at the center of the cell; the solid boundary conditions are met at the channel sides (this is simply so by construction) but not at the cell ends (therefore this is not a true standing wave, such as would be produced by a pair of oppositely-propagating plane waves in the absence of rotation); the current component in along-shore direction at the cell end walls is small (very small as the channel width increases) but does not fall to zero except at the center point; the wave elevation pattern progresses clockwise around the cell-center point (the amphidromic point P): an amphidromic point is also located at each cell corner.
Fig. 3.19 illustrates four phases in one oscillation cycle of the third cross channel mode ($m = 3$) of such a pseudo-standing wave. As before the motion has cell structure. For $t = 0$, each cell with horizontal wave lengths λ_x and λ_y is subdivided into four subcells in which the elevation has a high and a low, respectively, and velocities are longitudinal "to and fro" as shown in the figure. A quarter period later, $t = T/4$, locations of elevation highs and lows are shifted as shown, and the current is transverse and again "to and fro". At the

time $t = T/2$ the elevation and velocity distributions are again as shown for $t = 0$ but with highs and lows interchanged and velocity directions reversed; a similar behaviour is seen at $t = 3T/4$ when compared with the graph for $t = T/4$. As time progresses the current vectors rotate in the clockwise direction and particle paths for a harmonic oscillation are elliptical and are traced once within a period. Finally, as before, the side boundary conditions are met, but there is no line across the channel at which the longitudinal velocity would vanish at all points. These pseudo-standing POINCARÉ wave solutions can therefore not hold for a closed basin.

3.7 Reflection from the end of a channel wall

Neither the standing KELVIN waves, nor the standing POINCARÉ waves satisfy the no-through flux condition at a transverse wall. But a standing KELVIN wave solution plus a whole spectrum of standing POINCARÉ waves allows the no-through flux condition at a transverse wall to be met. The problem was first attacked by TAYLOR (1922) and further investigated by RAO (1966) and BROWN (1973).

The reflection of a KELVIN wave incident onto the end of a channel is solved by adding to this wave a reflected KELVIN wave of the same frequency and same wave number and an infinite number of POINCARÉ waves of the same frequency. The free amplitudes of these POINCARÉ waves are determined by the condition that there is no flux through the channel wall. If the incoming KELVIN wave has a frequency below the cut-off frequency, ω_c, of POINCARÉ waves, the POINCARÉ modes are all evanescent as one moves away from the end wall. Although they are important for the satisfaction of the boundary condition, their presence is felt only near the boundary. For this reason, this type of reflection is called *complete reflection*. *Fig. 3.20* shows the amphidromic system and the current ellipses in a rectangular basin closed at one end as determined by G. I. TAYLOR. The influence of the POINCARÉ waves is only seen near the channel end. When the frequency of the incoming wave is above the cut-off frequency then this incoming KELVIN wave is reflected by a reflected KELVIN and a series of POINCARÉ waves, the presence of which is recognized in the *entire* basin. Reflections are then called *incomplete*.

An example of the amphidrome modification near the closed end is shown for this case in *Fig. 3.21*. For higher frequencies, the amphidromic pattern diverges from that due to a pair of KELVIN waves, acquiring more and more lateral structure as more POINCARÉ modes are reflected from the closed end of the channel. *Fig. 3.21*, taken from BROWN shows the distortion of the amphidromes with increasing frequencies.

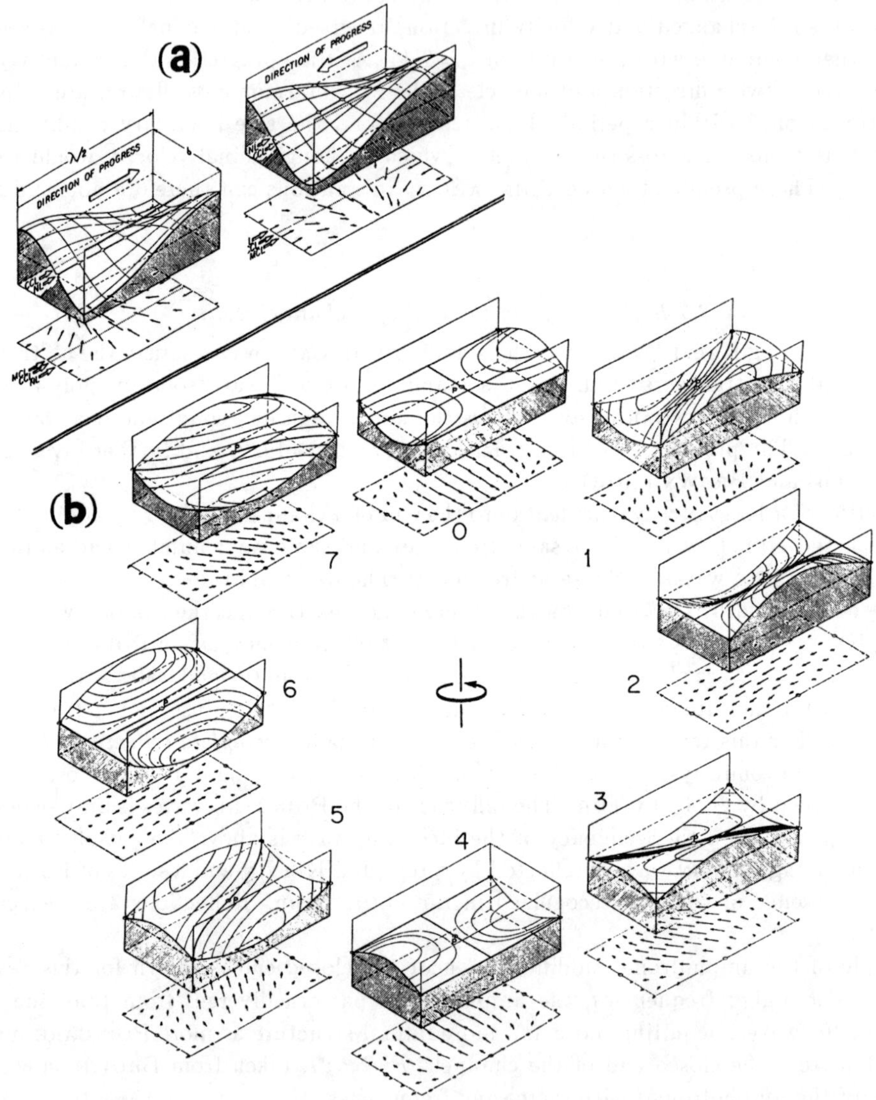

Figure 3.18: (From MORTIMER, 1977). Two oppositely-propagating but otherwise identical progressive POINCARÉ waves, from *Fig. 3.17* and shown here in (a), combine to produce the amphidromic "standing" POINCARÉ wave, half-wavelength cells of which are illustrated at 1/8-cycle intervals in (b). The combined wave, with amphidromic point P at the cell-center, is truly standing (with one node in this case) across but quasi-standing along the channel, as explained in the text. Wave topography can be viewed either as that of a water surface or as that of a "thermocline" interface in a two-layered model, in which case the horizontal current vectors (illustrated beneath each cell) are those in the lower layer.

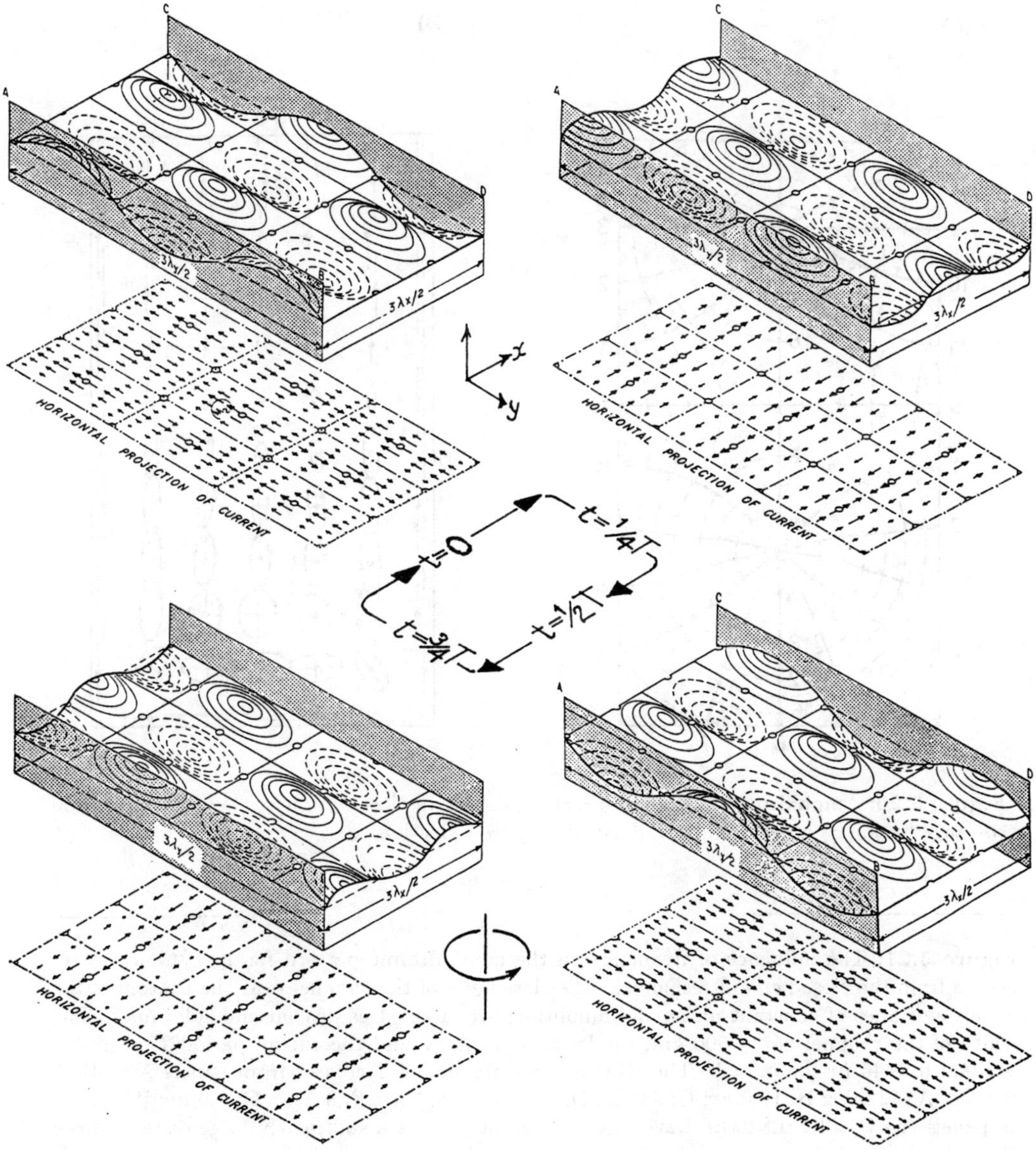

Figure 3.19: A "standing" POINCARÉ wave of cross-channel nodality three, i.e., a trinodal equivalent of *Fig.3.18b*, but with only four phases of the oscillation cycle shown. The legend for *Fig. 3.18* applies; and the progressive POINCARÉ wave component, contributing to this combination, is illustrated in *Fig. 3.17* (lower part) (from MORTIMER, 1980).

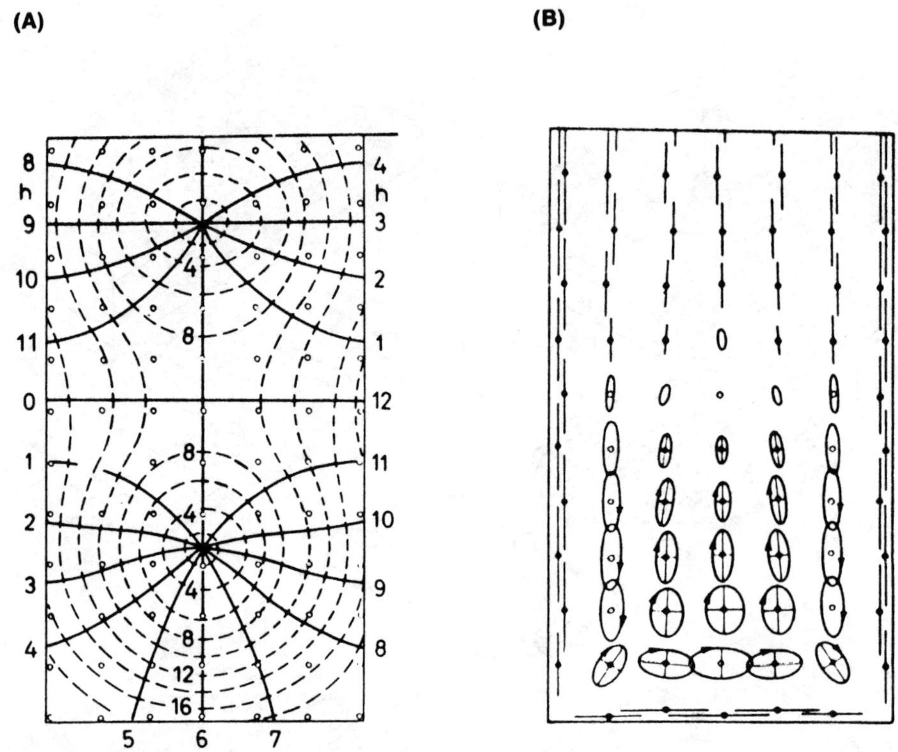

Figure 3.20: Amphidromic system in a rectangular basin closed at one end. (A) co-tidal and co-range lines; (B) current ellipses, illustrating pure reflection.

Figure 3.21: The influence of frequency on the amphidromic pattern for KELVIN waves reflected from the closed end of a channel. The closed end of the channel is at the bottom of the panels and lines of constant phase and amplitude are labelled as dashed and solid lines. The width of the channel is $l = 500\ km$, the latitude $54.46° N$ and the cut-off period $T_c = 2\pi/\omega_c$ is equal to 8.46 hr for $m = 1$. The KELVIN wave frequency decreases from panels A to H as follows: A: 12.0 hr; B: 10.0 hr; C: 9.0 hr; D: 8,6 hr; E: 8,1 hr; F: 8.0 hr; G: 7.0 hr; H: 6.0 hr. In panels A - D, all POINCARÉ waves are evanescent, and at a sufficiently large distance from the reflecting end the interference pattern becomes that of a pair of KELVIN waves travelling in opposite directions. In the last four panels (E - H) the $m = 1$ POINCARÉ mode propagates and the field far from the reflecting end consists of two KELVIN waves and a POINCARÉ wave. The asymmetry introduced by the reflected POINCARÉ wave is evident in the four panels (from BROWN, 1973).

Figure 3.21

4. Topographic Waves

4.1 Review of earlier work

In the last chapter the notions of first and second class waves were introduced. The former were said to be due to the action of the gravity force. These waves are therefore also called gravity waves. The latter are due to the rotation of the Earth, and cease to exist when the frame of reference is inertial. These waves are alternatively also termed ROSSBY-, *vortex-*, *geostrophic-* or *gyrational waves.* In a homogeneous fluid, subject to the rigid lid assumption and in the limit of the shallow water approximation the second class linearized waves are described by the differential equation

$$\begin{aligned} T[\Psi] &= \frac{\partial}{\partial t}E[\Psi] + \mathcal{J}\left[\Psi, \frac{f}{H}\right] = 0 \\ &= \frac{\partial}{\partial t}\nabla \cdot \left(\frac{\nabla\Psi}{H}\right) + \frac{\partial \Psi}{\partial x}\frac{\partial}{\partial y}\left(\frac{f}{H}\right) - \frac{\partial \Psi}{\partial y}\frac{\partial}{\partial x}\left(\frac{f}{H}\right) = 0 \end{aligned} \tag{4.1}$$

in which Ψ is the volume transport streamfunction introduced in (3.27) and H is the water depth. It was also demonstrated that equation (4.1) could be derived from the conservation law of potential vorticity (see equation (3.31) and the following derivation). Moreover, when the frame of reference is inertial ($f = 0$) $T[\Psi] = 0$ reduces to $\partial E[\Psi]/\partial t = 0$ which shows that in this case no second class waves can propagate. Thus, H = constant and variable f defines the planetary ROSSBY waves, while f = constant and H = variable defines the topographic (ROSSBY) waves (TW's); these only exist when the basins have variable bathymetry. We shall confine attention in this chapter to this latter case.

The interest in topographic waves in the ocean arose from the conspicuous long periodic wave signals that were observed along the continential shelves of our oceans. In lakes a considerable number of temperature and current observations in various lakes also disclosed pronounced oscillations with a characteristic period of a few days that could not be explained by (external or) internal gravity waves. POINCARÉ (1910) was the first to point out the existence of such long periodic oscillations in a rotating circular basin with parabolic depth profile, and LAMB gives the first analytic solution of these topographic waves. Much later, BIRCHFIELD and HICKIE (1977) considered the transient wind generation of the LAMB (1932) modes and demonstrated, how the gravity oscillations and the TW's modulate the seiche response by a slow rotation in the cyclonic sense of the pattern of coastal jets and return flow across the lake center. Free circular basin solutions for second class waves have been found by SAYLOR et al. (1980) and SAYLOR and MILLER (1983) and WENZEL (1978). An analytical method for determining the second class modes in an elliptical paraboloid was prescribed by BALL (1965), and MYSAK (1984), and JOHNSON (1987a) constructed solutions of TW's in an elliptical basin whose shoreline and depth contours form a family of confocal ellipses. Using a semi-analytic semi-numerical approach STOCKER and HUTTER (1986), (1987a,b) found solutions to the TW-equation in a rectangular basin with a symmetric bathymetric chart and could identify three typical wave modes of the TW-equation: (i) basin wide modes with large wavelengths, (ii) basin wide modes with shorter wavelengths and (iii) bay modes, in which the major part of the energy is concentrated in a localized (bay) region. From this study sprang a series of studies on localized TW's by STOCKER and JOHNSON[25].

Even though we treat here only the barotropic TW's, these waves exhibit a baroclinic coupling when they propagate in a stratified fluid. MYSAK et al. (1985) have shown that, in a two-layer basin of the size and depth scale of many intermontane lakes, barotropic TW's are only slightly affected by the interfacial motion. However, the interfacial oscillations are driven by the TW's. Thus, one can find the baroclinic TW mode from the solution of the barotropic TW-equation subject to the zero mass flux condition through the lake shores yet is able to observe the wave from temperature measurements in the thermocline.

4.2 Some observations and their interpretation

In this section we describe a series of long-periodic oscillations of temperature or current data indicating that the phenomenon which underlies these observations can probably be interpreted in terms of topographic waves. Our report is not in chronological order, but rather follows easiness in interpretation. The details of the theoretical models (assumptions, approximations etc.) which are used to explain the observations are postponed to the following section.

4.2.1 Lake Michigan

The analysis outlined here and the figures are due to SAYLOR, HUANG and REID (1980) and HUANG and SAYLOR (1982). During spring, summer and fall 1976 current meters were deployed in South Lake Michigan as indicated in *Fig. 4.1.* Most of these current meters were positioned along the eastern shore between Benton Harbor and Muskegon, however 10 current meters were also deployed along a straight line connecting opposite points along the shores between Racine and Holland. The bathymetry of the topography by a circular basin with radial profile dependence, see *Fig. 4.2*, may not be an oversimplification of the situation at hand.

Fig. 4.3 collects kinetic energy spectra of the Eastward (*Fig. 4.3a*) and Northward (*Fig. 4.3b*) velocity components (generally at the 25 m level) at the six stations indicated in *Fig. 4.1.* Vertical lines are drawn to accentuate the two conspicuous energy peaks at near inertial ($\approx$ 17 hours) and near 4-day periods. Thus, the lake responds distinctly at these periods. This is further substantiated by the graphs presented in *Fig. 4.4.* They show a progressive vector diagram of the hourly averaged low-pass filtered current velocities from the 32 m depth current meter at station 11 and coherence and phase difference between the East and North velocities from the same current meter.

[25]There is a vaste literature on TW's. A reference text may be LEBLOND and MYSAK (1980) who treat primarily waves in the open ocean. A rewiew, perhaps more adequate to the present topic is by STOCKER and HUTTER (1987a) and contains a large number of references pertinent to the propagation of TW's in closed or semi-closed basins. Relevant newer works are also by STOCKER and HUTTER (1986, 1987a, b, 1990), STOCKER and JOHNSON (1989, 1991, 1992) and JOHNSON (1987a, b, 1989), WILLMOTT and JOHNSON (1989) and others.

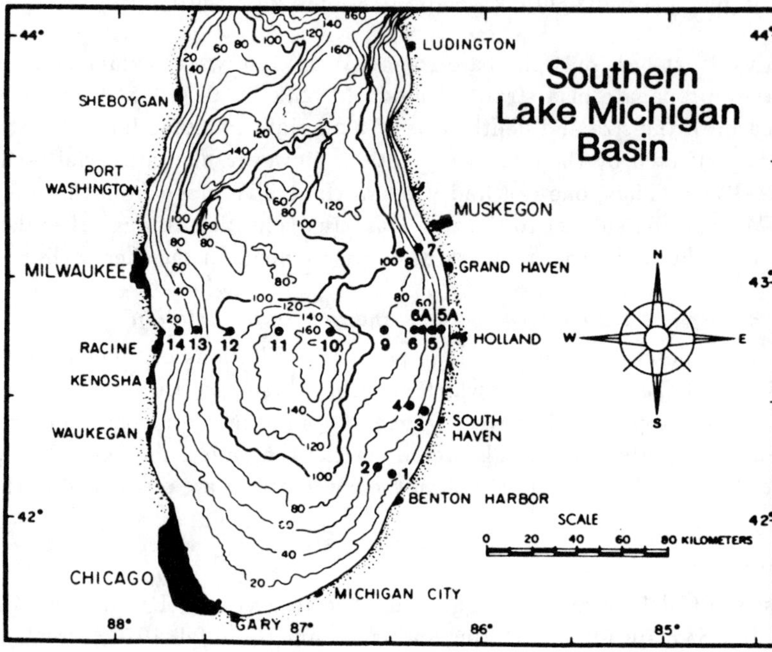

Figure 4.1: Location map showing the bathymetry of Southern Lake Michigan and the positions of the 16 current meter moorings (from SAYLOR et al.,(1980)).

The hodograph clearly reveals the oscillatory wave motion with a period of about 4 days, the rotation of the current vector being in a cyclonic (counterclockwise) direction for this station, which is in the center of the lake basin. Coherence and phase plots in *Fig. 4.4b* for the East (u)- and North (v)-components for one current meter at position 11 accentuate this, as they show two highly correlated signals centered at the two indicated frequencies.

At the 4-day period the u-component leads the v-component by a phase angle of 90° which corresponds to a cyclonic rotation; at the near-inertial period the phase angle of the u-component lags that of the v-component by approximately 90°, indicating an anti-cyclonic oscillation.

Further scrutiny of the data shows that (i) the 4-day period motions are clockwise rotations at all stations except 10 and 11 in the center of the lake, where they are counterclockwise and (ii) currents at on-shore positions are primarily along-shore while those at off-shore stations may have appreciable amplitudes for both, the East and North components.

These observations can be interpreted in terms of free TW's in a circular basin with nearly

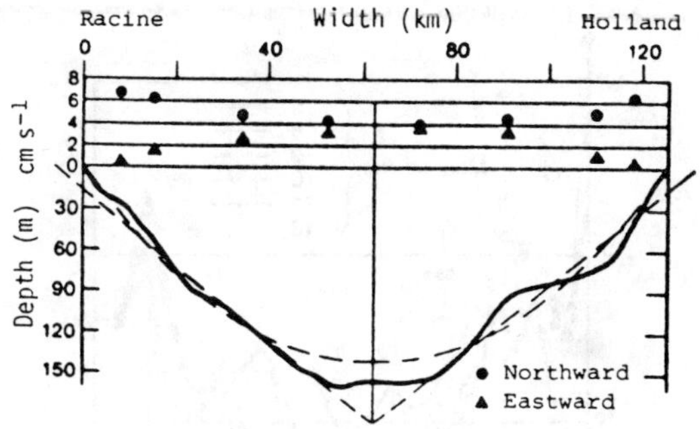

Figure 4.2: Average amplitudes of the north and east velocity components for rotational oscillations on the Southern Lake Michigan cross section during an episode of wave excitation, 1 - 15 July 1976. Basin approximations as a paraboloid or as an inverted cone that are referred to later are also shown (from SAYLOR et al., (1980)).

conical profiles. For a depth profile of the form

$$H = H_0 \left(1 - \left(\frac{r}{a}\right)^q\right) \quad , \; q > 0 \tag{4.2}$$

(H_0 is the maximum depth at the center, a is the radius of the circular basin and q a parameter characterizing the profile), SAYLOR, HUANG and REID (1980) deduce the frequency relation[26]

$$\frac{f}{\omega} = \frac{3m + 2q}{m} \; . \tag{4.3}$$

Here ω is the frequency, f the Coriolis parameter and $m = 1, 2, 3 \ldots$ the azimuthal mode number. Values for the frequencies or periods are summarized in *Table 4.1* and the streamline pattern of the vertically integrated transport for the lowest two modes is sketched in *Fig. 4.5.* The fundamental mode ($m = 1$) enjoys all properties of the observations mentioned abvove. In particular for the conical profile, its period is close to the observed 90 hour period. *Table 4.1* and equation (4.3), however, also show that for each mode the periods depend strongly on the topography, but not on the size of the basin (H_0 and a do not enter into the frequency relation; this was to be expected as it was shown in section 3 that the TW-equation is scale invariant, see text after (3.25)). Moreover, the same period arises for different modes and different bathymetries, indicating that the mode structure is very important in interpreting observations.

[26]Equation (4.3) is the frequency relation obtained when the TW-equation (4.1) is solved in a circular basin with bathymetry (4.2) and solutions are such that they have m azimuthal nodal lines, but no nodal circles other than the centre and the bounding circle (see *Fig. 4.5*).

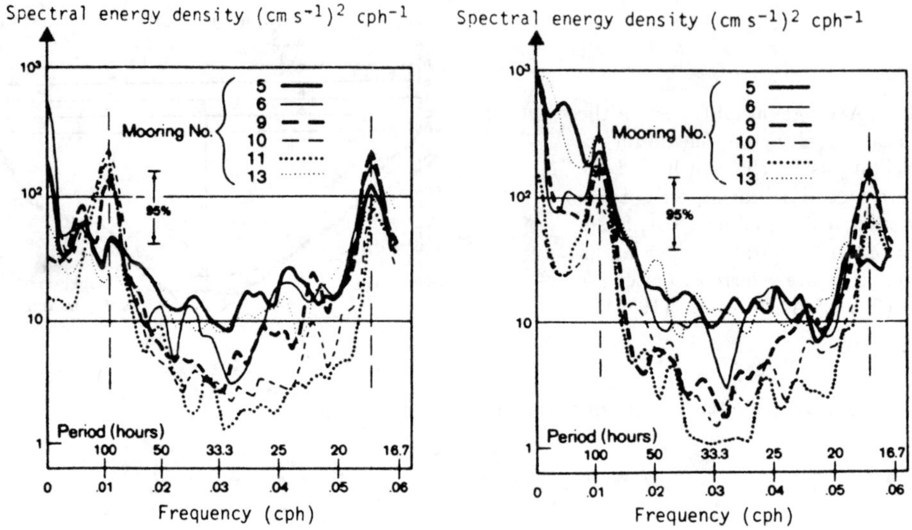

Figure 4.3: a) Kinetic energy spectra of the Eastward velocity component (u) recorded at the 25 m level at six stations in Southern Lake Michigan. Vertical lines are drawn to accentuate the two conspicuous energy peaks at near inertial and near 4-day periods. b) As in a), except for the Northward velocity component (v). (from SAYLOR et al., 1980)

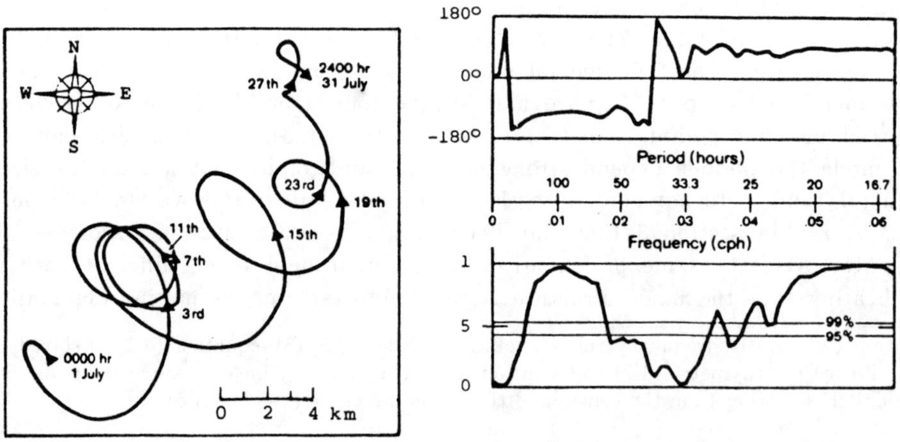

Figure 4.4: a) Hodograph of hourly averaged, lowpass filtered current velocities from the 32 m depth current meter at station 11 for the month of July 1976. b) Coherence (lower) and phase between the East and North velocities at the 32 m level from station 11. Two broad areas of high coherence highlight the cyclonic rotating 4-day oscillation and the anti-cyclonic near inertial oscillation. (from SAYLOR et al., 1980)

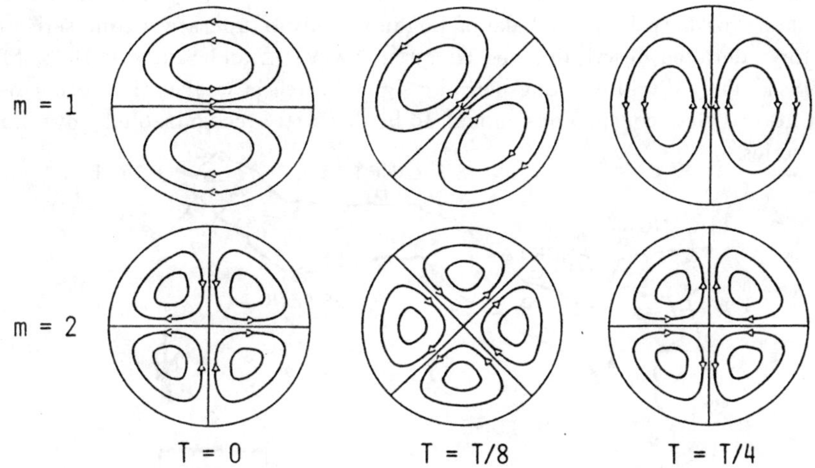

Figure 4.5: Schematic sketch of the mass transport streamline pattern of the fundamental and first "higher" mode plotted for three instances during a quarter period. During one cycle, the system of gyres rotates counterclockwise (on the Northern hemisphere) around the basin.

		$m = 1$		$m = 2$		$m = 3$	
Profile	q	f/w	T [hours]	f/w	T [hours]	f/w	T [hours]
	1	5	84,5	4	67,6	3	50,7
	2	7	118	5	84,5	4	67,6
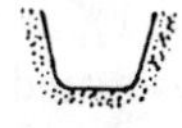	3	9	152	6	101	5	84,5
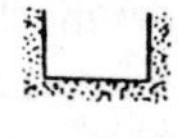	∞	0	∞	0	∞	0	∞

Table 4.1) Eigenfrequencies (-periods) of the first three TW-modes computed according to (4.3) using $f = 2\pi/16.9\ h$.

4.2.2 Lake of Lugano (North basin)

Southern Lake Michigan is a large and relatively shallow lake with a width of approximately 100 km, whereas Lake of Lugano is small (17 km long, approximately 2 km wide and 300 m deep). According to current thinking, existence of topographic waves should come as a surprise.

In summer 1979, thermistor chains and current meters were deployed at various locations in the stratified Northern basin of Lake of Lugano. Only temperature-time series could be analysed, (for a detailed description see HUTTER, SALVADÉ and SCHWAB 1983; MYSAK et al. 1983, 1985). They disclose a long-periodic signal which is weaker than other oscillation signals, but nevertheless conspicuous enough to be distinctly recognizable by eye in isotherm depth time series.

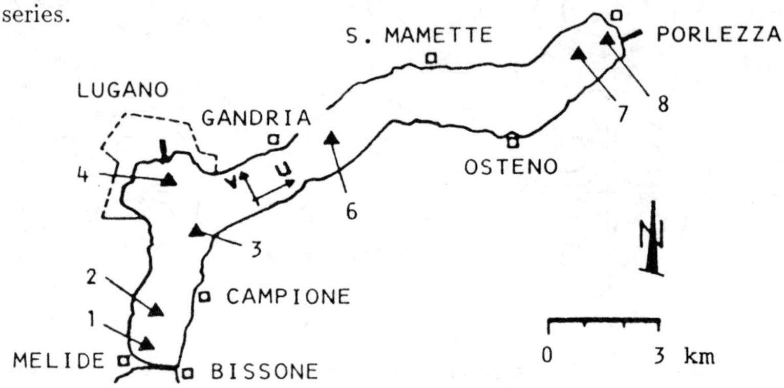

Figure 4.6: Northern Lake of Lugano with indicated positions of moorings.

Fig. 4.6 shows the map of Lake of Lugano (North basin) with indicated stations where the wind and temperature (generally at the lower portion of the metalimnion) were measured in July/August 1979. *Fig. 4.7* summarizes wind data at stations 4 and 7 (top) and isotherm-depth-time series at the stations 1 and 8, 6 and 4 (lower part). These time series "demonstrate a strong component of the motion with a period of perhaps 74 hours (marked by circles ⊗). To emphasize this wave the troughs of the isotherm depths have been brought into prominence by thick solid and dotted lines. The front arises first at station 4, propagates Southward, reaches station 1 approximately 7 hours later (indicated by the solid line connecting the troughs at Cassarate and Melide) and is reflected at the Southern end of the lake. The reflected wave passes through station 4 again (though split up into two smaller minima with an intermediate maximum, indicated by an arrow marked with an encircled 1), and after 37 hours, can be recognized at station 8 with a conspicuous minimum (heavy dotted lines). It appears later at Cassarate, Melide and as reflected wave at Cassarate (arrow marked by an encircled 2), Porlezza, etc.", (HUTTER, SALVADÉ and SCHWAB, 1983). The corresponding wave speed of approximately 12 cms^{-1} is substantially lower than the wave speed of the internal gravity wave of the two-layer model; neither can it be explained as a higher order baroclinic KELVIN wave of, say, a three-layer model. Furthermore, direct wind forcing can be excluded as a likely cause of excitation (MYSAK et al., 1985).

This description suggests that the 74h-signal corresponds to a basin wide motion, and a careful statistical analysis of the temperature-time series seems to corroborate that the phase of the oscillation with a suspected 74h-period propagates counterclockwise around the basin. *Fig. 4.8* shows variance spectra, coherence and phase difference of time-series pairs of the

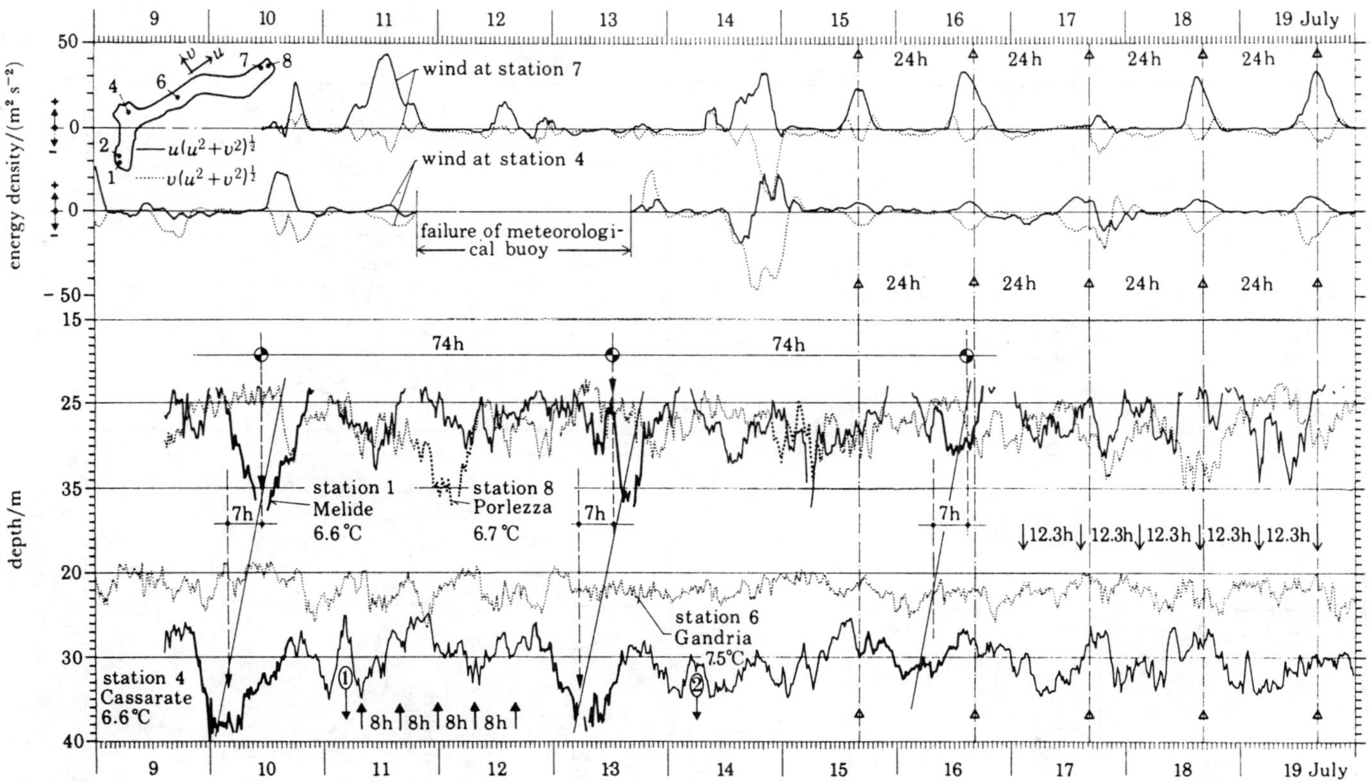

Fig. 4.7: Time series of 2 h means of the longitudinal and transverse components of the wind energy ($(u^2+v^2)^{\frac{1}{2}}u$, $(u^2+v^2)^{\frac{1}{2}}v$, with directions of u and v components as indicated in the insert in the upper left corner) at the stations 7 (Porlezza) and 4 (Cassarate) and unfiltered time series of selected isotherm depths at the stations 1 (Melide, top solid curve), 8 (Porlezza, dotted, superposed on the Melide curve) and 4 (Cassarate, solid curve). Components of the motion with conspicuous periods are marked by special symbols: troughs in heavy solid and dotted lines and marked by circles ⊕ are for 74 h; triangles for 24 h; arrows for 12 h or 8 h. The time shown is from 9 to 19 July 1979. (From Hutter *et al.* 1983.)

(Cortesy of the R. Soc. London)

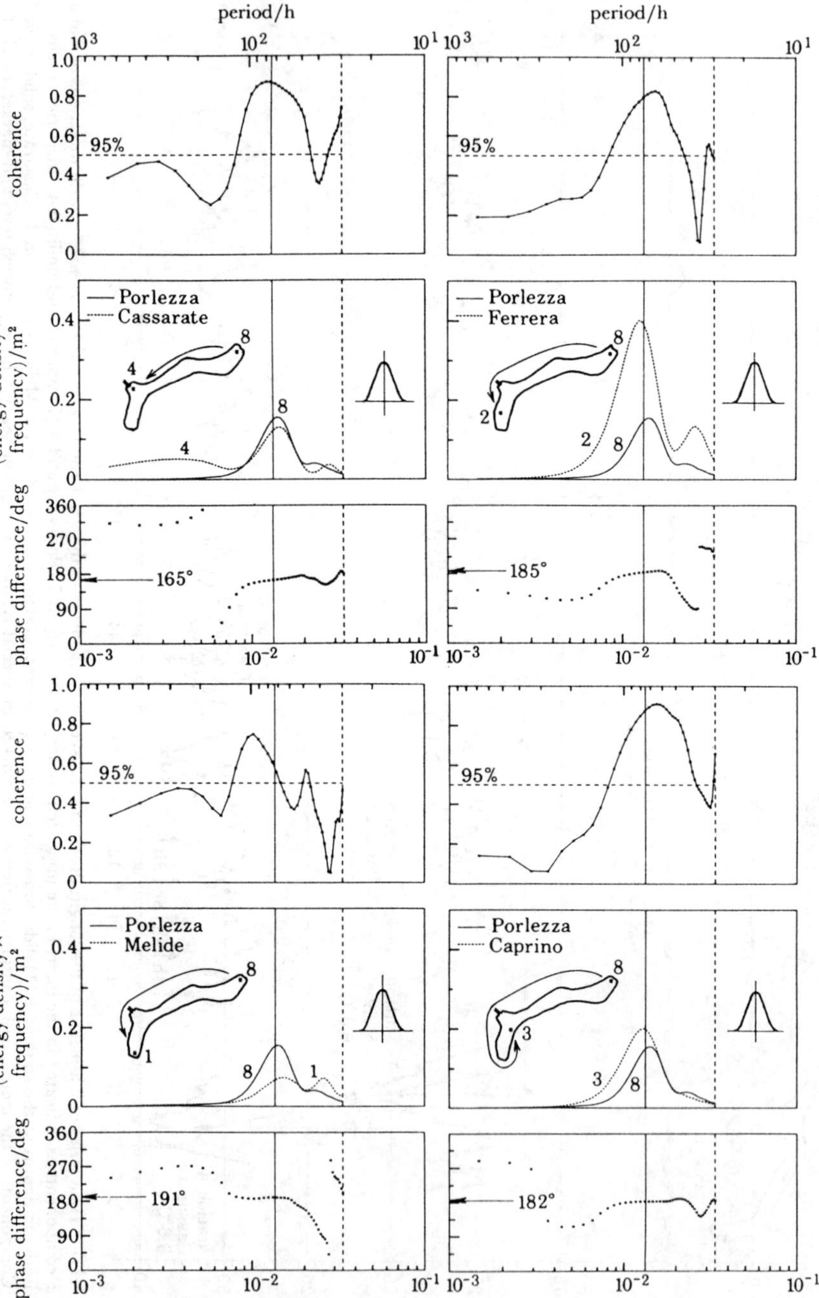
period/h
coherence
95%
Porlezza
Cassarate
Ferrera
Melide
Caprino
(energy density × frequency)/m²
phase difference/deg
165°
185°
191°
182°
frequency/cycles per hour

Figure 4.8: (see previous page) Variance spectra, coherence and phase difference of time-series pairs of the mean temperature-displacement function for an event of 43 days duration. Each column is subdivided into three subfigures where coherence c, variance and phase difference are plotted. The vertical broken line marks the cut-off period of 30 h and the horizontal broken lines in the coherence diagrams mark the 95% confidence limit of arctan c. The inset maps show the station pairs, the arrow indicating the direction of phase progression. The inserted window indicates the tapering of the spectra. The vertical solid line marks the period of 74 h, and the arrow indicates the phase difference between the time-series pair at that period (from MYSAK et al.,1985, Cortesy of the R. Soc. London).

mean temperature displacement function for an event of 43 days duration. The energy spectra of all mean temperature-displacement functions[27]) have a maximum between approximately 69 and 80 hours (with a mean of 74,5 hours), where coherence between data pairs is generally high. The maps inserted in the graphs of the energy spectra show the locations of the stations where the temperature measurements from the thermistor chains were taken, the arrow indicating the direction in which the phase is supposed to propagate. Phase differences between station pairs at the 74 hour period are indicated in *Fig. 4.8* by an arrow at the left margin of the corresponding phase difference.

MYSAK et al. (1985) explained the 74 h oscillation and the anticlockwise propagation of the phase as the baroclinic trace of a barotropically driven basin-wide TW. Such a model explains all the observations (except the "discrepancy" in the phase of the pair 3/8 of the mean temperature displacement function, which interrupts the anticlockwise increase of the phase difference)[28] in terms of the fundamental mode of TW's in an elliptical two-layer basin. The streamlines as constructed by JOHNSON (1986) are sketched in *Fig. 4.9.*
This interpretation of the observed long periodic signal (in the period range of 69 - 90 hours) is, however, incorrect. The TW-equation was namely approximately and numerically solved by the finite element technique (TRÖSCH, 1984) with results which do not at all support the interpretation in terms of basin wide modes. The numerical calculations with the finite element technique suggest that solutions in the 65 to 90 hour period range are generally *localized* to the narrow ends and the bay of Lugano. *Fig. 4.10* shows the streamline pattern of three such modes having periods of 68.5, 80.5 and 91 hours, respectively. In the following sections we will give arguments in support of this second interpretation.

Further examples of observations of TW-signals have been collected by STOCKER and HUTTER (1987a).

[27]The mean temperature-displacement function is representative of a mean isotherm depth-time series. Its exact definition is not relevant here and can be looked up in HUTTER et al. (1983).

[28]MYSAK et al. (1985, p. 52) list six arguments in support of the TW-model and only the above mentioned discrepancy in the phase relation against it.

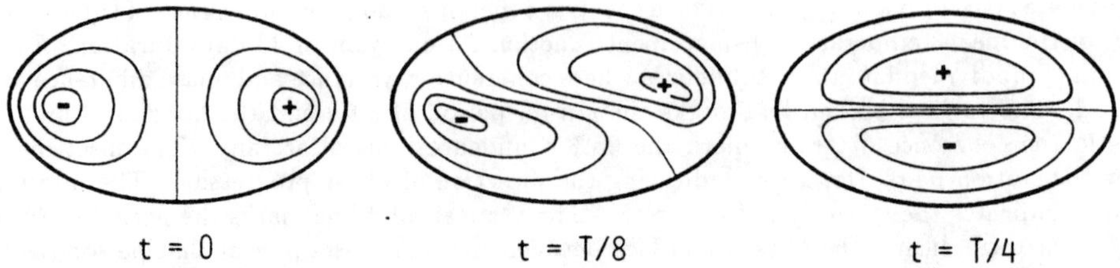

Figure 4.9: Contours of the mass transport streamfunction of the fundamental mode for the elliptical topographic waves (from JOHNSON, 1986).

4.3 Some known solutions of the TW-equation in various domains

In this and the following section we will construct explicit solutions for the TW-equation. The aim is to extract through this analysis the physical properties of TW's.

4.3.1 Circular basin with parabolic bottom

Following LAMB (1932, p. 212) we start our analysis of TW's in circular basins with equation (3.22), or better (3.24) which is its dimensionless form, or

$$\nabla \cdot \left(h \nabla \frac{\partial \zeta}{\partial t} \right) + \mathcal{J}\left[h, \zeta \right] - \left(\frac{L}{R} \right)^2 \mathcal{L} \frac{\partial \zeta}{\partial t} = 0 \ .$$

Here all quantities are dimensionless except L and R, a typical length and the ROSSBY radius, respectively. In polar coordinates this equation may be written as

$$\frac{\partial}{\partial r}\left(h \frac{\partial^2 \zeta}{\partial r \partial t} \right) + \frac{1}{r}\frac{\partial}{\partial \Theta}\left(\frac{h}{r}\frac{\partial^2 \zeta}{\partial \Theta \partial t} \right) + \frac{h}{r}\frac{\partial^2 \zeta}{\partial r \partial t} +$$

$$\frac{1}{r}\frac{\partial h}{\partial r}\frac{\partial \zeta}{\partial \Theta} - \frac{1}{r}\frac{\partial h}{\partial \Theta}\frac{\partial \zeta}{\partial r} - \left(\frac{L}{R} \right)^2 \mathcal{L}\frac{\partial \zeta}{\partial t} = 0 \ , \ 0 < r < 1 \ . \tag{4.4a}$$

The boundary conditions (no mass flux at the outer boundary, finiteness of ζ at the origin) are

$$\zeta = \text{ finite} \ , \quad \text{at } r = 0 \ , \quad \frac{\partial \zeta}{\partial r} = 0 \ , \quad \text{at } r = 1 \ . \tag{4.4b}$$

Consider a radial topography, $h = h(r)$ and assume an azimuthal wave solution of the form

$$\zeta = Z(r)\, e^{i(m\Theta - \sigma t)} \ , \tag{4.5}$$

travelling counterclockwise around the basin; $\sigma = \omega / f$ is the dimensionless frequency and m the azimuthal wavenumber. With these, the boundary value problem (4.4) assumes the

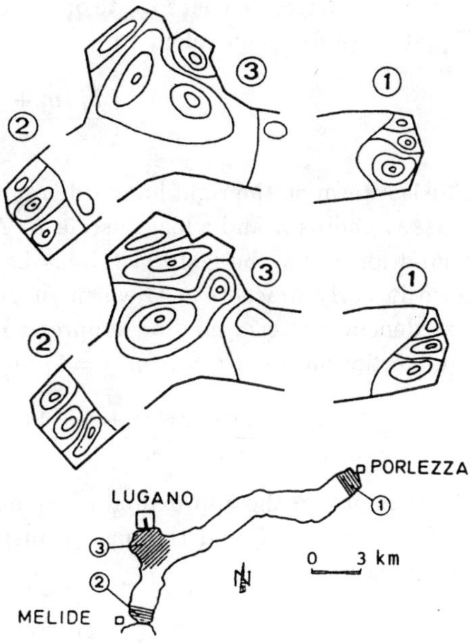

Figure 4.10: Three modes of long periodic waves in Lake of Lugano obtained by finite element technique. $T_1 = 68,5\ h$, $T_2 = 80,5\ h$ for $t = 0$ (left) and $T_3 = 91\ h$, $t = T/4$ (right) (from TRÖSCH, 1984 with alterations).

form

$$\begin{gathered}(hZ')' + \frac{h}{r}Z' - \left[\frac{m^2}{r^2}h + \frac{m}{\sigma}\frac{h'}{r} + \frac{1-\sigma^2}{(R/L)^2}\right] Z = 0\ ,\ 0 < r < 1\ , \\ Z = \text{ finite},\quad r = 0,\qquad Z' = 0,\quad r = 1.\end{gathered} \tag{4.6}$$

Primes denote differentiations with respect to r. From this two-point-boundary-value problem the solutions presented by LAMB (1932), WENZEL (1978) and SAYLOR, HUANG and REID (1980) can be obtained as special cases.
For the parabolic bottom profile,

$$h = 1 - r^2$$

$Z(r)$ can be expressed in terms of a hypergeometric polynomial (LAMB, 1932, p. 212; MILES and BALL, 1963; ABRAMOWITZ and STEGUN, 1972)

$$Z\,(r) = A_{mj} r^m F\left(m + j; m + 1; r^2\right)\ ,\quad m = 0,1,2,\ldots,\quad j = 1,2,3,\ldots, \tag{4.7}$$

in which A_{mj} is a free amplitude and σ satisfies the frequency relation

$$\frac{\sigma^2 - 1}{(R/L)^2} + \frac{2m}{\sigma} = 2\,[2j\,(m + j - 1) - m]\ , m = 0,1,2,\ldots,\ j = 1,2,3,\ldots. \tag{4.8}$$

This may be written as a third order polynomical equation, $P_3(\sigma) = 0$, i.e., the frequency occurs in third order which corresponds to three wave types, two first class and one second class wave. Here, we concentrate on second class waves and therefore exclude the case $m = 0$. Equation (4.8) is equivalent to

$$\frac{1}{\sigma} = \left\{ \frac{2j\,(m+j-1)}{m} - 1 \right\} - \frac{\sigma^2 - 1}{2m\,(R/L)^2} \ .$$

The last term on the right hand side represents the influence of the size effect via the external ROSSBY radius R and a length scale L. An order of magnitude for R is 500 km and an upper bound for L may be 200 km (Great Lakes), so $2(R/L)^2 \geq 12$. The minimum value of the term in curly brackets is 1, which suggests that the two first class modes entering via the size dependent term may be suppressed. Approximately, we may write, after neglection of the inertial motion ($\sigma = 1$ for $j = 1$) and transformation $n = j - 2$, $n = 0, 1 \ldots$

$$\frac{1}{\sigma} = \frac{2\,(n+2)\,(m+n+1)}{m} - 1 \ , \qquad m = 1, 2, 3, \ldots, \ n = 0, 1, 2, \ldots . \tag{4.9}$$

We thus obtain the approximate frequencies and periods of *Table 4.2.* The real parts of the surface elevation ζ and the mass transport streamfunction Ψ for the mode $(m, n) = (1, 0)$ are given by

$$\zeta\,(r, t) = Ar\left(1 - \tfrac{3}{2} r^2\right) \cos\left(\Theta - \sigma t\right) \ ,$$
$$\Psi\,(r, t) = Ar\,(1 - r^2)\left(1 - \sigma - \tfrac{3}{2}\,(1 - 3\sigma)\, r^2\right) \cos\left(\Theta - \sigma t\right) \ .$$

The streamlines of this solution and those of the $(1, 1)$- and $(2, 0)$-modes are sketched in *Fig. 4.11.* The periods are larger than 50.7 hours (2.1 days), but it is likely that only those modes with a very simple structure will be excited. The simplest fundamental mode has a period of 118 hours and consists of two basin-wide gyres, a cyclonic and an anticyclonic vortex. The entire system of gyres rotates counterclockwise (on the Northern hemisphere) around the basin. From this fact it follows that current vectors at mid basin positions rotate in the anticlockwise direction while those at near-shore positions rotate in the clockwise direction[29]. This is exactly the current pattern discovered by SAYLOR, HUANG and REID (1980) in Southern Lake Michigan, but the period of 118 hours is too large to fit the period of 100 hours inferred from measurements. They therefore studied the effect of the topography on the TW-solutions in a circular basin.

Table 4.2 also points at another peculiar property of TW's in enclosed basins. For all n the eigenfrequencies σ lie in a very narrow band between 0 and 1. This means that the eigenvalues are clustered close to one another, making their numerical seperation very difficult. This is the major reason why identification of TW wave structures in numerical solution procedures

[29] This result can be also be inferred if the radial and azimuthal compoments of the mass transport vector **V** are expressed in terms of the mass transport streamfunction

$$V_r = -\frac{1}{r}\frac{\partial \Psi}{\partial \Theta}, \qquad V_\Theta = \frac{\partial \Psi}{\partial r}$$

if those expressions are evaluated with the aid of the above expression for $r = 0$ and $r \sim 1$.

	$n=0$		$n=1$		$n=2$	
m	σ	$T[h]$	σ	$T[h]$	σ	$T[h]$
1	0,143	118	0,0588	287	0,0323	524
2	0,200	84,5	0,0909	186	0,0526	321
4	0,250	67,6	0,125	135	0,0769	220
∞	0,333	50,7	0,200	84,5	0,143	118

Table 4.2) Periods of TW's in a circular basin with parabolic bottom profile ignoring the size dependent term in (3.8) and computed with $f = 2\pi/16.9\ h$.

is so difficult. Needless to say that many modes must be mixed among themselves and with higher order baroclinic gravity waves in observed time series of temperature and/or velocity.

4.3.2 Circular basin with a power-law bottom profile

The influence of the bathymetric gradients on the periods of the topographic wave motion can be studied by selecting the profile (SAYLOR, HUANG and REID (1980)).

$$h(r) = (1 - r^q)\ ,\ 0 \le r \le 1\ ,\ q > 0\ . \tag{4.10}$$

Varying the exponent q yields an entire sequence of profile geometries with strong and weak topographic gradients. For $q = 1$ the radial depth profile is conical, for $q = 2$ it is parabolic, for $q > 2$ it becomes blunt and for $q \to \infty$ approaches constant depth. On the other hand, for $0 < q \le 1$ the profile has a vertex at the center and (except for $q = 1$) has a convex curvature. With $\Psi = \Psi(r)e^{i(m\Theta - \sigma t)}$, use of the dimensionless TW-equation in cylindrical coordinates

$$\frac{\partial}{\partial r}\left\{\frac{1}{h}\frac{\partial}{\partial r}\left(\frac{\partial \Psi}{\partial t}\right)\right\} + \frac{1}{r}\frac{\partial}{\partial \Theta}\left\{\frac{1}{h}\frac{1}{r}\frac{\partial}{\partial \Theta}\left(\frac{\partial \Psi}{\partial t}\right)\right\} + \frac{1}{r}\frac{\partial \Psi}{\partial r}\frac{\partial}{\partial \Theta}\left(\frac{1}{h}\right) - \frac{1}{r}\frac{\partial \Psi}{\partial \Theta}\frac{\partial}{\partial r}\left(\frac{1}{h}\right) = 0\ . \tag{4.11}$$

and a trial solution

$$\Psi(r) = A r^m h^2(r)\ ,$$

(which satisfies the boundary conditions) the depth profile must fulfil the differential equation

$$h'' + \frac{3m + 2 - m/\sigma}{2r} h' = 0\ . \tag{4.12}$$

The depth profile (4.10) is compatible with this provided that

$$\sigma = \frac{m}{3m + 2q}\ ,$$

which is the frequency-wavenumber relationship for the prescribed topographic profile that was already listed in (4.3). *Table 4.3* lists the frequencies for a sequence of topography parameters q and the wavenumbers $m = 1, 2$. The table indicates that the topography has a dominant effect on the periods. The solutions

$$\Psi(r) = A r^m (1 - r^q)\ ,\ \sigma = \frac{m}{3m + 2q}$$

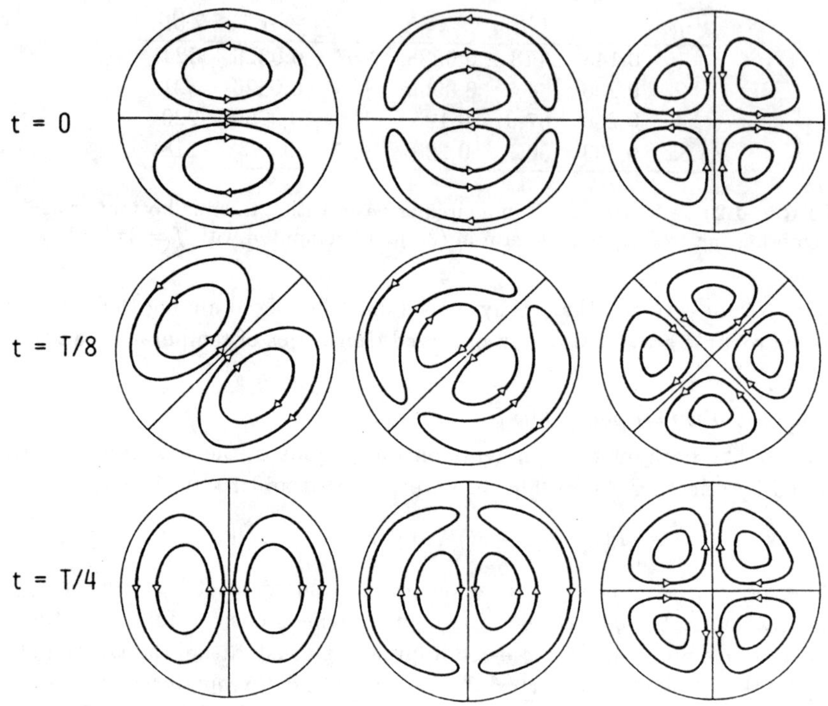

Figure 4.11: Contour lines of the mass transport streamfunction of the three modes with the simplest wave structure. The gyres rotate anticlockwise (on the Northern hemisphere) around the basin (from STOCKER and HUTTER, 1987a).

embrace all those motions whose streamfunction Ψ has no radial nodal circle. Hence, they contain in particular the solutions for the parabolic depth profile as shown in the left and right columns of *Fig. 4.11*.

4.3.3 Elliptic basin with parabolic bottom

Most lakes are long in one direction and not well approximated by circles. It is interesting, therefore, to see how the periods and mode structures of TW's depend on the aspect ratio (i.e. the width to length ratio) of the basins.

We consider the TW-equation in dimensionless form and in Cartesian coordinates

$$\frac{\partial}{\partial x}\left(h^{-1}\frac{\partial^2\Psi}{\partial x\partial t}\right)+\frac{\partial}{\partial y}\left(h^{-1}\frac{\partial^2\Psi}{\partial y\partial t}\right)-\left(\frac{\partial h^{-1}}{\partial x}\right)\frac{\partial\Psi}{\partial y}+\left(\frac{\partial h^{-1}}{\partial y}\right)\frac{\partial\Psi}{\partial x}=0\ , \quad \text{in } \mathcal{D}\ , \tag{4.13}$$

$$\Psi = 0\ , \qquad \text{on} \quad \partial\mathcal{D},$$

q	$m = 1$	$m = 2$
0,5	0,250	0,286
1	0,200	0,250
2	0,143	0,200
5	0,0769	0,125

Table 4.3) Topography effect on the dimensionless eigenfrequencies, σ, of the two first modes in the model of SAYLOR et al. (1980).

and choose a parabolic depth profile

$$h = \frac{1}{2}\left((1-a)\,x^2 + (1+a)\,y^2\right) - 1\,, \tag{4.14}$$

where $A(h) = \sqrt{2\,(h+1)\,/\,(1-a)}$ and $B(h) = \sqrt{2\,(h+1)\,/\,(1+a)}$ are the semi-axes of the elliptic depth-contours. These have all identical aspect ratios

$$r = \sqrt{\frac{B(h)}{A(h)}} = \sqrt{\frac{1-a}{1+a}}\,, \quad a = \sqrt{\frac{1-r^2}{1+r^2}}\,, \tag{4.15}$$

and the profile has a maximum depth $|h|_{max} = 1$. The basin is bounded by the zero depth-contour line, an ellipse with $A(0)$ and $B(0)$ as semi-axes.
The following analysis is due to BALL (1965). With (4.14) and the transformation $\psi = h^{-2}\Psi$ it is straightforward to show that (4.14) takes the form

$$\begin{aligned} &4\frac{\partial\psi}{\partial t} + 3\left[(1-a)\,x\frac{\partial^2\psi}{\partial x\partial t} + (1+a)\,y\frac{\partial^2\psi}{\partial y\partial t}\right] + h\left(\frac{\partial^3\psi}{\partial x^2\partial t} + \frac{\partial^3\psi}{\partial y^2\partial t}\right) + \\ &(1-a)\,x\frac{\partial\psi}{\partial y} - (1+a)\,y\frac{\partial\psi}{\partial x} = 0, \qquad \text{for } h < 0, \end{aligned} \tag{4.16}$$

$$\psi = \text{ finite}, \qquad \text{for } h = 0.$$

Note that the boundary condition $\Psi = 0$ along $\partial\mathcal{D}$ necessarily requires that ψ is bounded on $\partial\mathcal{D}$. The velocities are given by

$$u = -h^{-1}\frac{\partial\,(h^2\psi)}{\partial y} = -\frac{\partial h}{\partial y}\psi - h\frac{\partial\psi}{\partial y}\,, \qquad v = h^{-1}\frac{\partial\,(h^2\psi)}{\partial x} = 2\frac{\partial h}{\partial x}\psi + h\frac{\partial\psi}{\partial x}\,.$$

The advantage of the introduction of the streamfunction ψ is that (4.13) transforms into a differential equaton with the following special property. Suppose, ψ is an even (odd) polynomial of degree N[30], then the differential equation (4.16) generates again an even (odd) polynomial of the same degree.
Taking advantage of this fact, we consider first a polynomial with degree $N = 0$, i.e. a

[30]A polynomical of degree N in the variables x and y contains only terms of the form $x^k y^{N-k}$, for $k = 0, 1, 2, \ldots, N$, and no lower order terms.

constant ψ_{00} which obviously satisfies (4.16). This is a simple steady gyre with the velocity field

$$(u,v) = 2\psi_{00}\left(-(1+a)\,y, (1-a)\,x\right) \ ,$$

representing an elliptical rotation with constant vorticity $4\psi_{00}$. Maximal speeds are experienced along the shore-line.
More insight provides the choice of a homogeneous odd polynomial of degree $N = 1$, the linear BALL-mode:

$$\psi_1 = \psi_{10}\,(t) \cdot x + \psi_{01}\,(t) \cdot y \ . \tag{4.17}$$

Substitution into (4.16) yields the coupled system

$$(7-3a)\,\dot{\psi}_{10} + (1-a)\,\psi_{01} = 0, \qquad (7+3a)\,\dot{\psi}_{01} - (1+a)\,\psi_{10} = 0, \tag{4.18}$$

with $()^{\cdot} \equiv d/dt$. Assuming a harmonic time evolution $e^{-i\sigma t}$ for both coefficient functions, (4.18) allows nontrivial ψ_{10} and ψ_{01}, if and only if

$$\sigma^2 = \frac{1-a^2}{49-9a^2} \ . \tag{4.19}$$

This relation describes the dependence of the frequency on the aspect ratio parameter r (via a, see (4.15)). *Table 4.4* lists the periods obtained with (4.19). Obviously, $a = 0$ recovers the solution for the circle with parabolic bottom profile. Smaller a results in larger σ; consequently, the more elongated the ellipses become the larger will be the periods. In view of observational results for Lake of Lugano reported above this is unfortunate as this lake is long and narrow, and measurements point at oscillations with periods of 3-4 days. This is lower than the 118 hours obtained as a lower bound for the fundamental linear BALL mode.
The linearity of (4.17) implies that the line $\psi_1 = 0$ which separates vortices of different signs is a straight line that, owing to (4.18), rotates anticlockwise around the basin. *Fig. 4.12* shows the time evolution of this mode. The structure of the wave pattern, i.e. the number of gyres is conserved in the course of a wave cycle. This is in accord with the wave patterns found in the circular basin.

		linear		quadratic	
r	a	σ	$T[h]$	σ	$T[h]$
1,0	0	0,143	118	0,200	84,5
0,67	0,385	0,134	126	0,190	88,8
0,50	0,600	0,118	143	0,173	97,7
0,33	0,800	0,091	185	0,139	121
0,1	0,980	0,031	542	0,051	335
0	1,0	0	∞	0	∞

Table 4.4) Frequencies and periods of the linear and quadratic BALL-mode for various aspect ratios r. The periods are calculated with $f = 2\pi/16.9\ h$.

To obtain the next higher mode, we select an even polynomial of degree $N = 2$.

$$\psi_2 = \psi_{00}(t) + \psi_{20}(t)x^2 + \psi_{11}(t)xy + \psi_{02}(t)y^2 , \tag{4.20}$$

with the time dependent coefficient functions $\psi_{mn}(t)$. Equation (4.20) characterizes the quadratic BALL-mode. Substitution into (4.16) and equating equal powers of x and y, respectively, yields the homogeneous system of ordinary differential equations

$$\begin{aligned} (11 - 7a)\,\dot{\psi}_{20} + (1-a)\,\dot{\psi}_{02} + 2\,(1-a)\,\psi_{11} &= 0 , \\ 10\dot{\psi}_{11} + (1-a)\,\psi_{02} - (1+a)\,\psi_{20} &= 0 , \\ (11 + 7a)\,\dot{\psi}_{02} - (1+a)\,\dot{\psi}_{20} - 2\,(1+a)\,\psi_{11} &= 0 , \\ 2\dot{\psi}_{00} - \dot{\psi}_{20} - \dot{\psi}_{02} &= 0 , \end{aligned} \tag{4.21}$$

which allows periodic solutions proportional to $e^{-i\sigma t}$ provided that

$$\sigma\left(5\sigma^2\left(5 - 2a^2\right) - \left(1 - a^2\right)\right) = 0 .$$

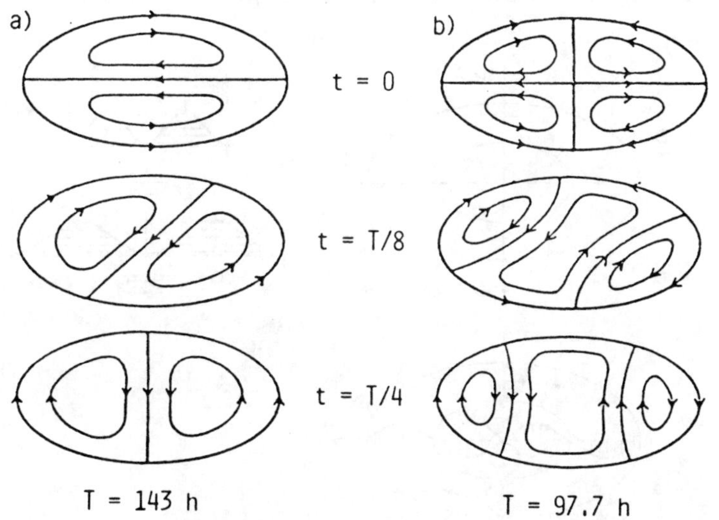

Figure 4.12: Mass transport streamline patterns for the "linear" (a) and "quadratic" (b) mode of the TW-equation in an elliptic basin with parabolic bottom profile (from BALL, (1965)).

Again, there is a steady solution $\sigma = 0$ and an oscillating solution with

$$\sigma^2 = \frac{1 - a^2}{5\,(5 - 2a^2)} . \tag{4.22}$$

Table 4.4 collects frequencies and periods for several aspect ratios r. For a fixed aspect ratio the periods of the quadratic mode are smaller than those of the fundamental linear mode. As (4.21) indicates, a steady solution must have

$$\psi_{11}^{st} = 0 \qquad \text{and} \qquad \psi_{02}^{st} = \frac{1+a}{1-a}\psi_{20}^{st} ,$$

and hence

$$\psi_2^{st} = \psi_{00} + A\left[(1-a)\,x^2 + (1+a)\,y^2\right] ,$$

where ψ_{00} and A are constants. $\psi_{00} \neq 0$, $A = 0$ recovers the simple steady gyre whereas $\psi_{00} = 0$, $A \neq 0$ yields the steady second order solution

$$\Psi_2^{st} = h^2 \psi_2^{st} = 2ah^2\,(h+1) \ . \tag{4.23}$$

This streamfunction vanishes along the boundary ($h = 0$) and at the center $(x, y) = (0, 0)$ and is positive otherwise; furthermore, its value is constant along similar ellipses and assumes a maximum value along the ellipse with $h = -2/3$ between the center and the shore line, *Fig. 4.13a.* The steady flow corresponding to the solution (4.23) is qualitatively indicated in *Fig. 4.13b.* An anticyclonic elliptical gyre in the center is surrounded by an elliptical ring of cyclonically rotating fluid. Oscillating solutions are obtained by constructing the eigenvector of (4.21) corresponding to the frequency given by (4.22). We quote BALL's result (real part)

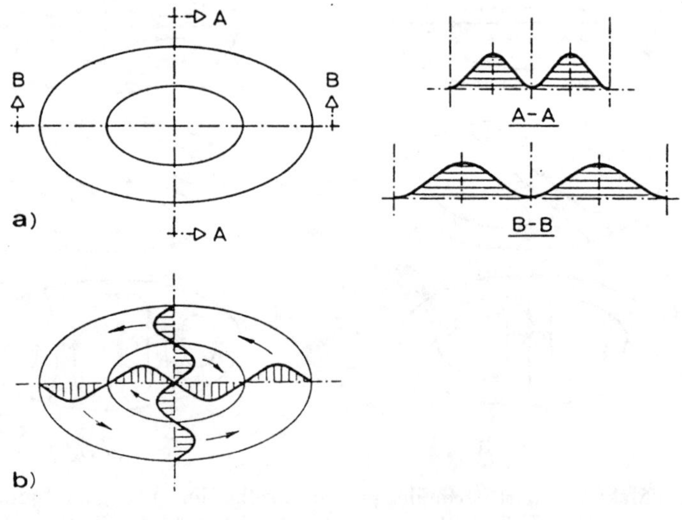

Figure 4.13: a) Distribution of the mass transport stream function and b) of the associated velocity field of the "quadratic" steady solution (4.23), (from STOCKER and HUTTER, 1987a).

$$\Psi_2 = h^2\psi_2 = Ah^2\left\{\left[(1+a)(3-2a)\,y^2 - (1-a)(3+2a)\,x^2 + a\right]\sin\sigma t \right.$$
$$\left. + \left[6\left(1-a^2\right)/5\sigma\right] xy\cos\sigma t\right\} \ . \tag{4.24}$$

For $t = 0$ the nodal lines $\Psi_2 = 0$ are the lines $x = 0$ and $y = 0$, whereas for $t > 0$ they are rotating hyperbolas (note that $0 < a < 1$). As illustrated in *Figure 4.12b* the wave pattern starts with four gyres of which the two positive vortices merge together building three gyres in the basin for most part within a period. The structure of this mode is therefore not conserved during the cycle. This is a new phenomenon due to the influence of the aspect ratio parameter a.

Solutions of polynomial degree higher than 2 have not been constructed so far mainly because computations become unwieldy. For this reason MYSAK (1985) and JOHNSON (1986) chose elliptical basins whose bathymetric contour lines follow confocal ellipses. The TW-equation may then be written down in elliptic coordinates. Choosing an exponential depth profile in the "radial" direction and a wave ansatz for azimuthally travelling waves then yields ordinary differential equations for the eigenfunctions. JOHNSON (1986) demonstrates that the streamline pattern is qualitatively similar to that obtained for a parabolic bottom, see e.g. *Fig. 4.9.* MYSAK, as oppposed to JOHNSON constructs solutions for an elliptic basin with an interior elliptic island. These are mathematically easier to construct than BALL's solutions. For a detailed analysis see STOCKER and HUTTER (1987a).

4.3.4 Topographic waves in infinite domains

Major developments in the understanding of second class waves were not advanced by solving the TW-equation in finite domains, but rather by studying these waves in infinite domains, such as channels, continental shelves, trenches, etc. Our aim is not to provide a complete account of the history and the availability of known solutions, for this the reader is directed to MYSAK (1980) and more recent literature. By presenting solutions to analytically accessible configurations we would rather like to work out the typical physical properties of the characteristics.

a) *Straight channel:*

Consider a straight, infinite channel with the Cartesian coordinate system as indicated in *Fig. 4.14.* With a depth profile $h(y)$ which is constant along the channel axis and a carrier-wave ansatz of the form

$$\Psi = \Psi(y)e^{i(kx-\sigma t)} \ , \quad \sigma = \omega/f$$

the TW-equation reduces to

$$\left(h^{-1}\Psi'\right)' - \left[\frac{k}{\sigma}\left(h^{-1}\right)' + k^2h^{-1}\right]\Psi = 0 \ , \tag{4.25}$$

in which $(\cdot)' = d(\cdot)/dy$. Equation (4.25) is subject to the familiar no-flux conditions

$$\Psi(0) = \Psi(r) = 0 \ .$$

Furthermore, at interior points we request that

$$[\![\Psi]\!] = 0 \ , \qquad [\![(\Psi' - k/\sigma\Psi)/h]\!] = 0 \ , \qquad \text{at} \quad y = s \ . \tag{4.26}$$

Here $[\![f(y)]\!]$ at $y = s$ denotes the jump of the quantity f defined by $[\![f(s)]\!] = \lim_{\varepsilon\to 0}\{f(s+\varepsilon) - f(s-\varepsilon)\}$. Equation $(4.26)_1$ means that the transport is continuous whereas $(4.26)_2$ follows by integrating (4.25) across the discontinuity:

$$\lim_{\varepsilon\to 0}\int_{s-\varepsilon}^{s+\varepsilon}\left(\left(h^{-1}\Psi'\right)' - \frac{k}{\sigma}\left(h^{-1}\right)'\Psi - k^2 h^{-1}\Psi\right)dy = 0\,,$$

$$\lim_{\varepsilon\to 0}\left\{\left[h^{-1}\Psi' - \frac{k}{\sigma}h^{-1}\Psi\right]_{s-\varepsilon}^{s+\varepsilon} + \int_{s-\varepsilon}^{s+\varepsilon}\left(\frac{k}{\sigma}h^{-1}\Psi' - k^2 h^{-1}\Psi\right)dy\right\} = 0\,.$$

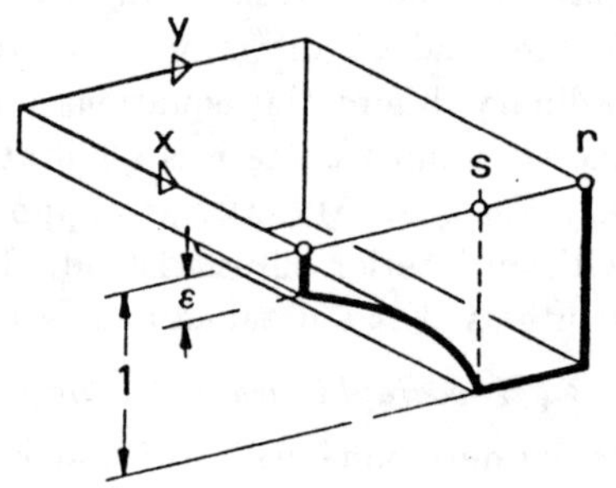

Figure 4.14: Infinite channel with onesided shelf. At the boundary points, 0, s, r the functions h and h' may not be continuous.

Because h, Ψ and Ψ' are all bounded at $y = s$ and h is nonzero at $y = s$ the last integral vanishes in the limit as $\varepsilon \to 0$. If Ψ and h suffer no jump discontinuity, then this implies

$$[\![\Psi]\!] = [\![\Psi']\!] = 0, \qquad \text{at} \quad y = s.$$

b) *Channel with one-sided topography:*

Consider the piecewise exponential depth profile (see *Fig. 4.14*)

$$h(y) = \begin{cases} \varepsilon e^{by} & , \quad 0 \le y \le s\,, \\ 1 & , \quad s \le y \le r\,, \end{cases} \qquad b = \frac{1}{s}\ln\frac{1}{\varepsilon} \tag{4.27}$$

It renders (4.25) an ordinary differential equation with constant coefficients. Subject to the boundary conditions $\Psi(0) = \Psi(r) = 0$. The solution is

$$\Psi(y) = \begin{cases} e^{by/2}\sin\lambda y & , \quad 0 \le y \le s\,, \\ A\sinh\left(k\left(y-r\right)\right) & , \quad s \le y \le r\,, \end{cases} \quad \lambda^2 = \frac{k}{\sigma}b - k^2 - \frac{b^2}{4}\,.$$

Here, the amplitude of $\Psi(y)$ for $0 \le y \le s$ has been set equal to unity. Evaluating (4.26) yields the *implicit dispersion relation*

$$\frac{1}{\lambda}\tan s\lambda = \frac{-1}{k\coth\left(k\left(r-s\right)\right) + b/2}\,, \qquad \lambda^2 = \frac{k}{\sigma}b - k^2 - \frac{b^2}{4}\,, \tag{4.28a, b}$$

and

$$A = \frac{e^{bs/2} \sin \lambda s}{\sinh (k (s - r))} \tag{4.28c}$$

of TW's in this infinite channel with one-sided topography. *Fig. 4.15* displays the left hand side and right hand side of (4.28a) as functions of λr. The left hand side is independent of σ whereas the right hand side is a double-valued relation of λ due to (4.28b). For frequencies lower than a critical value there exists a finite number of intersections (σ, λ) or (σ, k). This number increases stepwise with decreasing σ due to the periodicity of the tangent function. Note that λ^2 is only positive provided the signs of k and σ are identical. The dashed curve in *Fig. 4.15* shows the left hand side for $\lambda^2 < 0$, i.e. the tan-function is replaced by tanh; there are no intersections and hence no real pairs (σ, k) satisfying (4.28). This implies that for real waves phase propagation is into the positive x-direction for this configuration, which expresses that shelf waves on the Northern hemisphere are *right-bounded*, i.e., the shallow region of the shelf is to the right when looking into the direction of phase propagation.

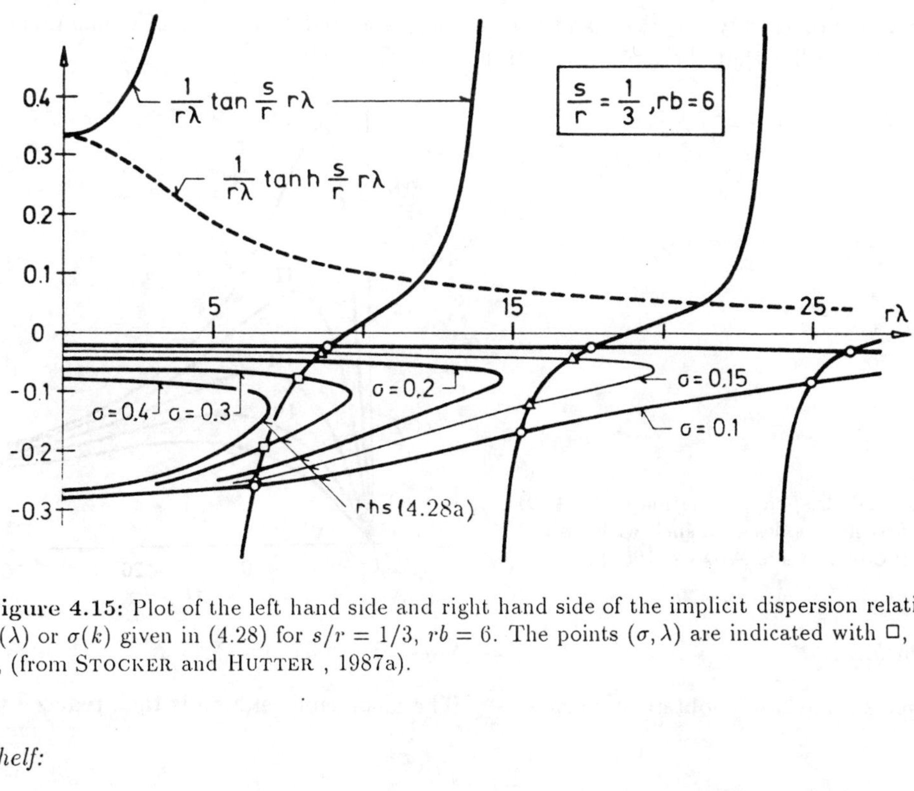

Figure 4.15: Plot of the left hand side and right hand side of the implicit dispersion relation $\sigma(\lambda)$ or $\sigma(k)$ given in (4.28) for $s/r = 1/3$, $rb = 6$. The points (σ, λ) are indicated with □, △, o, (from STOCKER and HUTTER , 1987a).

c) *Shelf:*

In the limit as $r \to \infty$ the depth profile (4.27) becomes the well-known exponential shelf.

Correspondingly the dispersion relation reads

$$\frac{1}{\lambda}\tan s\lambda = -\frac{1}{k+b/2}\,, \tag{4.29}$$

and the result of BUCHWALD and ADAMS (1968) is recovered. *Fig. 4.16* displays this dispersion relation for the first five modes. The shapes of these curves exhibit features which are typical of topographically trapped second-class waves: First, long shelf waves are non-dispersive, i.e. as $k \to 0$ $c_{gr} = \partial\sigma/\partial k \to c = \sigma/k > 0$. Phase and group velocity are the same, Second, when h'/h is bounded[31], for $y \in [0, \infty)$ then $c_{gr} < 0$ for some range of $k > 0$. For $k < k_0$, c and c_{gr} are both positive, and phase and energy propagate in the same direction; for $k > k_0$, c is still positive but c_{gr} is negative. Furthermore, $\sigma(k) \to 0$ as $k \to \infty$. These properties which hold in an infinite domain were proven by HUTHNANCE (1975) in a more general context.
The streamfunction in the shelf-wave limit takes the form

$$\begin{aligned}\Psi(y) &= e^{by/2}\sin\lambda y, & 0 \le y \le s,\\ \Psi(y) &= e^{bs/2}\sin\lambda s\cdot e^{-k(y-s)}, & s < y.\end{aligned} \tag{4.30}$$

It decays exponentially for $y > s$ and is essentially sinusoidal in the shelf domain. This is why we said earlier that these waves are trapped to the shelf.

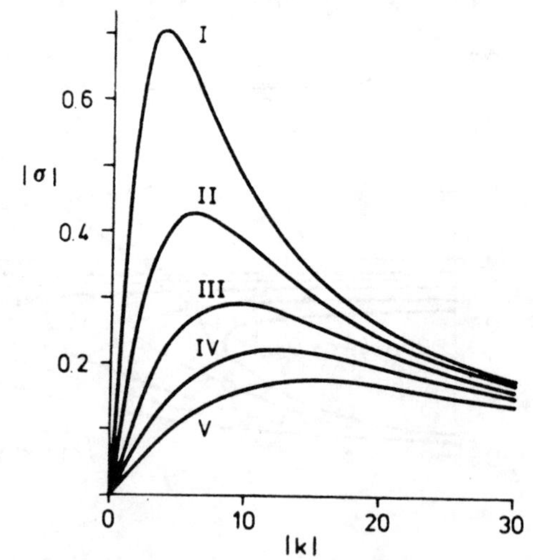

Figure 4.16: Dispersion relation $\sigma(k)$ (4.29) for the first five modes on a shelf with $b = 5.4$ (from BUCHWALD and ADAMS, 1968)

d) *Trench:*

A second simple case is obtained when $s \to r$. The dispersion relation is then reduced to

$$\tan r\lambda = 0\,,$$

[31] Because of its significance h'/h is often referred to as *slope parameter* $S = h'/h$.

whence

$$\sigma = \frac{kb}{k^2 + \left[\frac{b^2}{4} + \left(\frac{n\pi}{r}\right)^2\right]}, \qquad n = 1, 2, 3, \ldots, \tag{4.31}$$

where the integer n denotes the mode of the wave. The streamfunction of the n-th mode has $n-1$ nodes across the channel. Long waves in this channel are non-dispersive with phase and group velocity

$$c = c_{gr} = \frac{b}{\frac{b^2}{4} + \left(\frac{n\pi}{r}\right)^2}, \qquad \text{as} \quad k \to 0 \,. \tag{4.32}$$

For very short waves the frequency is inversely proportional to the wavenumber, $\sigma = b/k$, and phase and group velocity have opposite sign. The critical point (k_0, σ_0), where the group velocity vanishes is given by

$$(k_0, \sigma_0) = \left[\sqrt{\frac{b^2}{4} + \left(\frac{n\pi}{r}\right)^2}, b \Big/ \left(2\sqrt{\frac{b^2}{4} + \left(\frac{n\pi}{r}\right)^2}\right)\right] . \tag{4.33}$$

The critical frequency σ_0 strongly depends on the slope parameter $S = b = h'/h$.

e) *Single-step shelf:*

We now demonstrate that boundedness of h'/h is important for the existence of a vanishing group velocity at finite $k_0 < \infty$. To this end consider the profile

$$h(y) = \begin{cases} d \;, & 0 < y < s \,, \\ 1 \;, & s < y < r \end{cases} \,. \tag{4.34}$$

The profile was used by SEZAWA and KANAI (1939), SNODGRASS et al. (1962) and LARSEN (1969) to explain edge waves and trapped long waves. Clearly, since $h' = (1-d)\,\delta(y-s)$ vanishes everywhere in $y \in (0, r)$ except at $y = s$, where h' is infinite, TW's only exist because of this singularity.
With (4.34) the differential equation (4.25) reduces to $\Psi'' - k^2\Psi = 0$. Its solutions are

$$\Psi(y) = \begin{cases} \sinh ky \,, & 0 \le y \le s \,, \\ A \sinh k(y-r) \,, & s \le y \le r . \end{cases}$$

Inserting this into the matching conditions (4.26) yields the dispersion relation

$$\sigma = (1-d)\,\frac{\tanh ks \cdot \tanh k(r-s)}{d \tanh ks + \tanh k(r-s)} \,. \tag{4.35}$$

Fig. 4.17 displays $\sigma(k)$ for different values of d, and we notice that these are monotonically growing functions of k. Indeed

$$\sigma \to \frac{1-d}{1+d}, \text{ as } \quad k \to \infty, \qquad \sigma \propto k, \text{ as } \quad k \to 0, \tag{4.36}$$

and no finite critical wavenumber k_0 exists. Furthermore, there is only one single fundamental TW mode.

In reality channels have bathymetric gradients on both sides. From the results obtained so far, it can be concluded that in such a channel (e.g. with a parabolic depth profile) there will be TW's travelling along either sides of the channel each in a right-bounded way. The dispersion relation then consists of two branches (assume $\sigma > 0$), one for $k > 0$ representing those waves trapped to $y = 0$ and $k < 0$ for those trapped to $y = r$.

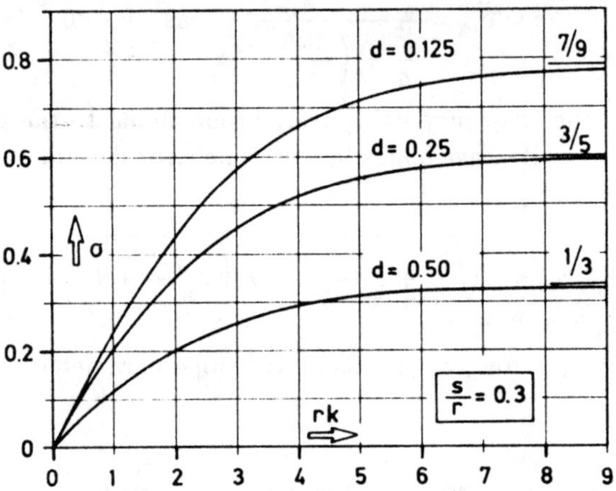

Figure 4.17: Graph of the dispersion relation (4.35) for $s/r = 0.3$ and 3 values of d. Note $\sigma\,(rk)$ is monotone.

4.4 TW's in rectangular basins

A detailed account on the theory of linear TW's in rectangular basins with a doubly symmetric bathymetric profile is too complicated and too long to be presented here. The reader is asked to consult the relevant literature (STOCKER and HUTTER, 1986, 1987 a,b; STOCKER, 1988). Here it may suffice to give a descriptive account of the subject matter.

The basic boundary value problem in dimensionless coordinates, after an exponential time dependent factor $e^{i\sigma t}$ has been split off, is

$$\begin{gathered} i\sigma\nabla\cdot\left(\nabla\frac{\Psi}{h}\right)+\mathcal{J}\left[\Psi,\frac{1}{h'}\right]=0, \quad \text{in } \mathcal{D}, \\ \Psi=0, \quad \text{on } \partial\mathcal{D}\,, \end{gathered} \tag{4.37}$$

in which σ is the dimensionless frequency and h the dimensionless depth, which we shall assume to be given in the form

$$h\,(s,n)=h_0\,(s)\left(1+\varepsilon-\left|\frac{2n}{B}\right|^q\right)\,,\quad 0<s<L\,,\quad -\frac{B}{2}<n<\frac{B}{2} \tag{4.38}$$

with constant ε and $q(0 < q < \infty)$. L and B denote the length and the width of the rectangle and s and n are the coordinates along the long and short axes of symmetry, respectively. We call ε the sidewall and q the topography parameter, see *Fig. 4.18*; $\varepsilon \neq 0$ prevents zero shore depth and $q > 0$ ($q < 0$) allows concave (convex) transverse depth profiles to be modelled. The function $h_0(s)$ describes the longitudinal variation of the depth thalweg profile which we shall choose as

$$h_0(s) = \eta + \sin^p(\pi s/L) \ , \ (p > 1) \ , \tag{4.39}$$

with constant parameters η and p, see *Fig. 4.18b*.
STOCKER and HUTTER (1986) presented a method which economically solves the boundary value problem (4.37) in a straight, infinite channel. The idea was to make use of the narrow, elongated shape of the bathymetry. The streamfunction was expanded in a complete function set $\{P_\alpha\}$, according to

$$\Psi(s,n) = \sum_{\alpha=1}^{2N} P_\alpha(s,n)\,\Psi_\alpha(s) \ , \tag{4.40}$$

thereby incorporating all functional dependence on n in the preselected $2N$ basis functions P_α (N symmetric and N skewsymmetric with respect to $n = 0$). These can be chosen for convenience, as trigonometric functions fulfilling the boundary conditions, $\Psi = 0$, along the shore[32]. Each basis function is weighted by a residue function $\Psi_\alpha(s)$, which remains to be determined. For a finite order of expansion, N, it is anticipated that (4.40) represents an approximation of the exact solution. With (4.38), (4.39) the two-dimensional problem (4.37) can be approximated by a system of ordinary differential equations for the residue functions. The system is obtained by a weighted integration of the two-dimensional equation (4.37) over the narrow coordinate n and using the expansion (4.40).

4.4.1 Dispersion relation for an infinite channel

We confine here attention to an infinite channel with $h_0(s) = 1$. Depth contour lines are then all parallel to the shore and the solution of the TW-equation amounts to solving a system of $2N$ homogeneous ordinary differential equations subject to homogeneous boundary conditions. Requiring that non-trivial solutions to this eigenvalue problem exist yields the dispersion relation $\sigma = \sigma(k)$.
For a first order model, $N = 1$, (this is a model which uses one symmetric and one skew-symmetric basis function of the form (4.40) and is of lowest possible order) we have plotted in

[32]When symmetric and skewsymmetric basis functions are chosen then (4.40) may be written as

$$\Psi(s,n) = \sum_{\alpha=1}^{N} P_\alpha^+(s,n)\,\Psi_\alpha^+(s) + \sum_{\alpha=1}^{N} P_\alpha^-(s,n)\,\Psi_\alpha^-(s) \ ,$$

where for trigonometric functions

$$P_\alpha^+(s,n) = \cos\left(\pi\left(\alpha - \tfrac{1}{2}\right)\tfrac{2n}{B(s)}\right) \ ,$$
$$P_\alpha^-(s,n) = \sin\left(\pi\alpha\tfrac{2n}{B(s)}\right), \qquad (\alpha = 1, 2, 3, \ldots, N) .$$

$B(s)$ is the width of the basin that may, strictly depend on s.

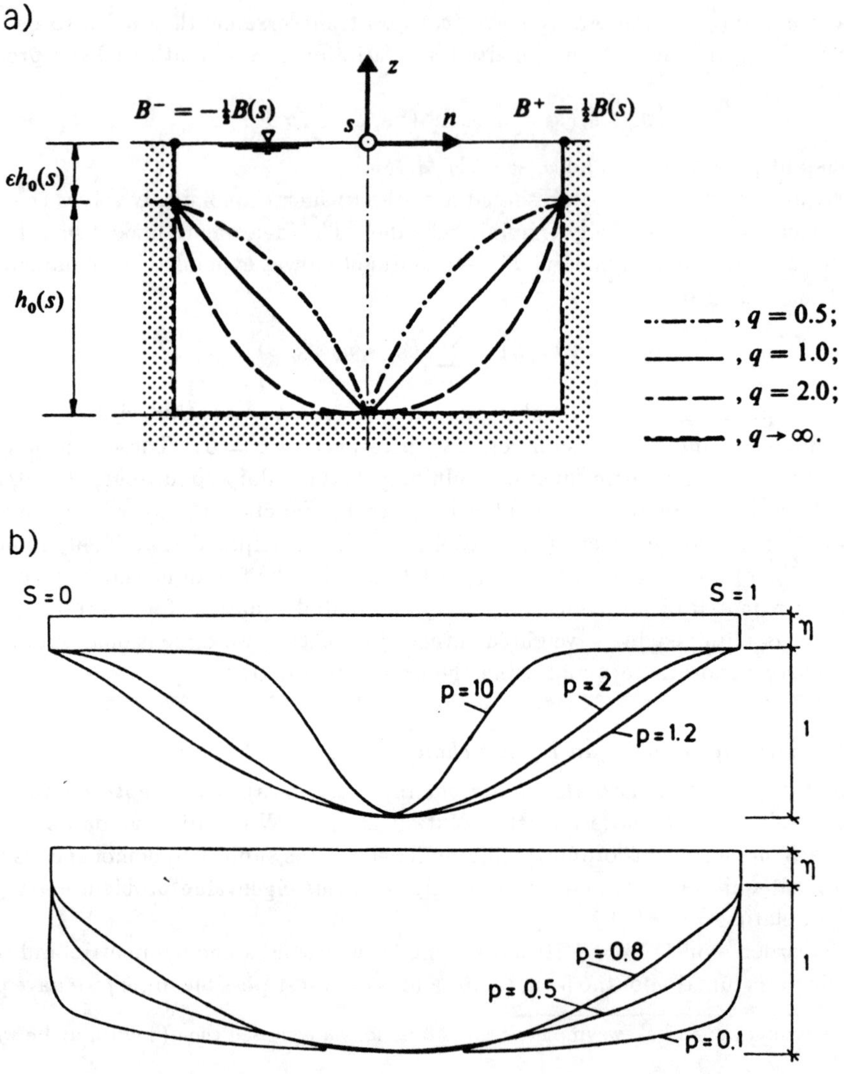

Figure 4.18: a) Cross-sectional depth profile of a symmetrical channel. b) Thalweg profiles (4.39) for different values of the exponent p. For $p > 1$ slopes at the channel ends are zero (top), when $p < 1$ they are infinite (bottom).

Fig. 4.19 this dispersion relation in a $(\Re k, \Im k, \sigma)$-coordinate system. Its graph is symmetric with respect to both axes $\Re(k) = 0$ and $\Im(k) = 0$. Three regimes 1, 2, 3 can be distinguished where the wavenumbers k take real, complex and purely imaginary values, respectively. In regime 1, all wavenumbers k are real and, therefore, represent physically possible channel solutions. Typically, for Rossby waves, for each frequency there exists a long and a short wave. It is worth noting that $\Re(k)$ can have both signs. This is due to the fact that h'/h changes sign across the channel, enabling propagation of TW's in both channel directions. In either case, as an effect of the Coriolis force, the structure of the wave on the Northern hemisphere is right-bound with respect to the direction of phase propagation. The structure of the streamfunction depends upon the frequency range. Small frequencies favour periodic patterns along the channel. A critical wavenumber exists at which no energy is transported along the channel (regime 1). Waves with intermediate frequencies (of order unity) have a mixed periodic-exponential structure and do not represent possible solutions
in an infinite channel; however they are candidates in semi-infinite channels and in rectangles (regime 2). At very large frequencies the solutions grow or decay exponentially. The union of these regimes will be called a *mode unit.*

A second order model, $N = 2$, furnishes eight wavenumbers to each frequency, and its dispersion relation consists of two interlocking mode units, see *Fig. 4.20.* Thus there are now two branches with real, complex and imaginary k, respectively. The relative size of the mode units and their spatial position within the (k, σ)-coordinate system depend crucially upon the topography. The cylindrical surface of the first-order model degenerates to a smaller bell-shaped surface, i.e. $|k|$ now depends on the frequency. The second mode unit forms an outer shell, which here has the form of a cone. Physically possible solutions in an infinite channel exist in regime 1 for both mode units and in regime 2 only for the first mode unit. The qualitative shape of the dispersion relation for an N-th-order model can be guessed from these figures. The modulus $|k|$ is plotted for a third-order model in *Fig. 4.21* demonstrating clearly the addition of the next mode unit.

As an illustration, *Fig. 4.22* shows the temporal evolution of the streamfunction of the 1st TW-mode in equidistant temporal steps through half a period, in a third-order model $N = 3$. Panel a) shows the streamfunction surface, panel b) the streamlines, details being defined in the figure caption. From these a counterclockwise rotation of the gyres can easily be inferred.

4.4.2 Three types of TW's

We investigate the spectrum of TW's in a basin with an aspect ratio $r = 0.5$, a parabolic cross-section ($q = 2.0$), and a thalweg varying as a $(\sin)^2$-function according to (4.39). *Fig. 4.23* displays a selection of modes from the spectrum. It is apparent that in the period range from 43 to 147 hours (corresponding to 45° latitude) a large variety of qualitatively different eigenmodes can be detected. We distinguish three types of eigen modes.

Type-1 is the well known modal pattern described earlier in the analytical models of TW's in enclosed basins. It is akin to BALL's solution and therefore called BALL-*type.* Both, the linear ($\sigma = 0,153$) and the quadratic ($\sigma = 0.211$) BALL mode occur in the spectrum, and additional eigenmodes are identified as type 1. All exact models existing so far, have shown qualitatively

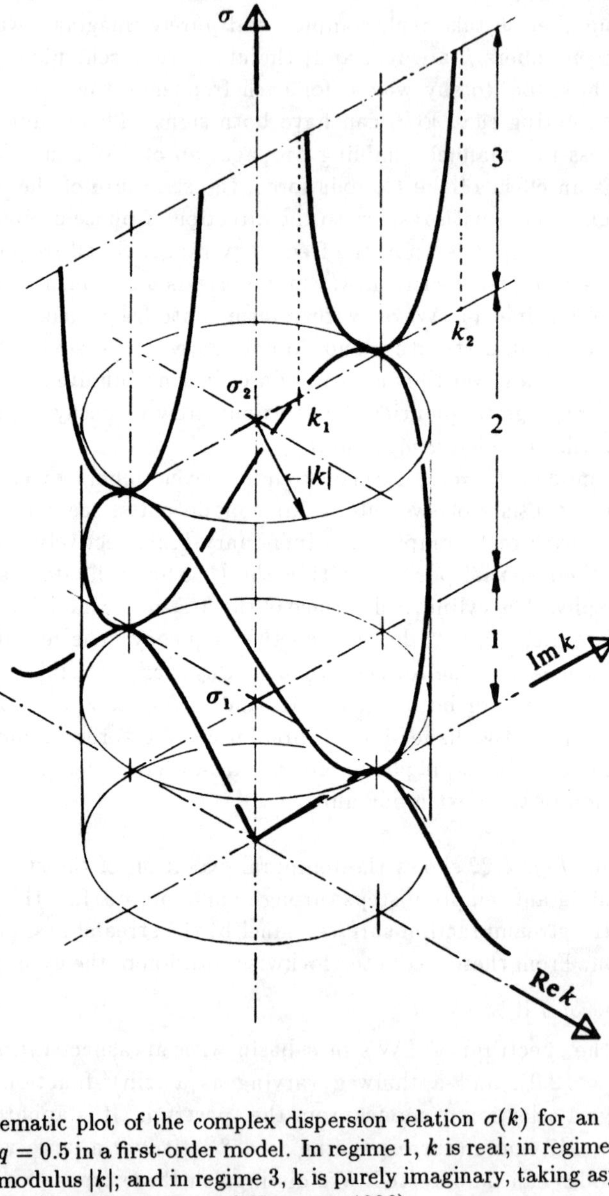

Figure 4.19: Schematic plot of the complex dispersion relation $\sigma(k)$ for an infinite channel with $\varepsilon = 0.05$ and $q = 0.5$ in a first-order model. In regime 1, k is real; in regime 2, it is complex with the constant modulus $|k|$; and in regime 3, k is purely imaginary, taking asymptotic values k_1 and k_2 for large σ (from STOCKER and HUTTER, 1986).

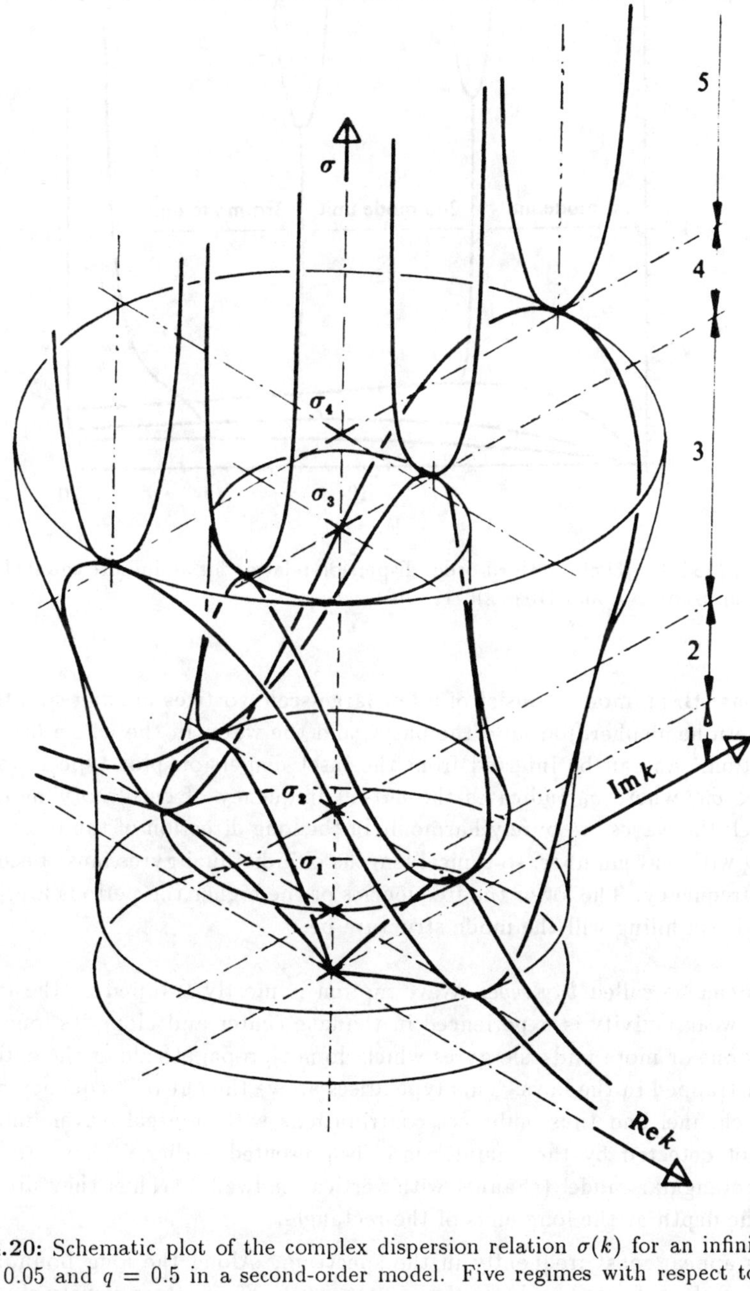

Figure 4.20: Schematic plot of the complex dispersion relation $\sigma(k)$ for an infinite channel with $\varepsilon = 0.05$ and $q = 0.5$ in a second-order model. Five regimes with respect to σ can be differentiated (from STOCKER and HUTTER, 1986).

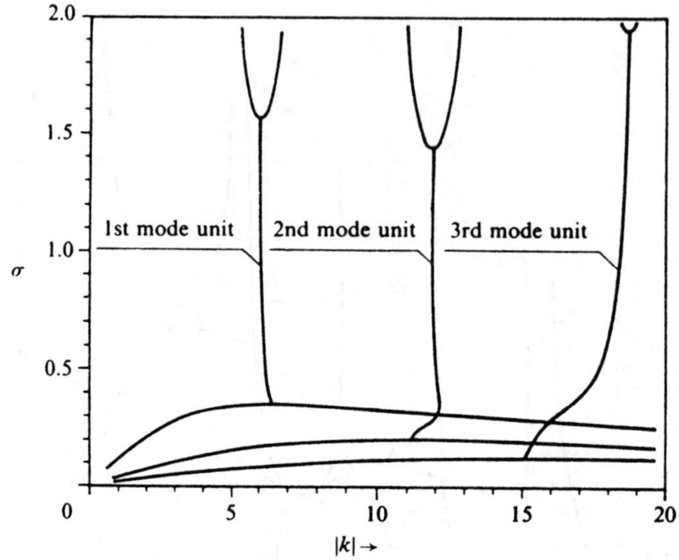

Figure 4.21: Modulus $|k|$ of the third-order dispersion relation for an infinite channel, $q = 0.5$; $\varepsilon = 0.05$, (from STOCKER and HUTTER, 1986).

similar solutions. BALL modes consist of a few large-scale vortices moving counterclockwise (on the Northern hemisphere) around the basin, and the water in the whole basin underlies this wave motion. As can be inferred from the inset dispersion plot type-1 waves have a frequency $\sigma < \sigma_0$, where σ_0 indicates the cut-off frequency of oscillatory motions in the infinite channel; the waves are purely harmonic in the long direction of the basin. Moreover, σ is increasing with wavenumber; so typical lengths of individual gyres must become smaller with growing frequency: The lower the frequencies or the higher the periods are, the simpler and the more basin filling will the mode structure be.

Type-2 waves can be called *bay-type*. Wave motion is mostly trapped at the long ends of the lake; very weak activity is experienced in the lake center and along its long sides. The patterns show one or more midscale gyres which do not propagate along the entire isobaths but are rather trapped in the bays. This type arises above the cut-off frequency σ_0 for TW's in an infinite channel and thus embraces contributions with nonreal wavenumbers. These waves were not detected by the analytic models presented earlier. They are also absent in a cruder rectangular model (channel with vertical endwalls). Thus they are due to the variation of the depth at the long ends of the rectangle.

Type-3 waves appear most frequently in the spectrum. Along the long boundaries of the basin a large number of small scale vortices is observed. The pattern is very similar to that found in straight infinite channels, see *Fig. 4.22*.
The modal structure and related properties of the different types can easily be explained

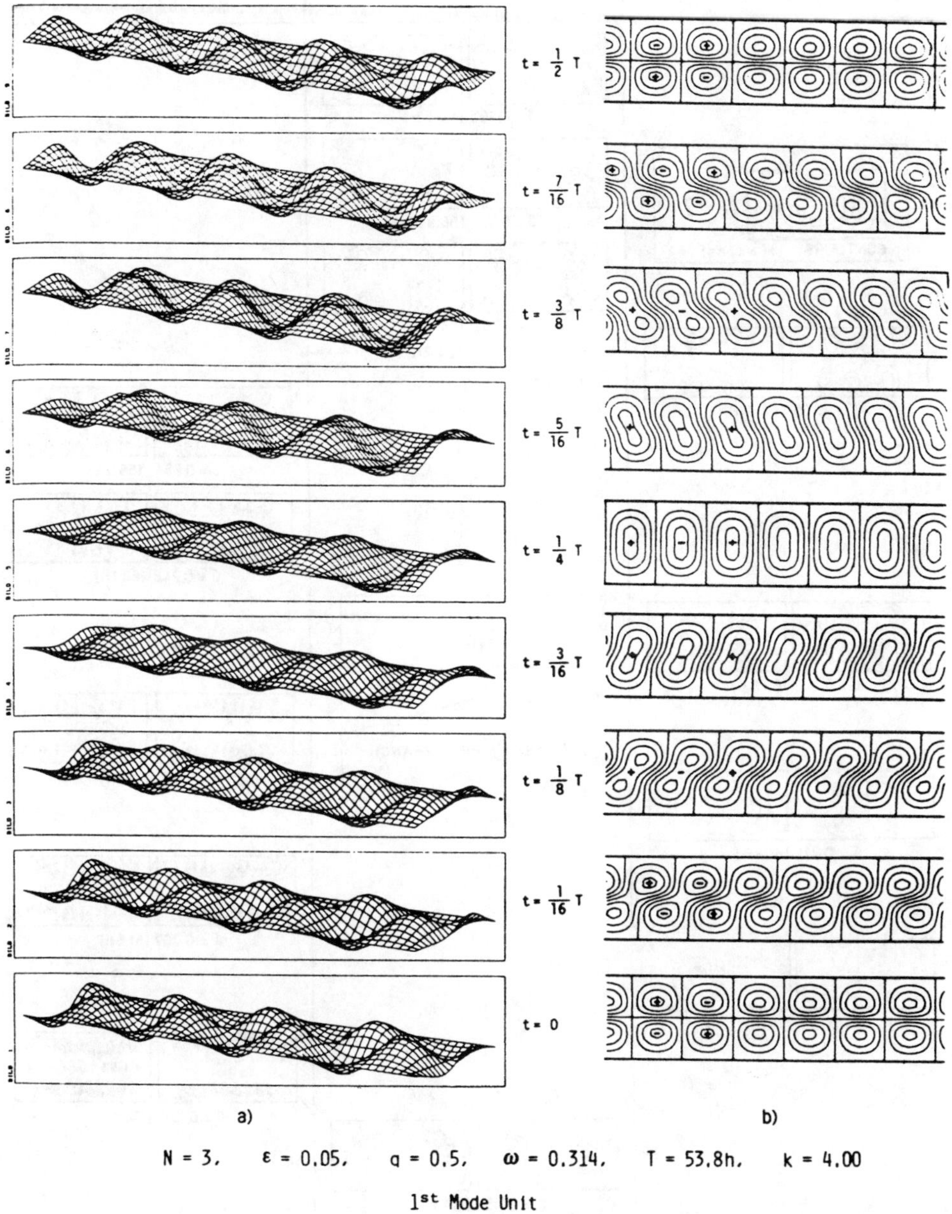

Figure 4.22: a) Time sequence of the streamfunction surface in steps of $1/16T$ in a channel $-1/2B \leq n \leq 1/2B$, $0 \leq s \leq 6Lr$ and aspect ratio $r = 1$. Note that the phase motion in the domains $n > 0$ and $n < 0$ is right bounded.
b) Time sequence of lines of constant Ψ relative to 90% of the maximum value at each time step. The cellular structure of cyclonic (+) and anticyclonic (-) vortices is clearly visible. (from STOCKER and HUTTER, 1987a)

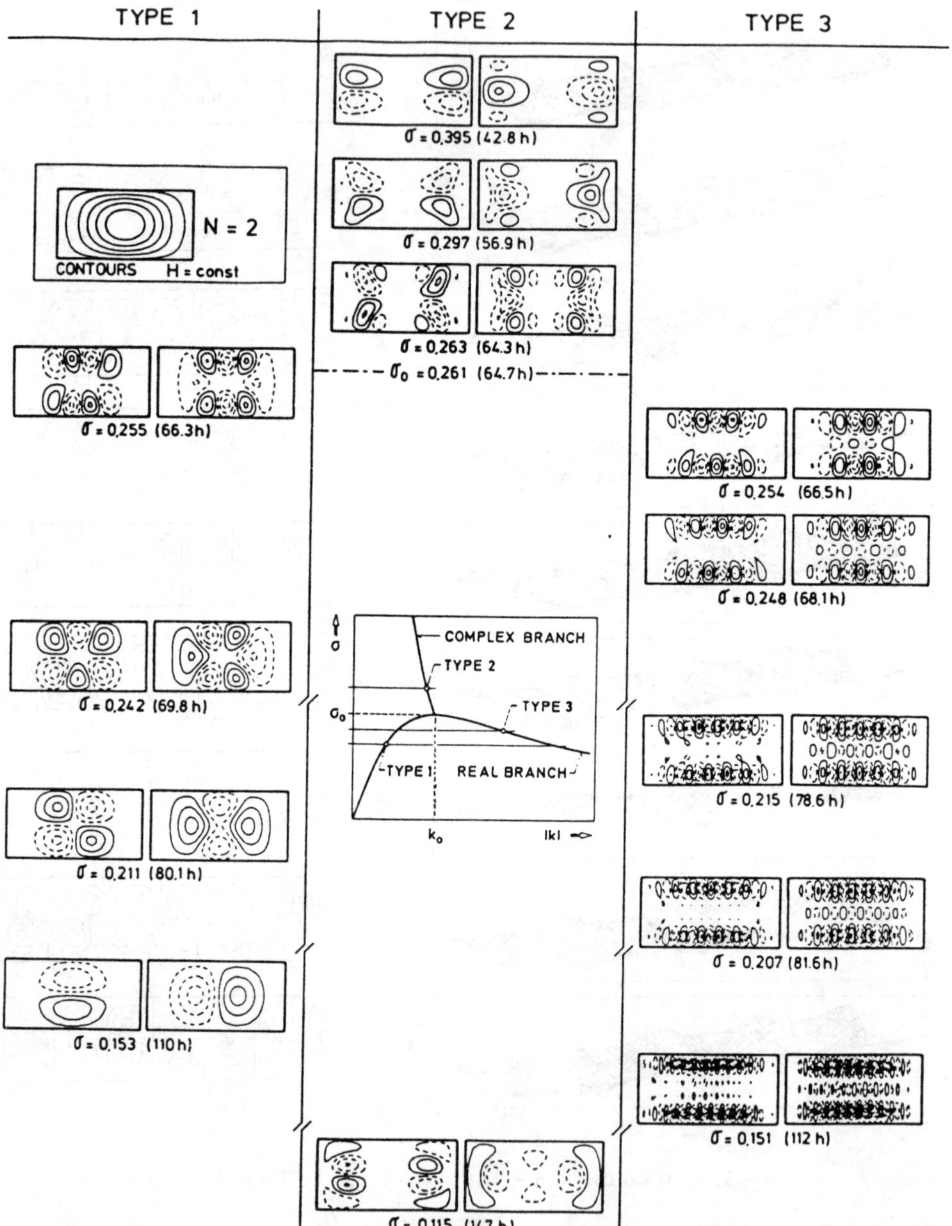

Figure 4.23: Selection through the spectrum of topographic wave modes in a second-order model. The eigenfrequencies increase towards the top of the figure. The contour lines of ψ are plotted for time $t = 0$ (left) and $t = T/4$ (right). Three types of solutions can be distinguished and cuts of the vertical lines indicate further modes not shown here. The inset displays the dispersion relation for the straight infinite channel (from STOCKER and HUTTER, 1987a).

with the help of the Rossby dispersion relation, see inset map in *Fig. 4.23*. For $\sigma < \sigma_0$ the wavenumbers are real and $\sigma(k)$ takes the familiar form with a critical point (σ_0, k_0) where the group velocity vanishes. Type-1 defines those (Ball) modes which become more structured as σ increases and for which the phase and energy propagate in the same direction. Type-3 defines the (channel) modes which are finely structured; the typical length of an individual gyre becomes here larger with growing frequency, and phase and energy propagate in opposite directions. For $\sigma > \sigma_0$ wavenumbers k are complex valued and waves are evanescent as one moves away from the ends.

The above analysis in based upon a second order model, $N = 2$, with two symmetric and two skew-symmetric basis functions. STOCKER and HUTTER (1986, 1987a,b) have also performed computations for $N = 3$ and $N = 4$ and demonstrate that the results will change quantitatively (frequencies will be slightly shifted and streamlines slightly changed) but qualitatively no new features were found. Moreover, STOCKER (1988) makes a careful analysis of wave reflection in a semi-infinite channel and shows that the reflection patterns at the wall of a semi-infinite gulf can be explained with the dispersion relation exhibiting the three mode types.

4.4.3 TW's in Lake of Lugano

Interpretation of the long-periodic signals of the order of 70 to 80 hours observed in Lake Lugano (Northern basin) was puzzling for a long time. MYSAK et al., (1985) propose that the wave motion is a fundamental BALL-mode. JOHNSON (1987b), using the same elliptical basin with a realistic aspect ratio, concludes, due to the strongly U-shaped transverse depth profiles, that it is rather a mode with a frequency close to the cut-off and a large along-axis wavenumber. A third interpretation arose when solving (4.37) by a finite element procedure, TRÖSCH (1984). In the period range 65 to 100 h rather small scale and bay-trapped wave patterns evolve. The nature and properties of the latter were not understood, but a qualitative comparison of these findings with the streamfunction patterns in the straight, semi-infinite channel brings clarification, see *Fig. 4.24*. For this comparison a highly idealized configuration is used. Effects of curvature of the thalweg are small (see, however, STOCKER and HUTTER, 1990) and need not be considered. Bay-modes are seen to be sustained independently in the three bays of Porlezza, Melide and Lugano. The basin is sufficiently elongated such that the coupling of the individual modes is negligible. At $T =$ 190 h a basin wide eigenmode occurs which can be related to a channel-mode with $\sigma < \sigma_0$.

This result is of importance particularly for elongated lakes with very steep transverse topography implying a small cutoff frequency. Let us estimate the frequency of the quadratic BALL-mode using the model of BALL (1965). With an aspect ratio of $1.5/17 = 0.088$ we obtain the estimate of $T = 360$ h. Recall that the transverse depth profiles are steeper than the parabolic of BALL's elliptic basin. Periods would have to be even longer. Measurements, however, clearly point at roughly 70 to 80 hours being close to, or rather above, the cutoff frequency of this channel-shaped basin. It is thus reasonable that the signal could be the trace of a bay-trapped topographic wave of one of the bays. In the idealized model only the topography parameters p and q determine the frequency of the bay-mode. Decreasing

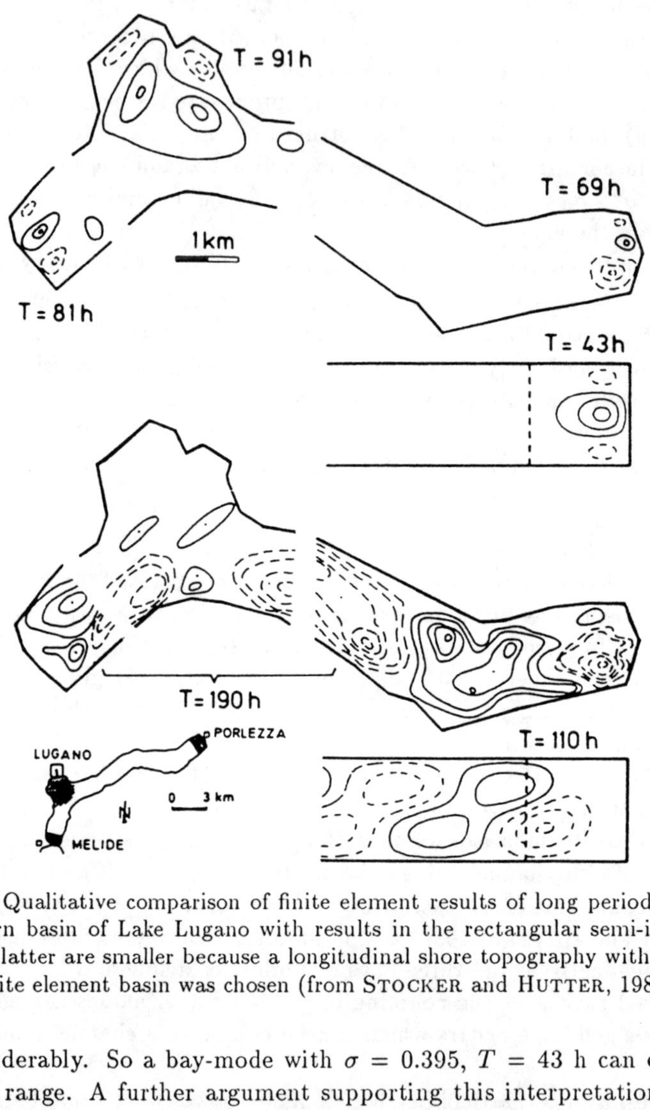

Figure 4.24: Qualitative comparison of finite element results of long periodic wave motion in the Northern basin of Lake Lugano with results in the rectangular semi-infinite channel. Periods of the latter are smaller because a longitudinal shore topography with a smaller slope than in the finite element basin was chosen (from STOCKER and HUTTER, 1987a).

p lowers σ considerably. So a bay-mode with $\sigma = 0.395$, $T = 43$ h can easily be brought into the correct range. A further argument supporting this interpretation is the fact that spectral peaks of temperature time series from Porlezza and Melide do not exactly coincide (see *Fig. 4.8*). This difference is likely be due to two independent bound modes at the two lake ends. Giving a final answer would however require data which would uncover the spatial structure much more clearly than can be done with the data at hand.

5. Closed Basins - Comparison with Observations

5.1 General remarks

The wave response in natural basins with variable bathymetry is a combination of gravity and topographic waves, so that in general the original shallow water equations in a stratified fluid ought to be solved. A separation of the vertical mode structure from the horizontal mode structure is however not possible, because this procedure requires constant depth over the entire domain. Because the three-dimensional problem with continuous stratification is too complex, the general procedure is to restrict considerations to simple approximations from which the barotropic and baroclinic mode response can be deduced. These approximations are one-, two- and, exceptionally, multi-layer models.

Analyses usually concentrate on separate studies of the gravity oscillations and, seldom, TW's. Clearly, such separate analyses are only justified when the periods of the first class and second class waves are sufficiently distant from one another, so that mutual interaction is unlikely. This is the case for small lakes for the barotropic and first baroclinic (two-layer model) motions, but is becoming questionable when higher order baroclinic gravity waves are excited.

As far as gravity waves are concerned, detailed studies that describe the behaviour of KELVIN and POINCARÉ waves are well known (MORTIMER, 1974) and have been treated earlier. According to these the rotation of the Earth can be ignored in the description of the structure of the wave motion if the ROSSBY radius of deformation

$$R = \begin{cases} \dfrac{1}{f}\sqrt{gH_{mean}}, & \text{barotropic mode}\ , \\ \dfrac{1}{f}\sqrt{g\dfrac{\Delta\rho}{\rho}H_{Eq}}, & \text{baroclinic mode} \end{cases} \tag{5.1}$$

is of the same order of magnitude as a typical horizontal extent (the width!) of the lake. For barotropic modes this requires basins of the dimension of the Great Lakes. For baroclinic motions a width greater than approximately 4 – 5 km makes rotational effects in KELVIN wave dynamics visible, and for higher order baroclinic oscillations a width of 1 – 2 km is sufficiently large. (These statements depend on density; our estimates are obtained with $\Delta\rho/\rho \approx 10^{-3}$). Unfortunately, signals of higher order baroclinic oscillations are generally rather weak and often covered by noise, or they are hardly separable from periods of topographic waves so that identification can be difficult.

5.2 A linear N-layer model

Except when large amplitude internal bores are analysed the linearized equations are sufficient to obtain the eigenmodes and their periods. Consider, *Fig. 5.1*, a baroclinic model with frozen-in stratification[33]; then,

$$\frac{\partial p}{\partial z} = -\rho_0(z) g \, . \tag{5.2}$$

If we approximate the real density structure by layers, integration of (5.2) followed by taking horizontal gradients yields

$$\begin{aligned} \frac{\nabla p^{i+1/2}}{\rho_0^{i+1/2}} &= \sum_{j=1}^{j=i} g_{i+1/2}^{j+1/2} \nabla(\zeta_j - \zeta_{j-1}) + g\nabla\zeta_0 \, , \\ g_{i+1/2}^{j+1/2} &= g\frac{\rho_0^{i+1/2} - \rho_0^{j-1/2}}{\rho_0^{i+1/2}} g \ll g \, . \end{aligned} \tag{5.3}$$

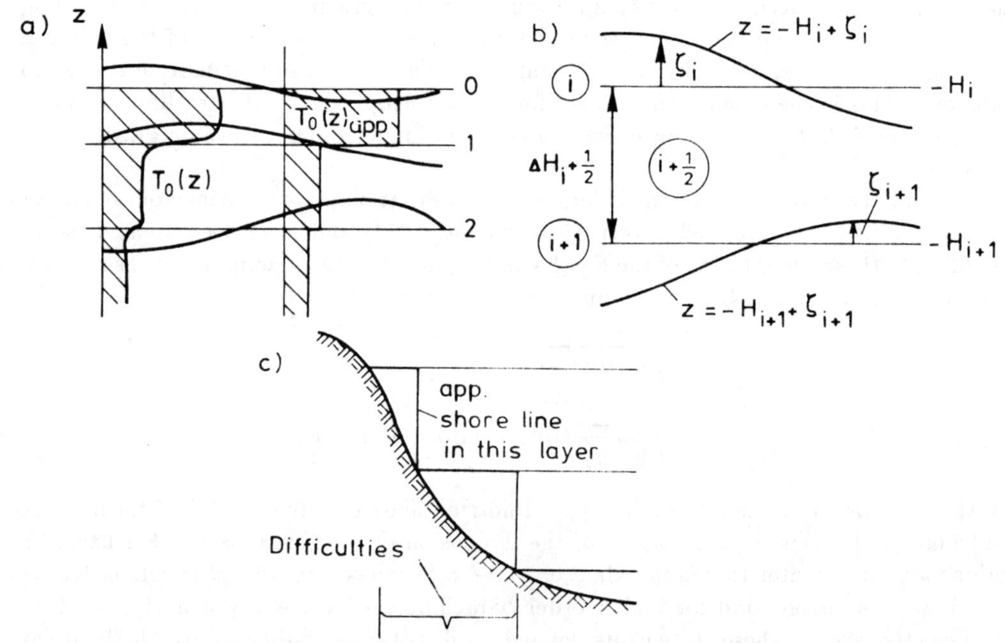

Figure 5.1: a) Ground state of the density profile $\rho_0(z)$ and its approximation by a step function with jumps at the infaces between the layers; shown are also in heavy lines the deformed interfaces. b) Definitions of layers, their variables in an approximate model according to equations (5.4). c) Definition of approximate shore lines in a layered model.

Interface surfaces are denoted by the index i, variables pertinent to the layer between the ith and $(i+1)^{\text{st}}$ interface are identified with the index $(i+\frac{1}{2})$.

33 By "frozen-in stratification" we mean a stratified ground state $\rho_0(z)$ which is not destroyed by the perturbed motion.

Similarly, vertical integration of the dynamic equations $(2.26)_{1,2,3}$ over deformed layers subjected to the kinematic boundary conditions $(2.33)_1$ and $(2.34)_1$ and imposing (5.3) leads to

$$\begin{aligned}
\frac{\partial \zeta_i}{\partial t} - \frac{\partial \zeta_{i+1}}{\partial t} + \nabla \cdot \left(\Delta H^{i+1/2} \cdot \mathbf{v}^{i+1/2} \right) &= 0 \, , \\
\frac{\partial u^{i+1/2}}{\partial t} - f v^{i+1/2} + \sum_{j=1}^{j=i} g_{i+1/2}^{j+1/2} \frac{\partial}{\partial x} (\zeta_j - \zeta_{j-1}) + g \frac{\partial \zeta_0}{\partial x} &= 0 \, , \\
\frac{\partial v^{i+1/2}}{\partial t} + f u^{i+1/2} + \sum_{j=1}^{j=i} g_{i+1/2}^{j+1/2} \frac{\partial}{\partial y} (\zeta_j - \zeta_{j-1}) + g \frac{\partial \zeta_0}{\partial y} &= 0 \, , \\
i = 0, 1, 2, \ldots, N-1, &
\end{aligned} \tag{5.4}$$

N being the number of layers. These equations incorporate both barotropic and baroclinic gravity waves as well as topographic ROSSBY waves. They must be complemented by the boundary condition of vanishing mass flow through the bottom surface. In the layer model this condition is fulfilled by requesting

$$\mathbf{v}^{i+1/2} \cdot \mathbf{N} = 0 \tag{5.5}$$

along the vertical "shore lines" of each layer as indicated in *Fig. 5.1.* This causes computational difficulties, and so (5.5) is generally satisfied at the inner-most approximate shore line. Obviously, this may make the computational domain considerably smaller than the real lake domain in areas of shallow water depths and may falsify the eigenperiods.

The number of layers that is generally used is two, one for the epilimnion and one for the hypolimnion, in rear exceptions three. MORTIMER (1952) achieved this way a better resolution of the metalimnion and could thus identify a higher order baroclinic mode in Windermere; SALVADÈ, ZAMBONI and BARBIERI (1988) traced such a higher baroclinic signal in the Northern basin of Lake of Lugano to the existence of a chemical density stratification at a depth much lower than the thermocline.

In a two-layer model, using a harmonic ansatz $\psi \rightarrow \psi e^{i\omega t}$ the above equations transform to an eigenvalue problem for the amplitude functions $\mathbf{v}_1$, $\mathbf{v}_2$, ζ_1, ζ_2, which in a real basin must be solved by numerical techniques. Computational problems for this system are, generally, enormous. It is, therefore advantageous to first eliminate the upper and lower layer velocities. The eigenvalue problem for the surface and the interface elevations ζ_1 and ζ_2 (*Fig. 3.12*) then becomes[34]

$$\begin{aligned}
&\left.\begin{aligned}
(\omega^2 - f^2)(\zeta_2 - \zeta_1) - \nabla \cdot (g H_1 \nabla \zeta_1) &= 0 \, , \\
(\omega^2 - f^2)\zeta_2 + \nabla \cdot \{ g h_2 \nabla \zeta_1 + g' H_2 \nabla (\zeta_2 - \zeta_1) \} &= 0 \, ,
\end{aligned}\right\} \quad (\mathbf{x}, \mathbf{y}) \in \mathcal{D}, \\
&\nabla \zeta_1 \cdot \mathbf{N} = 0, . \qquad \nabla \zeta_2 \cdot \mathbf{N} = 0, \qquad \text{on } (\mathbf{x}, \mathbf{y}) \in \partial \mathcal{D},
\end{aligned} \tag{5.6}$$

[34]We depart here from the notation in equations (5.4) and call the layer velocities $\mathbf{v}_1$ and $\mathbf{v}_2$ and the layer thicknesses H_1 and H_2, respectively; g' is the reduced gravity constant.

while the velocity amplitudes can be found a posteriori from

$$\begin{aligned}\mathbf{v}_1 &= \left(\omega^2 - f^2\right)^{-1} g\left(i\omega\nabla\zeta_1 - f\hat{\mathbf{k}}\times\nabla\zeta_1\right),\\ \mathbf{v}_2 &= \left(\omega^2 - f^2\right)^{-1}\left\{g\left(i\omega\nabla\zeta_1 - f\hat{\mathbf{k}}\times\nabla\zeta_1\right) + g'\left(i\omega\nabla\left(\zeta_2-\zeta_1\right) - f\hat{\mathbf{k}}\times\nabla\left(\zeta_2-\zeta_1\right)\right)\right\}.\end{aligned} \tag{5.7}$$

One can also deduce an eigenvalue problem for the velocities $\mathbf{v}_1$ and $\mathbf{v}_2$ instead. When barotropic and baroclinic gravity modes and TW modes have time scales which are far apart and basin dimensions are small in comparison to the internal ROSSBY radius, then the above equations may be solved with $f = 0$. This automatically eliminates all second class waves. Or one works with a two-layered equivalent depth model, in spite of the variable topography. In this case one starts from the equations

$$\frac{\partial\zeta_\alpha}{\partial t} + \nabla\cdot\mathbf{V}_\alpha = 0, \qquad \frac{\partial\mathbf{v}_\alpha}{\partial t} + f\hat{\mathbf{k}}\times\mathbf{V}_\alpha + gh_\alpha\nabla\zeta_\alpha = 0 \tag{5.8}$$

and may deduce the eigenvalue problem

$$\begin{aligned}\left(\omega^2 - f^2\right)\zeta_\alpha + \nabla\cdot\left(gh_\alpha\nabla\zeta_\alpha\right), &\qquad \text{in } \mathcal{D},\\ \nabla\zeta_\alpha\cdot\mathbf{N} = 0, &\qquad \text{on } \partial\mathcal{D},\end{aligned} \tag{5.9}$$

for the spatial part of the function ζ_α. The volume flux amplitude $\mathbf{V}_\alpha$ follows in a second step from

$$\mathbf{V}_\alpha = \frac{gh_\alpha}{\omega^2 - f^2}\left[i\omega\nabla\zeta_\alpha - f\hat{\mathbf{k}}\times\nabla\zeta_\alpha\right]. \tag{5.10}$$

For surface seiches equations (5.9) and (5.10) are exact; for internal gravity modes they have been shown to yield accurate thermocline excursion structures (BÄUERLE, 1981; HUTTER and SCHWAB, 1982). It has also successfully been used by KANARI (1975), HUTTER et al., (1983) STOCKER et al., (1987) and others.

Applying the rigid lid assumption to equation (5.4) ($\partial\zeta_1/\partial t$ ignored in equation $(5.4)_1$ for $i = 1$) filters out all surface gravity waves from the system. Adding the thus transformed mass balance equations yields in this case $\nabla\cdot(H_1\mathbf{v}_1 + H_2\mathbf{v}_2)$; thus, a streamfunction Ψ for the barotropic current exists such that

$$H_1\mathbf{v}_1 + H_2\mathbf{v}_2 = \hat{\mathbf{k}}\times\nabla\Psi .$$

Consequently , by invoking the rigid lid assumption it should be possible to deduce from (5.4) an eigenvalue problem involving Ψ and ζ_2, by which internal gravity seiches and barotropic and baroclinic topographic modes can be studied.

5.3 Seiches

A lake may be regarded as a dynamical system in which time series of external wind forces (and pressure fluctuations) are transferred into a spectral distribution of gravitational seiches (and topographic waves). Usually, the lowest few (~ 5) eigenmodes are excited to the extent that they can be identified through spectral analysis of time series of temperature, velocity and surface elevation recordings. These analyses are not so much important as models for the description of the water circulation; however they help to understand the physics of a system of otherwise rather obscure data. Seiche models are therefore primarily of diagnostic value.

Internal seiche, or more generally, internal wave analyses are also important for hydrobiological reasons. Since (i) the plankton and other micro-organisms tend to stay in the lower part of the thermocline, (ii) thermocline amplitudes are up to a few meters and (iii) internal wave periods are on the order of one to several days, the population dynamics of such species must be influenced by this hydrodynamical component if only because of the accompanied variation of the light intensity.

5.3.1 External, or surface seiches

These can be computed by the one-layer equation (5.8). In general, the Coriolis parameter may be assumed constant and often, namely in basins of the extent of 200 km or less the effects of the rotation of the Earth can be ignored. In large basins close to the Equator (within 3° latitude North or South, as e.g. for Lake Victoria or Lake Tanganyika) β-effects, i.e. a South-North variation of the Coriolis parameter should not be ignored. Neglecting the effects of the rotation of the Earth simplifies computations considerably; the matrix eigenvalue problem of the finite difference approximation of the one-layer equation (5.9) is not smaller than when $f \neq 0$, however, the TW-modes are excluded and only the gravity wave eigenvalues which are well separated must be determined. Identification of modes has proved to be much simpler[35]. We illustrate results for Lake of Lugano demonstrating a co-oscillation of the individual basins of a lake system, and Lake Superior illustrating the role of the rotation of the Earth.

Lake of Lugano, *Fig. 5.2*, lies at the Southern edge of the Swiss alps and consists of three basins. The North basin is L-shaped, possesses steep, nearly vertical shores and stretches from Porlezza to the glacial morene at Melide. Its length is 17 km, its width 1500 m and its mean and maximum depths are 170 and 288 m, respectively.

The South basin, essentially S-shaped, extends from Riva San Vitale to Agno. With its length of 19 km and its width of 1000 m it is more narrow and longer and with its 55 and 95 m mean and maximum depth it is shallower than the North basin. The appended lagoon of Ponte Tresa is fairly round, with a surface of 1 km^2 and mean and maximum depths of 33 and 55 m, respectively. The North and South basins are separated by a 50 m wide and 100 m long canal with a water depth of 5 to 6 m and a cross section of 200 m. A second

[35]It was mentioned in §4 when discussing topographic waves in enclosed basins that eigenvalues for these modes are clustered in a narrow band below the inertial frequency. This makes exact identification of particular modes extremely difficult. In particular cases one may not even be able to separate higher order baroclinic modes from topographic modes. Because the distribution of the frequencies of baroclinic gravity modes is less dense in general than that of TW's, elimination of the second class mode effects makes identification of baroclinic responses easier, let alone computations simpler and much faster.

canal, that connects the South basin with the lagoon at Ponte Tresa, is 500 m long and 40 to 50 m wide with a depth of 4 to 5 m. The only outflow of this lake system is at Ponte Tresa so that the mean flow must be from the North to the South basin and further to the lagoon at Ponte Tresa. Water can, however, also slosh forth and back between the three basins, and here we demonstrate that barotropic co-oscillations contribute to a significant exchange of water masses forth and back between the basins.

A 50×50 m mesh width was chosen for the finite difference approximation of the eigenvalue problem emerging from the one-layer equations. It resulted in a matrix eigenvalue problem with 20263 unknowns, of which the solution required special computational procedures, owing to its size. The canal at Melide was still only resolved by two elements this way.

The seiches of the individual basins were analyzed by HUTTER et al. (1982b). The first two modes of the isolated basins have periods of 13.8, 6.6 minutes (N-basin) and 27.6, 13.9 minutes (S-basin), respectively. The coupled system exhibits further, longer periods which are listed in *Table 5.1* (SALVADÈ and ZAMBONI, 1987). Perspective views of the surface elevations in the moment of maximum amplitudes are shown in *Fig. 5.3*. *Fig. 5.4* shows energy spectra of velocity-time series in the channel directions of current meters of position 8a (canal at Medile, *Fig. 5.4a*) and position L (canal at Lavena, *Fig. 5.4b*). Inspection of *Fig. 5.3* and *5.4* and *Table 5.1* reveals the following results.

- A variation of the depth in the channel at Melide affects only the fundamental mode with a period of app. 110 min. Depending on the water level this variation is also discernable in the measurements. The structure of the amplitudes of the surface elevations (*Fig. 5.3a*) shows that the currents in the South basin and the lagoon of Ponte Tresa are in phase and, together, out of phase with those in the North basin, causing substantial velocities only in the canal at Melide and somewhat smaller ones in the channel at Lavena. The computations have shown that a maximum amplitude of the surface elevation of only 2.5 mm gives rise to channel velocities as large as 10 cms^{-1}. This corresponds to a tremendous water mass transport through the channel.
- Mode 2 (*Fig. 5.3b*) represents the co-oscillation KPT of the South basin with the lagoon at Ponte Tresa; indeed, the oscillations in the two basins are out of phase, with significant velocities only in the channel at Lavena. A small co-oscillation in the North basin is discernable; however when computations are repeated with no flow through the Melide channel, no changes are observed in the period and the mode structure of the S-basin and the lagoon.
- Mode 3 is the fundamental seiche of the South basin as seen in *Fig. 5.3c*.

In large lakes or ocean basins (of the size of the Laurentian Great Lakes) rotation of the Earth will play a role, the criterion being that a typical horizontal distance is of the order of the external ROSSBY radius $R \sim 500$ km; in small lakes it can be ignored.

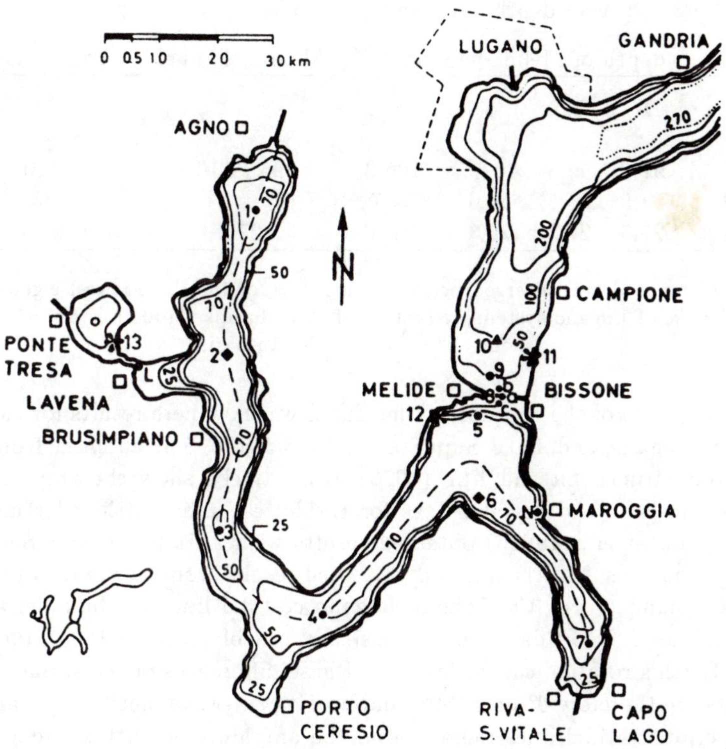

Figure 5.2: Bathymetric chart of the Lake of Lugano contoured in m and showing the entire Southern basin, part of the Northern basin and the lagoon at Ponte Tresa, and the connecting channels at Melide and Lavena. The entire lake system is shown in the inset. Symbols mark the positions of the following instruments:

- • Mooring with a current meter in the epi- and another in the hypolimnion and a thermistor chain encompassing the thermocline
- ◆ Meteorological buoys
- ▲ Positions with current meters, thermistor chains, and meteorological instruments
- ⋆ Seiche-limnigraphs

The dashed line in the Southern basin shows the thalweg of the latter. (from HUTTER, 1991)

Mode	Computed Periods [min] (mean depth 116,8 m)				Measured Periods [min]	
	depth of Melide channel [m]				Melide channel	Lavena channel
	6	5,5	5	4		
1	97,8	99,5	101,5	106,3	95 - 110	110
2	54,5	54,5	54,6	54,7		56
3	27,3	27,3	27,3	27,3	27,3	27,3

Table 5.1: Computed and observed periods in minutes for the three lowest eigenvalues of the seiches of the Lake of Lugano system, consisting of three basins. Mode 3 is also the fundamental mode of the isolated South basin.

To contrast the effects of the rotation of the Earth we show here results for the South basin of the Lake of Lugano and Lake Superior. *Fig. 5.5* and *5.6*, adopted from HUTTER et al., (1982b) and MORTIMER and FEE (1976), respectively, show the amphidromic system with co-range (dashed) and co-tidal lines for the lowest and fourth-order modes (Lake of Lugano). HUTTER et al., (1982b) obtained results with a finite element representation of the governing equations; MORTIMER and FEE used results from PLATZMAN (1972) based on KELVIN wave dynamics. For the Lake of Lugano co-tidal lines are bundled and elevations on two sides of the 90° co-tidal lines are basically out of phase (±180°), indicative of the fact that the Earth's rotation can be ignored. Phase differences in the surface amplitudes of any two points are therefore 0° or ±180°, and no KELVIN-wave mechanism can be observed. For Lake Superior, co-tidal lines radiate from the amphidrome with a wider spread; phase differences between elevations at two points vary according to this spreading and the wave travels counterclockwise around the amphidromic point with high amplitudes at the shores.

Limnigraphs erected along the shore at the points indicated by full triangles (*Fig. 5.5*) or arrows (*Fig. 5.6*) permitted identification of dominant oscillations from recorded time series using direct inspection by eye and spectral analysis. In *Fig. 5.7* energy spectra, coherence, and phase difference are shown for the limnigraph pair at the Lake of Lugano basin ends *Agno* and *Riva S. Vitale*. Evidently, energy maxima in the frequency spectrum coincide with the eigenperiods of the oscillations (27.8, 13.6, 9.4, 8.5 and 5.7 minutes for the five lowest modes). Coherence of those modes is high and phase differences are close to 0° and ±180°, depending on mode number, but consistent with expectation. Almost all these inferences can also be obtained from a comparative inspection of time series (*Fig. 5.8*). Such comparisons should be made whenever possible, because they provide easy access to the physical interpretation. For Lake Superior, *Fig. 5.6*, "observed" phases relative to 0° degree at the limnigraph position PI, shown in boxes, are compared with the computed phases outside the boxes; the two compare favourably as do the measured and the computed periods (7.89 h) and the periods and structure of higher order modes (the latter are not shown here).

There have been many more studies on surface seiches of which Lake of Lugano serves

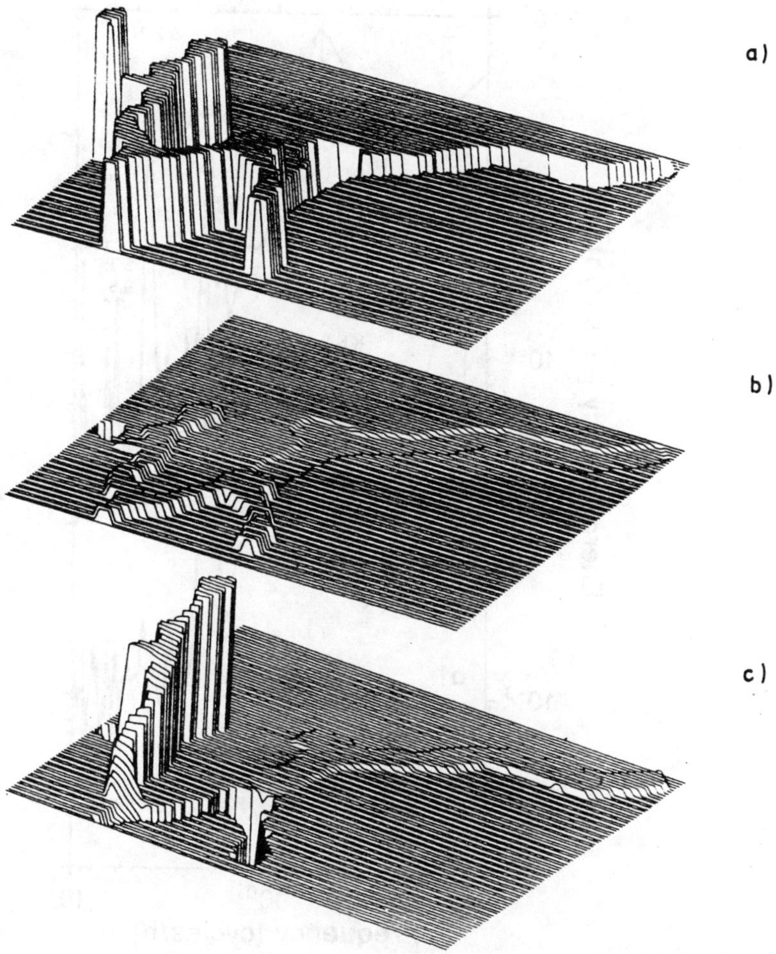

Figure 5.3: a) Perspective view of the surface elevation at the moment of largest amplitude of the coupled KNS oscillation of the Lake of Lugano with a period of app. 97 minutes. In the canal at Melide a nodal line arises, North and South basin (together with the Lagoon at Ponte Tresa) move out of phase. The largest amplitudes occur in the canal at Melide. b) same as a), but for the coupled KPT oscillation with a 54 minutes period. The largest velocities arise in the canal of Lavena. c) same as a), but for the fundamental mode of the South basin with a period of 27 minutes. The largest velocities arise in the middle of the South basin. For the definitions of KNS,KPT see figure legend 5.4. (from SALVADÈ and ZAMBONI, 1987, HUTTER, 1987b, with changes).

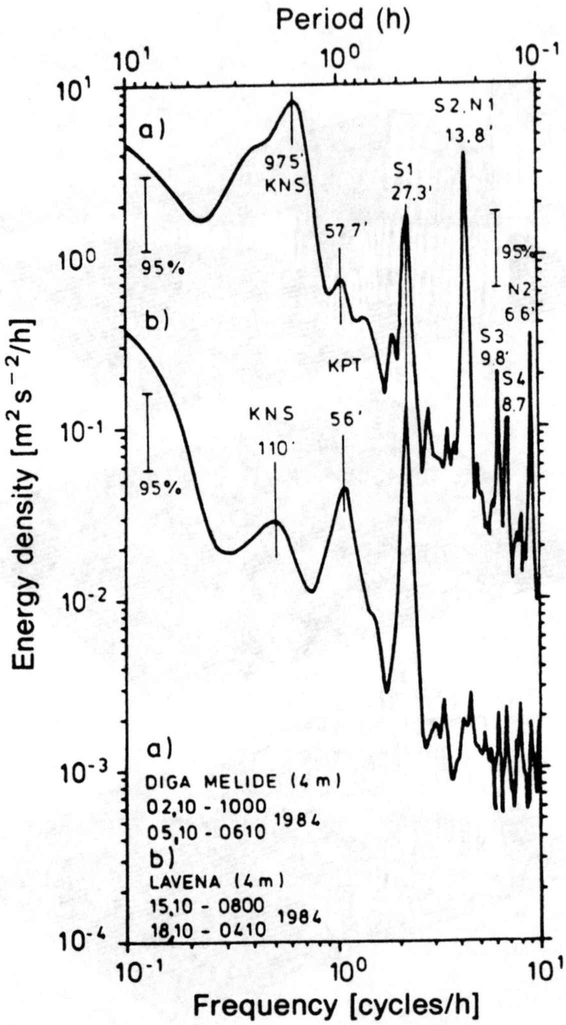

Figure 5.4: a) Energy spectrum of the current component at 4 m depth in the direction of the channel axis in position 8a at Melide for a 68 h-time series. Relative maxima are indicated with their periods (in min). S1 - S4 are the periods for the South basin alone; N1, N2 are those of the North basin alone. KNS denotes the period of the coupled oscillation of the North and South basin, KPT that of the lagoon at Ponte Tresa and the South basin. For both cases the interval of the time series is indicated. b) same as a) but for the current component in the long direction of the canal at Lavena. For both cases the duration of the time series is indicated (from HUTTER, 1987b).

only as an example. TISON and TISON (1969) collect the literature worldwide as of 1969. PLATZMAN and RAO (1964 a, b) study the seiches in Lake Erie, HAMBLIN and HOLLAN (1978), MÜHLEISEN and KURTH (1978), HOLLAN (1979, 1980) and HOLLAN et al. (1980) those of Lake Constance, HUTTER et al. (1982a) those of Lake of Zürich, MORTIMER and FEE (1976) those of Lakes Michigan and Superior, and SCHWAB and RAO (1977) those of Lake Huron. These works, typically, apply the full two-dimensional equations. Simpler, channel type, models have been developed by RAGGIO and HUTTER (1982 a, b, c) and HUTTER and RAGGIO (1982). Their simplest form, known as the Chrystal equations has been the basis for computations in the pre-computer time, see TISON and TISON (1969) and is still used by limnologists for smaller elongated lakes.

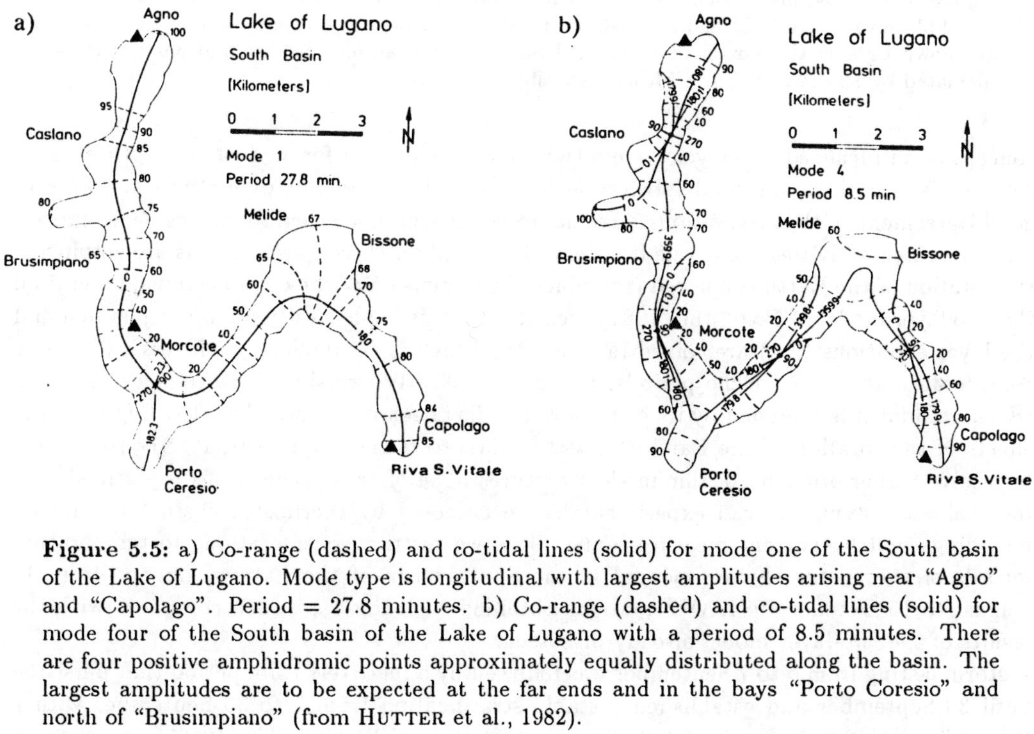

Figure 5.5: a) Co-range (dashed) and co-tidal lines (solid) for mode one of the South basin of the Lake of Lugano. Mode type is longitudinal with largest amplitudes arising near "Agno" and "Capolago". Period = 27.8 minutes. b) Co-range (dashed) and co-tidal lines (solid) for mode four of the South basin of the Lake of Lugano with a period of 8.5 minutes. There are four positive amphidromic points approximately equally distributed along the basin. The largest amplitudes are to be expected at the far ends and in the bays "Porto Coresio" and north of "Brusimpiano" (from HUTTER et al., 1982).

5.3.2 Internal seiches

Their analysis is important, both from a physical and biological point of view because internal waves form persistent motions to which the plankton population dynamics will respond. Rough analyses can be performed with a one layer equivalent-depth Defant-type channel model (MORTIMER, 1979; LEMMIN and MORTIMER, 1986); it ignores the rotation of the Earth, treats the lake as long and small and imposes CHARNEY's (1955) equivalent depth

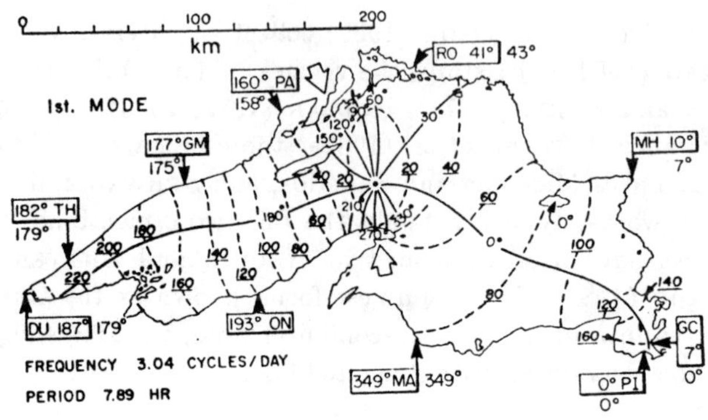

Figure 5.6: Phase progression of the first longitudinal surface mode in Lake Superior. Observed phases relative to 0° at PI, are shown in boxes and the phases computed by PLATZMAN are shown outside the boxes. The co-tidal lines and the amplitude distributions are those calculated by PLATZMAN (from MORTIMER and FEE, 1976).

concept of an idealized two-layer system (which is only correct for a constant depth model). Nevertheless, periods and mode structures for first order baroclinic oscillations are often in good agreement with observations. For more refined model calculations, using equations (5.4), research activities are at the frontiers. Two, seldom three, layer models are employed, the rotation of the Earth is ignored (to reduce the dimension of the computations, exception: the two layer model of BÄUERLE, 1981) even though this is physically often not justified and the layer equations (5.4) are solved in a domain which is common to all layers. This may considerably affect the seiche periods, as we shall shortly explain. A three-layer model in which each layer is bounded by its own shore line has been developed, but it still ignores the effects of the rotation of the Earth (SALVADÈ, ZAMBONI and BARBIERI, 1988). To be able to resolve higher order baroclinic mode structures none of these neglections is justified.
Internal wave dynamics can experimentally be detected by thermistor chain temperature recordings and by current measurements. Here we restrict considerations to temperature records of thermistor chains, moored in the South basin of the Lake of Lugano from 15 August to 15 October 1984. Observations obtained with these are then compared with the results of the two-layer model already discussed.
A storm lasting from 5 to 7 September was followed by a relatively calm period that persisted until 30 September and established a stable stratification from 7 to 30 September with a themocline at 12.2 m depth and a mean temperature of 12.3°C. The 12°C-isotherm now represents thermocline motion; and the 8°C-isotherm correspond to the upper edge of the hypolimnion. A detailed data analysis of poststorm motions was carried out by STOCKER et al. (1986). Here we restrict ourselves to a summary of the main features. *Figs. 5.9* and *5.10* yield the following inferences:

- The most conspicuous event in *Fig. 5.9* is the storm on 5 to 6 September, which was followed by a long interval of fairly calm weather. At positions 2 and 6 no daily wind

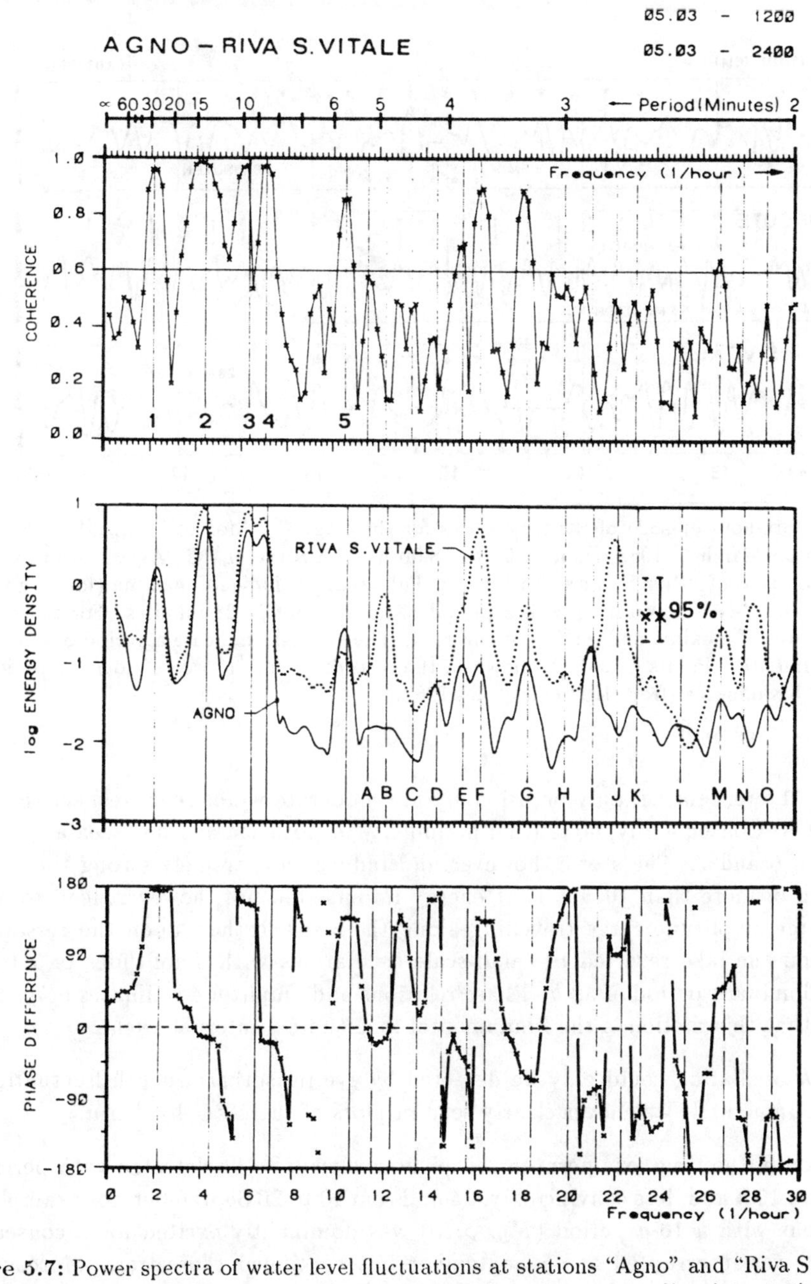

Figure 5.7: Power spectra of water level fluctuations at stations "Agno" and "Riva S. Vitale" (middle) for a twelve-hour episode on March 5, 1982, and corresponding interstation phase difference (bottom) and coherence (top) (from HUTTER et al., 1982b).

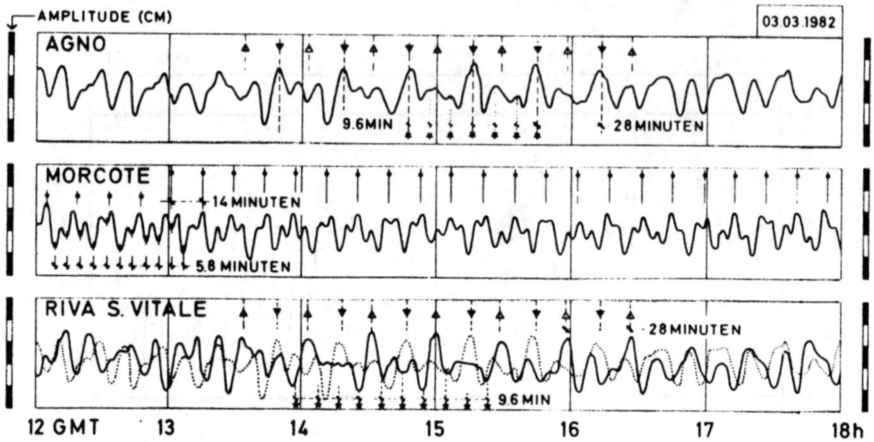

Figure 5.8: Six-hour episode of the time series for the "Agno", "Morcote" and "Riva S. Vitale" gauges on March 3, 1982, from 12.00 to 18.00 h. The curve of the "Agno" recording is superposed on that of "Riva S. Vitale" (dotted). Full triangles indicate maxima (for "Agno") and minima (for "Riva S. Vitale") with a period of approximately 28 minutes. For the open triangle this role of maxima and minima is reversed. The asterisks indicate a period of approximately 9.6 minutes. Points (· and \), shown in the recordings for "Morcote" identify periods with 14 and 5.8 minutes (from HUTTER et al., 1982b).

cycle can be recognized; only occasionally are moderate winds seen, respectively from NW and N. Consequently, no diurnal periodicity in isotherm depth is seen at positions 1, 2, 3, 4, 6 and 7. The storm, however, did induce an extremely strong thermocline excursion of more than 10 m. That energy impulse did not, however, lead to a non-linear large-amplitude surge travelling along the lake. Rather, upon the cessation of the storm, the lake reverted to small-scale oscillations with amplitudes of 1 to 2 m and predominant period of 38 h. Phase relations and amplitude estimates (*Fig. 5.12a*) suggest that this motion is the fundamental longitudinal internal seiche.

- The 38 h oscillation could only be detected by eye in isotherm depth fluctuations in the hypolimnion. It was never clearly seen in plots of the 12°C-Isotherm.

- Expansion of the time axis permits conspicuous signals to be detected with periods of 16, 12 and 10 h and, less convincingly, 24 h. From 10 to 20 September, for example, the oscillations with a 16-h period (*Fig. 5.10*) was dominantly excited for 5 consecutive days. It was interrupted at 15 September but continued thereafter. Moreover, at positions 2 and 5, brief episodes can be seen with 10- and 24-h periods. Alternatively, from 21 September until 1 October (*Fig. 5.11*), oscillations with a 12-h period prevailed; and an episode with a 10-h period on 25/26 September was relatively short. Further scrutiny between 17 and 24 September (not shown here) discloses a predominance of the 10-h oscillation.

Estimates of mean amplitudes and their variances are summarized in *Fig. 5.12* , *b*, *c* and *d*. Evidently, from *b*), with two nodal lines near positions 2 and 5 and 6, a second mode structure can be inferred. For the two episodes of 12$-h$ oscillation in *c*), third mode behaviour is suggested. There are three nodal lines; but the inferred position of the middle node is different in the two episodes. The amplitude distribution constructed for the 10-h oscillation again differs for the two events in *d*). There, an internal seiche of order four is the most likely interpretation.

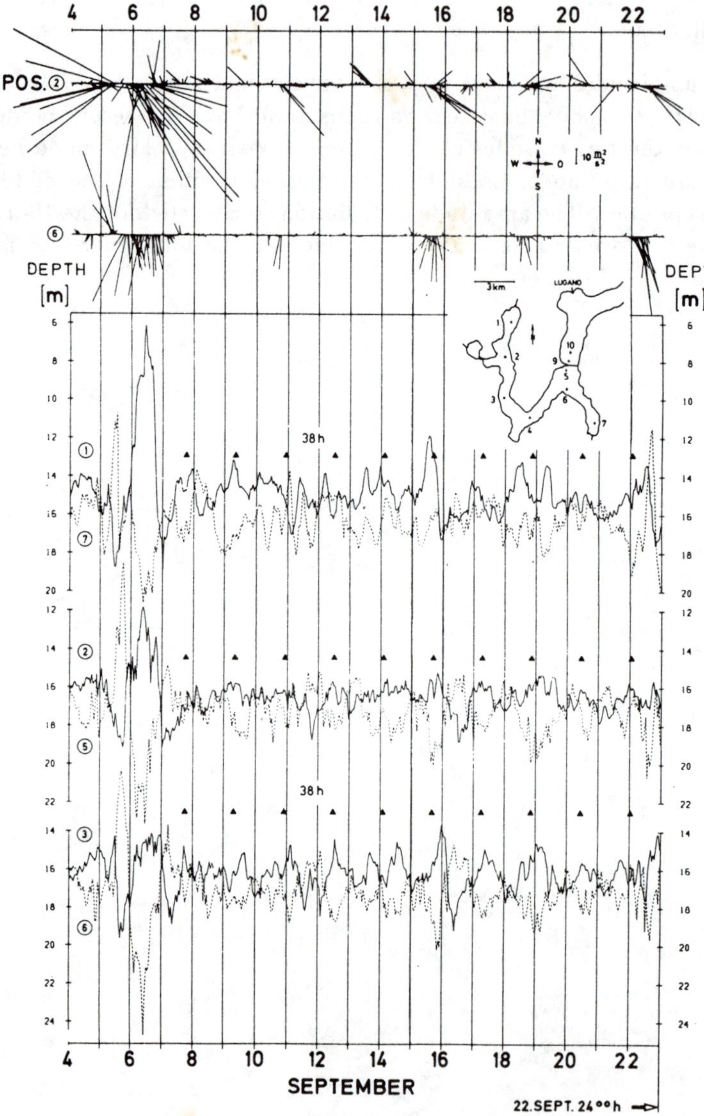

Figure 5.9: Wind speed squared $\left((u^2+v^2)^{\frac{1}{2}}(u,v)\right.$ in m^2s^{-2} as vector stick plots at positions 2,6) and fluctuations in depth of the $8°C$-isotherm at six positions (see inset) in the Southern basin, 4 to 22 September, 1984. Wind vectors and isotherm depths are hourly averaged; and the latter are plotted for the position pairs (1,7), (2,5) and (3,6). Note the dramatic storm of 5/6 September which generated very large thermoclinic excursions, which did not generate a non-linear surge. Oscillations with a period of $38h$ are marked by full triangles (from STOCKER et al., 1987).

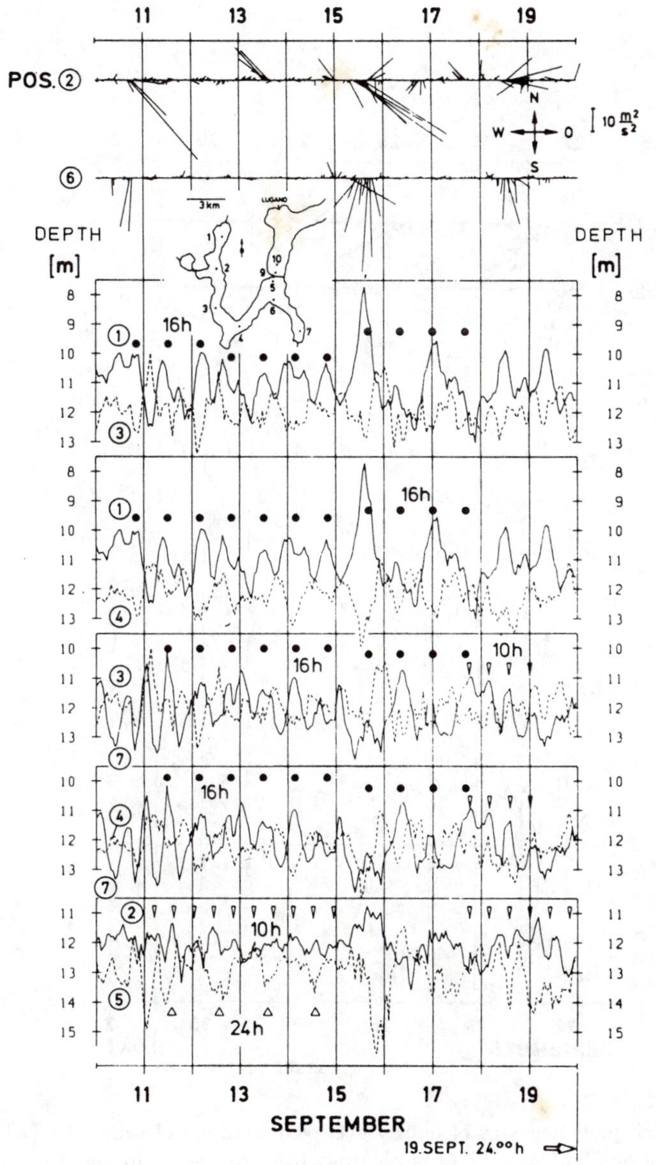

Figure 5.10: Stick plot diagrams of hourly averaged wind speed squared ($(u^2+v^2)^{\frac{1}{2}}(u,v)$ in m^2s^{-2} at positions 2,6) and $12°C$-isotherm depth fluctuations in the Southern basin, 10 to 20 September, 1984, plotted for the station pairs (1,3), (1,4), (3,7), (4.7) and (2,5). The dominant periodicities during oscillatory episodes are marked by symbols as in *Fig. 5.7* (from STOCKER et al., 1987).

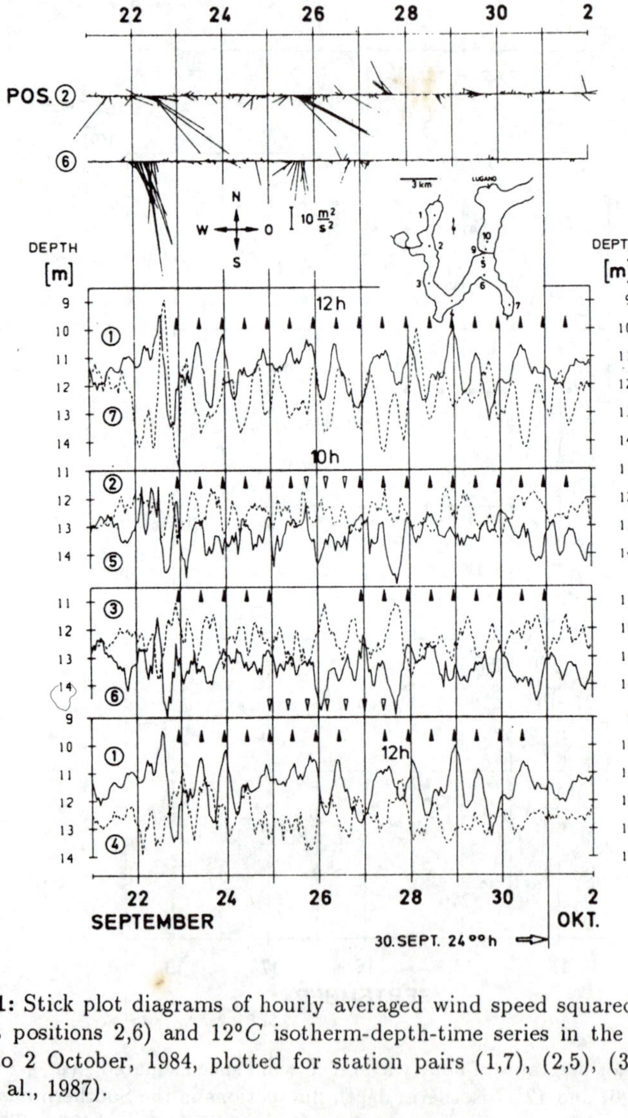

Figure 5.11: Stick plot diagrams of hourly averaged wind speed squared ($(u^2+v^2)^{\frac{1}{2}}(u,v)$ in m^2s^{-2} at positions 2,6) and $12°C$ isotherm-depth-time series in the Southern basin, 21 September to 2 October, 1984, plotted for station pairs (1,7), (2,5), (3,6) and (1,4) (from STOCKER et al., 1987).

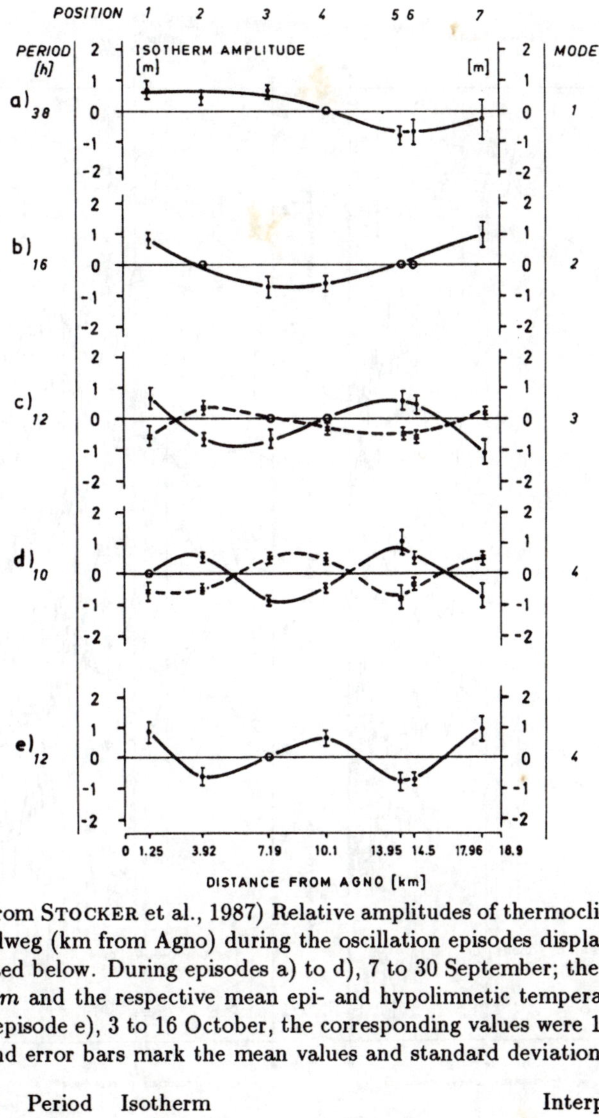

Figure 5.12: (from STOCKER et al., 1987) Relative amplitudes of thermocline deflection along the unrolled thalweg (km from Agno) during the oscillation episodes displayed in *Figs. 5.9* to *5.11* and tabulated below. During episodes a) to d), 7 to 30 September; the mean thermocline depth was 12, 2 m and the respective mean epi- and hypolimnetic temperatures were 16.4°C and 6.2°C. For episode e), 3 to 16 October, the corresponding values were 14.5 m, 13.9°C and 6.0°C. Points and error bars mark the mean values and standard deviations:

Graph	Period [h]	Isotherm [°C		Interpretation
a)	38	8	7 - 21 Sept.	mode 1
b)	16	12	10 - 14 Sept.	mode 2
c)	12	12	22 - 30 Sept. 8 - 10 Sept. (dashed)	mode 3
d)	10	8 12	24 - 27 Sept. 17 - 21 Sept. (dashed)	mode 4
e)	12	12	6 - 12 Oct.	mode 4

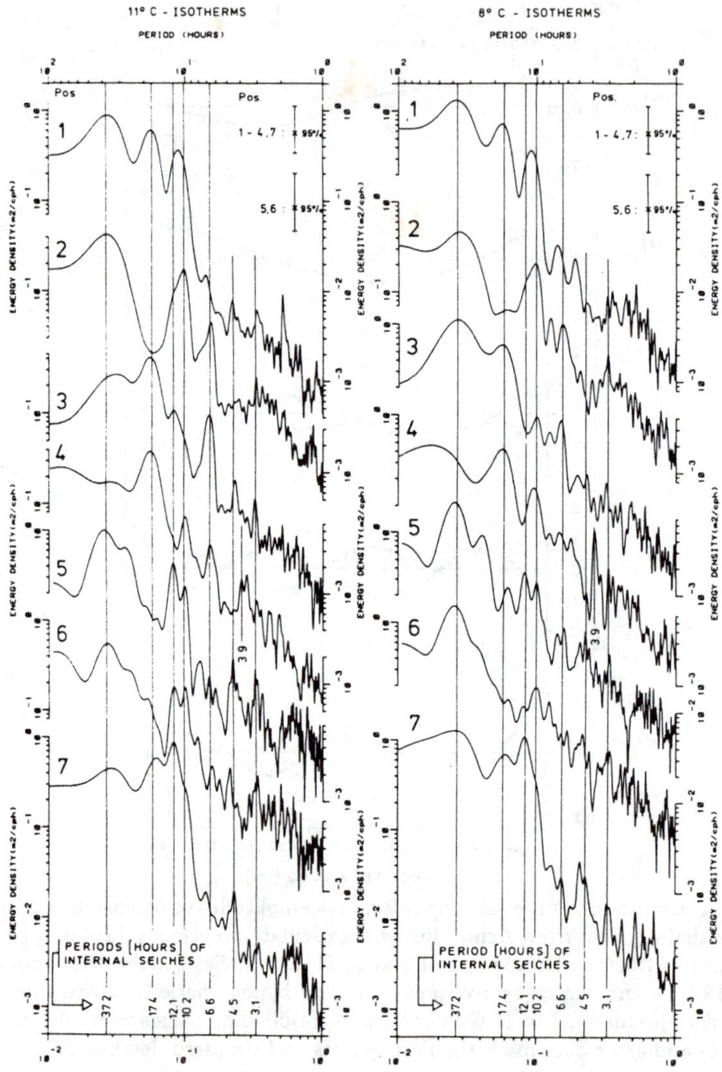

Figure 5.13: Energy spectra of fluctuations of 11°C and 8°C isotherm depth at positions 1 to 7 for the period 7 September 00.00 h to 2 October 00.00 h, 1984. Log energy density (m^2/cph) is plotted against log frequency (cph) or period (h). Vertical lines identify the periods of significant peaks. Confidence intervals are displayed at upper right (from STOCKER et al., 1987).

Mode number	Period [h]	Accuracy [h]
1	37,2	$\pm 1,3$
2	17,4	$\pm 0,7$
3	12,1	$\pm 0,2$
4	10,2	$\pm 0,2$
5	6,6	$\pm 0,3$
?	4,5	$\pm 0,2$
?	3,9	$\pm 0,2$
?	3,1	$\pm 0,2$

Table 5.2: Periods where spectral peaks of the isotherm depth energy density spectra arise in *Fig. 5.13*, their accuracy, and associated mode numbers.

Harmonic analysis:

Based on inspection alone the foregoing analysis is approximate; but energy spectra demonstrate that the deductions were correct.

In the left- and right-hand portions at *Fig. 5.13* we respectively present the energy spectra for the 11°C- and the 8°C-isotherm-depth fluctuations during the middle interval 7 September to 2 October. The main features of the applied spectral analysis are elimination of a linear trend, tapering with a "split cosine bell window" (10 % on each side) and smoothening of the periodogram with a modified Daniell procedure (BLOOMFIELD, 1976). With the thermocline at a mean depth of 12 m, the selected isotherms lie respectively about 1 and 4 m below it. In *Fig. 5.13* log energy density (m^2/cph) is plotted against log frequency (cph) or period (h). Vertical lines, centered on peaks rising above background "noise", are assigned periods and mode numbers as listed in *Table 5.2*. Mode number identification, from the model analysis described below, was also confirmed by cross-spectral analysis, but only up to the 5th mode. At higher mode orders the interseries coherence was too low. Furthermore, 7 records from moorings permit at most 6 nodes to be located.

The two-layered variable-depth model:

In order to model the internal free oscillations in the South basin mathematically, the stratification is approximated by two immiscible layers of fluid with densities ρ_1 and ρ_2 ($\rho_1 < \rho_2$). The model assumes that the lake domain is bordered by vertical walls such that the entire domain, where the two-layer model applies, consists of both, an epilimnion layer and a hypolimnion layer. The boundary of this domain is the intersection of the topography and the thermocline (the depth of discontinuity of the two layers) and will be referred to as "thermocline shore". Introduction of this "virtual" vertical wall ascertains that everywhere inside this shoreline two layers do exist. Clearly, the domain of the two-layered variable depth model (TVD-model) is smaller than the actual lake domain. This will make the periods of the internal gravity seiches somewhat smaller than if the full lake domain could be used, but for deep intermontane lakes the effects are negligible.

Interpretation:

The stratification of the lake in the TVD-model is described by the thermocline depth and the relative density difference $\varepsilon = (\rho_2 - \rho_1)/\rho_2$. From the temperature and resistivity measurements of 13 positions within the lake, taken on 24 September and thus representative of the period from 7 to 30 September the relative density difference was determined once via mean two-layer temperatures and once via mean two layer densities (*Fig. 5.14*). BÜHRER and AMBÜHL's (1975) equation of state was used. The two values for ε are

$$\varepsilon_I = 1{,}0184 \times 10^{-3}, \qquad \varepsilon_{II} = 1{,}168 \times 10^{-3},$$

and the interface of the two layers is at 12 m for both cases. *Table 5.3* assembles the computed periods of the 10 lowest modes ($n = 1, 2, \ldots, 10$) and compares these with our earlier estimates from inspection of the isotherm depth plots. The computed and observed periods show satisfactory agreement (maximum deviations of 15 %, mode 2, $\varepsilon = \varepsilon_I$). Interestingly, the computation tends to overestimate the periods when $\varepsilon = \varepsilon_I$, but does not consistently underestimate them when $\varepsilon = \varepsilon_{II} > \varepsilon_I$. An estimate of the periods, which the model would predict at different stratification, is given by the following approximate formula (see HORN et al., 1986)

$$T_I^2 \varepsilon_I h_I^* = T_{II}^2 \varepsilon_{II} h_{II}^* ,$$

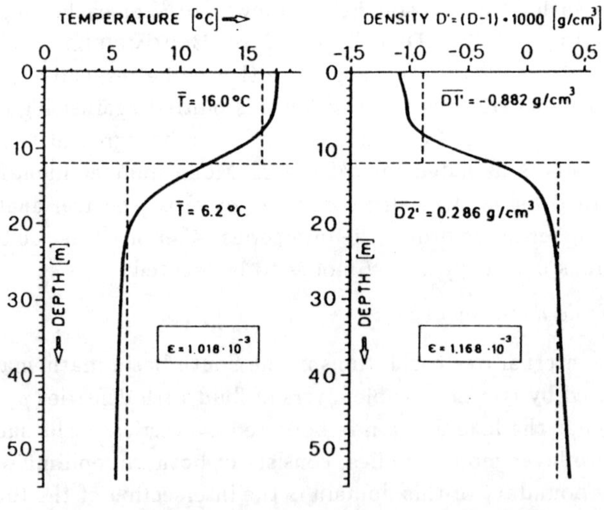

Figure 5.14: a) The temperature profile of 24 September, 1984, determined as the mean of profiles taken at 13 different positions in the South basin, and the two-layer approximation derived from it.
(b) Mean density profile of 21 September, 1984, determined from temperature and salinity profiles, and two-layer approximation derived from it (from STOCKER et al., 1987).

in which h_I^* and h_{II}^* are the equivalent depths in the two different stratifications, i.e.

$$h_I^* = \sqrt{\frac{h_1 h_2}{h_1 + h_2}}_I .$$

The reader can easily check that the computed periods in *Table 5.3* satisfy the above relation.

Mode	Computed TVD-model		Observed from isotherm depth time series	
n	$\varepsilon_I = 1{,}0184 \cdot 10^{-3}$	$\varepsilon_{II} = 1{,}169 \cdot 10^{-3}$	eye-fitted	spectral analysis
	T[h]	T[h]	12°, 8°(⋆); T[h]	11°, 8°(⋆); T[h]
1	37,3	34,8	38	37,2 (± 1,3)
2	19,2	18,0	16	17,4 (± 0,7)
3	12,8	11,9	12	12,1 (± 0,2)
4	11,5	10,8	10	10,2 (± 0,2)
5	7,2	6,7		6,6 (± 0,3)
6	6,3	5,9		
7	5,2	4,9		
8	5,0	4,7		4,5+ (± 0,2)
9	4,4	4,1		3,9+ (± 0,2)
10	4,2	4,0		3,1+ (± 0,2)

Table 5.3: Computed and measured periods of the internal seiches in stratified Southern Lake of Lugano (7 to 30 September).
(⋆) Indicates chosen isotherm for the isotherm-depth-time series.
(+) Not classified

The TVD model also permits distribution of interface displacement and near epi- and hypolimnetic velocities to be displayed for each mode. Displacement amplitudes for the first four modes, scaled in percent of the maximum amplitude in each case, are presented in *Fig. 5.15*, in which the transitions between positive (solid) and negative (dotted) contours correspond to nodes (heavy lines). "Observed" nodal lines, inferred from the above-described analysis of isotherm-depth fluctuations, are shown as heavy dashed lines. Given the uncertainty of the location of the latter, we must regard the agreement as satisfactory. The figure further discloses that the maximum deflections occur at Agno (modes 1 and 2) and at Riva S. Vitale (modes 3 and 4), but it can also be shown that, for mode 5, the maximum arises at a mid-lake position near Brusimpiano (*Fig. 5.2*). Along the shore in *Fig. 5.15* we also show in pairs the phases of the 11°C- and 8°C-isotherm-depth-time series at the respective mode period, except at those positions where amplitudes were too small to identify a phase. The expected jump of 180° when nodal lines are crossed is reasonably well achieved in all modes up to order five.

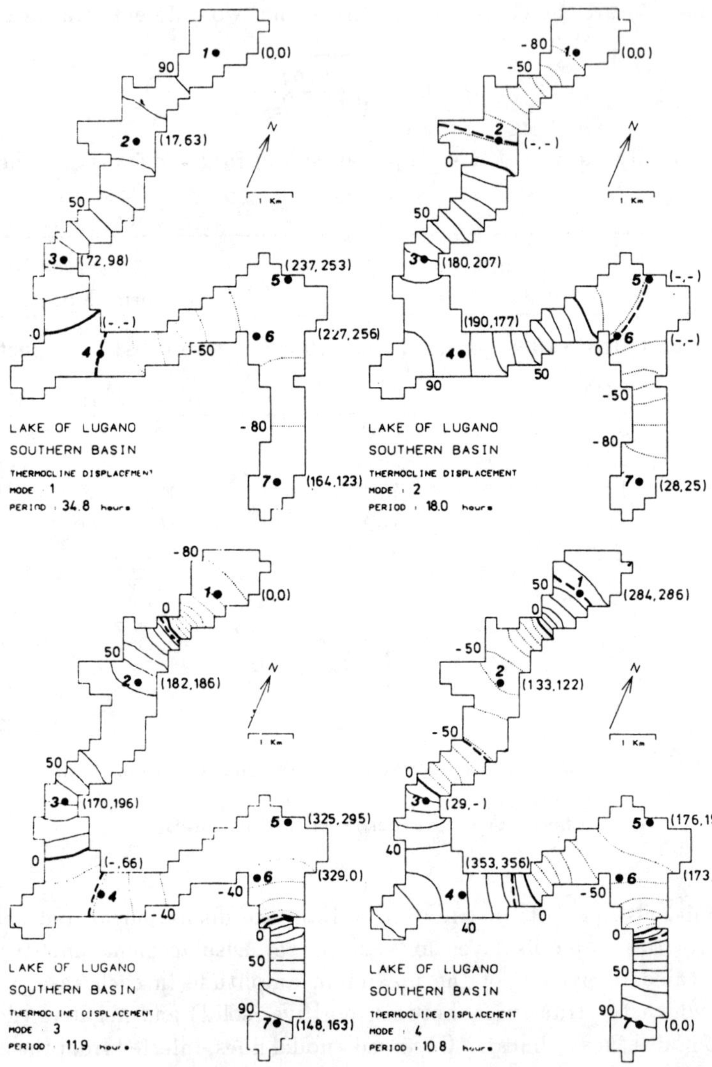

Figure 5.15: Distributions, predicted by the TVD model desccribed in the text, of interface elevations corresponding to the first four internal modes, relative to a maximum elevation of 100 in each case and with the upper and lower layer temperatures taken from *Fig. 5.14a*. Elevation contours (thin lines, solid positive, dotted negative) are drawn at 10 % intervals. Also shown are: mooring positions (bold italic numbers); computed periods; location of modes (heavy solid lines); and "observed" nodes (dashed heavy lines) and "observed" phase angles (bracketed paired numbers, respectively, derived from analysis of 8°C and 12°C-isotherm-depth fluctuations) relative to 0°C at position 1 (modes 1 to 3) and position 7 (mode 4) (from STOCKER et al., 1987).

Figs. 5.16 and *5.17* illustrate, for the first four modes, the distribution of the current fields in the upper and lower layers. Generally, the highest velocities are found in nodal regions and the lower currents are less and always opposed in direction to those in the upper layer. Exceptions occur. For instance the mode 1 velocities are highest where the basin has its smallest width. Comparison with *Fig. 5.15* shows that velocity maxima always coincide with locations where lines of constant amplitude of interface displacements are closest to each other. *Figs. 5.16* and *5.17* also help to identify the moorings at which certain modes can best be inferred from velocity measurements.

Current structure of the internal seiches:

Isotherm motions are coupled with the current field in the entire basin. Therefore, inferences based on analysis of isotherm depths should also be deducible from current meter records. Such reductions are, however, more difficult and could often only be made with knowledge of isotherm behaviour. One difficulty arises because the speed threshold of the Aanderaa current meters is about 2.5 cms^{-1}, while current directions are also reliably recorded at lower speed. Therefore, while complete records of direction were obtained from all moorings (except No. 3 which failed) and in both layers, speed was rarely recorded in the hypolimnion; and there were episodes, sometimes lasting several hours, when speed was neither recorded in the epilimnion. This ruled out spectral or other statistical analysis of the whole current time series. *Fig. 5.18a, b* shows the distributions of the water current directions in the epilimnion and hypolimnion, respectively. Evidently, current in the epilimnion is dominantly oriented along the thalweg; such a preference is not seen in the hypolimnion, except at position 5 (Melide) where the epilimnetic current is towards W, while the hypolimnion current is towards E. The reason for this behaviour is not clear, but it may be due to the increased thickness of the epilimnion close to the channel, inducing a horizontal pressure gradient, which drives the epi- and hypolimnetic currents in opposite directions.
Whereas the TVD model predicts currents for the first four modes, which are predominantly longitudinally-directed and always opposed in the two layers, the observed current directions show greater variability, particularly in the hypolimnion (*Fig. 5.18b*). To render visible any periodic behaviour which may be present in the current records, we have plotted (in *Fig. 5.19*) hourly averages of the longitudinal component of the current at positions 1, 2, 4, 6 and 7 for the interval 26 August to 5 September, i.e. that covered by the isotherm depth plots in *Fig. 5.20.* To assist comparison of the two figures we have repeated (in *Fig. 5.19*) the symbols which marked the presence of 24, 16, 12 and 10 h oscillations in *Fig. 5.20*, those symbols generally coincided with elevation maxima or minima, in *Fig. 5.19* they more often than not coincide with vanishing current. Such behaviour is to be expected from the model-predicted phase lag of one quarter cycle between maximum elevation and maximum current, and is further confirmed by the along-thalweg amplitudes of isotherm displacement (dashed lines) and epilimnetic current (solid lines) in *Fig. 5.20* for the interval 25 August to 5 September. Excitation of the 3rd mode is evident.

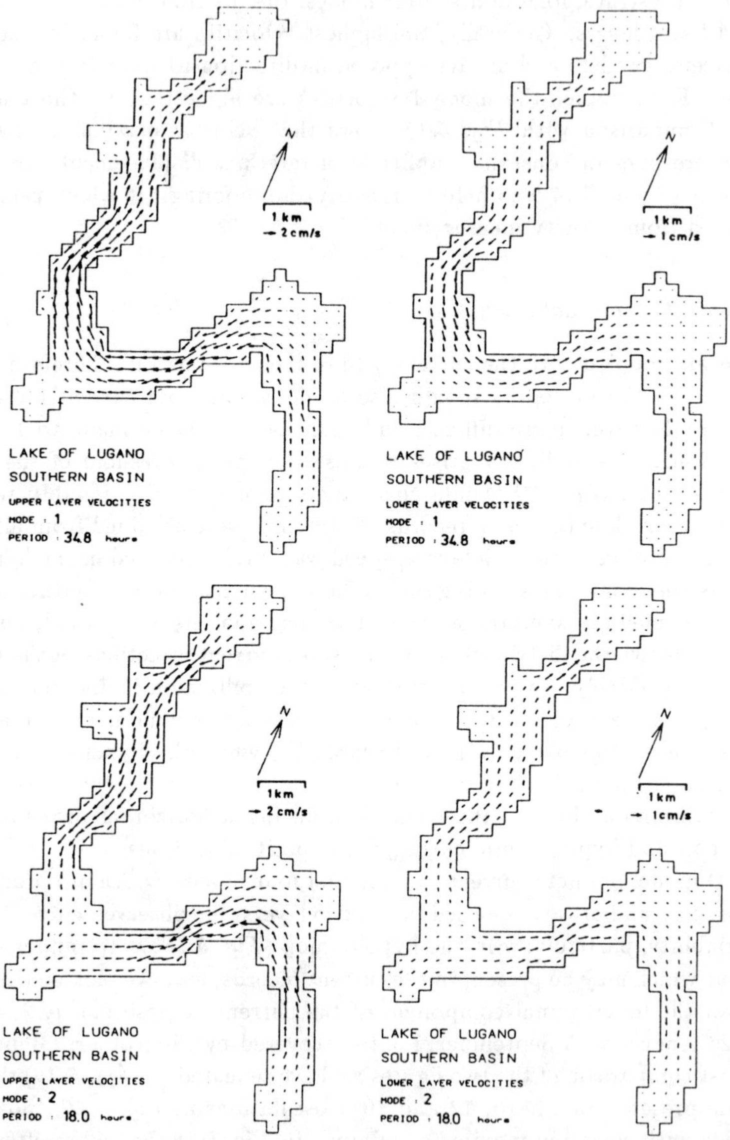

Figure 5.16: TVD-model-predicted current vectors in the upper and lower layer corresponding to the internal seiches of modes 1 and 2, the elevation distributions of which is shown in *Fig. 5.15.* Vector lengths indicate maximum velocities (speed scale as indicated in the inset), attained in each case one quarter-period after the elevation maximum, (from STOCKER et al., 1987).

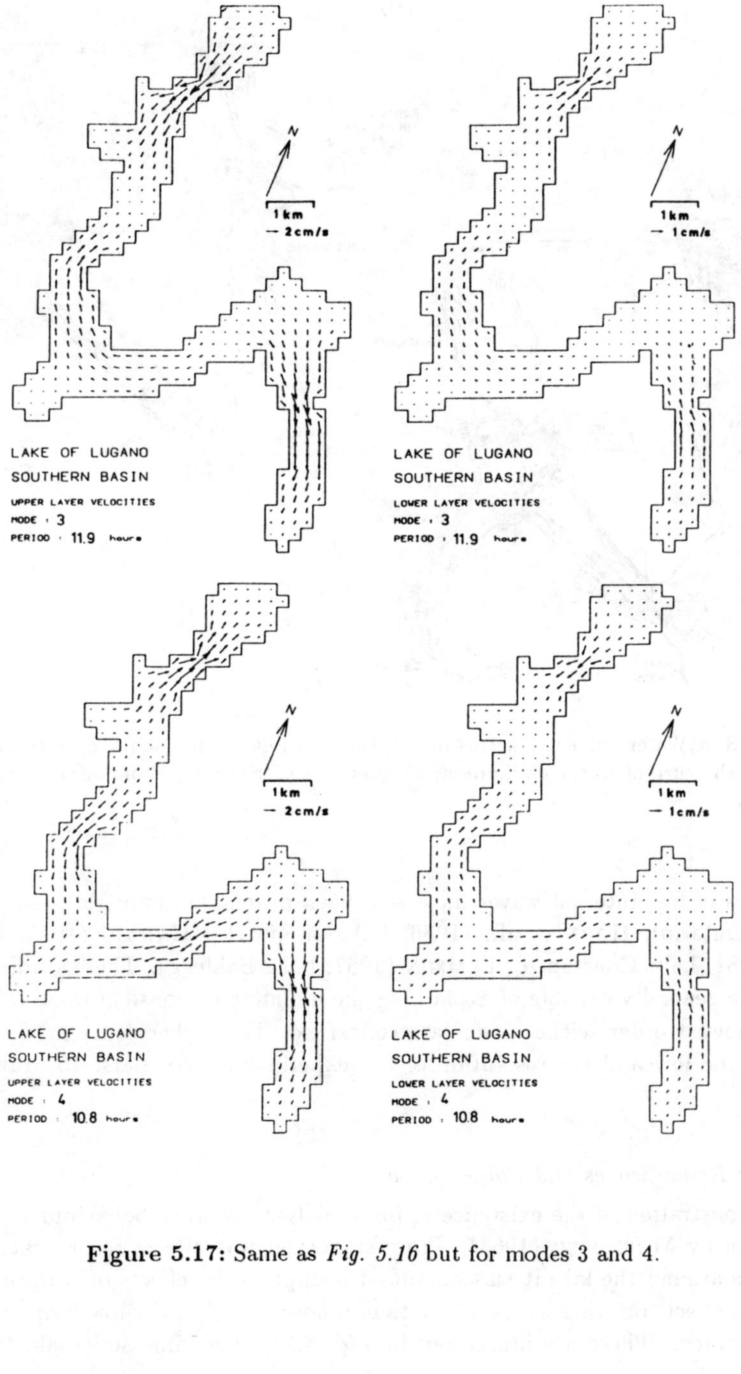

Figure 5.17: Same as *Fig. 5.16* but for modes 3 and 4.

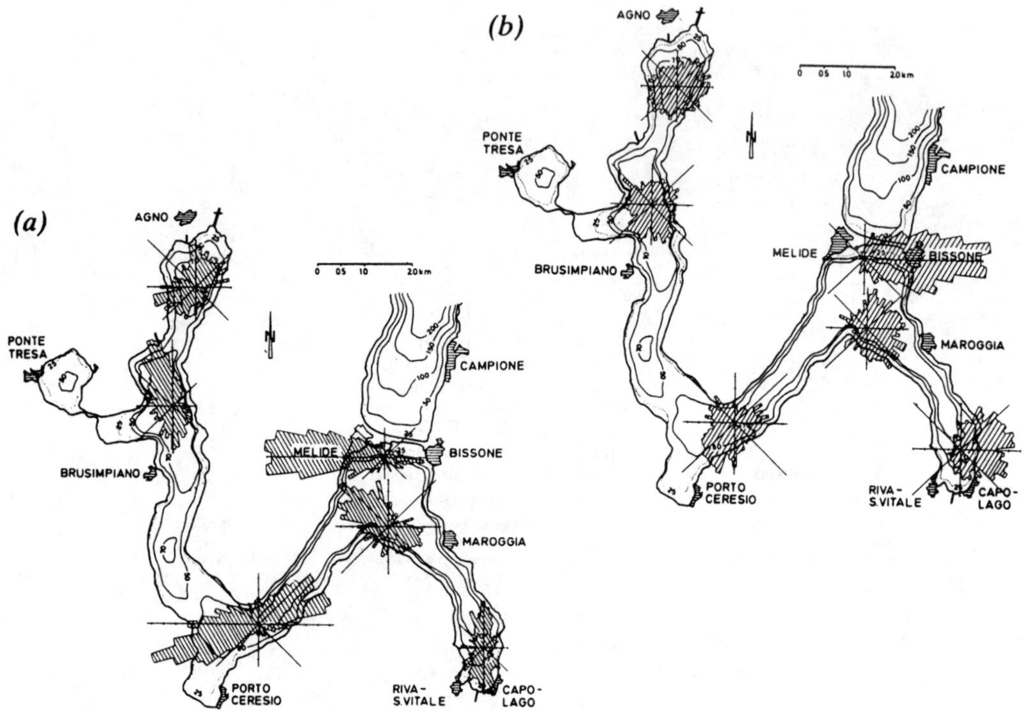

Figure 5.18: a)Water current rosettes illustrating the relative frequency of occurence of the direction of the current in the epilimnion, b) same as a) for the hypolimnion, (from STOCKER et al., 1987).

Similar studies of free internal waves have also been presented, among others, by SCHWAB (1977, Lake Ontario), HORN et al., (1986, Lake of Zürich), KANARI (1975, Lake Biwa), BÄUERLE (1981, Lake Constance), LEMMIN (1987, Lake Baldegg). These studies show that the models are generally capable of explaining the dynamic response of these lakes as far as the first few lowest order seiche modes are concerned. For higher order modes the signals are generally too low and the resolution of the experiments too coarse to draw conclusive inferences.

5.3.3 Internal Kelvin waves and Poincaré waves

The first demonstration of the existence of internal KELVIN wave behaviour has been done for Lac Léman by MORTIMER (1963). From smoothed water level records at eight limnigraph stations around the lake it was possible to suppress the effects of surface seiches and transient wind effects on time scales of less than 6 hours, so that the low-frequency response could be examined. These are illustrated in *Fig. 5.21*. The time series plotted from top

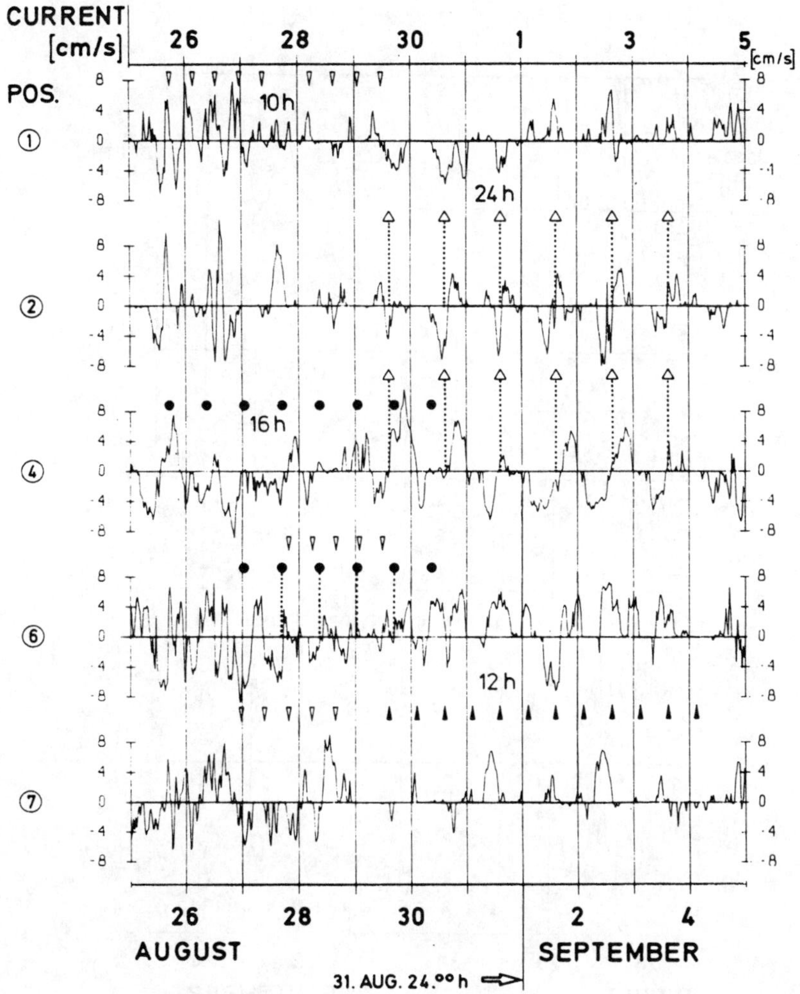

Figure 5.19: Along-thalweg components of the epilimnetic current in positions 1, 2, 4, 6 and 7. Positive direction is towards Agno. Symbols mark times when the $12°C$-isotherms were near minima in *Fig. 5.20*. In *Fig. 5.19* those symbols generally coincide with velocity zeros (see the 24 h-episodes of positions 2 and 4, from STOCKER et al., 1987).

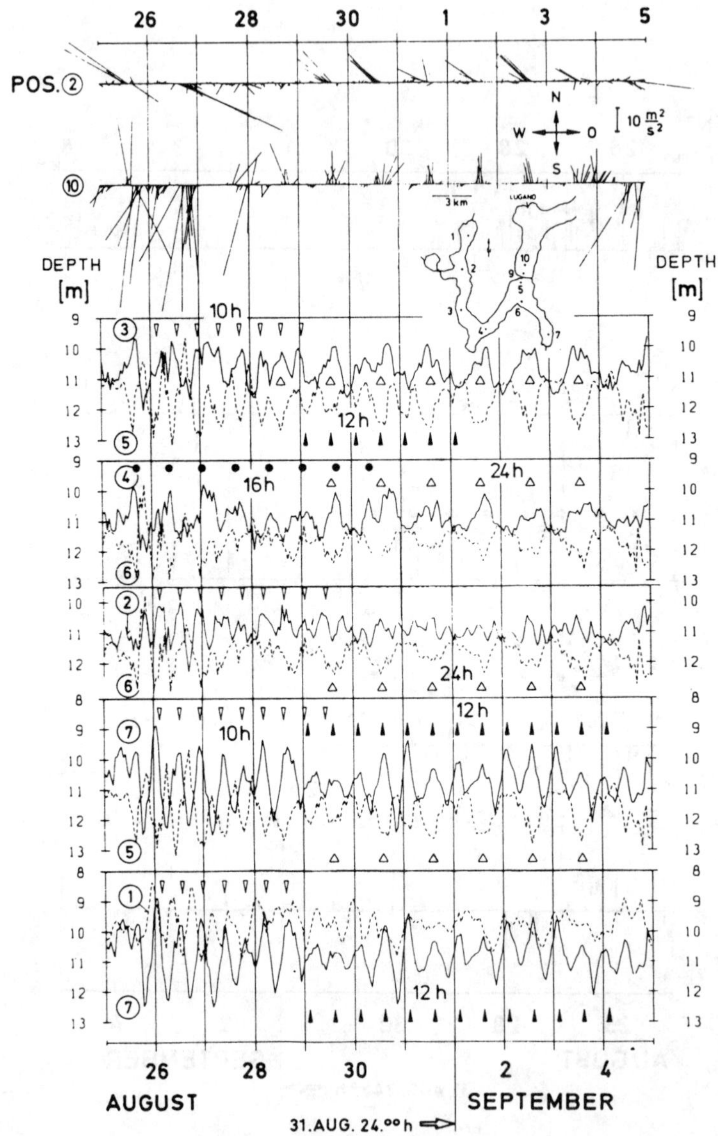

Figure 5.20: Wind speed squared ($\sqrt{u^2+v^2}(u,v)$ in m^2s^{-2} as vector stick plots at positions 2, 10), and fluctuations in depth of the $12°C$-isotherm of seven positions (see inset) in the Southern basin from 25 August to 5 September, 1984. Wind vector and isotherm depth plots are hourly averaged. The latter are grouped in selected pairs to display episodes of approximately out-of-phase oscillation. Such episodes are marked by symbols repeated at the appropriate periodicity (Δ 24 h, • 16 h, ∇ 10 h and Δ 12 h, from STOCKER et al., 1987).

to bottom consecutively in a counter-clockwise order around the lake disclose a cyclonically progressing wave with a period of about $3\frac{1}{2}$ days. In terms of a two-layer model a water level high at a particular station corresponds to a low of the interface excursion at the same position. This implies that for baroclinicity, water level and temperature records at the same position should be in-phase. Partial corroboration is provided in *Fig. 5.21* which also shows the Geneva waterworks intake temperature taken nearby Sécheron.
A detailed analysis of internal free oscillations in the Lake of Geneva on the basis of the above two-layered equivalent depth model but now including the rotational effects of the Earth has been given by BÄUERLE (1985).
Resonant POINCARÉ wave response can be detected with thermistor chain arrangements perpendicular to the long direction of the lake. It is clear that POINCARÉ waves are likely to be seen off-shore. For Lakes Ontario and Michigan repeated scans of vertical temperature distribution across the basin, made from railroad ferries on their regular crossings were analyzed and thermographs and current meter records taken in 1963 and during the international field year 1972 were analysed. The cross-lake standing wave behavior of the internal POINCARÉ wave in Lake Michigan is illustrated in *Fig. 5.22.* It shows the distribution of the $10°C$ isotherms as observed on five consecutive railroad ferry runs between Milwaukee and Muskegon from August, 19 - 20, 1950. It so happend that the ferry crossed the midpoint, steaming either east or west, at approximately eight hour intervals, i.e., close to half an inertial period, permitting a very clear phase change to be seen in the figure. The strong shore-trapped upwelling and downwelling was caused by an episode of northerly winds and caused the internal KELVIN- and POINCARÉ-type wave response. A detailed analysis of the data can be found in MORTIMER (1971, 1980).

5.3.4 β-effect close to the equator

Often, β-effects are hidden behind topographic effects for which reason we present results obtained with computations of a one-layer equivalent depth model of a rectangular basin with constant depth, stratified in two-layers. It is convenient to present results in non-dimensional form such that frequencies are scaled with the Merian frequency $\omega_M^2 = (\pi^2/L^2)\,/g'H_{Eq}$ where L is the length of the rectangle, g' the reduced gravity and H_{Eq} the equivalent depth. This gives rise to the introduction of the rotation number F and the dimensionless eigenfrequencies σ, defined by

$$F = \frac{f}{\omega_M}\ , \ \sigma_n = \frac{\omega_n}{\omega_M}\ . \tag{5.11}$$

The first describes the significance of the rotation of the Earth in comparison to the effects of the gravity, the second indicates how large the eigenfrequency is in a particular mode in comparison to the MERIAN frequency.
We present results for a square whose sides are parallel and perpendicular to the equator and which for $\beta \neq 0$ has a North-South extent from 0° to −2.5° latitude on the Southern hemisphere. *Fig. 5.23* shows co-range lines (interface amplitude distribution, dashed lines) and co-tidal lines (phases, solid lines) when the rotation number is 2 and 1, respectively, in the panels on the left when $\beta = 0$, in those on the right when at the center of the basin $\beta \neq 0$. In the (indicated) (1,0)-mode the nodal line would be a horizontal line through the

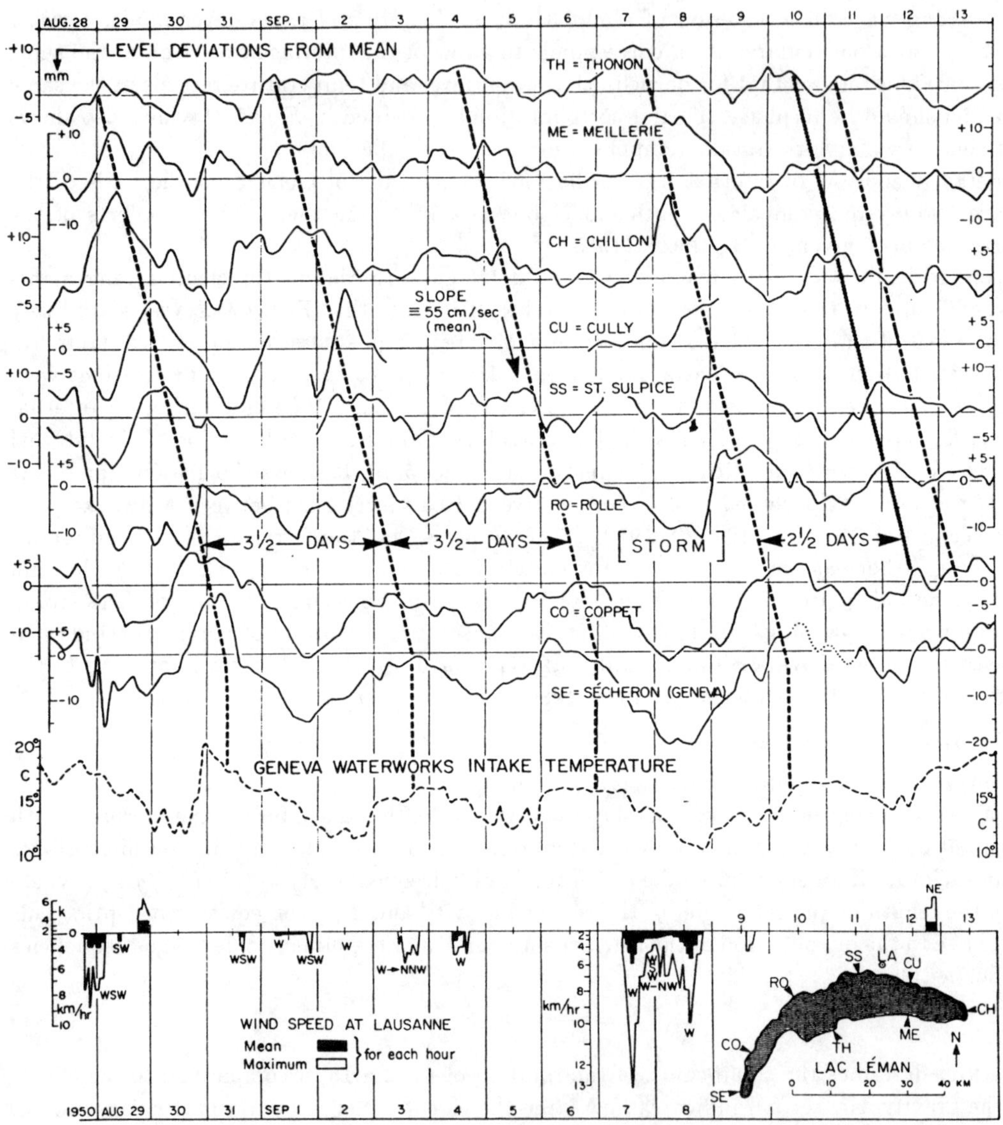

Figure 5.21: Lac Léman, 1950: Fluctuations of smoothed water levels at eight stations, arranged in counterclockwise order around the basin and compared with temperature at the City of Geneva waterworks intake and with wind observations at Lausanne (from MORTIMER, 1963).

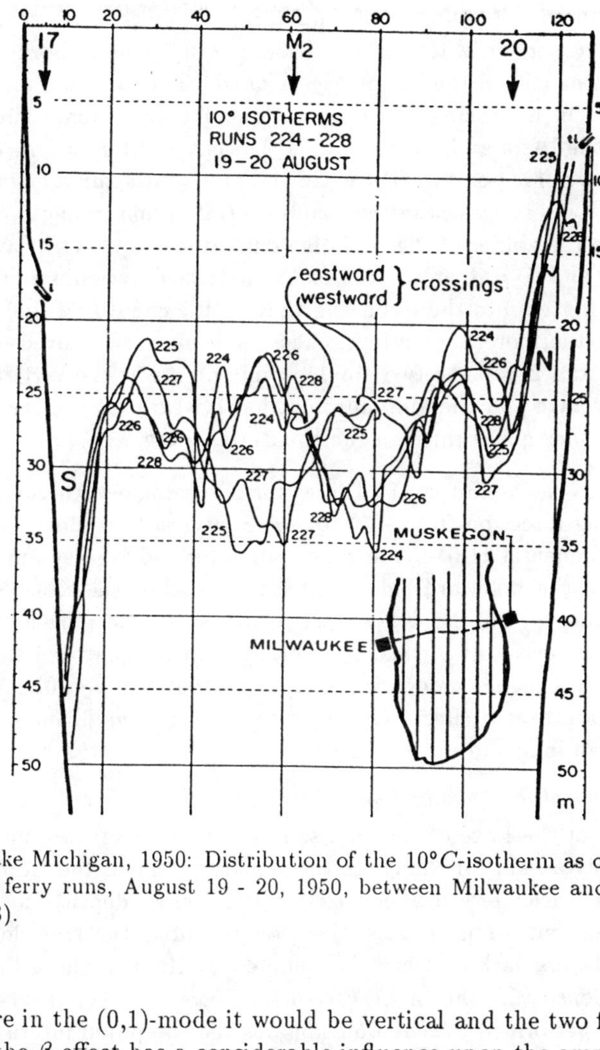

Figure 5.22: Lake Michigan, 1950: Distribution of the 10°C-isotherm as observed on 5 consecutive railroad ferry runs, August 19 - 20, 1950, between Milwaukee and Muskegon (from MORTIMER, 1963).

center of the square in the (0,1)-mode it would be vertical and the two frequencies σ would coalesce. Clearly, the β-effect has a considerable influence upon the amphidromic system:

- With $\beta = 0$ the co-range and co-tidal lines have a NS- and a EW-symmetry; for $\beta \neq 0$ the NS-symmetry is destroyed.
- In the (1,0)-mode and when $\beta = 0$ the elevation amplitude of the interface at the poleward shore is about twice as large as at the equatorial shore but the parity[36] of the amphidromy persists to be cyclonic for all rotation numbers F, that were studied.

[36]An amphidromy is said to have positive (negative) parity if the phase propagates cyclonically (anti-cyclonically) around the amphidromic point.

- For the (0,1)-mode the number of amphidromic points that occur depends upon the value of F irrespective whether $\beta = 0$ or $\beta \neq 0$. This is borne out very clearly in the lower two panels on the left of *Fig. 5.23* which show 1 and 5 amphidromic points, respectively. On increasing F from its value 1 four cyclonic mid-side amphidromic points enter the basin and move towards the centre. At $F = 1.74$ they nearly coalesce with the anticyclonic point at the centre and move out again along the diagonal leaving (when $F = 2$) a very weak (anticyclonic) central amphidromic point (lowest panel on the left). The remainder of the basin is dominated by KELVIN wave dynamics.
  When $\beta \neq 0$ and $F = 1$ at the center the poleward cyclonic amphidromic point has considerably moved into the basin and that on the equatorial side is far from entering the basin. Furthermore, the anticyclonic amphidromy has moved towards the equator. When $F = 2$ and $\beta \neq 0$ the two amphidromic points at the vertical sides have moved towards the centre and the amphidromy that is nearly at the centre when $F = 1$ has moved out of the square through the equatorial shore.

Fig. 5.24 displays co-range and co-tidal lines for a rectangle with aspect ratio 1:10 and a NS-extent in its long direction from $-3°$ to $-9°$ Southern latitude. For a rotation number $F = 6.85$, the fundamental (1,0)-mode shows some limited NS-asymmetry (higher density of co-range lines on the poleward side), but certainly classical KELVIN wave dynamics is essentially not violated. On the other hand, the (0,1)-mode (the finite transverse mode when $f = 0$, $\beta = 0$) shows a first trace of difference at the equatorial shore. This difference is even larger when yet higher modes are analyzed (BÄUERLE, unpublished).
These results indicate that β-effects are likely to be more significant in basins with aspect ratios close to 1 than long and narrow lakes.

5.3.5 Three layer model for Lake of Lugano (North basin)

In the North basin of the Lake of Lugano, salinity and temperature measurements suggest that there is not only an abrupt change in the density profile at the 10 m thermocline depth but equally also (however less pronounced) at the 100 m depth due to a change in the composition and concentration of salts. For computations the true density structure was replaced by three layers each of which is bounded by its own shore line and each having its own constant density as shown in *Fig. 5.1.* *Table 5.4* lists the conditions for which computations were performed. *Table 5.5* compares the computed internal seiche periods for the three layer model of which each layer has its own shoreline and a simpler two-layer model (layer depth at 10 m) with a common shore line for epi- and hypolimnion with periods as inferred from temperature (and velocity) time series through spectral analyses. Temperature spectra at various depths from two moorings are shown in *Fig. 5.25.* They indicate that motions with the 35 h and the 16.7 h periods are best seen at the 57 m depth (the lowest level where measurements were made), whereas the motion with the 25 h period has its largest amplitudes higher up, most likely close to the thermocline. Other observations and computations do corroborate this.
Interesting inferences can be drawn from the mode structures associated with the periods of *Table 5.5.* These may be summarized as follows:

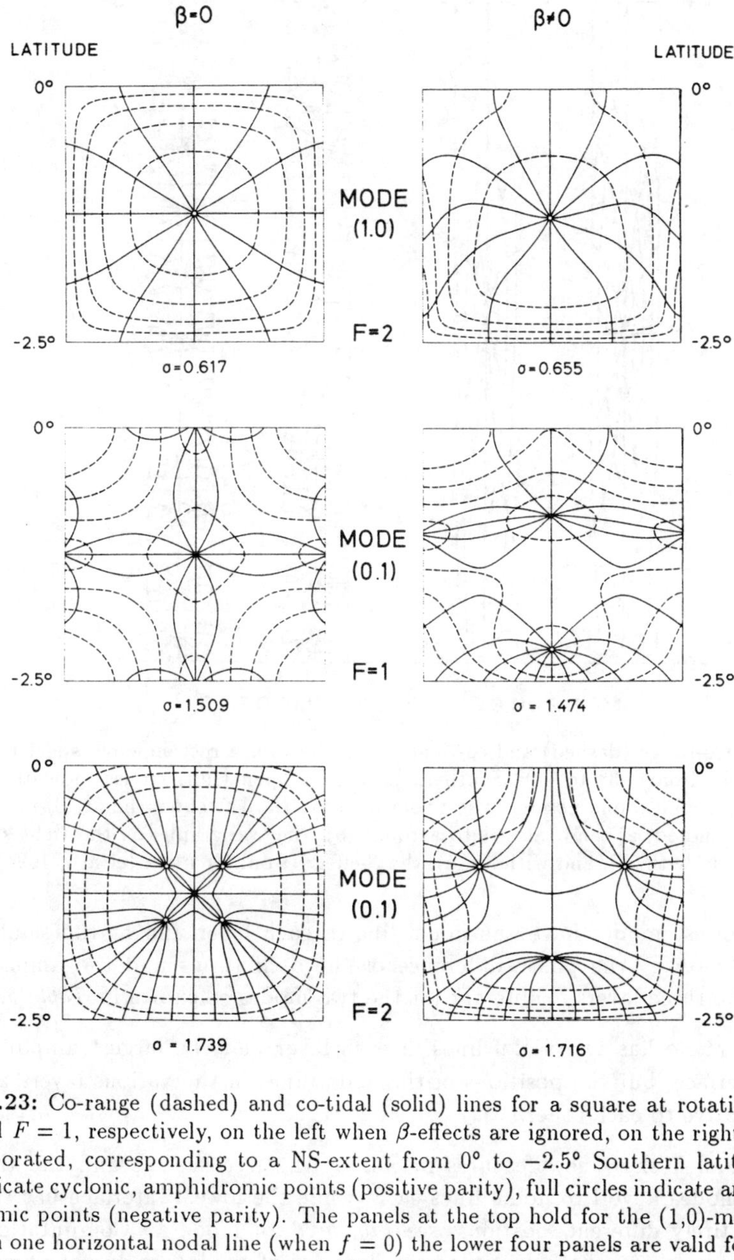

Figure 5.23: Co-range (dashed) and co-tidal (solid) lines for a square at rotation numbers $F = 2$ and $F = 1$, respectively, on the left when β-effects are ignored, on the right when they are incorporated, corresponding to a NS-extent from 0° to −2.5° Southern latitude. Open circles indicate cyclonic, amphidromic points (positive parity), full circles indicate anti-cyclonic, amphidromic points (negative parity). The panels at the top hold for the (1,0)-mode, i.e. the mode with one horizontal nodal line (when $f = 0$) the lower four panels are valid for the (0,1)-mode, i.e. the mode with one vertical nodal line (when $f = 0$). The dimensionless frequencies are also indicated (from HUTTER (et al.), 1991).

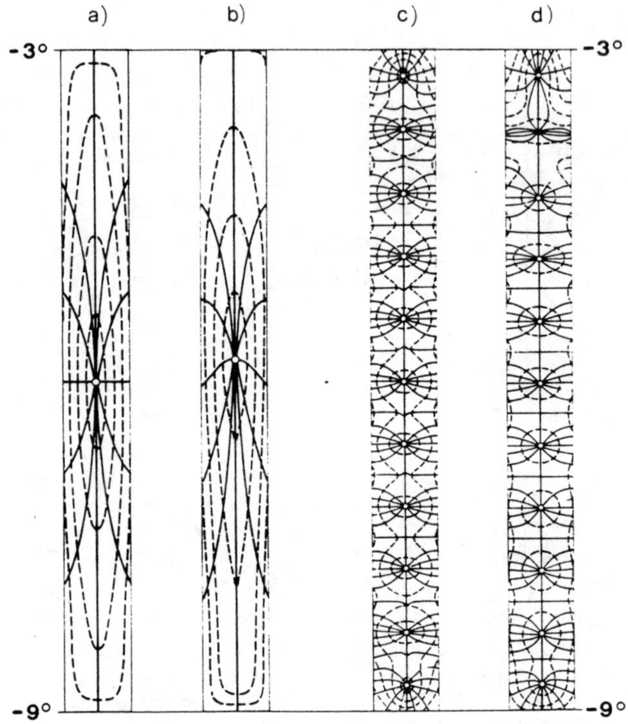

Figure 5.24: Co-range (dashed) and co-tidal (solid) lines for a rectangle of aspect ratio 1:10 and a NS-extent from $-3°$ to $-9°$ Southern latitude. Open (full) circles indicate cyclonic (anti-cyclonic) amphidromic points. The two graphs on the left show the (1, 0)-mode when the β-effect is ignored, a), and taken into account, b); the two graphs on the right show the (0, 1)-mode again with, d), and without, c), the β-effect (from HUTTER (et al.), 1991).

- The two lowest modes have one nodal line in each layer and at the same mid lake position; however, the phase difference of the oscillations and the amplitude ratio between the three layers is different for the two modes as shown in *Table 5.4*.

- The third mode has two nodal lines in each layer and the largest amplitude at the 100 m interface, but the positions of the nodal lines in the various layers are slightly shifted relative to each other.

- Modes 4 and 5 show new features insofar as the free surface and the thermocline elevation modes have two nodal lines each, while the lowest interface has three nodal lines at entirely different locations, see *Fig. 5.26* for mode 4. Computationally, this feature is a property of the fact that the domains of the layers do not agree with one another. The result implies that the phase difference of the oscillation between layers can no longer be defined globally but only locally and that the velocity distribution in the layer, can no longer be guessed from the mode structure of the interface excursions, compare *Figs. 5.26* and *5.27.* Of course, these results have important bearings on the design of field compaigns.

Layer	Depth [m]	Temperature [°C]	$(\rho - 1) \times 100$ [ρ in g/cm^3]	
1	0 - 10	21,0	-1,8210	$\varepsilon_{21} = 0,1904 \cdot 10^{-2}$
2	10 - 100	8,1	$8,289 \cdot 10^{-2}$	$\varepsilon_{31} = 0,2037 \cdot 10^{-2}$
3	variable	5,4	$2,161 \cdot 10^{-1}$	$\varepsilon_{32} = 0,1331 \cdot 10^{-3}$

Table 5.4: Parameters used for the three layer seiche model of the Northern basin of Lake of Lugano (from SALVADÈ et al., 1988)

Mode	Computed Periods in [h]		Observed periods	Phase structure	Number of nodal lines	Amplitude of interface surface
	3 layer	2 layer	in [h]	in layers (1/2/3)		[m]
1	35,1		35 - 38	+	1	0,0001
				−	1	0,1
				+	1	1,1
2	25,1	26,0	≈ 24	+	1	0,0002
				−	1	0,1
				−	1	0,2
3	16,7		15 - 17	+	2	0,0001
				−	2	0,1
				+	2	1,3
4	11,4	11,6	-	+	2	0,0002
				−	2	0,1
				?	3	0,3
5	11,0		-	+	2	0,0002
				−	2	0,1
				?	3	1,3

Table 5.5: Internal seiche periods of the North basin of Lake of Lugano in [h] as computed with the three layer model of SALVADÈ et al., (1988) and the two-layer model of HUTTER et al., (1983), compared with observed periods from temperature and velocity data of ANONYMOUS (1983). Phase structure and number of nodal lines of the surface elevation motion for the three layer model and corresponding amplitude scales are also shown. The symbol + − + means that the motions of the top and bottom surfaces are in phase with each other, but both are out of phase with the motion of the thermocline. Absolute values of the amplitudes of the interfaces are insignificant; only the ratios are important.

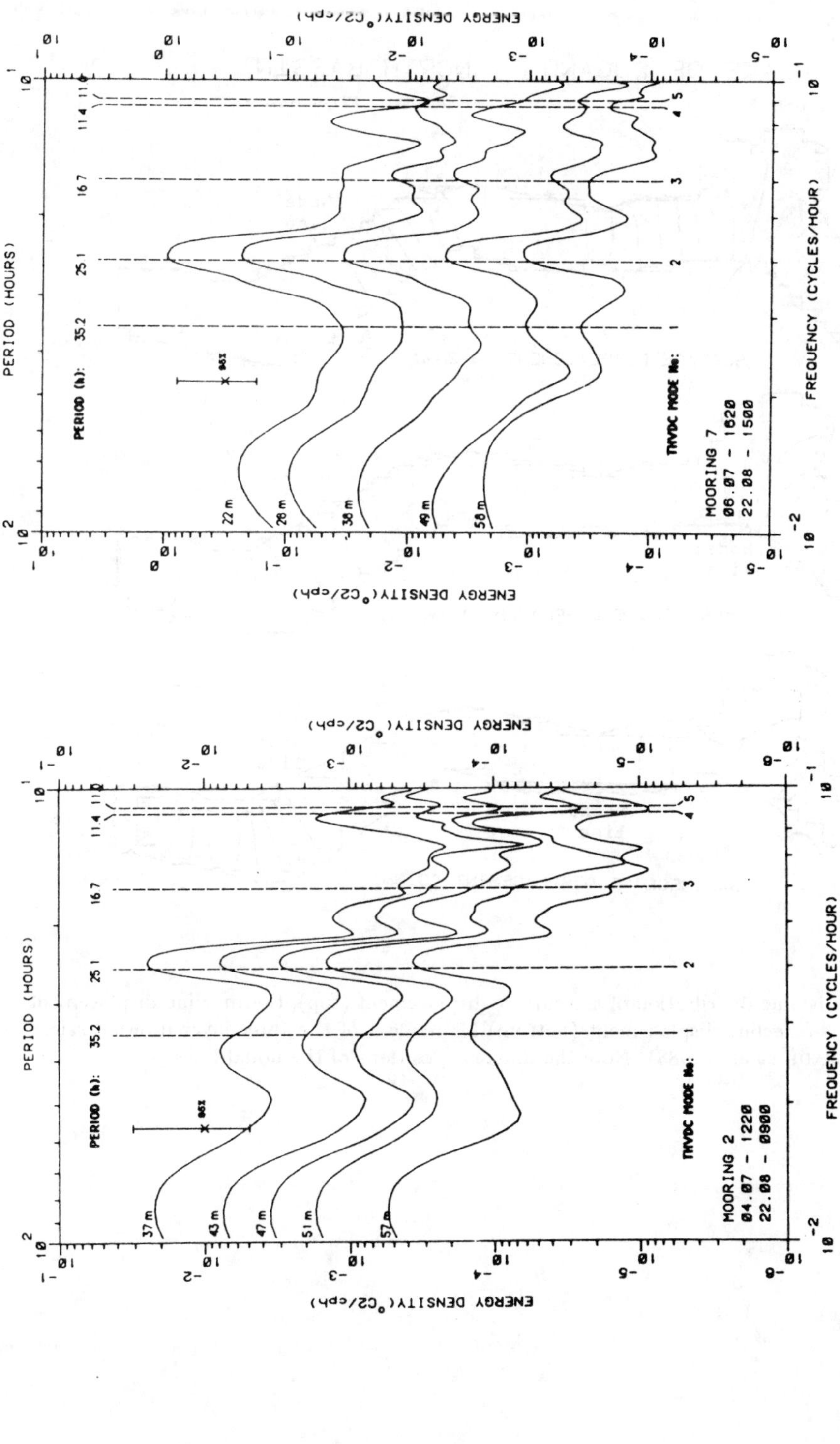

Fig. 5.25: a) Energy spectra of temperature time series at different depths plotted against frequency (cycles/hour) or period (hours) recorded at station 7902 from 4 July to 22 August 1979. Vertical lines indicate the periods computed with the three-layer internal seiche model of SALVADÈ et al., (1988). b) Same as in panel a), but for station 7907 from 6 July to 22 August 1979. The 35.1 *h* and the 16.7 *h* periods are only seen at the deeper thermistor recordings.

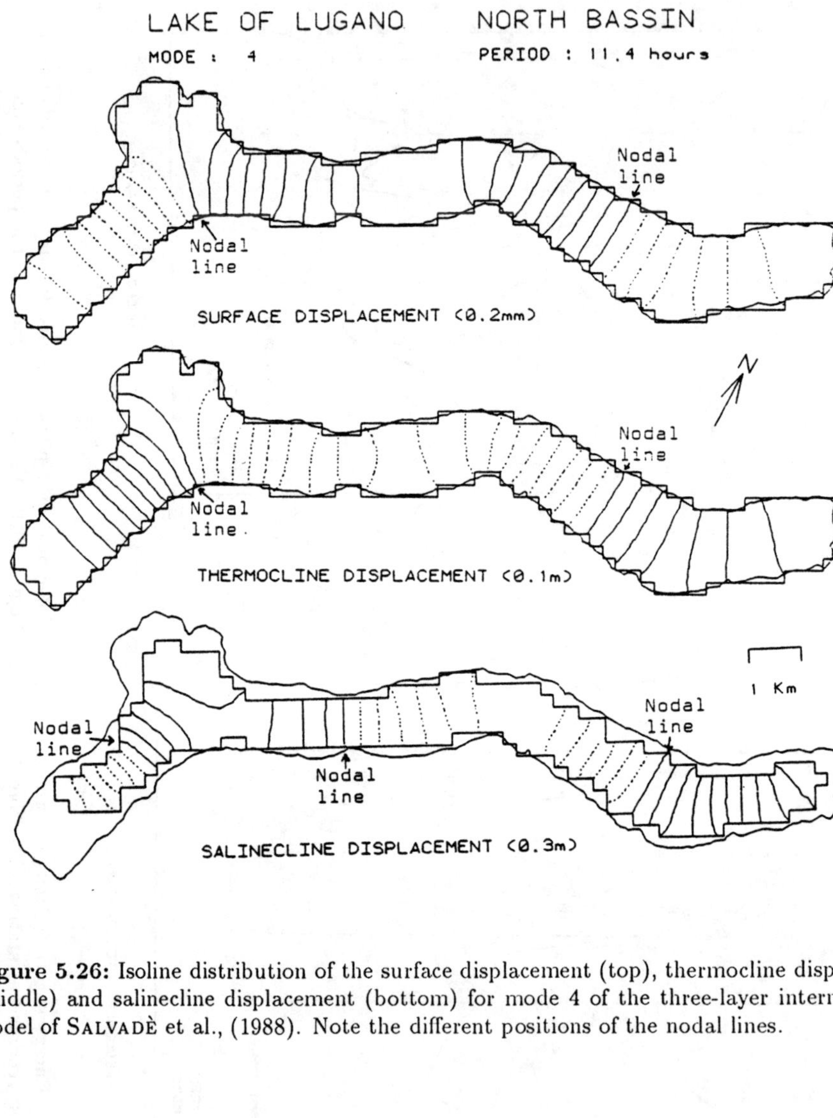

Figure 5.26: Isoline distribution of the surface displacement (top), thermocline displacement (middle) and salinecline displacement (bottom) for mode 4 of the three-layer internal seiche model of SALVADÈ et al., (1988). Note the different positions of the nodal lines.

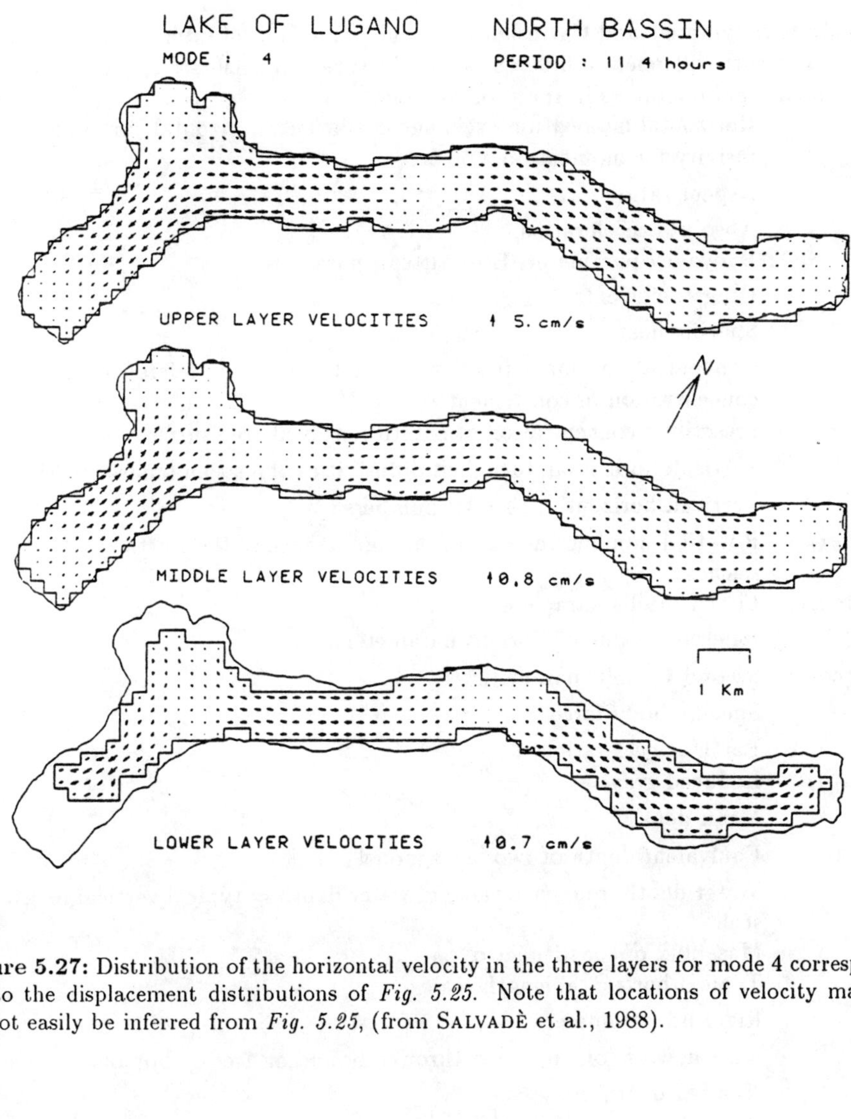

Figure 5.27: Distribution of the horizontal velocity in the three layers for mode 4 corresponding to the displacement distributions of *Fig. 5.25*. Note that locations of velocity maxima cannot easily be inferred from *Fig. 5.25*, (from SALVADÈ et al., 1988).

List of Symbols

In the main text symbolic and Cartesian tensor notation is used interchangingly. Operators grad, div, and curl will refer to differentiations in three-dimensional space, ∇, $\nabla\cdot$ and $\nabla\times$ are two dimensional operators in the f- or β-plane.

A	Horizontal momentum exchange coefficient, horizontal turbulent diffusivity for momentum
$\mathbb{A}$	Aspect ratio
$\mathbf{a}$	Absolute acceleration
$\mathbb{B}$	Buoyancy parameter, Boussinesq parameter
C, c	Drag coefficients
c_v	Specific heat
c_α, c_{α_0}	Concentration (mass fraction) of constituent α, reference value of concentration of constituent α .
c_α^f, c_α^H	prescribed concentration of constituent α at free surface and at base
D^V, D^H	Vertical, horizontal, turbulent diffusivities of heat and mass transport
$\mathbb{E}_V$, $\mathbb{E}_H$	Vertical, horizontal, Ekman numbers
$F = f/\omega_M$	Rotation-number, indicating the significance of the Earth rotation in seiches
$f = 2\Omega \sin\phi$	First Coriolis parameter
f_0	Reference value of Coriolis parameter
$\tilde{f} = 2\Omega \cos\phi$	Second Coriolis parameter
$\mathbf{f}$	Specific body force
$\mathbf{g}$, g	Earth's acceleration
$g' = g\Delta\rho/\rho$	Reduced gravity
H_{mean}	Mean depth
H_{E_q}	Equivalent depth of two-layer model
H, H_0	Water depth, reference value of water depth = typical vertical length scale
j_α	Mass flux of constituent α
L_0	Typical horizontal length scale
$\mathbb{L}$	Rigid lid parameter
M_α^f, M_α^H	mass flow of constituent α through bounding free or bottom surface
N	Number of tracers
p, p_{atm}	Pressure, atmospheric pressure
$\mathbb{P}_V$, $\mathbb{P}_A$	Vertical, horizontal Prandtl number
$\mathbf{q}$	Heat flux vector
Q^{atm}	Heat flow from the atmosphere
R	External, or internal Rossby radius

$\mathbb{R}_i$	RICHARDSON number
$\mathbb{R}_0 = \frac{V_0}{f_0 L_0}$	ROSSBY number
S	Salinity
T, T_0	Temperature, mean temperature
T_f, T_H	prescribed temperature functions at the free surface and at the base
T^*	Reference temperature = $4°C$
$T^H_{diffusion} = \frac{H_0^2}{D_V}$	Typical time scale for vertical diffusion
$\mathbf{t}$	Cauchy stress tensor (deviator)
$\mathbf{t}^*_{(n)}$	shear traction at a surface with unit normal vector $\mathbf{n}$
u, v, w	Cartesian components of the velocity vector
V_0	Typical velocity
$\mathbf{V}$	Volume transport, velocity amplitude
$\mathbf{v}$	Velocity vector
x, y, z	Cartesian coordinates, (x, y) forming the f- or β-plane, z pointing vertically
α	Index counting the tracer constituents
$\beta = df/dy$	North-South gradient of Coriolis parameter
ΔT_0	Temperature range $\cong 20°C$
$\varepsilon = 6.8 \cdot 10^{-6}[C^{-2}]$	Coefficient arising in the temperature dependence of the density anomaly, see (2.8)
	Density difference ratio in layer model, see *Table 4*
ζ, ζ_0	Free surface elevation, reference value of ζ
ϕ	Latitude angle in spherical coordinates
Ψ	Mass transport streamfunction
ν	Horizontal turbulent diffusivity of momentum
ν_0, ν_1	Coefficients arising in formula (2.31) for ν
π_α	Mass production of constituent α
Π_α	Production number
Ω	Angular velocity of the Earth
ω_M	Merian frequency
ρ, ρ^*	Mass density, reference density = density of water at $4°C$
$\rho_0(z)$	Steady vertical density profile
$\sigma = \rho - \rho_0/\rho_0$	Density anomaly
σ_0	Reference value for density anomaly
$\tau^w = \left(\tau_x^w, \tau_y^w\right)$	Wind shear stress vector
$\Theta = T - T^*$	Temperature difference
Θ_α	Dimensionless temperature, scaled mass fractions, $\alpha = 1, 2, \ldots, N+1$

Acknowledgement

The following institutions and journals have granted copy-right permission for figures reprinted in this article: *American Meteorological Society* for figures 4.1, 4.2, 4.3 and 4.4, reproduced from Saylor, Huang and Reed (1980), *Journal of Physical Oceanography. Journal of Fluid Mechanics* for fig. 4.12, reproduced from Ball (1965), fig. 3.5, reproduced from Rhines (1969), figs. 4.9, 4.20 and 4.21, reproduced from Stocker & Hutter (1986). *Aquatic Sciences* for figs. 5.2, 5.23, 5.24, 5.25, 5.26, 5.27, from Hutter et al. (1990), figs. 5.5, 5.7, 5.8 from Hutter et al. (1982b). *The Royal Society* for figs. 4.7 and 4.8 from Mysak et al. (1985). *Springer Verlag*, Berlin, etc. for figs. 4.5, 4.10, 4.11, 4.13, 4.15, 4.17, 4.18, 4.22, 4.23, 4.24 from Stocker and Hutter (1987a), for fig. 5.4 from Hutter (1987b). *Verlag Gebrüder Bornträger* for figs. 3.16 and 3.20 from Krauss (1966). *Gordon and Breach, Science Publishers*, S. A. for fig. 4.7 from Hutter et al. (1983) and fig. 4.9 from Johnson (1987).

References

ANONYMOUS. 1983. *Hydrologische Messungen mit verankerten Instrumenten im Nordbecken des Luganersees während der Monate Juli/August 1979.* Gemeinschaftsarbeit des Laboratorio di Fisica Terrestre, ICTS Lugano-Trevano, sowie der Gruppe " Physikalische Linologie" der VAW-ETH Zürich, Interner Bericht **I/68**, 328 pp.

ABRAMOWITZ, M. and STEGUN, I.A. 1972. *Handbook of mathematical functions.* Dover.

BALL, F.K. 1965. Second class motions of a shallow liquid. *J. Fluid Mech.*, **23**, 545-561.

BÄUERLE, E. 1981. *Die Eigenschwingungen abgeschlossener, zwei-geschichteter Wasserbecken bei variabler Bodentopographie.* Dissertation an der Christian Alberts Universität, Kiel.

BÄUERLE, E. 1985. Internal free oscillations in the Lake of Geneva. *Annales Geophysicae*, **3**, 199-206.

BIRCHFIELD, G.E. and HICKIE, B.P. 1977. The time dependent response of a circular basin of variable depth to a wind stress. *J. Phys. Oceanogr.*, **7**, 691-701.

BLOOMFIELD, P. 1976. *Fourier analysis of time series: An introduction.* John Wiley, New York.

BROWN, P.J. 1973. Kelvin wave reflection in semi-infinite canal. *J. Mar. Res.*, **31**, 1-10.

BUCHWALD, V.T. and ADAMS, J.K. 1968. The propagation of continental shelf waves. *Proc. Roy. Soc.*, **A305**, 235-250.

BÜHRER, H. and AMBÜHL, A.H. 1975. Die Einleitung von gereinigtem Abwasser in Seen. *Schweiz. Z. Hydrol.*, **37**, 347-369.

Charney, J.G. 1955. Generation of oceanic currents by wind. *J. Mar. Res.*, **14**, 477-498.

DEFANT, A. 1961. *Physical Oceanography.* Oxford University Press, Oxford.

DEFANT, F. 1953. Theorie der Seiches des Michigansees und ihre Abwandlung durch Wirkung der Corioliskraft. *Arch. Met. Geophys. Biokl.*, **A 6**, 218-241.

ERTEL, H. 1942. Ein neuer hydrodynamischer Wirbelsatz. *Meteorolog. Z.*, **59**, 277-281.

GALAZII, G.I. ed. 1969. *Baikal Atlas.* Limnol. Inst. Acad. Sci. USSR. Siberian Section, publ. by Govt. Dept. Geodasy and Cartography, Irkutsk and Moscow, 30 p.

GILL, A.E. 1982. *Atmosphere - Ocean Dynamics.* Academic Press, San Diego etc.

GONELLA, J.A. 1974. Effect d'écran d'une thermocline observationes faites sur la bonée-laboratoire. *La Houille Blanche*, **7/8**, 607-613.

GREENSPAN, H.P. 1968. *The Theory of Rotating Fluids.* Cambridge University Press.

GUSTAFSON, T. and KULLENBERG, B. 1936. Untersuchungen von Trägheitsströmungen in der Ostsee. *Svensk. Hydrgr.-Biol. Komm. Skr.* (New Series)*Hydrogr.*, **13**, 28 pp.

HAMBLIN, P.F. and HOLLAN, E. 1978. On the gravitational seiches of Lake Constance and their generation. *Schweiz. Z. Hydr.*, **40**, 119-154.

HEAPS, N.S. 1961. Seiches in a narrow lake, uniformly stratified in three layers. *Geophys. Suppl. J. Roy. Astron. Soc.*, **5**, 134-156.

HOLLAN, E. 1979. Hydrodynamische Modellrechnungen über die Eigenschwingungen des Bodensee-Obersees mit einer Deutung des Wasserwunders von Konstanz im Jahre 1549. *Schw. Verein für Geschichte des Bodensees und seiner Umgebung*, **97**, 157-192.

HOLLAN, E. 1980. *Die Eigenschwingungen des Bodensee-Obersees und eine Deutung des "Wasserwunders von Konstanz" im Jahre 1549.* Deutsches gewässerkundliches Jahrbuch, ISSN 0344-0788, Abflussjahr 1979.

HOLLAN, E., RAO, D.B. and BÄUERLE, E. 1980. Free surface oscillations in Lake Constance with an interpretation of the "Wonder of the rising water at Konstanz" 1549. *Arch. Met. Geophys. Biokl.*, **A 29**, 301-325.

HORN, W., MORTIMER, C.H. and SCHWAB, D.J. 1986. Internal wave dynamics of Lake of Zurich. *Limnol. Oceanogr.*, **31**, 1232-1254.

HUANG, J.C.K. and SAYLOR, J.H. 1982. Vorticity waves in a shallow basin. *Dyn. Atmos. Oceans.*, **6**, 177-196.

HUTTER, K. 1983. Strömungsdynamische Untersuchungen im Zürich- und im Luganersee. Ein Vergleich von Feldmessungen mit Resultaten theoretischer Modelle. *Schweiz. Z. Hydrol.*, **45**, 101-144.

HUTTER, K. 1984a. Mathematische Vorhersage von barotropen und baroklinen Prozessen im Zürich- und Luganersee. *Vierteljahrschrift der Naturforschenden Gesellschaft in Zürich.*, **129**, 51-92.

HUTTER, K. 1984b (Ed.). *Hydrodynamics of Lakes*, CISM Courses and Lectures No **286**, Springer Verlag Wien - New York.

HUTTER. K. 1987a. Hydrodynamic modeling of Lakes. Chapter 22 in: *Encyclopedia of Fluid Mechanics (ed. Cheremesinoff)*, Gulf Publ. Com. Houston, **6**, 897-998.

HUTTER, K. 1987b. Schwingungen in einem Seensystem: Der Luganersee. *Naturwissenschaften*, **74**, 405-414.

HUTTER, K. (et al.). 1991. Large scale water movements in lakes (with contributions by G. Salvadè, C. Spinedi, F. Zamboni and E. Bäuerle). *Aquatic Sciences*, **53**, 100-135.

HUTTER, K. and RAGGIO, G. 1982. A Chrystal-model describing gravitational barotropic motion in elongated lakes. *Arch. Met. Geophys. Biokl.*, **A 31**, 361-378.

HUTTER, K. and SCHWAB, D.J. 1982. *Baroclinic channel models*, Internal report No. 62, Laboratory of Hydraulics, Hydrology and Glaciology, ETH Zurich.

HUTTER, K., RAGGIO, G., BUCHER, C. and SALVADÈ, G. 1982a. The surface seiches of Lake of Zurich. *Schweiz. Z. Hydr.*, **44**, 423-454.

HUTTER, K., RAGGIO, G., BUCHER, C., SALVADÈ, G. and ZAMBONI, F. 1982b. The surface seiches of the Lake of Lugano. *Schweiz. Z. Hydr.*, **44**, 455-484.

HUTTER, K., SALVADÈ, G. and SCHWAB, D.J. 1983. On internal wave dynamics of the Lake of Lugano. *Geophys. Astrophys. Fluid Dyn.*, **27**, 299-336.

JOHNSON, E.R. 1987a. Topographic waves in elliptical basins. *Geophys. Astrophys. Fluid Dyn.*, **37**, 279-295.

JOHNSON, E.R. 1987b. A conformal mapping technique for topographic wave problems: semi-infinite channels and elongated basins. *J. Fluid Mech.*, **177**, 395-405.

JOHNSON, E.R. 1989. Topographic waves in open domains. Part I: Boundary conditions and frequency estimates. *J. Fluid Mech.*, **200**, 69-76.

KANARI, S. 1975. The long period internal waves in Lake Biwa. *Limnol. & Oceanogr.*, **20**, 544-553.

KRAUSS, W. 1966. Methoden und Ergebnisse der theoretischen Ozeanographie. *2. Intern. Wellen*, Borntraeger, Berlin-Stuttgart.

LAMB, H. 1932. *Hydrodynamics*, 6th ed., Cambridge University Press.

LARSEN, J.C. 1969. Long waves along a single-step topography in a semi-infinite uniformly rotating ocean. *J. Mar. Res.*, **27**, 1-6.

LEBLOND, P.H. and MYSAK, L.A. 1980. *Waves in the Ocean*, Elsevier Oceanography Series. Elsevier Scientific Publishing Company, Amsterdam-Oxford-New York.

LEMMIN, U. and MORTIMER, C.H. 1986. Tests of an extension to internal seiches of Defant's procedure for determination of surface seiche characteristics in real lakes. *Limnol. Oceanogr.*, **31**, 1207-1231.

LIGHTHILL, J. 1969. Dynamic response of the Indian Ocean to onset of the southwest monsoon. *Phil. Trans. Royal. Soc. London*, **A 265, 1159**, 45-92.

LIGHTHILL, J. 1978. *Waves in Fluids*, Cambridge University Press, Cambridge.

MARCHUK, G.I. and SARKISYAN, A.S. 1989. *Mathematical Modeling of Ocean Circulation*, Springer Verlag Berlin etc., 292 pp.

MILES, J.W. and BALL, F.K. 1963. On free-surface oscillations in a rotating paraboloid. *J. Fluid Mech.*, **17**, 257-266.

MORTIMER, C.H. 1952. Water movements in lakes during summer stratification; evidence from the distribution of temperature in Windermere. *Phil. Trans. Royal Soc. London*, **B 236**, 355-404.

MORTIMER, C.H. 1963. Frontiers in physical limnology with particular reference to long waves in rotating basins. In: *Proc. 5th Conf. Great Lakes Res., Great Lakes Res. Div. Univ. Mich.*, **9**, 9-42.

MORTIMER, C.H. 1971. *Large scale oscillatory motions and seasonal temperature changes in Lake Michigan and Lake Ontario.* Special Report No 12, Part **I**, 111 pp : Text, Part **II**, 106 pp : Illustrations, Center for Great Lakes studies, The University of Wisconsin-Milwaukee.

MORTIMER, C.H. 1974. Lake hydrodynamics. *Mitt. Int. Ver. Theor. Angew. Limnol.*, **20**, 124-197.

MORTIMER, C.H. 1975. Substantive corrections to SIL Communications (IVL Mitteilungen) Numbers 6 and 20, *Mitt. Int. Ver. Theor. Angew. Limnol.*, **19**, 60-72.

MORTIMER, C.H. 1977. Internal waves observed in Lake Ontario during the International Field Year for the Great Lakes (IFYGL), 1977: Descriptive theory and preliminary interpretations of near-inertial oscillations in terms of linear channel-wave models. *Special Report No 32, Center for Great Lakes Studies, The University of Winconsin-Milwaukee*, 122p.

MORTIMER, C.H. 1979. Strategies for coupling data collection and analysis with the dynamic modelling of lake motions. In: *Lake Hydrodynamics (eds.* W.H. GRAF & C.H. MORTIMER*), Elsevier, Amsterdam*, 183-222E.

MORTIMER, C.H. 1980. *Internal motion and related internal waves in Lake Michigan and Lake Ontario as responses to impulsive wind stresses.* Special Report No 37, Center for Great Lakes Studies, The University of Winconsin-Milwaukee, 192p.

MORTIMER, C.H. and FEE, E.J. 1976. Free surface oscillations and tides of Lakes Michigan and Superior, *Phil. Trans. Royal Soc. London*, **A 281**, 1-61.

MÜHLEISEN, R. and KURTH, W. 1978. Experimental investigations on the seiches of Lake Constance. *Schweiz. Z. Hydr.*, **40**.

MYSAK, L.A. 1980a. Recent advances in shelf wave dynamics. *Reviews of Geophysics and Space Physics*, **18**, 211-241.

MYSAK, L.A. 1980b. Topographically trapped waves. *Ann. Rev. Fluid Mech.*, **12**, 45-76.

MYSAK, L.A. 1984. Topographic waves in lakes, in *Hydrodynamic of Lakes*, CISM 286, K. HUTTER (Ed.), 81-128, Springer, Wien-New York.

MYSAK, L.A., SALVADÈ, G., HUTTER, K. and SCHEIWILLER, T. 1983. Lake of Lugano and topographic waves. *Nature*, **306**, 46-48.

MYSAK, L.A., SALVADÈ, G., HUTTER, K. and SCHEIWILLER, T. 1985. Topographic waves in elliptical basin, with application to the Lake of Lugano. *Phil. Trans. R. Soc. London*, **A 316**, 1-55.

OMAN, G. 1982. *Das Verhalten des geschichteten Zürichsees unter äußeren Windlasten.* Mitt. No. 60 der Versuchsanstalt für Wasserbau, Hydrologie und Glaziologie an der ETH Zürich, 185pp.

PHILLIPS, O.M. 1966. *The Dynamics of the Upper Ocean.* Cambridge University Press, Cambridge.

PLATZMANN, G.W. 1972. Two-dimensional free oscillation in natural basins. *J. Phys. Oceanogr.*, **2**, 117-138.

PLATZMANN, G.W. and RAO, D.B. 1964a. The free oscillations of Lake Erie. *Studies in Oceanography* (Hidaka Volume), Tokyo University Press, 359-382.

PLATZMANN, G.W. and RAO, D.B. 1964b. Spectra of Lake Erie water levels, *J. Geophys. Res.*, **69**, 225.

POINCARÉ, H. 1910. *Théorie des Marées. Seçons de Mécanique Celeste*, **3**, Gauthiers-Villars, Paris.

RAGGIO, G. and HUTTER, K. 1982a. An extended channel model for the prediction of motion in elongated homogeneous lakes. Part 1: Theoretical Introduction, *J. Fluid Mech.*, **121**, 231-255.

RAGGIO, G. and HUTTER, K. 1982b. An extended channel model for the prediction of motion in elongated homogeneous lakes. Part 2: First order model applied to ideal geometry; rectangular basin with flat bottom, *J. Fluid Mech.*, **121**, 257-281.

RAGGIO, G. and HUTTER, K. 1982b. An extended channel model for the prediction of motion in elongated homogeneous lakes. Part 3: Free oscillations in natural basins, *J. Fluid Mech.*, **121**, 283-299.

RAO, D.B. 1966. Free gravitational oscillations in rotating rectangular basins, *J. Fluid Mech.*, **24**, 523-555.

RHINES, P.B. 1969. Slow oscillations in an ocean of varying depth, *J. Fluid Mech.*, **37**, 161-205.

ROSSBY, C.G. 1936. Dynamics of steady ocean currents in the light of experimental fluid mechanics, *Pap. Phys. Oceanogr. Met.*, **5**, (1), 1-43.

ROSSBY, C.G. 1937. On the mutual adjustment of pressure and velocity distributions in certain simple current systems. I. *J. Mar. Res.*, **1**, 15-28.

ROSSBY, C.G. 1938a. On the mutual adjustment of pressure and velocity distributions in certain simple current systems. II. *J. Mar. Res.*, **2**, 239-263.

ROSSBY, C.G. 1938b. On temperature changes in the stratosphere resulting from shrinking and stretching. *Beitr. Phys. Freien Atmos.*, **24**, 53-59. ROSSBY, C.G. 1940. Planetary flow patterns in the atmosphere. *Q.J.R. Meteorol. Soc.*, **66**, Suppl., 68-97.

ROSSBY, C.G., et al. 1939. Relation between variations in the intensity of the zonal circulation of the atmosphere and the displacements of the semi-permanent centers of action. *J. Mar. Res.*, **2**, 38-55.

SALVADÈ, G. and ZAMBONI, F. 1987. External gravity oscillations of the coupled basins of the Lake of Lugano. *Annales Geophysicae*, **5 B**, 247-254.

SALVADÈ, G., ZAMBONI, F. and BARBIERI, A. 1988. Three-layer model of the North basin of the Lake of Lugano. *Annales Geophysicae*, **6**, 463-474.

SATO, G.K. and MORTIMER, C.H. 1975. *Lake currents and temperatures near the western shore of Lake Michigan.* Special Report No 22, Center for Great Lake Studies, The University of Wisconsin-Milwaukee.

SAYLOR, J.H., HUANG, J.S.K. and REID, R.O. 1980. Vortex modes in Southern Lake Michigan. *J. Phys. Oceanogr.*, **10**, 1814-1823.

SAYLOR, J.H. and MILLER, G.S. 1983. *Vortex Modes of Particular Great Lakes Basins*, Great Lakes Environmental Research Laboratory, Constribution No. 394, NOAA, Ann Arbor, Michigan.

SCHWAB, D.J. 1977. Internal free oscillations in Lake Ontario, *Limnol. & Oceanogr.* **22**, 700-708.

SCHWAB, D.J. and RAO, D.B. 1977. Gravitational oscillations of Lake Huron, Saginow Bay, Georgian Bay, and the North Channel, *J. Geophys. Research* **82**, 2105-2116.

SEZAWA, K. and KANAI, K. 1939. On shallow water waves transmitted in the direction parallel to a sea coast, with special reference to long-waves in heterogeneous media, *Bull. Earthquake Res. Inst. Tokyo*, **17**, 685-694.

SNODGRASS, F.E., MUNK, W.H. and MILLER, G.R. 1962. Long-period waves over California's continental borderland. I: Background spectra, *J. Mar. Res.*, **20**, 3-30.

STOCKER, T. 1988. A numerical study of topographic wave reflection in semi-infinite channels. *J. Phys. Oceanogr.*, **18**, 609-618.

STOCKER, T. and HUTTER, K. 1986. One-dimensional models for topographic Rossby waves in elongated basins on the f-plane, *J. Fluid Mech.*, **170**, 435-459.

STOCKER, T. and HUTTER, K. 1987a. *Topographic waves in channels and lakes on the f-plane*, Lecture Notes on Coastal and Estuarine Studies, **9**, Springer Verlag, Berlin etc.

STOCKER, T. and HUTTER, K. 1987b. Topographic Rossby waves in rectangular basins, *J. Fluid Mech.*, **185**, 107-120.

STOCKER, T. and HUTTER, K. 1990. Qualitative aspects of topographic waves in closed basins gulfs and channels, in: *Modeling Marine Systems*, **I**, (A.M. DAVIES, ed.), CRC-Press,

255-289.

STOCKER, T. and JOHNSON, E.R. 1989. Topographic waves in open domains. Part II. Bay modes and resonances. *J. Fluid Mech.*, **200**, 77-93.

STOCKER, T. and JOHNSON, E.R. 1991. The trapping and scattering of topographic waves by estuaries and headlands. *J. Fluid Mech.*, **222**, 501-524.

STOCKER, T. and JOHNSON, E.R. 1992. Transmission and reflection of shelf waves by esteraries and beadlands *J. Fluid Mech..*, to appear.

STOCKER, K., HUTTER, K., SALVADÈ, G., TRÖSCH, J. and ZAMBONI, F. 1987. Observations and analysis of internal seiches in the Southern basin of Lake Lugano. *Annales Geophysicae*, **5 B**, 553-568.

TAYLOR, G.I. 1922. Tidal oscillations in gulfs and rectangular basins. *Proc. London Math. Soc. Ser. 2*, **20**, 148-181.

TISON, L.J. and TISON, G. Jr. 1969. *Seiches et dénivellations causées par le vent dans les lacs, baies, estuaries.* Note Technique No. 102, Organisation Méteorologique Mondiale, Genève, Suisse.

TURNER, J.S. 1973. *Buoyancy Effects in Fluids.* Cambridge University Press.

TRÖSCH, J. 1984. Finite element calculation of topographic waves in lakes. *Proceedings of the 4th International Conference on Applied Numerical Modeling, Tainan, Taiwan.*

VAN DYCK, M. 1976. An album of fluid motions, Parabolic Press, Palo Alto.

WENZEL, M. 1978. *Interpretation der Wirbel im Bornholmbecken durch topographische Rossby Wellen in einem Kreisbecken.* Diplomarbeit, Christian Albrechts Universität Kiel, 52 pp.

WILLMOT, A.J. and JOHNSON, E.R. 1989. Topographic waves in a rotating stratified basin. *Geophys. Astrophys. Fluid Dynamics*, **45**, 71-87.

WITHAM, G.B. 1974. *Linear and Non-linear Waves.* Wiley, New York.

YIH, C.S. 1965. *Dynamics of Nonhomogeneous Fluids.* MacMillan, New York.

Continuum Description of Natural Slopes in Slow Movement

Kolumban Hutter
Institut für Mechanik, Technische Hochschule Darmstadt
Hochschulstraße 1, D - 6100 Darmstadt, Germany

A two-component mixture model is proposed that is capable of describing the slow motion of natural slopes which may creep (deform within their body) and slide along their basal surface under the effect of gravity. The field equations comprise the balance laws of mass and momentum and constitutive relations for the partial stresses of the fluid and the solid as well as the seepage force. The fluid is modelled as an incompressible perfect liquid and the soil as a viscous isotropic body in which the stress is rate dependent and may also depend on the porosity. The seepage force is described by Darcy's law. Explicit constitutive formulas are prescribed. The kinematic and dynamic boundary conditions for the fixed basal surface and the movable free and phreatic surfaces are stated. All these equations comprise a well posed boundary value problem for the moving mass of soil that is saturated with water.

The shallowness that usually prevails in such moving slides permits reduction and simplification of the governing equations that define this initial boundary value problem. This shallow slope approximation makes it possible to separate the determination of the stress and velocity distributions from the evolution of the movable surfaces. As for the material response, objective, rate-type constitutive relations are presented for the soil which is treated as a non-Newtonian viscous fluid matrix with an ideal interstitial fluid, whereby the effective stress is additively decomposed into a so-called quasi-elastic and a viscous component. The free phenomenological coefficient functions are calibrated with the aid of laboratory and field data. Computational results for real landslides are compared with field data, demonstrating the usefulness of the proposed theoretical model and approximations.

1. Introduction

Natural slopes in slow movement are often found, particularly in the Alpine arc. Their occurrence seems also to be increasing due to the increase of periods with intensive rainfall during the last few years. During such periods the phreatic surface of the groundwater may rise to an unusually high level, stay there for some time and thus trigger soil motion that is not present otherwise. Typical sliding or creeping areas are of the order of $10^6\,m^2$ and long term velocities (over 50 years, say) can be $1 - 20\,cma^{-1}$! However, short term fluctuations may be ten times as large or more, and on occasion cause catastrophic landslides. In these circumstances roads, buildings and entire villages may suffer damage, and loss of life may occur.

As such, the slow movement of natural slopes is a well known geological process, however, from a geomechanical point of view its analysis is still insufficiently understood. The difficulties in analyzing, modelling, and predicting such phenomena emerge from essentially four different sources:

- Most of these movements occur in isolated and remote areas (sometimes even as submarine events), so access to the sites is usually difficult.

- The movements, which are typically a few centimetres per year or less, are difficult to measure and the instrumentation is often complex and expensive. The measurements involve surface and aerial photogrammetry, borehole excavation, inclinmometry, piezometry and other geophysical techniques. Moreover, the larger the moving soil areas and the duration of motion are, the more complex and more expensive the surveys will be.

- To quantify the full time-dependent behaviour the constitutive response of the soil must be known, however, the material behaviour is presently still insufficiently understood. Laboratory tests are often far from being representative of long-term in situ behaviours (whose response times are several weeks, months or years), and long-term laboratory tests are expensive and difficult to control. On the other hand, field tests are often scarce and incomplete for an adequate test of a theoretical model.

- Slope movements are often strongly influenced by hydrological factors. When using any mathematical model that incorporates hydrological factors in predicting time-dependent behaviour of a slope, one is forced to prescribe hydrological boundary conditions. This is very hard, as it is fraught with uncertainties.

In spite of these and other difficulties, interest in this field has risen because more and more structures (such as roads, buildings touristic resorts, etc.) are being built in landslide-prone areas. They then form a potential danger, since slowly moving slopes often develop into more brutal or catastrophic landslides.

A multidisplinary four-year study of "landslide prone areas" ended in Switzerland in 1986 (see DUTI 1985). It led to the collection of a large amount of data for landslides of different geometrical, geological, and hydrogeological conditions. We shall draw heavily from this data set. Interesting studies concerning the movement of natural slopes have also been done by HAEFELI (1953, 1965). HUDER (1976). DYSLI and RECORDON (1981), ENGEL (1986), and VULLIET (1986). In France, similar studies were undertaken (see ANTOINE et al. 1971; BIAREZ and BOUCEK 1973), some of which are particularly concerned with the behaviour of an embankment built on an unstable and creeping slope (MORIN, 1979; BLONDEAU et al., 1983; CARTIER and POUGET, 1987; POUGET et al., 1985; BOURDEAU et al., 1988). Further studies from Germany (GUDEHUS et al., 1976; KOERNER, 1969), eastern Europe (MESCHYAN and BADALYAN, 1976; TER-MARTINOSYAN, 1971; TER-STEPHANIAN, 1963, 1965, 1980). Italy (Consiglio nazionale delle ricerche, 1982), Canada (TAVENAS and LEOUEIL,

1981; KENNEY and LAU, 1984), and the USA (NELSON and THOMPSON, 1974; NOBLE, 1973) demonstrate the worldwide occurrence of (time-dependent) landslide phenomena. The above list is, however, far from being exhausive.

Creeping (instability) phenomena tend to arise in the first (ten of) metres of the sedimentary superficial layers; these deposits are then of marine, lacustrine, or glacial origin. In rare cases, the slopes are formed of altered or schistous rocks, in which case the sliding masses may extend over several hundred metres in thickness (see HUDER, 1976). According to experience, the primary cause of the movement can often be attributed to glacier retreat (the slopes are no longer subject to the retaining pressure offered by the glacier) or to erosion of valleys by rivers (leading to increased slope angles). The actual "driving forces" are gravity, groundwater seepage, and, not to be forgotten, the additional weight of man-made structures.

Continuous creeping is not the usual stability problem of geotechnical analysis and it cannot be solved by a limit analysis approach (for a review of slope stability analysis see, for example, the book of CHOWDHURY, 1978). The considered slopes are neither still nor ruptured: *they move.* Conventional stability analyses (e. g. the method of slices) treat the soil as a rigid perfectly plastic body. They do not permit determination of displacement and velocity fields, but instead provide information regarding the safety factor of a stable mass of soil. Methods of analysis that permit determination of the velocity or displacement fields are still rare and even when solutions are proposed, it is for simple geometries (HAEFELI, 1953; STROGANOV, 1961; TER-STEPHANIAN, 1963) or simple material behavior (EMERY, 1978; TRUNK et al, 1986). Continuum models that are capable of simulating continuous slope movements do exist and may have various microstructural complexities: The soil may be treated as a *granular continuum* subject to quasistatic deformations; or it may be thought of as a mixture of two interpenetrating continua, a *solid matrix* and a *fluid filling the pore volume.* In this second model, the granular structure of the soil may be regarded as important, and then the model will be a mixture of interpenetrating constituent continua one of which at least will have granular structure. In the second case, the interaction of the interstitial water with the soil matrix is regarded as more significant than the microstructural pecularities of the soil, and so a *simple classical mixture concept* may be applied. In a yet even farther reaching simplification, one may also ignore the dynamic role of the water and treat the wet soil as a single constituent continuum with nonlinear rheological behaviour.

We shall employ here the classical mixture concept and demonstrate its usefulness by comparing results that have been obtained from theory with observations. Several continuum models that are capable of simulating continuous slope movements and were developed on these lines are:

1. the homogeneous thermoviscous body (HUTTER and VULLIET, 1985),

2. the two–phase elastoviscous mixture (VULLIET and HUTTER, 1988a; VULLIET and RECONDON, 1987),

3. a multilayer model (VULLIET and HUTTER, 1988b).

We also reviewed and developed constitutive models for creeping soils (VULLIET and HUTTER, 1988c,d). All of these models make use of rate-dependent constitutive responses, relationships which are somwhat alien to the geotechnical readership. Our studies have, however, demonstrated that a rate-dependent formulation is a competitive alternative to the classical approach. As will become apparent, more than just a single model is capable of reproducing the flow of soil in reasonable agreement with observations. The reason for this lies in the insufficiently detailed knowledge of the geometry and the varied material properties at a given site, facts which make an "exact" reproduction of the detailed motion of the soil impossible. On the other hand the continuing difficulty of collecting sufficiently detailed flow data in the field justifies restriction to simple theoretical models that are capable of predicting the gross features properly but may not resolve the details on the small scale level.

2. Descriptive features

Most creeping slopes exhibit deformation (or velocity) profiles similar to those given in *Fig.* 1. The examples of inclinometer measurements shown are taken from Swiss and French landslides; they are typical of all and may thus be representative of any other region.

An inclinometer curve gives the displacements of a plastic tube sealed in a borehole relative to its reference position at a previous time. Repeating the recordings permits determination of the deflection (displacement) of any point along the tube as a function of time. From these measurements, weakly, monthly, or yearly velocity profiles may be deduced. As a consequence, the inclinometer profiles given in *Fig.* 1 may be interpreted as velocity profiles.

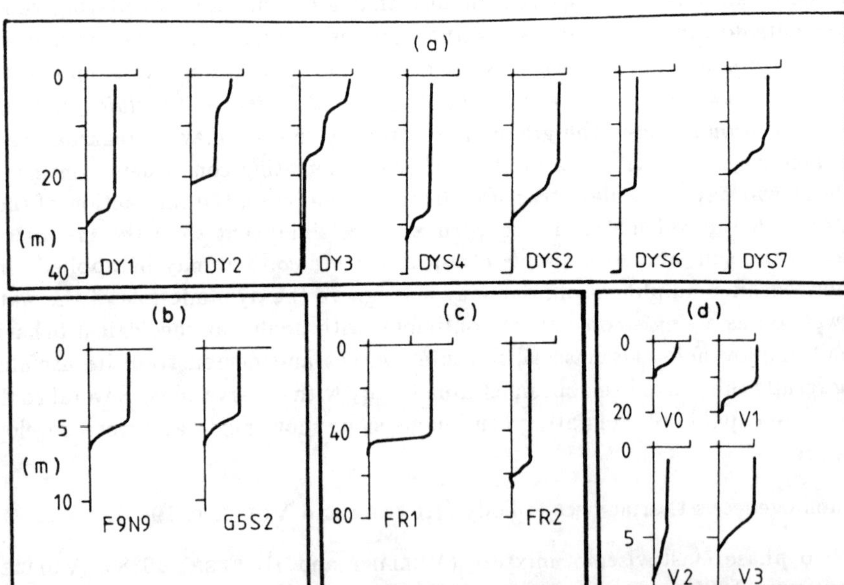

Fig. 1: Measured inclinometer curves for boreholes from four different landslides in Switzerland (CH) and France (F): (a) LeDay (CH); (b) Sallèdes (F); (c) La Frasse (CH); (d) Villaberney (CH).

It can be distinctly discerned from *Fig.* 1 that the deformations with depth are non-uniform. In general, the greater part of the displacements occurs in a narrow zone. In some cases (*Fig.* 1a) these zones occur at several depths, and the landslide may be regarded as a set of layers, one sliding over the other. Often, this sliding can be interpreted as the limit of a pronounced shear deformation in a thin layer. If its resolution is thought not to be of practical significance this shearing is mathematically more conveniently described as a jump in velocity. If this is done, the velocity vector at any location of the moving soil mass may be decomposed into two parts (see *Fig.* 2), $\boldsymbol{v} = \boldsymbol{v}_B + \boldsymbol{v}_S$, where $\boldsymbol{v}_B$ is the (relative) sliding velocity at the base (or interface) and $\boldsymbol{v}_S$ is the shear velocity due to the shear strain rates of the deforming mass. In some cases, when $\boldsymbol{v}_S$ is small in magnitude in comparison with $\boldsymbol{v}_B$

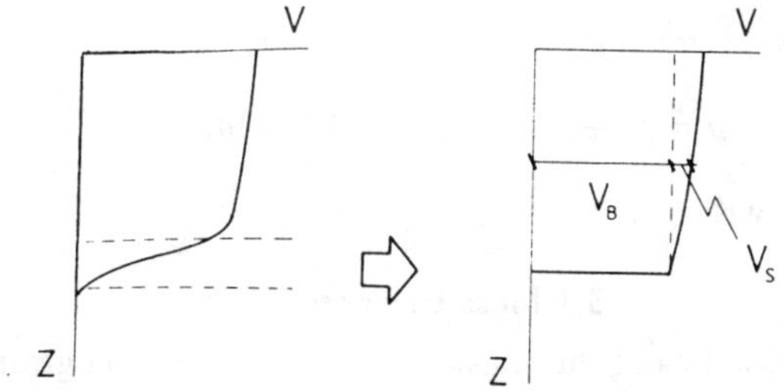

Fig. 2: Idealization of the sliding at the base, and decomposition of the velocity vector into sliding and shearing deformation.

the deformation can be described as the motion of a passively deforming body sliding over a fixed base (*Fig.* 1b,c), in others, $\boldsymbol{v}_B$ may be small in comparison to $\boldsymbol{v}_S$ (*Fig.* 1a, DYS2, DYS7) and the motion can be described as the result of creeping or shearing of a mass of granular material adhering to a rigid bed; in yet other situations both, sliding and shearing under creep contribute to the deformation. This situation prevails in *Fig.* 1a which shows inclinometer curves for the same landslide at various positions of the site. At places a sliding type deformation is dominant, at others it is the creep deformation.

It follows that under such circumstances a combined model must be employed for the entire slide under consideration.

3. Balance laws and their simplification

We shall base our formulation of the continuous deformation of the moving soil that is saturated with water on the basis of a two-phase mixture model. Mass and momentum balances of both constituents, one forming the solid phase, the other the liquid phase, will be regarded as significant; however thermodynamic considerations will only be taken into consideration insofar as the postulated constitutive relations must be in conformity with the second law of thermodynamics. Thermal processes will be ignored. This implies that the temperature does formally not enter the field equations. Such a procedure, of course, eliminates treatment of geothermal heat processes.

It is convenient, and certainly also realistic to assume the granules and the interstitial fluid to be density preserving. If the intrinsic densities of the mineral grains and the water are denoted by $\hat{\rho}_s$ and $\hat{\rho}_w$ respectively, then this assumption may be stated as

$$\hat{\rho}_s = \text{constant} , \quad \hat{\rho}_w = \text{constant} \tag{3.1}$$

with $\hat{\rho}_s = 2700\, kgm^{-3}$ and $\hat{\rho}_w = 1000\, kgm^{-3}$. Assuming the constituent bodies to be density preserving does not mean that the constituent bodies or the mixture be volume preserving. Indeed, let $\hat{\rho}_s$ and $\hat{\rho}_w$ be the constituent densities in the mixture and n the porosity (i. e. the volume of water per unit total volume); then

$$\rho_s = (1-n)\hat{\rho}_s , \quad \rho_w = n\hat{\rho}_w , \tag{3.2}$$

and the mixture density, ρ , is given by

$$\rho = \rho_s + \rho_w = n\hat{\rho}_w + (1-n)\hat{\rho}_s . \tag{3.3}$$

Clearly, ρ_s , ρ_w and ρ vary with n .

3.1 Mass Balances

Let $\boldsymbol{v}_w$ and $\boldsymbol{v}_s$ be the velocities of the constituents water and soil grains relative to a fixed coordinate system, respectively. With the help of (3.2) and (3.3) the constituent balance laws of mass

$$\frac{\partial \rho_\alpha}{\partial t} + \operatorname{div}(\rho_\alpha \boldsymbol{v}_\alpha) = 0 , \quad (\alpha = s, w) , \tag{3.4}$$

which are written here for conservative processes (zero mass production rates), take the form

$$\begin{aligned} \frac{\partial n}{\partial t} &+ \operatorname{div}(n\boldsymbol{v}_w) &= 0 , \\ -\frac{\partial n}{\partial t} &+ \operatorname{div}((1-n)\boldsymbol{v}_s) &= 0 , \end{aligned} \tag{3.5}$$

or after addition

$$\operatorname{div}\left[n\boldsymbol{v}_w + (1-n)\boldsymbol{v}_s\right] = 0 , \tag{3.6}$$

which may be interpreted as the balance law for the mixture as a whole. It is common in soil mechanics and soil hydrology to introduce the *filter velocity* as the porosity-weighted-difference velocity of the velocity of the interstitial water and the soil:

$$\boldsymbol{v}_F = n(\boldsymbol{v}_w - \boldsymbol{v}_s) . \tag{3.7}$$

Often this is also called the *filtration velocity of* DARCY. As the difference of two velocities it is an objective vector under Euclidean transformations and may therefore serve as a dependent or independent constitutive quantity in a material theory that obeys the principle of material objectivity. With (3.7), Equation (3.6) takes the form

$$\operatorname{div}(\boldsymbol{v}_F + \boldsymbol{v}_s) = 0 . \tag{3.8}$$

The sum of the filtration and soil velocities is solenoidal. We emphasize, however, that neither (3.6) nor (3.7) states that the mixture as a whole is incompressible, because in that case the barycentric mixture velocity

$$\rho \boldsymbol{v} := \rho_s \boldsymbol{v}_s + \rho_w \boldsymbol{v}_w \Longrightarrow \boldsymbol{v} = \frac{\rho_s}{\rho}\boldsymbol{v}_s + \frac{\rho_w}{\rho}\boldsymbol{v}_w \tag{3.9}$$

would have to be solenoidal which it obviously is not. Nevertheless the filtration velocity is a very useful concept as we shall soon see.

3.2 Balances of momenta

As is known from the general formulation of continuum mechanics of mixtures, there are several levels of complexity such theories can be based upon. In a purely diffusive model all constituent balance laws of mass are combined with the balance law of momenta for the mixture as a whole, and these equations are complemented by constitutive equations that close the system of equations. In the description of the deformation of water-saturated soil such a diffusive model is insufficient, because the cause of the deformation of the soil matrix and the flow of the interstitial fluid is gravity. Thus, the momentum balances of both constituents are important dynamical statements, and none of them can simply be replaced by a diffusive constitutive model. This is the physical essence behind DARCY's law. Nevertheless, a full dynamic treatment of the momentum balance laws is not needed. Indeed, it can be demonstrated by means of scaling arguments that the acceleration terms are, in general, much smaller than the forces balancing these equations. In other words, the momentum balances reduce to the force balances:

$$\begin{aligned} \operatorname{div}\boldsymbol{t}_s &+ \rho_s \boldsymbol{g} + \rho\boldsymbol{\beta} = \boldsymbol{0} , \\ \operatorname{div}\boldsymbol{t}_w &+ \rho_w \boldsymbol{g} - \rho\boldsymbol{\beta} = \boldsymbol{0} , \end{aligned} \tag{3.10}$$

in which $\boldsymbol{t}_s$ and $\boldsymbol{t}_w$ are the partial (constituent) CAUCHY stress tensors, $\boldsymbol{g}$ is the gravity vector and $\rho\boldsymbol{\beta}$ the seepage interaction force. In writing Equations (3.10) it was observed that the momentum production terms add up to zero; i. e. the volume force acted upon the solid by the interstitial fluid is equal and opposite to the force acted upon the fluid by the solid matrix. Moreover, by adding the two force balances (3.10),

$$\operatorname{div}\boldsymbol{t} + \rho\boldsymbol{g} = \boldsymbol{0} \tag{3.11}$$

is obtained as the force balance of the mixture as a whole in which

$$\boldsymbol{t} = \boldsymbol{t}_s + \boldsymbol{t}_w , \quad \boldsymbol{t} = \boldsymbol{t}^T \tag{3.12}$$

is the symmetric CAUCHY stress of the mixture. It is emphasized that (3.12) does not conform with the classical definition of the mixture stress,

$$\begin{aligned} \boldsymbol{t}^{\text{class}} &= \boldsymbol{t} + \sum_{\alpha} \rho_\alpha \boldsymbol{c}_\alpha \otimes \boldsymbol{c}_\alpha , \\ \boldsymbol{c}_\alpha &= \boldsymbol{v}_\alpha - \boldsymbol{v} , \qquad (\alpha = s, w) , \end{aligned} \tag{3.13}$$

which contains the sum of the diffusive fluxes of the constituent diffusive momenta ($\boldsymbol{c}_\alpha$ = diffusion velocity of constituent α). This term is missing in (3.12) because the acceleration terms were ignored in the momentum balances (3.10).

Equations (3.5) and (3.10) of the balance statements of mass and linear momenta are the basic dynamical statements of a purely mechanical (as opposed to thermodynamical) model and need be complemented by constitutive relations. Before we turn to these statements a few remarks are in order that clarify the approach taken there. Firstly, the partial stress in the water does not equal the pore pressure p_w [1]. In fact, it is the pore pressure that is measured piezometrically. Secondly, since we have assumed the constituents to be density preserving (but not necessarily also the mixture) our mixture model is subjected to a constraint, which induces a free pressure, in a similar fashion as with every incompressible single material. This pressure is the free pressure of the mixture and will be identified with the pore pressure p_w We will assume here that it is distributed among the two constituents according to the volume fraction present. Thus we may write

$$\begin{aligned} \boldsymbol{t}_w &= -np_w\mathbf{1} + \boldsymbol{\sigma}_w\,, \\ \boldsymbol{t}_s &= -(1-n)p_w\mathbf{1} + \boldsymbol{\sigma}_s\,, \\ \boldsymbol{t} &= -p_w\mathbf{1} + (\boldsymbol{\sigma}_w + \boldsymbol{\sigma}_s) = -p_w\mathbf{1} + \boldsymbol{\sigma}\,. \end{aligned} \tag{3.14}$$

The pore pressure p_w is a free scalar field that will be determined through the solution of the boundary value problem that will eventually be formulated for the variables under consideration; $\boldsymbol{\sigma}_w$ and $\boldsymbol{\sigma}_s$ are the extra stresses in the interstitial water and the soil matrix, respectively. It is for these variables that constitutive relations are formulated. Thirdly, we recall that, as a consequence of the balance law of momentum the total stress $\boldsymbol{t}$ (or the sum $\boldsymbol{\sigma}_w + \boldsymbol{\sigma}_s$) need be symmetric. At this level of development this symmetry is not required for $\boldsymbol{\sigma}_w$ and $\boldsymbol{\sigma}_s$ separately; one only must request that $\text{skew}(\boldsymbol{\sigma}_w) = -\text{skew}(\boldsymbol{\sigma}_s)$. However, if one of the partial stresses is symmetric (by assumption or through rational argumentation), then the other partial stress will necessarily be symmetric as well. This will be the case in the constitutive model we are about to develop.

[1] In soil mechanics this pore pressure is denoted by u . We deviate from this convention since u has also other meanings in this text.

4. A set of constitutive models for soil under slow movement

4.1 General remarks

As the interstitial water and the soil interact in a saturated soil, it is to be expected that both constituents will also influence the creeping properties according e. g. to the proportion of the two constituents that are present. In other words, the constitutive relation for the partial CAUCHY stress $\boldsymbol{t}_s$ of the soil will not only depend upon the (history of) deformation of the soil, but also on the properties of the interstitial fluid or at least on a variable that describes its presence.

One purpose in establishing a relationship between stress and (a measure of) deformation is to determine the stability of a soil mass. Limit-type formulas are adequate to describe these situations (TERZAGHI, 1939; DRUCKER & PRAGER, 1952). A stress measure approaches a limit value, when large or sudden changes of configuration arise. The deformation field or its time rate of change is usually not determined in these formulations. Of a different kind are rate-dependent formulas, which relate stress to the deformation more explicitly. They permit the determination of configurational changes of a soil mass under external forces and allow, at least in principle, the prediction of the deformation of natural slopes in slow movement. There is an abundant literature on experimental and theoretical investigations of this kind (for a excerpt of results see *Tables 1 & 2*). Usually, uni-, bi- and tri-axial compression tests and shear tests under various conditions are performed. Creep and stress curves are plotted and mathematical relations are deduced that functionally relate a strain rate measure with the stress, (perhaps time) and further phenomenological parameters.

To make these ideas a bit more precise, consider a shear test in which a layer of the material (soil plus water) is held between two plates and sheared by applying a pair of forces parallel to the plates as indicated in *Fig.* 3. An experiment that is performed under a HEAVISIDE type external force $\tau := \tau H(t)$ is referred to a creep experiment under constant stress.

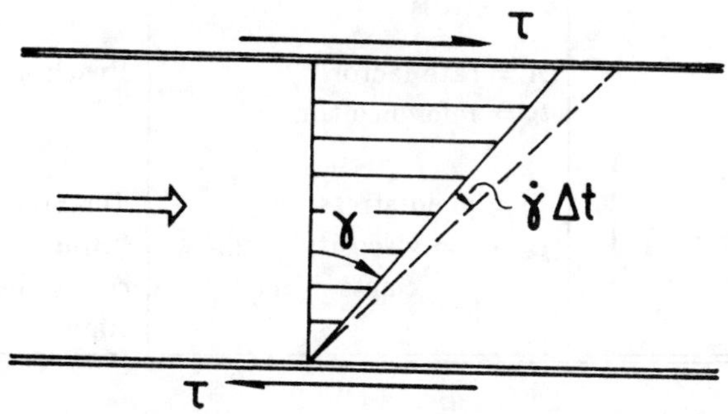

Fig. 3: Creep experiment under pure shear: Applied forces correspond to a shear stress, γ is the shear angle.

Formula (1)	Parameters (2)	Remarks and literature (3)
$\dot{\gamma}(t) = \frac{\tau}{2\mu}\exp\left(-\frac{G}{\mu}t\right)$	G = shear modulus; μ = dynamic viscosity	Kelvin model proposed by Calabresi (1970). Combinations of such models in series were proposed by Fedder (1972), Komamura and Huang (1974), and Murayama et al. (1974).
$\gamma(t) = \frac{\tau(t)}{G} + \int_{-\infty}^{t} K(t-s)\tau(s)ds$	G = shear modulus; K = creep function, often a combination of exponentials	Cakmak et al. (1970) and Krizek (1970).
$\dot{\gamma}(t) = A\exp\left(\alpha\frac{\tau}{\tau_r}\right)\left(\frac{t_1}{t}\right)^m$	A = rate factor, often temperature-dependent; α, m = exponents; t_1 = reference time; τ_r = yield (failure) shearing stress	Popular, but physically not objective formulation. Singh and Mitchell (1968), Hirst (1968), Kavazajian (1978) and Mesri et al. (1981). Orders of magnitude for A, α, m are collected by Vulliet (1986).
$\dot{\gamma}(t) = A\left(\frac{\tau}{\tau_r}\right)^m\left(\frac{t_1}{t}\right)^n$	A = rate factor; m, n = exponents (n may depend on τ); t_1 = reference time; τ_r = yield (failure) shearing stress	Feda et al. (1973).
$\dot{\gamma}(t) = A\frac{\tau}{t+t_o}$	A = rate factor; t_o = reference time	Pusch and Feltham (1980).
$\dot{\gamma}(t) = \left\{\begin{array}{ll} \frac{\tau-\tau_o}{\mu_o+\mu_1 t}, & \tau > \tau_o \\ 0, & \tau \leq \tau_o \end{array}\right\}$	τ_o = yield stress; μ_o, μ_1 = viscosity coefficients	Ter-Stephanian (1973). "Bingham-model" with viscosity that varies linearly with time.

Table 1: Flow laws – Explicit Functions of Time.

The relative displacement of the upper to the lower plate stipulates a linear distribution of the horizontal displacements and velocities. This motion can be quantified by recording the shear angle $\gamma(t)$ or its time rate of change $\dot{\gamma}(t)$ as functions of time. Because it is anticipated that under usual circumstances a slowly creeping soil quickly loses its memory, stress is related to strain rate, and a further dependence on strain is omitted.

Formula (1)	Parameters (2)	Remarks and literature (3)
$\dot{\gamma}(t) = \dfrac{\tau}{\mu}$	μ = Newtonian viscosity	Newtonian fluid model. Geuze (1953), Haefeli (1953) and Boucek (1964).
$\dot{\gamma}(t) = \left\{ \begin{array}{ll} \dfrac{\tau - \tau_o}{\mu}, & \tau > \tau_o \\ 0, & \tau < \tau_o \end{array} \right\}$	τ_o = yield stress μ = viscosity	Bingham body. Geuze and Tan Tjong-Kie (1953), Haefeli (1953), Stroganov (1961), Ter-Stephanian (1963), Yen (1969) and Trunk et al. (1986).
$\dot{\gamma}(t) = A\tau^m$ [a]	A = rate factor, often temperature-dependent via an Arrhenius relation;	Norton's law (1929). Barden (1965), Gudehus et al. (1976), Oka (1981) and Hutter and Vulliet (1985)
$\dot{\gamma}(t) = A\sinh(B_\tau)$ [b]	A = rate factor, often temperature-dependent via an Arrhenius relation; $B = C/T$, where C = constant and T = absolute temperature	Prandtl-Eyring model, see Prandtl (1928) and Glasstone et al. (1941). Murayama and Shibata (1961), Mitchell (1964), Christensen and Wu (1964), Andersland and Douglas (1970), and Gudehus et al. (1976)

[a] Polynomials $\dot{\gamma} = \sum_{j=1}^{N} A_j \tau^j$ are possible generalizations.

[b] For large τ, we may write $\gamma = (1/2)A\exp(B\tau)$, which is the logarithmic viscosity law. A similar alternative of the Prandtl-Eyring law, $\dot{\gamma} = A\sinh^n(B\tau)$, $n < 1$, was developed by Garofalo (1963). Neither one has been proposed in soil mechanics.

Table 2: Flow laws – Implicit functions of time.

With these prerequisites we may now properly esteem the relationships collected in *Tables 1 & 2*, which separate formulas that relate a strain rate measure with stress (and time). The formulas in *Table 1* are the result of a direct fit of experimental data. Often they are *ad-hoc* and do not permit deduction of physically objective constitutive relations. The second formulas (*Table 2*) do not explicitly involve time. These formulas are reductions to simple experimental situations of physically objective three-dimensional forms of constitutive relations of stress. They are possible reductions of adequate modelling schemes of the creeping flow of soil in natural slopes.

As a final remark to this introductory section we mention that in this present article we shall primarily confine attention to the rate dependent response. This tacitly implies that the model equations will be able to describe the creeping deformation of a mass of soil (saturated with water) but not the rupture of a stable mass of soil that is in the process of initiating its onsetting motion.

4.2 Material response for differential creep

In this and the following sections continuum mechanics will be used to postulate physically objective constitutive relationships for the three-dimensional constitutive relation of stress. Additional hypotheses will help to simplify the mathematical expressions further, so that uni-, bi-, or tri-axial compression experiments permit determination of material parameters. Two classes of soil models will be analysed: (1) an incompressible, viscous, isotropic (one component) soil with no interstitial fluid, which is representative of a soil whose moisture content hardly varies (and thus does not explicitly affect its creep response); and (2) a mixture of a solid (matrix) and an interstitial fluid, which is appropriate for saturated conditions.

4.2.1) Incompressible viscous model

Let $\boldsymbol{t} = \boldsymbol{\sigma}_s$ be the CAUCHY stress and

$$\boldsymbol{D} = \tfrac{1}{2}(\boldsymbol{L} + \boldsymbol{L}^T)\,, \quad \boldsymbol{L} = \mathrm{grad}\boldsymbol{v}$$

the stretching tensor as deduced from the velocity field $\boldsymbol{v} = \boldsymbol{v}_s$ of the soil. We now anticipate that a constitutive relation of the form

$$\boldsymbol{D} = \hat{\boldsymbol{D}}(\boldsymbol{t}) \tag{4.1}$$

will model creep response of deforming soil adequately. It follows from general principles of continuum mechanics (see e. g. TRUESDELL, 1977, 1978, MÜLLER, 1973, 1985) that the materially objective isotropic form of the constitutive relation (4.1) is given by

$$\begin{aligned} \boldsymbol{D} &= a\mathbf{1} + b\,\boldsymbol{t} + c\,\boldsymbol{t}^2\,, \\ a\,, b\,, c &= \mathrm{fcts}(I_{\boldsymbol{t}}\,, II_{\boldsymbol{t}}\,, III_{\boldsymbol{t}})\,, \end{aligned} \tag{4.2}$$

in which

$$I_{\boldsymbol{t}} = \mathrm{tr}\boldsymbol{t}\,, \quad II_{\boldsymbol{t}} = \tfrac{1}{2}\mathrm{tr}(\boldsymbol{t}^2)\,, \quad III_{\boldsymbol{t}} = \tfrac{1}{6}\mathrm{tr}(\boldsymbol{t}^3) \tag{4.3}$$

are three independent stress invariants, tr being the trace operator; furthermore, a, b and c are scalar valued coefficients and $\mathbf{1}$ is the unit tensor. The class (4.2), (4.3) of constitutive relations is known as a Reiner-Rivlin fluid.

Volume preservance imposes further restrictions. First, since the pressure is a free variable, we may without loss of generality assume that $\boldsymbol{t}$ is a deviator, so tr$\boldsymbol{t} = I_{\boldsymbol{t}} = 0$. Then it is easy to show that (4.2) takes the form [2]

$$\begin{aligned} \boldsymbol{D} &= b\,\boldsymbol{t} + c\left(\boldsymbol{t}^2 - \tfrac{2}{3}II_{\boldsymbol{t}}\boldsymbol{1}\right) , \\ b &= b(II_{\boldsymbol{t}}\,, III_{\boldsymbol{t}})\,, \quad c = c(II_{\boldsymbol{t}}\,, III_{\boldsymbol{t}})\,. \end{aligned} \tag{4.4}$$

This relation is still more general than the creep laws listed in *Table 2.* The simpler class is obtained from equation (4.4) by the following postulates:

- The stress deviator is collinear to the stretching tensor. This necessarily requires $c = 0$.
- The remaining scalar function b does not depend on the third stress deviator invariant, $III_{\boldsymbol{t}}\,:\, b = b(II_{\boldsymbol{t}})$.

These postulates imply that

$$\boldsymbol{D} = b(II_{\boldsymbol{t}})\,\boldsymbol{t}\,, \tag{4.5}$$

where b is an arbitrary function of its argument which must be determined from experiment. *Table 3* shows the form of this creep function, when the laws of Table 2 are considered.

Flow law (1)	$b(II_t)$ (2)	Remarks (3)
Newton	$1/\mu$	–
Bingham	$[1-\sqrt{II_{t_o}/II_t}]/\mu$	Applicable, if $II_t > II_{t_o}$.
Norton	$AII_t^{(m-1)/2}$	Also called power law.
Prandtl-Eyring	$[A\sinh(B\sqrt{II_t})]/\sqrt{II_t}$	$\to AB$, if $\sqrt{II_t}\to 0$.
Prandtl	$[A\exp(B\sqrt{II_t})]/\sqrt{II_t}$	$\to (B/\sqrt{II_t})$, if $\sqrt{II_t}\to 0$.

Table 3: Generalized creep response functions for various laws. II_t denotes the second stress deviator invariant.

4.2.2 Constitutive law for a saturated compressible soil

We now return to the original concept and regard the soil as a saturated mixture of a "solid" and an interstitial fluid. We use quotation marks, because our model for the granular matrix will be a very viscous fluid; we further suppose the water to be an ideal or perfect fluid

[2] For simple shear $u = u(y)\,,\ v = 0\,,\ w = 0\,,$ the stress that develops for a constitutive relation (4.4) is not simply the corresponding shear stress. Normal stresses are equally developed. In fact (4.4) imposes five additional relations that would have to hold in order that simple shear could exist for a general Reiner–Rivlin fluid.

which cannot sustain any viscous stress. With this supposition it follows that the partial stresses $\boldsymbol{\sigma}_s$ and $\boldsymbol{\sigma}_w$ are both symmetric. More specifically, since in a density preserving saturated soil, as defined in (3.1), the isotropic stress is absorbed in the pore-water pressure p_w, see Equations (3.14), and when elastic response is ignored we have $\boldsymbol{\sigma}_w = \mathbf{0}$. Thus, the mixture stress $\boldsymbol{\sigma}$ is given by

$$\boldsymbol{\sigma} = \boldsymbol{\sigma}_s + \boldsymbol{\sigma}_w = \boldsymbol{\sigma}_s \, . \tag{4.6}$$

We shall now take the stress $\boldsymbol{\sigma}$ to be of the constitutive class $\boldsymbol{\sigma} = \boldsymbol{\sigma}(n, \boldsymbol{D})$, where n equals the porosity as before and $\boldsymbol{D}$ is the stretching tensor of the deformation field of the solid phase.

We now suppose *viscotropy*, i. e. additive decomposition of the effects of n and $\boldsymbol{D}$ in the following manner:

$$\boldsymbol{\sigma}(n, \boldsymbol{D}) = \boldsymbol{\sigma}^e(n) + \boldsymbol{\sigma}^v(n, \boldsymbol{D}) \tag{4.7}$$

i. e., the constitutive part of the mixture stress has a contribution that depends merely upon the porosity and a second contribution which depends primarily on the "solid-matrix" deformation rate (and also on the porosity). Following common usage, we shall call $\boldsymbol{\sigma}^e(n)$ the elastic stress, however, this designation is inappropriate and misleading in general, because $\boldsymbol{\sigma}^e(n)$ need not be a "reversible stress". We shall use the term "elastic stress" in a very broad sense and absorb in $\boldsymbol{\sigma}^e$ all non-viscous effects. Terms like "pseudo-elastic", "quasi-elastic", and "elasto-plastic" may all be more appropriate substitutes. [3] The second term in (4.7), $\boldsymbol{\sigma}^v$, however, is justly referred to as viscous stress.

a) Elastic stress

Clearly, because $\boldsymbol{\sigma}^e = \boldsymbol{\sigma}^e(n)$ depends upon a scalar variable only, it must be isotropic so that

$$\boldsymbol{\sigma}^e(n) = -p^e(n)\mathbf{1} \, , \tag{4.8}$$

where $p^e(n)$ is an "elastic pressure". Explicitly, the p^e versus n relationship follows from an oedometric experiment or an isotropic compression test. In general, one plots void ratio $e = n/(1-n)$ against the logarithm of the spherical stress p^e . Loading and unloading follow separate curves. The compression curve may be written as

$$e = e_c - 0.435 C_c \ln \frac{p^e}{p^c} \tag{4.9}$$

where p^c is the *consolidation pressure*, e_c the *void ratio* corresponding to this consolidation pressure, and C_c the *compression index.*

[3] σ^e may for instance also contain rate independent contributions and describe plastic behaviour. For instance if

$$\sigma^e = a\mathbf{1} + b\frac{\boldsymbol{D}}{\sqrt{\operatorname{tr}(\boldsymbol{D}^2)}}$$

where a and b are scalars which may depend on n, then this stress depends on n and D but is rate independent. It would allow to model a Mohr–Coulomb type yield behaviour that includes cohesion. We shall be less general above.

Strictly speaking, Equation (4.9) is only valid for normally consolidated soils. Unloading, for which swelling arises, follows a different curve. p^c now represents the maximum compressive stress that was attained in a load cycle. Modelling these hysteretic effects accordingly would amount to establishing an elasto-plastic model similar to the famous "Cam–clay" of ROSCOE and BURLAND (1968) and SCHOFIELD and WROTH (1968). We shall not do it here and shall use the reversible law (4.9). In this sense the term "elastic stress" is appropriate.

b) Viscous stress

Many of the algebraic manipulations that are performed in the subsequent analysis are necessary, because the viscous stress $\boldsymbol{\sigma}^v$ is not written as a function of the stretching tensor but the inverse relationship, $\boldsymbol{D} = \boldsymbol{D}(n, \boldsymbol{\sigma}^v)$, is used instead. The reason is that creep tests stipulate a material behaviour that resembles a pseudo-plastic rather than a ***dilatant-fluid response.*** [4] When isotropic behaviour is imposed the constitutive relation for the viscous stress takes the form

$$\boldsymbol{D} = a\mathbf{1} + b\boldsymbol{\sigma}^v + c(\boldsymbol{\sigma}^v)^2 \tag{4.10}$$

with coefficients a, b, c , which may be regarded as functions of n and the stress invariants of $\boldsymbol{\sigma}^v$. By decomposing $\boldsymbol{\sigma}^v$ into an isotropic and a deviatoric contribution

$$\boldsymbol{\sigma}^v = -p^v\mathbf{1} + \boldsymbol{\sigma}', \quad (\mathrm{tr}\boldsymbol{\sigma}' = 0), \tag{4.11}$$

an alternative form of Equation (4.10) can be shown to be

$$\boldsymbol{E} = \left(\boldsymbol{D} - \tfrac{1}{3}\mathrm{tr}\boldsymbol{D}\mathbf{1}\right) = b\boldsymbol{\sigma}' + c\left\{(\boldsymbol{\sigma}')^2 - 2p^v\boldsymbol{\sigma}' - \tfrac{2}{3}II_{\boldsymbol{\sigma}'}\mathbf{1}\right\} \tag{4.12}$$

in which b and c are scalar valued functions of the porosity, n, the viscous pressure, p^v, and the second and third invariants of $\boldsymbol{\sigma}'$:

$$b = b(n, p^v, II_{\boldsymbol{\sigma}'}, III_{\boldsymbol{\sigma}'}), \quad c = c(n, p^v, II_{\boldsymbol{\sigma}'}, III_{\boldsymbol{\sigma}'}). \tag{4.13}$$

The coefficient a in Equation (4.12) has been eliminated by taking the trace of Equation (4.10). This trace equation yields the further statement, that

$$\mathrm{tr}\boldsymbol{D} = 3\left(a - bp^v + c(p^v)^2\right) + 2cII_{\boldsymbol{\sigma}'}. \tag{4.14}$$

Equations (4.12) and (4.14) comprise the complete constitutive relations for the viscous part of the granular stress. The two expressions are still rather complicated, and an experimental determination of the phenomenological coefficients a, b, c does not seem to be directly possible, because under usual circumstances there are not enough experimental data available

---

[4] In a pure shear experiment the shear rate $\dot{\gamma}$ and the shear stress τ may be related by the relationship

$$\dot{\gamma} = f(\tau)$$

such that

$$\lim_{\dot{\gamma}\to 0} \dot{\gamma} = \lim_{\tau\to 0} f(\tau) \Longrightarrow \tau^n.$$

A material with $0 < n < 1$ is called a dilatant fluid; if $n > 1$ then its response is denoted as pseudo-plastic; $n = 1$ characterizes Newtonian behaviour.

that would permit sufficient identification of these parameters. Thus, further simplifications are mandatory if only because of pragmatic reasons. Thus we impose the following additional assumptions:

- The stress deviator $\boldsymbol{\sigma}'$ and the *distorsion tensor* $\boldsymbol{E}$ are collinear. This implies that $c = 0$.

- The remaining phenomenological coefficients a and b do not depend upon the third deviator invariant $III_{\boldsymbol{\sigma}'}$.

These assumptions reduce Equations (4.12) and (4.14) to

$$\begin{aligned} \boldsymbol{E} &= \boldsymbol{D} - \tfrac{1}{3}(\mathrm{tr}\boldsymbol{D})\mathbf{1} = b\boldsymbol{\sigma}' , \\ \mathrm{tr}\boldsymbol{D} &= 3(a - bp^v) , \\ a &= a(n , p^v , II_{\boldsymbol{\sigma}'}) , \quad b = b(n , p^v , II_{\boldsymbol{\sigma}'}) . \end{aligned} \tag{4.15}$$

Consider pure shearing deformation for which the trace of $\boldsymbol{D}$ must vanish. Moreover it is feasable to assume that for such a special motion a viscous pressure cannot develop, implying $p^v = 0$. It follows from Equation $(4.15)_2$ under these circumstances that $a = 0$. It may therefore be advantageous to write

$$a = \tilde{a}p^v , \tag{4.16}$$

where $\tilde{a}$ is a function of the form shown in (4.15).

In summary: The constitutive model for the soil stress in a saturated binary mixture of granular material with interstitial water is given by

$$\boldsymbol{\sigma} = -(p^e + p^v)\mathbf{1} + \boldsymbol{\sigma}' = -p'\mathbf{1} + \boldsymbol{\sigma}' , \tag{4.17}$$

where p' is the effective pressure, p^e is given in (4.9) and p^v and $\boldsymbol{\sigma}'$, respectively, are related to the stretching tensor $\boldsymbol{D} = \boldsymbol{D}_s$ of the solid constituent according to

$$\begin{aligned} \mathrm{tr}\boldsymbol{D} &= 3(\tilde{a} - b)p^v , \\ \boldsymbol{E} &= \boldsymbol{D} - \tfrac{1}{3}(\mathrm{tr}\boldsymbol{D})\mathbf{1} = b\boldsymbol{\sigma}' , \end{aligned} \tag{4.18}$$

with

$$\tilde{a} = \tilde{a}(n , p^v , II_{\boldsymbol{\sigma}'}) , \quad b = b(n , p^v , II_{\boldsymbol{\sigma}'}) . \tag{4.19}$$

Here b^{-1} and $(\tilde{a}-b)^{-1}$ may be interpreted as an effective shear and bulk viscosity, respectively.

c) Results

The above constitutive relations can be transformed in part into slightly different forms by using the balance law of mass for the constituent soil, $(3.5)_2$, and by identifying $\mathrm{tr}\boldsymbol{D} = \mathrm{div}\boldsymbol{v}_s$. It is straightforward to show that this equation may also be written as

$$\mathrm{div}\boldsymbol{v}_s = \mathrm{tr}\boldsymbol{D} = \frac{1}{1-n}\frac{d^s n}{d^s t} = \frac{1}{1+e}\frac{d^s e}{dt} , \tag{4.20}$$

in which the time derivative d^s/dt is material with respect to the solid phase and $e := n/(1-n)$ is the void ratio. In view of Equations (4.9), (4.17) and (4.18), the materially dependent part of the pressure can thus be written as [5]

$$p' = \underbrace{p_c \exp\left(\frac{e_c - e}{0.435C_c}\right)}_{p^e} + \underbrace{\frac{1}{3(\tilde{a}-b)}\frac{1}{1+e}\frac{d^s e}{dt}}_{p^v} \; . \tag{4.21}$$

Two limiting cases are suggested by this formula.

Case A: The bulk viscosity $(\tilde{a}-b)^{-1}$ is moderate or small to the extent that $p^v \ll p^e$. In this case $p' = p^e$, and the volumetric creep rate takes the form

$$\mathrm{tr}\boldsymbol{D} = \frac{1}{1+e}\frac{d^s e}{dt} = -\frac{1}{1+e}0.435C_c\frac{d^s}{dt}(\ln(p')) \tag{4.22}$$

and Equations $(4.18)_1$ may be replaced by this relation. This equation demonstrates that in this case the rate of volume change is governed by the rate of change of the effective pressure

$$\frac{d^s e}{dt} = -0.435C_c\frac{d^s}{dt}(\ln(p')) \; . \tag{4.23}$$

Case B: When the bulk viscosity of the soil is large (i. e., when the soil is unconsolidated and compressible), then $p^v >> p^e$ and $p' \simeq p^v$. In this case, Equations (4.18) cannot be simplified. Whereas in *case A*, the coefficient b is the only phenomenological coefficient left to be determined by experiments, there are both, $\tilde{a}$ and b in this case that must be determined. How this is done will be demonstrated in the next section.

4.3: Determination of phenomenological parameters by means of laboratory and field experiments

Consider the cylindrical test specimen of *Fig.* 4.

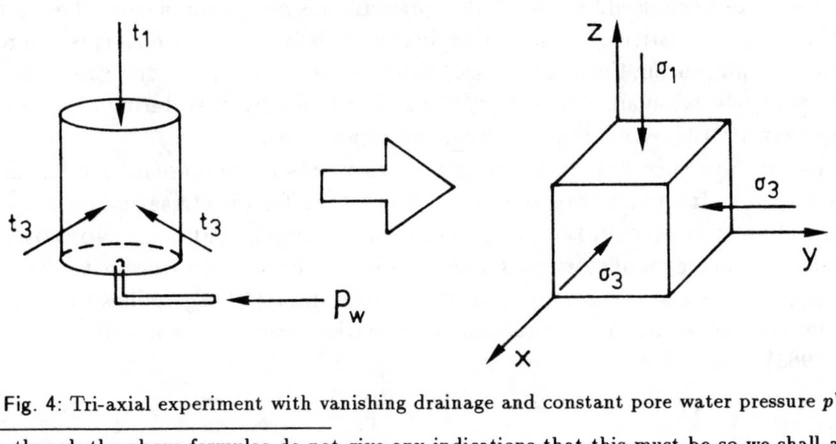

Fig. 4: Tri-axial experiment with vanishing drainage and constant pore water pressure p^w .

[5]Even though the above formulas do not give any indications that this must be so we shall assume that the bulk viscosity is positive. This implies $\tilde{a} - b > 0$.

It is sealed by rapping it in a rubber hull and subjecting it to tri-axial compression with two principle stresses, t_1, vertical and t_3 horizontal, and externally applied pore water pressure p^w. In the centre of the cylinder the state of stress is as shown on the right of *Fig.* 4, and in the coordinate system shown the stress and strain rate tensors have diagonal form

$$\begin{aligned} \operatorname{diag}\boldsymbol{t} &= (-t_3\,,-t_3\,,-t_1)\,, \\ \operatorname{diag}\boldsymbol{\sigma} &= (-t_3+p^w\,,-t_3+p^w\,,-t_1+p^w)\,, \\ \operatorname{diag}\boldsymbol{\sigma}' &= \operatorname{diag}\left(\boldsymbol{\sigma}-\tfrac{1}{3}(\operatorname{tr}\boldsymbol{\sigma})\mathbf{1}\right) = \tfrac{1}{3}(t_1-t_3)(1\,,1\,,-2)\,, \\ \operatorname{diag}\boldsymbol{D} &= (-\dot\epsilon_3\,,-\dot\epsilon_3\,,-\dot\epsilon_1)\,, \\ \operatorname{diag}\boldsymbol{E} &= \tfrac{1}{3}(\dot\epsilon_1-\dot\epsilon_3)(1\,,1\,,-2)\,. \end{aligned} \tag{4.24}$$

Here, the principal stretchings in the x, y and z directions are denoted by $-\dot\epsilon_3$, $-\dot\epsilon_3$ and $-\dot\epsilon_1$, respectively, the negative sign corresponding to contraction. Clearly,

$$\begin{aligned} II_{\boldsymbol{\sigma}'} &= \tfrac{1}{2}\operatorname{tr}(\boldsymbol{\sigma}')^2 = \tfrac{1}{3}(t_1-t_3)^2\,, \\ II_{\boldsymbol{E}} &= \tfrac{1}{2}\operatorname{tr}(\boldsymbol{E})^2 = \tfrac{1}{3}(\dot\epsilon_1-\dot\epsilon_3)^2\,. \end{aligned} \tag{4.25}$$

4.3.1 Tri-axial compression

a) Incompressible soil

If the soil is unsaturated but dry or only slightly moist, then p^w plays the role of the free pressure of an incompressible body. For such a case the constitutive relationship (4.5) was said to form the proper material response. Writing this constitutive relation down for any component of the stress and stretching tensors yields $b(II_{\boldsymbol{\sigma}'}) = D_{ij}/\sigma'_{ij}$ for any $i, j = 1, 2, 3$ (no sum). With Equations (4.24) and (4.25), this becomes

$$b(II_{\boldsymbol{\sigma}'}) = \frac{3\dot\epsilon_1}{2(t_1-t_3)}\,, \qquad II_{\boldsymbol{\sigma}'} = \tfrac{1}{3}(t_1-t_3)^2\,, \tag{4.26a}$$

where $2\dot\epsilon_3 = -\dot\epsilon_1$ has been used. Evidently the pressure has disappeared from these formulas; it does not affect the constitutive relation explicitly. This is no surprise as it is the result of our construction. Measuring the strain rate $\dot\epsilon_1$ and the stresses t_1 and t_2 determines b. To this end we plot $\frac{3}{2}\dot\epsilon_1/(t_1-t_3)$ against $(t_1-t_3)^2/3$ and then find b by postulating a mathematical relationship that fits the points inferred from the experiment.

Figs. 5 and 6 show such fits. In the former, data points are representative for an undisturbed clay specimen from San Francisco Bay, 32 minutes after the stress had been applied to the specimen (Singh & Mitchell, 1968). The dashed curve is a fit using Norton's power flow law with a power exponent $m = 3.5$ and a rate factor as indicated in the inset. The second example (*Fig.* 6), also taken from Singh & Mitchell (1968), applies to a remoulded clay, 1000 minutes after stress application. Our mathematical fit is a finite viscosity law (Hutter, 1983)

$$b = \frac{1}{\mu} + A(II_{\boldsymbol{\sigma}'})^{(m-1)/2} \tag{4.26b}$$

with coefficients μ , A and m as indicated in the figure.

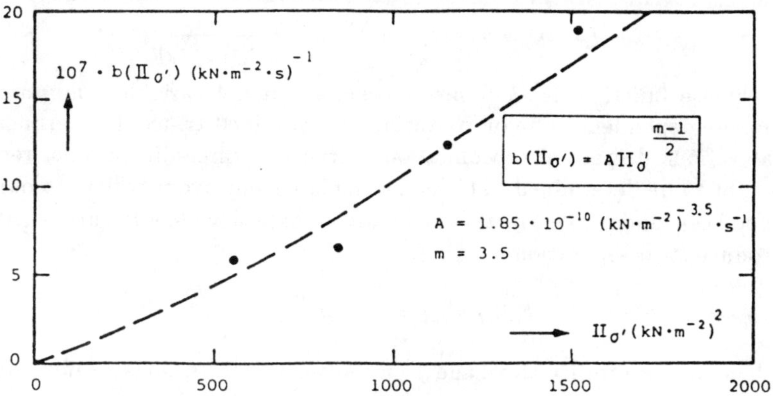

Fig. 5: Determination of the function $b(II_{\sigma'})$ for an undisturbed clay specimen from San Francisco Bay, 32 minutes after stress application. Data from Singh & Mitchell (1968).

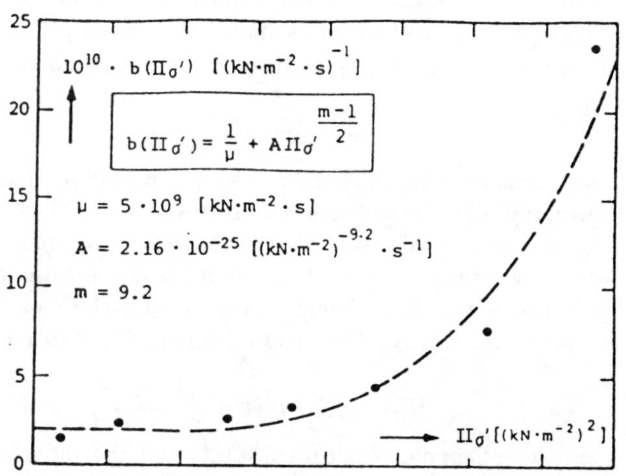

Fig. 6: Determination of the function $b(II_{\sigma'})$ for a remoulded clay, 1000 minutes after stress application. Data from Singh & Mitchell (1968).

b) Saturated binary mixture

According to §4.2b two limiting cases are possible; we first assume *case A*, i. e. small bulk viscosity. Thus the constitutive relations have the form of Equations (4.22) and $(4.18)_2$. In an undrained tri-axial experiment, the interstitial pressure is kept constant, $d^s p'/dt = 0$, so $\text{tr}\boldsymbol{D} = 0$: Only the deviatoric part of the stress-stretching relation remains to be determined.

Thus

$$b(\Pi_{\boldsymbol{\sigma}'}, n) = b\left(\tfrac{1}{3}(t_1 - t_3)^2\,,\, n\right) = \frac{3\dot{\epsilon}_1}{2(t_1 - t_3)}\,. \tag{4.27}$$

The creep response function depends now on one additional variable: the porosity, n. To find the complete dependence of b on its variables, undrained tri-axial experiments must be performed at various degrees of consolidation. Strictly, a three-dimensional representation $\Pi_{\boldsymbol{\sigma}'}$, n , b ought to be determined. This is a formidable endeavour which, to our knowledge has so far not been undertaken in this generality. Instead we conjecture a simplification, namely a product-form separation

$$b(\Pi_{\boldsymbol{\sigma}'}\,, n) = A(n)\, f(\Pi_{\boldsymbol{\sigma}'})\,, \tag{4.28}$$

in which A depends on porosity alone and f on the second stress deviator invariant. A power law

$$f(\Pi_{\boldsymbol{\sigma}'}) = (\Pi_{\boldsymbol{\sigma}'})^{(m-1)/2} \tag{4.29}$$

is a suitable candidate to match such experiments. The graphs in *Fig.* 5 and 6 serve again as examples, because they are applicable to a particular value of the porosity, n .

An *alternative procedure* is obtained as follows. Assume that the soil is normally consolidated and that there exists a one-to-one correspondence between porosity, n, and pressure, $p' = p^e$ for *case A*. In this case n can be replaced by p' in the phenomenological equation for b :

$$b = b(\Pi_{\boldsymbol{\sigma}'}\,, p') \tag{4.30}$$

We now introduce the pressure p' through the following qualitative analysis, the behaviour of which is typical for many soils: Under "normal" states of stress the creeping deformation of a mass of soil may be slow. If this state of stress, however, approaches a limiting value then the deformation accelerates and may at this limit assume a catastrophic value. What transpires from this is the notion of a "yield", even though the meaning of this word is not exactly as in the plasticity theory. The yield-criterion of DRUCKER & PRAGER can be written as

$$\Pi^r_{\boldsymbol{\sigma}'} = \left[k' - \alpha'(\mathrm{tr}\boldsymbol{\sigma}')\right]^2 \tag{4.31}$$

in which k' and α' are two phenomenological constants and the superscript r in $\Pi^r_{\boldsymbol{\sigma}'}$ is an identifier for conditions at yield (rupture). The meaning of (4.31) is that when this equation is satisfied at a particular point, then the deformation of the soil at that point is becoming large. The pressure p' enters (4.31) if (4.17) is substituted

$$\Pi^r_{\boldsymbol{\sigma}'} = \left[k' + 3\alpha' p'\right]^2\,. \tag{4.32}$$

Furthermore, k' and α' can be shown to be expressible in terms of the cohesion c' and the internal angle of friction, ϕ , of a soil obeying the MOHR-COULOMB yield criterion:

$$k' = \frac{6c'\cos\phi}{(3-\sin\phi)\sqrt{3}}\,, \quad \alpha' = \frac{2\sin\phi}{(3-\sin\phi)\sqrt{3}}\,. \tag{4.33}$$

We now write b in the form

$$b(\mathit{II}_{\boldsymbol{\sigma}'}, p') = A\frac{(\mathit{II}_{\boldsymbol{\sigma}'})^{(m-1)/2}}{(k' + 3\alpha' p')^m} = A\frac{(\mathit{II}_{\boldsymbol{\sigma}'})^{(m-1)/2}}{(\mathit{II}^r_{\boldsymbol{\sigma}'})^{m/2}} \tag{4.34}$$

with constant A (of dimension s^{-1}). This law expresses that, for a fixed deviatoric stress, the strain rate increases with decreasing spherical stress. This agrees qualitatively with experimental results as obtained for instance by TER-STEPHANIAN (1973) and others. With the aid of (4.25) and (4.26) it is easily shown that (4.34) takes the form

$$\dot{\epsilon}_1 = \frac{2}{\sqrt{3}}A\left(\frac{t_1 - t_3}{(t_1 - t_3)_r}\right)^m , \tag{4.35}$$

valid for the tri-axial experiment explained above. This is a univariate relationship. Its adequacy can be tested, if $\dot{\epsilon}_1$ is plotted against $(t_1 - t_3)/(t_1 - t_3)_r$ for different values of the consolidation stress p'_c . SINGH & MITCHELL (1968) have performed experiments in which it was varied and found that the postulate of univariability expressed in Equation (4.35) is well satisfied, as is the fit with a power law. For the conditions of *Fig.* 7 they found

$$A = 2.6 \cdot 10^{-5}\,[s^{-1}]\,, \quad m = 4.7\,.$$

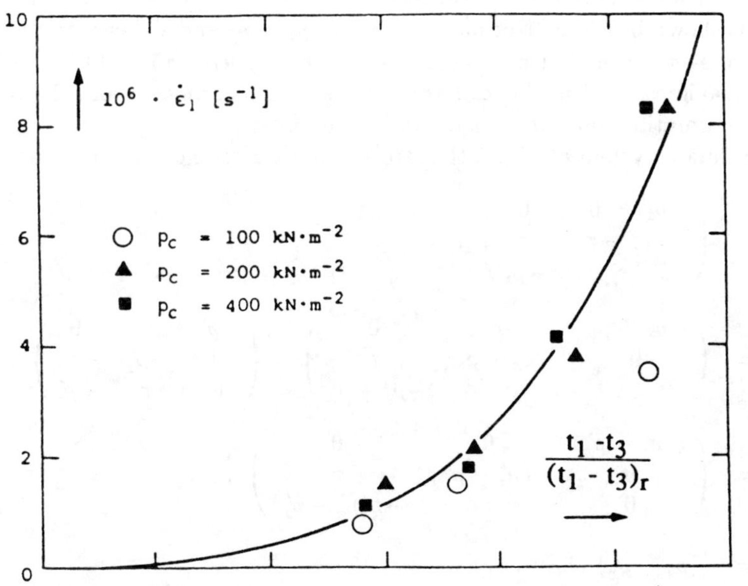

Fig. 7: Determination of the function $b(\mathit{II}_{\sigma'}, p')$. Data points are taken from Singh & Mitchell (1968).

We close this section by briefly outlining the procedure that would have to be taken when *case B* is considered for which the compressibility is large, $p' = p^v$. In this case tr$\boldsymbol{D}$ does not vanish and the full Equations (4.18), (4.19) apply. Together with the relations (4.24) and (4.25) these equations imply the relations

$$\begin{aligned} b(p', \mathit{II}_{\boldsymbol{\sigma}'}, n) &= \frac{\dot{\epsilon}_1 - \dot{\epsilon}_3}{t_1 - t_3}, \\ \tilde{a}(p', \mathit{II}_{\boldsymbol{\sigma}'}, n) &= \frac{\dot{\epsilon}_1 - \dot{\epsilon}_3}{t_1 - t_3} - \frac{\dot{\epsilon}_1 + 2\dot{\epsilon}_3}{3p'}, \end{aligned} \tag{4.36}$$

in which

$$p' = p^w + \tfrac{1}{3}(t_1 + 2t_3), \quad \mathit{II}_{\boldsymbol{\sigma}'} = \tfrac{1}{3}(t_1 - t_3)^2 . \tag{4.37}$$

In the laboratory p_w, t_1, t_3, $\dot{\epsilon}_1$ and $\dot{\epsilon}_3$ must be controlled and measured which determines all variables arising in (4.36) and (4.37) except n. So the coefficient functions a and b are only determined if an explicit dependence on n is ignored. An alternative would be to measure n both at the beginning and at the end of the experiment and to assign to n in (4.36) a mean value of the two. An exact determination is not possible in a tri-axial experiment in which drainage is not inhibited.

4.3.2 Shearing test "Geonor H 12"

In this test a shallow cylindrical specimen of a saturated soil that is sealed and subject to a constant pore water pressure p^w is sheared. The idealized state of stress at the centre of the specimen is shown in *Fig.* 8. Normal stresses p_z, $p_x = p_y$ and a shear stress τ are applied to the specimen as is a constant pore pressure p_w. The specimen is held by an inextensible mantle so that no horizontal strains can arise. Thus the shear angle γ and the vertical strain are the only deformation quantities that can be measured.

In the coordinate system of *Fig.* 8 the stresses and stretchings are given by

$$\begin{aligned} \boldsymbol{t} &= \begin{pmatrix} -p_x & 0 & 0 \\ 0 & -p_x & \tau \\ 0 & \tau & -p_z \end{pmatrix}, \\ \boldsymbol{\sigma} &= \begin{pmatrix} -p_x + p_w & 0 & 0 \\ 0 & -p_x + p_w & \tau \\ 0 & \tau & -p_z + p_w \end{pmatrix} = \begin{pmatrix} -p'_x & 0 & 0 \\ 0 & -p'_x & \tau \\ 0 & \tau & -p'_z \end{pmatrix}, \\ \boldsymbol{\sigma}' &= \begin{pmatrix} \frac{1}{3}(p'_z - p'_x) & 0 & 0 \\ 0 & \frac{1}{3}(p'_z - p'_x) & \tau \\ 0 & \tau & -\frac{2}{3}(p'_z - p'_x) \end{pmatrix}, \\ \mathit{II}_{\boldsymbol{\sigma}'} &= \tfrac{1}{3}(p'_z - p'_x)^2 + \tau^2 , \\ p' &= \tfrac{1}{3}(2p'_x + p'_z) \end{aligned} \tag{4.38}$$

as well as

$$
\begin{aligned}
\boldsymbol{D} &= \begin{pmatrix} 0 & 0 & 0 \\ 0 & 0 & \dot{\gamma}/2 \\ 0 & \dot{\gamma}/2 & -\dot{\epsilon}_z \end{pmatrix} , & \boldsymbol{E} &= \begin{pmatrix} \frac{1}{3}\dot{\epsilon}_z & 0 & 0 \\ 0 & \frac{1}{3}\dot{\epsilon}_z & \dot{\gamma}/2 \\ 0 & \dot{\gamma}/2 & -\frac{2}{3}\dot{\epsilon}_z \end{pmatrix} , \\
\mathrm{tr}\boldsymbol{D} &= -\dot{\epsilon}_z , & II_{\boldsymbol{E}} &= \tfrac{1}{3}\dot{\epsilon}_z^2 + \tfrac{1}{4}\dot{\gamma}^2 .
\end{aligned}
\tag{4.39}
$$

These expressions will now be used in the experimental determination of the creep response functions of the various models.

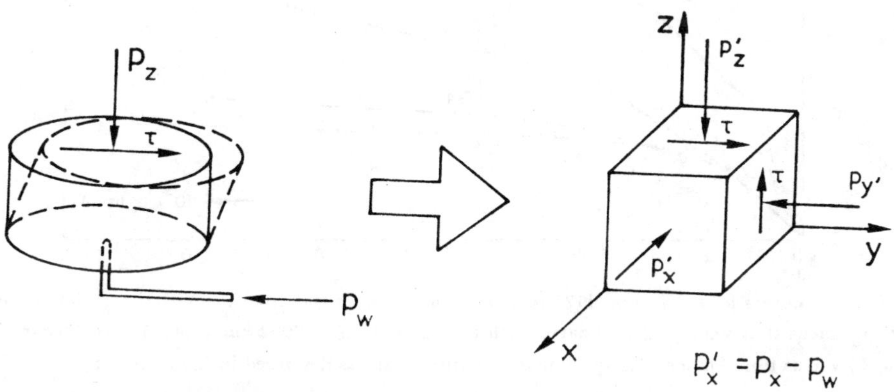

Fig. 8: Tri-axial state of stress as applied in the shear test experiment "Geonor H 12"

a) Incompressible soil

We ignore the interstitial fluid, i. e. assume dry conditions. Because of the incompressibility condition we have $\dot{\epsilon}_z = 0$ and so $\boldsymbol{E} = \boldsymbol{D}$. Thus, since $\boldsymbol{D} = b\boldsymbol{\sigma}'$, we may deduce the relationship

$$
b(II_{\boldsymbol{\sigma}'}) = \sqrt{\frac{II_{\boldsymbol{E}}}{II_{\boldsymbol{\sigma}'}}}
\tag{4.40}
$$

or in view of the relations (4.38) and (4.39), since $\dot{\epsilon}_z = 0$

$$
b\left(\tfrac{1}{3}(p'_z - p'_x)^2 + \tau^2\right) = \left\{ \frac{\frac{1}{4}\dot{\gamma}}{\frac{1}{3}(p'_z - p'_x)^2 + \tau^2} \right\}^{1/2} .
\tag{4.41}
$$

b) Compressible saturated mixture

If we consider *case A*, i. e., small bulk viscosity of the soil, and if the experiment is performed without drainage, then $\dot{\epsilon}_z = 0$, $\mathrm{tr}\boldsymbol{D} = 0$ and $d^s p/dt = 0$, according to (4.22). Thus (4.41) applies again, but b is now also a function of n : $b = b(II_{\boldsymbol{\sigma}'}, n)$.

For a clay from the Villarbeney landslide (DUTI, 1984), four specimens were preconsolidated with a normal stress of $200\,kPa$ and subsequently subjected to a shear test for determination of the rupture stress. It was found that $(\tau/p'_z)_r = 0.49$. Two of the specimens were subsequently exposed to a Geonor H12 test at a stress level $0.8\,(\tau/p'_z)_r$, two others at $0.6\,(\tau/p'_z)_r$. In the course of the experiment of seven days duration, a small increase of the interstitial stress was measured ($10-20\,kPa$). This is small enough that conditions of a constant state of stress are sufficiently satisfied.

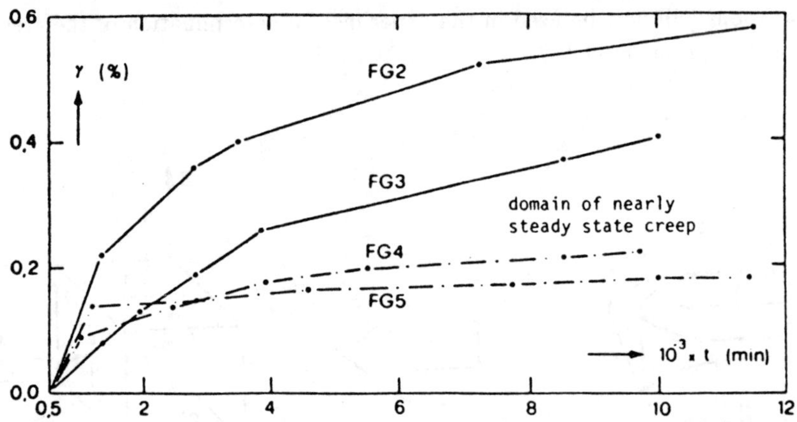

Fig. 9: Creep curves for a "Geonor H12" test under shearing for clay from the Villarbeney landslide for four specimens that were preconsolidated with a normal stress of 200 kPa. Plotted is the shear angle γ (in %) versus time for the four specimens. Conditions are as described in the main text.

Fig. 9 shows the creep curves (shear angle versus time) for the four samples. Even though a trend is not entirely conclusive from *Fig.* 9, we may nevertheless suppose that the creep curves approach steady slopes for large time and that this steady creep has been reached at $t = (1-1.2)\times 10^4$ minutes . In *Fig.* 10, where $\dot{\gamma}$ is plotted against τ/τ_r , this steady–state creep is represented by the four full circles. The solid line corresponds to the power law

$$\begin{aligned} \dot{\gamma} &= \tilde{A}\left(\frac{\tau}{\tau_r}\right)^m , \\ \tilde{A} &= 1.17\times 10^{-8}[s^{-1}]\,, \quad m = 4.8 \end{aligned} \tag{4.42}$$

where A and m were computed by a regression analysis with $r^2 = 0.85$. It remains to connect the formulas (4.42) with an explicit relation for the creep response function. For the choice Eq. (4.42) and with the experimental conditions $p'_z = p'_y = p'_x$ (TAN TJONG-KIE, 1981; KAVAZANJIAN and MITCHELL, 1984), we easily deduce that

$$A = \tfrac{1}{2}\tilde{A} = 5.85\times 10^{-9}[s^{-1}]\,, \quad m = 4.8\,. \tag{4.43}$$

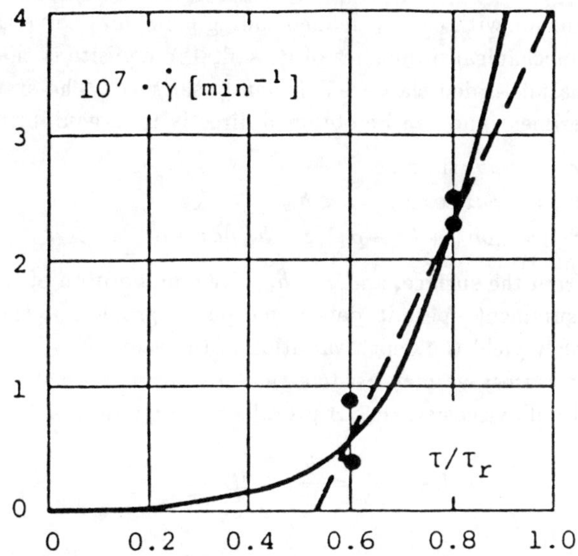

Fig. 10: Shear strain rate as a function of stress level. Solid circles are for the four specimens of *Fig.* 9, their strain rate values corresponding to the asymptotic slopes of the curves in *Fig.* 9 for large time. The solid line corresponds to the power law fit (4.42), the dashed line to a Bingham fluid behaviour.

The dashed line in *Fig.* 10 corresponds to the flow law of a BINGHAM body with yield stress $(\tau/\tau_r) = 0.5$. Clearly, with only four experimental points, no one is preferable over the other.

For *case B*, i. e., large compressibility, Equations (4.18) apply. From these, it follows that

$$\begin{aligned} b(p', II_{\boldsymbol{\sigma}'}, n) &= \sqrt{\frac{II_{\boldsymbol{E}}}{II_{\boldsymbol{\sigma}'}}}\,, \\ \tilde{a}(p', II_{\boldsymbol{\sigma}'}, n) &= \frac{\mathrm{tr}\boldsymbol{D}}{3p'} + b(p', II_{\boldsymbol{\sigma}'}, n)\,. \end{aligned} \tag{4.44}$$

If we substitute in these the relevant expressions of Equations (4.38) and (4.39) then these expressions become

$$\begin{aligned} b(p', II_{\boldsymbol{\sigma}'}, n) &= \left[\frac{\frac{1}{3}\dot{\epsilon}_z^2 + \frac{1}{4}\dot{\gamma}^2}{\frac{1}{3}(p'_z - p'_x)^2 + \tau^2}\right]^{1/2}, \\ \tilde{a}(p', II_{\boldsymbol{\sigma}'}, n) &= b(p', II_{\boldsymbol{\sigma}'}, n) - \frac{\dot{\epsilon}_z}{2p' + p'_z}\,. \end{aligned} \tag{4.45}$$

An experimental determination of these trivariate functions is still missing.

4.3.3 In situ flow tests

A long shallow landslide with clearly defined sliding plane may serve as a test site for the determination of the mechanical properties of its soil. Prerequisite is that the assumptions of an infinitely long parallel-sided slab are sufficiently satisfied. The state of stress is then that of a Geonor experiment and can be obtained directly from equilibrium considerations

$$\begin{aligned} \tau &= \rho g z \sin\alpha\,, \\ p' &= \rho g z \cos\alpha\,, \quad z < h_w\,, \\ p' &= [\rho h_w + (\rho - \rho_w)(z - h_w)]g\cos\alpha\,, \quad z > h_w \end{aligned} \tag{4.46}$$

where z is measured from the surface, and $z = h_w$ gives the position of the phreatic surface.

Inclinometer measurements permit determination of profiles of the shear angle $\gamma(t)$ through integration; they yield the depth variation of the velocity.

With regard to pressures, we may assume that p_x and p_y are equal. With the further supposition of a small bulk viscosity, *case A* prevails and Equation (4.4) applies and reduces to

$$b(II_{\boldsymbol{\sigma}'}, n) = \frac{\dot{\gamma}}{2\tau}\,, \quad II_{\boldsymbol{\sigma}'} = \tau^2\,. \tag{4.47}$$

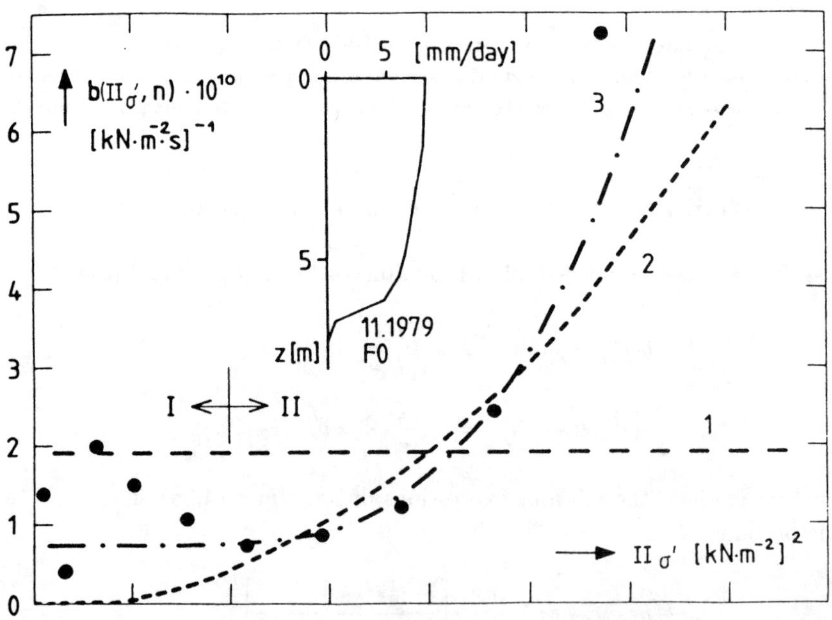

Fig. 11: Determination of $b(II_{\sigma'}, n)$ by means of in-situ measurements for the landslide Villarbeney, Switzerland. Full circles are obtained from measurements at 10 different depths of one borehole. The curves 1, 2, 3 are fits according to Equation (4.26) with coefficients as listed in Table 4. The inset shows the velocity profile in one borehole from which measurements are inferred.

The procedure is now to compute b from the right-hand side of Equation (4.47) and, similarly, to record $II_{\sigma'}$ and n . Apart from the inclinometer measurements, porosity measurements are thus needed.

As an example, consider the landslide at Villarbeney, Switzerland. It has been active for many generations, and detailed investigations on the flow were collected and interpreted by LMS (1980), DYSLI and RECORDON (1981) and DUTI (1985). The geometric characteristics of this landslide are as follows: the length is $750\,m$, the width $700\,m$ in average, the depth $10\,m$, and the mean slope angle is 15° . Only profile FO, which lies in the central, most active portion of this slide, will be treated herein. The mean values of the geotechnical characteristics in the profile are $w = 16{,}8\%$, $w_p = 18.2\%$, $w_L = 41{,}2\%$, $n = 0.3$, and the clay fraction is 32 %.

Because in this example the variation of n is small, an n-dependence of b was not sufficiently observable. *Fig.* 11 shows values of b plotted against $II_{\sigma'}$ and a sketch of the velocity profile. The ten points are obtained from measurements at ten different depths. A fair scatter in their position makes the postulation of a mathematical formula for b difficult. Three choices are indicated in the figure. Curve 1 sets b equal to a constant mean value, its coefficient of variation being 104 %. This law would correspond to Newtonian behaviour. Another possibility, curve 2, is a power law, but it would reasonably represent material behavior in this landslide only in the range II of the figure. A better fit is obtained with a finite viscosity law of form of Equation (4.16). Values of μ , m and A are collected in *Table 4*; V_μ = the coefficient of variation; r^2 = the correlation coefficient; and in the dimension of A, the asterisk has the value of m . Obviously curve 3 provides the best fit.

Curve number (*Fig.* 11) (1)	μ $(kN \cdot m^{-2} \cdot s^{-1}$ (2)	m (3)	A $(kN \cdot m^{-2} \cdot s^{-1}$ (4)	Remarks (5)
1	$5.3 \cdot 10^9$	–	0	$V_\mu = 104\%$
2	∞	5.3	$6.7 \cdot 10^{-17}$	$r^2 = 0.83$
3	$1.4 \cdot 10^{10}$	9.6	$1.2 \cdot 10^{-23}$	$r^2 = 0.97$

Table 4: Values of μ , m , A (Eq. 4.26) in *Fig.* 11

4.3.4 Concluding remarks

It seems to be appropriate to summarize the findings of paragraph 4.3. Constitutive models for the stress tensor of a creeping soil are presented. The soil is treated either as a non-Newtonian, very viscous fluid or a saturated mixture of soil and water, in which water is an inviscid fluid, while the soil is treated as a non-Newtonian viscous fluid. Rate-type constitutive relations of the REINER-RIVLIN type are reduced by the additional postulates that: (1) Stretching $\boldsymbol{D}$ cannot depend on $\boldsymbol{t}^2$ except through its invariant; and (2) invariants of $\boldsymbol{t}^3$ cannot enter such a relationship. It is then demonstrated that

- For the one-component model, there is only a single scalar-valued function b, Equation (4.5) which must be determined experimentally to close the system.

- For the saturated mixture, the "elastic" stress contribution is governed by Equation (4.9), whereas the "viscous" stress is given by, at most, two scalar-valued functions, $\tilde{a}$ and b (Equations (4.18), (4.19)).

When the bulk viscosity is ignored $\tilde{a}$ no longer needs to be determined. Proposed mathematical relations of these functions are calibrated with laboratory data using tri-axial experiments and shear tests with the Geonor H12 test cells. In one case, field data were used for this fit. All these demonstrate the usefulness of the procedure.

A specialization of this concept for the time-dependent sliding of a moving soil mass along a basal surface is possible. Viscous-type sliding relations giving the relative sliding velocity as a function of the state of stress at the interface will be given below, as will particular applications of these concepts to real landslides.

4.4 Application of the model in the context of visco–plasticity

So far it was assumed that creep occurs at all stress levels; one exception was the BINGHAM law in *Table 3*, where the strain rate remains zero below a "yield", $II_{\boldsymbol{\sigma}'} < II_{\boldsymbol{\sigma}'_0}$. A simple way to incorporate such plastic effects is to use an extension of PERZYNA's theory of visco-plasticity (PERZYNA, 1966; CORMEAU, 1976) in which the strain rate $\boldsymbol{D}(= \boldsymbol{D}_s)$ of the soil is decomposed into an elastic part $\boldsymbol{D}^e$ and a viscoplastic part $\boldsymbol{D}^{vp}$,

$$\boldsymbol{D} = \boldsymbol{D}^e + \boldsymbol{D}^{vp} \tag{4.48}$$

and where the elastic strain rate is related to the objective stress derivative (JAUMANN derivative)

$$\begin{aligned} \boldsymbol{\sigma}^{\nabla} &= \frac{d^s\boldsymbol{\sigma}}{dt} + \boldsymbol{\omega}\boldsymbol{\sigma} - \boldsymbol{\sigma}\boldsymbol{\omega} , \\ \boldsymbol{\omega} &:= \tfrac{1}{2}(\boldsymbol{L} - \boldsymbol{L}^T) \end{aligned} \tag{4.49}$$

for instance via the linear relation

$$\boldsymbol{\sigma}^{\nabla} = \boldsymbol{C}_e[\boldsymbol{D}^e] , \tag{4.50}$$

or in view of isotropy and with (4.17), $\boldsymbol{\sigma} = -p'\mathbf{1} + \boldsymbol{\sigma}'$

$$\boldsymbol{D}^e = \frac{1}{3K}\frac{d^s p'}{dt}\mathbf{1} + \frac{1}{2G}\boldsymbol{\sigma}'^{\nabla} . \tag{4.51}$$

In the above, $\boldsymbol{\omega}$ is the spin tensor, $\boldsymbol{C}_e$ the elastic constitutive matrix, a fourth rank tensor, and K and G the constant bulk and shear moduli, respectively; $\boldsymbol{\sigma}'$ is the stress deviator.

In these formulations, the visco-plastic stretching $\boldsymbol{D}^{vp}$ is given by

$$\boldsymbol{D}^{vp} = \gamma < \phi(F) > \frac{\partial Q}{\partial \boldsymbol{\sigma}'} , \tag{4.52}$$

where $\phi(F)$ is a scalar-valued yield function [6], Q a scalar-valued visco-plastic potential and γ the so-called fluidity paramter which may be a function of stress. Moreover, the bracket

[6] In other words, $F(\sigma) = 0$ defines a closed surface in the stress space. The interior of $F(\sigma) = 0$ is given by all those σ for which $F(\sigma) < 0$. The exterior is defined by $F(\sigma) \geq 0$. The origin, $\sigma = 0$, lies inside this surface.

expression $<>$ defines a yield condition,

$$< \phi(F(\boldsymbol{\sigma})) >= \begin{cases} \phi(F(\boldsymbol{\sigma})) & , \quad \text{for} \quad F > 0 \,, \\ 0 & , \quad \text{for} \quad F \leq 0 \,. \end{cases} \tag{4.53}$$

i. e., the visco-plastic stretching equals zero for a state of stress which lies inside the yield surface (in the stress space).

4.4.1 Material without yield

If we choose

$$Q \equiv F \equiv \sqrt{II_{\boldsymbol{\sigma}'}} \,, \tag{4.54}$$

then visco-plastic strain rate occurs at any stress level. Indeed, with

$$\frac{\partial Q}{\partial \boldsymbol{\sigma}'} = \frac{1}{2\sqrt{II_{\boldsymbol{\sigma}'}}} \boldsymbol{\sigma}' \tag{4.55}$$

we may write the visco-plastic strain rate in the form

$$\boldsymbol{D}^{vp} = \frac{\gamma < \phi(\sqrt{II_{\boldsymbol{\sigma}'}}) >}{2\sqrt{II_{\boldsymbol{\sigma}'}}} \boldsymbol{\sigma}' \,. \tag{4.56}$$

This formula indicates that $\boldsymbol{D}_{vp}$ must be a deviator, because $\boldsymbol{\sigma}'$ is one. In the special case that the elastic strain is ignored, $\boldsymbol{\sigma}^{\triangledown} = 0$, or $K = G \to \infty$, Equation (4.56) agrees with $(4.18)_2$ provided that

$$b(II_{\boldsymbol{\sigma}'}) = \frac{\gamma < \phi(\sqrt{II_{\boldsymbol{\sigma}'}}) >}{2\sqrt{II_{\boldsymbol{\sigma}'}}} \,. \tag{4.57}$$

A possible dependence of b on the porosity, n, could be incorporated in a dependence of γ upon n.

4.4.2 Material with yield and non-associative visco-plasticity

For any well behaved yield function $F(\boldsymbol{\sigma}') = 0$ and for a von Mises-type visco-plastic potential

$$Q = II_{\boldsymbol{\sigma}'} - c \,, \tag{4.58}$$

where c is a constant, we easily deduce

$$\frac{\partial Q}{\partial \boldsymbol{\sigma}'} = \frac{\partial Q}{\partial II_{\boldsymbol{\sigma}'}} \boldsymbol{\sigma}' = \boldsymbol{\sigma}' \tag{4.59}$$

from (4.52)

$$\boldsymbol{D}^{vp} = \gamma < \phi(F(\boldsymbol{\sigma}')) > \boldsymbol{\sigma}' \,, \tag{4.60}$$

which is again deviatoric. With the elastic contributions as given in (4.51), we therefore see that the total stretching may be written as

$$\begin{aligned} \text{tr}\boldsymbol{D} &= \frac{1}{K} \frac{d^s p'}{dt} \,, \\ \boldsymbol{E} &= \frac{1}{2G} \boldsymbol{\sigma}'^{\triangledown} + \gamma < \phi(F(\boldsymbol{\sigma}')) > \boldsymbol{\sigma}' \,. \end{aligned} \tag{4.61}$$

$\boldsymbol{E}$ is the distortion tensor, the deviatoric part of $\boldsymbol{D}$, as before. If we compare this with our earlier results in *case A*, see Equations $(4.18)_1$ and (4.22), for a saturated compressible soil, it can be recognized that the two laws are very similar for constant creep tests ($\boldsymbol{\sigma}'^{\triangledown} = 0$) but not equivalent. An other case for which similar results are obtained is the limit $G \to \infty$. A difference that remains is that b in Equation (4.18), may depend on the porosity, n, while (4.61) does not.

To make the comparison complete, let

$$\gamma < \phi(F) >= A\frac{F^{m-1}}{(II_{\boldsymbol{\sigma}'} - F)^m} \tag{4.62}$$

with

$$F = \sqrt{II_{\boldsymbol{\sigma}'}} - k' - 3\alpha' p' \, , \tag{4.63}$$

where k' and α' can be expressed in terms of the cohesion, c', and the internal angle of friction, ϕ, as given in (4.33). Equation (4.63) corresponds to a PRAGER & DRUCKER yield criterion. Equation (4.62) expresses that two different points in the stress deviator space, positioned at the same distance from the yield surface $F = 0$ will not generate the same strain rate, except if identical mean pressures are applied.

5. On Darcy's law

DARCY's law in its classical interpretation expresses the fact that due to the friction of the interstitial fluid along the pore boundaries mechanical energy is lost (and through this dissipation transferred to heat). From a continuum mechanical point of view DARCY's law is a constitutive postulate for the percolation force $\boldsymbol{\beta}$.

Let us *motivate* (not derive!) this law in such a way as it is classically usually done. The energy loss is expressed as a pressure loss by postulating that the pressure gradient be proportional to the filter velocity.

$$\text{grad} p_w = -\frac{\mu}{k^*}\boldsymbol{v}_F = -\frac{\mu}{k^*}n(\boldsymbol{v}_w - \boldsymbol{v}_s) \, . \tag{5.1}$$

Here, μ is the dynamic viscosity with dimension $[Nsm^{-2}] = [kgm^{-1}s^{-1}]$ and k^* the permeability with dimension $[m^2]$. Considering the balance law of momentum for the constituent water, see Equation (3.10) in which $\boldsymbol{t}_w = -np_w\mathbf{1}$ (ideal perfect fluid assumption for constituent water), we have

$$\text{grad}(np_w) = -\rho_w \boldsymbol{g} + \rho\boldsymbol{\beta} \, , \tag{5.2}$$

or upon ignoring the specific gravity and the gradient of n , grad$n$

$$n\,\text{grad}(p_w) = -\rho\boldsymbol{\beta} \, . \tag{5.3}$$

Combination of (5.1) and (5.3) motivates the material equation

$$\rho\boldsymbol{\beta} = \frac{n^2\mu}{k^*}(\boldsymbol{v}_w - \boldsymbol{v}_s) = \frac{n\mu}{k^*}\boldsymbol{v}_F \, . \tag{5.4}$$

This is DARCY's *law*. Accordingly, the percolation force is assumed to be proportional to the filter velocity and to the porosity with constant of proportionality μ/k^* . Most likely, a material scientist or a continuum mechanician would have started from a postulate like (5.4). We shall now simply adopt (5.4) as the constitutive postulate for the percolation force and free ourselves from its motivation.

Substituting DARCY's law into the force balances (3.10), imposing (3.14) and assuming that $\boldsymbol{\sigma}_w = \mathbf{0}$ $(\boldsymbol{\sigma}_s = \boldsymbol{\sigma})$, yields

$$\begin{aligned} -\text{grad}((1-n)p_w) \quad &+ \text{div}\boldsymbol{\sigma} + \rho_s \boldsymbol{g} + \frac{n^2\mu}{k^*}(\boldsymbol{v}_w - \boldsymbol{v}_s) = \mathbf{0}\,, \\ -\,\text{grad}(np_w) \quad &+ \rho_w \boldsymbol{g} - \frac{n^2\mu}{k^*}(\boldsymbol{v}_w - \boldsymbol{v}_s) = \mathbf{0}\,, \end{aligned} \tag{5.5}$$

or, after addition

$$-\,\text{grad}p_w + \text{div}\boldsymbol{\sigma} + \rho\boldsymbol{g} = \mathbf{0}\,, \tag{5.6}$$

which is the force balance of the two component mixture. The two mass balance statements (3.5) and the force balances (5.5) constitute eight scalar equations, which, upon substitution of a constitutive relation for the stress $\boldsymbol{\sigma}$, contain eight independent field variables, n, p_w, $\boldsymbol{v}_s$ and $\boldsymbol{v}_w$. So, they form field equations for these field variables; however, equations (3.5) and (5.5) may not constitute the ideal combination of equations for the construction of solutions. Such an alternative set is e. g. obtained if (3.5) is combined with (5.6) and one of the Equations (5.5).

Consider the force balance for the constituent water $(5.5)_2$ and assume that $\|\text{grad}n/n\| \ll \|\text{grad}p_w/p_w\|$. Then it is straightforward to show that $(5.5)_2$ is approximately expressible as

$$\boldsymbol{v}_F = -\underbrace{\frac{k^*\hat{\rho}_w g}{\mu}}_{k}\,\text{grad}\,\underbrace{(z + \frac{p_w}{\hat{\rho}_w g})}_{h}\,, \tag{5.7}$$

where z is the vertical coordinate, or

$$\boldsymbol{v}_F = -k\,\text{grad}h\,, \tag{5.8}$$

in which h is usually referred to as *piezometric head* and k (with dimension $[ms^{-1}]$) is DARCY's *permeability*. Equation (5.8) is a very popular version of DARCY's law , often introduced without any notion of a binary mixture concept. Our derivation shows that it is obtained by combining the original DARCY law with the force balance of constituent water and by ignoring grad$n$ terms. Alternative field equations to (3.5) and (5.5) are now (3.5) combined with (5.6) and (5.8).

Consider two limiting cases. (a) If water and soil move with the same speed then $\boldsymbol{v}_F = \mathbf{0}$; equation (5.7) then implies

$$\text{grad}p_w = \hat{\rho}_w \boldsymbol{g}\,. \tag{5.9}$$

The pore water pressure is hydrostatic in this case; more specifically, the pore water pressure does not "see" the soil at all. Clearly, such conditions prevail if both constituents are at rest.

Combining (5.9) with the force balance of the mixture as a whole, (5.6), yields

$$\operatorname{div}\boldsymbol{\sigma} + (\rho - \hat{\rho}_w)\boldsymbol{g} = \mathbf{0} \,. \tag{5.10}$$

The stresses of the mixture are equilibrated as if a buoyance force would act on the gravel that is fully submerged in water. (b) Suppose that the velocity of the soil is much smaller than that of the interstitial fluid: $\|\boldsymbol{v}_s\| \ll \|\boldsymbol{v}_w\|$. Thus $\boldsymbol{v}_F \simeq n\boldsymbol{v}_w$ and (5.8) takes the form

$$n\boldsymbol{v}_w = -k \operatorname{grad} h \,. \tag{5.11}$$

When this is combined with the mass balance law for the constituent water, $(3.5)_1$, then

$$\frac{\partial n}{\partial t} = \operatorname{div}(k \operatorname{grad} h) \tag{5.12}$$

which demonstrates that DARCY's law gives rise to a diffusive mechanism. The field equations that must be solved in this case are

$$\begin{aligned} \frac{\partial n}{\partial t} \quad &- \quad \operatorname{div}(k \operatorname{grad} h) &= 0 \,, \\ \frac{\partial n}{\partial t} \quad &- \quad \operatorname{div}((1-n)\boldsymbol{v}_s) &= 0 \,, \\ -\operatorname{grad} p_w \quad &+ \quad \operatorname{div}\boldsymbol{\sigma} + \rho\boldsymbol{g} &= 0 \,. \end{aligned} \tag{5.13}$$

in which $h = z + p_w/(\hat{\rho}_w g)$ and $\boldsymbol{\sigma}$ is given by a constitutive relation in terms of n and $\boldsymbol{D}_s$. The set (5.13) is then a system of five partial differential equations for n , p_w and $\boldsymbol{v}_s$. Once these fields are determined, the groundwater velocity follows from Equation (5.11).

6. Boundary and transition conditions

As is evident from paragraph 2 the moving and deforming soil that may be saturated with water in its lower part and water-free or at most moist in its upper part contains a number of distinct surfaces across which not all field variables can be continuously differentiable. We thus refer to the surfaces as *singular surfaces.* The *free surface* (*Fig.* 12), bounding the soil against the outer atmosphere is such a surface, as is the *base* or *basal surface* and the *water table* of the groundwater, that is commonly referred to as *phreatic surface.* However, as is stipulated by the inclinometer measurements shown in *Fig.* 1, there may also exist internal *sliding surfaces*, or internal layers across which the shearing deformation is so large that the layer may be collapsed to a sliding surface. We shall not assume this latter case to arise and refer the reader to VULLIET & HUTTER (1988a).

In general, two different kind of statements must hold at a singular surface, (1) the *kinematic surface condition* that describes the evolution of the surface through time and (2) *dynamic statements* that correspond to the surface balance laws and express jump properties of the involved field variables.

Let $\boldsymbol{u}$ be the velocity vector with which the singular surface $F(\boldsymbol{x},t) = 0$ moves; it need to coincide neither with $\boldsymbol{v}_s$ nor with $\boldsymbol{v}_w$. Then, the kinematic surface equation is the statement

$$\frac{DF}{Dt} := \frac{\partial F}{\partial t} + (\operatorname{grad} F) \cdot \boldsymbol{u} = 0 \,, \tag{6.1a}$$

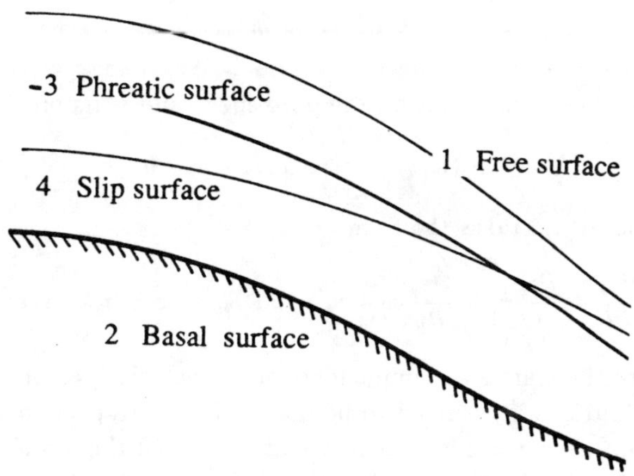

Fig. 12: Boundary and internal surfaces in a deforming mass of soil. (1) Free surface, (2) basal surface (rock–bed or a different sediment base), (3) phreatic surface, (4) internal sliding surface. Below the phreatic surface the soil is saturated, above it, it is not.

in which D/Dt expresses time derivative for an observer that is moving with the surface; it need not be a material motion. [7] Moreover, let f be any field quantity that may experience a jump discontinuity across a singular surface $F(\boldsymbol{x},t)=0$. Then we define $[\![f]\!] := f^+ - f^-$, where $f^\pm$ are the values of f when the singular surface is approached from the positive and the negative side, respectively, the positive side of the surface being defined as that part of the body into which the unit normal vector, $\boldsymbol{n}$, of the surface points. With these definitions, the jump conditions of mass and momenta for constituent α read

$$\begin{aligned} [\![\rho_\alpha(\boldsymbol{v}_\alpha - \boldsymbol{u})\cdot\boldsymbol{n}_F]\!] &= 0\,, \\ [\![\boldsymbol{t}_\alpha\boldsymbol{n}_F]\!] - [\![\rho_\alpha\boldsymbol{v}_\alpha \otimes (\boldsymbol{v}_\alpha - \boldsymbol{u})\boldsymbol{n}_F]\!] &= \boldsymbol{0}\,, \end{aligned} \tag{6.2}$$

where in our case $\alpha = s$ (soil) and $\alpha = w$ (water).

The laws (6.1) and (6.2) are the basis for all boundary and transition conditions.

[7] Let $\boldsymbol{w}$ be any velocity field (of a certain material point) defined on the singular surface. Then (6.1a) may equally be written as

$$\frac{\partial F}{\partial t} + (\text{grad}F)\cdot\boldsymbol{w} = ||\text{grad}F||(\boldsymbol{w}-\boldsymbol{u})\cdot\underbrace{\frac{\text{grad}F}{||\text{grad}F||}}_{\boldsymbol{n}_F} = ||\text{grad}F||\,\underbrace{(\boldsymbol{w}-\boldsymbol{u})\cdot\boldsymbol{n}_F}_{Q_\perp}\,, \tag{6.1b}$$

where $\boldsymbol{n}_F$ is the unit normal vector of the surface. Its direction defines the positive side of the surface. The left hand side defines the time derivative following the motion described by w. $Q_\perp$ is the diffusive flow of the material point moving with velocity w , relative to the motion of the singular surface. Thus (6.1b) may take the alternative form

$$\frac{d^w F}{dt} = ||\text{grad}F||Q_\perp \quad . \tag{6.1c}$$

Special cases are, if the surface is material with respect to the grains, $\boldsymbol{u} = \boldsymbol{v}_s$ and $\boldsymbol{w} = \boldsymbol{v}_w$.

6.1 Free surface

This surface is material with respect to the soil. Water may be present and percolate through the soil, however, it does not fill the pore space and is ignored. With

$$F_s(\boldsymbol{x},t) := z_s(x,y,t) - z = 0$$

and $\boldsymbol{u} = \boldsymbol{v}_s$, Equation (6.1) takes the form

$$\frac{\partial z_s}{\partial t} + \frac{\partial z_s}{\partial x} v_x^s + \frac{\partial z_s}{\partial y} v_y^s - v_z^s = 0\,, \quad \text{at} \quad z = z_s(x,y,t)\,, \tag{6.3}$$

where v_x^s , v_y^s , v_z^s are the Cartesian components of the velocity vector $\boldsymbol{v}_s$.

Because the free surface is assumed to be material with respect to the grains, the jump condition of mass $(6.2)_1$ for $\alpha = s$ is an empty statement and the jump condition of momentum implies $\boldsymbol{t}_s \boldsymbol{n}_s = \boldsymbol{0}$, if it is assumed that the surface traction from the atmospheric side vanishes (vanishing atmospheric pressure, no wind shear stress). With

$$\boldsymbol{n}_s = \left(-\frac{\partial z_s}{\partial x}, -\frac{\partial z_s}{\partial y}, 1\right) / N_s\,, \quad N_s := \left\{1 + \left(\frac{\partial z_s}{\partial x}\right)^2 + \left(\frac{\partial z_s}{\partial y}\right)^2\right\}^{1/2} \tag{6.4}$$

the traction condition takes the usual Cartesian form

$$\left.\begin{array}{rcrcrcl} -t_{xx}^s \dfrac{\partial z_s}{\partial x} & - & t_{xy}^s \dfrac{\partial z_s}{\partial y} & + & t_{xz} & = & 0\,, \\ -\,t_{xy}^s \dfrac{\partial z_s}{\partial x} & - & t_{yy}^s \dfrac{\partial z_s}{\partial y} & + & t_{yz} & = & 0\,, \\ -\,t_{xz}^s \dfrac{\partial z_s}{\partial x} & - & t_{yz}^s \dfrac{\partial z_s}{\partial y} & + & t_{zz} & = & 0\,, \end{array}\right\} \text{at} \quad z = z_s(x,y,t)\,. \tag{6.5}$$

If the landslide is shallow and the free surface slowly varying in the x- and y- directions such that $\partial z_s / \partial x$ and $\partial z_s / \partial y$ may be negligible, then (6.5) simply requires the horizontal shear stress components and the surface normal stress to vanish.

6.2 Rigid bed

We suppose the moving and deforming landslide to be bounded from below by a rigid, undeformable bed that may be permeable to the constituent water but impermeable for the grains. Thus the basal surface, defined by

$$F_B := z - z_B(x,y) = 0 \tag{6.6}$$

is material with respect to the grains and the kinematic condition takes the form

$$\frac{\partial z_s}{\partial x} v_x^s + \frac{\partial z_s}{\partial y} v_y^s - v_z^s = 0\,, \text{ at} \quad z = z_B(x,y)\,. \tag{6.7}$$

This is only one kinematic equation. It expresses the tangency of the velocity vector $\boldsymbol{v}_s$ to the basal surface: $\boldsymbol{v}_s \cdot \boldsymbol{n}_B = 0$. Written in terms of the water velocity, it takes the form

$$\frac{\partial z_s}{\partial x} v_x^w + \frac{\partial z_s}{\partial y} v_y^w - v_z^w = Q_\perp^B N^B \ , \ N_B := \left\{ 1 + \left(\frac{\partial z_B}{\partial x} \right)^2 + \left(\frac{\partial z_B}{\partial y} \right)^2 \right\}^{1/2} \tag{6.8}$$

according to Equation (6.1c) in the footnote to formula (6.1a), $Q_\perp^B$ is the *infiltration rate* into the rock bed that must be a prescribed function of position and time, if the water transport through the rock is not modelled by other means. When $Q_\perp^B = 0$ then the rock bed is impermeable for the constituent water and $\boldsymbol{v}_w$ is tangential to the base : $\boldsymbol{v}_w \cdot \boldsymbol{n}_B = 0$. The general form of Equation (6.8) is simply

$$\boldsymbol{v}_w \cdot \boldsymbol{n}_B = Q_\perp^B \ , \ \text{at} \quad z = z_B(x,y) \tag{6.9}$$

and is particularly transparent in this form. When the piezometric head is used as a basic variable then owing to the definition of $Q_\perp^B = (\boldsymbol{v}_w - \boldsymbol{v}_s) \cdot \boldsymbol{n}_B$ (see (6.1b)) and $(\boldsymbol{v}_w - \boldsymbol{v}_s) = \boldsymbol{v}_F / n$ we may write

$$nQ_\perp^B = \boldsymbol{v}_F \cdot \boldsymbol{n}_B = -k \,\mathrm{grad} h_B \ ,$$

or

$$Q_\perp^B = -\frac{k}{n} \frac{\partial h}{\partial n_B} \ , \ \text{at} \quad z = z_B(x,y) \ . \tag{6.10}$$

The infiltration rate (or the drainage rate) is proportional to the normal derivative of the piezometric head and inversely proportional to the porosity. Often, condition (6.10) is replaced by prescribing the piezometric head, or the pore water pressure,

$$\left. \begin{array}{rcl} h & = & h_B(x,y,t) \ , \\ \text{or} \qquad p_w & = & p_w^B(x,y,t) \ , \end{array} \right\} \ \text{at} \quad z = z_B(x,y) \ . \tag{6.11}$$

These boundary conditions would replace the kinematic statements derived above, but they are not equivalent to them and are easily interpretable.

We begin the derivation of the dynamic boundary conditions by writing the jump conditions of mass, (6.2), as

$$[\rho_\alpha (\boldsymbol{v}_\alpha - \boldsymbol{v}_s) \cdot \boldsymbol{n}_B]^- = [\rho_\alpha (\boldsymbol{v}_\alpha - \boldsymbol{v}_s) \cdot \boldsymbol{n}_B]^+ \ , \tag{6.12}$$

where the negative (positive) sign refers to the rock bed (soil–water mixture). If $\alpha = s$, then $\boldsymbol{v}_\alpha^- = \boldsymbol{v}_s$ and (6.12) becomes trivial. On the other hand, for $\alpha = w$, $\boldsymbol{v}_\alpha^- = \boldsymbol{v}_w$. Defining the positive bracket in (6.12) by M_w^B which equals the flow of water mass per unit area of the basal surface, and recognizing that the left hand side of (6.12) equals $\rho_w^- Q_\perp^B = n\hat{\rho}_w Q_\perp^B$, Equation (6.12) becomes

$$Q_\perp^B = \frac{M_w^B}{n\hat{\rho}_w} \ , \ \text{at} \quad z = z_B(x,y) \ . \tag{6.13}$$

It follows that the jump conditions of mass have not led to any new equation. It simply resulted in an alternative interpretation of the infiltration rate $Q_\perp^B$. The variable (of the two) that can physically most easily be quantified is M_w with dimension $[kgm^{-2}s^{-1}]$; once it is specified, the infiltration rate (with dimension $[ms^{-1}]$ follows from (6.13).

The second set of dynamic statements that must be fulfilled at the bed are the momentum jump conditions $(6.2)_2$. These condition are not helpful here, because they only lead to a connection between the tractions at the base on either side of the basal surface. Because the tractions on the rock bed side are not needed, these conditions are superfluous.

Nevertheless a further dynamic condition is still needed because we have expressed above the tangency of the velocity $\boldsymbol{v}_s$ to the basal surface but not requested the *no–slip* boundary condition. Actually we could request this and would then simply have

$$\boldsymbol{v}_s = \mathbf{0} \text{ , at } \quad z = z_B(x,y) \, . \tag{6.14}$$

This condition, which is kinematic in nature and makes (6.7) obsolete because it is identically satisfied, has dynamic meaning. Actually vertical profiles of in-plane velocities of creeping soils indicate that the main deformation often occurs in a thin layer close to the bottom of the moving and deforming mass (Fig. 1). In a simplified attempt this thin layer can be shrunk to zero and the substantial shearing replaced by a viscous sliding law. In this law the tangential velocity $\boldsymbol{v}_s$ may be functionally related to the state of stress at the base, viz.,

$$\boldsymbol{v}_s = \hat{\boldsymbol{v}}_s(\boldsymbol{\tau}_B \, , p_\perp^B) \, , \text{ at } \quad z = z_B(x,y) \, , \tag{6.15a}$$

where $\boldsymbol{\tau}_B$ is the shear traction at the bed and $p_\perp^B$ the traction normal to the basal surface at the bed. If $\boldsymbol{v}_s$ and $\boldsymbol{\tau}_B$ are antiparallel which is natural to suppose, then

$$\boldsymbol{v}_s = -f\left(||\boldsymbol{\tau}_B|| \, , p_\perp^B\right) \boldsymbol{\tau}_B \tag{6.15b}$$

with $f > 0$ and

$$\begin{aligned} \boldsymbol{\tau}_B &= \boldsymbol{t}\boldsymbol{n}_B - (\boldsymbol{n}_B \cdot \boldsymbol{t}\boldsymbol{n}_B)\boldsymbol{n}_B = \boldsymbol{\sigma}'\boldsymbol{n}_B - (\boldsymbol{n}_B \cdot \boldsymbol{\sigma}'\boldsymbol{n}_B) \, , \\ p_\perp^B &= \boldsymbol{n}_B \cdot \boldsymbol{t}\boldsymbol{n}_B \, . \end{aligned} \tag{6.15c}$$

This law automatically satisfies the tangency condition $\boldsymbol{n}_B \cdot \boldsymbol{v}_s = 0$. $\boldsymbol{t}$ is the total mixture stress. Any isotropic contribution contained in it does, however, not contribute to the shear traction. The law (6.15) is a viscous-type sliding law of which the functional relations for the coefficient $f > 0$ will be discussed in section 7.

6.3 Transition conditions at the phreatic surface

In general, the upper free surface of the ground water is neither material for the soil grains nor for the water. The phreatic surface, as it is alternatively called, is material with respect to the constituent water only if no water percolates from the free surface through the unsaturated soil above it. If

$$F_w = z - z_w(x,y,t) = 0 \tag{6.16}$$

defines the phreatic surface then

$$\frac{\partial z_w}{\partial t} + \frac{\partial z_w}{\partial x} v_x^w + \frac{\partial z_w}{\partial y} v_y^w - v_z^w = Q_\perp^w N_w \,,$$
$$N_w = \left\{ 1 + \left(\frac{\partial z_w}{\partial x} \right)^2 + \left(\frac{\partial z_w}{\partial y} \right)^2 \right\}^{1/2} \tag{6.17}$$

is the corresponding kinematic equation where $Q_\perp^w > 0$ is the *accretion rate* perpendicular to the surface. In writing down (6.17) we have set, in (6.1b), $\boldsymbol{w} = \boldsymbol{v}_w$ and $(\boldsymbol{v}_w - \boldsymbol{u}) \cdot \boldsymbol{n}_w = -Q_\perp^w$ (note the negative sign, which makes a flow of percolating water from above positive). The accretion rate function must be prescribed in a model that does not describe the water flow through the domain of unsaturated soil. In periods without rain, $Q_\perp^w = 0$; during heavy rainfall episode $Q_\perp^w$ is best set equal to the endrained surface water, but this does not free the hydrologist from separating the total rainfall from surface runoff. We take here the position that $Q_\perp^w(x, y, t)$ is known.

Next, consider the dynamic conditions (6.2). Without introducing further additional assumptions these jump conditions are rather complicated and do not provide much physical insight. The jump condition of mass for the constituent water allows deduction of the relation

$$Q_\perp^w = \frac{M_w^a}{\hat{\rho}_w n} \,, \text{ at } \quad z = z_w(x, y, t) \,, \tag{6.18}$$

in which M_w^a is the *accretion mass rate* from above. Better insight is gained if the following additional assumptions are imposed:

- The porosity does not jump across the phreatic surface: $[\![n]\!] = 0$.
- In the momentum jump condition the velocity jump $[\![\boldsymbol{v}_\alpha]\!]$ is negligibly small.

The first of these conditions appears as natural if the motion of the phreatic surface relative to the soil matrix is considered. It does not seem realistic that the presence of the water surface when crossing a solid matrix will suddenly change the soil's porosity. From $(6.2)_1$ then follows

$$[\![n]\!] = 0 \implies [\![\boldsymbol{v}_s \cdot \boldsymbol{n}_w]\!] = 0 \,, \text{ at } \quad z = z_w(x, y, t) \,, \tag{6.19}$$

and $[\![(\boldsymbol{v}_w - \boldsymbol{u}) \cdot \boldsymbol{n}]\!] = 0$. When the porosity suffers no jump, there can neither be a jump in the normal velocity of the soil, nor can there be a jump in volume flow of water. These conditions are physically plausible.

The second of the above conditions implies that

$$[\![\boldsymbol{t}_\alpha \, \boldsymbol{n}_w]\!] - [\![\boldsymbol{v}_\alpha]\!] \rho_\alpha (\boldsymbol{v}_\alpha - \boldsymbol{u}) \cdot \boldsymbol{n}_w \simeq [\![\boldsymbol{t}_\alpha \, \boldsymbol{n}_w]\!] = \boldsymbol{0} \,. \tag{6.20}$$

Therefore, the peculiar tractions are continuous across the phreatic surface. Since there is no water above it, $\boldsymbol{t}_w^+ = \boldsymbol{0}$, implying $p_w \boldsymbol{n} = \boldsymbol{0} \Rightarrow p_w = 0$. So,

$$\left.\begin{aligned} p_w &= 0 \,, \\ [\![\boldsymbol{t}_s \, \boldsymbol{n}_w]\!] = [\![-(p' + p^v) \boldsymbol{n}_w + \boldsymbol{\sigma}' \, \boldsymbol{n}_w]\!] &= \boldsymbol{0} \,, \end{aligned}\right. \quad \text{at} \quad z = z_w(x, y, t) \,. \tag{6.21}$$

In summary, Equations (6.19) and (6.21) together define the jump conditions at the phreatic surface.

7. Viscous-type sliding laws

7.1 Remarks on objective laws

We return to the problem of sliding of the soil mass along its basal surface, as briefly touched upon in paragraph 6.2, Equations (6.15). Here in this section we shall propose various possible constitutive laws. Before turning to these, note that such relationships must – as any other constitutive relation – be materially objective statements. Equations (6.15*a*), however, do not appear in an objective form, and so one might be tempted to reject them as admissible constitutive relationships. It is easily demonstrated that $\boldsymbol{\tau}_B$ and $p_\perp^B$ defined in (6.15c) transform under Euclidean transformations as an objective vector and scalar, respectively, since $\boldsymbol{t}$ and $\boldsymbol{n}$ transform as an objective second rank tensor and as a vector, respectively. If then $\hat{\boldsymbol{v}}_s(\cdot)$ in (6.15a) or $\boldsymbol{f}(\cdot)$ in (6.15b) transform under Euclidean transformations as an objective vector and scalar respectively, then it follows that the variables on the left of (6.15a,b) ought to transform as vectors under the Euclidean transformation group. As velocity vectors it appears that they do not. This (spurious) inconsistency is resolved by recognizing that the sliding law should be written as $[\![\boldsymbol{v}_s]\!] = -\hat{\boldsymbol{v}}_s(\boldsymbol{\tau}_B, p_\perp^B)$. Thus, the variable on the left is the difference velocity $\boldsymbol{v}_s^{\text{bed}} - \boldsymbol{v}_s^{\text{sole}}$, and in our case $\boldsymbol{v}_s^{\text{bed}} = \boldsymbol{0}$. This difference velocity is in fact an objective vector under Euclidean transformations. Incidentally, in the above we have used the terms "sole" and "bed" to denote the moving upper portion and the rigid immobile lower part of the sliding interface, respectively. In what follows we shall use the notation $\boldsymbol{v}_s^{\text{sole}} = \boldsymbol{v}_B$.

7.2 Mathematical sliding laws

The drag force along a sliding plane can be described in several ways. Dry solid friction is one popular possibility and is expressed in terms of a solid friction angle. This is the approach preferred in limit analysis and does not permit determination of the velocity field, since a yield-type behaviour is postulated. The sliding velocity $\boldsymbol{v}_B$ is expressed as the step-function

$$\begin{aligned} |\boldsymbol{v}_B| &= 0\,, && \text{for} \quad |\tau_B| < c + \sigma_B \tan\phi\,, \\ |\boldsymbol{v}_B| &= \text{undetermined}\,, && \text{for} \quad |\tau_B| \geq c + \sigma_B \tan\phi\,, \end{aligned} \tag{7.1}$$

where τ_B and σ_B are the shear stress and normal stress at the base, c is the cohesion, and ϕ is the angle of friction. These parameters are chosen according to the type of processes that are analyzed (see, for example, LAMBE and WHITMAN, 1963; SKEMPTON, 1964, 1985; CHOWDHURY, 1978). For short–term (undrained) behaviour $c = S_u$ is the undrained shear strength and $\phi = 0$; for long–term (drained) conditions $c = c'$ and $\phi = \phi'$ are the cohesion intercept and friction angle based on effective stresses; for residual conditions (i. e., large displacements) the cohesion c is usually set equal to zero and $\phi = \phi_{\text{res}}$ is the residual angle of friction. Note that for drained conditions (7.1) is expressed in terms of the *effective* normal stress $\sigma'_B = (-p_\perp^B - p_w)$, p_w being the pore-water pressure acting at the base. TAVENAS and LEROUEIL (1981) discuss the influence of creep deformations on the strength parameters.

We will now demonstrate that the resistivity exerted on the sole by the bed is effectively a viscous fluid resistance and that (7.1) is a limiting case of a more general formulation.

Glaciologists (WEETMAN, 1964; LLIBOUTRY, 1975; FOWLER, 1981) who were confronted with similar behaviour on temperate glaciers postulated the law

$$\boldsymbol{v}_B = -f(\cdot)\boldsymbol{\tau}_B \tag{7.2}$$

where $\boldsymbol{v}_B$ is the excess velocity of the sole over that of the bed, $\boldsymbol{\tau}_B$ is the shear traction within the sliding surface, and $f(\cdot)$ a sliding coefficient (the inverse of viscosity), which may depend on additional parameters that are related to the local physics in the vicinity of the sliding surface. Relation (7.2) agrees with (6.15b).

Of primary importance is the stress dependence of $f(\cdot)$, which will be confirmed by examples later. Three types of relations will be presented here, for cases where the velocity depends on (1) the shear traction only, (2) the shear *and* normal effective stresses, and (3) the ratio of shear traction to a function of normal effective pressure.

7.2.1 $\boldsymbol{\tau}_B$ dependence

The simplest form is given by the well-known WEETMAN's law or power law from glaciology, where the sliding velocity in one dimension is written

$$|\boldsymbol{v}_B| = B|\boldsymbol{\tau}_B|^n\,, \quad n \geq 1 \tag{7.3}$$

and B, n are phenomenological constants. The generalized form of (7.3) is found by substituting in (7.2) the following:

$$f(\cdot) = B\tau_B^{n-1}\,, \quad n \geq 1\,, \tag{7.4}$$

where

$$\tau_B = \sqrt{\boldsymbol{\tau}_B \cdot \boldsymbol{\tau}_B} \tag{7.5}$$

is the norm of the basal shear stress vector defined in (6.15b).

A variant of (7.4), appropriate for stick slip, is as follows: $\boldsymbol{v}_B$ vanishes, and thus the no–slip condition prevails as long as the shear stress is smaller than a yield value, τ_0. For larger shear tractions viscous sliding is permitted. In this case the function f in (7.2) must depend on the "excessive stress", $\tau_B - \tau_0$, and on τ_B, a possible form being

$$\begin{aligned} f(\cdot) &= 0\,, & &\text{if} \quad \tau_B \leq \tau_0\,, \\ f(\cdot) &= B\frac{(\tau_B - \tau_0)^n}{\tau_B} = B\left(1 - \frac{\tau_0}{\tau_B}\right)^n \tau_B^{n-1}\,, & &\text{if} \quad \tau_B > \tau_0\,. \end{aligned} \tag{7.6}$$

where B and n are again constants. Equation (7.6) reduces to WEERTMAN's law (7.4) for $\tau_0 = 0$. A qualitative representation of its behaviour is given in *Fig.* 13 a, where the velocity contours in the τ, σ plane are horizontal lines, as can be inferred from the general form of the velocity – shear stress plot.

The above types of equations may be used for incompressible soils, or soils under undrained conditions. These conditions are not likely to prevail in the long-term behaviour of landslides or for slopes strongly influenced by hydrological factors. Inclusion of these processes requires introduction of the effective stress σ' in the sliding law.

7.2.2 τ_B and σ'_B dependence

One way to introduce the normal effective pressure σ'_B at the interface is to postulate that τ_0 is a function of σ'_B . For example, assuming a MOHR-COULOMB expression we have

$$\tau_0 = c' + \sigma'_B \tan\phi \,. \tag{7.7}$$

This suggests introduction of the yield function

$$F(\tau_B) \equiv \tau_B - c' - \sigma'_B \tan\phi \tag{7.8}$$

with $F(\tau_0) = 0$. Here c' is the cohesion, and ϕ is the angle of internal friction defining the *viscous-sliding limit of the interface*, τ_0. These parameters are not necessarily the same as those in the limit analysis and may differ from those defining the strength of the interface, in the plastic sense. Examples will be provided later, they show that sliding occurs at stress levels lower than the strength of the interface.

With (7.7) after substitution into (7.6) we have

$$\begin{aligned} f(\cdot) &= 0\,, && \text{if} \quad \tau_B \le \tau_0(\sigma'_B)\,, \quad \text{or} \quad F \le 0\,, \\ f(\cdot) &= B\left(1 - \frac{c' - \sigma'_B \tan\phi}{\tau_B}\right)^n \tau_B^{n-1}\,, && \text{if} \quad \tau_B \le \tau_0(\sigma'_B)\,, \quad \text{or} \quad F > 0\,. \end{aligned} \tag{7.9}$$

With $\sigma'_B = -p_\perp^B - p_w$ it is readily seen that the pore-water pressure p_w enhances the sliding velocity. Obviously, this relation is more general than either (7.4) or (7.6). It reduces to (7.6) by setting $\phi = 0$ and to (7.4) by imposing $c = 0$, $\phi = 0$. A qualitative representation is given in *Fig.* 13b. Note that the velocity contours are parallel to the viscous-sliding limit line.

The MOHR-COULOMB relation (7.8) for the viscous-sliding limit is not the only possible choice of a sliding law dependent on σ'_B. If a different yield criterion, $F(\sigma'_B, \tau_B) = 0$, is used (7.9) becomes

$$\begin{aligned} f(\cdot) &= 0\,, && \text{if} \quad F \le 0\,, \\ f(\cdot) &= B\frac{F^n}{\tau_B}\,, && \text{if} \quad F > 0\,. \end{aligned} \tag{7.10}$$

For large values of the exponent n all these sliding laws tend to the soil friction law (7.1).

7.2.3 $\tau_B/\tau_0(\sigma'_B)$ dependence

VULLIET (1986) proposed a *power sliding law* in which the sliding velocity is not a function of the distance to a viscous-sliding limit as in (7.6) or *Fig.* 13b, but a function of the ratio of the shear stress τ_B to the viscous-sliding limit $\tau_0(\sigma'_B)$. We define the stress level s as

$$s := \frac{\tau_B}{\tau_0(\sigma'_B)} \,. \tag{7.11}$$

Thus $|\boldsymbol{v}_B| = |\text{fct}(s)|$. Once again, adopting (7.2) using the MOHR-COULOMB criterion for τ_0 and postulating a power law relation for velocity gives the following expression for $f(\cdot)$:

$$f(\cdot) = B\frac{\tau_B^{n-1}}{(c' + \sigma'_B \tan\phi)^n}\,, \tag{7.12a}$$

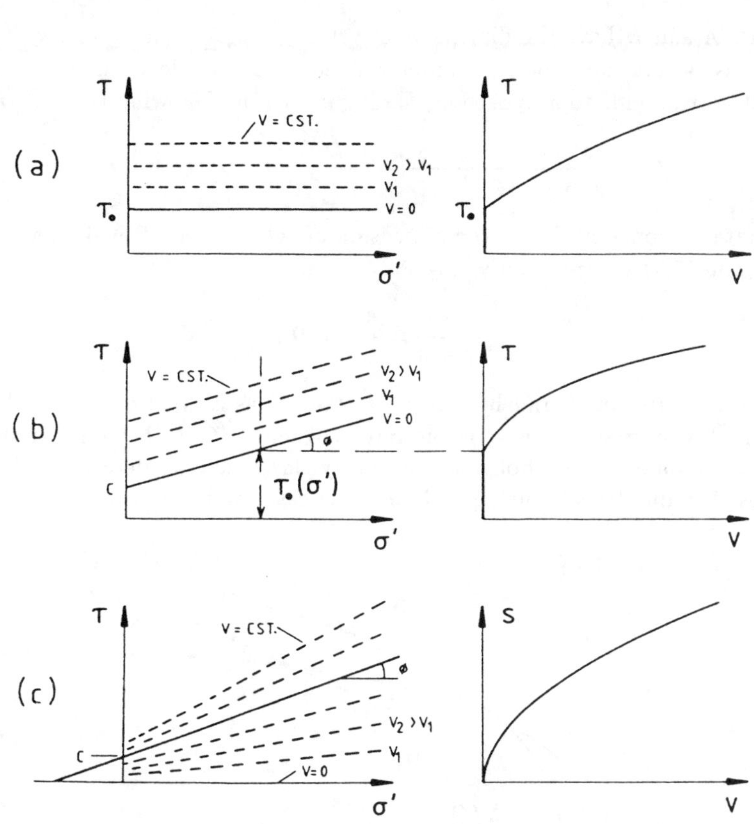

Fig. 13: Schematic representation of the sliding laws. Velocity contours in the (σ', τ) plane (left) and velocity versus shear stress or stress level (right). σ' is the effective normal stress. a) as according to (7.2) and (7.6); b) as according to (7.2) and (7.9); c) as according to (7.2) and (7.12a).

which for cohesionless interfaces reduces to

$$f(\cdot) = B\frac{\tau_B^{n-1}}{(\sigma_B)^n} \; . \tag{7.12b}$$

The schematic representation of the sliding law (7.12a) is shown in *Fig.* 13c. We see that the velocity contours in the τ, σ' plane are not parallel, and that the velocity is a monotonically increasing function of the stress level s. Moreover, this law predicts a nonzero sliding velocitiy as soon as $\tau_B > 0$.

Fig. 14 illustrates the response performance for different choices of $f(s)$, and plots v_B versus s . Curve (a) is obtained by substituting (7.12b) into (7.2). Curve (b) is linear in the stress ratio and has a yield value; the corresponding *"linear" sliding law with yield* is

$$f(\cdot) = \frac{B}{c' + \sigma'_B \tan\phi} - \frac{A}{\tau_B} \, , \quad \text{if } s > A/B \, . \tag{7.13}$$

The constants A and B have the dimension of velocity and are shown in *Fig.* 14. Curve (c) is a limiting case where the velocity is undetermined for a stress level equal to or higher than 1 ($s \geq 1$). It corresponds to a hyperbolic sliding law of the following type:

$$f(\cdot) = \frac{C}{c' + \sigma_B' \tan\phi - \tau_B} - \frac{CD}{\tau_B}\,, \quad \text{if } s > D/(1+D)\,, \tag{7.14}$$

where the material constant C has the dimension of velocity and D is dimensionless. Using (7.11), (7.7) and (7.3) this relation yields

$$\frac{s}{1-s} = \frac{1}{C}v + D\,. \tag{7.15}$$

These parameters are thus found by plotting $s/(1-s)$ as a function of v and using linear regression; in this representation D is the intercept and $1/C$ is the slope of the regression line, and $D = 0$ implies a hyperbolic law without yield. Below we determine the coefficients of the proposed sliding laws by using real landslide cases and in situ measurements.

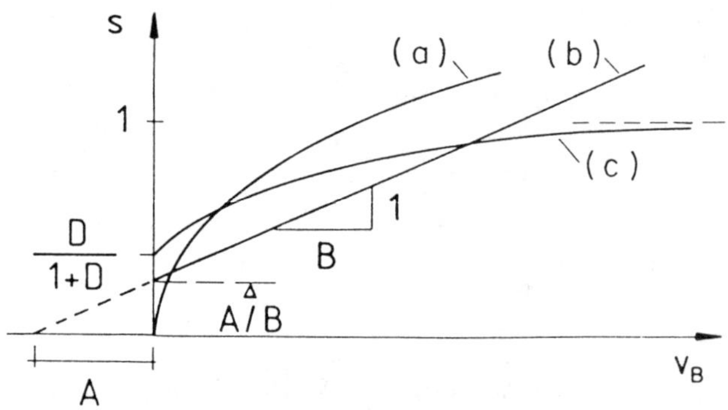

Fig. 14: Relation between sliding velocity and stress level s: a) power law according to (7.2) and (7.12a); b) "linear" law with yield according to (7.2) and (7.13); c) hyperbolic law with yield according to (7.2) and (7.14).

8. Scaling and asymptotic development

The equations derived in the previous sections comprise a complete system and open a variety of alternatives to study the effects of the various constitutive postulates. Resolution is not easy if the problem is tried to be solved in its full three-dimensionality. An initial value problem for porosity, n, porewater pressure, p_w, and the velocity fields of the soil, $\boldsymbol{v}_s$, and the interstitial water $\boldsymbol{v}_w$ with two evolving surfaces – the free upper surface and the phreatic surface – and fixed basal surface, has to be solved. Simplification is almost mandatory if explicit insight into the physical behaviour is to be gained. A great deal of help is provided by the recognition that natural slopes have peculiar geometric and kinematic characteristics. The thicknesses of the moving and deforming masses are usually small when compared with the longitudinal and transversal dimensions, and movement is mainly from high to low altitudes.

To incorporate this *shallowness* into the equations the latter are non-dimensionalized by introducing typical scales for the dimensional fields. Lengths in the x-, y- and z-directions are scaled with a characteristic length $[L_x]$, width $[L_y]$ and depth $[H]$, Fig. 15. If dimensionless velocities, stresses, etc. are also introduced by scaling them with reference values, then the governing equations can be written in dimensionless form with dimensionless coefficients two of which are

$$\epsilon_x = \frac{[H]}{[L_x]}, \quad \epsilon_y = \frac{[H]}{[L_y]}, \quad \text{with } \eta = \frac{\epsilon_y}{\epsilon_x}. \tag{8.1}$$

The dimensionless equations are listed in the Appendix. Study of more than 40 landslides in Switzerland has shown that ϵ_x is of the order of $0.01 - 0.16$ and ϵ_y of the order of $0.004 - 0.21$, see *Table 5*. This suggests that in the scaled field equations and boundary and transition conditions, terms involving ϵ_x (and possibly ϵ_y) could be ignored and limiting equations be deduced.

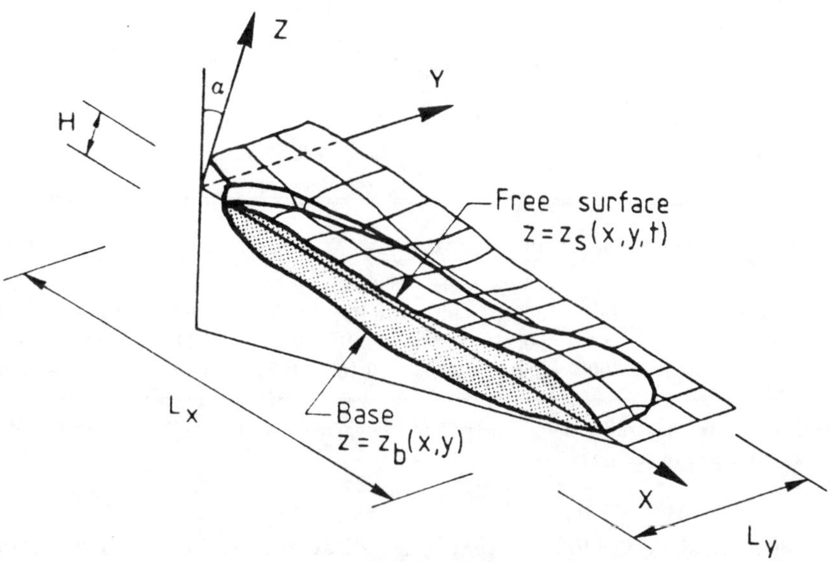

Fig. 15: Three dimensional view and section of a landslide, showing the typical dimensions and the (inclined) axis of the Cartesian coordinate system. x is in the direction of steepest descent, y is horizontal and z perpendicular to the (x, y)-plane; α is the slope angle and L_x, L_y and H are typical scales for length, width and depth. The phreatic surface is not shown.

For example $\epsilon_y \tan\alpha$ will be ignored where α is the mean slope angle, but terms involving the factor $\epsilon_y \cot\alpha$ are retained. If the equations are so reduced, a new approximate set of equations is obtained (the so-called lowest order shallow slope approximation). It is characterized by the following properties:

- The diagonal components of the deviatoric stress tensor as well as the in-plane shear stress τ_{xy} are zero.
- Provided the initial distribution of the porosity, n, and the z-components of the water velocity are known, the remaining stress components p, τ_{xz} and τ_{yz} and the velocity components u_s, v_s, w_s of the soil, and u_w, v_w of the interstitial water for the corresponding geometric configuration may be computed by simple quadrature rules.
- Using the continuity equations these solutions can be improved by iterating on the porosity, n, and the velocity component w_w.
- With these quantities determined, the new free and phreatic surfaces can be computed by forward stepping in time.

The details of this reduction are somewhat tedious and will not be presented here, see VULLIET (1986), it is, however, evident that the stress and velocity computations are separate from those of the domain geometry. Essential to this separation is the scaling that is outlined in the Appendix.

9. Illustrations

Typical calculations for a strictly parallel-sided slab and a study on the variation of the hydraulic conditions are given by VULLIET & HUTTER (1988a).

	α (deg)	$[H]$ (m)	$[L_x]$ (m)	$[U]$ (cm/y)	ϵ_x	ϵ_y	η	$\epsilon_x \cot\alpha$	$\epsilon_y \cot\alpha$
Mean value	17.1	64	1400	28	0.048	0.068	1.144	0.157	0.234
Standard deviation	5.7	96	1300	92	0.035	0.045	1.265	0.119	0.161
Minimum value	8.4	8	90	1	0.01	0.004	0.11	0.020	0.020
Maximum value	34.9	700	6000	500	0.16	0.21	5.60	0.53	0.640

Table 5: Statistical data concerning the geometry and average sliding velocity for 40 Swiss landslides (parameters as explained in the text).

Here we consider three sites; the first is situated in Le Day near Vallorbe in Switzerland, the second is in Sallèdes near Clermond-Ferrand in France and the third is in La Frasse in the Swiss alps, 60 km east of Lausanne.

9.1 Le Day landslide

This landslide is situated in the Swiss Jura, and affects the international Lausanne - Paris railway line (see *Fig.* 16). It is composed of silty-clay moraines and rhodanian glacio-lacuster deposits, with up to 50 % by weight less than 2 μm; the clay fraction is mainly micas, chlorites, and smectites. The mean downslope angle is 17° and the landslide area is approximately 700 × 1000 m, over a mean depth of 25 m. The average water content and

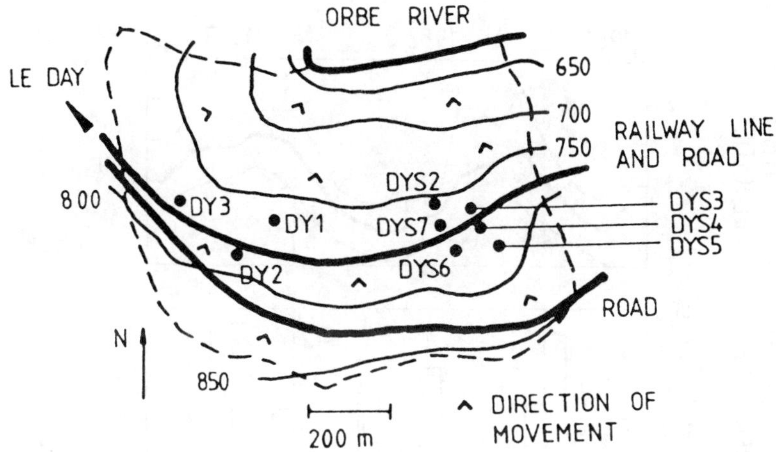

Fig. 16: Le Day landslide. Schematic situation map of the site and position of the boreholes; compare also *Fig. 1a.*

consistency indices are $w = 20.5\%$, $w_L = 42\%$, $I_p = 24\%$. The mean strength parameters are $\phi_u = 13^o$ and $c_u = 25\,kPa$ (direct shear test). These quantities are given for boreholes DY1, DY2, DY3, and close to the sliding surface. The sliding mass covers a carstic calcareous formation that strongly influences the hydrological environment, so that the sliding surface may be considered to be a pervious boundary.

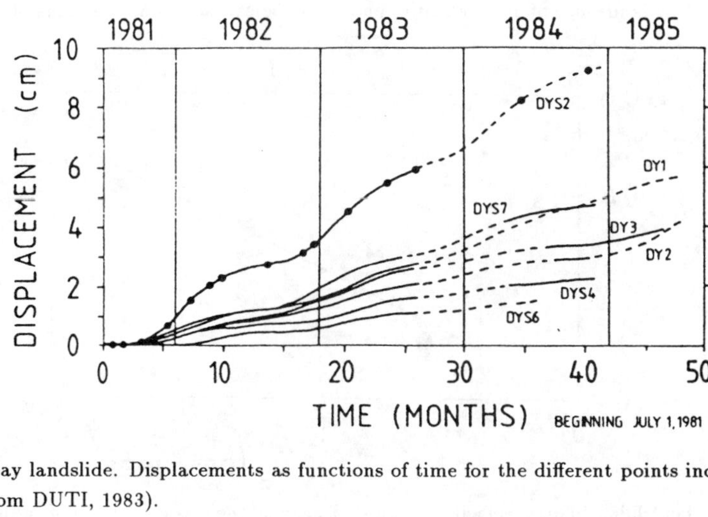

Fig. 17: Le Day landslide. Displacements as functions of time for the different points indicated in *Fig.* 16 (Partly from DUTI, 1983).

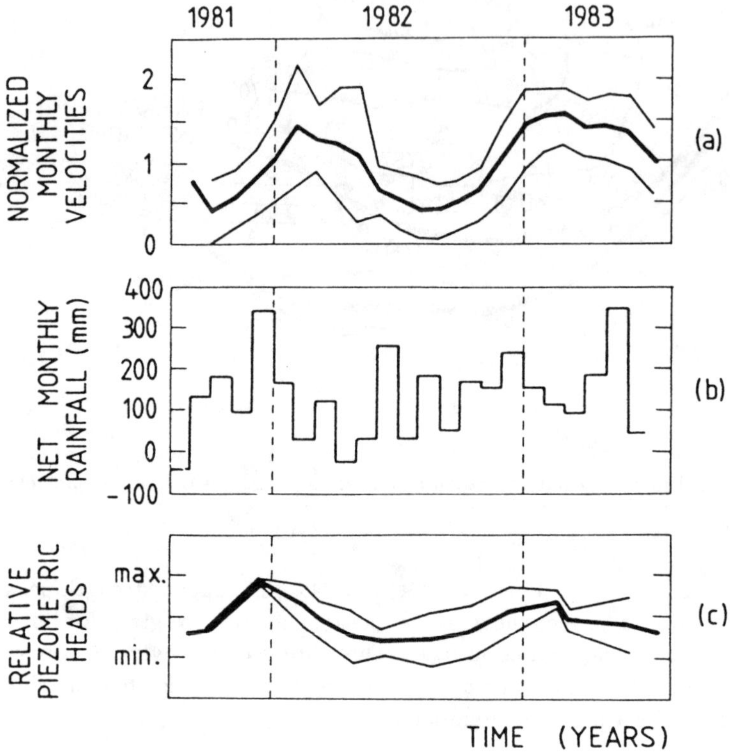

Fig. 18: Le Day landslide. Sliding velocities and hydrological factors as functions of time.

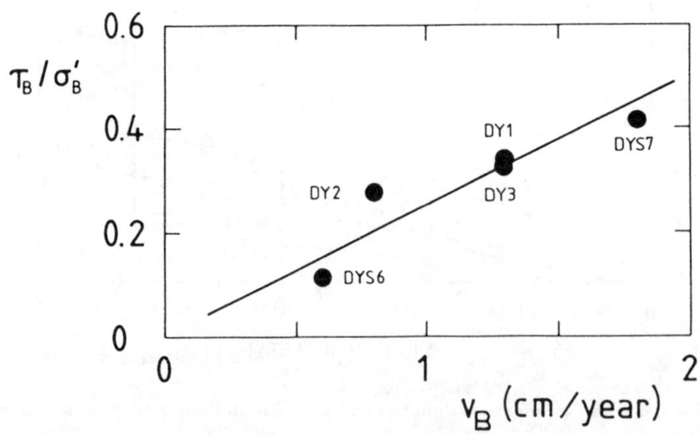

Fig. 19: Le Day landslide. Sliding velocity v_B plotted against the stress ratio for five different bore holes (full circles) and a best linear fit.

Fig. 16 gives a schematic situation map of the landslide, as well as the positions of the boreholes, which are all used for inclinometer measurements. Some displacement profiles are given in *Fig.* 18a, showing that the greater part of the deformation is concentrated in a narrow zone, close to the bottom. The displacements as functions of time are given in *Fig.* 17 for the years 1981 – 1985. It can be seen that the mean velocity over this period of time (the trend) is of the order of $1-3cm/year$, and that seasonal fluctuations are superimposed. This is more significant if the monthly velocities (normalized with the average velocity) are plotted as a function of time, as shown in *Fig. 18a*.

The heavy line in *Fig. 18a* represents the mean monthly velocity over the seven measurement points (or boreholes), and the thin lines are one standard deviation apart. This indicates that the periods of maximum velocity occur in late winter, but are not reached at the same time everywhere on the site; the minimum velocities are reached in summer.

For the same period of time, the net rainfall is represented in *Fig. 18b*. Direct correlation with *Fig. 18a* is not readily apparent, but the peaks of net rainfall do not necessarily have a direct influence on the movements because they are smoothed by the runoff, which is not taken into account here.

Finally, in *Fig. 18c*, the mean variation of the piezometric head close to the sliding surface (averaged over six micropiezometers) is seen. Noting, however, the relative dispersion of these measures, a good fit exists between sliding velocities (*Fig. 18a*) and water heads (*Fig. 18c*). This landslide gives a good illustration of the qualitative action of the groundwater on the movements of a natural slope in which the periods of high velocity are associated with high piezometric levels. The identification of a sliding surface is possible except for drilling hole DYS2 (see *Fig. 1a*). We estimated the basal stresses from the formulas

$$(\sigma_B, \tau_B) = \rho g H (\cos\alpha, \sin\alpha) , \tag{9.1}$$

where $\rho g = 20kN/m^3$, H is the depth, and α is the local inclination angle of the free surface (see *Table 6*). The pore-water pressures were obtained from piezometric measurements. *Table 6* lists these values and the deduced effective pressures, shear stresses, and measured sliding velocities in five boreholes of the site.

Borehole No.	$\nu_{B,mes}$ $(mm/year)$	H (m)	α (deg)	$\tau_{B,est}$ (kPa)	$\sigma_{B,est}$ (kPa)	$p_{w,mes}$ (kPa)	σ'_B (kPa)	$\tau_B \sigma'_B$
DY1	13	26.5	13	119.2	516.4	165	315.4	0.34
DY2	8	20	9	62.5	395.1	168	227.1	0.28
DY3	13	14.85	11	56.7	291.5	117	174.5	0.32
DYS6	6	22	4	30.7	438.9	158	280.9	0.11
DYS7	18	15.35	16	84.6	295.1	93	202.1	0.42

NOTE: Subscript "mes" denotes a measured value and "est" an estimated value.

Table 6: Le Day landslide. Sliding velocity, depth of the sliding plane, surface inclination, and stress at the base.

Fig. 19 shows the relation v_B versus τ_B/σ'_B with the points corresponding to these boreholes indicated by solid dots. This suggests a power law for the sliding velocity of the type of (7.12b) with the material constants $n = 1$ and $B' = 1.27 \times 10^{-9} ms^{-1}$.

9.2 Sallèdes landslide

This site has been selected by the Laboratoire central des ponts et chausées (Paris, France) as an experimental site to study the impact of a road embankment on an already unstable slope. Information on it can be found in CARTIER and POUGET (1987), MORIN (1979), POUGET et al. (1985), POUGET and CARTIER (1988), and BOURDEAU et al. (1988). The superficial layers belong to the upper Oligocene (chattien) and consist of marl of which the first $1 - 10\,m$ constitute the unstable mass. This moving layer is subject to groundwater seepage, with a groundwater level that is affected by strong seasonal fluctuations (with $1 - 5\,m$ amplitudes). As would be expected, instabilities are related to the position of the groundwater level.

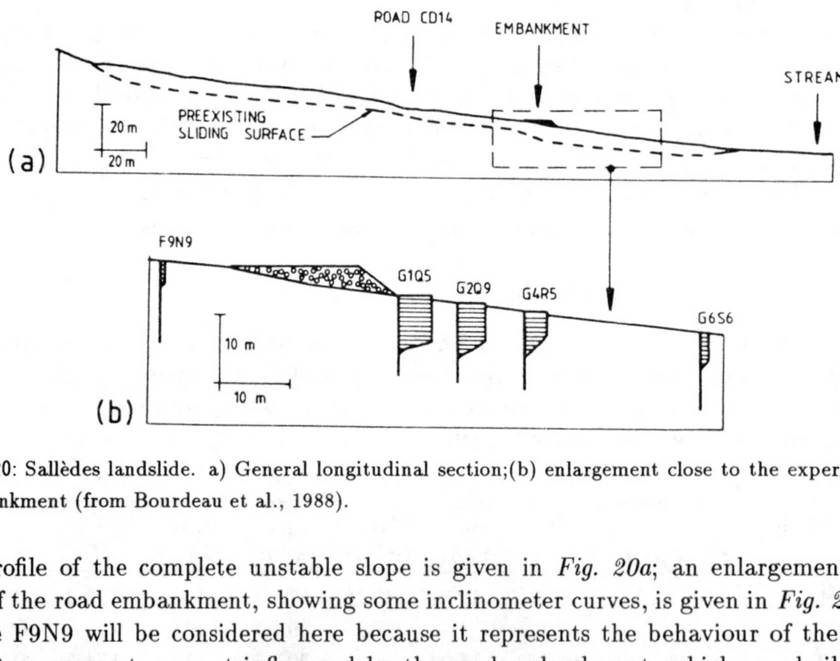

Fig. 20: Sallèdes landslide. a) General longitudinal section;(b) enlargement close to the experimental embankment (from Bourdeau et al., 1988).

A profile of the complete unstable slope is given in *Fig. 20a*; an enlargement in the region of the road embankment, showing some inclinometer curves, is given in *Fig. 20b*. The borehole F9N9 will be considered here because it represents the behaviour of the natural slope. Its movement was not influenced by the road embankment, which was built in two steps, first in March and then in September 1980. *Fig. 21a* shows the surface displacements as functions of time for the different indicated points and *Fig. 21b* gives the corresponding piezometric levels close to the sliding surface. Note the correlation between piezometry and displacements. (For more details see the cited literature).

The following geotechnical data were obtained from POUGET and CARTIER (1988): The average specific weight γ is $18\,kNm^{-1}$, the consistency parameters are $w = 30 - 55\%$,

$w_L = 55 - 85\%$, $I_p = 20 - 70\%$. The "clay fraction" is $30 - 70\% < 2\mu m$ with an important contribution of smectites. Strength parameters are variable, but POUGET and CARTIER conclude that one can retain $c' = 20\,kPa$ and $\phi' = 21^o$ for the first $3\,m$, and $c' = 15\,kPa$ and $\phi' = 25^o$ between 3 and $5\,m$. On the pre-existing sliding surface, the values are $c_{\text{res}} = 0.9^o \leq \phi_{\text{res}} \leq 13^o$.

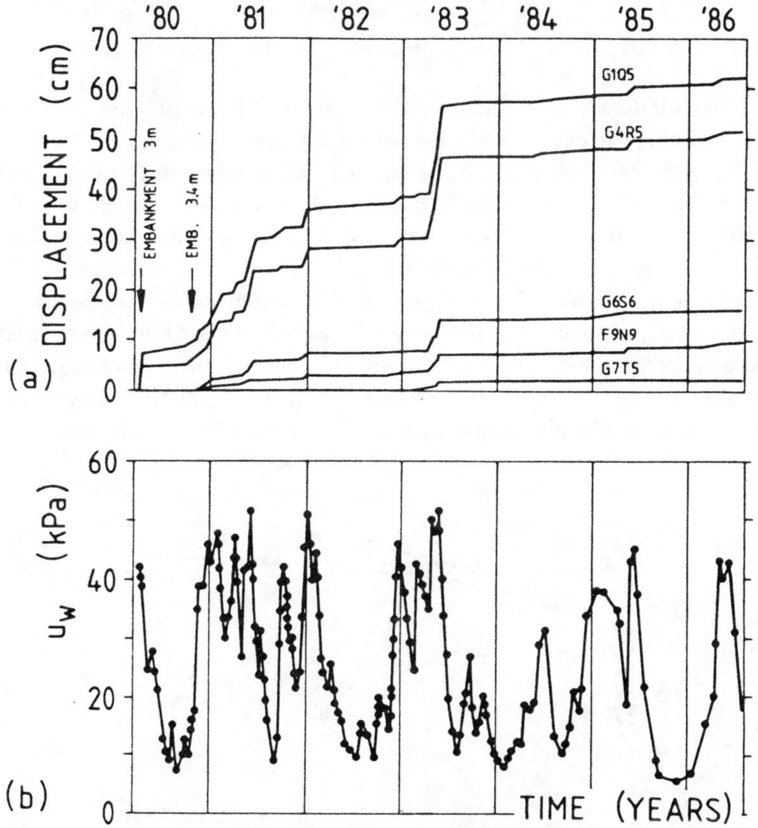

Fig. 21: Sallèdes landslide. a) Displacement versus time for different points indicated in *Fig. 20b*; b) porewater pressure at the sliding surface versus time (from Pouget & Cartier, 1988).

Fig. 22 presents the sliding velocity v_B as a function of the stress ratio s (7.11). Here the points do not correspond to different locations on the site but to the same borehole (F9N9) and different hydrological conditions. Evidently, the variation of the position of the groundwater level conspicuously affects the movements of the $6\,m$ thick sliding mass. Several forms of the sliding law are possible and will be proposed below:

Curve a) of *Fig.* 22 represents the power sliding law (7.12a) with $n = 15$, $B = 2.16 \times 10^{-8} ms^{-1}$, $c = 0$, $\phi = 9.5^o$, and a coefficient of correlation of 0.84. For the sake of comparison with the average values given in "Le Day" it is straightforward to pass from (7.12a) to (7.12b), as $c = 0$. We then obtain the values $n = 15$ and $B' = 9.56 \times 10^3 ms^{-1}$. This value of the exponent n is very high in comparison with the value in the previous case (Le Day); it reflects the pseudoplastic behavior of this landslide, that is, vanishing velocities up to $s = 0.7$, small velocities between this level and $s = 0.9$, and very high velocities for stresses above this level. A better correlation may be obtained if the exponent n is postulated to be a function of the stress level, $n = n(s)$. With the simple linear relation

$$n = as ,$$

where $a = 21.5$ and using a sliding coefficient $B = 3.11 \times 10^{-8} ms^{-1}$ the coefficient of correlation is 0.93 and the curve would lie between curves a) and c) in *Fig.* 22.

Curve b) in *Fig.* 22 is obtained by a bilinear sliding law (15), with two yield limits indicated in the figure. In this case the parameters are $A = 0$, $B = 0$ for $s \leq 0.71$; $A = 9.95 \times 10^{-9} ms^{-1}$, $B = 1.40 \times 10^{-8} ms^{-1}$ for $0.71 < s < 0.945$; $A = 1.46 \times 10^{-6} ms^{-1}$, $B = 1.54x10^{-6} ms^{-1}$ for $s > 0.945$; $c = 0$; $\phi = 9.5^o$.

For this landslide, the power law with constant exponent does not offer the best fit. The practical use of the hyperbolic law is questionable, as it leads to an undetermined velocity for $s \geq 1$; the fit, however, is good. Finally, the "linear" law with two yield limits (curve b) seems to be both simple and accurate. However, from a numerical point of view piecewise-defined relations are not welcome; a power law with varying exponent is preferred.

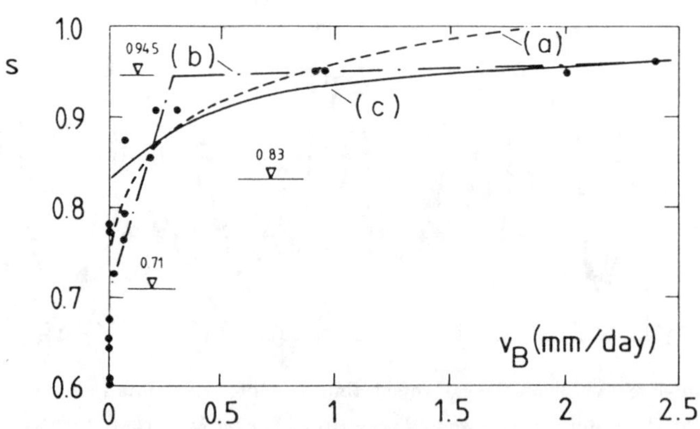

Fig. 22: Sallèdes landslide. Sliding velocity v_B plotted against the stress level s for the point F9N9 and different hydrological conditions (full dots). The curves are best fits corresponding to a) power law, b) piecewise linear law, c) hyperbolic law with yield.

9.3 La Frasse

The La Frasse landslide lies in the Swiss alps, 60 km east of Lausanne. It covers an area 2000 m long by 400 m wide; it is about 60 m thick and has a mean slope angle of approximately 10°. A situation map is given in *Fig.* 23.

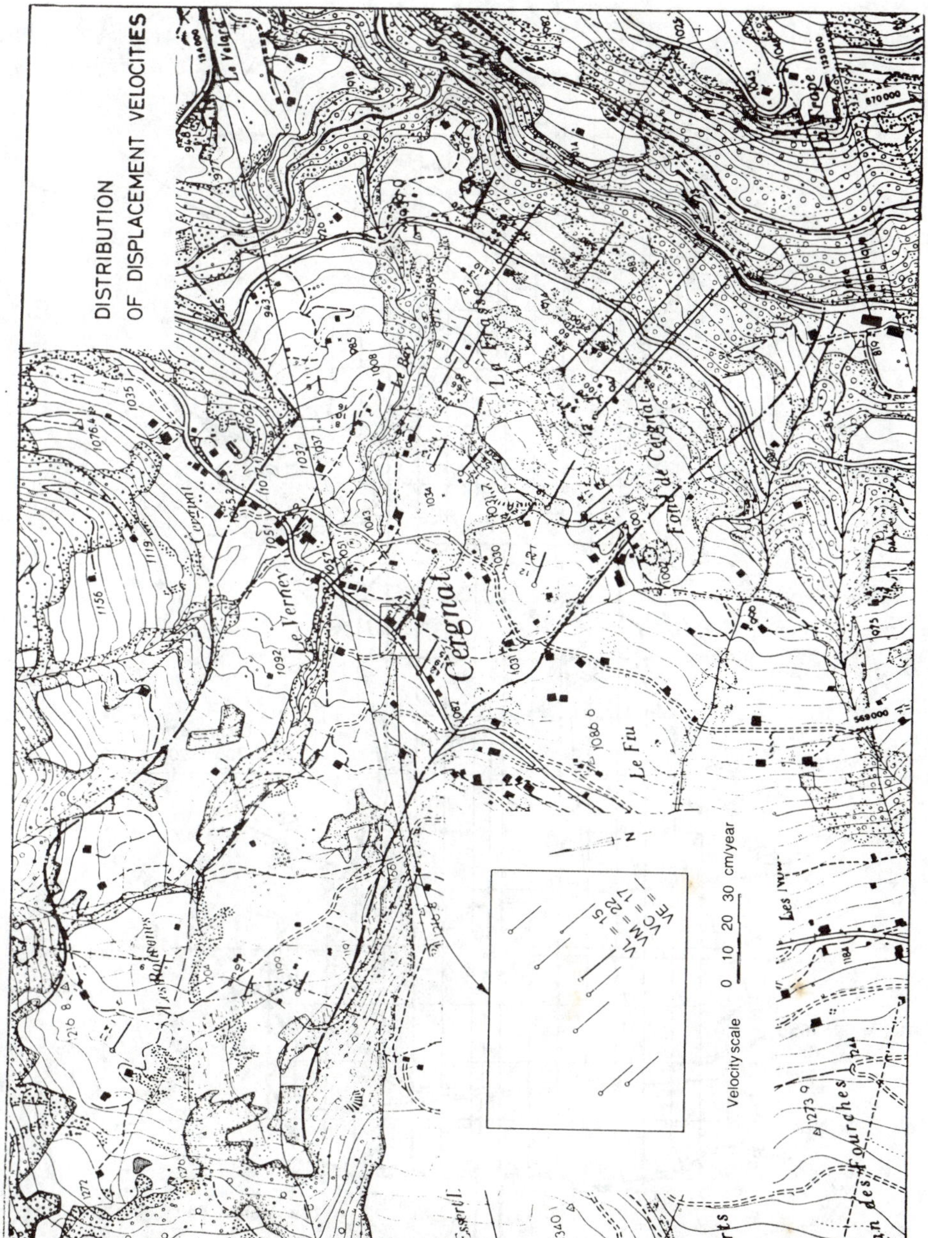

Fig. 23: La Frasse landslide: map showing the level lines, houses, roads, and some measured velocity vectors.

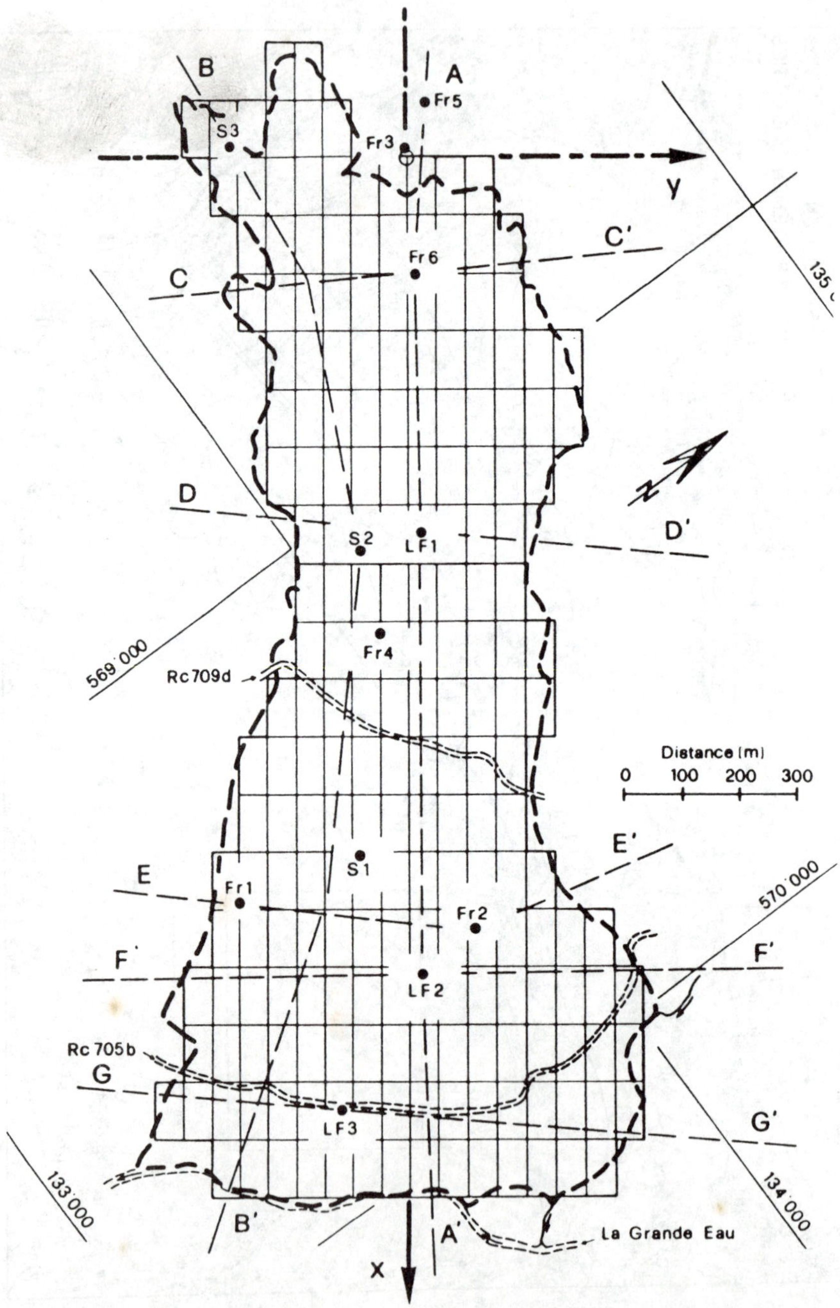

Fig. 24: La Frasse landslide: computational grid, borehole sites, coordinate system x, y and cross section B–B.

The geology consists mainly of clayed flisch and marl from the Eocene period. Further details of this landslide are given by DUTI (1986).

La Frasse has been intensely studied and instrumented so that the geometry of the free surface is known and the locations of the phreatic and basal surfaces have been determined for some points. The complete three-dimensional shape of the base has been constructed by successive interpolation between known points, taking into account the general geology of the site, available geophysical results and so on. The phreatic surface being close to the free surface, it was assumed that the two surfaces coincide everywhere. Here, the long term (120 years) surface velocity field was used, which was obtained by comparing old and new maps (ENGEL, 1986). From inclinometer readings it could be inferred that the total sliding motion overrides differential creep.

For computation, the domain of the landslide was divided into a rectangular grid with mesh size of $\Delta x = 100\,m$ and $\Delta y = 50\,m$; the total number of nodes amounted to 259 (*Fig.* 24). Perspective views of the free and basal surfaces are given in *Fig.* 25. The heavy lines in these figures indicate the roads Aigle-Le Sépey and Le Sépey-Leysin, which cross the region of the landslide.

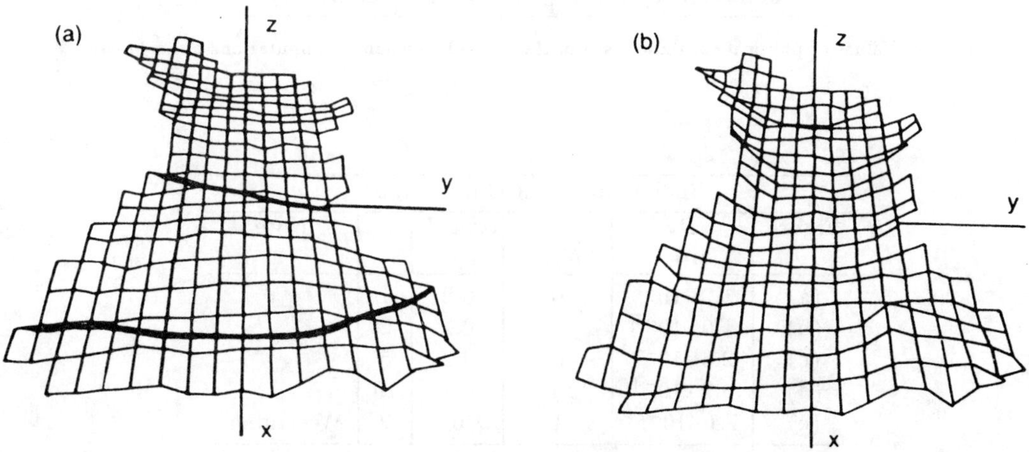

Fig. 25: La Frasse: a) perspective view of the free surface; b) perspective view of the base.

A non-linear sliding law (Eq. (7.12a)) was employed with phenomenological constants as listed in *Table 7*. A preliminary two-dimensional computation along the line B–B in *Fig.* 24 was then performed (ignoring differential shear and transverse dependencies), and surface velocities alone B–B were computed by varying the coefficients B for exponents $n = 1$ and $n = 2$ so as to obtain the best fit with the observations. This optimization gave the values shown in *Table 7* and the corresponding optimal fits in *Fig.* 26. The free parameters in the sliding law having been determined, a full three-dimensional computation was performed to obtain the long term velocity field for the given configuration of the landslide.

Computation	B : m/s	n
1	10^{-7}	1
2	$2.6 \cdot 10^{-6}$	2

Table 7: Values of the phenomenological coefficients B and n in the sliding law for the two-dimensional computation in *Fig.* 26.

Quantity	Value	Dimension
Slope angle	$\alpha = 13$	degrees, ^{o}
Intrinsic density		
(a) of the soil	$\hat{\rho}_s = 2.7$	t/m^3
(a) of the water	$\hat{\rho}_w = 1.0$	t/m^3
Characteristic length	$[L_x] = 2000$	metres, m
Characteristic width	$[L_y] = 600$	metres, m
Characteristic depth	$[H] = 50$	metres, m

Table 8: Values of physical constants used in the three-dimensional computations of La Frasse.

Sliding dominant (Eqn. 7.12b)					
Figure	B : m/s	c' : kN/m^2	$\tan\phi'$	n	Note
(a)	$5.2 \cdot 10^{-8}$	0	0.2	1	
(b)	$1.0 \cdot 10^{-7}$	0	0.2	2	
(c)	$2.6 \cdot 10^{-6}$	0	0.2	5	
(d)	$5.2 \cdot 10^{-4}$	0	0.2	10	
(e)	$7.1 \cdot 10^{-14}$	1	0	2	Weertman
Differential creep dominant (Eqn. 4.34)					
Figure	A : s^{-1}	k' : kN/m^2	α'	m	Note
(f)	$7.1 \cdot 10^{-11}$	1	0	1	Newton
(g)	$2.1 \cdot 10^{-8}$	0	0.2	5	
(h)	$4.2 \cdot 10^{-23}$	1	0	5	Norton

Table 9: Values of the parameters used in the computations with which *Fig.* 28 was constructed.

Some physical constants are given in *Table* 8. A comparison between the computed and measured surface velocities is given in *Fig.* 27, the agreement between the two is satisfactory. The directions of the velocity vectors are well reproduced as are their magnitudes in most cases.

In the course of the study of this landslide the influence of a variation in the parameters of the sliding law and the flow law have been investigated. Results are summarized in *Fig.* 28.Corresponding phenomenological parameters are collected in *Table* 9. In *Fig. 28a-e* the sliding conditions are varied while the creep law is held fixed with a coefficient A so small that differential creep is negligible. In *Fig. 28 f-h* the creep law is varied, and the no-slip condition is imposed.

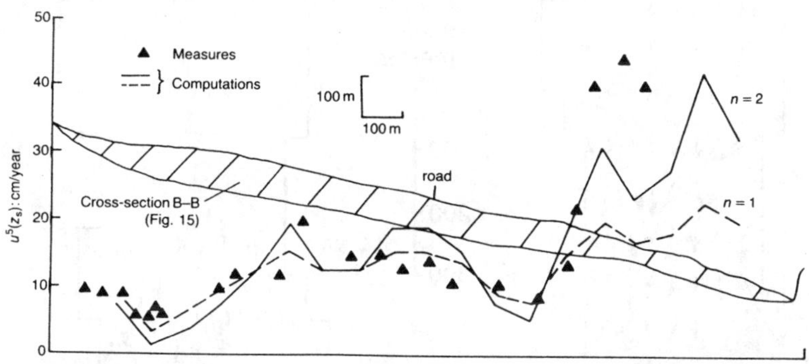

Fig. 26: La Frasse: Illustration of results from two–dimensional computations along B–B (Fig. 25). Surface velocity as a function of the distance x along B–B. Comparison between computed and measured velocities. The computations are based on Eq. (7.12a) using values from *Table* 7.

The result of *Fig. 28a* corresponds to the case $n = 1$ in *Fig.* 26, in this full three - dimensional computation the surface velocities are too high. In *Fig. 28b-d* the influence of the sliding exponent n is studied, with values of 2, 5 and 10, respectively. In these cases the effective pressure enters the sliding law. *Fig. 28e* shows the flow pattern when a WEERTMAN-type sliding law is used (WEERTMAN, 1964). This law is independent of pressure and contains only a shear stress dependence. The upper part of the sliding motion is poorly reproduced and velocities in the lower part are too large in general.

The influence of the creep laws is shown in *Fig. 28 f-h*. In these cases, the velocity profiles do not match the inclinometer measurements (*Fig.* 1). Thus, what follows is a theoretical study. In *Fig. 28f*, Newtonian behaviour applies (i. e. the body is linearly viscous). In the next case a high order power fluid is applied in which flow enhancement due to effective pressure is accounted for (*Fig. 28g*). When this latter effect is ignored (*Fig. 29h*) Norton's flow law (with exponent $n = 5$) is obtained. Comparison of *Fig. 28g* and *28 h* demonstrates the influence of the effective pressure.

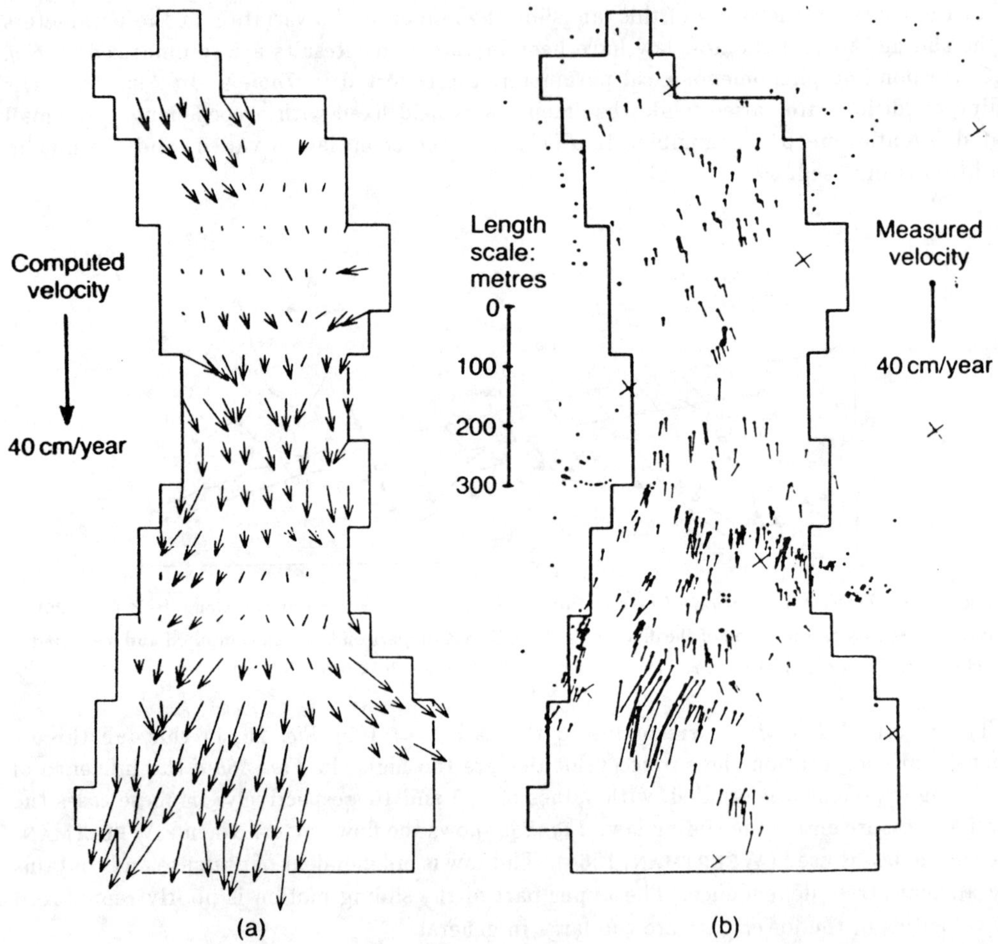

Fig. 27: La Frasse: a) results from three-dimensional computation Eq. (7.12 a) $n = 2$; b) measured velocity field (note the different velocity scales).

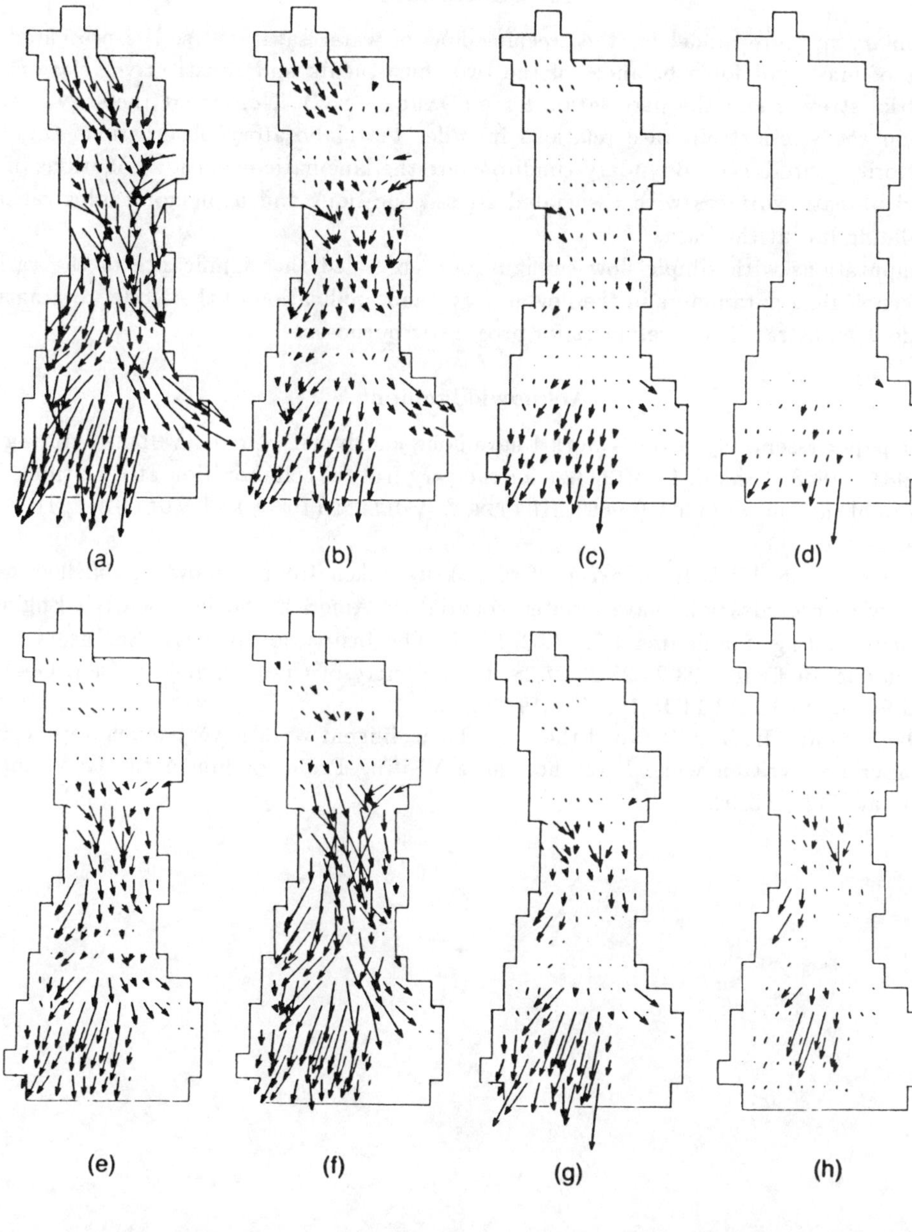

Fig. 28: La Frasse: a) – e) influence of the variation in sliding conditions; f) – h) and creep law, when Eq. (7.12b) is used in computations of the sliding velocity and Eq. (4.34) is used in evaluation of the creep law. Numerical values of phenomenological coefficients are as in *Table* 9.

10. Conclusions

A binary mixture model for the creeping flow of water saturated soil is presented. It consists of mass and force balances of the two constituents and constitutive relations for the partial stresses and the percolation force (DARCY's law). Non-linear rheology became crucial in the stress-strain rate relations in order that laboratory observations could be satisfactorily reproduced. Boundary conditons are the kinematic evolution equations of the free and phreatic surfaces with associated stress conditions and a phenomenological non-linear sliding law at the base.

Computations with simple flow configurations revealed the significance of the various phenomenological parameters in the flow process, and application of the model to a natural landslide demonstrated its usefulness for prognostic purposes.

Acknowledgement

This paper essentially reviews work I have done jointly with Dr. L. VULLIET during the years 1984 – 1988. I thank L. VULLIET for the very frutiful collaboration at that time. The work is published in VULLIET (1986), HUTTER & VULLIET (1985) and VULLIET & HUTTER (1988 a, b, c, d, e).

The figures of this article, except figure 3 are taken from our own published work. The following organizations have granted copyrights: American Society of Civil Engineers, J. Geotech. Eng. for figures 4,5,6,7,8,9,10,11; The Institution of Civil Engineers (UK), Géotechnique for figures 23,24,25,26,27,28; Can. Society of Civil Engineers, Can. Geotech. Journal for figures1,2,13,14,15,16,17,18,19,20,21,22.

Mrs. Danner and Mr. Staudt typed the text. Their diligent work is very much appreciated. This paper was written while I was holding a Visiting Professorship in the Geographical Institute at ETH, Zürich.

Appendix I. Scaled equations

We shall illustrate how the governing equations are non-dimensionalized. To this end, orders of magnitude are introduced for each variable which are typical for the processes under consideration.

The scalings are:

$$
\begin{aligned}
x &= [L_x]\bar{x}\,, \quad & y &= [L_y]\bar{y}\,, \quad & z &= [H]\bar{z}\,, \\
u^s &= [U_s]\bar{u}^s\,, \quad & v^s &= [V_s]\bar{v}^s\,, \quad & w^s &= [W_s]\bar{w}^s\,, \\
t &= [T]\bar{t}\,, \quad & p &= \tilde{\rho} g[H]\cos\alpha\,\bar{p}\,,
\end{aligned}
$$

$$
\begin{aligned}
&(t_{xx}\,, t_{xy}\,, t_{xz}\,\ldots) = \tilde{\rho} g[H]\sin\alpha(\bar{t}_{xx}\,, \bar{t}_{xy}\,, \bar{t}_{xz}\,\ldots) \\
&u^w = [U_w]\bar{u}^w\,, \quad v^w = [V_w]\bar{v}^w\,, \quad w^w = [W_s]\bar{w}^w\,, \\
&p_w = \hat{\rho}_w g[H]\cos\alpha\,\bar{p}_w\,, \qquad Q = [W_w]\bar{Q}\,.
\end{aligned}
\tag{A.1}
$$

Variables carrying an overhead bar are dimensionless and quantities in brackets are the scales. Moreover, the velocity components in the x-, y- and z- directions are denoted by u, v, w with indices s and w depending upon whether they are those of the soil or water, respectively. We shall write

$$
\epsilon_x = \frac{[H]}{[L_x]}\,, \quad \epsilon_y = \frac{[H]}{[L_y]}\,, \quad \eta = \frac{\epsilon_y}{\epsilon_x} = \frac{[L_y]}{[L_x]}
\tag{A.2}
$$

and implement the order relations

$$
\frac{[L_x]}{[L_y]} = \frac{[U_s]}{[V_s]} = \frac{[U_w]}{[V_w]} = \frac{1}{\eta}\,.
\tag{A.3}
$$

If the definitions and choices, respectively

$$
\begin{aligned}
\kappa &= \rho/\hat{\rho}_w\,, & \kappa_s &= \hat{\rho}_s/\hat{\rho}_w\,, \\
[U_w] &= k\sin\alpha\,, \\
\zeta &= [W_w]/[U_w]\,, & \xi &= [W_s]/[U_s]\,, \\
\beta &= [U_s]/[U_w]\,, & [T] &= [L_x]/[U_w]\,,
\end{aligned}
\tag{A.4}
$$

are used, then after lengthy but straightforward manipulations the following dimensionless equations can be deduced:

Mass balances

$$
\begin{aligned}
-\beta^{-1}\frac{\partial n}{\partial t} + \frac{\partial[(1-n)\bar{u}^s]}{\partial\bar{x}} + \frac{\partial[(1-n)\bar{v}^s]}{\partial\bar{y}} + \xi\epsilon_x^{-1}\frac{\partial[(1-n)\bar{w}^s]}{\partial\bar{z}} &= 0\,, \\
\frac{\partial n}{\partial t} + \frac{\partial(n\bar{u}^w)}{\partial\bar{x}} + \frac{\partial(n\bar{v}^w)}{\partial\bar{y}} + \xi\epsilon_x^{-1}\frac{\partial(n\bar{w}^w)}{\partial\bar{z}} &= 0\,.
\end{aligned}
\tag{A.5}
$$

Normalized force balances

$$
\begin{aligned}
\epsilon_x \cot\alpha \left(n\frac{\partial(\bar{p}_w)}{\partial \bar{x}} - \kappa\frac{\partial(\bar{p})}{\partial \bar{x}} \right) + \kappa\frac{\partial t_{xz}}{\partial \bar{z}} + (1-n)\kappa_s + n^2(\bar{u}^w - \beta\bar{u}^s) &= 0\,, \\
\epsilon_y \cot\alpha \left(n\frac{\partial(\bar{p}_w)}{\partial \bar{y}} - \kappa\frac{\partial(\bar{p})}{\partial \bar{y}} \right) + \kappa\frac{\partial t_{yz}}{\partial \bar{z}} + \eta n^2(\bar{v}^w - \beta\bar{v}^s) &= 0\,, \\
\cot\alpha \left(n\frac{\partial(\bar{p}_w)}{\partial \bar{z}} - \kappa\frac{\partial(\bar{p})}{\partial \bar{z}} \right) - (1-n)\kappa_s \cot\alpha + n^2(\zeta\bar{w}^w - \zeta\beta\bar{w}^s) &= 0\,, \\
1 - \epsilon_x \cot\alpha \frac{\partial(\bar{p}_w)}{\partial \bar{x}} - n(\bar{u}^w - \beta\bar{u}^s) &= 0\,, \\
\epsilon_y \cot\alpha \frac{\partial(\bar{p}_w)}{\partial \bar{y}} + \eta n(\bar{v}^w - \beta\bar{v}^s) &= 0\,, \\
\frac{\partial(\bar{p}_w)}{\partial \bar{z}} + 1 + n\tan\alpha(\zeta\bar{w}^w - \zeta\beta\bar{w}^s) &= 0\,.
\end{aligned}
\tag{A.6}
$$

Constitutive relations

$$
\begin{aligned}
\bar{t}_{xx} = \bar{t}_{yy} = \bar{t}_{zz} = \bar{t}_{xy} &= 0\,, \\
\frac{\partial \bar{u}^s}{\partial \bar{z}} &= 2\boldsymbol{b}(\mathit{\Pi}_{\bar{\boldsymbol{t}}}, n)\bar{t}_{xz}\,, \\
\frac{\partial \bar{v}^s}{\partial \bar{z}} &= \frac{2}{\eta}\boldsymbol{b}(\mathit{\Pi}_{\bar{\boldsymbol{t}}}, n)\bar{t}_{yz}\,, \\
\frac{\partial \bar{u}^s}{\partial \bar{x}} + \frac{\partial \bar{v}^s}{\partial \bar{y}} + \xi\epsilon_x^{-1}\frac{\partial \bar{w}^s}{\partial \bar{z}} &= -3C(1-n)\frac{\kappa\dot{\bar{p}} - \dot{\bar{p}}_w}{\kappa\bar{p} - \bar{p}_w}\,,
\end{aligned}
\tag{A.7}
$$

with

$$
\begin{aligned}
\boldsymbol{b}(\mathit{\Pi}_{\bar{\boldsymbol{t}}}, n) &= \frac{b((\tilde{\rho}g[H]\sin\alpha)^2 \mathit{\Pi}_{\bar{\boldsymbol{t}}}, n)}{b((\tilde{\rho}g[H]\sin\alpha)^2, n)}\,, \\
[U_s] &= [H]^2\tilde{\rho}g\sin\alpha\, b((\tilde{\rho}g[H]\sin\alpha)^2, n)\,, \\
C &= \frac{0.435}{3}C_c\,, \\
(\;)^{\cdot} &= \beta^{-1}\frac{\partial(\;)}{\partial \bar{t}} + \frac{\partial(\;)}{\partial \bar{x}}\bar{u}^s + \frac{\partial(\;)}{\partial \bar{y}}\bar{v}^s + \xi\epsilon_x^{-1}\frac{\partial(\;)}{\partial \bar{z}}\bar{w}^s\,.
\end{aligned}
\tag{A.8}
$$

Boundary conditions at the free surface

$$\begin{aligned}
\beta^{-1}\frac{\partial \bar{z}_s}{\partial \bar{t}} + \frac{\partial \bar{z}_s}{\partial \bar{x}}\bar{u}^s + \frac{\partial \bar{z}_s}{\partial \bar{y}}\bar{v}^s - \xi\epsilon_x^{-1}\bar{w}^s &= 0\,, \\
\epsilon_x \cot\alpha(-\kappa\bar{p} + \bar{p}_w)\frac{\partial \bar{z}_s}{\partial \bar{x}} - \kappa\bar{t}_{xz} &= 0\,, \\
-\,\epsilon_y \cot\alpha(\kappa\bar{p} - \bar{p}_w)\frac{\partial \bar{z}_s}{\partial \bar{y}} - \kappa\bar{t}_{yz} &= 0\,, \\
\kappa\bar{p} - \bar{p}_w &= 0\,,
\end{aligned} \tag{A.9}$$

where $\bar{z} = \bar{z}_s(\bar{x}, \bar{y}, \bar{t})$

Transition conditions at the phreatic surface

$$\begin{aligned}
\frac{\partial \bar{z}_w}{\partial \bar{t}} + \frac{\partial \bar{z}_w}{\partial \bar{x}}\bar{u}^w + \frac{\partial \bar{z}_w}{\partial \bar{y}}\bar{v}^w - \xi\epsilon_x^{-1}(\bar{w}^w + \bar{Q}) &= 0\,, \\
\bar{p}_w &= 0\,, \\
[\![n]\!] &= 0\,, \\
[\![\bar{t}_{xz}]\!] &= 0\,, \\
[\![\bar{t}_{yz}]\!] &= 0\,, \\
[\![\kappa\bar{p} - \bar{p}_w]\!] &= 0\,,
\end{aligned} \tag{A.10}$$

where $\bar{z} = \bar{z}_s(\bar{x}, \bar{y}, \bar{t})$. As before, $[\![\ ]\!]$ denotes the difference of the bracketed quantity.

Boundary conditions at the base

$$\begin{aligned}
0 &= \frac{\partial \bar{z}_B}{\partial \bar{x}}\bar{u}^s + \frac{\partial \bar{z}_B}{\partial \bar{y}}\bar{v}^s - \bar{w}^s\,, \\
\bar{u}^s &= C_B F_B(\bar{\tau}_B\,, \bar{\sigma}'_B)\bar{\tau}_{xz}\,, \\
\bar{v}^s &= \tfrac{C_B}{\eta} F_B(\bar{\tau}_B\,, \bar{\sigma}'_B)\bar{\tau}_{yz}\,, \\
\frac{\partial \bar{p}_w}{\partial \bar{z}} + 1 &= \frac{-q}{\cos\alpha} \quad \text{(if flux } q \text{ is prescribed)}\,, \\
\bar{p}_w &= \frac{h - (z\cos\alpha - x\sin\alpha + a_2)}{[H]\cos\alpha} \quad \text{(if potential } h \text{ is prescribed)}\,,
\end{aligned} \tag{A.11}$$

where $\bar{z} = \bar{z}_B(\bar{x}, \bar{y})$, and in which

$$\begin{aligned} C_B &= \tilde{\rho} g[H] \sin\alpha F_B(\tilde{\rho} g[H] \sin\alpha \, , \, \tilde{\rho} g[H] \sin\alpha)/[U_s] \, , \\ \boldsymbol{F}_B(\bar{\tau}'_B \, , \, \bar{\sigma}'_B) &= \frac{F_B((\tilde{\rho} g[H] \sin\alpha)\bar{\tau}_B \, , \, (\tilde{\rho} g[H] \sin\alpha)\bar{\sigma}_B)}{F_B((\tilde{\rho} g[H] \sin\alpha)1 \, , \, (\tilde{\rho} g[H] \sin\alpha)1)} \, , \end{aligned} \tag{A.12}$$

with

$$\bar{\sigma}'_B = \left(\bar{\sigma}'^2_{B_x} + \bar{\sigma}'^2_{B_y} + \bar{\sigma}'^2_{B_z}\right)^{1/2} \quad , \quad \bar{\tau}_B = \left(\bar{\tau}^2_{B_x} + \bar{\tau}^2_{B_y} + \bar{\tau}^2_{B_z}\right)^{1/2} . \tag{A.13}$$

Appendix II. Notation

The following symbols are used in this paper:

$A, \bar{A}$	rate factors
$a, b, c, \tilde{a}$	coefficients in constitutive relation of REINER-RIVLIN fluid
B	constant in the PRANDTL-EYRING model; material constant (sliding function)
b	creep response function
C_c	compression index
c'	cohesion
$\boldsymbol{D}$	stretching tensor
$\boldsymbol{D}^e, \boldsymbol{D}^{vp}$	elastic, viscoplastic stretching
$I_{\boldsymbol{D}}, II_{\boldsymbol{D}}, III_{\boldsymbol{D}}$	invariants of stretching tensor
$\boldsymbol{E}\boldsymbol{D} - \frac{1}{3}(\mathrm{tr}\boldsymbol{D})1$	stretching tensor, distortion tensor
$II_{\boldsymbol{E}}$	second invariant of $\boldsymbol{E}$
e, e_c	void ratios
f	sliding function
$F0$	equation of an arbitrary singular surface
$F(\tau_B, \sigma'_B)$	sliding function
G	shear modulus
g	gravity
g_i	gravity vector
$[H]$	characteristic depth
h	piezometric head
I_p	plasticity index
k	coefficient of permeability $[ms^{-1}]$
k'	DRUCKER-PRAGER coefficient
$[L_x]$	characteristic length
$[L_y]$	characteristic width
m, n	exponents, material constants

n	porosity
$\boldsymbol{n}$	normal vector
p, p'	pressures
p^e, p^c, p^v	elastic, consolidation, viscous pressures
p_w	pore-water pressure
$Q_\perp, Q_\perp^B, Q_\perp^w$	diffusive flows through singular surface
Q	visco-plastic potential
T	absolute temperature
$[T]$	characteristic time
t, t_0, t_1	time, reference times
$\boldsymbol{t}$	Cauchy stress
$\boldsymbol{t}_s, \boldsymbol{t}_w$	stress tensors
$II_{\boldsymbol{t}}$	second invariant of $\boldsymbol{t}$
$\boldsymbol{u}$	velocity of a singular surface
$[U_s], [V_s], [W_s]$	characteristic soil velocities in direction x, y, z
$[U_w], [V_w], [W_w]$	charcteristic water velocities in direction x, y, z
$\boldsymbol{u}_B^s$	sliding velocity of the soil at the base
$\boldsymbol{v}^F$	filtration velocity (DARCY)
$\boldsymbol{v}^s$	velocity of the soil grains $\boldsymbol{v}^s(u^s, v^s, w^s)(v_x^s, v_y^s, v_z^s)$
$\boldsymbol{v}^w$	velocity of the water $\boldsymbol{v}^s(u^w, v^w, w^w)(v_x^w, v_y^w, v_z^w)$
$\boldsymbol{w}$	velocity vector of unspecified particle
$z_B(x, y)$	position of the base
$z_s(x, y, t)$	position of the free surface
$z_w(x, y, t)$	position of the phreatic surface
α	slope angle, inclination angle
α'	DRUCKER-PRAGER parameter
β	velocity ratio $[W_w]/[U_w]$
$\boldsymbol{\beta}$	seepage force
γ	fluidity parameter, shear angle
$\dot{\gamma}$	shear rate
ϵ_x	aspect ratio $[H]/[L_x]$
ϵ_y	aspect ratio $[H]/[L_y]$
$\dot{\epsilon}_1, \dot{\epsilon}_2, \dot{\epsilon}_3$	strain rates
ζ	velocity ratio $[W_w]/[U_w]$
η	aspect ratio $[L_y]/[L_x]$
κ	density ratio mixture/water
κ_s	density ratio soil/water
μ, μ_0, μ_1	dynamic viscosities $[kN/m^2]$
ξ	velocity ratio $[W_s]/[U_s]$
ρ	total bulk mass density
$\tilde{\rho}$	mean total bulk mass density (for the scalings)
ρ_s	partial density of the soil grains

ρ_w	partial density of the water
$\hat{\rho}_s$	intrinsic density of the soil grains
$\hat{\rho}_w$	intrinsic density of the water
$\boldsymbol{\sigma}, \boldsymbol{\sigma}^s, \boldsymbol{\sigma}^w, \boldsymbol{\sigma}'$	total, "solid", water, effective stress tensors
$\boldsymbol{\sigma}^e, \boldsymbol{\sigma}^v, \boldsymbol{\sigma}_r$	elastic, viscous, rupture (failure) stress tensors
σ'_B	effective normal stress at the base
τ, τ_r, τ_0	shear stress tensors
$\boldsymbol{\tau}_B$	shear traction at the base
ϕ, ϕ_f	internal angles of friction
ϕ_r	residual angle of friction .

Operators

diag$(\cdot)$	diagonal of a matrix
div$(\cdot)$	divergence
tr$(\cdot)$	trace .

References

ANDERSLAND, O. B., and DOUGLAS, A. G. (1970). "Soil deformation rates and activiation energies." *Geotechnique*, **20**(1), 1 – 16

ANTOINE, P., BIAREZ, J., DESVARREUX, P., and MOUGIN, J.-P. (1971). "Les problèmes posés par la stabilité des pentes dans les régions montageneuses." *Géologie alpine*, **47**, 5 – 24

BARDEN, L. (1965). "Consolidation of clay with non-linear viscosity." *Geotechnique*, **15**(9), 345 – 362

BIAREZ, J., and BOUCEK, B. (1973). "Viscoplasticité de l'argile in situ et en laboratoire." *Proceedings, 8th Int. Conf. on Soil Mechanics and Foundation Engineering*, Moscow, Vol. **1.1**, 51 – 56.

BLONDEAU, F., MORIN, P., and POUGET, P. (1983). "Comportement d'un remblai construit jusqu' à la rupture sur un versant naturel, Site expérimental de Sallèdes." *Rapport de recherche LPC*, No. 126.

BJERRUM, L., and LANDVA, A. (1966). "Direct simple-shear test on a Norwegian quick clay." *Geotechnique*, **16**(1), 1 – 20

BOURDEAU, P. L., RECORDON, E., VULLIET, L., POUGET, P., PILOT, G, and DELMAS, PH. (1988). "La stabilité des versants: Évolutions récentes du calcul." *Proceedings, 5th Int. Symposium on Landslides*, Lausanne, Switzerland.

CACMAK, A. S., RICHARDS, R., and COLE, D. H. (1970). "Shift functions for hydrorheologically simple clays." *Proc., 5th Int. Congress on Rheology*, Tokyo, Japan, 491 – 499.

CALABRESI, G. (1970). "Creep di argille indisturbate in prove di taglio di Lunga durate" (with English abstract). *Estratto dalla rivista Italiana di Geotechnica*, No. **1** (in Italian).

CARTIER, G., and POUGET, P. (1987). "Corrélation entre la pluviométrie et les déplacements de pentes instables." *Proceedings, 9th European Conference on Soil Mechanics and Foundation Engineering*, Dublin

CHOWDHURY, R. N. (1978). "Slope analysis." *In Developments in geotechnical engineering, Vol.* **22**, Elsevier, Amsterdam, The Netherlands.

CONSIGLIO NAZIONALE DELLE RICHERCHE (1982). Progetto finalizzato "Convervazione del suolo." *Atti del convegno conclusivo*, Roma, Italy

CHRISTENSEN, R. W., and WU, T. H. (1964). "Analysis of clay deformation as a rate process." *J. Soil Mech. Found. Div., ASCE*, **90**(SM6), 125 – 157.

COURMEAU, I. C. (1976). "*Visco-plasticity and plasticity in the finite element method.* Thesis in the University of Wales, Swansea.

DRUCKER, D. C., and PRAGER, W. (1952). "Soil mechanics and plastic analysis or limit design." *Q. J. Appl. Math.*, **10**(2), 157 - 165.

DUTI (1983). *Étude pluridisciplinaire du glissement du Day. Secteur des Gaudennes, secteur de Praz-Mattey.* École polytechnique fédérale de Lausanne, Switzerland.

DUTI (1985). *"Projet d'École Détection et Utilisation des Terrains Instables."* Rapport d'activité a fin 1983, EPFL, Lausanne, Switzerland.

DYSLI, M., and RECORDON, E. (1981). "Fluage des formations argileuses alpines." Proc., 10th *Int. Conf. on Soil Mechanics and Foundation Engineering*, Stockholm, Sweden, Vol. **3**, 395 - 400.

EMERY, J. J. (1978). "Simulation of slope creep." *Dev. in Geotech. Eng. Vol.* **14A**, *Rockslides and Avalanches*, 1, B. Voight editor, Elsevier, Amsterdam.

ENGEL, T. (1986). *"Nouvelles méthodes de messure et d'analyse pour l'étude des mouvements du sol en terrains instables."* Thèse No. 601, Ècole polytechnique fédérale de Lausanne, Switzerland.

FEDA, J., KAMENOV, B., and KLABLENA, P. (1973). "Investigation of creep and structure of clayed materials." *Proc., 8th ICSMFE, Moscow, USSR, Vol.* **1**, 119 - 128.

FEDDER, D. (1972). "Zustandsgleichung eines Lehmes, ermittelt mit Hilfe einer neuentwickelten Versuchsapparatur." *Proc., 6th Int. Congress on Rheology*, Lyon, France, 1372 - 1379.

FOWLER, A. C. (1981). "A theoretical treatment of the sliding of glaciers in the absence of cravitation." *Philosophical Transactions of the Royal Society of London*, **298**, 637 - 685.

GAROFALO, F. (1963). "An empirical relation defining the stress dependence of minimum creep rate in metals." *Trans., Metal Soc., ASME*, **227**, 351 - 355.

GEUZE, E. C. W. A., and TAN T. JONG-KIE (1950). "The shearing properties of soils." *Geotechnique*, **2**, 141 - 161.

GEUZE, E. C. W. A. 1953). "Laboratory investigations, including compaction tests, Improvement of soil properties." *Proc., 3rd Int. Conf. on Soil Mechanics and Foundation Engineering, Zürich, Switzerland, Vol.* **3**, 119 - 122.

GLASSTONE, S., LAIDLER, K. J., and EYRING, H. (1941). "*The theory of rate processes.*" McGraw Hill, New York, N. Y.

GUDEHUS, G., KOLYMBAS, D., and LEINENKUGEL, H.-J. (1976). "Zeitverhalten von Böschungen und Einschnitten in weichem und steifem Ton." *Proc., 6th European Conf. on Soil Mechanics and Foundation Engineering, Wien, Austria, Vol.* **1**, 51 – 58.

HAEFELI, R. (1953). "Creep problems in soils, snow and ice." *Procedings, 3rd Int. Conf. on Soil Mechanics and Foundation Engineering, Zürich, Switzerland, Vol.* **3**, 238 – 251.

HAEFELI, R. (1965). "Creep and progressive failure in snow, soil, rock, and ice." *Proceedings, 6th Int. Conf. on Soil Mechanics and Foundation Engineering, Montreal, Canada, Vol.* **3**, 134 – 148.

HIRST, J. T. (1968). "*The influence of compositional factors on the stress-strain-time behavior of soils.*" Thesis presented to the University of California at Berkely, Calif., in partial fulfillment of the requirements for a degree.

HUDER, J. (1976). "Creep in Bundner Schist", *Laurits Bjerrum Memorial Volume, Norwegian Geotechnical Institute, Oslo, Norway.*

HUTTER, K. (1983). "*Theoretical glaciology*". Terra Scientific Publishing Company, Tokyo, Japan

HUTTER, K., and SAVAGE, S. B. (1988). "Avalanche dynamics: the motion of a finite mass of gravel down a mountain side." *Proceedings, 5th Int. Symposium on Landslides, Lausanne, Switzerland.*

HUTTER, K., and VULLIET, L. (1985). "Gravity-driven slow creeping flow of a thermoviscous body at elevated temperatures." *J. Therm. Stresses*, **8**, 99 – 138.

KAVAZANJIAN, E. (1978). "*A generalized approach to the prediction of the stress-strain-time behavior of soft clay.*" Thesis presented to the University of California, at Berkeley, Calif., in partial fulfillment of the requirements for a degree.

KAVAZANJIAN, E., and MITCHELL, J. K. (1984). "Time dependence of lateral earth pressure." *J. Geotech. Engrg., ASCE*, **110**(4), 530 – 533.

KENNEY, T. C., and LAU, K. C. (1984). "Temporal changes of groundwater pressure in a natural slope of nonfissured clay." *Canadian Geotechnical Journal*, **21**, 138 – 146

KOERNER, H. J. (1969). "Kinematische Betrachtung zum Rankineschen Spannungszustand in der geneigten, kriechenden Schicht." *Felsmechanik und Ingenieurgeologie, Suppl. Vol.*, **33** – 54.

KOMAMURA, F., and HUANG, R. J. (1974). "New rheological model for soil behaviour." *J. Geotech. Engrg., ASCE*, **100** (GT7), 807 – 824.

KRIZEK, R. J. (1970). "Constitutive behavior of clay soils." *Proc., 5th Int. Congress on Rheology, Tokyo, Japan*, 469 – 489.

LAMBE, T. C., and WHITEMAN, R. V. (1969) "*Soil Mechanics*", Wiley and Sons, N. Y.

LLIBOUTRY, L. A. (1975). "Loirs de glissement dun glacier sans cavitation". *Annales de géophysique*, **31**, 207-226.

LLIBOUTRY, L. A. (1979) "Local friction laws for glaciers: A critical review and new openings". *J. Glaciology*, **23**, 67-95.

LMS (1980) Glissement de Villarbeney. 2ème rapport intermédiaire, Lab. de Mec. des Sobs, Étude GX59, Août 1980, EPEL, Lausanne.

MESCHYAN, S. R. and BADALYAN, R. G. (1976). "Regularity of creep of clays and deformation of slopes." *Proceedings 6th European Conference on Soil Mechanics and Foundation Engineering, Vienna, Austria*, **1**, 71 – 74.

MESRI, G., et al. (1981). "Shear stress–strain–time behavior of clays." *Geotechnique*, **31**(4), 537 – 552.

MITCHELL, J. K. (1964). "Shearing resistance of soils as a rate process." *J. Soil Mech. and Found. Div., ASCE*, **90**(SM1), 29 – 61.

MORIN, P. (1979). "*Étude du comportement avant rupture d'un remblai expérimental sur versant à Sallèdes.*" Thésis 'Ecole nationale des ponts et chausées, Paris, France.

MÜLLER, I. (1973). "*Thermodynamik, Die Grundlagen der Materialtheorie.*" Bertelsmann Universitätsverlag, Düsseldorf, West Germany.

MÜLLER,I. (1985) "*Thermodynamics*, Pitman Advanced Publishing Program, Boston · London · Melbourne.

MURAYAMA, S. and SHIBATA, T., (1961). "Rheological characteristics of clays". Proc. 5th ICSMFE, Paris.

MURAYAMA, S. (1983). "Formulation of stress-strain-time behaviour of soils under deviatoric stress condition.' *Soils Found.*, **23**(2), 43 – 57.

MURAYAMA, S., SEKIGUCHI, H., and UEDA, T. (1974). "A study of the stress–strain-time behavior of saturated clays based on a theory of nonlinear viscoelasticity.' *Soils Found.*, **14**(2), 19 – 33.

MURAYAMA, S., SHIBATA, T. (1961). "Rheological characteristics of clays." *Proc., 5th Int. Conf. on Soil Mechanics and Foundation Engineering, Paris, France.*

NELSON, J. D., and THOMPSON, E. G. (1974). "Creep failure of slopes in clay shale." *Extraordinary Symposium Soil Engineering and Engineering Geology, Idaho, USA*, 177 – 195

NELSON, J. D., and THOMPSON, E. G. (1977). "A theory of creep failure in overconsolidated clay." *Journal of the Geotechnical Engineering Division, ASCE*, **103**, 1281 - 1294.

NOBLE, H. L. (1973). "Residual strength and landslides in clay and shale." *Journal of the Soil Mechanics and Foundations Division, ASCE* **99**(SM9), 705 - 719.

NORTON, F. H. (1929). "*The creep of steel at high temperature.*" McGraw Hill, New York, N. Y.

OKA, F. (1981). "Prediction of time-dependent behavior of clay." *Proc. 10th Int. Conf. on Soil Mechanics and Foundation Engineering, Stockholm, Sweden*, **1**, 215 - 218.

PERZYNA, P. (1966). "Fundamental problems in plasticity". *Adv. in Appl. Mech.* **9**, 243-377.

POUGET, P., and CARTIER, G. (1988). "Étude du comportement d'un remblai construit sur un versant instable. Le remblai B de Sallèdes (Puy de Dôme)." *Rapport de recherche LPC*, No. **151**.

POUGET, P., CARTIER, G. and PILOT, G. (1985). "Comportement d'un remblai constuit sur un sit instable." *Proceedings, 11th Int. Conference on Soil Mechanics and Foundation Engineering, San Francisco*, **4**, 2345 - 2348.

PRANDTL, L. (1928). "Ein Gedankenmodell zur kinematischen Theorie der festen Körper." *Ztschr. Angew. Math. u. Mech.*, **8**, Heft 2, 85 - 105.

PUSCH, R., and FELTHAM, P. (1980). "A stochastic model of the creep of soils." *Geotechnique*, **30**(4), 497 - 506.

ROSCOE, K.H., and BURLAND, J.B., (1968). "On the generalized stress-strain behaviour of "wet" clay, in: *Engineering Plasticity* (ed J. Heyman and P.A. Leckie) Cambridge University Press, pp 535-609.

SAVAGE, S. B. and HUTTER, K. (1989). "The motion of a finite mass of granular material down a rough incline." *Journal of Fluid Mechanics*, **199**, 177 - 215

SINGH, A., and MITCHELL, J. K. (1968). "General stress-strain-time function for soils." *J. Soil Mech. Found. Div., ASCE*, **94**(SM1), 21 - 46.

SKEMPTON, A. W., (1964). "Long-term stability of clay slopes", *Geotechnique*, **14** (2), 77-102.

SKEMPTON, A. W., (1985). "Residual strength of clays in landslides, folded strata and the laboratory", *Geotechnique*, **35** (1), 3-18.

STROGANOV, A. S. (1961). "Visco–plastic flow of soils." *Proc., 5th Int. Conf. on Soil Mechanics and Foundation Engineering, Paris, France*, **1**, 721 - 726.

TAN, TJ.–K. (1981). "Time dependent lateral pressures and Poisson's ratio measurement." *Proc., 10th Int. Conf. on Soil Mechanics and Foundation Engineering, Stockholm, Sweden*, **1**, 797 – 800.

TAVENAS, F., and LEROUEIL, S. (1981). "Creep and failure of slopes in clays." *Canadian Geotechnical Journal*, **18**, 106 – 120.

TER-MARTIROSYAN, Z. G. (1971). "Creep of an inclined bed under variable pore pressure." Translated from *Osnovaniya, Fundamentry i Mekhanika Gruntov*, No. **4**, 8 – 10.

TER-STEPHANIAN, G. (1963). "On the long-term stability of slopes." *Publ. No. 52, Norwegian Geotech. Inst., Oslo, Norway*, 1 – 13.

TER-STEPHANIAN, G. (1965). "In situ determination of the rheological characteristics of soils on slopes." *Proceedings, 6th Int. Conference on Soil Mechanics and Foundation Engineering, Montreal*, **2**, 575 – 577.

TER-STEPHANIAN, G. (1973). "New rheological model of creep of a clay at shear." Reported at the *Symposium on the Theory of Landslide Processes, May, Dilijan.*

TER-STEPHANIAN, G. (1980). "Creep on natural slopes and cuttings. State-of-the-art Report." *Proceedings, Int. Symposium on Landslides, New Dehli*, **2**, 95 – 108.

TERZAGHI, C., (1931). "The static rigidity of plastic clays". *J. of Rheology* **2** (3), 253-262.

TRUESDELL, C. (1977). "*Rational continuum mechanics.*" Vol. 1, Academic Press, New York, N. Y.

TRUESDELL, C. (1978). "*Rational continuum mechanics.*" Vol. 2, Academic Press, New York, N. Y.

TRUNK, F. J., DENT, J. D., and LANG, T. E. (1986). " Computer modeling of large rock slides." *J. Geotechn. Engrg., ASCE*, **111**(3), 348 – 360.

VULLIET, L. (1986). "*Modélisation des pentes naturelles en mouvement.*" Thèse No. 635, Ecole Polytechnique Fédérale de Lausanne, Lausanne, Switzerland.

VULLIET, L., and HUTTER, K. (1988 a). "Continuum model for natural slopes under slow movement." *Géotechnique*, **38**, 199 – 217.

VULLIET, L., and HUTTER, K. (1988 b). "A multi-layer, multi-sliding surface model for three-dimensional creeping slopes." *Proceedings, 5th Int. Symposium on Landslides, Lausanne, Switzerland.*

VULLIET, L., and HUTTER, K. (1988 c). "Some constitutive laws for creeping soils and for rate-dependent sliding at interfaces." *Proceedings, 6th Int. Conference on Numerical Methods in Geomechanics, Innsbruck, Austria.*

VULLIET, L., and HUTTER, K. (1988 d). "Set of constitutive models for soils under slow movement." *Journal of Geotechnical Engineering, ASCE*, **114**(9), 1022 – 1041.

VULLIET, L., and HUTTER, K. (1988 e). "Viscous-type sliding laws for landslides." *Can. Geotech. J.* **25**, 467 – 477.

VULLIET, L., and RECORDON, E. (1987). "The effect of groundwater seepage on the movement of natural slopes." *Proceedings, 9th European Conference on Soil Mechanics and Foundation Engineering, Dublin.*

WEERTMAN, J. (1964). "The theory of glacier sliding", *J. Glaciology*, **5**, 287-303.

WEERTMAN, J. (1979). "The unsolved general glacier sliding problem", *J. Glaciology*, **23**, 97-115.

YEN, B. C. (1969). "Stability of slopes undergoing creep deformation." *J. Soil Mech. Found. Div. ASCE*, **95** (SM4), 1075 – 1096.

Constitutive Equations for Granular Materials in Geomechanical Context

Wolfgang Ehlers
Institut für Mechanik, Technische Hochschule Darmstadt
Hochschulstraße 1, D-6100 Darmstadt, Germany

This lecture outlines the possibilities of porous media theories in describing granular materials in geomechanical context as, for example, saturated or unsaturated soils or granular rocks, etc. In the present investigations, porous media theories are referred to as classical mixture theories extended by the concept of volume fractions. This approach, assuming statistically distributed and superimposed continua with internal interactions, implies the diverse field functions of both the porous solid matrix and the respective pore contents to be represented by average functions of the macroscale.

In particular, fluid-saturated or empty porous solid materials are taken into consideration including elastic and elasto-plastic constitutive equations for the solid matrix combined with convenient constitutive relations for the viscous pore fluids and the respective interaction mechanisms between the constituents. The different findings, exhibiting the model description of compressible and incompressible phases, respectively, are based on the concept of second-grade materials, thus making use of the most natural framework in constitutive modelling for multiphase media.

1. Introduction

Brittle and granular solid matrices, especially, geomaterials, generally fall into the category of porous media, that is, porous solid materials saturated by an arbitrary number of fluids (liquids and/or gases). The reader who is interested in the foundations of porous media theories (continuum theories of mixtures extended by the concept of volume fractions) is referred to, for example, DRUMHELLER (1978), BOWEN (1980, 1982, 1984), DE BOER & EHLERS (1986b, 1988, 1990a, b), EHLERS (1989a–d, 1991, 1992), DE BOER *et al.* (1991), DE BOER & LADE (1991) or EHLERS (in prep.). Furthermore, a broad review of the different approaches to classical mixture theories (heterogeneously composed continua with internal interactions) can be taken from ATKIN & CRAINE (1976a, b), BOWEN (1976), HUTTER (1977), DRUMHELLER (1978), BEDFORD & DRUMHELLER (1983), DE BOER & EHLERS (1986b) or EHLERS (1989b).

As was shown by the author (EHLERS, 1989a), the concept of porous media theories allows for the description of both saturated and empty matrix materials, thus offering a well-founded theoretical background for a lot of branches in engineering practice, such as geomechanics (soil and rock mechanics as well as glacier and rock ice mechanics), oil production industries, mining industries, sintering technologies, biomechanics, etc.

Usually, porous media models are employed by the assumption of an immiscible mixture of a porous solid material saturated by an arbitrary number of fluids. In assuming a statistical distribution of the single constituents through the control space (superimposed continua), all mechanical and thermodynamical quantities are described via average functions of the macroscale. The individual average functions must then be understood as the result of an averaging procedure relating functions from the microscale towards the macroscale, compare, for example, HASSANIZADEH & GRAY (1979a, b), NIGMATULIN (1979), BACHMAT & BEAR (1986) or DE BOER *et al.* (1991).

Proceeding from porous media theories, the general constitutive framework must be based on the assumption of second-grade materials (MÜLLER, 1968; CROSS, 1973; WANG, 1973), thus making use of the most natural framework in constitutive modelling for multiphase media, additionally avoiding so-called "simple" results as were concluded from several former approaches to our subject, compare, for example, the different publications of Biot (cited in DE BOER & EHLERS, 1988). However, in describing compressible or incompressible porous media models with elastic or viscoelastic phases, the basic constitutive approach to materials of second grade and the thermodynamical restrictions governing the respective constitutive equations are generally well known, whereas the use of hybrid or mixed mode models, respectively (herewith we mean incompressible matrix materials saturated by compressible pore fluids or, vice versa, compressible solid matrices saturated by incompressible pore liquids), is still under study and needs further clarification. Apart from these general remarks on constitutive modelling for porous media, it should be noted that, for a long time, only a very few papers were available taking into account elasto-plastic solid matrices, compare, for example, MORLAND (1972), KOJIĆ & CHEATHAM (1974a, b), PRÉVOST (1981), DE BOER & KOWALSKI (1983) or DE BOER & EHLERS (1986a). Most of these articles, however, are restricted to geometrical linear (small strain) theories or proceed from simplified thermodynamical approaches, in some extend combined with other rigorous simplifications. However, it was only very recenty that a general approach to porous media elastoplasticity based on second-grade materials could be discussed and published by the author (EHLERS, 1989a, 1991, 1992). This approach, consequently making use of geometrically finite (large strain) theories will be used as the basic framework for our present investigations in porous media elastoplasticity.

This lecture outlines the possibilities of porous media theories in describing fluid-saturated and empty granular materials in geomechanical context, thus contributing to a better understanding of soil mechanical problems of civil engineering type like general elasto-plastic deformations of soils or coupled solid-fluid motions occuring, for instance, in computing the setting of buildings on liquid-saturated soil, the problem of groundwater and/or pollutant flow in deformable matrix materials or the flow of oil and/or gas in natural reservoirs. Furthermore, the theory can easily be used for the description of glacier or rock ice consolidation and related problems, if required.

In particular, after a short preliminary introduction to the fundamentals of porous media theories and classical mixture theories, respectively, the general problem of constitutive modelling based on certain combinations of "mixture" materials is dicussed. However, since a general arbitrariness of material combinations usually leads to rather lengthy and complex formulæ, we will restrict our consideration to the different combinations of two-phase models, taking into consideration, within the framework of porous media theories, one porous solid matrix material (compressible or incompressibe, elastic or elastoplastic) saturated by one compressible or incompressible viscous fluid. These models naturally allow for the description of a lot of geomechanical materials such as liquid- or gas-saturated granular soils or rocks as well as temperate ice defined as an immiscible mixture of an ice matrix filled with melt water.

Subsequent to these general remarks, the following sections concern, by use of the concept of Lagrangian multipliers, four typical two-phase media combinations, that is, one compressible model consisting of two compressible constituents, two hybrid models consisting of one compressible and one incompressible constituent, and, finally, one incompressible model with, of course, incompressible constituents. Furthermore, from the second-grade character of general heterogeneous media, finite elastoplasticity for the solid constituent is introduced via the multiplicative decomposition of both the first and the second solid deformation gradients. This concept compatible with an additive decomposition of strain and strain rate measures (Lie derivatives) was recently pointed out by the author (EHLERS, 1989a) and subsequently succesfully applied to porous media elastoplasticity (EHLERS, 1989b, 1991, 1992). For simplicity and convenience for the reader, the different findings in porous media theories are outlined excluding mass exchanges between the constituents and, in addition, assuming the model to be governed by a single temperature function.

Within the framework of a consistent thermodynamical approach to porous media theories, the concept of second-grade elastoplasticity requires the plastic solid deformation gradients to be incorporated among the list of constitutive variables. Following this, the plastic part of the solid deformation gradient must be treated in the sense of an internal state variables, that is, (as is usual in plasticity) a constitutive equation (flow rule) must be formulated for the plastic rate of deformation tensor. Apart from the full set of thermodynamical restrictions and basic constitutive relations for the model under discussion, the above procedure reveals a restriction for the plastic dissipation including a term which can be interpreted as the back-stress tensor of kinematically hardening second-grade solid materials (EHLERS, 1989a). Thus, a finite and load history dependent constitutive law can be used to govern the multiaxial Bauschinger effect. In addition, as a consequence of the multiplicative decomposition, the plastic response (yield function and flow rule) must be given with respect to the plastic intermediate configuration of the solid constituent. However, by use of adequate push-forward and pull-back transformations, the plastic response can be transferred from the intermediate configuration to the actual configuration or to the reference configuration of the matrix material, if required.

Concerning applications of the general theory to practical geomechanical problems as were described above, it is convenient to simplify the general model by use of a "principle of constituent separation" (EHLERS, 1989a, b). Following this, finite elasticity of porous media is discussed by the introduction of a new elasticity law characterized by an extension of the neo-Hookean model towards the finite compressible range. In particular, this elasticity law includes, within the scope of compressible solid constituents, two compressibilities which will be called "phase (or material) compressibility" due to the compressibility of the grains of the granular matrix material and "configuration (or structural) compressibility" due to volume porosity changes. Obviously, in dealing with incompressible matrix materials, the "material compressibility" naturally degenerates to a Lagrangian multiplier; additionally, compare EHLERS (1989b, 1991). Concerning the plastic range of the porous solid matrix, one proceeds from the concept of "constituent plastic phase (or material) incompressibility", that is, we assume in formal analogy to plasticity theories for non-porous single solid continua that the matrix material itself behaves incompressible under purely plastic loading conditions. Following this, the description of porous media plasticity does not depend on the fact whether the solid matrix material is compressible or incompressible. Furthermore, the plastic response is based on a non-associated flow rule for the plastic rate of deformation tensor and a constitutive equation for the back-stress tensor together with a convenient yield criterion (EHLERS, in prep.) which is defined as an extension of de Boer's failure function (DE BOER, 1988) towards the finite kinematically hardening range of the porous solid material.

The above constitutive model (including constitutive equations for the viscous pore fluid and the interaction mechanisms between the constituents) can be utilized to describe ductile matrices as well as brittle and granular materials. Both the elastic and the plastic responses can easily be applied to empty skeletons as well.

Throughout this paper, direct tensor notation is used; some additional information concerning certain types of tensor operations and derivatives is offered in the Appendix.

2. Preliminaries

Proceeding from general porous media theories, the present section offers a brief review of kinematics, balance laws and the entropy principle for mixtures together with the concept of volume fractions. In addition, the basic constitutive postulates for the second-grade medium under discussion are outlined. Concerning the different notions and several further information, the reader is referred to DE BOER & EHLERS (1986b) and EHLERS (1989a, b, 1991). In what follows, all introduced functions are assumed to be sufficiently smooth in space-time.

Considering a fluid-saturated porous medium as an immiscible mixture of constituents φ^i with particles X^i ($i = S$: solid phase; $i = F$: fluid phases), a macroscopic formulation implies a model of superimposed continua (compare, Figure 1), where at any time t each spatial point of the current configuration is simultaneously occupied by particles X^i of constituents φ^i. These particles proceed from different reference positions $\mathbf{X}_i$ at time t_o. Thus, each constituent is assigned its own motion function, viz.:

$$\mathbf{x} = \chi_i(\mathbf{X}_i, t). \tag{2.1}$$

The volume fractions

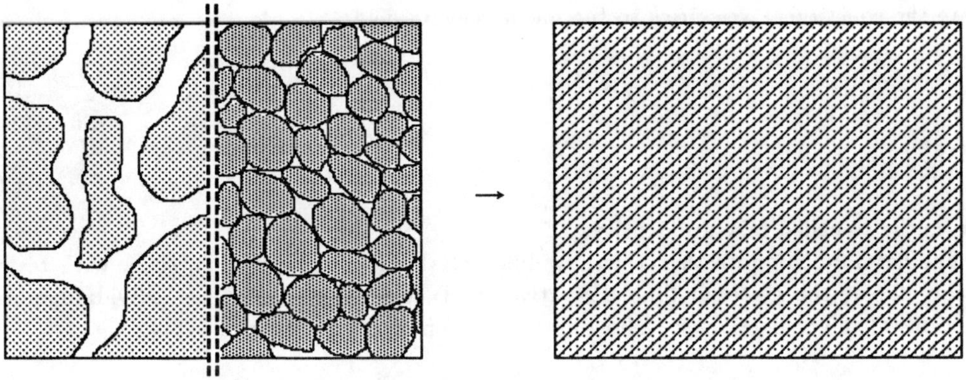

Figure 1: Geometrical (left) and "smeared" (right) representation of a porous medium.

$$n^i = n^i(\mathbf{x}, t) \tag{2.2}$$

are defined as the local ratios of the constituent volumes v^i with respect to the bulk volume v. Associated with each φ^i is an effective (or material) density ρ^{iR} which is defined as the mass of φ^i per unit of v^i and a partial (or bulk) density ρ^i defined as the mass of φ^i per unit of v. The density functions are related by

$$\rho^i = n^i \rho^{iR}. \tag{2.3}$$

Constituent incompressibility, as it is partly assumed for the present investigations, implies the effective densities to be constant during deformation, that is:

$$\rho^{iR} = \text{const.} \tag{2.4}$$

From eqns (2.3) and (2.4), it is immediately concluded that constituent incompressibility does not cause macroscopic incompressibility of φ^i since the bulk density functions can still change through changes in volume fractions.

From eqn (2.1), each constituent is assigned its own velocity and acceleration fields,

$$\overset{\prime}{\mathbf{x}}_i = \frac{\partial \chi_i(\mathbf{X}_i, t)}{\partial t}, \qquad \overset{\prime\prime}{\mathbf{x}}_i = \frac{\partial^2 \chi_i(\mathbf{X}_i, t)}{\partial t^2}. \tag{2.5}$$

By use of the inverse motion functions, an equivalent representation of the above equations is

$$\overset{\prime}{\mathbf{x}}_i = \overset{\prime}{\mathbf{x}}_i\,(\mathbf{x}, t), \qquad \overset{\prime\prime}{\mathbf{x}}_i = \overset{\prime\prime}{\mathbf{x}}_i\,(\mathbf{x}, t). \tag{2.6}$$

The diffusion velocities

$$\mathbf{u}_i = \overset{\prime}{\mathbf{x}}_i - \dot{\mathbf{x}} \tag{2.7}$$

relate the constituent velocities to the mean velocity field,

$$\dot{\mathbf{x}} = \frac{1}{\rho}(\rho^S \overset{\prime}{\mathbf{x}}_S + \rho^F \overset{\prime}{\mathbf{x}}_F), \tag{2.8}$$

where

$$\rho = \rho^S + \rho^F \tag{2.9}$$

defines the so-called mixture density.

Furthermore, suppose that Γ is any differentiable function of $(\mathbf{x}, t)$, then, with respect to eqn $(2.6)_1$, the material time derivatives of Γ corresponding to the individual motion functions of the respective φ^i (following the motion of φ^i) are defined via

$$(\Gamma)'_i = \frac{\partial \Gamma}{\partial t} + \operatorname{grad} \Gamma \cdot \overset{\prime}{\mathbf{x}}_i \,. \tag{2.10}$$

Additionally,

$$\dot{\Gamma} = \frac{\partial \Gamma}{\partial t} + \operatorname{grad} \Gamma \cdot \dot{\mathbf{x}} \,, \tag{2.11}$$

where, obviously,

$$(\Gamma)'_i - \dot{\Gamma} = \operatorname{grad} \Gamma \cdot \mathbf{u}_i \,. \tag{2.12}$$

From eqn (2.1), the deformation gradient of φ^i and its inverse are

$$\mathbf{F}_i = \operatorname{Grad}_i \mathbf{x} \,, \qquad \mathbf{F}_i^{-1} = \operatorname{grad} \mathbf{X}_i \,. \tag{2.13}$$

In the above equations, the symbols "grad" and "Grad_i" mean partial differentiation with respect to the spatial position $\mathbf{x}$ and the reference position $\mathbf{X}_i$ of φ^i, respectively. Finally, the material and the spatial velocity gradients of φ^i read:

$$\begin{aligned} (\mathbf{F}_i)'_i &= \operatorname{Grad}_i \overset{\prime}{\mathbf{x}}_i \,, \\ \mathbf{L}_i &= \operatorname{grad} \overset{\prime}{\mathbf{x}}_i \,. \end{aligned} \tag{2.14}$$

Excluding mass exchanges between the solid and the fluid phases, the concept of volume fractions and the constituent balance equations are given with respect to DE BOER & EHLERS (1986b) and EHLERS (1989b, 1991):

Concept of volume fractions:

$$n^S + n^F = 1 \,. \tag{2.15}$$

Balance of mass:

$$(\rho^i)'_i + \rho^i \operatorname{div} \overset{\prime}{\mathbf{x}}_i = 0 \,. \tag{2.16}$$

Balance of momentum:

$$\operatorname{div} \mathbf{T}^i + \rho^i(\mathbf{b}^i - \overset{\prime\prime}{\mathbf{x}}_i) + \mathbf{p}^i = \mathbf{0}. \tag{2.17}$$

Balance of moment of momentum:

$$\begin{aligned} \mathbf{T}^i &= \mathbf{T}^{Ti} - \mathbf{M}^i , \\ \mathbf{M}^S + \mathbf{M}^F &= \mathbf{0} . \end{aligned} \tag{2.18}$$

Balance of energy:

$$\begin{aligned} \rho^i(\varepsilon_i)'_i = -\mathbf{p}^i \cdot \overset{\prime}{\mathbf{x}}_i + \mathbf{T}^i \cdot \mathbf{L}_i + \rho^i r^i - \operatorname{div}\mathbf{q}^i + e^i , \\ e^S + e^F = 0 . \end{aligned} \tag{2.19}$$

In these relations, $\mathbf{p}^i$, $\mathbf{M}^i$, and e^i are the supply terms of momentum, moment of momentum and energy representing the transfers to φ^i caused by the respective other constituents occupying $\mathbf{x}$ at time t. For the binary model under consideration, the momentum supplies may also be interpreted as the interaction forces (or the drag forces) per unit of bulk volume. The quantities $\mathbf{T}^i$, ε^i, r^i, and $\mathbf{q}^i$ are the partial Cauchy stresses, the internal energy densities, the external heat supplies, and the heat influx vectors of φ^i; "div" is the divergence operator corresponding to "grad", the symbol $(\ldots)^T$ defines the transpose of the respective tensorial object. For several applications, it is sufficient to use the body force density $\mathbf{b}$ instead of the individual functions $\mathbf{b}^i$. Then, it is understood that

$$\mathbf{b} = \mathbf{b}^S = \mathbf{b}^F . \tag{2.20}$$

Finally, assuming the present model to be governed by one single temperature function,

$$\theta = \theta^S = \theta^F , \tag{2.21}$$

the entropy inequality reads, compare EHLERS (1989b):

$$\begin{aligned} -\rho^S(\psi^S)'_S - \rho^F(\psi^F)'_F - \dot{\theta}(\rho^S\eta^S + \rho^F\eta^F) - \mathbf{p}^F \cdot (\overset{\prime}{\mathbf{x}}_F - \overset{\prime}{\mathbf{x}}_S) + \\ + \mathbf{T}^S \cdot \mathbf{L}_S + \mathbf{T}^F \cdot \mathbf{L}_F - \frac{1}{\theta}\mathbf{h} \cdot \operatorname{grad}\theta \geq 0. \end{aligned} \tag{2.22}$$

In this inequality, ψ^i and η^i are the partial free energy and entropy densities which are connected with the internal energy functions via the Legendre transform

$$\psi^i = \varepsilon^i - \theta\,\eta^i , \tag{2.23}$$

whereas

$$\mathbf{h} = \mathbf{q}^S + \mathbf{q}^F + \theta\,(\rho^S\,\eta^S\,\mathbf{u}_S + \rho^F\,\eta^F\,\mathbf{u}_F) \tag{2.24}$$

is one of the possible influx vectors of the multiphase medium under consideration.

Concerning the skew-symmetric moment of momentum supply terms $\mathbf{M}^i$ introduced to mixture theories by KELLY (1964), it was shown by HASSANIZADEH & GRAY (1979b) for mixtures with immiscible constituents that $\mathbf{T}^i$ must be symmetric as far as the individual φ^i are microscopically non-polar. Thus, for a fluid-saturated binary medium with non-polar constituents, it is sufficient to conclude that

$$\mathbf{T}^i = \mathbf{T}^{Ti} \qquad \text{and} \qquad \mathbf{M}^i = \mathbf{0}\,. \tag{2.25}$$

Moreover, concerning the quantities $(\rho^i)'_i$ incorporated into eqn (2.16), it should be noted that

$$(\rho^i)'_i = n^i(\rho^{iR})'_i + (n^i)'_i\,\rho^{iR}, \tag{2.26}$$

where eqn (2.3) was used. Thus, it is easily concluded from eqns (2.4) and (2.16) that, for any incompressible φ^i,

$$(n^i)'_i + n^i \operatorname{div} \overset{\prime}{\mathbf{x}}_i = 0\,, \tag{2.27}$$

that is, the balance of mass equations reduce to balance equations for the volume fractions n^i.

3. Basic constitutive theory

3.1 General framing

From the preceding relations, the general model is defined by the balance equations (2.15)–(2.19) together with the entropy inequality (2.22). In addition, it is easily concluded that, within the framework of general continuum mechanics, the model must be completed by a unique set of constitutive postulates which, from the principle of equipresence, must be assumed in the first place to depend on a common set of independent state variables, e.g. $\mathcal{S}$.

Thus, if one considers the basic model to consist of the porous solid matrix φ^S saturated by one viscous fluid φ^F, the following constitutive postulates hold:

$$\begin{aligned} &(\psi^i, \eta^i, \mathbf{T}^i, \mathbf{q}^i, \mathbf{p}^F, (n^F)'_F) = \mathcal{R}(\mathcal{S}_c)\,, \\ &\mathcal{S}_c = \{\theta,\ \operatorname{grad}\theta,\ n^F,\ \operatorname{grad} n^F,\ \rho^F,\ \operatorname{grad}\rho^F, \\ &\qquad \mathbf{F}_S,\ \operatorname{Grad}_S \mathbf{F}_S,\ \overset{\prime}{\mathbf{x}}_F - \overset{\prime}{\mathbf{x}}_S,\ \mathbf{D}_F\}. \end{aligned} \tag{3.1}$$

In $(3.1)_2$, the second-grade character of the model is reflected by including the gradients of n^F, ρ^F and $\mathbf{F}_S$ into the set $\mathcal{S}_c$. Furthermore, since in the above equations, there is no explicit inclusion of the reference positions $\mathbf{X}_i$ of constituents φ^i, it is easily seen from WANG (1973) that the set $\mathcal{S}_c$ defines a homogeneous binary second-grade medium, where

$$\operatorname{Grad}_i n^i_{oi} = \mathbf{0}, \qquad \operatorname{Grad}_i \rho^{iR}_{oi} = \mathbf{0}. \tag{3.2}$$

Therein, the quantities n^i_{oi} and ρ^{iR}_{oi} denote the volume fractions and effective density functions in the reference positions of φ^i.

Concerning the above relations, it must be noted that the constitutive postulates $(3.1)_1$ define a non-polar compressible binary model. Non-polarity is easily seen from the fact that, on the one hand, the relation $(3.1)_1$ does not include the skew-symmetric moment of momentum supply terms defined by eqn (2.18), whereas, on the other hand, the compressible model requires a constitutive postulate for the fluid volume fraction rate $(n^F)'_F$; a quantity which, in the compressible model, is not balanced by eqn (2.16). Following this, the fluid volume fraction, also occurring in the set $\mathcal{S}_c$, must be understood as an internal state variable of the general deformation process. The reader who is interested in further details concerning the postulates (3.1) is referred to EHLERS (1989b), where, within the framework of porous media theories, an extended discussion on the general problem of constitutive modelling can be found.

Furthermore, note in passing that, from eqn (2.15) together with the definition of the material time derivative (2.10),

$$\begin{aligned} n^S &= 1 - n^F, \\ \operatorname{grad} n^S &= -\operatorname{grad} n^F, \\ (n^S)'_S &= -(n^F)'_F + \operatorname{grad} n^F \cdot (\overset{\prime}{\mathbf{x}}_F - \overset{\prime}{\mathbf{x}}_S), \end{aligned} \tag{3.3}$$

that is, the solid volume fraction and its gradient are uniquely determined by the respective fluid quantities, whereas the solid volume fraction rate is given by the constitutive postulate for the fluid volume fraction rate together with other variables of the thermodynamical process. Thus, no constitutive equation for $(n^S)'_S$ must be given. Additionally, the "dependent quantities," n^S and $\operatorname{grad} n^S$, cannot be understood to contribute to the set $\mathcal{S}_c$. Moreover, the above relations make clear that, instead of using the fluid volume fraction terms to contribute to the modelling of the constitutive assumptions (3.1), an equivalent solution can be obtained by alternatively making use of the respective solid volume fraction terms.

However, in the case of incompressible models or hybrid models, respectively, certain modifications of the basic relations described above are necessary:
Firstly, in considering a binary medium, where a compressible solid matrix is saturated by an incompressible pore liquid (hybrid model of first type), it follows from eqns (2.3), (2.4), (2.27) and $(3.2)_2$ that

$$\begin{aligned} (n^F)'_F &= -n^F (\mathbf{D}_F \cdot \mathbf{I}), \\ \rho^F &= n^F \rho^{FR}, \\ \operatorname{grad} \rho^F &= \rho^{FR} \operatorname{grad} n^F, \end{aligned} \tag{3.4}$$

that is, the quantities ρ^F and $\operatorname{grad} \rho^F$ are not longer independent but are now uniquely determined by n^F and $\operatorname{grad} n^F$. Thus, ρ^F and $\operatorname{grad} \rho^F$ must be dropped from the set $\mathcal{S}_c$. In addition, it is seen from eqn $(3.4)_1$ that, for the hybrid model of first type, it is not necessary to construct a constitutive equation for $(n^F)'_F$. From this remark, it is directly concluded that, in this case, n^F is not an internal but is now an external state variable. Furthermore, since incompressibility of any constituent gives rise to a certain constraint to be incorporated into the entropy inequality of the model, a combination of eqns $(3.3)_3$ and $(3.4)_1$ yields

$$\lambda^{FR} [(n^S)'_S - n^F \operatorname{div} \overset{\prime}{\mathbf{x}}_F - \operatorname{grad} n^F \cdot (\overset{\prime}{\mathbf{x}}_F - \overset{\prime}{\mathbf{x}}_S)] = 0, \tag{3.5}$$

where λ^{FR} is a Lagrangian multiplier.

Following the above remarks, the constitutive postulates for the hybrid model of first type combine to

$$
\begin{aligned}
&(\psi^i\,, \eta^i\,, \mathbf{T}^S\,, \mathbf{T}^F + n^F \lambda^{FR}\,\mathbf{I}\,, \mathbf{q}^i\,, \mathbf{p}^F - \lambda^{FR}\,\mathrm{grad}\,n^F) = \mathcal{R}\,(\mathcal{S}_1)\,,\\
&\qquad \mathcal{S}_1 = \{\theta,\ \mathrm{grad}\,\theta,\ n^F,\ \mathrm{grad}\,n^F,\\
&\qquad\qquad \mathbf{F}_S,\ \mathrm{Grad}_S\,\mathbf{F}_S,\ \overset{\prime}{\mathbf{x}}_F - \overset{\prime}{\mathbf{x}}_S,\ \mathbf{D}_F\,\}.
\end{aligned}
\tag{3.6}
$$

Additionally, the entropy inequality (2.22) reads

$$
\begin{aligned}
-\rho^S(\psi^S)'_S - \rho^F(\psi^F)'_F - \dot{\theta}(\rho^S\eta^S + \rho^F\eta^F) - (\mathbf{p}^F - \lambda^{FR}\,\mathrm{grad}\,n^F)\cdot(\overset{\prime}{\mathbf{x}}_F - \overset{\prime}{\mathbf{x}}_S) + \\
+\,\mathbf{T}^S\cdot\mathbf{L}_S + (\mathbf{T}^F + n^F\lambda^{FR}\,\mathbf{I})\cdot\mathbf{L}_F - \frac{1}{\theta}\mathbf{h}\cdot\mathrm{grad}\,\theta \geq 0,
\end{aligned}
\tag{3.7}
$$

where the constraint (3.5) was used.

Secondly, in considering a binary medium, where an incompressible solid matrix is saturated by a compressible pore fluid (hybrid model of second type), one now obtains from eqns (2.3), (2.4), (2.27) and $(3.2)_2$ that

$$
\begin{aligned}
(n^S)'_S &= n^S\,[\mathrm{div}\,(\overset{\prime}{\mathbf{x}}_F - \overset{\prime}{\mathbf{x}}_S)\ -\ (\mathbf{D}_F\cdot\mathbf{I})],\\
\rho^S &= n^S\,\rho^{SR},\\
\mathrm{grad}\,\rho^S &= \rho^{SR}\,\mathrm{grad}\,n^S.
\end{aligned}
\tag{3.8}
$$

Thus, with the aid of eqns (3.3), the relations

$$
\begin{aligned}
n^F &= 1 - n^S_{oS}(\det\mathbf{F}_S)^{-1},\\
\mathrm{grad}\,n^F &= (1-n^F)\,\mathbf{F}_S^{T-1}(\mathbf{F}_S^{T-1}\mathrm{Grad}_S\,\mathbf{F}_S)^{\underline{1}}\,,\\
(n^F)'_F &= -\,(1-n^F)\,[\mathrm{div}\,(\overset{\prime}{\mathbf{x}}_F - \overset{\prime}{\mathbf{x}}_S)\ -\ (\mathbf{D}_F\cdot\mathbf{I})]\ +\ \mathrm{grad}\,n^F\,\cdot\,(\overset{\prime}{\mathbf{x}}_F - \overset{\prime}{\mathbf{x}}_S)
\end{aligned}
\tag{3.9}
$$

hold, where

$$
n^S = n^S_{oS}(\det\mathbf{F}_S)^{-1}
\tag{3.10}
$$

to be obtained by integration of eqn (2.27) together with

$$
\mathrm{grad}\,n^S = -n^S\,\mathbf{F}_S^{T-1}(\mathbf{F}_S^{T-1}\mathrm{Grad}_S\,\mathbf{F}_S)^{\underline{1}}
\tag{3.11}
$$

have been used. Note in passing that eqn (3.11) directly follows from eqn (3.10) by differentiation, wherein the symbol $(\ \ldots\ )^{\underline{1}}$ defines a contraction of the arguments in parentheses towards a vector. From eqns (3.8)–(3.11), it is concluded, for the hybrid model of second type, that the quantities n^F and $\mathrm{grad}\,n^F$ included into the set $\mathcal{S}_c$ are not longer independent but are uniquely determined by $\mathbf{F}_S$ and $\mathrm{Grad}_S\,\mathbf{F}_S$. Thus, these variables must be dropped from $\mathcal{S}_c$. Furthermore, it is not necessary to formulate a constitutive equation for $(n^F)'_F$, compare eqn $(3.9)_3$. Moreover, in analogy to eqn (3.5), the constraint reflecting constituent incompressibility of φ^S is given by

$$
\lambda^{SR}\,[-n^S\,\mathrm{div}\,\overset{\prime}{\mathbf{x}}_S + (n^F)'_F - \mathrm{grad}\,n^F\,\cdot\,(\overset{\prime}{\mathbf{x}}_F - \overset{\prime}{\mathbf{x}}_S)] = 0,
\tag{3.12}
$$

where λ^{SR} is the Lagrangian multipier.

Thus, from the above relations, the constitutive postulates for the hybrid model of second type combine to

$$\begin{aligned}(\psi^i, \eta^i, \mathbf{T}^S + n^S \lambda^{SR}\, \mathbf{I}, \mathbf{T}^F, \mathbf{q}^i, \mathbf{p}^F - \lambda^{SR}\, \mathrm{grad}\, n^F) = \mathcal{R}(\mathcal{S}_2),\\ \mathcal{S}_2 = \{\theta,\ \mathrm{grad}\,\theta,\ \rho^F,\ \mathrm{grad}\,\rho^F,\\ \mathbf{F}_S,\ \mathrm{Grad}_S\,\mathbf{F}_S,\ \overset{\prime}{\mathbf{x}}_F - \overset{\prime}{\mathbf{x}}_S,\ \mathbf{D}_F\}.\end{aligned} \tag{3.13}$$

Additionally, the entropy inequality (3.7) must be modified towards

$$\begin{aligned}-\rho^S(\psi^S)'_S - \rho^F(\psi^F)'_F - \dot{\theta}(\rho^S\eta^S + \rho^F\eta^F) - (\mathbf{p}^F - \lambda^{SR}\, \mathrm{grad}\, n^F)\cdot(\overset{\prime}{\mathbf{x}}_F - \overset{\prime}{\mathbf{x}}_S) +\\ +(\mathbf{T}^S + n^S\lambda^{SR}\,\mathbf{I})\cdot\mathbf{L}_S + \mathbf{T}^F\cdot\mathbf{L}_F - \frac{1}{\theta}\mathbf{h}\cdot\mathrm{grad}\,\theta \geq 0.\end{aligned} \tag{3.14}$$

Therein, the constraint (3.12) was used.

Apart from the compressible and the hybrid two-phase media, we finally discuss the incompressible model defined by an incompressible solid matrix saturated by an incompressible viscous pore liquid. In this case, it is immediately concluded from the preceding considerations (in particular, compare eqns (3.4) and (3.9)) that both the liquid volume fractions and volume fraction gradients, n^F and $\mathrm{grad}\, n^F$, as well as the partial liquid density function and its gradient, ρ^F and $\mathrm{grad}\,\rho^F$, are uniquely determined by the first and the second solid deformation gradients, $\mathbf{F}_S$ and $\mathrm{Grad}_S\,\mathbf{F}_S$. Thus, the above mentioned terms do not contribute to the set of state variables. Furthermore, by addition of the constraint functions (3.5) and (3.12), one obtains with the aid of eqn (2.27) that

$$\lambda^R\,[-n^S\,\mathrm{div}\,\overset{\prime}{\mathbf{x}}_S - n^F\,\mathrm{div}\,\overset{\prime}{\mathbf{x}}_F - \mathrm{grad}\,n^F\cdot(\overset{\prime}{\mathbf{x}}_F - \overset{\prime}{\mathbf{x}}_S)] = 0, \tag{3.15}$$

where, obviously,

$$\lambda^R = \lambda^{FR} + \lambda^{SR}. \tag{3.16}$$

Based on these remarks, one directly concludes the constitutive postulates for the incompressible model to yield

$$\begin{aligned}(\psi^i, \eta^i, \mathbf{T}^i + n^i\lambda^R\,\mathbf{I}, \mathbf{q}^i, \mathbf{p}^F - \lambda^R\,\mathrm{grad}\,n^F) = \mathcal{R}(\mathcal{S}_{in}),\\ \mathcal{S}_{in} = \{\theta,\ \mathrm{grad}\,\theta,\ \mathbf{F}_S,\ \mathrm{Grad}_S\,\mathbf{F}_S,\ \overset{\prime}{\mathbf{x}}_F - \overset{\prime}{\mathbf{x}}_S,\ \mathbf{D}_F\}.\end{aligned} \tag{3.17}$$

Finally, the corresponding modification of the entropy inequality reads:

$$\begin{aligned}-\rho^S(\psi^S)'_S - \rho^F(\psi^F)'_F - \dot{\theta}(\rho^S\eta^S + \rho^F\eta^F) - (\mathbf{p}^F - \lambda^R\,\mathrm{grad}\,n^F)\cdot(\overset{\prime}{\mathbf{x}}_F - \overset{\prime}{\mathbf{x}}_S) +\\ +(\mathbf{T}^S + n^S\lambda^R\,\mathbf{I})\cdot\mathbf{L}_S + (\mathbf{T}^F + n^F\lambda^R\,\mathbf{I})\cdot\mathbf{L}_F - \frac{1}{\theta}\mathbf{h}\cdot\mathrm{grad}\,\theta \geq 0.\end{aligned} \tag{3.18}$$

The four different two-phase models described above deliver a lot of possibilities for the general modelling of fluid-saturated solid skeletons within the framework of porous media theories. However, concerning the validity of the different models, especially, with respect to the development of constitutive equations for granular materials in geomechanical context, one easily comes to realize that those model types including an incompressible solid matrix exhibit the most realistic character and thus, of course, the greatest practical importance.

3.2 Second-grade elastoplasticity

Within the scope of second-grade solid materials, the general response is affected by the first and the second solid deformation gradients, that is:

$$\mathcal{R} = \mathcal{R}\,(\,\dots, \mathbf{F}_S, \overset{3}{\mathbf{G}}_S, \dots), \tag{3.19}$$

where the third-order tensor $\overset{3}{\mathbf{G}}_S$ is defined via

$$\overset{3}{\mathbf{G}}_S = \operatorname{Grad}_S \mathbf{F}_S. \tag{3.20}$$

Concerning the representation (3.19), CROSS (1973) has shown that the pair $(\mathbf{F}_S, \overset{3}{\mathbf{G}}_S)$ is an element in the second-grade linear group $\mathcal{L}$ defined by

$$\mathcal{L} = \{(\mathbf{H}, \overset{3}{\mathbf{C}})\}. \tag{3.21}$$

Therein, $\mathbf{H}$ is a nonsingular linear transformation and $\overset{3}{\mathbf{C}}$ is a third-order tensor with the defining property

$$\overset{3}{\mathbf{C}} = (\overset{3}{\mathbf{C}})^{\overset{23}{T}}. \tag{3.22}$$

In the above relation, the symbol $(\overset{3}{\dots})^{\overset{23}{T}}$ indicates the transposition of $(\overset{3}{\dots})$ with respect to the second and the third base vectors or the indices 2 and 3, respectively.

Following this, finite elastoplasticity for the solid constituent results in a multiplicative decomposition of the pair $(\mathbf{F}_S, \overset{3}{\mathbf{G}}_S)$ into elastic and plastic parts, that is:

$$(\mathbf{F}_S, \overset{3}{\mathbf{G}}_S) = (\mathbf{F}_{Se}, \overset{3}{\mathbf{G}}_{Se}) \circ (\mathbf{F}_{Sp}, \overset{3}{\mathbf{G}}_{Sp}). \tag{3.23}$$

The symbol $(\dots, \overset{3}{\dots}) \circ (\dots, \overset{3}{\dots})$, the composition, is the binary operator defining the group. Since the composition rule is induced by the chain rule (CROSS, 1973), it it easily proved that

$$\begin{aligned} \mathbf{F}_S &= \mathbf{F}_{Se}\,\mathbf{F}_{Sp}, \\ \overset{3}{\mathbf{G}}_S &= [(\overset{3}{\mathbf{G}}_{Se}\,\mathbf{F}_{Sp})^{\underline{3}\,\overset{23}{T}}\,\mathbf{F}_{Sp}]^{\underline{3}\,\overset{23}{T}} + (\mathbf{F}_{Se}\,\overset{3}{\mathbf{G}}_{Sp})^{\underline{3}}, \end{aligned} \tag{3.24}$$

where $(\dots)^{\underline{3}}$ defines a contraction of the arguments in parantheses towards a third-order tensor. For convenience for the reader, eqn $(3.24)_2$, in a simple index notation, reads

$$(\mathbf{G}_S)_{aBC} = (\mathbf{G}_{Se})_{a\alpha\beta}\,(\mathbf{F}_{Sp})_{\alpha B}\,(\mathbf{F}_{Sp})_{\beta C} + (\mathbf{F}_{Se})_{a\alpha}\,(\mathbf{G}_{Sp})_{\alpha BC}, \tag{3.25}$$

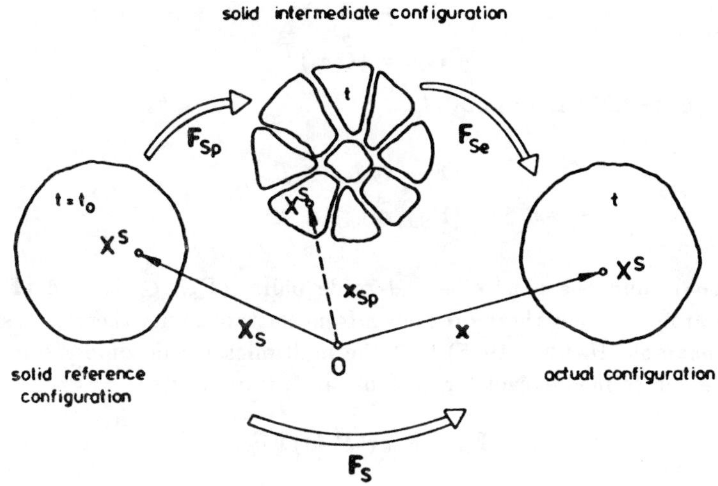

Figure 2: Solid configurations.

small Latin subscripts belonging to the actual, small Greek subscripts to the intermediate and capital Latin subscripts to the reference configuration of φ^S.

In addition, it is commonly known from elastoplasticity of non-porous single solid continua (materials of first grade) that the concept of the multiplicative decomposition of deformation gradients is connected with the suggestion of a stress-free intermediate configuration (compare Figure 2), where the purely plastic state of deformation is frozen into the memory of the material; additionally, the temperature function is pulled back to its reference value. However, in the case of general deformation processes, this intermediate configuration is incompatible with the existence of partial derivatives when these functions are considered in the sense of usual field functions. Thus, the quantities

$$\begin{aligned} \mathbf{F}_{Se} &= \operatorname{grad}_{Sp} \mathbf{x}\,, & \overset{3}{\mathbf{G}}_{Se} &= \operatorname{grad}_{Sp} \mathbf{F}_{Se}\,, \\ \mathbf{F}_{Sp} &= \operatorname{Grad}_S \mathbf{x}_{Sp}\,, & \overset{3}{\mathbf{G}}_{Sp} &= \operatorname{Grad}_S \mathbf{F}_{Sp} \end{aligned} \tag{3.26}$$

must be understood as "point functions" only defined in the particular subregions of the intersected plastic intermediate configuration. On the other hand, within the case of homogeneous deformation processes, the field character of the above relations generally holds. Furthermore, note in passing that $\mathbf{x}_{Sp}$ defines the position of particles X^S in the intermediate configuration of φ^S and "grad_{Sp}" means partial differentiation with respect to $\mathbf{x}_{Sp}$.

Moreover, it is seen from eqns (3.26) that

$$\begin{aligned} \overset{3}{\mathbf{G}}_{Se} &= (\overset{3}{\mathbf{G}}_{Se})^{\overset{23}{T}}, \\ \overset{3}{\mathbf{G}}_{Sp} &= (\overset{3}{\mathbf{G}}_{Sp})^{\overset{23}{T}} \end{aligned} \tag{3.27}$$

and thus, from eqns (3.21) and (3.22),

$$\begin{aligned} (\mathbf{F}_{Se}, \overset{3}{\mathbf{G}}_{Se}) &\in \mathcal{L}, \\ (\mathbf{F}_{Sp}, \overset{3}{\mathbf{G}}_{Sp}) &\in \mathcal{L}; \end{aligned} \tag{3.28}$$

that is, the elastic and the plastic second-grade pairs, $(\mathbf{F}_{Se}, \overset{3}{\mathbf{G}}_{Se})$ and $(\mathbf{F}_{Sp}, \overset{3}{\mathbf{G}}_{Sp})$, are elements in the group. It is furthermore known from elastoplastcity theories of single continua (compare, for example HAUPT, 1985) that the multiplicative decomposition is not unique but can only be determined except for an arbitrary rotation, viz.:

$$\mathbf{F}_S = \mathbf{F}_{Se} (\hat{\mathbf{Q}}^T \hat{\mathbf{Q}}) \mathbf{F}_{Sp} . \tag{3.29}$$

Therein, $\hat{\mathbf{Q}}$ is a proper orthogonal linear transformation corresponding to the intermediate configuration of φ^S. Note in passing that terms belonging to this intermediate configuration are characterized by the topscript $(\hat{\ldots})$. From the above arguments, it is easily concluded that the rotation of the plastic intermedite configuration generally remains unspecified. Following this, the arbitrary rotation for materials of second grade must be introduced by the pair

$$(\hat{\mathbf{Q}}, \overset{3}{\mathbf{0}}) \in \mathcal{O}_p \in \mathcal{L}, \tag{3.30}$$

where $\mathcal{O}_p$ defines the proper orthogonal subgroup of $\mathcal{L}$. From the character of $\mathcal{O}_p$, it is easily proved that

$$(\hat{\mathbf{Q}}, \overset{3}{\mathbf{0}})^T \circ (\hat{\mathbf{Q}}, \overset{3}{\mathbf{0}}) = (\mathbf{I}, \overset{3}{\mathbf{0}}), \tag{3.31}$$

where the pair $(\mathbf{I}, \overset{3}{\mathbf{0}})$ is the identity element in the group.

Combining eqns (3.23) and (3.31), one finds that

$$(\mathbf{F}_S, \overset{3}{\mathbf{G}}_S) = (\tilde{\mathbf{F}}_{Se}, \overset{3}{\tilde{\mathbf{G}}}_{Se}) \circ (\tilde{\mathbf{F}}_{Sp}, \overset{3}{\tilde{\mathbf{G}}}_{Sp}), \tag{3.32}$$

where

$$\begin{aligned} (\tilde{\mathbf{F}}_{Se}, \overset{3}{\tilde{\mathbf{G}}}_{Se}) &= (\mathbf{F}_{Se}, \overset{3}{\mathbf{G}}_{Se}) \circ (\hat{\mathbf{Q}}, \overset{3}{\mathbf{0}})^T, \\ (\tilde{\mathbf{F}}_{Sp}, \overset{3}{\tilde{\mathbf{G}}}_{Sp}) &= (\hat{\mathbf{Q}}, \overset{3}{\mathbf{0}}) \circ (\mathbf{F}_{Sp}, \overset{3}{\mathbf{G}}_{Sp}) . \end{aligned} \tag{3.33}$$

In addition, the composition rule yields

$$\begin{aligned} \tilde{\mathbf{F}}_{Se} &= \mathbf{F}_{Se} \hat{\mathbf{Q}}^T, & \overset{3}{\tilde{\mathbf{G}}}_{Se} &= [(\overset{3}{\mathbf{G}}_{Se} \hat{\mathbf{Q}}^T)^{\underline{3}\,\overset{23}{T}} \hat{\mathbf{Q}}^T]^{\underline{3}\,\overset{23}{T}}, \\ \tilde{\mathbf{F}}_{Sp} &= \hat{\mathbf{Q}} \mathbf{F}_{Sp}, & \overset{3}{\tilde{\mathbf{G}}}_{Sp} &= (\hat{\mathbf{Q}} \overset{3}{\mathbf{G}}_{Sp})^{\underline{3}} . \end{aligned} \tag{3.34}$$

Furthermore, by use of the polar decomposition theorem,

$$\begin{aligned} \mathbf{F}_S &= \mathbf{R}_S \mathbf{U}_S = \mathbf{V}_S \mathbf{R}_S\,, \\ \mathbf{F}_{Se} &= \mathbf{R}_{Se} \mathbf{U}_{Se} = \mathbf{V}_{Se} \mathbf{R}_{Se}\,, \\ \mathbf{F}_{Sp} &= \mathbf{R}_{Sp} \mathbf{U}_{Sp} = \mathbf{V}_{Sp} \mathbf{R}_{Sp}\,. \end{aligned} \tag{3.35}$$

Thus, it is seen from eqns (3.33) and (3.34) that only eqn $(3.35)_1$ is unique while eqns $(3.35)_2$ and $(3.35)_3$ depend on $\hat{\mathbf{Q}}$. Moreover, the definition of right and left Cauchy-Green tensors via

$$\begin{aligned} \mathbf{C}_S &= \mathbf{F}_S^T \mathbf{F}_S\,, & \mathbf{B}_S &= \mathbf{F}_S \mathbf{F}_S^T\,, \\ \hat{\mathbf{C}}_{Se} &= \mathbf{F}_{Se}^T \mathbf{F}_{Se}\,, & \mathbf{B}_{Se} &= \mathbf{F}_{Se} \mathbf{F}_{Se}^T\,, \\ \mathbf{C}_{Sp} &= \mathbf{F}_{Sp}^T \mathbf{F}_{Sp}\,, & \hat{\mathbf{B}}_{Sp} &= \mathbf{F}_{Sp} \mathbf{F}_{Sp}^T \end{aligned} \tag{3.36}$$

reveals the same undeterminancy insofar as only the right Cauchy-Green tensors $\mathbf{C}_S$ and $\mathbf{C}_{Sp}$ relative to the reference configuration of φ^S and the left Cauchy-Green tensos $\mathbf{B}_S$ and $\mathbf{B}_{Se}$ of the actual configuration are unique, whereas the elastic quantity $\hat{\mathbf{C}}_{Se}$ and the plastic quantity $\hat{\mathbf{B}}_{Sp}$ correspond to the intermediate configuration and are thus influenced by $\hat{\mathbf{Q}}$.

Within the framework of the above definition and comments, the multiplicative decomposition of the pair $(\mathbf{F}_{Se}\,, \overset{3}{\mathbf{G}}_{Se})$ is compatible with an additive decomposition of strain measures, where, in the present article, several elements of the so-called "first family" of strain tensors are introduced. Concerning the notion "first family," the reader is referred to the work of HAUPT & TSAKMAKIS (1989). Within the scope of general elasto-plastic deformations, this first family can be introduced via the Lagrangian strain $\mathbf{E}_S$, the plastic Lagrangian strain $\mathbf{E}_{Sp}$ and the elastic strain measure $\hat{\mathbf{\Gamma}}_{Se}$, viz:

$$\begin{aligned} \mathbf{E}_S &= \tfrac{1}{2}(\mathbf{C}_S - \mathbf{I}), \\ \mathbf{E}_{Sp} &= \tfrac{1}{2}(\mathbf{C}_{Sp} - \mathbf{I}), \\ \hat{\mathbf{\Gamma}}_{Se} &= \tfrac{1}{2}(\hat{\mathbf{C}}_{Se} - \mathbf{I}). \end{aligned} \tag{3.37}$$

Furthermore, it is easily seen from Figure 2 and eqns (3.36) and (3.37) that the multiplicative decomposition implies the geometrical concepts of elasticity and plasticity to be defined by the "distances" between the different solid configurations. Concerning the first family of strain tensors, these "distances" must be understood as the differences betweeen the squares of material line elements of the different configurations. In particular, elasticity is defined by the "distance" between the current configuration and the intermediate configuration of φ^S, whereas plasticity is characterized by the "distance" between this intermediate configuration and the reference configuration of φ^S. However, by pulling $\hat{\mathbf{\Gamma}}_{Se}$ back with $\mathbf{F}_{Sp}^T(\,\hat{\ldots}\,)\mathbf{F}_{Sp}$, one immediately obtains the elastic strain measure defined in terms of the solid reference configuration, namely, as the difference between $\mathbf{E}_S$ and $\mathbf{E}_{Sp}$:

$$\mathbf{E}_{Se} = \mathbf{F}_{Sp}^T \hat{\mathbf{\Gamma}}_{Se} \mathbf{F}_{Sp} = \mathbf{E}_S - \mathbf{E}_{Sp}\,. \tag{3.38}$$

Then, by pushing the Lagrangian strains $\mathbf{E}_S$, $\mathbf{E}_{Se}$ and $\mathbf{E}_{Sp}$ forward with $\mathbf{F}_{Sp}^{T-1}(\ldots)\mathbf{F}_{Sp}^{-1}$ or with $\mathbf{F}_S^{T-1}(\ldots)\mathbf{F}_S^{-1}$, respectively, one easily concludes to the strain measures $\hat{\mathbf{\Gamma}}_S$, $\hat{\mathbf{\Gamma}}_{Sp}$ and

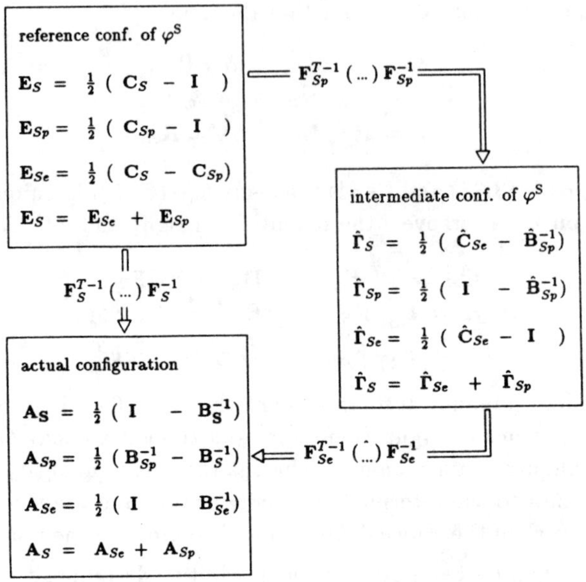

Table 1: Lagrangian and Almansian strain tensors of φ^S.

$\hat{\mathbf{\Gamma}}_{Se}$ corresponding to the intermediate configuration of φ^S and to the Almansian strains $\mathbf{A}_S$, $\mathbf{A}_{Sp}$ and $\mathbf{A}_{Se}$ belonging to the actual configuration. For convenience for the reader, the above strain measures of the different configurations are summarized in Table 1. Note in passing that, by use of a tensor representation given with respect to the natural basis systems of the different configurations, the strain tensors listed in Table 1 exhibit contravariant vector bases. Thus, the respective push-forward and, vice versa, pull-back transformations can be understood in the sense of "contravariant transports".

For several purposes in finite elastoplasticity, it is convenient to use Karni-Reiner strain measures instead of Lagrangian or Almansian strains (concerning the notion "Karni-Reiner strain," the reader is referred to MACVEAN, 1968). Proceeding from the polar decompositions (3.35) together with eqns (3.37), the Karni-Reiner strain tensors can be introduced by the following push-forward rotations:

$$\begin{aligned} \mathbf{K}_S &= \mathbf{R}_S \, \mathbf{E}_S \, \mathbf{R}_S^T \,, \\ \hat{\mathbf{K}}_{Sp} &= \mathbf{R}_{Sp} \, \mathbf{E}_{Sp} \, \mathbf{R}_{Sp}^T \,, \\ \mathbf{K}_{Se} &= \mathbf{R}_{Se} \, \hat{\mathbf{\Gamma}}_{Se} \, \mathbf{R}_{Se}^T \,. \end{aligned} \tag{3.39}$$

By use of eqns (3.36), an equivalent representation of the definitions (3.39) is given by

$$\begin{aligned} \mathbf{K}_S &= \tfrac{1}{2}(\mathbf{B}_S - \mathbf{I}), \\ \hat{\mathbf{K}}_{Sp} &= \tfrac{1}{2}(\hat{\mathbf{B}}_{Sp} - \mathbf{I}), \\ \mathbf{K}_{Se} &= \tfrac{1}{2}(\mathbf{B}_{Se} - \mathbf{I}). \end{aligned} \tag{3.40}$$

reference conf. of φ^S

$$\overset{R}{\mathbf{K}}_S = \tfrac{1}{2}\,(\,\mathbf{I} - \mathbf{C}_S^{-1})$$
$$\overset{R}{\mathbf{K}}_{Sp} = \tfrac{1}{2}\,(\,\mathbf{I} - \mathbf{C}_{Sp}^{-1})$$
$$\overset{R}{\mathbf{K}}_{Se} = \tfrac{1}{2}\,(\,\mathbf{C}_{Sp}^{-1} - \mathbf{C}_S^{-1})$$
$$\overset{R}{\mathbf{K}}_S = \overset{R}{\mathbf{K}}_{Se} + \overset{R}{\mathbf{K}}_{Sp}$$

$\Longrightarrow \mathbf{F}_{Sp}\,(\dots)\,\mathbf{F}_{Sp}^T \Longrightarrow$

intermediate conf. of φ^S

$$\hat{\mathbf{K}}_S = \tfrac{1}{2}\,(\,\hat{\mathbf{B}}_{Sp} - \hat{\mathbf{C}}_{Se}^{-1})$$
$$\hat{\mathbf{K}}_{Sp} = \tfrac{1}{2}\,(\,\hat{\mathbf{B}}_{Sp} - \mathbf{I}\,)$$
$$\hat{\mathbf{K}}_{Se} = \tfrac{1}{2}\,(\,\mathbf{I} - \hat{\mathbf{C}}_{Se}^{-1})$$
$$\hat{\mathbf{K}}_S = \hat{\mathbf{K}}_{Se} + \hat{\mathbf{K}}_{Sp}$$

$\Downarrow \mathbf{F}_S\,(\dots)\,\mathbf{F}_S^T$ (reference → actual) $\qquad \Longleftarrow \mathbf{F}_{Se}\,(\hat{\dots})\,\mathbf{F}_{Se}^T$ (intermediate → actual)

actual configuration

$$\mathbf{K}_S = \tfrac{1}{2}\,(\,\mathbf{B}_S - \mathbf{I}\,)$$
$$\mathbf{K}_{Sp} = \tfrac{1}{2}\,(\,\mathbf{B}_S - \mathbf{B}_{Se})$$
$$\mathbf{K}_{Se} = \tfrac{1}{2}\,(\,\mathbf{B}_{Se} - \mathbf{I}\,)$$
$$\mathbf{K}_S = \mathbf{K}_{Se} + \mathbf{K}_{Sp}$$

Table 2: Karni-Reiner strain tensors of φ^S.

Thus, the Karni-Reiner strains prove to yield the negative values of the Piola and Finger strain measures which are the basic elements in a so-called "second family of strain tensors". For further information concerning this second family, the reader is again referred to HAUPT & TSAKMAKIS (1989). From the definition of the above strain measures, it is furthermore seen that $\mathbf{K}_S$ and $\mathbf{K}_{Se}$ are defined with respect to the actual configuration and are thus independent of the arbitrary rotation, while $\hat{\mathbf{K}}_{Sp}$ corresponds to the intermediate configuration of φ^S and, hence, is influenced by $\hat{\mathbf{Q}}$. In order to complete the set of Karni-Reiner strain tensors, the push-forward transformation of $\hat{\mathbf{K}}_{Sp}$ with $\mathbf{F}_{Se}\,(\,\hat{\dots}\,)\,\mathbf{F}_{Se}^T$ yields

$$\mathbf{K}_{Sp} = \mathbf{F}_{Se}\,\hat{\mathbf{K}}_{Sp}\,\mathbf{F}_{Se}^T = \mathbf{K}_S - \mathbf{K}_{Se}\,. \tag{3.41}$$

Then, by pulling the Karni-Reiner strains $\mathbf{K}_S$, $\mathbf{K}_{Sp}$ and $\mathbf{K}_{Se}$ back with $\mathbf{F}_{Se}^{-1}\,(\,\dots)\,\mathbf{F}_{Se}^{T-1}$ or with $\mathbf{F}_S^{-1}\,(\,\dots)\,\mathbf{F}_S^{T-1}$, respectively, one obtains the strain measures $\hat{\mathbf{K}}_S$, $\hat{\mathbf{K}}_{Sp}$ and $\hat{\mathbf{K}}_{Se}$ of the intermediate configuration of φ^S and the solid reference configuration strains $\overset{R}{\mathbf{K}}_S$ $\overset{R}{\mathbf{K}}_{Sp}$ and $\overset{R}{\mathbf{K}}_{Se}$ (compare Table 2).

Finally, note in passing that the Karni-Reiner strains included into Table 2, in contrast to the Lagrangian and Almansian strain measures of Table 1, exhibit a covariant representation with respect to their natural vector bases. Thus, the above pull-back and corresponding push-forward transformations can be understood as "covariant transports".

As was shown by the author (EHLERS, 1989a, b), the multiplicative decomposition of deformation gradients implies the existence of only two proper spatial velocity gradients, namely $\mathbf{L}_S$ and $\hat{\mathbf{L}}_{Sp}$, the former relative to the actual and the latter relative to the plastic intermediate configuration or φ^S (in addition, compare PAULUN, 1980):

$$\begin{aligned} \mathbf{L}_S &= (\mathbf{F}_S)'_S\,\mathbf{F}_S^{-1}, \\ \hat{\mathbf{L}}_{Sp} &= (\mathbf{F}_{Sp})'_S\,\mathbf{F}_{Sp}^{-1}. \end{aligned} \tag{3.42}$$

Their symmetric and skew-symmetric parts are:

$$\begin{aligned} \mathbf{D}_S &= \tfrac{1}{2}(\mathbf{L}_S + \mathbf{L}_S^T), & \hat{\mathbf{D}}_{Sp} &= \tfrac{1}{2}(\hat{\mathbf{L}}_{Sp} + \hat{\mathbf{L}}_{Sp}^T), \\ \mathbf{W}_S &= \tfrac{1}{2}(\mathbf{L}_S - \mathbf{L}_S^T), & \hat{\mathbf{W}}_{Sp} &= \tfrac{1}{2}(\hat{\mathbf{L}}_{Sp} - \hat{\mathbf{L}}_{Sp}^T). \end{aligned} \tag{3.43}$$

In the above representations, $\mathbf{D}_S$ characterizes the solid deformation rate of the actual configuration, whereas $\hat{\mathbf{D}}_{Sp}$ denotes the purely plastic solid deformation rate corresponding to the intermediate configuration of φ^S. In addition, $\mathbf{W}_S$ and $\hat{\mathbf{W}}_{Sp}$ are the total and the plastic solid spin tensors, the former defined in terms of the actual configuration, the latter with respect to the plastic intermediate configuration.

Following the concept of Lie derivatives or Oldroyd derivatives, respectively, one comes to realize that the above solid rate of deformation tensors can be expressed via

$$\begin{aligned} \mathbf{D}_S &= (\mathbf{A}_S)^{\vartriangle} = (\mathbf{K}_S)^{\triangledown}, \\ \hat{\mathbf{D}}_{Sp} &= (\hat{\mathbf{\Gamma}}_{Sp})_p^{\vartriangle} = (\hat{\mathbf{K}}_{Sp})_p^{\triangledown}. \end{aligned} \tag{3.44}$$

Therein, the symbols $(\,\dots)^{\vartriangle}$ and $(\,\dots)^{\triangledown}$ define the "upper" und the "lower" Lie derivatives of φ^S corresponding to the current configuration:

$$\begin{aligned} (\,\dots)^{\vartriangle} &= (\,\dots)'_S + \mathbf{L}_S^T(\,\dots) + (\,\dots)\mathbf{L}_S, \\ (\,\dots)^{\triangledown} &= (\,\dots)'_S - \mathbf{L}_S(\,\dots) - (\,\dots)\mathbf{L}_S^T; \end{aligned} \tag{3.45}$$

the symbols $(\hat{\dots})_p^{\vartriangle}$ and $(\hat{\dots})_p^{\triangledown}$ denote the respective derivatives of the plastic intermediate configuration:

$$\begin{aligned} (\hat{\dots})_p^{\vartriangle} &= (\hat{\dots})'_S + \hat{\mathbf{L}}_{Sp}^T(\hat{\dots}) + (\hat{\dots})\hat{\mathbf{L}}_{Sp}, \\ (\hat{\dots})_p^{\triangledown} &= (\hat{\dots})'_S - \hat{\mathbf{L}}_{Sp}(\hat{\dots}) - (\hat{\dots})\hat{\mathbf{L}}_{Sp}^T. \end{aligned} \tag{3.46}$$

From the above considerations, it is easily concluded that, by use of a representation of the different strain measures with respect to the natural basis systems of the different configurations, the concept of Lie derivatives means nothing else than taking the material time derivative of the respective tensorial components, keeping the contravariant bases (upper Lie derivative) or the covariant bases (lower Lie derivative) fixed. Thus, concerning the different strain measures of Table 1 and Table 2, the concept of Lie derivatives implies a concept of frame-indifferent strain rates which are related to each other by the same push-forward and pull-back transformations as are the respective strain tensors. For convenience for the reader, the strain rates are summarized in Table 3 and Table 4.

Finally, note in passing that the concept of Karni-Reiner strain rates included into Table 4 differs from the definition originally proposed by HAUPT & TSAKMAKIS (1989).

$$(\dots)^{\triangle} = (\dots)'_S + \mathbf{L}_S^T (\dots) + (\dots)\, \mathbf{L}_S$$

$$(\hat{\dots})^{\triangle}_p = (\hat{\dots})'_S + \hat{\mathbf{L}}_{Sp}^T (\hat{\dots}) + (\hat{\dots})\, \hat{\mathbf{L}}_{Sp}$$

reference conf. of φ^S

$$(\mathbf{E}_S)'_S = (\mathbf{E}_{Se})'_S + (\mathbf{E}_{Sp})'_S$$

$$(\mathbf{E}_{Sp})'_S$$

$$(\mathbf{E}_{Se})'_S$$

reference conf. → intermediate conf.: $\mathbf{F}_{Sp}^{T-1} (\dots)\, \mathbf{F}_{Sp}^{-1}$

reference conf. → actual configuration: $\mathbf{F}_S^{T-1} (\dots)\, \mathbf{F}_S^{-1}$

intermediate conf. of φ^S

$$(\hat{\mathbf{\Gamma}}_S)^{\triangle}_p = (\hat{\mathbf{\Gamma}}_{Se})^{\triangle}_p + (\hat{\mathbf{\Gamma}}_{Sp})^{\triangle}_p$$

$$\hat{\mathbf{D}}_{Sp} = (\hat{\mathbf{\Gamma}}_{Sp})^{\triangle}_p = \tfrac{1}{2} (\hat{\mathbf{L}}_{Sp} + \hat{\mathbf{L}}_{Sp}^T)$$

$$(\hat{\mathbf{\Gamma}}_{Se})^{\triangle}_p$$

intermediate conf. → actual configuration: $\mathbf{F}_{Se}^{T-1} (\hat{\dots})\, \mathbf{F}_{Se}^{-1}$

actual configuration

$$\mathbf{D}_S = (\mathbf{A}_S)^{\triangle} = \tfrac{1}{2} (\mathbf{L}_S + \mathbf{L}_S^T)$$

$$(\mathbf{A}_{Sp})^{\triangle}$$

$$(\mathbf{A}_{Se})^{\triangle}$$

$$(\mathbf{A}_S)^{\triangle} = (\mathbf{A}_{Se})^{\triangle} + (\mathbf{A}_{Sp})^{\triangle}$$

Table 3: Lagrangian and Almansian strain rates of φ^S.

$$(\dots)^{\nabla} = (\dots)'_S - \mathbf{L}_S^T (\dots) - (\dots)\, \mathbf{L}_S$$

$$(\hat{\dots})^{\nabla}_p = (\hat{\dots})'_S - \hat{\mathbf{L}}_{Sp}^T (\hat{\dots}) - (\hat{\dots})\, \hat{\mathbf{L}}_{Sp}$$

reference conf. of φ^S

$$(\overset{R}{\mathbf{K}}_S)'_S = (\overset{R}{\mathbf{K}}_{Se})'_S + (\overset{R}{\mathbf{K}}_{Sp})'_S$$

$$(\overset{R}{\mathbf{K}}_{Sp})'_S$$

$$(\overset{R}{\mathbf{K}}_{Se})'_S$$

reference conf. → intermediate conf.: $\mathbf{F}_{Sp} (\dots)\, \mathbf{F}_{Sp}^T$

reference conf. → actual configuration: $\mathbf{F}_S (\dots)\, \mathbf{F}_S^T$

intermediate conf. of φ^S

$$(\hat{\mathbf{K}}_S)^{\nabla}_p = (\hat{\mathbf{K}}_{Se})^{\nabla}_p + (\hat{\mathbf{K}}_{Sp})^{\nabla}_p$$

$$\hat{\mathbf{D}}_{Sp} = (\hat{\mathbf{K}}_{Sp})^{\nabla}_p = \tfrac{1}{2} (\hat{\mathbf{L}}_{Sp} + \hat{\mathbf{L}}_{Sp}^T)$$

$$(\hat{\mathbf{K}}_{Se})^{\nabla}_p$$

intermediate conf. → actual configuration: $\mathbf{F}_{Se} (\hat{\dots})\, \mathbf{F}_{Se}^T$

actual configuration

$$\mathbf{D}_S = (\mathbf{K}_S)^{\nabla} = \tfrac{1}{2} (\mathbf{L}_S + \mathbf{L}_S^T)$$

$$(\mathbf{K}_{Sp})^{\nabla}$$

$$(\mathbf{K}_{Se})^{\nabla}$$

$$(\mathbf{K}_S)^{\nabla} = (\mathbf{K}_{Se})^{\nabla} + (\mathbf{K}_{Sp})^{\nabla}$$

Table 4: Karni-Reiner strain rates of φ^S.

3.3 Constitutive framing

Within the framework of porous media elastoplasticity, the second-grade character of the model implies the general response (3.19) to be replaced by the assumption that

$$\mathcal{R} = \mathcal{R}\,(\,\dots\,,\mathbf{F}_{Se}\,,\overset{3}{\mathbf{G}}_{Se}\,,\mathbf{F}_{Sp}\,,\overset{3}{\mathbf{G}}_{Sp}\,,\,\dots\,). \tag{3.47}$$

In writing eqn (3.47), it is understood that the "elastic variables," $\mathbf{F}_{Se}$ and $\overset{3}{\mathbf{G}}_{Se}$, are taken as independent external state variables for a given geometry of the plastic intermediate configuration, whereas, as a consequence of the path-dependence in plasticity, the "plastic variables," $\mathbf{F}_{Sp}$ and $\overset{3}{\mathbf{G}}_{Sp}$, must be interpreted in the sense of internal state variables. Thus, the set $\mathcal{R}$ must be extended by an additional constitutive equation (fow rule) for the purely plastic rate of deformation tensor $\hat{\mathbf{D}}_{Sp}$ or for any related plastic strain rate. Note in passing that it is not necessary to formulate a constitutive equation for the plastic spin tensor $\hat{\mathbf{W}}_{Sp}$; this term naturally remains unspecified, namely, in the same sense as the plastic rotation. Finally, it should be noted that, to describe the effect of kinematical hardening, a constitutive equation for the back-stress tensor $\hat{\mathbf{Y}}^S$ of the solid matrix is needed, in general.

From the preceding considerations, the constitutive assumptions (3.1) and (3.47) combine to yield

$$\begin{aligned}
&(\psi^i\,,\eta^i\,,\mathbf{T}^i\,,\mathbf{q}^i\,,\mathbf{p}^F\,,(n^F)'_F\,) = \mathcal{R}\,(\mathcal{S}_{cp})\,,\\
&\hat{\mathbf{D}}_{Sp} \;=\; \hat{\mathbf{D}}_{Sp}\,(\mathcal{S}_{cp}\,,\,\dots\,)\,, \quad \hat{\mathbf{Y}}_S \;=\; \hat{\mathbf{Y}}_S\,(\mathcal{S}_{cp}\,,\,\dots\,)\,,\\
&\mathcal{S}_{cp} = \{\theta,\ \operatorname{grad}\theta,\ n^F,\ \operatorname{grad} n^F,\ \rho^F,\ \operatorname{grad}\rho^F,\\
&\qquad \mathbf{F}_{Se}\,,\overset{3}{\mathbf{G}}_{Se}\,,\mathbf{F}_{Sp}\,,\overset{3}{\mathbf{G}}_{Sp}\,,\ \overset{\prime}{\mathbf{x}}_F - \overset{\prime}{\mathbf{x}}_S,\ \mathbf{D}_F\,\}.
\end{aligned} \tag{3.48}$$

In addition, the plastic response is governed by a yield function which must be understood as a secondary condition for the respective thermodynamical process.

However, the constitutive model defined by eqns (3.48) governs the compressible two-phase model consisting of an elasto-plastic solid matrix saturated by a compressible viscous pore fluid. Thus, in order to describe elasto-plastic deformations of hybrid models or incompressible models, respectively, the above equations must be modified.

In particular, for the hybrid model of first type, one obtains the following relations by combination of eqns (3.6) and (3.47)

$$\begin{aligned}
&(\psi^i\,,\eta^i\,,\mathbf{T}^S\,,\mathbf{T}^F + n^F\lambda^{FR}\,\mathbf{I}\,,\mathbf{q}^i\,,\mathbf{p}^F - \lambda^{FR}\operatorname{grad} n^F) = \mathcal{R}\,(\mathcal{S}_{1p})\,,\\
&\hat{\mathbf{D}}_{Sp} \;=\; \hat{\mathbf{D}}_{Sp}\,(\mathcal{S}_{1p}\,,\,\dots\,)\,, \quad \hat{\mathbf{Y}}_S \;=\; \hat{\mathbf{Y}}_S\,(\mathcal{S}_{1p}\,,\,\dots\,)\,,\\
&\mathcal{S}_{1p} = \{\theta,\ \operatorname{grad}\theta,\ n^F,\ \operatorname{grad} n^F,\\
&\qquad \mathbf{F}_{Se}\,,\overset{3}{\mathbf{G}}_{Se}\,,\mathbf{F}_{Sp}\,,\overset{3}{\mathbf{G}}_{Sp}\,,\ \overset{\prime}{\mathbf{x}}_F - \overset{\prime}{\mathbf{x}}_S,\ \mathbf{D}_F\,\}.
\end{aligned} \tag{3.49}$$

Furthermore, it is then easily seen that hybrid models of second type are governed by

$$
\begin{aligned}
&(\psi^i\,,\eta^i\,,\mathbf{T}^S + n^S\lambda^{SR}\,\mathbf{I}\,,\mathbf{T}^F\,,\mathbf{q}^i\,,\mathbf{p}^F - \lambda^{SR}\,\mathrm{grad}\,n^F) = \mathcal{R}\,(\mathcal{S}_{2p})\,,\\
&\hat{\mathbf{D}}_{Sp} \;=\; \hat{\mathbf{D}}_{Sp}\,(\mathcal{S}_{2p}\,,\,\ldots)\,, \quad \hat{\mathbf{Y}}_S \;=\; \hat{\mathbf{Y}}_S\,(\mathcal{S}_{2p}\,,\,\ldots)\,,\\
&\mathcal{S}_{2p} = \{\theta,\ \mathrm{grad}\,\theta,\ \rho^F,\ \mathrm{grad}\,\rho^F,\\
&\qquad \mathbf{F}_{Se}\,,\overset{3}{\mathbf{G}}_{Se}\,,\mathbf{F}_{Sp}\,,\overset{3}{\mathbf{G}}_{Sp}\,,\ \overset{\prime}{\mathbf{x}}_F - \overset{\prime}{\mathbf{x}}_S,\ \mathbf{D}_F\,\},
\end{aligned}
\tag{3.50}
$$

where eqns (3.13) and (3.47) have been used.

Finally, with the aid of eqns (3.17) and (3.47), the constitutive assumptions for the incompressible model combine to

$$
\begin{aligned}
&(\psi^i\,,\eta^i\,,\mathbf{T}^i + n^i\lambda^R\,\mathbf{I}\,,\mathbf{q}^i\,,\mathbf{p}^F - \lambda^R\,\mathrm{grad}\,n^F) = \mathcal{R}\,(\mathcal{S}_{inp})\,,\\
&\hat{\mathbf{D}}_{Sp} \;=\; \hat{\mathbf{D}}_{Sp}\,(\mathcal{S}_{inp}\,,\,\ldots)\,, \quad \hat{\mathbf{Y}}^S \;=\; \hat{\mathbf{Y}}^S\,(\mathcal{S}_{inp}\,,\,\ldots)\,,\\
&\mathcal{S}_{inp} = \{\theta,\ \mathrm{grad}\,\theta,\ \mathbf{F}_{Se}\,,\overset{3}{\mathbf{G}}_{Se}\,,\mathbf{F}_{Sp}\,,\overset{3}{\mathbf{G}}_{Sp}\,,\ \overset{\prime}{\mathbf{x}}_F - \overset{\prime}{\mathbf{x}}_S,\ \mathbf{D}_F\,\}.
\end{aligned}
\tag{3.51}
$$

Moreover, to obtain thermodynamical restrictions from the dissipation principle, the entropy inequalities of the respective models, the relations (2.22), (3.7), (3.14) and (3.18), must be modified towards elasto-plastic deformations. In particular, from the assumption of non-polar constituents, compare eqns (2.25), the solid stress-power W^S yields

$$
\begin{aligned}
W^S &= \boldsymbol{\tau}^S \cdot \mathbf{L}_S \;=\; \boldsymbol{\tau}^S \cdot \mathbf{D}_S\,,\\
\boldsymbol{\tau}^S &= \det \mathbf{F}_S\,\mathbf{T}^S\,,
\end{aligned}
\tag{3.52}
$$

where $\boldsymbol{\tau}^S$ is the Kirchhoff stress tensor (weighted Cauchy stress tensor) of the solid material. Furthermore, the solid rate of deformation tensor $\mathbf{D}_S$ must be split into elastic and plastic parts, where, from Table 3 and Table 4, two principally different possibilities seem to be available, at first glance. However, since the solid stress-power must be universally invariant, it is evident that

$$
W^S \;=\; \boldsymbol{\tau}^S \cdot \mathbf{D}_S \;=\; \hat{\boldsymbol{\tau}}^S \cdot (\hat{\boldsymbol{\Gamma}}_S)_p^{\triangle} \;=\; \mathbf{S}^S \cdot (\mathbf{E}_S)'_S\,,
\tag{3.53}
$$

where Table 3 together with the covariant push-forward and pull-back transformations between the Kirchhoff stress $\boldsymbol{\tau}^S$ of the current configuration, the Kirchhoff stress $\hat{\boldsymbol{\tau}}^S$ of the plastic intermediate configuration and the Piola-Kirchhoff stress $\mathbf{S}^S$ of the solid reference configuration have been used, compare Table 5.

Note in passing that the back-stress tensors $\mathbf{Y}^S$, $\hat{\mathbf{Y}}^S$ and $\mathbf{Z}^S$ corresponding to the different solid configurations must be related to each other by use of the same covariant push-forward and pull-back transformations as are used for the different stress measures. For convenience for the reader, both the stress rates and the back stress rates are included into Table 5, where the rates corresponding to the intermediate and the actual configurations are defined within the framework of lower Lie derivatives.

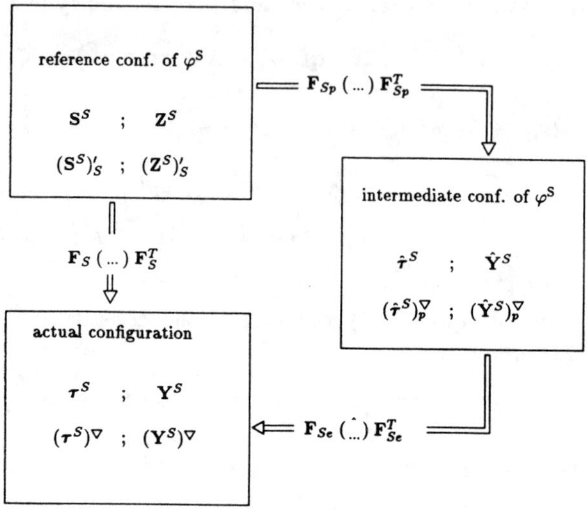

Table 5: Kirchhoff stress tensors, back-stress tensors, Kirchhoff stress rates and back-stress rates of φ^S.

From the preceding considerations, it is immediately concluded that $\mathbf{D}_S$ must be additively decomposed with respect to Table 3. Thus, one obtains

$$\begin{aligned} W^S &= W^{Se} + W^{Sp}, \\ W^{Se} &= \boldsymbol{\tau}^S \cdot (\mathbf{A}_{Se})^{\triangle} = \hat{\boldsymbol{\tau}}^S \cdot (\hat{\boldsymbol{\Gamma}}_{Se})^{\triangle}_p = \mathbf{S}^S \cdot (\mathbf{E}_{Se})'_S, \\ W^{Sp} &= \boldsymbol{\tau}^S \cdot (\mathbf{A}_{Sp})^{\triangle} = \hat{\boldsymbol{\tau}}^S \cdot (\hat{\boldsymbol{\Gamma}}_{Sp})^{\triangle}_p = \mathbf{S}^S \cdot (\mathbf{E}_{Sp})'_S. \end{aligned} \tag{3.54}$$

Hence, in the case of elasto-plastic deformations, the entropy inequality (2.22) of the compressible two-phase model reads

$$\begin{aligned} &-\rho^S(\psi^S)'_S - \rho^F(\psi^F)'_F - \dot{\theta}(\rho^S\eta^S + \rho^F\eta^F) - \mathbf{p}^F \cdot (\acute{\mathbf{x}}_F - \acute{\mathbf{x}}_S) + \\ &\quad + \mathbf{T}^S \cdot [(\mathbf{A}_{Se})^{\triangle} + (\mathbf{A}_{Sp})^{\triangle}] + \mathbf{T}^F \cdot \mathbf{D}_F - \frac{1}{\theta}\mathbf{h} \cdot \operatorname{grad}\theta \geq 0. \end{aligned} \tag{3.55}$$

Analogously, the entropy inequality (3.7) of the hybrid model of first type yields

$$\begin{aligned} &-\rho^S(\psi^S)'_S - \rho^F(\psi^F)'_F - \dot{\theta}(\rho^S\eta^S + \rho^F\eta^F) - (\mathbf{p}^F - \lambda^{FR}\operatorname{grad} n^F) \cdot (\acute{\mathbf{x}}_F - \acute{\mathbf{x}}_S) + \\ &\quad + \mathbf{T}^S \cdot [(\mathbf{A}_{Se})^{\triangle} + (\mathbf{A}_{Sp})^{\triangle}] + (\mathbf{T}^F + n^F\lambda^{FR}\,\mathbf{I}) \cdot \mathbf{D}_F - \frac{1}{\theta}\mathbf{h} \cdot \operatorname{grad}\theta \geq 0, \end{aligned} \tag{3.56}$$

whereas the hybrid model of second type is governed by the modification of inequality (3.14) towards

$$-\rho^S(\psi^S)'_S - \rho^F(\psi^F)'_F - \dot{\theta}(\rho^S\eta^S + \rho^F\eta^F) - (\mathbf{p}^F - \lambda^{SR}\,\mathrm{grad}\,n^F)\cdot(\overset{\prime}{\mathbf{x}}_F - \overset{\prime}{\mathbf{x}}_S) +$$
$$+(\mathbf{T}^S + n^S\lambda^{SR}\,\mathbf{I})\cdot[(\mathbf{A}_{Se})^{\triangle} + (\mathbf{A}_{Sp})^{\triangle}] + \mathbf{T}^F\cdot\mathbf{D}_F - \frac{1}{\theta}\mathbf{h}\cdot\mathrm{grad}\,\theta \geq 0. \quad (3.57)$$

Furthermore, one obtains

$$-\rho^S(\psi^S)'_S - \rho^F(\psi^F)'_F - \dot{\theta}(\rho^S\eta^S + \rho^F\eta^F) - (\mathbf{p}^F - \lambda^{R}\,\mathrm{grad}\,n^F)\cdot(\overset{\prime}{\mathbf{x}}_F - \overset{\prime}{\mathbf{x}}_S) +$$
$$+(\mathbf{T}^S + n^S\lambda^{R}\,\mathbf{I})\cdot[(\mathbf{A}_{Se})^{\triangle} + (\mathbf{A}_{Sp})^{\triangle}] + (\mathbf{T}^F + n^F\lambda^R\,\mathbf{I})\cdot\mathbf{D}_F - \frac{1}{\theta}\mathbf{h}\cdot\mathrm{grad}\,\theta \geq 0 \quad (3.58)$$

as the modified version of inequality (3.18) governing the incompressible model.

Finally, it should be noted that the stress tensors, the back-stress tensors and the different stress and back-stress rates included into Table 5 (lower Lie derivatives) together with the strain measures of Table 1 and the strain rates of Table 3 (upper Lie derivatives), yield a system of dual variables. Concerning the notion "dual variable," the reader is referred to HAUPT & TSAKMAKIS (1989). In addition, since the transport mechanisms between the quantities of Table 5 must be understood in the sense of covariant transformations, the finite character of the model requires, by use of the natural basis systems of the different configurations, both the finite and the rate type constitutive equations of the respective stress and back-stress tensors to reflect a covariant basis representation (concept of covariance). Following this, the dualism of stress, back-stress and strain measures is guaranteed throughout the deformation process.

4. General constitutive modelling

4.1 Preliminary remarks

The basic constitutive approach to binary models in porous media theories as was discussed in the preceding section is sufficient for the determination of the respective sets of response functions, say $\mathcal{R}$. However, it is very well known that these response functions cannot be chosen arbitrarily; instead, they must be compatible with the second law of thermodynamics. Thus, one easily comes to realize that constitutive equations must reflect the restrictions imposed on the different models by the dissipation principle. In particular, thermodynamical restrictions for the binary elasto-plastic models under consideration result from the individual versions of the entropy inequality combined with the basic constitutive assumptions for the respective materials.

Within the present considerations, we proceed from the assumption of plastic incompressibility of the matrix material, that is, we assume in analogy to plasticity theories for single continua that the matrix material itself, in the case of purely plastic deformations, exhibits a constant material density function, while the bulk density can change due to volume fraction changes. Thus, it follows from

$$\rho^S = \rho^S_{oS}\,(\det\mathbf{F}_S)^{-1} \quad (4.1)$$

to be obtained by integration of the mass balance equation of φ^S, eqn (2.16), together with the multiplicative decomposition of deformation gradients defined by eqns (3.24), that

$$\begin{aligned}\rho^S &= \hat{\rho}^S\,(\det\mathbf{F}_{Se})^{-1},\\ \hat{\rho}^S &= \rho^S_{oS}\,(\det\mathbf{F}_{Sp})^{-1}.\end{aligned} \quad (4.2)$$

Furthermore, one concludes from

$$\hat{\rho}^S = \hat{n}^S \, \hat{\rho}^{SR} \tag{4.3}$$

that constituent plastic incompressibility results in

$$\hat{n}^S = n^S_{oS} \, (\det \mathbf{F}_{Sp})^{-1}, \quad \hat{\rho}^{SR} = \rho^{SR}_{oS} = \text{const.} \tag{4.4}$$

In the above equations, the topscript $(\hat{\ldots})$ again characterizes terms of the plastic intermediate configuration of φ^S. Moreover, one obtains by "differentiation" of eqn $(4.4)_1$ with respect to $\mathbf{x}_{Sp}$[1] that

$$\text{grad}_{Sp} \, \hat{n}^S = -\hat{n}^S \, \mathbf{F}^{T-1}_{Sp} (\mathbf{F}^{T-1}_{Sp} \, \overset{3}{\mathbf{G}}_{Sp})^{\underline{1}}, \tag{4.5}$$

where eqn $(3.2)_1$ has been used.

However, it must be noted that eqns (4.4) and (4.5) are valid in both cases compressible and incompressible matrices. Thus, it is proved that the argument given in the introduction to the present article holds, namely the remark that, by use of the concept of constituent plastic incompressibility, the description of the plastic response of porous media does not depend on the remaining compressibility properties of the matrix itself.

Within the framework of porous media theories, the general procedure of deriving thermodynamical restrictions from the entropy inequalities of the respective two-phase models proves to be rather laborious, producing lots of lengthy formulæ that are clearly beyond the scope of the present article. Thus, only the main results are listed below. In addition, the reader who is interested in further details concerning the technical procedure is referred to EHLERS (1989b).

In the remainder of this section, we will continue the discussion for the four different binary media under consideration, separately.

4.2 The compressible two-phase model

(a) General results

The thermodynamocal restrictions governing the constitutive equations of the compressible fluid-saturated elasto-plasic solid matrix result from the entropy inequality (3.55) combined with the constitutive assumptions (3.48). In particular, the following set of statements is obtained from a linear expansion of the entropy inequality about the thermodynamical minimum state defined by

$$\begin{aligned} \mathcal{S}_{cp} \to \mathcal{S}_{cp0} = \{ & \theta, \, \text{grad}\,\theta = \mathbf{0}, \, n^F, \, \text{grad}\, n^F, \, \rho^F, \, \text{grad}\, \rho^F, \\ & \mathbf{F}_{Se}, \overset{3}{\mathbf{G}}_{Se}, \mathbf{F}_{Sp}, \overset{3}{\mathbf{G}}_{Sp}, \, \mathbf{x}'_F - \mathbf{x}'_S = \mathbf{0}, \, \mathbf{D}_F = \mathbf{0} \}. \end{aligned} \tag{4.6}$$

[1] Recall from the discussion included into Subsection 3.1 that the directional derivative with respect to $\mathbf{x}_{Sp}$ is only defined in the particular subsections of the intersected plastic intermediate configuration.

Using standard arguments (BOWEN, 1976), combined with several symmetry and skew-symmetry conditions (EHLERS, 1989b), is it easily proved that the following relations hold:

$$\begin{aligned} \psi^i &= \xi^i\,(\theta,\, n^F,\, \rho^F,\, \mathbf{F}_{Se}, \mathbf{F}_{Sp}), \\ \eta^S &= -\frac{\partial \psi^S}{\partial \theta}, \\ \eta^F &= -\frac{\partial \psi^F}{\partial \theta}; \end{aligned} \tag{4.7}$$

$$\mathbf{q}^S + \mathbf{q}^F = -\,\beta_\vartheta \,\mathrm{grad}\,\theta + \theta\,\alpha_\vartheta\,(\overset{\prime}{\mathbf{x}}_F - \overset{\prime}{\mathbf{x}}_S); \tag{4.8}$$

$$\begin{aligned} \mathbf{T}^S &= \left(\rho^S \frac{\partial \psi^S}{\partial \mathbf{F}_{Se}} + \rho^F \frac{\partial \psi^F}{\partial \mathbf{F}_{Se}}\right) \mathbf{F}^T_{Se}, \\ \mathbf{T}^F &= -\rho^F \left(\rho^S \frac{\partial \psi^S}{\partial \rho^F} + \rho^F \frac{\partial \psi^F}{\partial \rho^F}\right) \mathbf{I} + \mathbf{T}^F_E, \\ \mathbf{T}^F_E &= 2\,\mu^F\,\mathbf{D}_F + \nu^F\,(\mathbf{D}_F \cdot \mathbf{I})\,\mathbf{I}; \end{aligned} \tag{4.9}$$

$$\begin{aligned} \mathbf{p}^F = {} & \rho^S \frac{\partial \psi^S}{\partial n^F}\,\mathrm{grad}\,n^F + \rho^S \frac{\partial \psi^S}{\partial \rho^F}\,\mathrm{grad}\,\rho^F - \mathbf{F}^{T-1}_{Se}\left(\rho^F \frac{\partial \psi^F}{\partial \mathbf{F}_{Se}}\;\overset{3}{\mathbf{G}}_{Se}\right)\overset{1}{} \\ & - \mathbf{F}^{T-1}_{Se}\,\mathbf{F}^{T-1}_{Sp}\left(\rho^F \frac{\partial \psi^F}{\partial \mathbf{F}_{Sp}}\;\overset{3}{\mathbf{G}}_{Sp}\right)\overset{1}{} + \mathbf{p}^F_E, \end{aligned} \tag{4.10}$$

$$\mathbf{p}^F_E = -\,\alpha_\vartheta\,\mathrm{grad}\,\theta - \mathbf{S}_V\,(\overset{\prime}{\mathbf{x}}_F - \overset{\prime}{\mathbf{x}}_S).$$

In the above relations, the quantities $\mathbf{T}^F_E$ and $\mathbf{p}^F_E$ are defined as the "extra" fluid stresses and the "extra" momentum supply, both terms representing the dissipative parts of $\mathbf{T}^F$ and $\mathbf{p}^F$, respectively. Furthermore, in the case of isotropic permeability, the permeability tensor $\mathbf{S}_V$ reduces to

$$\mathbf{S}_V = \frac{(n^F)^2\,\gamma^{FR}}{k^F}\,\mathbf{I}, \tag{4.11}$$

where γ^{FR} is the effective specific weight of the pore fluid and k^F is the coefficient of permeability. In addition,

$$\alpha_\vartheta\,,\,\beta_\vartheta\,,\,\mu^F,\,\nu^F,\,k^F = \mathcal{F}\,(S_{cp0}). \tag{4.12}$$

Therein, α_ϑ characterizes the entropy coupling parameter for the present single temperature model (EHLERS, 1989a, b); β_ϑ is the coefficient of thermal conductivity for the two-phase model under consideration. The quantities μ^F and ν^F define the macroscopic shear and bulk viscosity parameters of the fluid phase. Within the present single temperature model, the relations $(4.7)_{2,3}$ are implied by the existence of the Legendre transform, eqn (2.23). Apart from the above results, the entropy inequality yields two further dissipation mechanisms concerning the constitutive postulates to be defined for the fluid volume fraction rate $(n^F)'_F$ and for the purely plastic rate of deformation tensor $\hat{\mathbf{D}}_{Sp}$ or for any related plastic strain rate, as for example $(\mathbf{A}_{Sp})^{\triangle}$ (compare, Table 3), viz.:

$$- (n^F)'_F \left(\rho^S \frac{\partial \psi^S}{\partial n^F} + \rho^F \frac{\partial \psi^F}{\partial n^F} \right) \geq 0\,, \tag{4.13}$$

$$\begin{aligned} (\mathbf{A}_{Sp})^{\triangle} \cdot \big[\mathbf{T}^S - \mathbf{F}_{Se} & \left(\rho^S \frac{\partial \psi^S}{\partial \mathbf{F}_{Sp}} + \rho^F \frac{\partial \psi^F}{\partial \mathbf{F}_{Sp}} \right) \mathbf{F}^T_{Sp} \mathbf{F}^T_{Se} \\ + (\mathbf{F}_{Se} \mathbf{F}^T_{Se} - \mathbf{I}) & \left(\rho^S \frac{\partial \psi^S}{\partial \mathbf{F}_{Se}} + \rho^F \frac{\partial \psi^F}{\partial \mathbf{F}_{Se}} \right) \mathbf{F}^T_{Se} \big] \geq 0\,. \end{aligned} \tag{4.14}$$

However, note in passing that the dissipation mechanisms (4.13) and (4.14) must be understood in the sense of sufficient conditions.

In order to obtain further results, one concludes from the principle of material frame-indifference in accordance with the dualism between stress and strain measures, compare Section 3.2, that (Ehlers, 1989b)

$$\psi^i (\theta, n^F, \rho^F, \mathbf{F}_{Se}, \mathbf{F}_{Sp}) \;\rightarrow\; \psi^i (\theta, n^F, \rho^F, \mathbf{E}_{Se}, \mathbf{E}_{Sp}), \tag{4.15}$$

that is, the functional dependence of the free energy densities, ψ^S and ψ^F, on the variables $\mathbf{F}_{Se}$ and $\mathbf{F}_{Sp}$ must be expressible by the variables $\mathbf{E}_{Se}$ and $\mathbf{E}_{Sp}$. In addition, it is seen from eqn (3.36) and Table 1, that

$$\begin{aligned} \mathbf{E}_{Se} &= \mathbf{E}_{Se} (\mathbf{F}_{Se}\,, \mathbf{F}_{Sp})\,, \\ \mathbf{E}_{Sp} &= \mathbf{E}_{Sp} (\mathbf{F}_{Sp})\,. \end{aligned} \tag{4.16}$$

Thus, the chain rule yields (additionally, compare the Appendix)

$$\begin{aligned} \frac{\partial \psi^i}{\partial \mathbf{F}_{Se}} &= \mathbf{F}_{Se} \mathbf{F}_{Sp} \frac{\partial \psi^i}{\partial \mathbf{E}_{Se}} \mathbf{F}^T_{Sp}\,, \\ \frac{\partial \psi^i}{\partial \mathbf{F}_{Sp}} &= (\mathbf{F}^T_{Se} \mathbf{F}_{Se} - \mathbf{I}) \mathbf{F}_{Sp} \frac{\partial \psi^i}{\partial \mathbf{E}_{Se}} + \mathbf{F}_{Sp} \frac{\partial \psi^i}{\partial \mathbf{E}_{Sp}}\,. \end{aligned} \tag{4.17}$$

By use of the Kirchhoff stress tensor $\boldsymbol{\tau}^S = \det \mathbf{F}_S \, \mathbf{T}^S$, compare eqn (3.52), eqns $(4.9)_1$ and $(4.17)_1$ combine to

$$\boldsymbol{\tau}^S = \det \mathbf{F}_S \; \mathbf{F}_S \left(\rho^S \frac{\partial \psi^S}{\partial \mathbf{E}_{Se}} + \rho^F \frac{\partial \psi^F}{\partial \mathbf{E}_{Se}} \right) \mathbf{F}^T_S. \tag{4.18}$$

Alternatively,

$$\boldsymbol{\tau}^S = \det \mathbf{F}_S \left(\rho^S \frac{\partial \psi^S}{\partial \mathbf{A}_{Se}} + \rho^F \frac{\partial \psi^F}{\partial \mathbf{A}_{Se}} \right) \tag{4.19}$$

holds, where Table 1 has been used.

Similarly, with the aid of eqns (4.17), the dissipation mechanism (4.14) reads

$$(\mathbf{A}_{Sp})^{\triangle} \cdot [\mathbf{T}^S - \mathbf{F}_S \left(\rho^S \frac{\partial \psi^S}{\partial \mathbf{E}_{Sp}} + \rho^F \frac{\partial \psi^F}{\partial \mathbf{E}_{Sp}} \right) \mathbf{F}_S^T] \geq 0 \,. \tag{4.20}$$

Therein, as was pointed by the author (EHLERS, 1989a, b), the second term in square brackets may be interpreted to represent the constitutive functional for a finite representation of the back-stress tensor of the kinematically hardening second-grade solid material. In particular, it is suggested that

$$\mathbf{Y}^S = \det \mathbf{F}_S \; \mathbf{F}_S \left(\rho^S \frac{\partial \psi^S}{\partial \mathbf{E}_{Sp}} + \rho^F \frac{\partial \psi^F}{\partial \mathbf{E}_{Sp}} \right) \mathbf{F}_S^T . \tag{4.21}$$

Equivalently, one obtains

$$\mathbf{Y}^S = \det \mathbf{F}_S \left(\rho^S \frac{\partial \psi^S}{\partial \mathbf{A}_{Sp}} + \rho^F \frac{\partial \psi^F}{\partial \mathbf{A}_{Sp}} \right) . \tag{4.22}$$

However, in any case, the path-dependence in plasticity yields the values of ψ^i and $\mathbf{Y}^S$ to depend strongly on the plastic loading history.

The following remarks concern the case of single solid continua, where an inequality of type (4.20) was firstly suggested by GREEN & NAGHDI (1965). In this article, Green and Naghdi gave the interpretation that $\mathbf{Y}^S$ "*... may be regarded as a point in stress space*," where the yield function is "*star-shaped with respect to this point.*" Later, PHILLIPS & EISENBERG (1966) showed that $\mathbf{Y}^S$ "*which is associated with the state of plastic strain must always be within or on the yield surface.*" In addition, $\mathbf{Y}^S$ was interpreted as the "*thermodynamical reference stress.*" In a subsequent paper, PHILLIPS (1974) concluded that $\mathbf{Y}^S$ is probably located at the center of the isothermal yield function. This center, in a modern terminology, is defined by the back-stress tensor. Further remarks can be made on a paper by CABOCHE (1983) who assumed the solid free energy to depend on an internal state variable $\boldsymbol{\alpha}$ rather than to depend on the plastic strain measure $\mathbf{E}_{Sp}$. By use of the variable $\boldsymbol{\alpha}$, Chaboche concluded the back-stress tensor to be characterized by the gradient of ψ^S into the direction of $\boldsymbol{\alpha}$. Thus, by identification of $\boldsymbol{\alpha}$ with the internal state variable $\mathbf{E}_{Sp}$, the above interpretation and the interpretation of Chaboche coincide.

However, by use of the different push-forward and pull-back transformations of strain, stress and back-stress measures (compare Table 1, Table 3 and Table 5), each solid configuration (reference, plastic intermediate and actual) is assigned a constitutive relation for the solid stress and back-stress tensors which can be related to each other by covariant transport mechanisms, compare Table 6.

reference conf. of φ^S

$$\mathbf{S}_E^S = \det \mathbf{F}_S \left(\rho^S \frac{\partial \psi^S}{\partial \mathbf{E}_{Se}} + \rho^F \frac{\partial \psi^F}{\partial \mathbf{E}_{Se}} \right)$$

$$\mathbf{Z}^S = \det \mathbf{F}_S \left(\rho^S \frac{\partial \psi^S}{\partial \mathbf{E}_{Sp}} + \rho^F \frac{\partial \psi^F}{\partial \mathbf{E}_{Sp}} \right)$$

$\Longrightarrow \mathbf{F}_{Sp}\,(\ldots)\,\mathbf{F}_{Sp}^T \Longrightarrow$

intermediate conf. of φ^S

$$\hat{\boldsymbol{\tau}}_E^S = \det \mathbf{F}_S \left(\rho^S \frac{\partial \psi^S}{\partial \hat{\boldsymbol{\Gamma}}_{Se}} + \rho^F \frac{\partial \psi^F}{\partial \hat{\boldsymbol{\Gamma}}_{Se}} \right)$$

$$\hat{\mathbf{Y}}^S = \det \mathbf{F}_S \left(\rho^S \frac{\partial \psi^S}{\partial \hat{\boldsymbol{\Gamma}}_{Sp}} + \rho^F \frac{\partial \psi^F}{\partial \hat{\boldsymbol{\Gamma}}_{Sp}} \right)$$

$\Longleftarrow \mathbf{F}_{Se}\,(\ldots)\,\mathbf{F}_{Se}^T \Longleftarrow$

$\Downarrow \mathbf{F}_S\,(\ldots)\,\mathbf{F}_S^T$ (from reference conf.)

actual configuration

$$\boldsymbol{\tau}_E^S = \det \mathbf{F}_S \left(\rho^S \frac{\partial \psi^S}{\partial \mathbf{A}_{Se}} + \rho^F \frac{\partial \psi^F}{\partial \mathbf{A}_{Se}} \right)$$

$$\mathbf{Y}^S = \det \mathbf{F}_S \left(\rho^S \frac{\partial \psi^S}{\partial \mathbf{A}_{Sp}} + \rho^F \frac{\partial \psi^F}{\partial \mathbf{A}_{Sp}} \right)$$

Table 6: Basic relations between stress tensors and back-stress tensors of φ^S.

Furthermore, it is easily proved with the aid of eqns (3.52) and (4.21) that the plastic dissipation mechanism, inequality (4.20), is universally invariant. Thus, one obtains with the condition $\det \mathbf{F}_S \geq 0$:

$$(\mathbf{A}_{Sp})^{\triangle} \cdot (\boldsymbol{\tau}^S - \mathbf{Y}^S) = (\hat{\boldsymbol{\Gamma}}_{Sp})_p^{\triangle} \cdot (\hat{\boldsymbol{\tau}}^S - \hat{\mathbf{Y}}^S) = (\mathbf{E}_{Sp})'_S \cdot (\mathbf{S}^S - \mathbf{Z}^S) \geq 0\,. \quad (4.23)$$

From the preceding relations together with the balance equations defined in Section 2, the basic approach to compressible fluid-saturated elasto-plastic porous solid matrices is properly defined. Finally, note in passing that, for the single temperature model under consideration, it was not necessary to find constitutive equations for the energy supply terms e^i, since the temperature variation for the whole system must be calculated from the sum of the energy balance equations (2.19).

Concerning constitutive equations for the free energy functionals ψ^i and the fluid volume fraction rate $(n^F)'_F$ together with a convenient yield criterion and a flow rule for the purely plastic rate of deformation tensor $\hat{\mathbf{D}}_{Sp}$, the reader is referred to the following sections.

(b) Simplified constitutive model

Considering applications of the above theory to certain problems of applied engineering practice, it would appear that some simplifications are necessary. Since the relations for the fluid stress tensor, eqn $(4.9)_2$, the momentum supply term, eqn $(4.10)_1$, and the solid stress and back-stress tensors (compare Table 6) depend on the "choice" of the free energy functionals ψ^i, it seems to be reasonable to introduce simplifications with respect to these quantities. By use of a "principle of constituent separation" (EHLERS, 1989a) implying each free energy functional of a given constituent to be independent of those variables belonging to the respective other constituent, it follows with the aid of the relation (4.15) that the general thermodynamical restriction, eqn $(4.7)_1$, can be simplified towards

$$\begin{aligned} \psi^S(\theta, n^F, \rho^F, \mathbf{F}_{Se}, \mathbf{F}_{Sp}) &\rightarrow \psi^S(\theta, n^S, \mathbf{E}_{Se}, \mathbf{E}_{Sp}), \\ \psi^F(\theta, n^F, \rho^F, \mathbf{F}_{Se}, \mathbf{F}_{Sp}) &\rightarrow \psi^F(\theta, n^F, \rho^F), \end{aligned} \tag{4.24}$$

where, in the relation $(4.24)_1$, n^F was substituted by n^S via $n^S = 1 - n^F$. The quantity n^F or n^S, respectively, obviously represents the volume fraction variable of both the solid and the fluid materials, thus exhibiting the necessary kinematic coupling mechanism between the constituents. Following this, it is seen that "constituent separation" does not include "kinematic separation" for the compressible two-phase model which includes four independent kinematic variables, two solid variables ($\mathbf{F}_{Se}$ and $\mathbf{F}_{Sp}$), one fluid variable (ρ^F) and the coupling variable (n^F or n^S, respectively). Thus, one obtains from Table 5 and eqns $(4.9)_2$, $(4.10)_1$, (4.18) and (4.21) that

$$\begin{aligned} \mathbf{T}^F &= -(\rho^F)^2 \frac{\partial \psi^F}{\partial \rho^F} \mathbf{I} + \mathbf{T}_E^F, \\ \mathbf{p}^F &= \rho^S \frac{\partial \psi^S}{\partial n^F} \operatorname{grad} n^F + \mathbf{p}_E^F, \\ \boldsymbol{\tau}^S &= \det \mathbf{F}_S \ \mathbf{F}_S \left(\rho^S \frac{\partial \psi^S}{\partial \mathbf{E}_{Se}}\right) \mathbf{F}_S^T, \\ \hat{\mathbf{Y}}^S &= \det \mathbf{F}_S \ \mathbf{F}_{Sp} \left(\rho^S \frac{\partial \psi^S}{\partial \mathbf{E}_{Sp}}\right) \mathbf{F}_{Sp}^T, \end{aligned} \tag{4.25}$$

where further representations of eqns $(4.25)_3$ and $(4.25)_4$ can easily be obtained with the aid of Table 6.

In addition to the above results, it should be noted that the dissipation mechanism governing the fluid volume fraction rate, inequality (4.13), remains unchanged. Thus, the relation

$$-(n^F)'_F \left(\rho^S \frac{\partial \psi^S}{\partial n^F} + \rho^F \frac{\partial \psi^F}{\partial n^F}\right) \geq 0 \tag{4.26}$$

still holds.

Considering the representations of $\mathbf{T}^F$ and $\boldsymbol{\tau}^S$, several convenient modifications are possible. For instance, it is seen from the relation $(4.24)_2$ that a function $\psi^F(\theta, n^F, \rho^F)$ may be given by the representation[2]

$$\psi^F(\theta, n^F, \rho^F) \rightarrow \psi^F(\theta, n^F, \rho^{FR}). \tag{4.27}$$

[2] more precisely, one would have to distinguish between the two functions $\psi^F(\theta, n^F, \rho^F)$ and $\tilde{\psi}^F(\theta, n^F, \rho^{FR})$ by use of an additional tilde.

Thus, it is concluded from the chain rule that

$$\frac{\partial \psi^F}{\partial \rho^F} = \frac{1}{\rho^{FR}} \frac{\partial \psi^F}{\partial n^F} + \frac{1}{n^F} \frac{\partial \psi^F}{\partial \rho^{FR}}, \tag{4.28}$$

where the definition (2.3) was used. Given eqn (4.28), the fluid stress tensor, eqn $(4.25)_1$, yields

$$\mathbf{T}^F = -n^F \left(\rho^{FR} n^F \frac{\partial \psi^F}{\partial n^F} + (\rho^{FR})^2 \frac{\partial \psi^F}{\partial \rho^{FR}} \right) \mathbf{I} + \mathbf{T}^F_E. \tag{4.29}$$

Therein, the terms

$$\begin{aligned} p^{FR}_{\text{con.}} &= \rho^{FR} n^F \frac{\partial \psi^F}{\partial n^F}, \\ p^{FR}_{\text{mat.}} &= (\rho^{FR})^2 \frac{\partial \psi^F}{\partial \rho^{FR}} \end{aligned} \tag{4.30}$$

are defined as the fluid configuration pressure, $p^{FR}_{\text{con.}}$, and the fluid phase (or material) pressure, $p^{FR}_{\text{mat.}}$, respectively (additionally, compare BAER & NUNZIATO, 1986).

Concerning further representations of the constitutive postulates for the solid Kirchhoff stress tensor $\boldsymbol{\tau}^S$, eqn $(4.25)_3$, one obtains from the chain rule that

$$\frac{\partial \psi^S}{\partial \mathbf{E}_{Se}} = 2\, \mathbf{F}_{Sp}^{-1} \frac{\partial \psi^S}{\partial \hat{\mathbf{C}}_{Se}} \mathbf{F}_{Sp}^{T-1}, \tag{4.31}$$

where, from Table 1, $\mathbf{E}_{Se} = \mathbf{E}_{Se}(\hat{\mathbf{C}}_{Se}, \mathbf{F}_{Sp})$ has been used. In addition, it is convenient to proceed with the following multiplicative split of $\mathbf{F}_{Se}$, viz.:

$$\begin{aligned} \mathbf{F}_{Se} &= \mathbf{F}_{Se\,sph} \overline{\mathbf{F}}_{Se}, \\ \mathbf{F}_{Se\,sph} &= (\det \mathbf{F}_{Se})^{1/3}\, \mathbf{I}. \end{aligned} \tag{4.32}$$

Therein, $\mathbf{F}_{Se\,sph}$ is the spherical part of $\mathbf{F}_{Se}$, whereas $\overline{\mathbf{F}}_{Se}$, satisfying $\det \overline{\mathbf{F}}_{Se} = 1$, represents the respective volume-preserving part.

Thus, it is seen that the elastic state variable, $\mathbf{E}_{Se}$, included into the free energy functional ψ^S can be expressed via

$$\mathbf{E}_{Se} = \mathbf{E}_{Se}(\overline{\mathbf{C}}_{Se}, \det \mathbf{F}_{Se}, \mathbf{F}_{Sp}), \tag{4.33}$$

where, from eqns (3.36) and (4.32),

$$\hat{\mathbf{C}}_{Se} = (\det \mathbf{F}_{Se})^{2/3}\, \overline{\mathbf{C}}_{Se}, \qquad \overline{\mathbf{C}}_{Se} = \overline{\mathbf{F}}^T_{Se} \overline{\mathbf{F}}_{Se} \tag{4.34}$$

has been used.

Given the above formulæ, one now obtains from the chain rule that

$$\begin{aligned} \boldsymbol{\tau}^S = \det \mathbf{F}_S \Big[& 2\rho^S (\det \mathbf{F}_{Se})^{-2/3} \Big(\mathbf{F}_{Se} \frac{\partial \psi^S}{\partial \overline{\mathbf{C}}_{Se}} \mathbf{F}^T_{Se} \Big)^D \\ & + \rho^S \det \mathbf{F}_{Se} \frac{\partial \psi^S}{\partial \det \mathbf{F}_{Se}} \mathbf{I} \Big], \end{aligned} \tag{4.35}$$

eqn (4.35) thus substituting eqn $(4.25)_3$.

Furthermore, it is easy to show with the aid of eqns (4.1)–(4.3) that

$$\det \mathbf{F}_{Se} \, \frac{\partial \psi^S}{\partial \det \mathbf{F}_{Se}} = -\rho^S \, \frac{\partial \psi^S}{\partial \rho^S} \,, \tag{4.36}$$

and thus, in analogy to eqn (4.28),

$$\frac{\partial \psi^S}{\partial \rho^S} = \frac{1}{\rho^{SR}} \frac{\partial \psi^S}{\partial n^S} + \frac{1}{n^S} \frac{\partial \psi^S}{\partial \rho^{SR}} \,. \tag{4.37}$$

Following this, eqns (4.35) and (4.37) finally combine to

$$\begin{aligned} \boldsymbol{\tau}^S &= \det \mathbf{F}_S \, [\, 2\, \rho^S (\det \mathbf{F}_{Se})^{-2/3} (\, \mathbf{F}_{Se} \, \frac{\partial \psi^S}{\partial \overline{\mathbf{C}}_{Se}} \, \mathbf{F}^T_{Se} \,)^D \\ &\quad - n^S \left(\rho^{SR} \, n^S \, \frac{\partial \psi^S}{\partial n^S} + (\rho^{SR})^2 \, \frac{\partial \psi^S}{\partial \rho^{SR}} \right) \mathbf{I} \,] \,. \end{aligned} \tag{4.38}$$

Therein, in analogy to eqns (4.30),

$$\begin{aligned} p^{SR}_{\text{con.}} &= \rho^{SR} \, n^S \, \frac{\partial \psi^S}{\partial n^S} \,, \\ p^{SR}_{\text{mat.}} &= (\rho^{SR})^2 \, \frac{\partial \psi^S}{\partial \rho^{SR}} \end{aligned} \tag{4.39}$$

define the solid configuration (matrix or structural) pressure, $p^{SR}_{\text{con.}}$, and the solid phase (or material) pressure, $p^{SR}_{\text{mat.}}$, respectively.

However, concerning the different representations of the back-stress tensor, $\mathbf{Y}^S$ or $\hat{\mathbf{Y}}^S$, respectively, one obtains by use of the same procedure as was relating $\boldsymbol{\tau}^S$ from eqn $(4.25)_3$ to eqn (4.35) that

$$\begin{aligned} \hat{\mathbf{Y}}^S &= \det \mathbf{F}_S \, [\, 2\, \rho^S (\det \mathbf{F}_{Sp})^{-2/3} (\, \mathbf{F}_{Sp} \, \frac{\partial \psi^S}{\partial \overline{\mathbf{C}}_{Sp}} \, \mathbf{F}^T_{Sp} \,)^D \\ &\qquad + \rho^S \det \mathbf{F}_{Sp} \, \frac{\partial \psi^S}{\partial \det \mathbf{F}_{Sp}} \, \mathbf{I} \,] \,, \end{aligned} \tag{4.40}$$

where eqn (3.36) and Table 1 and Table 6 together with the multiplicative split

$$\begin{aligned} \mathbf{F}_{Sp} &= \mathbf{F}_{Sp\,sph} \, \overline{\mathbf{F}}_{Sp} \,, \\ \mathbf{F}_{Sp\,sph} &= (\det \mathbf{F}_{Sp} \,)^{1/3} \, \mathbf{I} \,. \end{aligned} \tag{4.41}$$

have been used. In eqn (4.41), analogously to (4.32), $\mathbf{F}_{Sp\,sph}$ is the spherical part of $\mathbf{F}_{Sp}$, whereas $\overline{\mathbf{F}}_{Sp}$, satisfying $\det \overline{\mathbf{F}}_{Sp} = 1$, represents the respective volume-preserving part.

However, concerning the dissipation mechanism governing the volume fraction rate of the fluid constituent, it is easily seen that the relations (4.26), (4.30) and (4.39) combine to

$$(n^F)'_F \, (p^{SR}_{\text{con.}} - p^{FR}_{\text{con.}}) \geq 0 \,, \tag{4.42}$$

where

$$\frac{\partial \psi^S}{\partial n^F} = -\frac{\partial \psi^S}{\partial n^S} \tag{4.43}$$

to be obtained with the aid of (2.15) has been used. Thus, it is finally concluded from the above dissipation mechanism, that

$$(n^F)'_F = \gamma_c \, (p^{SR}_{\text{con.}} - p^{FR}_{\text{con.}}) \, , \tag{4.44}$$

where

$$\gamma_c = \gamma_c(\mathcal{S}_c) \geq 0 \, , \tag{4.45}$$

is a material function, governing the dependence of the volume fraction rate, $(n^F)'_F$, by the difference of the solid and the liquid configuration pressure terms.

Further possible simplifications of the theory can be introduced in analogy to a suggestion originally proposed by GREEN & NAGHDI (1965). In the present context, this simplification reads

$$\psi^S(\theta, n^S, \mathbf{E}_{Se}, \mathbf{E}_{Sp}) = \psi^{Se}(\theta, n^S, \mathbf{E}_{Se}) + \psi^{Sp}(\mathbf{E}_{Sp}), \tag{4.46}$$

where ψ^S from $(4.24)_1$ was used. Hence, assuming the plastic deformation history to be included into the functional ψ^{Sp}, the free energy ψ^{Se} can be interpreted as the thermoelastic potential function. Finally, note in passing that the solid volume fraction n^S was not included into the functional ψ^{Sp} what, by the way, is directly concluded from the remarks given in Subsection 4.1.

However, it is seen from the above definitions and comments, that the simplified constitutive model must be understood as the most convenient tool for applications to compressible fluid-saturated elasto-plastic solid matrices.

4.3 The hybrid model of first type

(a) General results

The thermodynamical restrictions governing the constitutive equations of the hybrid model of first type (compressible elasto-plastic solid matrix saturated by an incompressible viscous pore liquid) result from the entropy inequality (3.56) together with the constitutive assumptions (3.49). In analogy to the preceding section, the following set of statements is obtained from a linear expansion of the entropy inequality about the thermodynamical minimum state which is now defined by

$$\begin{aligned} \mathcal{S}_{1p} \to \mathcal{S}_{1p0} = \{ & \theta, \operatorname{grad}\theta = \mathbf{0}, n^F, \operatorname{grad} n^F, \\ & \mathbf{F}_{Se}, \overset{3}{\mathbf{G}}_{Se}, \mathbf{F}_{Sp}, \overset{3}{\mathbf{G}}_{Sp}, \mathbf{x}'_F - \mathbf{x}'_S = \mathbf{0}, \mathbf{D}_F = \mathbf{0} \}. \end{aligned} \tag{4.47}$$

Proceeding from the same arguments as were used to conclude to eqns (4.7)–(4.10), it can be shown that the following relations hold:

$$
\begin{aligned}
\psi^i &= \xi^i\,(\theta,\, n^F,\, \mathbf{F}_{Se}, \mathbf{F}_{Sp}), \\
\eta^S &= -\frac{\partial \psi^S}{\partial \theta}\,, \\
\eta^F &= -\frac{\partial \psi^F}{\partial \theta}\,;
\end{aligned}
\tag{4.48}
$$

$$
\mathbf{q}^S + \mathbf{q}^F \;=\; -\,\beta_\vartheta\,\operatorname{grad}\theta \,+\, \theta\,\alpha_\vartheta\,(\overset{\prime}{\mathbf{x}}_F \,-\, \overset{\prime}{\mathbf{x}}_S); \tag{4.49}
$$

$$
\begin{aligned}
\mathbf{T}^S &= \left(\rho^S\,\frac{\partial \psi^S}{\partial \mathbf{F}_{Se}} + \rho^F\,\frac{\partial \psi^F}{\partial \mathbf{F}_{Se}}\right)\mathbf{F}^T_{Se}, \\
\mathbf{T}^F &= -n^F\,\lambda^{FR}\,\mathbf{I} - n^F\left(\rho^S\,\frac{\partial \psi^S}{\partial n^F} + \rho^F\,\frac{\partial \psi^F}{\partial n^F}\right)\mathbf{I} + \mathbf{T}^F_E\,, \\
\mathbf{T}^F_E &= 2\,\mu^F\,\mathbf{D}_F \,+\, \nu^F\,(\,\mathbf{D}_F\cdot\mathbf{I}\,)\,\mathbf{I}\,;
\end{aligned}
\tag{4.50}
$$

$$
\begin{aligned}
\mathbf{p}^F &= \lambda^{FR}\operatorname{grad} n^F \,+\, \rho^S\,\frac{\partial \psi^S}{\partial n^F}\operatorname{grad} n^F \,-\, \mathbf{F}^{T-1}_{Se}\left(\rho^F\,\frac{\partial \psi^F}{\partial \mathbf{F}_{Se}}\;\overset{3}{\mathbf{G}}_{Se}\right)\overset{1}{} \\
&\quad -\,\mathbf{F}^{T-1}_{Se}\,\mathbf{F}^{T-1}_{Sp}\left(\rho^F\,\frac{\partial \psi^F}{\partial \mathbf{F}_{Sp}}\;\overset{3}{\mathbf{G}}_{Sp}\right)\overset{1}{} \,+\, \mathbf{p}^F_E\,, \\
\mathbf{p}^F_E &= -\,\alpha_\vartheta\,\operatorname{grad}\theta \,-\, \mathbf{S}_V\,(\overset{\prime}{\mathbf{x}}_F \,-\, \overset{\prime}{\mathbf{x}}_S)\,.
\end{aligned}
\tag{4.51}
$$

Note in passing that, in the case of isotropic permeability, the permeability tensor $\mathbf{S}_V$ is again governed by eqn (4.11). Furthermore, the material parameter functions of the hybrid model of first type depend on the set $\mathcal{S}_{1po}$, viz.:

$$
\alpha_\vartheta\,,\, \beta_\vartheta\,,\, \mu^F,\, \nu^F,\, k^F \;=\; \mathcal{F}\,(\mathcal{S}_{1po})\,. \tag{4.52}
$$

Apart from the above results, the entropy inequality contains two further mechanisms which must be understood in the sense of sufficient conditions:

$$
-\,\lambda^{FR}\,(n^S)'_S \geq 0\,, \tag{4.53}
$$

$$
(\mathbf{A}_{Sp})^{\triangle}\cdot\,[\mathbf{T}^S - \mathbf{F}_S\left(\rho^S\,\frac{\partial \psi^S}{\partial \mathbf{E}_{Sp}} + \rho^F\,\frac{\partial \psi^F}{\partial \mathbf{E}_{Sp}}\right)\mathbf{F}^T_S\,] \geq 0\,. \tag{4.54}
$$

It is seen from the above relations that the inequality (4.53) restricts the values of the Lagrangian multiplier λ^{FR}, whereas the inequality (4.54), obviously identical with the relation

(4.20) of the compressible model, again represents the dissipation mechanism governing the plastic rate of deformation tensor. However, note in passing that the above representation of the dissipation mechanism, inequality (4.54), proceeds from the principle of material frame-indifference leading to

$$\psi^i\,(\theta, n^F, \mathbf{F}_{Se}, \mathbf{F}_{Sp}) \;\rightarrow\; \psi^i\,(\theta, n^F, \mathbf{E}_{Se}, \mathbf{E}_{Sp}), \tag{4.55}$$

and, additionally, makes use of eqns (4.16) and (4.17). For the hybrid model of first type, it is furthermore concluded that the equations (4.18), (4.19), (4.21) and (4.22) governing the Kirchhoff solid stress tensor and the back-stress tensor of the compressible model still hold. Hence, for example,

$$\boldsymbol{\tau}^S = \det \mathbf{F}_S \;\mathbf{F}_S \left(\rho^S \frac{\partial \psi^S}{\partial \mathbf{E}_{Se}} + \rho^F \frac{\partial \psi^F}{\partial \mathbf{E}_{Se}}\right) \mathbf{F}_S^T, \tag{4.56}$$

$$\mathbf{Y}^S = \det \mathbf{F}_S \;\mathbf{F}_S \left(\rho^S \frac{\partial \psi^S}{\partial \mathbf{E}_{Sp}} + \rho^F \frac{\partial \psi^F}{\partial \mathbf{E}_{Sp}}\right) \mathbf{F}_S^T. \tag{4.57}$$

Following this, it is easily seen with the aid of Table 3 and Table 6 that the universal invariance of the plastic dissipation mechanism is guaranteed throughout the deformation process, additionally, compare the relations (4.23). Thus,

$$(\mathbf{A}_{Sp})^{\triangle} \cdot (\boldsymbol{\tau}^S - \mathbf{Y}^S) = (\hat{\boldsymbol{\Gamma}}_{Sp})^{\triangle}_{p} \cdot (\hat{\boldsymbol{\tau}}^S - \hat{\mathbf{Y}}^S) = (\mathbf{E}_{Sp})'_S \cdot (\mathbf{S}^S - \mathbf{Z}^S) \geq 0\,. \tag{4.58}$$

From the preceding relations, the general constitutive approach to the hybrid model of first type is properly defined.

(b) Simplified hybrid model of first type

As was described in Subsection 4.2 (b), the simplified constitutive models make use of the "principle of constituent separation" implying each free energy functional of a given constituent to be independent of the variables of the respective other constituent. Thus, it is concluded by combination of the relations $(4.48)_1$ and (4.55) that

$$\begin{aligned} \psi^S\,(\theta, n^F, \mathbf{F}_{Se}, \mathbf{F}_{Sp}) &\;\rightarrow\; \psi^S\,(\theta, n^S, \mathbf{E}_{Se}, \mathbf{E}_{Sp}), \\ \psi^F\,(\theta, n^F, \mathbf{F}_{Se}, \mathbf{F}_{Sp}) &\;\rightarrow\; \psi^F\,(\theta, n^F), \end{aligned} \tag{4.59}$$

where n^F or $n^S = 1 - n^F$, respectively, again represents the volume fraction variable of both the solid and the liquid materials acting upon the constituents in the sense of a kinematic coupling mechanism between the phases of the hybrid model under consideration. Following this, one obtains from eqns $(4.50)_2$, $(4.51)_1$, (4.56) and (4.57) that

$$\begin{aligned}
\mathbf{T}^F &= -n^F\,[\,(\lambda^{FR} + \rho^S \frac{\partial \psi^S}{\partial n^F}) + \rho^{FR} n^F \frac{\partial \psi^F}{\partial n^F}\,]\,\mathbf{I} + \mathbf{T}^F_E\,,\\
\mathbf{p}^F &= (\lambda^{FR} + \rho^S \frac{\partial \psi^S}{\partial n^F})\,\mathrm{grad}\, n^F + \mathbf{p}^F_E\,,\\
\boldsymbol{\tau}^S &= \det \mathbf{F}_S\ \mathbf{F}_S \left(\rho^S \frac{\partial \psi^S}{\partial \mathbf{E}_{Se}}\right) \mathbf{F}^T_S,\\
\hat{\mathbf{Y}}^S &= \det \mathbf{F}_S\ \mathbf{F}_{Sp} \left(\rho^S \frac{\partial \psi^S}{\partial \mathbf{E}_{Sp}}\right) \mathbf{F}^T_{Sp}.
\end{aligned} \tag{4.60}$$

Additionally, by use of eqns $(4.39)_1$ and (4.43), the fluid stress tensor reads

$$\mathbf{T}^F = -n^F\,[\,(\lambda^{FR} - p^{SR}_{\mathrm{con.}}) + p^{FR}_{\mathrm{con.}}\,]\,\mathbf{I} + \mathbf{T}^F_E\,. \tag{4.61}$$

By comparison of the fluid stress tensors of the compressible and the hybrid model of first type, eqns (4.29) and (4.61), it is observed that the phase pressure term, $p^{FR}_{\mathrm{mat.}}$, included into eqn (4.29) is substituted, concerning the above relation, by the Lagrangian multiplier λ^{FR} minus the solid configuration pressure, $p^{SR}_{\mathrm{con.}}$.

Apart from the above results, further modifications of the solid Kirchhoff stress tensor, eqn $(4.60)_3$, and the back-stress tensor, eqn $(4.60)_4$, respectively, can directly be taken from the preceding compressible model. Within the framework of the simplified constitutive theory, this fact is easily concluded from the identical representations of the solid free energy functionals, compare the relations $(4.24)_1$ and $(4.59)_1$. Thus, eqns (4.35), (4.38) and (4.40) hold, viz.:

$$\begin{aligned}
\boldsymbol{\tau}^S &= \det \mathbf{F}_S\,[\,2\,\rho^S (\det \mathbf{F}_{Se})^{-2/3} (\mathbf{F}_{Se} \frac{\partial \psi^S}{\partial \overline{\mathbf{C}}_{Se}} \mathbf{F}^T_{Se})^D + \rho^S \det \mathbf{F}_{Se} \frac{\partial \psi^S}{\partial \det \mathbf{F}_{Se}}\,\mathbf{I}\,],\\
\boldsymbol{\tau}^S &= \det \mathbf{F}_S\,[\,2\,\rho^S (\det \mathbf{F}_{Se})^{-2/3} (\mathbf{F}_{Se} \frac{\partial \psi^S}{\partial \overline{\mathbf{C}}_{Se}} \mathbf{F}^T_{Se})^D\\
&\qquad - n^S \left(\rho^{SR} n^S \frac{\partial \psi^S}{\partial n^S} + (\rho^{SR})^2 \frac{\partial \psi^S}{\partial \rho^{SR}}\right) \mathbf{I}\,],\\
\hat{\mathbf{Y}}^S &= \det \mathbf{F}_S\,[\,2\,\rho^S (\det \mathbf{F}_{Sp})^{-2/3} (\mathbf{F}_{Sp} \frac{\partial \psi^S}{\partial \overline{\mathbf{C}}_{Sp}} \mathbf{F}^T_{Sp})^D + \rho^S \det \mathbf{F}_{Sp} \frac{\partial \psi^S}{\partial \det \mathbf{F}_{Sp}}\,\mathbf{I}\,].
\end{aligned} \tag{4.62}$$

Finally, in discussing the simplified constitutive theory of elasto-plastic hybrid models of first type, the separation of the functional ψ^S from (4.59) into thermoelastic and plastic parts yields

$$\psi^S(\theta, n^S, \mathbf{E}_{Se}, \mathbf{E}_{Sp}) = \psi^{Se}(\theta, n^S, \mathbf{E}_{Se}) + \psi^{Sp}(\mathbf{E}_{Sp}), \tag{4.63}$$

where eqn (4.63) obviously equals the respective relation of the compressible elasto-plastic two-phase model, eqn (4.46).

4.4 The hybrid model of second type

(a) General results

The thermodynamical restrictions governing the constitutive equations of the hybrid model of second type (incompressible elasto-plastic solid matrix saturated by a compressible viscous pore fluid) result from the entropy inequality (3.57) together with the constitutive assumptions (3.50). Thus, similarly to the preceding section, the following set of statements is obtained from a linear expansion of the entropy inequality about the thermodynamical minimum state defined by

$$\begin{aligned} \mathcal{S}_{2p} \rightarrow \mathcal{S}_{2p\,0} \;\; = \;\; & \{\theta,\ \operatorname{grad}\theta = \mathbf{0},\ \rho^F,\ \operatorname{grad}\rho^F, \\ & \mathbf{F}_{Se},\ \overset{3}{\mathbf{G}}_{Se},\ \mathbf{F}_{Sp},\ \overset{3}{\mathbf{G}}_{Sp},\ \mathbf{x}'_F - \mathbf{x}'_S = \mathbf{0},\ \mathbf{D}_F = \mathbf{0}\}. \end{aligned} \qquad (4.64)$$

Proceeding from the same arguments as were used in the two preceding sections, one easily obtains the following relations :

$$\begin{aligned} \psi^i &= \xi^i\,(\theta,\ \rho^F,\ \mathbf{F}_{Se}, \mathbf{F}_{Sp}), \\ \eta^S &= -\frac{\partial \psi^S}{\partial \theta}, \\ \eta^F &= -\frac{\partial \psi^F}{\partial \theta}; \end{aligned} \qquad (4.65)$$

$$\mathbf{q}^S + \mathbf{q}^F = -\,\beta_\vartheta \operatorname{grad}\theta + \theta\,\alpha_\vartheta\,(\mathbf{x}'_F - \mathbf{x}'_S)\,; \qquad (4.66)$$

$$\begin{aligned} \mathbf{T}^S &= -n^S\,\lambda^{SR}\,\mathbf{I} + \left(\rho^S\,\frac{\partial \psi^S}{\partial \mathbf{F}_{Se}} + \rho^F\,\frac{\partial \psi^F}{\partial \mathbf{F}_{Se}}\right)\mathbf{F}^T_{Se}, \\ \mathbf{T}^F &= -\rho^F\left(\rho^S\,\frac{\partial \psi^S}{\partial \rho^F} + \rho^F\,\frac{\partial \psi^F}{\partial \rho^F}\right)\mathbf{I} + \mathbf{T}^F_E, \\ \mathbf{T}^F_E &= 2\,\mu^F\,\mathbf{D}_F + \nu^F\,(\mathbf{D}_F\cdot\mathbf{I})\,\mathbf{I}; \end{aligned} \qquad (4.67)$$

$$\begin{aligned} \mathbf{p}^F = \lambda^{SR} \operatorname{grad} n^F + \rho^S\,\frac{\partial \psi^S}{\partial \rho^F}\operatorname{grad}\rho^F - \mathbf{F}^{T-1}_{Se}\left(\rho^F\,\frac{\partial \psi^F}{\partial \mathbf{F}_{Se}}\ \overset{3}{\mathbf{G}}_{Se}\right)\underset{\sim}{1} \\ -\,\mathbf{F}^{T-1}_{Se}\,\mathbf{F}^{T-1}_{Sp}\left(\rho^F\,\frac{\partial \psi^F}{\partial \mathbf{F}_{Sp}}\ \overset{3}{\mathbf{G}}_{Sp}\right)\underset{\sim}{1} + \mathbf{p}^F_E, \end{aligned} \qquad (4.68)$$

$$\mathbf{p}^F_E = -\,\alpha_\vartheta \operatorname{grad}\theta - \mathbf{S}_V\,(\mathbf{x}'_F - \mathbf{x}'_S)\,.$$

As was just pointed out above, the permeability tensor $\mathbf{S}_V$, in the case of isotropic permeability, is again governed by eqn (4.11). Furthermore, the material parameter functions included into the hybrid model of second type depend on the set $\mathcal{S}_{2po}$, viz.:

$$\alpha_\vartheta\,,\ \beta_\vartheta\,,\ \mu^F\,,\ \nu^F\,,\ k^F\ =\ \mathcal{F}\,(\mathcal{S}_{2po})\,. \tag{4.69}$$

In addition to the above results, the entropy inequality contains two further mechanisms which, again, must be understood in the sense of sufficient conditions:

$$-\,\lambda^{SR}\,(n^F)'_F\ \geq\ 0\,, \tag{4.70}$$

$$(\mathbf{A}_{Sp})^\triangle\,\cdot\,[\,\mathbf{T}^S_E\,-\,\mathbf{F}_S\left(\rho^S\,\frac{\partial\psi^S}{\partial\mathbf{E}_{Sp}}\,+\,\rho^F\,\frac{\partial\psi^F}{\partial\mathbf{E}_{Sp}}\right)\mathbf{F}^T_S\,]\ \geq\ 0\,. \tag{4.71}$$

Concerning the inequality (4.71),

$$\mathbf{T}^S_E\ =\ \mathbf{T}^S\,+n^S\,\lambda^{SR}\,\mathbf{I} \tag{4.72}$$

defines one possible version of the extra stress tensor of the incompressible solid matrix material. Furthermore, it is seen that the inequality (4.70) restricts the values of the Lagrangian multiplier λ^{SR}, whereas the inequality (4.71) represents the plastic dissipation mechanism for the hybrid model of second type. However, recall that the latter inequality proceeds from the principle of material frame-indifference including

$$\psi^i\,(\theta,\rho^F,\mathbf{F}_{Se},\mathbf{F}_{Sp})\ \rightarrow\ \psi^i\,(\theta,\,\rho^F,\,\mathbf{E}_{Se},\mathbf{E}_{Sp}), \tag{4.73}$$

where, additionally, eqns (4.16) and (4.17) have been used. For the hybrid model of second type, it is furthermore concluded by combination of eqn $(4.67)_1$ and (4.72) that

$$\begin{aligned}\boldsymbol{\tau}^S\ &=\ -n^S_{oS}\,\lambda^{SR}\,\mathbf{I}\,+\,\boldsymbol{\tau}^S_E\,,\\ \boldsymbol{\tau}^S_E\ &=\ \det\mathbf{F}_S\ \mathbf{F}_S\left(\rho^S\,\frac{\partial\psi^S}{\partial\mathbf{E}_{Se}}\,+\,\rho^F\,\frac{\partial\psi^F}{\partial\mathbf{E}_{Se}}\right)\mathbf{F}^T_S,\end{aligned} \tag{4.74}$$

where $\boldsymbol{\tau}^S_E$ is one version of the solid Kirchhoff extra stress tensor,

$$\boldsymbol{\tau}^S_E\ =\ \det\,\mathbf{F}_S\,\mathbf{T}^S_E, \tag{4.75}$$

and n^S_{oS} is defined via eqn (3.10). Furthermore, it follows from inequality (4.71) in formal analogy to the conclusion leading to the back-stress tensor $\mathbf{Y}^S$ included into eqn (4.21) that we can write

$$\mathbf{Y}^S\ =\ \det\mathbf{F}_S\ \mathbf{F}_S\left(\rho^S\,\frac{\partial\psi^S}{\partial\mathbf{E}_{Sp}}\,+\,\rho^F\,\frac{\partial\psi^F}{\partial\mathbf{E}_{Sp}}\right)\mathbf{F}^T_S. \tag{4.76}$$

Finally, in contrast to preceding subsections which are governed by a compressible matrix material, the plastic dissipation mechanism (compare, for example, (4.58)) now reads:

$$(\mathbf{A}_{Sp})^\triangle\,\cdot\,(\,\boldsymbol{\tau}^S_E\,-\,\mathbf{Y}^S\,)\,=\,(\hat{\boldsymbol{\Gamma}}_{Sp})^\triangle_p\,\cdot\,(\,\hat{\boldsymbol{\tau}}^S_E\,-\,\hat{\mathbf{Y}}^S\,)\,=\,(\mathbf{E}_{Sp})'_S\,\cdot\,(\,\mathbf{S}^S_E\,-\,\mathbf{Z}^S\,)\,\geq\,0\,. \tag{4.77}$$

Therein, the extra stress tensors, $\boldsymbol{\tau}_E^S$, $\hat{\boldsymbol{\tau}}_E^S$ and $\mathbf{S}_E^S$, are related to each other by the same transport mechanisms as are the corresponding total stress tensors, $\boldsymbol{\tau}^S$, $\hat{\boldsymbol{\tau}}^S$ and $\mathbf{S}^S$. Following this, it is easily seen with the aid of Table 3 and Table 6 that the universal invariance of the plastic dissipation mechanism is again guaranteed throughout the deformation process.

From the preceding relations, the general constitutive approach to the hybrid model of second type is properly defined.

(b) Simplified hybrid model of second type

In analogy to the Subsections 4.2 (b) and 4.3 (b), the "principle of constituent separation" implying each free energy functional of a given constituent to be independent of the variables of the respective other constituent yields by combination of the relations $(4.65)_1$ and (4.73):

$$\begin{aligned} \psi^S(\theta, \rho^F, \mathbf{F}_{Se}, \mathbf{F}_{Sp}) &\rightarrow \psi^S(\theta, \mathbf{E}_{Se}, \mathbf{E}_{Sp}), \\ \psi^F(\theta, \rho^F, \mathbf{F}_{Se}, \mathbf{F}_{Sp}) &\rightarrow \psi^F[\theta, \rho^F, n^S(\hat{\mathbf{C}}_{Se}, \mathbf{C}_{Sp})], \end{aligned} \tag{4.78}$$

where, from eqns (3.10) and (3.36) together with the multiplicative split (3.26),

$$\begin{aligned} n^S &= n_{oS}^S (\det \mathbf{F}_{Se})^{-1} (\det \mathbf{F}_{Sp})^{-1} \\ &= n_{oS}^S (\det \hat{\mathbf{C}}_{Se})^{-1/2} (\det \mathbf{C}_{Sp})^{-1/2} \end{aligned} \tag{4.79}$$

has been used to represent the volume fraction variable or the kinematic coupling variable, respectively, of both the solid and the fluid materials. Following this, one concludes from eqns $(4.67)_2$, $(4.68)_1$, (4.74) and (4.76) that

$$\begin{aligned} \mathbf{T}^F &= -(\rho^F)^2 \frac{\partial \psi^F}{\partial \rho^F} \mathbf{I} + \mathbf{T}_E^F, \\ \mathbf{p}^F &= \lambda^{SR} \operatorname{grad} n^F - \mathbf{F}_{Se}^{T-1} \Big(\rho^F \frac{\partial \psi^F}{\partial \mathbf{F}_{Se}} \overset{3}{\mathbf{G}}_{Se} \Big) \overset{1}{} \\ &\quad - \mathbf{F}_{Se}^{T-1} \mathbf{F}_{Sp}^{T-1} \Big(\rho^F \frac{\partial \psi^F}{\partial \mathbf{F}_{Sp}} \overset{3}{\mathbf{G}}_{Sp} \Big) \overset{1}{} + \mathbf{p}_E^F, \\ \boldsymbol{\tau}_E^S &= \det \mathbf{F}_S \, (2 \rho^S \mathbf{F}_{Se} \frac{\partial \psi^S}{\partial \hat{\mathbf{C}}_{Se}} \mathbf{F}_{Se}^T - n^S \rho^F \frac{\partial \psi^F}{\partial n^S} \mathbf{I}), \\ \hat{\mathbf{Y}}^S &= \det \mathbf{F}_S \, (2 \rho^S \mathbf{F}_{Sp} \frac{\partial \psi^S}{\partial \mathbf{C}_{Sp}} \mathbf{F}_{Sp}^T - n^S \rho^F \frac{\partial \psi^F}{\partial n^S} \mathbf{I}). \end{aligned} \tag{4.80}$$

Note in passing that to obtain the eqns $(4.80)_3$ and $(4.80)_4$ from eqns (4.74) and (4.76), the following relations have been used:

Firstly, with the aid of Table 1 and eqns (3.36), it is found by differentiation in analogy to eqn (4.31) that

$$\begin{aligned} \frac{\partial \psi^i}{\partial \mathbf{E}_{Se}} &= 2\,\mathbf{F}_{Sp}^{-1}\,\frac{\partial \psi^i}{\partial \hat{\mathbf{C}}_{Se}}\,\mathbf{F}_{Sp}^{T-1}, \\ \frac{\partial \psi^i}{\partial \mathbf{E}_{Sp}} &= 2\,\frac{\partial \psi^i}{\partial \mathbf{C}_{Sp}}. \end{aligned} \tag{4.81}$$

Additionally, by use of eqn (4.79), the chain rule yields

$$\begin{aligned} \frac{\partial n^S}{\partial \hat{\mathbf{C}}_{Se}} &= -\tfrac{1}{2}\,n^S\,\hat{\mathbf{C}}_{Se}^{-1}, \\ \frac{\partial n^S}{\partial \mathbf{C}_{Sp}} &= -\tfrac{1}{2}\,n^S\,\mathbf{C}_{Sp}^{-1}. \end{aligned} \tag{4.82}$$

Secondly, differentiation of ψ^F from (4.78) combined with eqns (4.81) and (4.82) leads to

$$\begin{aligned} \frac{\partial \psi^F}{\partial \hat{\mathbf{C}}_{Se}} &= \frac{\partial \psi^F}{\partial n^S}\,\frac{\partial n^S}{\partial \hat{\mathbf{C}}_{Se}} = -\tfrac{1}{2}\,n^S\,\frac{\partial \psi^F}{\partial n^S}\,\hat{\mathbf{C}}_{Se}^{-1}, \\ \frac{\partial \psi^F}{\partial \mathbf{C}_{Sp}} &= \frac{\partial \psi^F}{\partial n^S}\,\frac{\partial n^S}{\partial \mathbf{C}_{Sp}} = -\tfrac{1}{2}\,n^S\,\frac{\partial \psi^F}{\partial n^S}\,\mathbf{C}_{Sp}^{-1}. \end{aligned} \tag{4.83}$$

Now, given eqns (4.81)–(4.83), one easily concludes the relations $(4.80)_3$ and $(4.80)_4$ to hold.

Further representations of the solid stress measures can be obtained with the aid of eqns (4.74) and $(4.80)_3$, viz.:

$$\begin{aligned} \boldsymbol{\tau}^S &= \det \mathbf{F}_S\,[\,-n^S(\,\lambda^{SR} + \rho^F\,\frac{\partial \psi^F}{\partial n^S}\,)\,\mathbf{I} + 2\,\rho^S\,\mathbf{F}_{Se}\,\frac{\partial \psi^S}{\partial \hat{\mathbf{C}}_{Se}}\,\mathbf{F}_{Se}^T\,] \\ &= \det \mathbf{F}_S\,[\,-n^S(\,\lambda^{SR} + \rho^F\,\frac{\partial \psi^F}{\partial n^S}\,)\,\mathbf{I} \\ &\quad + 2\,\rho^S\,(\det \mathbf{F}_{Se})^{-2/3}\,(\,\mathbf{F}_{Se}\,\frac{\partial \psi^S}{\partial \overline{\mathbf{C}}_{Se}}\,\mathbf{F}_{Se}^T\,)^D + \rho^S \det \mathbf{F}_{Se}\,\frac{\partial \psi^S}{\partial \det \mathbf{F}_{Se}}\,\mathbf{I}\,], \end{aligned} \tag{4.84}$$

Therein, eqns (4.32)–(4.34) have been used. For the incompressible solid material under consideration, an alternative representation of the above equation is

$$\begin{aligned} \boldsymbol{\tau}^S = {} & \det \mathbf{F}_S\,[\,-n^S(\,\lambda^{SR} + \rho^F\,\frac{\partial \psi^F}{\partial n^S}\,)\,\mathbf{I} \\ & + 2\,\rho^S\,(\det \mathbf{F}_{Se})^{-2/3}\,(\,\mathbf{F}_{Se}\,\frac{\partial \psi^S}{\partial \overline{\mathbf{C}}_{Se}}\,\mathbf{F}_{Se}^T\,)^D - n^S\,p_{\text{con.}}^{SR}\,\mathbf{I}\,], \end{aligned} \tag{4.85}$$

where, as was pointed out in eqn (4.39), the term

$$p_{\text{con.}}^{SR} = \rho^{SR}\,n^S\,\frac{\partial \psi^S}{\partial n^S} \tag{4.86}$$

again represents the configuration (or structural) pressure of the matrix. Note in passing that to obtain the above result, eqn (4.79) combined with eqns (4.36) and (4.37) has been used. In analogy to the preceding considerations, the back-stress tensor can be expressed via

$$\begin{aligned}\hat{\mathbf{Y}}^S &= \det \mathbf{F}_S \,[\, 2\,\rho^S\,(\det\,\mathbf{F}_{Sp})^{-2/3}\,(\,\mathbf{F}_{Sp}\,\frac{\partial \psi^S}{\partial \overline{\mathbf{C}}_{Sp}}\,\mathbf{F}^T_{Sp}\,)^D \\ &\quad + (\,\rho^S \det \mathbf{F}_{Sp}\,\frac{\partial \psi^S}{\partial \det \mathbf{F}_{Sp}} + n^S\,p^{FR}_{\text{con.}}\,)\,\mathbf{I}\,]\,. \end{aligned} \tag{4.87}$$

Therein, as was pointed out in eqn $(4.30)_1$,

$$\begin{aligned} p^{FR}_{\text{con.}} &= n^F\,\rho^{FR}\,\frac{\partial \psi^F}{\partial n^F} \\ &= -\,\rho^F\,\frac{\partial \psi^F}{\partial n^S} \end{aligned} \tag{4.88}$$

represents the fluid configuration pressure term. Note in passing that to obtain the above result, $n^F = 1 - n^S$ has been used.

However, concerning the representation $(4.82)_1$ for the fluid stress tensor, it follows from the same arguments as were leading from $(4.25)_1$ to (4.29) that

$$\begin{aligned} \mathbf{T}^F &= -n^F \left(\rho^{FR}\,n^F\,\frac{\partial \psi^F}{\partial n^F} + (\rho^{FR})^2\,\frac{\partial \psi^F}{\partial \rho^{FR}}\right)\mathbf{I} + \mathbf{T}^F_E \\ &= -n^F\,(\,p^{FR}_{\text{con.}} + p^{FR}_{\text{mat.}}\,)\,\mathbf{I} + \mathbf{T}^F_E\,. \end{aligned} \tag{4.89}$$

Apart from the above results, further representations for the interaction force $\mathbf{p}^F$ can be found. In particular, one obtains by differentiation that

$$\begin{aligned} \frac{\partial \psi^F}{\partial \mathbf{F}_{Se}} &= -\,\frac{\partial \psi^F}{\partial n^S}\,n^S\,\mathbf{F}^{T-1}_{Se}\,, \\ \frac{\partial \psi^F}{\partial \mathbf{F}_{Sp}} &= -\,\frac{\partial \psi^F}{\partial n^S}\,n^S\,\mathbf{F}^{T-1}_{Sp}\,, \end{aligned} \tag{4.90}$$

and thus,

$$\begin{aligned} -\,\mathbf{F}^{T-1}_{Se}\left(\rho^F\,\frac{\partial \psi^F}{\partial \mathbf{F}_{Se}}\,\overset{3}{\mathbf{G}}_{Se}\right)^{\underline{1}} &- \mathbf{F}^{T-1}_{Se}\,\mathbf{F}^{T-1}_{Sp}\left(\rho^F\,\frac{\partial \psi^F}{\partial \mathbf{F}_{Sp}}\,\overset{3}{\mathbf{G}}_{Sp}\right)^{\underline{1}} = \\ &= \rho^F\,\frac{\partial \psi^F}{\partial n^S}\,n^S\,\mathbf{F}^{T-1}_{Se}\,(\,\mathbf{F}^{T-1}_{Se}\,\overset{3}{\mathbf{G}}_{Se}\,)^{\underline{1}} + \rho^F\,\frac{\partial \psi^F}{\partial n^S}\,n^S\,\mathbf{F}^{T-1}_{Sp}\,(\,\mathbf{F}^{T-1}_{Sp}\,\overset{3}{\mathbf{G}}_{Sp}\,)^{\underline{1}} \\ &= \rho^F\,\frac{\partial \psi^F}{\partial n^S}\,n^S\,[\,(\det\,\mathbf{F}_{Se})^{-1}\,\mathrm{grad}\,(\det\,\mathbf{F}_{Se}) + (\det\,\mathbf{F}_{Sp})^{-1}\,\mathrm{grad}\,(\det\,\mathbf{F}_{Sp})\,]\,. \end{aligned} \tag{4.91}$$

where, as a rule (compare, EHLERS, 1989 b),

$$\begin{aligned} (\,\mathbf{F}^{T-1}_{Se}\,\overset{3}{\mathbf{G}}_{Se}\,)^{\underline{1}} &= (\det\,\mathbf{F}_{Se})^{-1}\,\mathrm{grad}\,(\det\,\mathbf{F}_{Se})\,, \\ (\,\mathbf{F}^{T-1}_{Sp}\,\overset{3}{\mathbf{G}}_{Sp}\,)^{\underline{1}} &= (\det\,\mathbf{F}_{Sp})^{-1}\,\mathrm{grad}\,(\det\,\mathbf{F}_{Sp}) \end{aligned} \tag{4.92}$$

has been used. Furthermore,

$$\begin{aligned} \operatorname{grad} n^F &= -\operatorname{grad} n^S \\ &= -n^S_{oS} \operatorname{grad}[(\det \mathbf{F}_{Se})^{-1} (\det \mathbf{F}_{Sp})^{-1}] \\ &= n^S [(\det \mathbf{F}_{Se})^{-1} \operatorname{grad}(\det \mathbf{F}_{Se}) \\ &\qquad + (\det \mathbf{F}_{Sp})^{-1} \operatorname{grad}(\det \mathbf{F}_{Sp})] \end{aligned} \tag{4.93}$$

holds. By comparison of eqns (4.91) and (4.93), it is immediately seen that

$$\begin{aligned} -\mathbf{F}^{T-1}_{Se} \left(\rho^F \frac{\partial \psi^F}{\partial \mathbf{F}_{Se}} \overset{3}{\mathbf{G}}_{Se}\right) \mathbf{1} - \mathbf{F}^{T-1}_{Se} \mathbf{F}^{T-1}_{Sp} \left(\rho^F \frac{\partial \psi^F}{\partial \mathbf{F}_{Sp}} \overset{3}{\mathbf{G}}_{Sp}\right) \mathbf{1} = \\ = \rho^F \frac{\partial \psi^F}{\partial n^S} \operatorname{grad} n^F, \end{aligned} \tag{4.94}$$

and thus, from eqn $(4.80)_2$,

$$\mathbf{p}^F = (\lambda^{SR} + \rho^F \frac{\partial \psi^F}{\partial n^S}) \operatorname{grad} n^F + \mathbf{p}^F_E. \tag{4.95}$$

Comparing the results of the hybrid models of first and second type, one comes to realize that, for instance, the structures of the momentum supply terms, given by eqns $(4.60)_2$ and (4.95), directly correspond to each other, except for the fact that, in the first model, $\mathbf{p}^F$ includes the Lagrangian multiplier of the incompressible pore liquid combined with the solid configuration pressure term, whereas, in the second model, these terms are substituted by the Lagrangian multiplier of the incompressible solid matrix combined with the fluid configuration pressure. In addition, the same terms are included into the fluid stress tensor $\mathbf{T}^F$ corresponding to the incompressible liquid of the first model and into the solid stress tensor $\boldsymbol{\tau}^S$ corresponding to the incompressible solid matrix of the second model, compare, for example, eqns $(4.60)_1$ and (4.85).

Finally, in discussing the simplified constitutive theory of elasto-plastic hybrid models of second type, the separation of the functional ψ^S into thermoelastic and plastic parts yields

$$\psi^S(\theta, \mathbf{E}_{Se}, \mathbf{E}_{Sp}) = \psi^{Se}(\theta, \mathbf{E}_{Se}) + \psi^{Sp}(\mathbf{E}_{Sp}), \tag{4.96}$$

where the relation $(4.78)_1$ has been used.

4.5 The incompressible two-phase model

(a) General results

For the incompressible two-phase model under consideration (incompressible elasto-plastic solid matrix saturated by an incompressible viscous pore liquid), the thermodynamical restrictions result from the entropy inequality (3.58) together with the constitutive assumptions

(3.51). Furthermore, the following thermodynamical restrictions are obtained from a linear expansion of the entropy inequality about the thermodynamical minimum state defined by

$$\begin{aligned} \mathcal{S}_{inp} \rightarrow \mathcal{S}_{inp\,0} \;=\; \{\theta,\ \operatorname{grad}\theta &= \mathbf{0}\,, \\ \mathbf{F}_{Se}\,, \overset{3}{\mathbf{G}}_{Se}\,, \mathbf{F}_{Sp}\,, \overset{3}{\mathbf{G}}_{Sp}\,, \overset{\prime}{\mathbf{x}}_F - \overset{\prime}{\mathbf{x}}_S &= \mathbf{0}\,,\ \mathbf{D}_F = \mathbf{0}\}. \end{aligned} \tag{4.97}$$

Using the same arguments as was done above, one easily concludes to

$$\psi^i = \xi^i\,(\theta,\, \mathbf{F}_{Se}, \mathbf{F}_{Sp}),$$

$$\begin{aligned} \eta^S &= -\frac{\partial \psi^S}{\partial \theta}\,, \\ \eta^F &= -\frac{\partial \psi^F}{\partial \theta}\,; \end{aligned} \tag{4.98}$$

$$\mathbf{q}^S + \mathbf{q}^F = -\,\beta_\vartheta \operatorname{grad}\theta + \theta\,\alpha_\vartheta\,(\overset{\prime}{\mathbf{x}}_F - \overset{\prime}{\mathbf{x}}_S)\,; \tag{4.99}$$

$$\begin{aligned} \mathbf{T}^S &= -n^S \lambda^R\,\mathbf{I} + \mathbf{T}^S_E\,, \\ \mathbf{T}^F &= -n^F \lambda^R\,\mathbf{I} + \mathbf{T}^F_E\,; \end{aligned} \tag{4.100}$$

$$\begin{aligned} \mathbf{T}^S_E &= \left(\rho^S\,\frac{\partial \psi^S}{\partial \mathbf{F}_{Se}} + \rho^F\,\frac{\partial \psi^F}{\partial \mathbf{F}_{Se}}\right) \mathbf{F}^T_{Se}\,, \\ \mathbf{T}^F_E &= 2\,\mu^F\,\mathbf{D}_F + \nu^F\,(\mathbf{D}_F \cdot \mathbf{I}\,)\,\mathbf{I}\,; \end{aligned} \tag{4.101}$$

$$\begin{aligned} \mathbf{p}^F &= \lambda^R \operatorname{grad} n^F - \mathbf{F}^{T-1}_{Se}\left(\rho^F\,\frac{\partial \psi^F}{\partial \mathbf{F}_{Se}}\;\overset{3}{\mathbf{G}}_{Se}\right)\overset{1}{\mathbf{\;}} \\ &\quad - \mathbf{F}^{T-1}_{Se}\,\mathbf{F}^{T-1}_{Sp}\left(\rho^F\,\frac{\partial \psi^F}{\partial \mathbf{F}_{Sp}}\;\overset{3}{\mathbf{G}}_{Sp}\right)\overset{1}{\mathbf{\;}} + \mathbf{p}^F_E\,, \\ \mathbf{p}^F_E &= -\,\alpha_\vartheta \operatorname{grad}\theta - \mathbf{S}_V\,(\overset{\prime}{\mathbf{x}}_F - \overset{\prime}{\mathbf{x}}_S)\,. \end{aligned} \tag{4.102}$$

Therein, in the case of isotropic permeability circumstances, the permeability tensor $\mathbf{S}_V$ is again governed by eqn (4.11). Furthermore, the material parameter functions included into the incompressible model depend on the set $\mathcal{S}_{ip\,0}$, viz.:

$$\alpha_\vartheta\,,\ \beta_\vartheta\,,\ \mu^F,\ \nu^F,\ k^F = \mathcal{F}\,(\mathcal{S}_{inp\,0})\,. \tag{4.103}$$

However, the entropy inequality moreover yields

$$(\mathbf{A}_{Sp})^{\triangle} \cdot [\mathbf{T}_E^S - \mathbf{F}_S \left(\rho^S \frac{\partial \psi^S}{\partial \mathbf{E}_{Sp}} + \rho^F \frac{\partial \psi^F}{\partial \mathbf{E}_{Sp}}\right) \mathbf{F}_S^T] \geq 0\,, \tag{4.104}$$

which must again be understood in the sense of a sufficient condition.

Recall, for convenience, that the inequality (4.104) was obtained with the aid of the principle of material frame-indifference leading to

$$\psi^i\,(\theta, \mathbf{F}_{Se}, \mathbf{F}_{Sp}) \;\rightarrow\; \psi^i\,(\theta,\, \mathbf{E}_{Se}, \mathbf{E}_{Sp}). \tag{4.105}$$

Within the framework of the incompressible two-phase model, it is furthermore concluded by combination of eqns (4.74) and $(4.100)_1$ that

$$\begin{aligned} \boldsymbol{\tau}^S &= -n_{oS}^S\, \lambda^R\, \mathbf{I} + \boldsymbol{\tau}_E^S\,, \\ \boldsymbol{\tau}_E^S &= \det \mathbf{F}_S\ \mathbf{F}_S \left(\rho^S \frac{\partial \psi^S}{\partial \mathbf{E}_{Se}} + \rho^F \frac{\partial \psi^F}{\partial \mathbf{E}_{Se}}\right) \mathbf{F}_S^T, \end{aligned} \tag{4.106}$$

where, additionally, eqns (4.16) and (4.17) have been used. As was pointed out in the preceding subsection, $\boldsymbol{\tau}_E^S$ is the solid Kirchhoff extra stress tensor, compare eqn (4.75). Furthermore, it is evident that

$$\mathbf{Y}^S = \det \mathbf{F}_S\ \mathbf{F}_S \left(\rho^S \frac{\partial \psi^S}{\partial \mathbf{E}_{Sp}} + \rho^F \frac{\partial \psi^F}{\partial \mathbf{E}_{Sp}}\right) \mathbf{F}_S^T. \tag{4.107}$$

Finally, compare the relations (4.77), the plastic dissipation mechanism reads:

$$(\mathbf{A}_{Sp})^{\triangle} \cdot (\boldsymbol{\tau}_E^S - \mathbf{Y}^S) = (\hat{\boldsymbol{\Gamma}}_{Sp})_p^{\triangle} \cdot (\hat{\boldsymbol{\tau}}_E^S - \hat{\mathbf{Y}}^S) = (\mathbf{E}_{Sp})_S' \cdot (\mathbf{S}_E^S - \mathbf{Z}^S) \geq 0\,, \tag{4.108}$$

where, again, the extra stress tensors, $\boldsymbol{\tau}_E^S$, $\hat{\boldsymbol{\tau}}_E^S$ and $\mathbf{S}_E^S$, are related to each other by the same transport mechanisms as are the corresponding total stress tensors, $\boldsymbol{\tau}^S$, $\hat{\boldsymbol{\tau}}^S$ and $\mathbf{S}^S$.

From the preceding relations, the general constitutive approach to the incompressible two-phase model is properly defined.

(b) Simplified constitutive model

In analogy to the arguments given in EHLERS (1989a, b), the "principle of constituent separation" and the statements $(4.98)_1$ and (4.105) now combine to

$$\begin{aligned} \psi^S\,(\theta, \mathbf{F}_{Se}, \mathbf{F}_{Sp}) &\;\rightarrow\; \psi^S\,(\theta, \mathbf{E}_{Se}, \mathbf{E}_{Sp}), \\ \psi^F\,(\theta, \mathbf{F}_{Se}, \mathbf{F}_{Sp}) &\;\rightarrow\; \psi^F\,(\theta). \end{aligned} \tag{4.109}$$

Note in passing that, within the framework of the simplified incompressible binary model, there is no kinematic coupling variable included into the representations (4.109); namely, as a result of the fact that the incompressible model does only contain one type of kinematic variables, $\mathbf{E}_{Se}$ and $\mathbf{E}_{Sp}$, both corresponding to the skeleton material. Thus, the following relations hold to substitute eqns $(4.102)_1$, $(4.106)_2$ and (4.107), viz.:

$$
\begin{aligned}
\mathbf{p}^F &= \lambda^R \operatorname{grad} n^F + \mathbf{p}_E^F, \\
\boldsymbol{\tau}_E^S &= \det \mathbf{F}_S \, (2\,\rho^S \mathbf{F}_{Se} \frac{\partial \psi^S}{\partial \hat{\mathbf{C}}_{Se}} \mathbf{F}_{Se}^T), \\
\hat{\mathbf{Y}}^S &= \det \mathbf{F}_S \, (2\,\rho^S \mathbf{F}_{Sp} \frac{\partial \psi^S}{\partial \mathbf{C}_{Sp}} \mathbf{F}_{Sp}^T).
\end{aligned}
\tag{4.110}
$$

Note in passing that to obtain eqns $(4.110)_2$ and $(4.110)_3$, use was made of eqns (4.81).

Further representations of the solid stress tensor can be obtained in direct analogy to eqns (4.85) and (4.85):

$$
\begin{aligned}
\boldsymbol{\tau}_E^S &= 2\,\rho_{oS}^S \, \mathbf{F}_{Se} \frac{\partial \psi^S}{\partial \hat{\mathbf{C}}_{Se}} \mathbf{F}_{Se}^T \\
&= 2\,\rho_{oS}^S (\det \mathbf{F}_{Se})^{-2/3} (\mathbf{F}_{Se} \frac{\partial \psi^S}{\partial \overline{\mathbf{C}}_{Se}} \mathbf{F}_{Se}^T)^D + \rho_{oS}^S \det \mathbf{F}_{Se} \frac{\partial \psi^S}{\partial \det \mathbf{F}_{Se}} \mathbf{I}.
\end{aligned}
\tag{4.111}
$$

An alternative representation of the above equation is given by

$$
\boldsymbol{\tau}_E^S = 2\,\rho_{oS}^S (\det \mathbf{F}_{Se})^{-2/3} (\mathbf{F}_{Se} \frac{\partial \psi^S}{\partial \overline{\mathbf{C}}_{Se}} \mathbf{F}_{Se}^T)^D - n_{oS}^S \, p_{\text{con.}}^{SR} \, \mathbf{I},
\tag{4.112}
$$

where $p_{\text{con.}}^{SR}$ again represents the configuration pressure of the matrix. In analogy to the preceding considerations, the back-stress tensor reads

$$
\hat{\mathbf{Y}}^S = 2\,\rho_{oS}^S (\det \mathbf{F}_{Sp})^{-2/3} (\mathbf{F}_{Sp} \frac{\partial \psi^S}{\partial \overline{\mathbf{C}}_{Sp}} \mathbf{F}_{Sp}^T)^D + \rho_{oS}^S \det \mathbf{F}_{Sp} \frac{\partial \psi^S}{\partial \det \mathbf{F}_{Sp}} \mathbf{I}.
\tag{4.113}
$$

Finally, in discussing the simplified constitutive theory of incompressible two-phase models, the separation of the functional ψ^S into thermoelastic and plastic parts yields

$$
\psi^S(\theta, \mathbf{E}_{Se}, \mathbf{E}_{Sp}) = \psi^{Se}(\theta, \mathbf{E}_{Se}) + \psi^{Sp}(\mathbf{E}_{Sp}),
\tag{4.114}
$$

where the relation $(4.109)_1$ has been used.

5. The solid constituent: Finite elasticity

5.1 Preliminary remarks

The present section concerns the discussion of finite elasticity laws for the solid constituent of the different binary media under consideration, thus describing the elasticity properties of compressible models, hybrid models of both types and incompressible models, respectively. However, concerning the representation of the solid matrix included into these models, we obviously have to handle two different solid materials, namely, compressible and incompressible ones. As was pointed out in the preceding section, the matrix material exhibits, in the compressible range, two compressibilities governed by the material and the configuration pressure terms, respectively, whereas the incompressible matrix is governed by only one compressiblity belonging to the matrix pressure.

However, it it seen from the simplified versions of the compressible model and the hybrid model of first type, compare eqns $(4.25)_1$ and $(4.60)_3$, that the basic representations of the Kirchhoff stress tensor of the compressible solid material read

$$\begin{aligned} \boldsymbol{\tau}^S &= \det \mathbf{F}_S \, (\rho^S \, \mathbf{F}_S \, \frac{\partial \psi^S}{\partial \mathbf{E}_{Se}} \, \mathbf{F}_S^T) \\ &= \det \mathbf{F}_S \, (2 \, \rho^S \, \mathbf{F}_{Se} \, \frac{\partial \psi^S}{\partial \hat{\mathbf{C}}_{Se}} \, \mathbf{F}_{Se}^T), \end{aligned} \tag{5.1}$$

where eqn (4.31) together with the multiplicative split $(3.21)_1$ has been used. Further representations of the above equation can be obtained from the chain rule and the transport mechanisms between the different strain tensors exhibited by Table 1.

Concerning the incompressible matrices included into the hybrid model of second type, one obtains from eqn $(4.80)_3$ that

$$\tilde{\boldsymbol{\tau}}_E^S = \det \mathbf{F}_S \, (2 \, \rho^S \, \mathbf{F}_{Se} \, \frac{\partial \psi^S}{\partial \hat{\mathbf{C}}_{Se}} \, \mathbf{F}_{Se}^T - n^S \, \rho^F \, \frac{\partial \psi^F}{\partial n^S} \, \mathbf{I}\,), \tag{5.2}$$

that is, the above version of the extra stress (marked by an additional tilde) is affected by the fluid configuration pressure, additionally compare eqn (4.88). Thus, in only considering the matrix properties, we redefine the extra stress tensor via

$$\begin{aligned} \boldsymbol{\tau}_E^S &= \det \mathbf{F}_S \, (\rho^S \, \mathbf{F}_S \, \frac{\partial \psi^S}{\partial \mathbf{E}_{Se}} \, \mathbf{F}_S^T) \\ &= \det \mathbf{F}_S \, (2 \, \rho^S \, \mathbf{F}_{Se} \, \frac{\partial \psi^S}{\partial \hat{\mathbf{C}}_{Se}} \, \mathbf{F}_{Se}^T), \end{aligned} \tag{5.3}$$

where

$$\tilde{\boldsymbol{\tau}}_E^S = \boldsymbol{\tau}_E^S - n_{oS}^S \, \rho^F \, \frac{\partial \psi^F}{\partial n^S} \, \mathbf{I} \tag{5.4}$$

to be obtained with the aid of eqn (3.10) has been used. Note in passing that the present definition and eqn $(4.74)_1$ combine to

$$\boldsymbol{\tau}^S = -n_{oS}^S \, (\lambda^{SR} + \rho^F \, \frac{\partial \psi^F}{\partial n^S}) \, \mathbf{I} + \boldsymbol{\tau}_E^S . \tag{5.5}$$

Furthermore, it is seen from eqns $(4.100)_1$ and $(4.110)_2$ that the incompressible model yields

$$\boldsymbol{\tau}^S = -n_{oS}^S \, \lambda^R \, \mathbf{I} + \boldsymbol{\tau}_E^S , \tag{5.6}$$

where $\boldsymbol{\tau}_E^S$ is again governed by eqn (5.3). Thus, it is easily concluded that the solid extra stresses of the incompressible matrix materials and the above stress tensors of the compressible solid matrices can be obtained from the same basic relations, although the included individual functionals ψ^S are different concerning to the respective elasticity properties under consideration.

Following this, we proceed by firstly introducing a finite elasticity law for compressible materials which, then, can be simplified towards incompressible matrices. These elasticity laws proceed from the introduced free energy split into elastic and plastic parts, compare eqns (4.46), (4.63), (4.96) and (4.114), where, for the compressible matrix materials included into the compressible two-phase model and into the hybrid model of first type, it was found that

$$\psi^{Se} \rightarrow \psi_c^{Se} = \psi_c^{Se}(\theta,\, n^S,\, \mathbf{E}_{Se}), \tag{5.7}$$

whereas the incompressible matrix materials included into the hybrid model of second type and into the incompressible model is based on

$$\psi^{Se} \rightarrow \psi_{in}^{Se} = \psi_{in}^{Se}(\theta,\, \mathbf{E}_{Se}). \tag{5.8}$$

In particular, the following definition of finite elasticity proceeds from the assumption of hyperelastic porous solid matrices, where, furthermore, use is made of on an extension of the neo-Hookean model (RIVLIN, 1948) towards the compressible range. From the definition of hyperelasticity (compare WANG & TRUESDELL, 1973, III.4), the relations

$$\begin{aligned} \psi_c^{Se}(\theta,\, n^S,\, \mathbf{E}_{Se}) &= \bar{\psi}^{Se}(\theta) + \tilde{\psi}_c^{Se}(n^S,\, \mathbf{E}_{Se}),\\ \psi_{in}^{Se}(\theta,\, \mathbf{E}_{Se}) &= \bar{\psi}^{Se}(\theta) + \tilde{\psi}_{in}^{Se}(\mathbf{E}_{Se}) \end{aligned} \tag{5.9}$$

hold and thus, from eqns (5.1) and (5.3),

$$\begin{aligned} \boldsymbol{\tau}^S &= \rho_{oS}^S\, \mathbf{F}_S\, \frac{\partial \tilde{\psi}_c^{Se}}{\partial \mathbf{E}_{Se}}\, \mathbf{F}_S^T,\\ \boldsymbol{\tau}_E^S &= \rho_{oS}^S\, \mathbf{F}_S\, \frac{\partial \tilde{\psi}_{in}^{Se}}{\partial \mathbf{E}_{Se}}\, \mathbf{F}_S^T. \end{aligned} \tag{5.10}$$

Note again that the geometrical concept of elasticity (compare Subsection 3.2) was defined by the "distance" between the plastic intermediate configuration of φ^S and the current configuration. Hence, it is seen from

$$\mathbf{E}_{Se} = \mathbf{E}_{Se}(\hat{\mathbf{C}}_{Se}, \mathbf{F}_{Sp}) = \tfrac{1}{2}\, \mathbf{F}_{Sp}^T(\hat{\mathbf{C}}_{Se} - \mathbf{I})\mathbf{F}_{Sp} \tag{5.11}$$

(compare Table 1), that the above definition of elasticity requires the plastic variables to be frozen at the particular instant. Following this, one must conclude the different stored energy functions to yield

$$\begin{aligned} \tilde{\psi}_c^{Se}(n^S,\, \mathbf{E}_{Se}) &= \tilde{\psi}_c^{Se}(n^S,\, \hat{\mathbf{C}}_{Se},\, \mathbf{F}_{Sp})\Big|_{\mathbf{F}_{Sp}\ \text{fixed}},\\ \tilde{\psi}_{in}^{Se}(\mathbf{E}_{Se}) &= \tilde{\psi}_{in}^{Se}(\hat{\mathbf{C}}_{Se},\, \mathbf{F}_{Sp})\Big|_{\mathbf{F}_{Sp}\ \text{fixed}}. \end{aligned} \tag{5.12}$$

In addition, the symmetry group for the solid constituent yields

$$\begin{aligned}
\tilde{\psi}_c^{Se} &= \tilde{\psi}_c^{Se}(n^S, \overline{\mathrm{I}}_e, \overline{\mathrm{II}}_e, J_e, \mathbf{F}_{Sp})\Big|_{\mathbf{F}_{Sp}\ \text{fixed}}, \\
\tilde{\psi}_{in}^{Se} &= \tilde{\psi}_{in}^{Se}(\overline{\mathrm{I}}_e, \overline{\mathrm{II}}_e, J_e, \mathbf{F}_{Sp})\Big|_{\mathbf{F}_{Sp}\ \text{fixed}}, \qquad (5.13) \\
J_e &= \det \mathbf{F}_{Se} = (\mathrm{III}_e)^{1/2},
\end{aligned}$$

where $\overline{\mathrm{I}}_e$ and $\overline{\mathrm{II}}_e$ are the principal invariants of the right elastic Cauchy-Green tensor $\overline{\mathbf{C}}_{Se}$ or the left elastic Cauchy-Green tensor $\overline{\mathbf{B}}_{Se}$ defined by

$$\begin{aligned}
\overline{\mathbf{C}}_{Se} &= (\det \mathbf{F}_{Se})^{-2/3}\, \hat{\mathbf{C}}_{Se}, \\
\overline{\mathbf{B}}_{Se} &= (\det \mathbf{F}_{Se})^{-2/3}\, \mathbf{B}_{Se},
\end{aligned} \qquad (5.14)$$

additionally, compare eqns (3.36), (4.32) and (4.34).

From the above definitions and comments, the basic concept of finite porous media elasticity is properly defined.

5.2 Compressible matrix materials

The present invesigations in finite elasticity for compressible matrix materials proceed from an extension of the neo-Hookean model towards the compressible range, where, in particular, use is made of a representation of Flory type (FLORY, 1961), viz.:

$$\boldsymbol{\tau}^S = \mu^S\, \overline{\mathbf{B}}_{Se}^D - P^S\, \mathbf{I}, \qquad (5.15)$$

μ^S is the macroscopic shear modulus. Following this, it is the main idea to substitute the unspecified function P^S included into the above representation by convenient pressure terms.

However, it follows by combination of eqns (4.31) and $(5.10)_1$ that

$$\begin{aligned}
\boldsymbol{\tau}^S &= 2\, \rho_{oS}^S\, \mathbf{F}_S\, \frac{\partial \tilde{\psi}_c^{Se}}{\partial \hat{\mathbf{C}}_{Se}}\, \mathbf{F}_S^T \\
&= 2\, \rho_{oS}^S\, J_e^{-2/3} \Big(\mathbf{F}_S\, \frac{\partial \tilde{\psi}_c^{Se}}{\partial \overline{\mathbf{C}}_{Se}}\, \mathbf{F}_S^T \Big)^D + \rho_{oS}^S\, J_e\, \frac{\partial \tilde{\psi}_c^{Se}}{\partial J_e}\, \mathbf{I},
\end{aligned} \qquad (5.16)$$

where, to obtain the above result,

$$\frac{\partial \tilde{\psi}_c^{Se}}{\partial \hat{\mathbf{C}}_{Se}} = \Big(\frac{\partial \overline{\mathbf{C}}_{Se}}{\partial \hat{\mathbf{C}}_{Se}} \Big)^T \frac{\partial \tilde{\psi}_c^{Se}}{\partial \overline{\mathbf{C}}_{Se}} + \frac{\partial \tilde{\psi}_c^{Se}}{\partial J_e}\, \frac{\partial J_e}{\partial \hat{\mathbf{C}}_{Se}} \qquad (5.17)$$

together with

$$\begin{aligned}
\frac{\partial J_e}{\partial \hat{\mathbf{C}}_{Se}} &= \tfrac{1}{2}\, J_e\, \hat{\mathbf{C}}_{Se}^{-1}, \\
\frac{\partial \overline{\mathbf{C}}_{Se}}{\partial \hat{\mathbf{C}}_{Se}} &= J_e^{-2/3} \big(\overset{4}{\mathbf{I}} - \tfrac{1}{3}\, \hat{\mathbf{C}}_{Se} \otimes \hat{\mathbf{C}}_{Se}^{-1} \big)
\end{aligned} \qquad (5.18)$$

has been used; $\overset{4}{\mathbf{I}}$ is the fourth-order identity. Note in passing that the multiplicative split of $\hat{\mathbf{C}}_{Se}$ into a spherical part and a volume-preserving part naturally results in a decoupling of the volume-preserving variables, $\bar{\mathrm{I}}_e$ and $\bar{\mathrm{II}}_e$, and the variable J_e.

Furthermore, it is seen from eqn $(5.13)_1$ that

$$\frac{\partial \tilde{\psi}_c^{Se}}{\partial \overline{\mathbf{C}}_{Se}} = \left(\frac{\partial \tilde{\psi}_c^{Se}}{\partial \bar{\mathrm{I}}_e} + \bar{\mathrm{I}}_e \frac{\partial \tilde{\psi}_c^{Se}}{\partial \bar{\mathrm{II}}_e} \right) \mathbf{I} - \frac{\partial \tilde{\psi}_c^{Se}}{\partial \bar{\mathrm{II}}_e} \overline{\mathbf{C}}_{Se} . \tag{5.19}$$

From eqn (5.19), it is now easily concluded by combination of the above representation together with the neo-Hookean model of Flory type that

$$\begin{aligned} \rho_{oS}^S \frac{\partial \tilde{\psi}_c^{Se}}{\partial \bar{\mathrm{I}}_e}\bigg|_{\mathbf{F}_{Sp}\ \text{fixed}} &= \tfrac{1}{2} \mu^S \bigg|_{\mathbf{F}_{Sp}\ \text{fixed}} , \\ \frac{\partial \tilde{\psi}_c^{Se}}{\partial \bar{\mathrm{II}}_e} &= 0. \end{aligned} \tag{5.20}$$

Thus, the following statement holds:

$$\rho_{oS}^S \tilde{\psi}_c^{Se} = [\tfrac{1}{2} \mu^S (\bar{\mathrm{I}}_e - 3) + \rho_{oS}^S \mathrm{U}_c (n^S, J_e, \mathbf{F}_{Sp})]\Big|_{\mathbf{F}_{Sp}\ \text{fixed}} . \tag{5.21}$$

However, concerning the representation of the function U_c, an alternative form is given by

$$\mathrm{U}_c (n^S, J_e, \mathbf{F}_{Sp})]\Big|_{\mathbf{F}_{Sp}\ \text{fixed}} = \mathrm{U}_c (n^S, \rho^{SR}, \mathbf{F}_{Sp})]\Big|_{\mathbf{F}_{Sp}\ \text{fixed}} , \tag{5.22}$$

where the independent state variables n^S and J_e have been substituted by the alternative variables n^S and ρ^{SR} via

$$n^S \rho^{SR} = \hat{n}^S \hat{\rho}^{SR} J_e^{-1} , \tag{5.23}$$

compare eqns (4.1)–(4.4). Note in passing that the solid volume fraction $\hat{n}^S$ and the effective denisity function $\hat{\rho}^{SR}$ of the plastic intermediate configuration of φ^S are kept fixed at a given plastic state of deformation. Thus, an alternative representation of the stored energy function of the compressible matrix material is

$$\rho_{oS}^S \tilde{\psi}_c^{Se} = [\tfrac{1}{2} \mu^S (\bar{\mathrm{I}}_e - 3) + \rho_{oS}^S \mathrm{U}_c (n^S, \rho^{SR}, \mathbf{F}_{Sp})]\Big|_{\mathbf{F}_{Sp}\ \text{fixed}} . \tag{5.24}$$

Given the above results, one concludes from eqns (5.16) together with (5.21) or (5.24), respectively, that

$$\boldsymbol{\tau}^S = \mu^S J_e^{-2/3} \mathbf{B}_{Se}^D + \rho_{oS}^S J_e \frac{\partial \mathrm{U}_c}{\partial J_e}\bigg|_{\mathbf{F}_{Sp}\ \text{fixed}} \mathbf{I} . \tag{5.25}$$

Proceeding from the above relations, one comes to realize in formal analogy to the description of the compressible range of porous media included into Ehlers (1989b, 1991) that a convenient representation is of U_c is given by

$$\rho_{oS}^S \mathrm{U}_c \Big|_{\mathbf{F}_{Sp}\ \text{fixed}} = k_{\text{con.}}^S \left(\frac{\hat{n}^S}{n^S}\right)^c \frac{c \ln\left(\dfrac{\hat{n}^S}{n^S}\right) - 1}{c^2} + \frac{k_{\text{con.}}^S}{c^2} + \tfrac{1}{2} k_{\text{mat.}}^S \left(\ln \frac{\hat{\rho}^{SR}}{\rho^{SR}} \right)^2 , \tag{5.26}$$

where $k^S_{\text{con.}}$ is the configuration compression modulus, $k^S_{\text{mat.}}$ is the material or the phase compression modulus, and c is a material number governing, in the framework of incompressible matrices, the so-called configuration pressure of compaction corresponding to that deformation state where all the pore spaces are closed and the materials transits to non-porous material properties. However, it is seen from the above mathematical representation of the function U_c that, in the case of vanishing c, the first two terms of the right hand side tend towards "$\frac{0}{0}$". Thus, given $c = 0$, the alternative function $\bar{\text{U}}_c$ can be expressed via

$$\rho^S_{oS}\,\bar{\text{U}}_c\Big|_{\mathbf{F}_{Sp}\ \text{fixed}} = \tfrac{1}{2}\,k^S_{\text{con.}}\left(\ln\frac{\hat{n}^S}{n^S}\right)^2 + \tfrac{1}{2}\,k^S_{\text{mat.}}\left(\ln\frac{\hat{\rho}^{SR}}{\rho^{SR}}\right)^2. \tag{5.27}$$

Combination of eqns (5.25) and (5.26) yields

$$\begin{aligned}\boldsymbol{\tau}^S &= 2\,\mu^S J_e^{-2/3}\,\mathbf{K}^D_{Se}\\ &\quad+[\,k^S_{\text{con.}}\,(\frac{\hat{n}^S}{n^S})^c\ln(\frac{\hat{n}^S}{n^S}) + k^S_{\text{mat.}}\ln(\frac{\hat{\rho}^{SR}}{\rho^{SR}})\,]\,\mathbf{I}\,,\end{aligned} \tag{5.28}$$

whereas, from eqns (5.25) and (5.27), one obtains

$$\begin{aligned}\boldsymbol{\tau}^S &= 2\,\mu^S J_e^{-2/3}\,\mathbf{K}^D_{Se}\\ &\quad+[\,k^S_{\text{con.}}\ln(\frac{\hat{n}^S}{n^S}) + k^S_{\text{mat.}}\ln(\frac{\hat{\rho}^{SR}}{\rho^{SR}})\,]\,\mathbf{I}\,,\end{aligned} \tag{5.29}$$

where, from eqn (3.36) and Table 1,

$$2\,\mathbf{K}^D_{Se} = \mathbf{B}^D_{Se} \tag{5.30}$$

and, from the chain rule,

$$\rho^S_{oS}\,J_e\,\frac{\partial \text{U}_c}{\partial J_e}\Big|_{\mathbf{F}_{Sp}\ \text{fixed}} = -\left(n^S\rho^S_{oS}\,\frac{\partial \text{U}_c}{\partial n^S} + \rho^{SR}\rho^S_{oS}\,\frac{\partial \text{U}_c}{\partial \rho^{SR}}\right) \tag{5.31}$$

have been used. Furthermore, it is easily seen that the Kirchhoff stress tensors from eqns (5.28) and (5.29) directly correspond to each other, except that, in eqn (5.29), $c = 0$. From the definition of the configuration and the matrix pressure terms, compare eqns (4.38) and (4.39), it is seen that

$$\begin{aligned}p^{SR}_{\text{con.}} &= -\frac{1}{n^S\det\mathbf{F}_S}\,k^S_{\text{con.}}\,(\frac{\hat{n}^S}{n^S})^c\ln(\frac{\hat{n}^S}{n^S})\,,\\ p^{SR}_{\text{mat.}} &= -\frac{1}{n^S\det\mathbf{F}_S}\,k^S_{\text{mat.}}\ln(\frac{\hat{\rho}^{SR}}{\rho^{SR}})\,.\end{aligned} \tag{5.32}$$

However, for several applications, it may be impossible to obtain experimental data on the configuration and the matrix compression moduli, separately. In considering these circumstances, it is usually suggested that

$$k^S_{\text{mat.}} = k^S_{\text{con.}} = k^S\,, \tag{5.33}$$

and thus, from eqn (5.29) by use of $c = 0$,

$$\boldsymbol{\tau}^S = 2\,\mu^S J_e^{-2/3}\,\mathbf{K}_{Se}^D + k^S \ln J_e\,\mathbf{I}, \tag{5.34}$$

where eqn (5.25) has been used.

From the preceding considerations, the different versions of finite elasticity for compressible porous solid matrices are properly defined. Finally, since finite elasticity must include geometrically linearized theories defined by

$$\begin{aligned} \mathbf{F}_{Se} &\rightarrow \mathbf{R}_{Se}\,, \\ \mathbf{U}_{Se} \rightarrow \mathbf{I}, &\text{ resp. } \mathbf{V}_{Se} \rightarrow \mathbf{I}, \end{aligned} \tag{5.35}$$

compare eqns (3.35), a linearization of the finite law about the state of small elastic strains must yield the Hookean law. Note in passing that this linearization must obviously be carried out by use of the finite law (5.34), since the other versions contain material parameters not occurring in linearized theories. In particular, a linearization with respect to the plastic intermediate configuration implies

$$\hat{\boldsymbol{\tau}}^S{}_{\text{lin.}} = \hat{\boldsymbol{\tau}}^S\Big|_{\hat{\boldsymbol{\Gamma}}_{Se}=\mathbf{0}} + \frac{\partial \hat{\boldsymbol{\tau}}^S}{\partial \hat{\boldsymbol{\Gamma}}_{Se}}\Big|_{\hat{\boldsymbol{\Gamma}}_{Se}=\mathbf{0}}\,\hat{\boldsymbol{\Gamma}}_{Se}\,. \tag{5.36}$$

Therein, by pull-back of eqn (5.34), compare Table 5,

$$\begin{aligned} \hat{\boldsymbol{\tau}}^S &= \mu^S J_e^{-2/3}\,[\,(\mathbf{I} - \tfrac{1}{3}(\,\hat{\mathbf{C}}_{Se}\cdot\mathbf{I}\,)\,\hat{\mathbf{C}}_{Se}^{-1}\,] + k^S \ln J_e\,\hat{\mathbf{C}}_{Se}^{-1}\,, \\ \hat{\boldsymbol{\tau}}^S\Big|_{\hat{\boldsymbol{\Gamma}}_{Se}=\mathbf{0}} &= \mathbf{0} \end{aligned} \tag{5.37}$$

hold, where $\mathbf{F}_{Se} = \mathbf{R}_{Se}$, satisfying $\hat{\boldsymbol{\Gamma}}_{Se} = \mathbf{0}$, $\hat{\mathbf{C}}_{Se} = \mathbf{I}$ and $J_e = 1$, has been used. Furthermore, the elastic tangent operator included into eqn (5.36) reads

$$\begin{aligned} \overset{4}{\hat{\mathbf{C}}}_e &= \frac{\partial \hat{\boldsymbol{\tau}}^S}{\partial \hat{\boldsymbol{\Gamma}}_{Se}} = 2\,\frac{\partial \hat{\boldsymbol{\tau}}^S}{\partial \hat{\mathbf{C}}_{Se}} \\ &= k^S(\,\hat{\mathbf{C}}_{Se}^{-1}\otimes\hat{\mathbf{C}}_{Se}^{-1}\,) - 2\,k^S \ln J_e(\,\hat{\mathbf{C}}_{Se}^{-1}\otimes\hat{\mathbf{C}}_{Se}^{-1}\,)^{\overset{23}{T}} \\ &\quad - \tfrac{2}{3}\,J_e^{-2/3}\mu^S[\,(\,\mathbf{I}\otimes\hat{\mathbf{C}}_{Se}^{-1}\,) - \tfrac{1}{3}(\,\hat{\mathbf{C}}_{Se}\cdot\mathbf{I}\,)(\,\hat{\mathbf{C}}_{Se}^{-1}\otimes\hat{\mathbf{C}}_{Se}^{-1}\,) \\ &\qquad + (\,\hat{\mathbf{C}}_{Se}^{-1}\otimes\mathbf{I}\,) - (\,\hat{\mathbf{C}}_{Se}\cdot\mathbf{I}\,)(\,\hat{\mathbf{C}}_{Se}^{-1}\otimes\hat{\mathbf{C}}_{Se}^{-1}\,)^{\overset{23}{T}}\,]. \end{aligned} \tag{5.38}$$

Following this, it is concluded that

$$\overset{4}{\hat{\mathbf{C}}}_e\Big|_{\hat{\mathbf{C}}_{Se}=\mathbf{I}} = k^S(\mathbf{I}\otimes\mathbf{I}) + 2\,\mu^S[\,(\mathbf{I}\otimes\mathbf{I})^{\overset{23}{T}} - \tfrac{1}{3}(\mathbf{I}\otimes\mathbf{I})\,], \tag{5.39}$$

where

$$(\,\mathbf{I}\otimes\mathbf{I}\,)^{\overset{23}{T}} = \overset{4}{\mathbf{I}} \tag{5.40}$$

defines the fourth-order identity. Thus, by combination of eqns (5.36)–(5.38), the linearized elasticity law of Hookean type yields

$$\hat{\tau}^S_{\text{lin.}} = 2\,\mu^S\,\hat{\Gamma}^D_{Se} + k^S\,(\hat{\Gamma}_{Se}\cdot\mathbf{I})\,\mathbf{I}. \tag{5.41}$$

Considering the above result, it is obvious that another possibility to linearize the finite law, eqn (5.34), is given by a linearization with respect to the actual configuration. Following this, an alternative version of eqn (5.36) is

$$\tau^S_{\text{lin.}} = \tau^S\Big|_{\mathbf{A}_{Se}=\mathbf{0}} + \frac{\partial\tau^S}{\partial\mathbf{A}_{Se}}\Big|_{\mathbf{A}_{Se}=\mathbf{0}}\,\mathbf{A}_{Se}\,, \tag{5.42}$$

where

$$\tau^S\Big|_{\mathbf{A}_{Se}=\mathbf{0}} = \mathbf{0} \tag{5.43}$$

to be obtained with the aid of $\mathbf{F}_{Se} = \mathbf{R}_{Se}$ has been used. However, it can be shown that the elastic tangent operator corresponding to the actual configuration is related from $\overset{4}{\hat{\mathbf{C}}}_e$ by a covariant fourth-order push-forward transformation, viz.:

$$\begin{aligned}
\overset{4}{\mathbf{C}}_e &= (\mathbf{F}_{Se}\otimes\mathbf{F}_{Se})^{\overset{23}{T}}\;\overset{4}{\hat{\mathbf{C}}}_e\;(\mathbf{F}^T_{Se}\otimes\mathbf{F}^T_{Se})^{\overset{23}{T}}\\
&= k^S\,(\mathbf{I}\otimes\mathbf{I}) - 2\,k^S\,\ln J_e\,(\mathbf{I}\otimes\mathbf{I})^{\overset{23}{T}}\\
&\quad - \tfrac{2}{3}\,J_e^{-2/3}\,\mu^S\,[\,(\mathbf{B}_{Se}\otimes\mathbf{I}) - \tfrac{1}{3}(\mathbf{B}_{Se}\cdot\mathbf{I})(\mathbf{I}\otimes\mathbf{I})\\
&\qquad + (\mathbf{I}\otimes\mathbf{B}_{Se}) - (\mathbf{B}_{Se}\cdot\mathbf{I})(\mathbf{I}\otimes\mathbf{I})^{\overset{23}{T}}\,],
\end{aligned} \tag{5.44}$$

and hence,

$$\overset{4}{\mathbf{C}}_e\Big|_{\mathbf{B}_{Se}=\mathbf{I}} = k^S\,(\mathbf{I}\otimes\mathbf{I}) + 2\,\mu^S[\,(\mathbf{I}\otimes\mathbf{I})^{\overset{23}{T}} - \tfrac{1}{3}(\mathbf{I}\otimes\mathbf{I})\,]. \tag{5.45}$$

Following this, one easily concludes to

$$\tau^S_{\text{lin.}} = 2\,\mu^S\,\mathbf{A}^D_{Se} + k^S\,(\mathbf{A}_{Se}\cdot\mathbf{I})\,\mathbf{I}, \tag{5.46}$$

eqn (5.46) thus defining the linearized elasticity law of Hookean type given with respect to the actual configuration. From eqns (5.41) and (5.46), it is easily concluded from the transport mechanisms of strain and stress measures that the above linearizations violate the general concept of covariance. Thus, the linearized elasticity laws can only be related to each other if the covariant and the contravariant transport mechanisms coincide, namely, within the range of vanishing elastic stretch, that is, $\mathbf{F}_{Se} = \mathbf{R}_{Se}$.

5.3 Incompressible matrix materials

As was pointed out in Subsection 5.1, the description of finite elasticity for incompressible porous matrix materials is governed by the same basic relations as are used within the general compressible range. Thus, one obtains in formal analogy to eqn (5.16) that

$$\begin{aligned} \boldsymbol{\tau}_E^S &= 2\,\rho_{oS}^S\,\mathbf{F}_S\,\frac{\partial\tilde{\psi}_{in}^{Se}}{\partial\hat{\mathbf{C}}_{Se}}\,\mathbf{F}_S^T \\ &= 2\,\rho_{oS}^S\,J_e^{-2/3}\,(\,\mathbf{F}_S\,\frac{\partial\tilde{\psi}_{in}^{Se}}{\partial\overline{\mathbf{C}}_{Se}}\,\mathbf{F}_S^T\,)^D + \rho_{oS}^S\,J_e\,\frac{\partial\tilde{\psi}_{in}^{Se}}{\partial J_e}\,\mathbf{I}\,, \end{aligned} \tag{5.47}$$

where, eqns (5.3) and $(5.13)_2$ have been used.

Furthermore, following the same arguments as were given by eqns (5.19)–(5.25), we easily accept that

$$\rho_{oS}^S\,\tilde{\psi}_{in}^{Se} = [\,\tfrac{1}{2}\,\mu^S\,(\,\bar{\mathrm{I}}_e - 3) + \rho_{oS}^S\,\mathrm{U}_{in}(\,J_e,\,\mathbf{F}_{Sp}\,)\,]\Big|_{\mathbf{F}_{Sp}\ \mathrm{fixed}}\,, \tag{5.48}$$

where an alternative expression for U_i is given by

$$\mathrm{U}_{in}(\,J_e,\,\mathbf{F}_{Sp}\,)\,]\Big|_{\mathbf{F}_{Sp}\ \mathrm{fixed}} = \mathrm{U}_{in}(\,n^S,\,\mathbf{F}_{Sp}\,)\,]\Big|_{\mathbf{F}_{Sp}\ \mathrm{fixed}}\,. \tag{5.49}$$

Therein, the independent state variable J_e has been substituted by the alternative variable n^S via

$$n^S = \hat{n}^S\,J_e^{-1}, \quad \rho^{SR} = \hat{\rho}^{SR}, \tag{5.50}$$

compare eqns (4.1)–(4.4). Note in passing that the solid volume fraction $\hat{n}^S$ of the plastic intermediate configuration of φ^S is again kept fixed at a given plastic state of deformation. Thus, another representation of the above stored energy function of the incompressible matrix material is

$$\rho_{oS}^S\,\tilde{\psi}_{in}^{Se} = [\,\tfrac{1}{2}\,\mu^S\,(\,\bar{\mathrm{I}}_e - 3) + \rho_{oS}^S\,\mathrm{U}_{in}(\,n^S,\,\mathbf{F}_{Sp}\,)\,]\Big|_{\mathbf{F}_{Sp}\ \mathrm{fixed}}\,. \tag{5.51}$$

From these results, one immediately concludes to

$$\boldsymbol{\tau}_E^S = \mu^S\,J_e^{-2/3}\,\mathbf{B}_{Se}^D + \rho_{oS}^S\,J_e\,\frac{\partial\mathrm{U}_{in}}{\partial J_e}\Big|_{\mathbf{F}_{Sp}\ \mathrm{fixed}}\,\mathbf{I}\,, \tag{5.52}$$

where the second term can easily be substituted by

$$\rho_{oS}^S\,J_e\,\frac{\partial\mathrm{U}_{in}}{\partial J_e}\Big|_{\mathbf{F}_{Sp}\ \mathrm{fixed}} = -\,n^S\,\rho_{oS}^S\,\frac{\partial\mathrm{U}_{in}}{\partial n^S}\Big|_{\mathbf{F}_{Sp}\ \mathrm{fixed}}\,, \tag{5.53}$$

additionally, compare eqn (5.31).

In comparison with the relations (5.26) and (5.27), one comes to realize that the function U_{in} corresponding to the compressible range of the porous matrix materials under consideration (herewith, we mean the compressiblitity properties of the pore structure) can be expressed via

$$\rho_{oS}^S \, \mathrm{U}_{in} \Big|_{\mathbf{F}_{Sp} \text{ fixed}} = k_{\text{con.}}^S \, (\frac{\hat{n}^S}{n^S})^c \, \frac{c \ln(\frac{\hat{n}^S}{n^S}) - 1}{c^2} + \frac{k_{\text{con.}}^S}{c^2} \, . \tag{5.54}$$

Alternatively, given $c = 0$,

$$\rho_{oS}^S \, \bar{\mathrm{U}}_{in} \Big|_{\mathbf{F}_{Sp} \text{ fixed}} = \tfrac{1}{2} \, k_{\text{con.}}^S \, \left(\ln \frac{\hat{n}^S}{n^S} \right)^2 . \tag{5.55}$$

Therein, $k_{\text{con.}}^S$ again defines the configuration compression modulus. However, note in passing that, obviously, there is no matrix compression modulus to be included into the functions U_{in} or $\bar{\mathrm{U}}_{in}$, respectively.

Given eqns (5.54) and (5.55), it is not difficult to find that

$$\boldsymbol{\tau}_E^S = 2 \, \mu^S \, J_e^{-2/3} \, \mathbf{K}_{Se}^D + k_{\text{con.}}^S \, (\frac{\hat{n}^S}{n^S})^c \, \ln(\frac{\hat{n}^S}{n^S}) \, \mathbf{I} \, , \tag{5.56}$$

or, proceeding from $c = 0$,

$$\boldsymbol{\tau}_E^S = 2 \, \mu^S \, J_e^{-2/3} \, \mathbf{K}_{Se}^D + k_{\text{con.}}^S \, \ln(\frac{\hat{n}^S}{n^S}) \, \mathbf{I} \, . \tag{5.57}$$

Furthermore, it is evident that only the configuration pressure term,

$$p_{\text{con.}}^{SR} = - \frac{1}{n^S \, \det \mathbf{F}_S} \, k_{\text{con.}}^S \, (\frac{\hat{n}^S}{n^S})^c \, \ln(\frac{\hat{n}^S}{n^S}) \, , \tag{5.58}$$

occurs in the above equations, whereas the matrix pressure is naturally excluded from $\boldsymbol{\tau}_E^S$. Thus, for the incompressible matrix,

$$k_{\text{con.}}^S = k^S \, . \tag{5.59}$$

Following this, one easily finds

$$\boldsymbol{\tau}_E^S = 2 \, \mu^S \, J_e^{-2/3} \, \mathbf{K}_{Se}^D + k_{\text{con.}}^S \, J_e{}^c \, \ln J_e \, \mathbf{I} \tag{5.60}$$

to hold, additionally, compare EHLERS (1991), where finite elasticity for incompressible matrices was discussed by a slightly different elasticity law that could be understood as an extension of the Simo-Pister law towards porous media theories, compare SIMO & PISTER (1984).

However, by use of the same linearization procedures as were discussed within the description of compressible matrices, it is easily concluded that

$$\begin{aligned} \hat{\boldsymbol{\tau}}^S{}_{\text{lin.}} &= \hat{\boldsymbol{\tau}}^S \Big|_{\hat{\boldsymbol{\Gamma}}_{Se} = \mathbf{0}} + \frac{\partial \hat{\boldsymbol{\tau}}^S}{\partial \hat{\boldsymbol{\Gamma}}_{Se}} \Big|_{\hat{\boldsymbol{\Gamma}}_{Se} = \mathbf{0}} \hat{\boldsymbol{\Gamma}}_{Se} \\ &= 2 \, \mu^S \, \hat{\boldsymbol{\Gamma}}_{Se}^D + k_{\text{con.}}^S \, (\hat{\boldsymbol{\Gamma}}_{Se} \cdot \mathbf{I}) \, \mathbf{I} \end{aligned} \tag{5.61}$$

or, respectively,

$$\begin{aligned} \boldsymbol{\tau}^S_{E\,\mathrm{lin.}} &= \boldsymbol{\tau}^S{}_{\mathrm{lin.}}\Big|_{\mathbf{A}_{Se}=\mathbf{0}} + \frac{\partial \boldsymbol{\tau}^S}{\partial \mathbf{A}_{Se}}\Big|_{\mathbf{A}_{Se}=\mathbf{0}} \mathbf{A}_{Se} \\ &= 2\,\mu^S\,\mathbf{A}^D_{Se} + k^S_{\mathrm{con.}}\,(\,\mathbf{A}_{Se}\cdot\mathbf{I}\,)\,\mathbf{I}\,. \end{aligned} \tag{5.62}$$

Equation (5.61) represents the linearization with respect to the plastic intermediate configuration, whereas eqn (5.62) linarizes the finite law by considering the geometry of the actual configuration.

The following discussion concerns the experimental evidence of the material parameter c. To explain this number, consider the range of purely spherical stresses which can be introduced with the aid of the multiplicative split of $\mathbf{F}_{Se}$ defined by eqn (4.32), that is,

$$\mathbf{F}_{Se} = J_e{}^{1/3}\,\overline{\mathbf{F}}_{Se}\,. \tag{5.63}$$

From eqns (5.60) and (5.62), the hydrostatic part of the solid extra stress tensor $\boldsymbol{\tau}^S_E$ yields by use of $\overline{\mathbf{F}}_{Se} = \mathbf{I}$:

$$\begin{aligned} \boldsymbol{\tau}^S_{E\,sph} &= -\,P^S\,\mathbf{I}\,, \\ P^S &= -\,k^S_{\mathrm{con.}}\,J_e{}^c\,\ln\,J_e\,\mathbf{I}\,. \end{aligned} \tag{5.64}$$

In this place, note again that macroscopic compressibility of an incompressible porous solid is, without exception, only due to variations in porosity. Thus, the point of compaction (closed pore spaces) cannot be exceeded, that is, beyond this point, the matrix material behaves incompressible. Since, within the range of purely elastic deformations ($\mathbf{F}_S \rightarrow \mathbf{F}_{Se} = n^S_{oS}/n^S$), the point of compaction is defined by $n^S \rightarrow 1$ and $P^S \rightarrow \overset{*}{P}{}^S$, where $\overset{*}{P}{}^S$ defines the material depending pressure of compaction (EHLERS, 1989b, 1991), it is easily concluded that

$$\overset{*}{P}{}^S = -\,k^S_{\mathrm{con.}}\,(n^S_{oS})^c\,\ln\,n^S_{oS}\,, \tag{5.65}$$

where eqn (5.50) together with $\hat{n}^S = n^S_{oS}$ has been used. Following this, it is obvious that c must be interpreted as the control parameter for $\overset{*}{P}{}^S$, compare Figure 3.

6. The solid constituent: Finite plasticity

6.1 Preliminary remarks

In the preceding sections, a general discussion on four different types of two-phase media was carried out, where no principle difference has been made concerning the skeleton material properties, except for the fact that the matrix material itself could be considered to be either compressible or incompressible. Furthermore, a plastic predeformation was included into the discussion by allowing, within the general framework of finite elastic solid deformations, for any value of a prescribed plastic state of deformation.

Considering constitutive equations for granular materials in geomechanical context, it is very well known that, within the majority of problems, the range of elastic deformations is usually small, whereas the plastic deformations exhibit considerable values. Therefore, a finite plasticity theory is necessary to govern the material response. However, it is

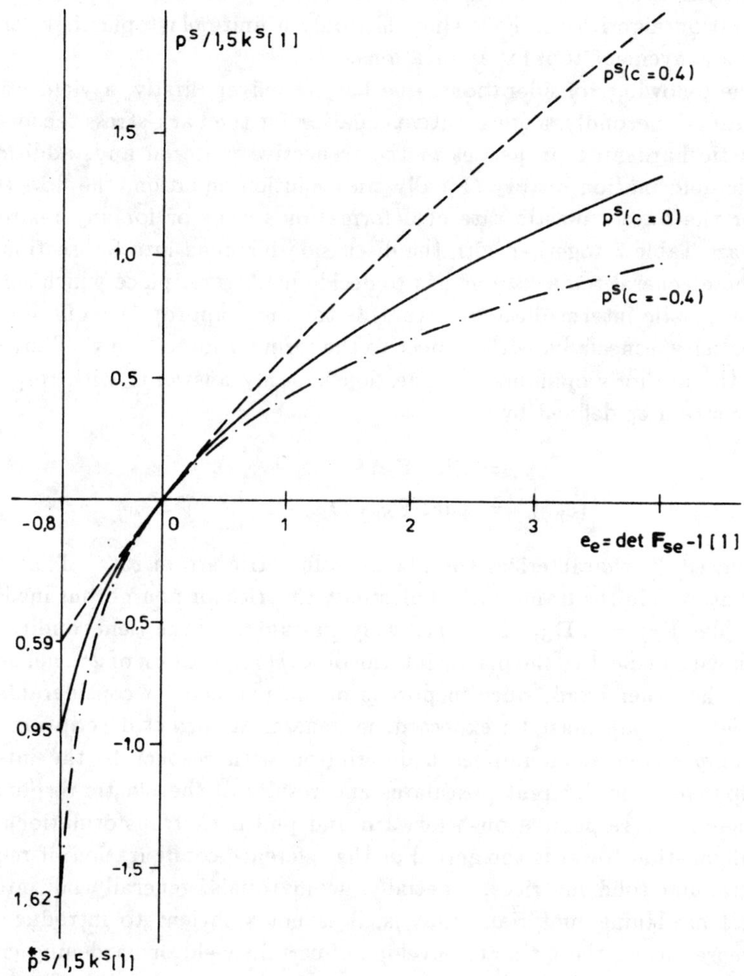

Figure 3: Hydrostatic stress P^S versus volumetric strain e_e of a purely elastic empty solid matrix ($n_{oS}^S = 0.2$).

known that a growing amount of numerical strategies in finite elastoplasticity proceeds from predictor-corrector schemes, thus making use of finite elastic deformations, at least within the numerical iteration algorithms. Following this, only a finite elastoplasticity theory can be understood as a convenient tool for applications.

Concerning the following considerations, one has to deliver, firstly, a yield criterion to frame the elastic range. Secondly, a constitutive equation for the back-stress tensor is needed to reflect the plastic hardening properties of the respective material and, additionally, to include the plastic deformation history. Finally, an evolution equation, the flow rule, must be formulated for the purely plastic rate of deformation tensor or for any related plastic strain rate, compare Table 3 together with the discussion included into Subsection 3.3.

Apart from these general remarks, one has to decide in the first place which solid configuration (reference, plastic intermediate or actual) is the most appropriate one for the basic description of plasticity, especially, with respect to the formulation of the yield criterion and the fow rule. In the author's opinion, this question is easily answered with respect to the plastic volumetric strain e_p defined by

$$\begin{aligned} e_p &= \det \mathbf{F}_{Sp} - 1\,, \\ (e_p)'_S &= \det \mathbf{F}_{Sp}\,(\hat{\mathbf{D}}_{Sp} \cdot \mathbf{I}), \end{aligned} \tag{6.1}$$

where, in addition, $(e_p)'_S$ characterizes the plastic volumetric strain rate. Thus, it is seen from eqns (6.1) that, within the framework of plasticity theories for non-porous media, plastic incompressibility ($\det \mathbf{F}_{Sp} = 1$; $\hat{\mathbf{D}}_{Sp} \cdot \mathbf{I} = 0$) is easily guaranteed if the yield condition and the flow rule are given with respect to the plastic intermediate configuration of φ^S, hence ensuring $\hat{\mathbf{D}}_{Sp} = \hat{\mathbf{D}}^D_{Sp}$. On the other hand, since in porous media plasticity a considerable value of the plastic volumetric strain must be expected, in general, a correct description of plastic dilatation and compression again implies a description with respect to the intermediate configuration. However, the different postulates and results of the plastic response can be transferred by use of the respective push-forward and pull-back transformations from the intermediate configuration towards the actual or the reference configuration, if required.

Brittle and granular solid matrices, especially, geomaterials, generally fall into the category of frictional hardening materials, that is, it is not sufficient to introduce a failure function alone, representing the ultimate envelope of certain yield or hardening criteria, respectively, but the respective yield function itself must be properly defined. Furthermore, it is known that ultimate envelopes exhibit an open cone in the principal stress space (the opening is directed towards the mean pressure), thus representing an unsensivity of failure states versus pure pressure (grain crunching excluded), whereas yield and hardening functions are characterized by an explicit pressure dependence, thus implying the yield and hardening surfaces to represent a smooth and closed shape in stress space. These types of yield surfaces are usually known as single surface yield criteria, in contrast to the different cap models, where the open cone of the failure surface is closed by use of another yield function, the cap. However, since cap models show certain disadvantages (some comments on classsical cap models can be taken from DESAI, 1989), it proves to be very convenient to proceed, in geomaterials terminology, from single surface yield criteria, thus describing such phenomena like phase change properties and brittle-ductile transition behaviour (compare, DESAI, 1990). In proposing convenient single surface models, two further requirements for the representation of the respective criteria should be considered, namely, (a) the possibility of only using one single mathematical function for the description of yielding, hardening (isotropic and/or kinematic) and failure, and (b) the inclusion of the smallest possible number of parameters to successfully represent the material response.

Meeting the above requirements, the present section concerns a new 7-parameter single surface yield criterion (EHLERS, in prep.) for brittle and granular matrices, especially, geomaterials, where, by use of a convexity condition in the deviatoric plane, the general 7-parameter form is easily reduced to a 6-parameter representation. The section furthermore contains a non-associated flow rule for the plastic strain rates and, in the framework of kinematical hardening materials, a constitutive equation for the back-stress (translation) tensor. The different findings are given with respect to geometrical nonlinear (finite) theories naturally including the usual small strain approaches.

6.2 Definition of a seven-parameter single surface yield criterion

(a) The necessary set of parameters

In the framework of geomaterials, the different yield functions generally meet the basic requirements of porous media plasticity, that is, the shapes of these functions, in the octahedral plane and in the hydrostatic plane or the triaxial plane, respectively, are principally characterized as is shown in Figure 4 and Figure 5, additionally, compare LADE (1979), DE BOER (1988) or KIM & LADE (1988). Following this, both the yield function and the failure condition must show, in the octahedral plane, the shape of a rounded triangle, compare Figure 4. This typical shape, which sometimes becomes more distinct near failure, is necessary for the description of triaxial compression, simple shear and extension tests.

Thus, concerning the necessary number of material parameters or material functions, respectively, to be included into a general single surface criterion, it is easily concluded that, firstly, one needs at least one or better two material parameters to govern the variation of the yield radius $\hat{r}$ with varying characterizing yield angles $\hat{\Theta}$. In using the latter possibility, one of the two parameters is responsible for the "smoothness" of the rounded triangle's shape, whereas the other one defines the ratio $\hat{r}_c/\hat{r}_t$ of the compression and the extension radii. Secondly, considering the hydrostatic plane, compare Figure 5, the closed shape (tear drop) must be governed by five further parameters. In particular, two parameters are necessary to specify the radius $\hat{r}_o(\hat{\Theta})$ and the tangential angle $\hat{\varphi}_o(\hat{\Theta})$ at vanishing hydrostatic pressure ($\hat{I} = 0$), for example, in the simple shear regime ($\hat{\Theta} = 0° : \hat{r}_{os}, \hat{\varphi}_{os}$), compare Figure 5 (b). Note in passing that herewith and in the remainder of this section, the notion "hydrostatic pressure" defines that part of the total hydrostatic pressure acting on the two-phase medium that influences solid plasticity. In particular, it follows from the concept of constituent plastic incompressibility, that this pressure necessarily is the solid configuration pressure.

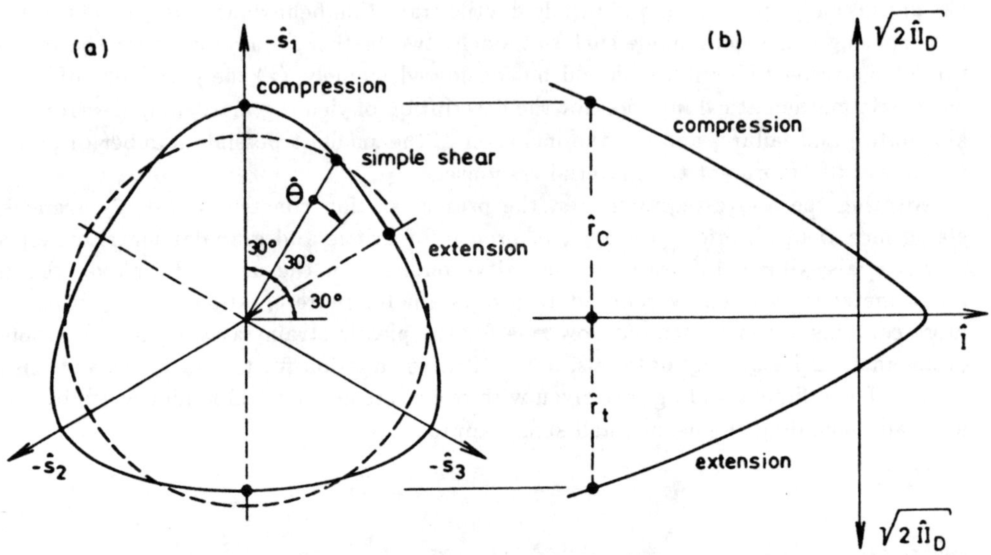

Figure 4: Characteristic shape of the general yield function in (a) octahedral plane and (b) hydrostatic plane (section). $\hat{I}$, $\hat{II}_D$: first and deviatoric second principal invariants of the stress state (positive sign means tension); $\hat{s}_1$, $\hat{s}_2$, $\hat{s}_3$: principal stresses.

In addition, one must introduce three parameters to independently control the length of the yield surface and the location and the amplitude of its maximum; this maximum usually characterizes the transition point between brittle and ductile behaviour (phase change[3]).

Following the above argumentation, one comes to realize that a general yield function for brittle and granular solid materials should proceed from at least seven material parameters, where, in the case of vanishing cohesion ($\hat{r}_o = 0$), the necessary number of parameters reduces to six. Furthermore, a basic requirement on any general yield function should be its reductability to simpler shapes. In particular, the mathematical formulation of the yield function should not only include the open cone of the ultimate envelope by reducing the necessary number of parameters from seven to five or from six to four, respectively, but it should also naturally include simpler yield surfaces like, for instance, the well-known Drucker-Prager criterion (compare, DRUCKER & PRAGER, 1952) or the Green criterion for porous metals (GREEN, 1972) and, of course, the classical von Mises type conditions for single-phase metals without voids. In any case, the different reductions should be obtainable by simply replacing selected material parameters by zero.

Finally, it should be noted that the total number of material parameters can be reduced by one, if one makes use of a convexity condition for the shape of the yield surface in the deviatoric plane. In this case, the two material parameters governing the variation of $\hat{r}$ with

[3]In geomaterials plasticity, the notion phase change addresses the transition of the material from brittle to ductlie behaviour in the high pressure regime and has nothing to do with the introduction of mass supply ("phase change") terms included into classical mixture theories.

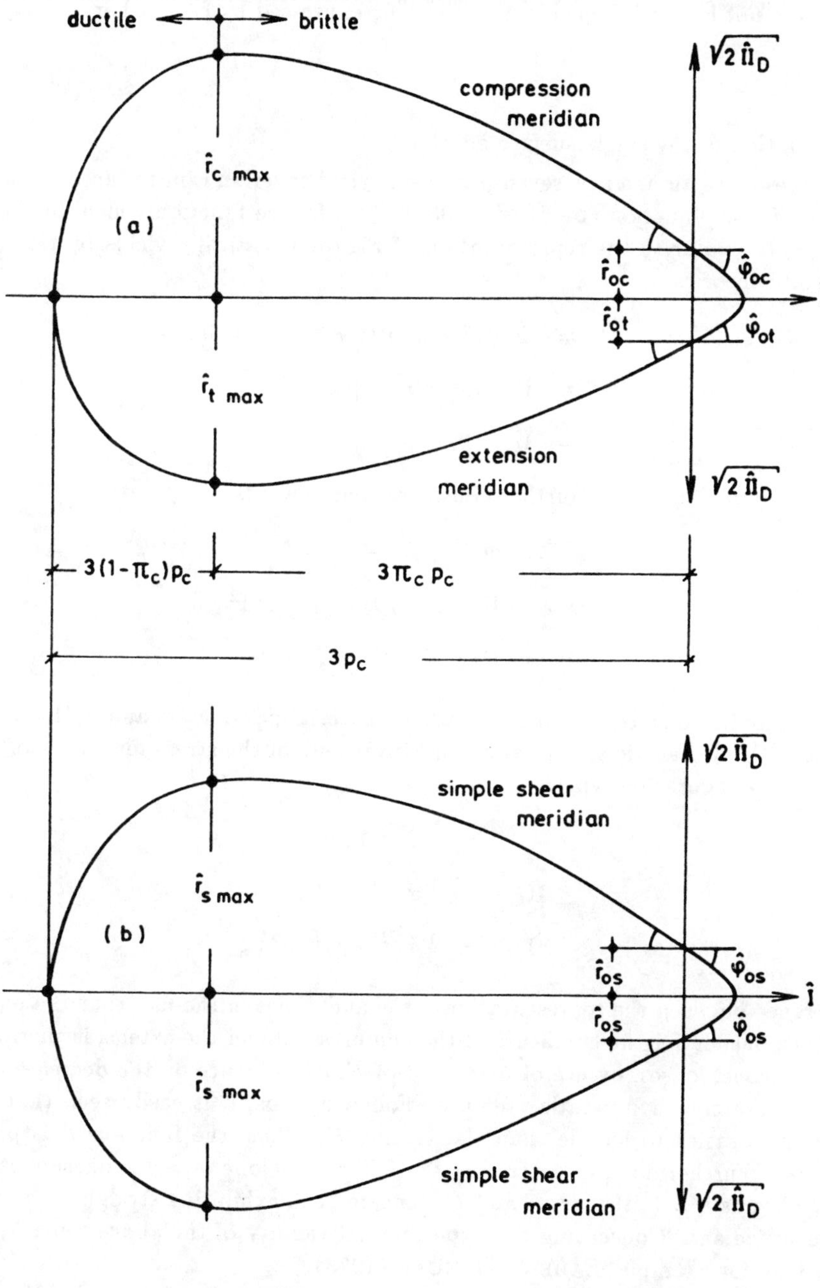

Figure 5: Characteristic shape of the general yield function in the hydrostatic plane (full representation). (a) compression and extension meridians ($\hat{\Theta} \mp 30°$). (b) simple shear meridians ($\hat{\Theta} = 0°$).

varying $\hat{\Theta}$ are not longer independent but can be expressed by each other.

(b) Definition of the single surface envelope

In the present section, a new seven-parameter yield criterion is introduced as an extension of de Boer's failure function (DE BOER, 1988). This failure function which, in the principal stress space, is similar to the representation of Figure 4, basically yields by transformation towards finite theories,

$$\begin{aligned} \bar{F} &= \bar{\Phi}^{1/2}(1+\gamma\bar{\vartheta})^{1/3} + \beta\,\hat{\mathrm{I}} - \kappa = 0\,, \\ \bar{\Phi} &= \hat{\mathrm{II}}_D + \tfrac{1}{2}\alpha^2\hat{\mathrm{I}}^2\,, \\ \bar{\vartheta} &= \hat{\mathrm{III}}_D/\bar{\Phi}^{3/2}\,, \end{aligned} \tag{6.2}$$

and may be simplified for our further considerations towards

$$\begin{aligned} \tilde{F} &= \tilde{\Phi}^{1/2} + \beta\,\hat{\mathrm{I}} - \kappa = 0\,, \\ \tilde{\Phi} &= \hat{\mathrm{II}}_D\,(1+\gamma\hat{\vartheta})^{2/3} + \tfrac{1}{2}\,\alpha^2\,\hat{\mathrm{I}}^2\,, \\ \hat{\vartheta} &= \hat{\mathrm{III}}_D/\hat{\mathrm{II}}_D{}^{3/2}. \end{aligned} \tag{6.3}$$

In both representations, α, β, γ and κ are four material parameters and $\hat{\mathrm{I}}\,, \hat{\mathrm{II}}_D$ and $\hat{\mathrm{III}}_D$ are the first and the deviatoric second and third invariants of the stress tensor $\hat{\tau}^S$ of the plastic intermediate configuration, viz.:

$$\begin{aligned} \hat{\mathrm{I}} &= \hat{\tau}^S \cdot \mathbf{I}\,, \\ \hat{\mathrm{II}}_D &= \tfrac{1}{2}\,\hat{\tau}^{SD} \cdot \hat{\tau}^{SD}\,, \\ \hat{\mathrm{III}}_D &= \tfrac{1}{3}\,\hat{\tau}^{SD} \cdot \hat{\tau}^{SD}\,\hat{\tau}^{SD}\,. \end{aligned} \tag{6.4}$$

The difference between the representations of $\bar{F}$ and $\tilde{F}$ lies in the fact that the parameter α, which is responsible for the curvature of the compression and the extension meridians of the open failure cone, looses, by use of $\tilde{F}$ instead of $\bar{F}$, its influence on the decrease of the ratio $\hat{r}_c/\hat{r}_t$ with increasing hydrostatic pressure. Following this, it is easily seen that the failure condition $\tilde{F}$ is easier to handle than the version $\bar{F}$. Thus, the function $\tilde{F}$ is preferred for what follows. Furthermore, in describing the failure envelope of, e.g., cohesionless soils, one generally obtains $\alpha = 0$, that is, $\bar{F}$ and $\tilde{F}$ coincide both exhibiting straight compression and extension meridians. Concerning the experimental validity of the above remark, the reader is referred to, for example, LADE & DUNCAN (1973).

The present modifications of de Boer's failure condition toward the general seven-parameter "single surface" yield function are given as follows,

$$\begin{aligned} F &= \hat{\Phi}^{1/2} + \beta\,\hat{\text{I}} + \epsilon\,\hat{\text{I}}^2 - \kappa = 0\,, \\ \hat{\Phi} &= \hat{\text{II}}_D\,(1+\gamma\,\hat{\vartheta})^m + \tfrac{1}{2}\alpha^2\,\hat{\text{I}}^2 + \delta^2\,\hat{\text{I}}^4\,, \\ \hat{\vartheta} &= \hat{\text{III}}_D/\hat{\text{II}}_D{}^{3/2}, \end{aligned} \tag{6.5}$$

thus extending the failure envelope $\tilde{F}$ by introducing the term $\epsilon\,\hat{\text{I}}^2$ [4] and modifying the function $\tilde{\Phi}$ by addition of the term $\delta^2\,\hat{\text{I}}^4$. Finally, the fixed power $(\ldots)^{2/3}$ is substituted by the material parameter m (additionally compare the shape function F_S of the HISS model included into DESAI, 1990). By use of the Reuss variables

$$\begin{aligned} \hat{r} &= \sqrt{2\,\hat{\text{II}}_D}\,, \\ \hat{\Theta} &= \tfrac{1}{3}\,\arcsin\left(\frac{\sqrt{27}}{2}\,\frac{\hat{\text{III}}_D}{\hat{\text{II}}_D{}^{3/2}}\right), \end{aligned} \tag{6.6}$$

where $\hat{r}$ is the yield radius and $\hat{\Theta}$ the characterizing deviatoric yield angle, compare Figure 4 (a), the representation of $F = F(\hat{\text{I}}\,,\,\hat{\text{II}}_D\,,\,\hat{\text{III}}_D)$ can be replaced by $F = F(\hat{\text{I}}\,,\,\hat{r}\,,\,\hat{\Theta})$, viz.:

$$\begin{aligned} F &= \hat{\Phi}^{1/2} + \beta\,\hat{\text{I}} + \epsilon\,\hat{\text{I}}^2 - \kappa = 0\,, \\ \hat{\Phi} &= \tfrac{1}{2}\,\hat{r}^2(1+\gamma\,\hat{\vartheta})^m + \tfrac{1}{2}\,\alpha^2\,\hat{\text{I}}^2 + \delta^2\,\hat{\text{I}}^4, \\ \hat{\vartheta} &= \frac{2}{\sqrt{27}}\,\sin(3\hat{\Theta}). \end{aligned} \tag{6.7}$$

Another convenient form of the yield envelope is given by solving eqn $(6.7)_1$ with respect to $\hat{r}$. Thus, one obtains with the aid of eqns (6.6) that

$$\begin{aligned} \hat{r}\,(\hat{\Theta},\hat{\text{I}}\,) = \sqrt{2}\,\Bigg\{\Bigg[&(\epsilon^2-\delta^2)\,\hat{\text{I}}^4 + 2\,\beta\,\epsilon\,\hat{\text{I}}^3 + (\beta^2 - \tfrac{1}{2}\,\alpha^2 - 2\,\epsilon\,\kappa)\,\hat{\text{I}}^2 \\ &-2\,\beta\,\kappa\,\hat{\text{I}} + \kappa^2\Bigg]\left[1 + \frac{2}{\sqrt{27}}\,\gamma\,\sin(3\hat{\Theta})\right]^{-m}\Bigg\}^{1/2}. \end{aligned} \tag{6.8}$$

It is easily seen that the above representation can be expressed via

$$\hat{r}\,(\hat{\Theta},\hat{\text{I}}\,) = \mathcal{F}_h(\,\hat{\text{I}}\,)\,\mathcal{F}_d(\hat{\Theta})\,, \tag{6.9}$$

where

$$\begin{aligned} \mathcal{F}_h &= \sqrt{2}\,[(\epsilon^2-\delta^2)\,\hat{\text{I}}^4 + 2\,\beta\,\epsilon\,\hat{\text{I}}^3 + (\beta^2 - \tfrac{1}{2}\,\alpha^2 - 2\,\epsilon\,\kappa)\,\hat{\text{I}}^2 - 2\,\beta\,\kappa\,\hat{\text{I}} + \kappa^2\,]^{1/2}\,, \\ \mathcal{F}_d &= [1 + \frac{2}{\sqrt{27}}\,\gamma\,\sin(3\hat{\Theta})]^{-m/2} \end{aligned} \tag{6.10}$$

[4] note in passing that a term $\epsilon\,\hat{\text{I}}^2$ was firstly used by Dresenkamp in his thesis of 1987 (DRESENKAMP, 1987).

are the shape functions of the yield envelope in the hydrostatic plane, $\mathcal{F}_h = \mathcal{F}_h(\hat{\mathrm{I}})$, and in the deviatoric plane, $\mathcal{F}_d = \mathcal{F}_d(\hat{\Theta})$, respectively. Following this, it is concluded that $\hat{r}_s = \mathcal{F}_h$ represents the yield function in the simple shear regime. Consequently, the multiplier $\mathcal{F}_d$ governs, in the octahedral plane, the typical deviation from a circle toward the compression and the extension radii, $\hat{r}_c$ and $\hat{r}_t$, additionally compare Figure 4 and Figure 5. From the above representation, one comes to realize that the five material parameters included into $\mathcal{F}_h$, α, β, δ, ϵ, and κ, are the governing parameters of the hydrostatic plane and can thus be taken, for example, from triaxial shear tests, whereas the remaining two parameters included into $\mathcal{F}_d$, γ and m, are responsible, in the deviatoric plane, for the smoothness of the rounded triangle's shape, γ, and for the ratio $\hat{r}_c/\hat{r}_t$ of the compression and the extension radii, m. In addition to the above remarks and comments, it should be noted that, in the principal stress space, the invariants $\hat{\mathrm{I}}$, $\hat{\mathrm{II}}_D$ and $\hat{\mathrm{III}}_D$ and furthermore the Reuss variables $\hat{r}$ and $\hat{\Theta}$ are given in case of compression, simple shear and extension tests by:

compression ($\hat{s}_1 < \hat{s}_2 = \hat{s}_3$):

$$\left.\begin{aligned} \hat{\mathrm{I}}_c &= \hat{s}_1 + 2\,\hat{s}_3 \\ \hat{\mathrm{II}}_{Dc} &= \tfrac{1}{3}\,(\hat{s}_1 - \hat{s}_3)^2 \\ \hat{\mathrm{III}}_{Dc} &= \tfrac{2}{27}\,(\hat{s}_1 - \hat{s}_3)^3 \end{aligned}\right\} \begin{aligned} &\hat{r}_c = \sqrt{\tfrac{2}{3}}\,(\hat{s}_3 - \hat{s}_1) \\ &\hat{\Theta}_c = -30^\circ \end{aligned} \tag{6.11}$$

simple shear ($\hat{s}_2 = \dfrac{\hat{s}_1 + \hat{s}_3}{2}$):

$$\left.\begin{aligned} \hat{\mathrm{I}}_s &= \tfrac{3}{2}\,(\hat{s}_1 + \hat{s}_3) \\ \hat{\mathrm{II}}_{Ds} &= \tfrac{1}{4}\,(\hat{s}_1 - \hat{s}_3)^2 \\ \hat{\mathrm{III}}_{Ds} &= 0 \end{aligned}\right\} \begin{aligned} &\hat{r}_s = \tfrac{1}{2}\sqrt{2}\,(\hat{s}_3 - \hat{s}_1) \\ &\hat{\Theta}_s = 0 \end{aligned} \tag{6.12}$$

extension ($\hat{s}_1 > \hat{s}_2 = \hat{s}_3$):

$$\left.\begin{aligned} \hat{\mathrm{I}}_t &= \hat{s}_1 + 2\,\hat{s}_3 \\ \hat{\mathrm{II}}_{Dt} &= \tfrac{1}{3}\,(\hat{s}_1 - \hat{s}_3)^2 \\ \hat{\mathrm{III}}_{Dt} &= \tfrac{2}{27}\,(\hat{s}_1 - \hat{s}_3)^3 \end{aligned}\right\} \begin{aligned} &\hat{r}_t = \sqrt{\tfrac{2}{3}}\,(\hat{s}_1 - \hat{s}_3) \\ &\hat{\Theta}_t = 30^\circ \end{aligned} \tag{6.13}$$

Therein, $\hat{s}_1$, $\hat{s}_2$ and $\hat{s}_3$ are the principal values of the intermediate Kirchhoff stress tensor, $\hat{\tau}^S$; positive values mean "tension".

However, the general seven-parameter single surface yield function is not only appropriate for the description of first yielding, but it can also be used to describe both isotropic and kinematic hardening as well as failure. In the case of first yielding, the included set of parameters,

$$\mathcal{P} = \{\alpha, \beta, \gamma, \delta, \epsilon, m, \kappa\}, \tag{6.14}$$

yields

$$\mathcal{P} \rightarrow \mathcal{P}_o = \{\alpha, \beta, \gamma, \delta, \epsilon, m, \kappa\}_o, \tag{6.15}$$

whereas, in the case of isotropic hardening, we have

$$\mathcal{P} = \mathcal{P}(p^D, p^V). \tag{6.16}$$

Therein, p^D and p^V define the deviatoric and the volumetric strain pathes with respect to the intermediate configuration via

$$\begin{aligned} p^D &= \int_{(t)} \sqrt{\tfrac{2}{3}\hat{\mathbf{D}}_{Sp}^D \cdot \hat{\mathbf{D}}_{Sp}^D}\, Dt, \\ p^V &= \int_{(t)} \tfrac{1}{3}\hat{\mathbf{D}}_{Sp} \cdot \mathbf{I}\, Dt. \end{aligned} \tag{6.17}$$

Furthermore, in using kinematic hardening, one principally proceeds from $\mathcal{P} = \mathcal{P}_o$ and additionally substitutes the stress invariants $\hat{\mathrm{I}}$, $\hat{\mathrm{II}}_D$ and $\hat{\mathrm{III}}_D$ as well as the corresponding Reuss variables $\hat{r}$ and $\hat{\Theta}$ by

$$\begin{aligned} \hat{\mathrm{I}}_k &= (\hat{\boldsymbol{\tau}}^S - \hat{\mathbf{Y}}^S) \cdot \mathbf{I}, \\ \hat{\mathrm{II}}_{kD} &= \tfrac{1}{2}(\hat{\boldsymbol{\tau}}^{SD} - \hat{\mathbf{Y}}^{SD}) \cdot (\hat{\boldsymbol{\tau}}^{SD} - \hat{\mathbf{Y}}^{SD}), \\ \hat{\mathrm{III}}_{kD} &= \tfrac{1}{3}(\hat{\boldsymbol{\tau}}^{SD} - \hat{\mathbf{Y}}^{SD}) \cdot (\hat{\boldsymbol{\tau}}^{SD} - \hat{\mathbf{Y}}^{SD})(\hat{\boldsymbol{\tau}}^{SD} - \hat{\mathbf{Y}}^{SD}), \end{aligned} \tag{6.18}$$

and thus, analogously to eqns (6.6),

$$\begin{aligned} \hat{r}_k &= \sqrt{2\,\hat{\mathrm{II}}_{kD}}, \\ \hat{\Theta}_k &= \tfrac{1}{3} \arcsin\left(\frac{\sqrt{27}}{2} \frac{\hat{\mathrm{III}}_{kD}}{\hat{\mathrm{II}}_{kD}{}^{3/2}}\right). \end{aligned} \tag{6.19}$$

Finally, the open cone of the ultimate envelope is naturally included into the general seven-parameter condition by introducing

$$\mathcal{P} = \mathcal{P}_u = \{\alpha, \beta, \gamma, m, \kappa\}_u, \tag{6.20}$$

where $\epsilon_u = \delta_u = 0$ has been used.

Since finite plasticity implies the yield function to be represented in terms of the plastic intermediate configuration, it is obviously necessary to transform the different material parameters or material functions, respectively, from that configuration, where they were measured (in general, that is the reference configuration), toward the intermediate configuration. With the aid of this parameter transformation, one easily avoids any drifting effect on plastic state variables, respectively, material parameters.

(c) Discussion of material parameters

The octahedral plane (γ, m): In the octahedral plane, the general yield function representation at constant hydrostatic stress, $\hat{\mathrm{I}} = \text{const.}$, is given by

$$\hat{r}\,(\hat{\Theta}, \hat{\mathrm{I}})\Big|_{\hat{\mathrm{I}} = \text{const.}} = \hat{r}_s\,(\hat{\mathrm{I}})\Big|_{\hat{\mathrm{I}} = \text{const.}} \mathcal{F}_d\,(\hat{\Theta})\,, \tag{6.21}$$

where

$$\hat{r}_s\,(\hat{\mathrm{I}}) = \mathcal{F}_h\,(\hat{\mathrm{I}}) \tag{6.22}$$

has been used, compare eqns (6.9) and (6.10). Thus, at any constant $\hat{\mathrm{I}}$, we have

$$\hat{r}\,(\hat{\Theta}) = \hat{r}_s\,[1 + \frac{2}{\sqrt{27}}\,\gamma \sin(3\hat{\Theta})]^{-m/2} \tag{6.23}$$

and, additionally, in the cases of triaxial compression, simple shear and extension:

$$\begin{aligned} \hat{r}_c &= \hat{r}\,(\hat{\Theta} = -30°) = \hat{r}_s\,(1 - \frac{2}{\sqrt{27}}\gamma)^{-m/2}\,, \\ \hat{r}_s &= \hat{r}\,(\hat{\Theta} = 0°)\,, \\ \hat{r}_t &= \hat{r}\,(\hat{\Theta} = 30°) = \hat{r}_s\,(1 + \frac{2}{\sqrt{27}}\gamma)^{-m/2}\,. \end{aligned} \tag{6.24}$$

Thus, the ratio between the compression and the extension radii is governed by

$$\hat{r}_c/\hat{r}_t = \left(\frac{1 + \dfrac{2}{\sqrt{27}}\gamma}{1 - \dfrac{2}{\sqrt{27}}\gamma} \right)^{m/2}\,. \tag{6.25}$$

However, it is seen from different test results on brittle and granular materials (compare, for example Lade & Duncan (1973), Lade (1979) and Desai & Salami (1987)) that, although the ratio $\hat{r}_c/\hat{r}_t$ varies with respect to the diverse specimens and the various test programs, one observes the shapes of the deviatoric yield surfaces to approach, as far as possible, a triangular representation with, of course, rounded corners. Following this, one comes to realize that one obtains the best model representation of this type of possible yield surfaces, if one makes use of a convexity condition with respect to the deviatoric plane, compare Figure 6.

Convexity in the deviatoric plane is guaranteed, if the yield function satisfies the condition

$$K = \frac{\hat{r}^2 + 2(\partial\hat{r}/\partial\hat{\Theta})^2 - \hat{r}(\partial^2\hat{r}/\partial\Theta^2)}{[\hat{r}^2 + (\partial\hat{r}/\partial\hat{\Theta})^2]^{3/2}} \geq 0\,. \tag{6.26}$$

Therein, at constant $\hat{\mathrm{I}}$, the yield radius $\hat{r}$ and its first and second derivatives with respect to

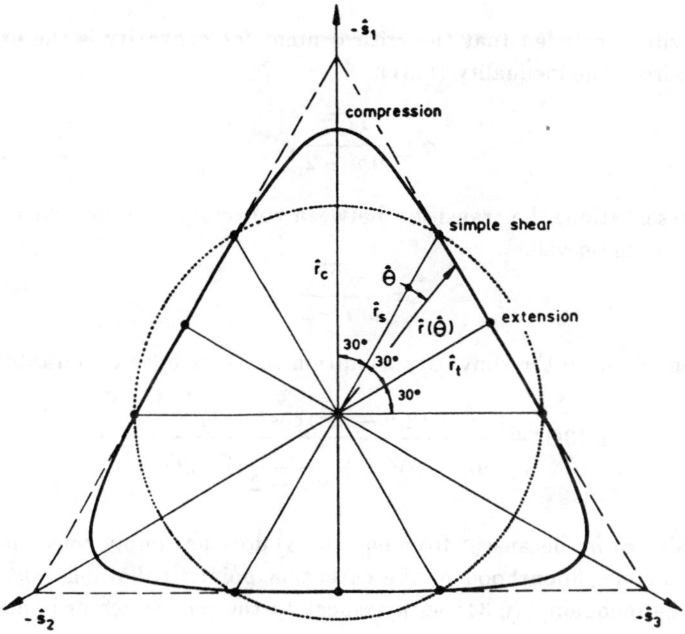

Figure 6: Convex shape of the deviatoric yield function ($\gamma = \overset{*}{\gamma}$, $m = \overset{*}{m}$) with maximum ratio $\hat{r}_c/\hat{r}_t$.

$\hat{\Theta}$ are given by eqn (6.23) and by

$$\begin{aligned}
\frac{\partial \hat{r}}{\partial \hat{\Theta}} &= -\hat{r}_s \frac{3m}{\sqrt{27}} \gamma \cos(3\hat{\Theta}) [1 + \frac{2}{\sqrt{27}} \gamma \sin(3\hat{\Theta})]^{-(2+m)/2}, \\
\frac{\partial^2 \hat{r}}{\partial \hat{\Theta}^2} &= \hat{r}_s \{ \frac{9m}{\sqrt{27}} \gamma \sin(3\hat{\Theta}) [1 + \frac{2}{\sqrt{27}} \gamma \sin(3\hat{\Theta})]^{-(2+m)2} \\
&\quad + \frac{2m(2+m)}{6} [\gamma \cos(3\hat{\Theta})]^2 [1 + \frac{2}{\sqrt{27}} \gamma \sin(3\hat{\Theta})]^{-(4+m)/2} \} .
\end{aligned} \tag{6.27}$$

Furthermore, because of

$$[\hat{r}^2 + (\partial \hat{r}/\partial \hat{\Theta})^2]^{3/2} > 0 , \tag{6.28}$$

the convexity condition (6.26) reduces to

$$\mathcal{C} = \hat{r}^2 + 2(\partial \hat{r}/\partial \hat{\Theta})^2 - \hat{r}(\partial^2 \hat{r}/\partial \hat{\Theta}^2) \geq 0 , \tag{6.29}$$

or to

$$\frac{\partial^2 \hat{r}}{\partial \hat{\Theta}^2} \leq \hat{r} + \frac{2}{\hat{r}} \left(\frac{\partial \hat{r}}{\partial \hat{\Theta}} \right)^2 , \tag{6.30}$$

respectively. Thus, one finds

$$\frac{9m-2}{\sqrt{27}} \gamma \sin(3\hat{\Theta}) + \frac{(2m - m^2) [\gamma \cos(3\hat{\Theta})]^2}{3 [1 + \frac{2}{\sqrt{27}} \gamma \sin(3\hat{\Theta})]} \leq 1 . \tag{6.31}$$

Moreover, it is easily concluded that the critical point for convexity is the extension point ($\hat{\Theta} = 30°$), where, from the inequality (6.31),

$$\gamma \leq \frac{\sqrt{27}}{9m - 2} . \tag{6.32}$$

In the above representation, the transition between convexity and concavity at $\hat{\Theta} = 30°$ is governed by the maximum value

$$\overset{*}{\gamma} = \frac{\sqrt{27}}{9m - 2} . \tag{6.33}$$

However, insertion of $\overset{*}{\gamma}$ into the convexity condition (6.31) does not guarantee

$$\sin(3\hat{\Theta}) + \frac{9\,(2m - m^2)\cos^2(3\hat{\Theta})}{(9m - 2)^2\,[1 + \frac{2}{9m - 2}\sin(3\hat{\Theta})]} \leq 1 \tag{6.34}$$

to hold for any value of m, because $\overset{*}{\gamma}$ from eqn (6.33) does not imply convexity or concavity of $\hat{r}\,(\hat{\Theta})$ in a certain neighbourhood of the extension point. Following this, an additional investigation of the inequality (6.34) with respect to the parameter m is necessary. From this investigation, one concludes to an overall convex representation of the deviatoric yield surface, if

$$m \geq \overset{*}{m} = \frac{6}{11} , \tag{6.35}$$

compare Figure 6.

With the aid of $\overset{*}{\gamma}$ from eqn (6.33), one obtains the ratio between the compression and the extension radii, eqn (6.25), to be substituted by

$$\hat{r}_c / \hat{r}_t = \left(\frac{1 + \dfrac{2}{9m - 2}}{1 - \dfrac{2}{9m - 2}} \right)^{m/2} . \tag{6.36}$$

Thus, the possible range of values $\hat{r}_c / \hat{r}_t$ is restricted to

$$1 \leq \hat{r}_c / \hat{r}_t \leq 1.584 , \tag{6.37}$$

where

$$\infty \geq m \geq \overset{*}{m} \tag{6.38}$$

has been used. Note in passing that the range of possible values $\hat{r}_c / \hat{r}_t$ defined above is sufficient to represent the usual model situations, compare, for example LADE & DUNCAN (1973), LADE (1979) or the different tests cited by DE BOER & DRESENKAMP (1989).

However, by combination of eqns (6.21)–(6.23), (6.33) and (6.35), the general seven-parameter yield surface reduces to the following six-parameter representation:

$$\begin{aligned}\hat{r}\,(\hat{\Theta},\hat{\mathrm{I}}) &= \hat{r}_s(\hat{\mathrm{I}})\,[1+\frac{2}{9m-2}\,\sin(3\hat{\Theta})\,]^{-m/2},\\ \hat{r}_s(\hat{\mathrm{I}}) &= \sqrt{2}\,[\,(\epsilon^2-\delta^2)\,\hat{\mathrm{I}}^4+2\,\beta\,\epsilon\,\hat{\mathrm{I}}^3\\ &\quad+(\beta^2-\tfrac{1}{2}\alpha^2-2\,\epsilon\,\kappa)\,\hat{\mathrm{I}}^2-2\,\beta\,\kappa\,\hat{\mathrm{I}}+\kappa^2\,]^{1/2},\\ m &\geq \overset{*}{m}=6/11\,.\end{aligned} \tag{6.39}$$

Therein, the deviatoric material parameter, m, is the governing parameter of the compression/extension ratio.

The hydrostatic plane (α, β, δ, ϵ, κ): In the hydrostatic plane, it is sufficient to investigate the yield function (6.39) at constant $\hat{\Theta}$, for example, in the simple shear regime, $\hat{\Theta}=0°$. Thus,

$$\hat{r}\,(\hat{\Theta},\hat{\mathrm{I}})\Big|_{\hat{\Theta}=0°} = \hat{r}_s(\hat{\mathrm{I}})\,, \tag{6.40}$$

compare eqn $(6.39)_2$. For the following investigations, it is furthermore convenient to use the representations

$$\begin{aligned}\hat{r}_s(\hat{\mathrm{I}}) &= (a_0+a_1\,\hat{\mathrm{I}}+a_2\,\hat{\mathrm{I}}^2+a_3\,\hat{\mathrm{I}}^3+a_4\,\hat{\mathrm{I}}^4)^{1/2},\\ \frac{\partial\hat{r}_s(\hat{\mathrm{I}})}{\partial\hat{\mathrm{I}}} &= \tfrac{1}{2}\,\hat{r}_s^{-1}\,(a_1+2a_2\,\hat{\mathrm{I}}+3a_3\,\hat{\mathrm{I}}^2+4a_4\,\hat{\mathrm{I}}^3)\,,\\ \frac{\partial^2\hat{r}_s(\hat{\mathrm{I}})}{\partial\hat{\mathrm{I}}^2} &= \tfrac{1}{2}\,\hat{r}_s^{-1}\,[(2a_2+6a_3\,\hat{\mathrm{I}}+12a_4\,\hat{\mathrm{I}}^2)\\ &\quad-\tfrac{1}{2}\,\hat{r}_s^{-2}(a_1+2a_2\,\hat{\mathrm{I}}+3a_3\,\hat{\mathrm{I}}^2+4a_4\,\hat{\mathrm{I}}^3)^2]\,,\end{aligned} \tag{6.41}$$

where

$$\begin{aligned}a_0 = 2\,\kappa^2,\quad a_1 &= -4\,\beta\,\kappa\,,\quad a_2 = 2\,(\beta^2-\tfrac{1}{2}\alpha^2-2\,\epsilon\,\kappa)\,,\\ a_3 &= 4\,\beta\,\epsilon\,,\quad a_4 = 2\,(\epsilon^2-\delta^2)\,.\end{aligned} \tag{6.42}$$

Concerning the above definition of parameters, $a_0 - a_4$, it is easily concluded that a_0 is governed by κ, a_1 by β, a_2 by α, a_3 by ϵ and a_4 by δ. The determination of $a_0 - a_4$ with respect to representative test results is obtained by the following set of boundary conditions, compare Figure 5 (b):

$$\begin{aligned}&\hat{r}_s(\hat{\mathrm{I}}=0)=\hat{r}_{os}\,, &&\frac{\partial\hat{r}_s}{\partial\hat{\mathrm{I}}}(\hat{\mathrm{I}}=0)=\tan\hat{\varphi}_{os}\,,\\ &\hat{r}_s(\hat{\mathrm{I}}=-3\,\pi_c\,p_c)=\hat{r}_{s\,\max}\,, &&\frac{\partial\hat{r}_s}{\partial\hat{\mathrm{I}}}(\hat{\mathrm{I}}=-3\,\pi_c\,p_c)=0\,,\\ &\hat{r}_s(\hat{\mathrm{I}}=-3\,p_c)=0\,.\end{aligned} \tag{6.43}$$

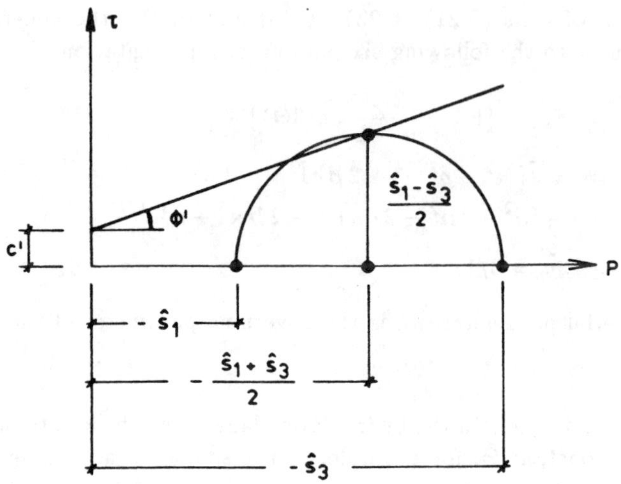

Figure 7: Effective shear parameters c' and ϕ' from classical Mohr-Coulomb theory ($c' = c\cos\phi\,,\ \phi' = \arctan(\sin\phi)$).

Therein, p_c is the natural consolidation pressure, π_c defines the ratio between the pressures corresponding to the curve's maximum and length. From eqns $(6.42)_{1,2}$, one finds in comparison with the well-known Mohr-Coulomb failure theory that

$$\begin{aligned} a_0 &= (\hat{r}_{os})^2 = 2\,(c\cos\phi)^2\,, \\ a_1 &= 2\,\hat{r}_{os}\,\tan\hat{\varphi}_{os} = -\tfrac{4}{3}\sin\phi\;c\cos\phi\,, \end{aligned} \tag{6.44}$$

and thus, from eqns $(6.42)_{1,2}$,

$$\begin{aligned} \kappa = c\cos\phi\,,\quad \hat{r}_{os} &= \sqrt{2}\,\kappa\,, \\ \beta = \tfrac{1}{3}\sin\phi\,,\quad \hat{\varphi}_{os} &= \arctan\,(-\tfrac{\sqrt{2}}{3}\sin\phi)\,; \end{aligned} \tag{6.45}$$

ϕ is the angle of internal friction, and c is the cohesion, compare Figure 7.

Based on the above result, the determination of the material parameters a_2, a_3 and a_4 reduces to the matrix equation

$$\underline{a} = \underline{L}^{-1}\underline{r}\,, \tag{6.46}$$

where

$$\underline{a}^T = [\, a_2 \quad a_3 \quad a_4 \,]\,,$$

$$\underline{L} = \begin{bmatrix} (-3\,p_c)^2 & (-3\,p_c)^3 & (-3\,p_c)^4 \\ (-3\,\pi_c\,p_c)^2 & (-3\,\pi_c\,p_c)^3 & (-3\,\pi_c\,p_c)^4 \\ -6\,\pi_c\,p_c & 3\,(-3\,\pi_c\,p_c)^2 & 4\,(-3\,\pi_c\,p_c)^3 \end{bmatrix}, \tag{6.47}$$

$$\underline{r} = \begin{bmatrix} -\,a_0 + \quad 3\,p_c\,a_1 \\ -\,a_0 + \quad 3\,\pi_c\,p_c\,a_1 + \quad \hat{r}^2_{s\,\max} \\ -a_1 \end{bmatrix}.$$

Finally, the convexity condition

$$\frac{\partial^2 \hat{r}_s(\hat{\mathrm{I}})}{\partial \hat{\mathrm{I}}^2} \le 0 \tag{6.48}$$

must hold for all values of $\hat{\mathrm{I}}$ defining the hydrostatic shape of the yield function.

Furthermore, within the scope of cohesionless materials where $c = 0$, the parameters a_0 and a_1 vanish, and one obtains from eqns (6.41) and (6.42)

$$\begin{aligned} \hat{r}_s(\hat{\mathrm{I}}) &= \hat{\mathrm{I}}\,(\bar{a}_2 + \bar{a}_3\,\hat{\mathrm{I}} + \bar{a}_4\,\hat{\mathrm{I}}^2)^{1/2}\,, \\ \frac{\partial \hat{r}_s(\hat{\mathrm{I}})}{\partial \hat{\mathrm{I}}} &= \frac{2\bar{a}_2 + 3\bar{a}_3\,\hat{\mathrm{I}} + 4\bar{a}_4\,\hat{\mathrm{I}}^2}{2\,(\bar{a}_2 + \bar{a}_3\,\hat{\mathrm{I}} + \bar{a}_4\,\hat{\mathrm{I}}^2)^{1/2}}\,, \end{aligned} \tag{6.49}$$

where

$$\bar{a}_2 = 2\,(\bar{\beta}^2 - \tfrac{1}{2}\bar{\alpha}^2), \quad \bar{a}_3 = 4\,\bar{\beta}\,\bar{\epsilon}, \quad \bar{a}_4 = 2\,(\bar{\epsilon}^2 - \bar{\delta}^2)\,. \tag{6.50}$$

From eqn $(6.49)_2$, one finds with the aid of the boundary condition $(6.43)_2$ that

$$\bar{a}_2 = (\tan\varphi_{os})^2\,, \tag{6.51}$$

and thus, in still requiring the Mohr-Coulomb theory to hold as a special case of our yield function,

$$\bar{\alpha} = 0, \qquad \bar{\beta} = \tfrac{1}{3}\,\sin\phi\,. \tag{6.52}$$

This result does not only yield a coupling between the boundary conditions $(6.43)_{3-5}$, but it also opens the possibility to express the consolidation pressure and the ratio π_c as functions of $\bar{\beta}$, $\bar{\delta}$ and $\bar{\epsilon}$:

$$\begin{aligned} p_c &= \frac{\bar{\beta}}{3\,(\bar{\epsilon} + \bar{\delta})}\,, \\ \pi_c &= \frac{1}{4\,(\bar{\epsilon} - \bar{\delta})}\,(3\,\bar{\epsilon} - \sqrt{8\,\bar{\delta}^2 + \bar{\epsilon}^2}\,)\,. \end{aligned} \tag{6.53}$$

Note in passing that the above result was obtained from eqns $(6.43)_{4,5}$. Consequently, in the case of $\bar{\alpha} = 0$, the curve's maximum depends on p_c and π_c.

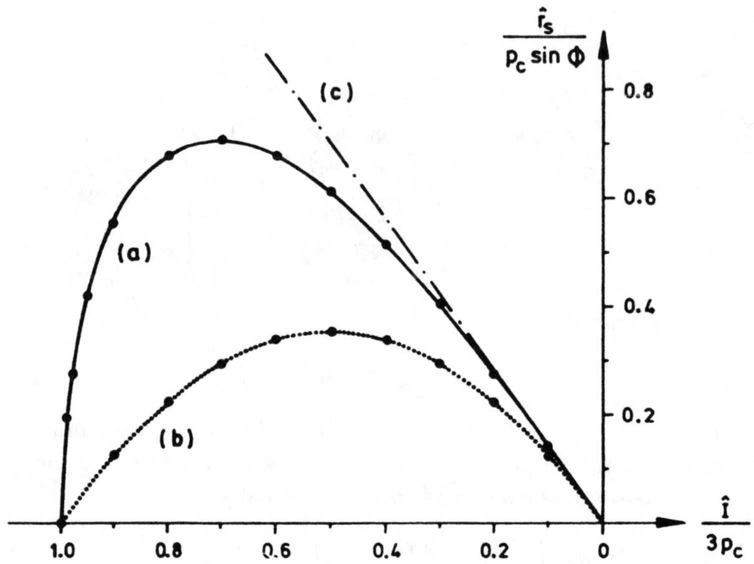

Figure 8: Hydrostatic yield envelope for cohesionless materials: (a) $\alpha = \epsilon = \kappa = 0$, (b) $\alpha = \delta = \kappa = 0$, (c) $\alpha = \delta = \epsilon = \kappa = 0$.

(d) Transition of the general yield function to simpler shapes

Based on the six-parameter representation of the general yield function, eqns (6.39), several simplifications are possible, especially, with respect to the hydrostatic range, where the need for the introduction of such simplifications often results from the lack of experimental information, necessary to fit the five general boundary conditions, defined by eqns (6.43). However, a minimum requirement on the closed hydrostatic shape of the yield envelope exhibiting a vertical tangent at $\hat{I} = -3\,p_c$ is the dependence on the parameters β, δ and κ, where β and κ can be defined via eqns $(6.45)_{1,3}$. Note in passing that a substitution of δ by ϵ would change the tear drop representation towards a simple parabola, compare Figure 8.

In the case of cohesionless materials, one obtains the simplest version of the general yield envelope (compare Figure 8 (a)) via

$$\begin{aligned} \hat{r}\,(\hat{\Theta}, \hat{I}) &= \hat{r}_s(\hat{I})\,[1 + \frac{2}{9m-2}\sin(3\hat{\Theta})]^{-m/2}, \\ \hat{r}_s(\hat{I}) &= \sqrt{2}\,(\bar{\beta}^2\,\hat{I}^2 - \bar{\delta}^2\,\hat{I}^4)^{1/2}, \end{aligned} \tag{6.54}$$

where, from eqns $(6.45)_3$ and (6.53),

$$\bar{\beta} = \tfrac{1}{3}\sin\phi, \quad \bar{\delta} = \frac{\sin\phi}{9p_c}, \quad \pi_c = \tfrac{1}{2}\sqrt{2}. \tag{6.55}$$

Therein, the shape parameter $\bar{\delta}$ is governed by the angle of internal friction, ϕ, and the consolidation pressure, p_c. Furthermore, the failure envelope corresponding to the yield surface (6.54) reads

$$\begin{aligned} \hat{r}_u(\hat{\Theta}, \hat{\mathrm{I}}) &= \hat{r}_{su}(\hat{\mathrm{I}})[1 + \frac{2}{9m-2}\sin(3\hat{\Theta})]^{-m/2}, \\ \hat{r}_{su}(\hat{\mathrm{I}}) &= \sqrt{2}\bar{\beta}\,|\,\hat{\mathrm{I}}\,|, \end{aligned} \tag{6.56}$$

where, in addition, $\bar{\delta} = 0$ has been used, compare Figure 8 (c).

However, apart from these results, the general yield condition contains further simplifications directly leading to well-known representations, which are included into the published literature. In particular, by use of $\delta = 0$ and $m = 2/3$, one obtains from eqns (6.39) that

$$\begin{aligned} \hat{r}(\hat{\Theta}, \hat{\mathrm{I}}) &= \hat{r}_s(\hat{\mathrm{I}})[1 + \tfrac{1}{2}\sin(3\hat{\Theta})]^{-1/3}, \\ \hat{r}_s(\hat{\mathrm{I}}) &= \sqrt{2}\,[\epsilon^2\,\hat{\mathrm{I}}^4 + 2\,\beta\,\epsilon\,\hat{\mathrm{I}}^3 + (\beta^2 - \tfrac{1}{2}\alpha^2 - 2\,\epsilon\,\kappa)\,\hat{\mathrm{I}}^2 - 2\,\beta\,\kappa\,\hat{\mathrm{I}} + \kappa^2]^{1/2}. \end{aligned} \tag{6.57}$$

Except that the starting point to obtain eqns (6.57) was given by eqns (6.3) instead of eqns (6.2), the above form of the yield surface corresponds to DE BOER & DRESENKAMP (1989) including Dresenkamp's convexity condition $\gamma = \overset{*}{\gamma} = \frac{3}{4}\sqrt{3}$ (DRESENKAMP, 1987). Furthermore, by additionally assuming $\epsilon = 0$, the yield function (6.57) reduces to

$$\begin{aligned} \hat{r}(\hat{\Theta}, \hat{\mathrm{I}}) &= \hat{r}_s(\hat{\mathrm{I}})[1 + \tfrac{1}{2}\sin(3\hat{\Theta})]^{-1/3}, \\ \hat{r}_s(\hat{\mathrm{I}}) &= \sqrt{2}\,[(\beta^2 - \tfrac{1}{2}\alpha^2)\,\hat{\mathrm{I}}^2 - 2\,\beta\,\kappa\,\hat{\mathrm{I}} + \kappa^2]^{1/2}, \end{aligned} \tag{6.58}$$

where these equations prove to represent the simplified de Boer's function (6.3). Further modifiations concern the special case, where $m = \infty$ and thus $\hat{r}_c = \hat{r}_s = \hat{r}_t$, compare eqns (6.36)–(6.38). In this case, the use of $\alpha = \delta = \epsilon = 0$ leads to

$$\hat{r}(\hat{\Theta}, \hat{\mathrm{I}}) = \hat{r}_s(\hat{\mathrm{I}}), \quad \hat{r}_s(\hat{\mathrm{I}}) = \sqrt{2}(\kappa - \beta\,\hat{\mathrm{I}}), \tag{6.59}$$

or, equivalently, to

$$\sqrt{\hat{\mathrm{II}}_D} + \beta\,\hat{\mathrm{I}} - \kappa = 0\,, \tag{6.60}$$

where eqn $(6.6)_1$ has been used. Obviously, these equations are nothing else than the well-known Drucker-Prager criterion (DRUCKER & PRAGER, 1952). Note in passing that the simplified yield functions (6.57)–(6.60) are principally characterized as open failure curves; that is, the validity of these functions, when they are used as yield or hardening conditions, respectively, is restricted to the range of moderate hydrostatic pressures, where no pressure influence on the hydrostatic yield limit occurs. Moreover, because of its circular shape in the deviatoric plane, the Drucker-Prager criterion can only be expected to yield senseful results within the simple shear regime.

Moreover, the general yield function contains the case of

$$\begin{aligned} \hat{r}(\hat{\Theta}, \hat{\mathrm{I}}) &= \hat{r}_s(\hat{\mathrm{I}}) \\ \hat{r}_s(\hat{\mathrm{I}}) &= \sqrt{2}(\kappa^2 - \tfrac{1}{2}\alpha^2\,\hat{\mathrm{I}}^2)^{1/2}, \end{aligned} \tag{6.61}$$

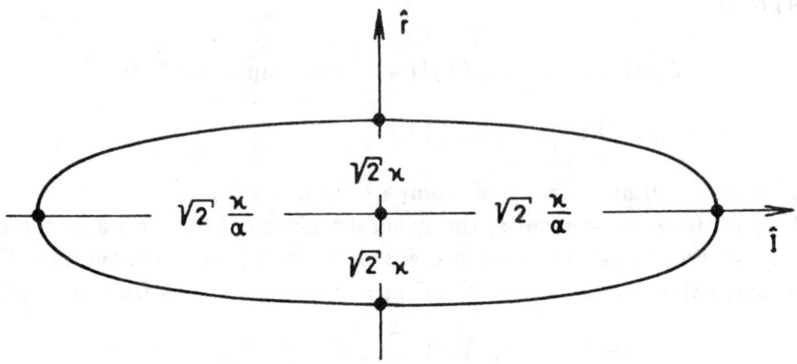

Figure 9: Yield function for ductile solids (hydrostatic plane).

or, equivalently,

$$(\,\text{II}_D + \tfrac{1}{2}\alpha^2\,\hat{\text{I}}^2)^{1/2} - \kappa = 0\,. \qquad (6.62)$$

These equations, which are obtained from eqns (6.39) by use of $\beta = \delta = \epsilon = 0$ and, additionally, $m = \infty$, directly correspond to Green's yield condition for ductile media (GREEN, 1972) representing an ellipsoid in stress space, compare Figure 9.

Finally, with the aid of $\alpha = 0$, eqn (6.62) reduces to the well-known yield function of von Mises type, namely,

$$\hat{\text{II}}_D - \kappa^2 = 0\,. \qquad (6.63)$$

(e) Two curve fitting examples

Deviatoric curve fitting: To set an example for a possible curve fitting process with respect to the octahedral shape of the yield function, consider Figure 10, where a typical deviatoric failure envelope is given. The experimental data of LADE & DUNCAN (1973) carried out at dense Monterey No. 0 Sand show the ratios

$$r_c/r_t = 1.544, \quad r_c/r_s = 1.385, \quad r_t/r_s = 0.897\,. \qquad (6.64)$$

Recall from eqns (6.37) and (6.39) that the ultimate ratios including a convex representation of the deviatoric yield shape are

$$\begin{aligned} \max \hat{r}_c/\hat{r}_t &= 1.584,\\ \max \hat{r}_c/\hat{r}_s &= 1.373,\\ \min \hat{r}_t/\hat{r}_s &= 0.867, \end{aligned} \qquad (6.65)$$

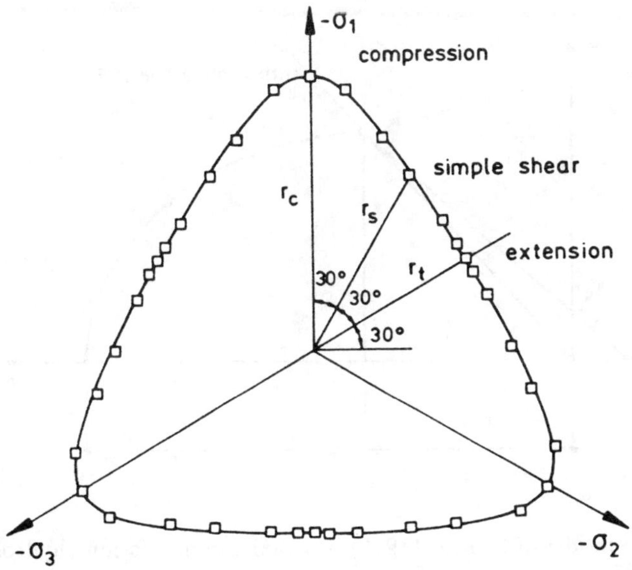

Figure 10: Deviatoric failure shape of dense Monterey No. 0 Sand (taken from LADE & DUNCAN, 1973).

thus exhibiting a sufficient agreement with Lade and Duncan's experiments. Furthermore, if one understands $\hat{r}_s$ as the reference value for the deviatoric curve, a comparison of eqns (6.64) and (6.65) leads to

$$\hat{r}_c/r_c = 0.99, \quad \hat{r}_t/r_t = 0.97, \tag{6.66}$$

that is, by use of $m = \overset{*}{m}$, one obtains an underestimation of 1% for the compression value r_c and of 3% for the extension value r_t.

However, other experiments may show more rounded shapes of the deviatoric envelope, thus leading to values $m > \overset{*}{m} = 6/11$, additionally compare Figure 13.

Hydrostatic curve fitting: Compression and simple shear tests on consolidated clay carried out by THAMM (1974) and adjusted to cap model yield surfaces by SCHAD (1979) exhibit shear parameters

$$c = 11\,\text{kN/m}^2, \quad \phi = 21°, \tag{6.67}$$

and thus, from eqns $(6.45)_{1,3}$:

$$\kappa = 10.27\,\text{kN/m}^2, \quad \beta = 0.1195. \tag{6.68}$$

Furthermore, by use of the critical state line concept, (SCHAD, 1979, Figure 4.8), Schad defines the pressure $\pi_c\, p_c$ corresponding to the maximum value of the compression meridian,

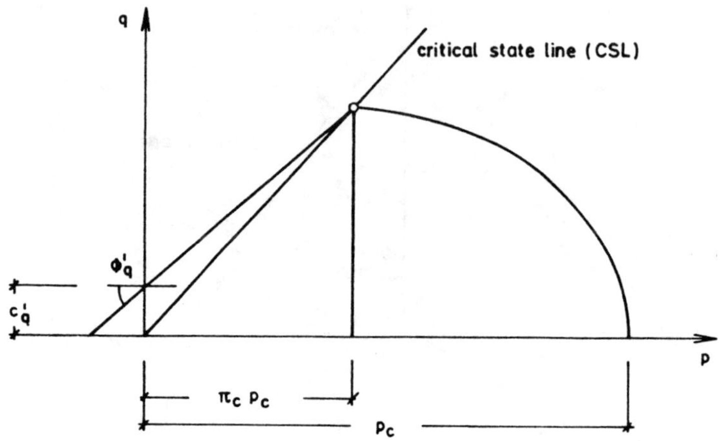

Figure 11: Cap model of SCHAD (1979) in triaxial compression (hydrostatic plane).

$q_{\max} = -(\hat{s}_1 - \hat{s}_3)_{\max}$:

$$\begin{aligned} \pi_c\, p_c &= 103.19 \text{ kN/m}^2, \\ q_{\max} &= \sqrt{\tfrac{3}{2}}\, \hat{r}_{c\max} = 107.32 \text{ kN/m}^2. \end{aligned} \tag{6.69}$$

Thus, one obtains the corresponding values

$$\begin{aligned} \tau_{\max} &= \tfrac{1}{2}\sqrt{2}\, \hat{r}_{s\max} = 47.25 \text{ kN/m}^2, \\ \hat{r}_c/\hat{r}_s &= 1.311, \end{aligned} \tag{6.70}$$

where SCHAD (1979, Figure 4.7) has been used. Finally, with the aid of Schad's cap model, compare Figure 11, Schad obtains the consolidation pressure, p_c, to yield

$$p_c = 258 \text{ kN/m}^2. \tag{6.71}$$

Principally, eqns (6.68)–(6.71) define the five boundary conditions from eqns (6.43), which are necessary for the determination of the five material parameters of the hydrostatic yield curve.

Unfortunately, several of the above values do not result from experimental data but only from Schad's calculations satisfying his cap model problem. Insertion of these values into our general yield function would necessarily lead to concave parts of the yield envelope, what is easily concluded by comparison of Figure 5 and Figure 11. Thus, the following curve fitting process is based on the boundary conditions (6.68), (6.70) and (6.71) substituting the condition $(6.69)_1$ by a numerical convexity control. By use of the relations (6.41)–(6.43), the

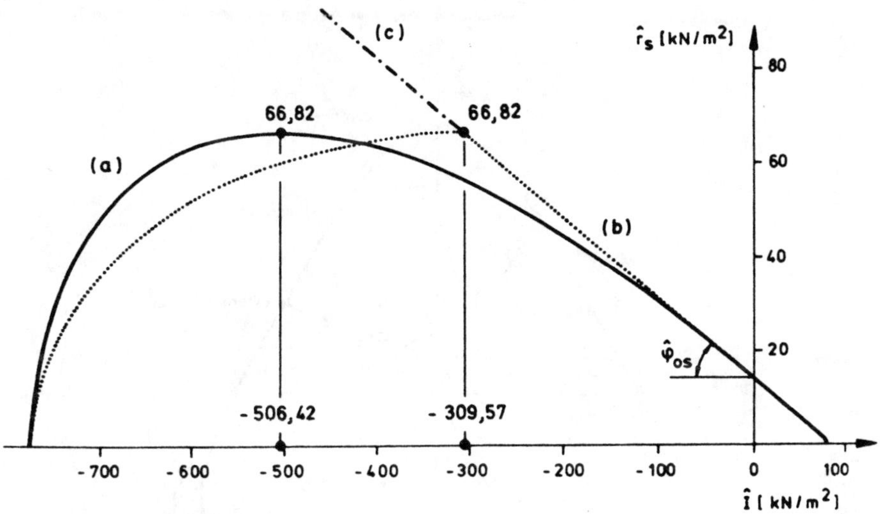

Figure 12: (a) Calculated single surface model with material parameters from eqns (6.68), (6.71) and (6.72). (b) Schad's cap model with material parameters corresponding to eqns (6.68)–(6.71). (c) Associated failure curve.

numerical process easily yields

$$\begin{aligned} \alpha &= -0.128\,430, \\ \epsilon &= 6.208\,342 \cdot 10^{-5}\ \mathrm{m^2/kN}, \\ \delta &= 1.100\,137 \cdot 10^{-4}\ \mathrm{m^2/kN}, \end{aligned} \tag{6.72}$$

additionally proving the convexity condition (6.48) to hold throughout the envelope's length defined by $\hat{I}_{01} \geq \hat{I} \geq \hat{I}_{02}$, where

$$\hat{I}_{01} = 82.56\ \mathrm{kN/m^2}, \quad \hat{I}_{02} = -774.0\ \mathrm{kN/m^2}. \tag{6.73}$$

Therein, $\hat{I}_{01}$ results from numerical computations, whereas $\hat{I}_{02} = -3\,p_c$ can directly be taken from eqn (6.71).

Figure 12 exhibits the above results for the present single surface model, curve (a), in comparison with Schad's cap model, curve (b). Furthermore, curve (c) represents the corresponding failure envelope, which directly results from the general yield function by use of the material parameters $\kappa_u = \kappa$ and $\beta_u = \beta$ from eqns (6.68) and, additionally, $\alpha_u = \epsilon_u = \delta_u = 0$. Note in passing that $\alpha_u \neq 0$ as was taken into account with eqn (6.20) would lead to slightly curved meridians of the hydrostatic failure envelope, compare, for example, de Boer & Dresenkamp (1989). However, the single surface envelope given by curve (a) of Figure 12 must be understood as the yield condition corresponding to first yielding, that is, in the hardening regime, isotropic hardening would imply a growing of the envelope, whereas kinematic hardening allows for a transition of curve (a) in stress space.

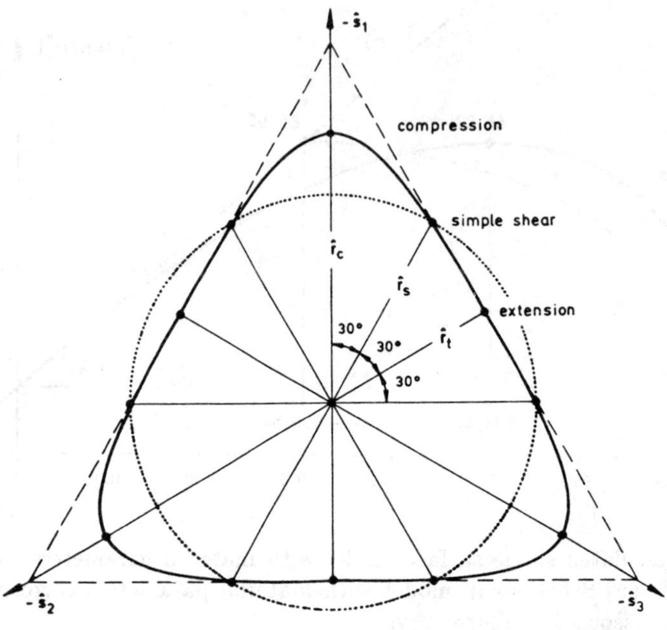

Figure 13: Deviatoric shape corresponding to the calculated single surface model at $\hat{\mathrm{I}} = -506.42\,\mathrm{kN/m^2}$. Material parameter $m = 0.593\,5$.

Finally, to complete the present example, it is concluded with the aid of eqn $(6.39)_1$ that the ratio $\hat{r}_c/\hat{r}_s = 1.311$ from eqn $(6.70)_2$ yields the deviatoric parameter

$$m = 0.593\,5. \tag{6.74}$$

The corresponding deviator curve as is shown in Figure 13 includes the ratios

$$\hat{r}_t/\hat{r}_s = 0.870, \quad \hat{r}_c/\hat{r}_t = 1.507, \tag{6.75}$$

and thus, at the hydrostatic curve's maximum defined by $\hat{\mathrm{I}} = -506.42\,\mathrm{kN/m^2}$,

$$\hat{r}_c = 87.60\,\mathrm{kN/m^2}, \quad \hat{r}_s = 66.82\,\mathrm{kN/m^2}, \quad \hat{r}_t = 58.14\,\mathrm{kN/m^2}. \tag{6.76}$$

From the above curve fitting procedures, one easily concludes the seven-parameter envelope to exhibit exactly the correct number of parameters, which is necessary to satisfy the respective boundary condition of the yield curve. In addition, recall that the use of two material parameters characterizing the deviatoric plane, γ and m, made it possible to construct a precisely convex deviator envelope, nevertheless allowing for a variation of the ratio $\hat{r}_c/\hat{r}_t$. The remaining five material parameters, α, β, δ, ϵ and κ, are sufficient to successfully describe the hydrostatic yield curve in all relevant representations, first yielding, isotrop or kinematic hardening, respectively, as well as failure for both cohesion and cohesionless frictional materials.

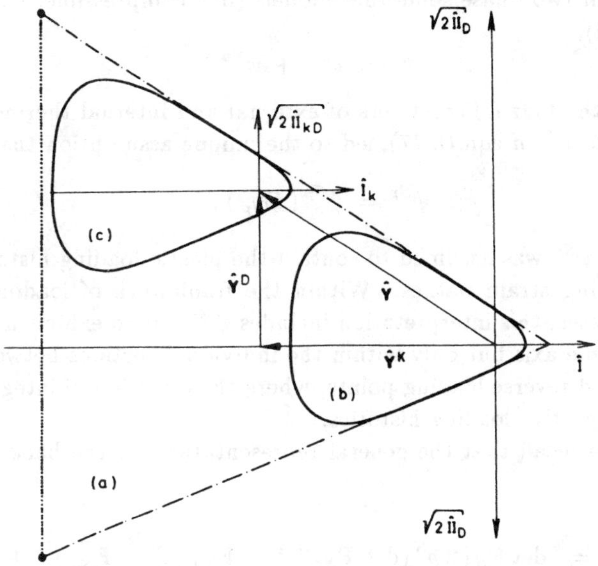

Figure 14: Schematic representation of kinematic hardening. (a) failure envelope, (b) first yielding and (c) initial yield function shifted by kinematic hardening.

6.3 Back-stress tensor and non-associated flow rule

(a) Back-stress tensor

Concerning the derivation of constitutive equations for the back-stress tensor of kinematically hardening porous solid matrices, lots of basic ideas are available from the published literature. Unfortunately, most of these possibilities are restricted to usual metal plasticity including the concept of plastic incompressibility for non-porous media. Thus, it is easily recognized that these back-stress laws do not apply to porous media plasticity, since, although we proceed from constituent plastic incompressibility, large volume changes may occur, namely, as a consequence of porosity variations induced by plastic dilatation or compression. Furthermore, concerning the finite representation of the back-stress tensor, one has to distinguish between its deviatoric part producing a translation of the yield surface in the deviatoric plane and its hydrostatic part representing the shifter on the hydrostatic axis, compare Figure 14.

Following this, the material behaviour in these two principally different directions must be caught by two different material functions depending on the plastic loading history through both the deviatoric and the hydrostatic plastic strain pathes. Recall from Section 3 and Section 4 that this history was assumed to be included into the functional ψ^S of the matrix material, where, from eqns (4.46), (4.63), (4.96) and (4.114) corresponding to the four different combinations of two-phase solid-fluid models (one compressible, two hybrid and one incompressible model),

$$\psi^S = \psi^{Se} + \psi^{Sp} \tag{6.77}$$

was used. However, the four different sets of external and internal thermodynamic process variables included into ψ^S of eqn (6.77), led to the unique assumption that

$$\psi^{Sp} = \psi^{Sp}(\mathbf{E}_{Sp}), \tag{6.78}$$

where the functional ψ^{Sp} was assumed to contain the plastic loading history by integration over the plastic loading strain pathes. Within the framework of loading, unloading and reverse loading processes, this interpretation includes ψ^{Sp} not to exhibit a smooth representation on the total time axis but only within the individual sections between the respective loading, unloading and reverse loading points, where the direction of integration is uniquely prescribed by the respective loading histories.

Following this, we recall that the general representation for the back-stress tensor, eqn (4.87), was given by

$$\begin{aligned}\hat{\mathbf{Y}}^S &= \det \mathbf{F}_S [\, 2\,\rho^S (\det \mathbf{F}_{Sp})^{-2/3} (\, \mathbf{F}_{Sp} \frac{\partial \psi^{Sp}}{\partial \overline{\mathbf{C}}_{Sp}} \mathbf{F}^T_{Sp}\,)^D + \\ &\quad + (\, \rho^S \det \mathbf{F}_{Sp} \frac{\partial \psi^{Sp}}{\partial \det \mathbf{F}_{Sp}} + n^S\, p^{FR}_{\text{con.}}\,)\, \mathbf{I}\,],\end{aligned} \tag{6.79}$$

where the fluid configuration pressure term defined by eqn (4.88) only occurs in the case of the hybrid model of second type, additionally compare eqn (4.40) for the compressible model, eqn $(4.63)_3$ for the hybrid model of first type and eqn (4.113) for the incompressible model. Thus, for the following considerations, we simplify the above representation by use of $p^{FR}_{\text{con.}} = 0$. Furthermore, since constituent plastic incompressibility implies each of the four models to be governed by

$$\begin{aligned}\hat{n}^S &= n^S_{oS}\,(\det \mathbf{F}_{Sp})^{-1}, \\ \det \mathbf{F}_{Sp} \frac{\partial \psi^{Sp}}{\partial \det \mathbf{F}_{Sp}} &= -\,\hat{n}^S \frac{\partial \psi^{Sp}}{\partial \hat{n}^S},\end{aligned} \tag{6.80}$$

compare eqns (4.3) and (4.4), it is easily seen from eqns (4.78) and (4.79), that the constitutive functional for $\hat{\mathbf{Y}}^S$ does not depend on the choice of the basic constitutive model.

Following the considerations included into EHLERS (1991), one obtains with the aid of the concept of material symmetry that

$$\begin{aligned}&\psi^{Sp}(\mathbf{E}_{Sp}) \rightarrow \psi^{Sp}(\mathbf{C}_{Sp}), \\ &\psi^{Sp} = \psi^{Sp}(\overline{\mathrm{I}}_p, \overline{\mathrm{II}}_p, J_p), \\ &J_p = \det \mathbf{F}_{Sp} = (\mathrm{III}_p)^{1/2},\end{aligned} \tag{6.81}$$

where $\overline{\mathrm{I}}_p$ and $\overline{\mathrm{II}}_p$ are the principal invariants of the right plastic Cauchy-Green tensor $\overline{\mathbf{C}}_{Sp}$ or the left plastic Cauchy-Green tensor $\overline{\mathbf{B}}_{Sp}$ defined by

$$\begin{aligned} \overline{\mathbf{C}}_{Sp} &= (\det \mathbf{F}_{Sp})^{-2/3}\,\mathbf{C}_{Sp}\,, \\ \overline{\mathbf{B}}_{Sp} &= (\det \mathbf{F}_{Sp})^{-2/3}\,\hat{\mathbf{B}}_{Sp}\,. \end{aligned} \tag{6.82}$$

Within the framework of finite theories, a possible set-up for the functional ψ^{Sp} is

$$\psi^{Sp}(\overline{\mathrm{I}}_p, \overline{\mathrm{II}}_p, J_p) \rightarrow \psi^{Sp}(\overline{\mathrm{I}}_p, J_p)\,. \tag{6.83}$$

Thus, eqns (6.79) and (6.83) combine to

$$\hat{\mathbf{Y}}^S = 2\,\rho_{oS}^S\,J_p^{-2/3}\,\frac{\partial\psi^{Sp}}{\partial\overline{\mathrm{I}}_p}\,\hat{\mathbf{B}}_{Sp}^D + \rho_{oS}^S\,J_p\,\frac{\partial\psi^{Sp}}{\partial J_p}\,\mathbf{I}\,. \tag{6.84}$$

Finally, by use of

$$\hat{\mathbf{B}}_{Sp}^D = 2\,\hat{\mathbf{K}}_{Sp}^D \tag{6.85}$$

(compare Table 2), the following version of the back-stress tensor holds:

$$\begin{aligned} \hat{\mathbf{Y}}_S &= \bar{c}\,J_p^{-2/3}\,\hat{\mathbf{K}}_{Sp}^D + \tilde{c}\ln J_p\,\mathbf{I}\,, \\ \bar{c} &= 4\,\rho_{oS}^S\,\frac{\partial\psi^{Sp}}{\partial\overline{\mathrm{I}}_p}\,, \\ \tilde{c} &= \rho_{oS}^S\,(\ln J_p)^{-1}\,\frac{\partial\psi^{Sp}}{\partial J_p}\,. \end{aligned} \tag{6.86}$$

Therein, as was discussed above, $\bar{c}$ and $\tilde{c}$ are functions of the deformation history.

However, for several purposes, it may be more convenient to proceed from the back-stress rate rather than from the back-stress tensor itself. Proceeding from the concept of Lie derivatives, the rate equation (lower Lie derivative) of eqn $(6.86)_1$ reads:

$$\begin{aligned} (\hat{\mathbf{Y}}^S)_p^{\nabla} &= (\bar{c})_S'\,J_p^{-2/3}\,\hat{\mathbf{K}}_{Sp}^D + (\tilde{c})_S'\ln J_p\,\mathbf{I} \\ &\quad + \bar{c}\,J_p^{-2/3}[(\hat{\mathbf{K}}_{Sp}^D)_p^{\nabla} - \tfrac{2}{3}(\hat{\mathbf{D}}_{Sp}\cdot\mathbf{I})\,\hat{\mathbf{K}}_{Sp}^D] \\ &\quad + \tilde{c}[(\hat{\mathbf{D}}_{Sp}\cdot\mathbf{I})\,\hat{\mathbf{I}} - 2\ln J_p\,\hat{\mathbf{D}}_{Sp}]\,, \end{aligned} \tag{6.87}$$

where

$$(\hat{\mathbf{K}}_{Sp}^D)_p^{\nabla} = \tfrac{1}{3}[(\hat{\mathbf{B}}_{Sp}\cdot\mathbf{I})\,\hat{\mathbf{D}}_{Sp} - (\hat{\mathbf{B}}_{Sp}\cdot\hat{\mathbf{D}}_{Sp})\,\mathbf{I}] \tag{6.88}$$

to be obtained from Table 2 and Table 4 has been used.

(b) Non-associated flow rule

It is very well known from experiment that an associated flow rule, where the general yield function equals the plastic potential, leads to extremely unrealistic results, especially with respect to the plastic volumetric strain; dilatation or compression. Thus, one must apply a non-associated flow rule by either assuming the existence of a plastic potential Q different from the yield surface F or by directly proposing a constitutive equation for the plastic rate of deformation tensor. However, although the first possibility may be more in accordance with classical metal plasticity, it must be considered that the existence of a plastic potential was originally coupled with the non-existence of plastic volume changes, compare VON MISES (1928, p. 163 and p. 181). Thus, any determination of a function $Q \neq F$, apart from the usual Λ-factor, means nothing else than an integration of the plastic strain rate with respect to the representative stress state. Hence, the second possibility proves to be more natural and will therefore be preferred in what follows.

In analogy to EHLERS (1991, 1992), the present non-associated flow rule is introduced via

$$\begin{aligned} \hat{\mathbf{D}}_{Sp} &= \tfrac{1}{2}\,(\,\hat{\mathrm{II}}_{kD}\,)^{-1/2}\ \overset{4}{\boldsymbol{\Lambda}}\ [\,(\,\hat{\boldsymbol{\tau}}^{SD} - \hat{\mathbf{Y}}^{SD}) + \zeta\,\hat{\mathrm{I}}_k\,\mathbf{I}\,]\,, \\ \overset{4}{\boldsymbol{\Lambda}} &= \Lambda\,(\hat{\mathbf{Q}}_p \otimes \hat{\mathbf{Q}}_p)^{\overset{23}{T}}\,, \\ \zeta &= \sqrt{\tfrac{2}{3}}\ \frac{(\,\hat{\mathrm{II}}_{kD}\,)^{1/2}}{\hat{\mathrm{I}}_k}\ \tan\nu_p\,. \end{aligned} \tag{6.89}$$

In the above flow rule characterizing kinematically hardening materials, the proper orthogonal tensor $\hat{\mathbf{Q}}_p$ may be expressed by (DE BOER, 1988)

$$\hat{\mathbf{Q}}_p = \mathbf{e} \otimes \mathbf{e} + (\,\mathbf{I} - \mathbf{e} \otimes \mathbf{e})\,\cos\xi_p + \mathbf{e} \times \mathbf{I}\,\sin\xi_p\,, \tag{6.90}$$

where, for our present investigations, ξ_p defines the deviatoric flow angle admitting non-coaxiality between $\hat{\mathbf{D}}^D_{Sp}$ and $\hat{\boldsymbol{\tau}}^{SD} - \hat{\mathbf{Y}}^{SD}$; $\mathbf{e}$ is a unit vector parallel to the rotation axis which is assumed to be directed perpendicular to the octahedral plane. However, in case of coaxiality, $\overset{4}{\boldsymbol{\Lambda}}$ obviously reduces to

$$\overset{4}{\boldsymbol{\Lambda}} = \Lambda\ \overset{4}{\mathbf{I}}\,. \tag{6.91}$$

Therein, the fourth-order identity is defined by eqn (5.40). Furthermore, the invariants $\hat{\mathrm{I}}_k$ and $\hat{\mathrm{II}}_{kD}$ included into eqns (6.89) can be taken from eqns $(6.18)_{1,2}$.

Considering isotropically instead of kinematically hardening properties, it is easily seen that the flow rule (6.89) simply reduces to

$$\begin{aligned} \hat{\mathbf{D}}_{Sp} &= \tfrac{1}{2}\,(\,\hat{\mathrm{II}}_{D}\,)^{-1/2}\ \overset{4}{\boldsymbol{\Lambda}}\ [\,\hat{\boldsymbol{\tau}}^{SD} + \zeta\,\hat{\mathrm{I}}\,\mathbf{I}\,]\,, \\ \overset{4}{\boldsymbol{\Lambda}} &= \Lambda\,(\hat{\mathbf{Q}}_p \otimes \hat{\mathbf{Q}}_p)^{\overset{23}{T}}\,, \\ \zeta &= \sqrt{\tfrac{2}{3}}\ \frac{(\,\hat{\mathrm{II}}_{D}\,)^{1/2}}{\hat{\mathrm{I}}}\ \tan\nu_p\,. \end{aligned} \tag{6.92}$$

However, in using either eqns (6.89) or eqns (6.92) , the hydrostatic parameter function ζ is defined by use of the plastic dilatation angle ν_p, compare Figure 15. Furthermore, from the definition of the dilatation angle, one obtains

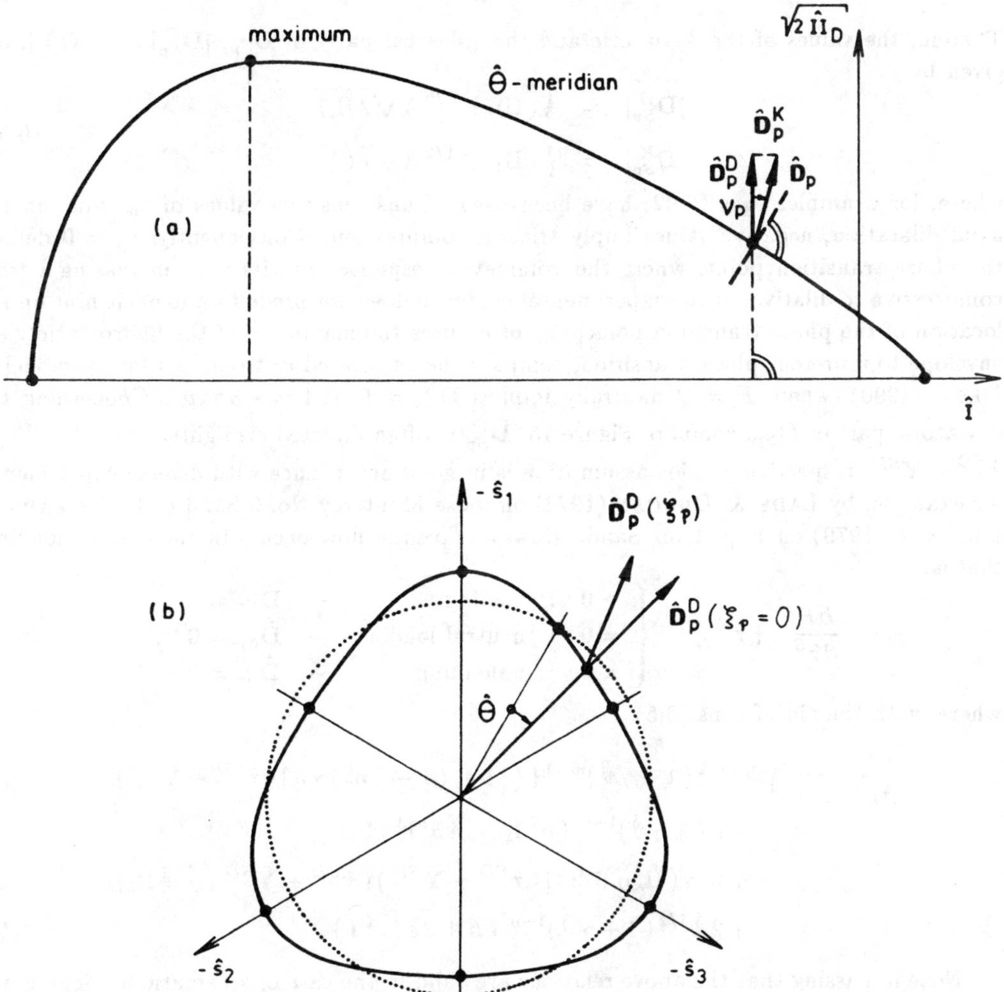

Figure 15: Plastic rate of deformation tensor in principal stress space; (a) hydrostatic plane and (b) deviatoric plane.

$$\tan \nu_p = \frac{|\mathbf{D}_{Sp}^K|}{|\mathbf{D}_{Sp}^D|} \operatorname{signum} \hat{\mathrm{I}} . \tag{6.93}$$

Therein, the values of the deviatoric and the spherical parts of $\hat{\mathbf{D}}_{Sp}$, $|\hat{\mathbf{D}}_{Sp}^D|$ and $|\hat{\mathbf{D}}_{Sp}^K|$, are given by

$$\begin{aligned} |\hat{\mathbf{D}}_{Sp}^D| &= \tfrac{1}{2}(\hat{\mathrm{II}}_D)^{-1/2} \Lambda \sqrt{2\,\hat{\mathrm{II}}_D}\,, \\ |\hat{\mathbf{D}}_{Sp}^K| &= \tfrac{1}{2}(\hat{\mathrm{II}}_D)^{-1/2} \Lambda \sqrt{3}\,\zeta\,|\,\hat{\mathrm{I}}\,|\,, \end{aligned} \tag{6.94}$$

where, for example, eqns (6.92) have been used. Thus, positive values of ν_p result in triaxial dilatation, negative values imply triaxial compression. Consequently, $\nu_p = 0$ defines the phase transition point, where the volumetric response transits with increasing $\hat{\mathrm{I}}$ from compressive to dilative. If the experimental evidence does not predict an explicit hint on the location of the phase transition point, one often uses the maximum of the hydrostatic yield envelope to represent phase transition, compare the associated isotropic hardening model of DESAI (1990), where $F = Q$ naturally implies $\hat{\mathbf{D}}_{Sp}^K = \mathbf{0}$ at $\hat{\mathrm{I}} = -3\,\pi_c\,p_c$. Concerning the deviatoric part of $\hat{\mathbf{D}}_{Sp}$, compare Figure 15, $\hat{\mathbf{D}}_{Sp}^D$ is often directed straightly towards $\hat{\boldsymbol{\tau}}^{SD}$ or $\hat{\boldsymbol{\tau}}^{SD} - \hat{\mathbf{Y}}^{SD}$, respectively. This assumption is in good accordance with diverse experiments, for example, by LADE & DUNCAN (1973) on loose Monterey No. 0 Sand or by YAMADA & ISHIHARA (1979) on Fuji River Sand. However, plastic flow occurs in the case of loading, that is,

$$\frac{\partial F}{\partial \hat{\boldsymbol{\tau}}^S} \cdot (\hat{\boldsymbol{\tau}}^S)_p^{\nabla} \begin{cases} > 0 & : \text{loading} & \rightarrow \hat{\mathbf{D}}_{Sp} \neq \mathbf{0}\,, \\ = 0 & : \text{neutral loading} & \rightarrow \hat{\mathbf{D}}_{Sp} = \mathbf{0}\,, \\ < 0 & : \text{unloading} & \rightarrow \hat{\mathbf{D}}_{Sp} = \mathbf{0}\,, \end{cases} \tag{6.95}$$

where, with the aid of eqns (6.5),

$$\begin{aligned} \frac{\partial F}{\partial \hat{\boldsymbol{\tau}}^S} = \tfrac{1}{2}\hat{\Phi}^{-1/2}(1+\gamma\hat{\vartheta})^{m-1}\{\{[1+(1-\tfrac{3}{2}m)\gamma\hat{\vartheta}](\hat{\boldsymbol{\tau}}^{SD} - \hat{\mathbf{Y}}^{SD}) \\ +(1+\gamma\hat{\vartheta})^{1-m}(\alpha^2\,\hat{\mathrm{I}}_k + 4\,\delta^2\,\hat{\mathrm{I}}_k^3)\,\mathbf{I}\} \\ +m\,\gamma\,(\hat{\mathrm{II}}_{kD})^{-1/2}[(\hat{\boldsymbol{\tau}}^{SD} - \hat{\mathbf{Y}}^{SD})(\hat{\boldsymbol{\tau}}^{SD} - \hat{\mathbf{Y}}^{SD}) - \tfrac{2}{3}\,\hat{\mathrm{II}}_{kD}\,\mathbf{I}] \\ +2\,\hat{\Phi}^{1/2}(1+\gamma\hat{\vartheta})^{1-m}(\beta + 2\,\varepsilon\,\hat{\mathrm{I}}_k)\,\mathbf{I}\}. \end{aligned} \tag{6.96}$$

Note in passing that the above relations are valid in the case of kinematic hardening and can easily be reduced to isotropic hardening conditions with $\hat{\mathbf{Y}}^S = \mathbf{0}$.

Finally, the Λ-factor can either be obtained from a consistency condition, $(F)'_S = 0$, or directly from F itself. The former ("classical") approach results in an incremental procedure, the latter approach includes the definition of a so-called "consistent tangent operator" combined, in numerical computations, with a predictor-corrector scheme.

7. Concluding remarks

In the present article, constitutive equations for granular materials in geomechanical context were discussed based on porous media theories. Proceeding from the consideration that granular porous solid materials can be saturated by both liquids and gases, one had to describe a macroscopic model of binary media consisting of compressible and/or incompressible constituents.

Meeting the above requirements, a unified approach to porous media theories was presented by use of the four principally different types of two-phase media, the compressible model, the two hybrid models and the incompressible model. In particular, not only the general constitutive considerations were presented, but we also introduced the "simplified constitutive models" as the most convenient tools for applications.

Following this, elasticity and plasticity properties of the granular solid constituent were taken into consideration. Concerning the elastic range, these considerations led to a new description of finite porous media elasticity based on a four-parameter elasticity law including two compressibility terms, the material or the phase compressibility and the configuration or the structural compressibility, respectively. Recall that the material compressibility, in describing incompressible matrix materials, degenerated to a Lagrangian multiplier. Furthermore, finite porous media plasticity including first yielding, isotropic and kinematic hardening as well as failure was discussed on the basis of a new seven-parameter "single surface" yield criterion combined with a non-associated flow rule and a constitutive functional for the finite back-stress (translation) tensor. Most of this elasticity and plasticity material was worked out in the last month and has never been published before.

In analogy to the matrix description, the pore fluid could also be shown to be equipped with two pressure terms, a material and a configuration pressure, where the material pressure, within the framework of incompressible liquids, also degenerated to a Lagrangian multiplier.

However, concerning applications of the presented theory within numerical algorithms, it must be noted that at least two different systems of differential equations have to be solved which, as a result of the momentum balance equations, are coupled by the momentum supply term or the interaction force, respectively. By use of incompressible constituents, an additional equation, the volume continuity, occurs for the determination of the Lagrangian multiplier. Finally, in considering thermomechanical processes, additional couplings arise from the energy balance equation which is necessary for the determination of the temperature variation.

8. Appendix

The following definitions and rules can easily be proved by use of usual tensor calculus.

8.1 Fourth-order tensors

Let $\mathbf{A}, \mathbf{B}, \mathbf{C}, \mathbf{D}$ be arbitrary nonsingular linear transformations (second-order tensors). Then, a fourth-order tensor $\overset{4}{\mathbf{A}}$ can be defined which has the following characteristics:

$$
\begin{aligned}
\overset{4}{\mathbf{A}} &= (\mathbf{A} \otimes \mathbf{B})^{\overset{14}{T}} = (\mathbf{B}^T \otimes \mathbf{A}^T)^{\overset{23}{T}}, \\
\overset{4}{\mathbf{A}}{}^T &= [(\mathbf{A} \otimes \mathbf{B})^{\overset{14}{T}}]^T = (\mathbf{A}^T \otimes \mathbf{B}^T)^{\overset{14}{T}}, \\
\overset{4}{\mathbf{A}}{}^{-1} &= [(\mathbf{A} \otimes \mathbf{B})^{\overset{14}{T}}]^{-1} = (\mathbf{A}^{-1} \otimes \mathbf{B}^{-1})^{\overset{14}{T}}.
\end{aligned} \tag{8.1}
$$

Therein, the symbol $\overset{4}{(\ldots)}{}^{\overset{ik}{T}}$ denotes the exchange of the i–th and k–th basis systems or a transposition with respect to the indices i and k, respectively. In addition,

$$\overset{4}{(\ldots)}{}^{T} = [\overset{4}{(\ldots)}{}^{\overset{13}{T}}]^{\overset{24}{T}} . \tag{8.2}$$

Given eqn $(8.1)_1$, the following rules hold:

$$\begin{aligned} (\mathbf{A}\otimes\mathbf{B})^{\overset{14}{T}}(\mathbf{C}\otimes\mathbf{D})^{\overset{14}{T}} &= (\mathbf{CA}\otimes\mathbf{DB})^{\overset{14}{T}} , \\ (\mathbf{A}\otimes\mathbf{B})^{\overset{14}{T}}(\mathbf{C}\otimes\mathbf{D}) &= (\mathbf{B}^T\mathbf{CA}\otimes\mathbf{D}) , \\ (\mathbf{A}\otimes\mathbf{B})(\mathbf{C}\otimes\mathbf{D})^{\overset{14}{T}} &= (\mathbf{A}\otimes\mathbf{DBC}^T) , \end{aligned} \tag{8.3}$$

and

$$(\mathbf{A}\otimes\mathbf{B})^{\overset{14}{T}}\mathbf{C} = \mathbf{B}^T\mathbf{CA} . \tag{8.4}$$

By definition of a fourth-order tensor $\overset{4}{\mathbf{B}}$ provided with the properties

$$\begin{aligned} \overset{4}{\mathbf{B}} &= (\mathbf{A}\otimes\mathbf{B})^{\overset{13}{T}} , \\ \overset{4}{\mathbf{B}}{}^{T} &= [(\mathbf{A}\otimes\mathbf{B})^{\overset{13}{T}}]^T = (\mathbf{B}\otimes\mathbf{A})^{\overset{13}{T}} , \\ \overset{4}{\mathbf{B}}{}^{-1} &= [(\mathbf{A}\otimes\mathbf{B})^{\overset{13}{T}}]^{-1} = (\mathbf{B}^{T-1}\otimes\mathbf{A}^{T-1})^{\overset{13}{T}} , \end{aligned} \tag{8.5}$$

it is easily proved in connection with the definition $(8.1)_1$ that

$$\begin{aligned} (\mathbf{A}\otimes\mathbf{B})^{\overset{13}{T}}(\mathbf{C}\otimes\mathbf{D})^{\overset{13}{T}} &= (\mathbf{D}^T\mathbf{A}\otimes\mathbf{CB}^T)^{\overset{14}{T}} , \\ (\mathbf{A}\otimes\mathbf{B})^{\overset{14}{T}}(\mathbf{C}\otimes\mathbf{D})^{\overset{13}{T}} &= (\mathbf{CA}\otimes\mathbf{B}^T\mathbf{D})^{\overset{13}{T}} , \\ (\mathbf{A}\otimes\mathbf{B})^{\overset{13}{T}}(\mathbf{C}\otimes\mathbf{D})^{\overset{14}{T}} &= (\mathbf{DA}\otimes\mathbf{BC}^T)^{\overset{13}{T}} , \\ (\mathbf{A}\otimes\mathbf{B})^{\overset{13}{T}}(\mathbf{C}\otimes\mathbf{D}) &= (\mathbf{BC}^T\mathbf{A}\otimes\mathbf{D}) , \\ (\mathbf{A}\otimes\mathbf{B})(\mathbf{C}\otimes\mathbf{D})^{\overset{13}{T}} &= (\mathbf{A}\otimes\mathbf{CB}^T\mathbf{D}) , \end{aligned} \tag{8.6}$$

and

$$(\mathbf{A}\otimes\mathbf{B})^{\overset{13}{T}}\mathbf{C} = \mathbf{BC}^T\mathbf{A} . \tag{8.7}$$

In addition,

$$(\overset{4}{\mathbf{C}}\overset{4}{\mathbf{D}})^T = \overset{4}{\mathbf{D}}{}^T\,\overset{4}{\mathbf{C}}{}^T \tag{8.8}$$

holds where $\overset{4}{\mathbf{C}}$ and $\overset{4}{\mathbf{D}}$ are arbitrary tensors of fourth order.

8.2 Tensor derivatives with respect to tensors of second order

Let $\mathbf{A}, \mathbf{B}, \mathbf{C}$ be arbitrary second-order tensors. Then, the following derivation rules hold:

$$\frac{\partial(\mathbf{AB})}{\partial\mathbf{A}} = (\mathbf{B}\otimes\mathbf{I})^{\overset{14}{T}}, \tag{8.9}$$

$$\frac{\partial(\mathbf{AB})}{\partial\mathbf{B}} = (\mathbf{I}\otimes\mathbf{A}^T)^{\overset{14}{T}}, \tag{8.10}$$

$$\frac{\partial(\mathbf{AA})}{\partial\mathbf{A}} = (\mathbf{A}\otimes\mathbf{I})^{\overset{14}{T}} + (\mathbf{I}\otimes\mathbf{A}^T)^{\overset{14}{T}}, \tag{8.11}$$

$$\frac{\partial(\mathbf{A}^T\mathbf{A})}{\partial\mathbf{A}} = (\mathbf{A}\otimes\mathbf{I})^{\overset{13}{T}} + (\mathbf{I}\otimes\mathbf{A})^{\overset{14}{T}}, \tag{8.12}$$

$$\frac{\partial(\mathbf{AA}^T)}{\partial\mathbf{A}} = (\mathbf{A}^T\otimes\mathbf{I})^{\overset{14}{T}} + (\mathbf{I}\otimes\mathbf{A})^{\overset{13}{T}}, \tag{8.13}$$

$$\frac{\partial(\mathbf{ABC})}{\partial\mathbf{B}} = (\mathbf{C}\otimes\mathbf{A}^T)^{\overset{14}{T}}, \tag{8.14}$$

$$\frac{\partial\mathbf{A}^{-1}}{\partial\mathbf{A}} = -(\mathbf{A}^{-1}\otimes\mathbf{A}^{T-1})^{\overset{14}{T}}. \tag{8.15}$$

In addition, it is known that

$$\frac{\partial\mathbf{A}}{\partial\mathbf{A}} = (\mathbf{I}\otimes\mathbf{I})^{\overset{23}{T}} = \overset{4}{\mathbf{I}}, \tag{8.16}$$

$$\frac{\partial(\mathbf{A}\cdot\mathbf{I})\mathbf{I}}{\partial\mathbf{A}} = (\mathbf{I}\otimes\mathbf{I}), \tag{8.17}$$

$$\frac{\partial\mathbf{A}^T}{\partial\mathbf{A}} = (\mathbf{I}\otimes\mathbf{I})^{\overset{13}{T}}, \tag{8.18}$$

where $\overset{4}{\mathbf{I}}$ characterizes the fourth-order identity.

8.3 Example

To set an example for the convenience of the above definitions and rules, it is proved that eqn $(4.17)_2$ holds. Given

$$\begin{aligned}
\psi^i &= \psi^i(\theta, \mathbf{E}_{Se}, \mathbf{E}_{Sp}), \\
\mathbf{E}_{Se} &= \mathbf{E}_{Se}(\mathbf{F}_{Se}, \mathbf{F}_{Sp}) = \mathbf{F}_{Sp}^T\hat{\mathbf{\Gamma}}_{Se}\mathbf{F}_{Sp}, \qquad \hat{\mathbf{\Gamma}}_{Se} = \hat{\mathbf{\Gamma}}_{Se}(\mathbf{F}_{Se}), \\
\mathbf{E}_{Sp} &= \mathbf{E}_{Sp}(\mathbf{F}_{Sp}) = \tfrac{1}{2}(\mathbf{F}_{Sp}^T\mathbf{F}_{Sp} - \mathbf{I}),
\end{aligned} \tag{8.19}$$

one obtains from the chain rule:

$$\frac{\partial\psi^i}{\partial\mathbf{F}_{Sp}} = \left(\frac{\partial\mathbf{E}_{Se}}{\partial\mathbf{F}_{Sp}}\right)^T\frac{\partial\psi^i}{\partial\mathbf{E}_{Se}} + \left(\frac{\partial\mathbf{E}_{Sp}}{\partial\mathbf{F}_{Sp}}\right)^T\frac{\partial\psi^i}{\partial\mathbf{E}_{Sp}}. \tag{8.20}$$

Therein, with the help of eqns $(8.1)_2$, $(8.5)_2$, and (8.12),

$$\begin{aligned} \frac{\partial \mathbf{E}_{Sp}}{\partial \mathbf{F}_{Sp}} &= \tfrac{1}{2}\,[(\mathbf{F}_{Sp}\otimes\mathbf{I})^{\overset{13}{T}} + (\mathbf{I}\otimes\mathbf{F}_{Sp})^{\overset{14}{T}}]\,, \\ \left(\frac{\partial \mathbf{E}_{Sp}}{\partial \mathbf{F}_{Sp}}\right)^T &= \tfrac{1}{2}\,[(\mathbf{I}\otimes\mathbf{F}_{Sp})^{\overset{13}{T}} + (\mathbf{I}\otimes\mathbf{F}_{Sp}^T)^{\overset{14}{T}}]\,. \end{aligned} \tag{8.21}$$

Furthermore, with the abbreviation

$$\mathbf{A} = \hat{\mathbf{\Gamma}}_{Se}\mathbf{F}_{Sp}\,, \tag{8.22}$$

the chain rule yields

$$\frac{\partial \mathbf{E}_{Se}}{\partial \mathbf{F}_{Sp}} = \frac{\partial(\mathbf{F}_{Sp}^T\mathbf{A})}{\partial \mathbf{F}_{Sp}^T}\frac{\partial \mathbf{F}_{Sp}^T}{\partial \mathbf{F}_{Sp}} + \frac{\partial(\mathbf{F}_{Sp}^T\mathbf{A})}{\partial \mathbf{A}}\frac{\partial \mathbf{A}}{\partial \mathbf{F}_{Sp}}\,. \tag{8.23}$$

Thus, by use of eqns (8.9), (8.10), and (8.18):

$$\frac{\partial \mathbf{E}_{Se}}{\partial \mathbf{F}_{Sp}} = (\hat{\mathbf{\Gamma}}_{Se}\mathbf{F}_{Sp}\otimes\mathbf{I})^{\overset{14}{T}}(\mathbf{I}\otimes\mathbf{I})^{\overset{13}{T}} + (\mathbf{I}\otimes\mathbf{F}_{Sp})^{\overset{14}{T}}(\mathbf{I}\otimes\hat{\mathbf{\Gamma}}_{Se})^{\overset{14}{T}}\,. \tag{8.24}$$

Then, with the help of $(8.1)_2$, $(8.3)_1$, $(8.5)_2$ and $(8.6)_2$,

$$\begin{aligned} \frac{\partial \mathbf{E}_{Se}}{\partial \mathbf{F}_{Sp}} &= (\hat{\mathbf{\Gamma}}_{Se}\mathbf{F}_{Sp}\otimes\mathbf{I})^{\overset{13}{T}} + (\mathbf{I}\otimes\hat{\mathbf{\Gamma}}_{Se}\mathbf{F}_{Sp})^{\overset{14}{T}}\,, \\ \left(\frac{\partial \mathbf{E}_{Se}}{\partial \mathbf{F}_{Sp}}\right)^T &= (\mathbf{I}\otimes\hat{\mathbf{\Gamma}}_{Se}\mathbf{F}_{Sp})^{\overset{13}{T}} + (\mathbf{I}\otimes\mathbf{F}_{Sp}^T\hat{\mathbf{\Gamma}}_{Se})^{\overset{14}{T}}\,. \end{aligned} \tag{8.25}$$

Now, (8.20), $(8.21)_2$, and $(8.25)_2$ combine to

$$\frac{\partial \psi^i}{\partial \mathbf{F}_{Sp}} = \hat{\mathbf{\Gamma}}_{Se}\mathbf{F}_{Sp}\left(\frac{\partial \psi^i}{\partial \mathbf{E}_{Se}}\right)^T + \hat{\mathbf{\Gamma}}_{Se}\mathbf{F}_{Sp}\frac{\partial \psi^i}{\partial \mathbf{E}_{Se}} + \tfrac{1}{2}\left[\mathbf{F}_{Sp}\left(\frac{\partial \psi^i}{\partial \mathbf{E}_{Sp}}\right)^T + \mathbf{F}_{Sp}\frac{\partial \psi^i}{\partial \mathbf{E}_{Sp}}\right], \tag{8.26}$$

where eqns (8.4) and (8.7) have been used.

Finally, since the second-order tensors $\partial\psi^i/\partial\mathbf{E}_{Se}$ and $\partial\psi^i/\partial\mathbf{E}_{Sp}$ are obviously symmetric, one concludes with the help of Table 1 that

$$\frac{\partial \psi^i}{\partial \mathbf{F}_{Sp}} = (\mathbf{F}_{Se}^T\mathbf{F}_{Se} - \mathbf{I})\,\mathbf{F}_{Sp}\frac{\partial \psi^i}{\partial \mathbf{E}_{Se}} + \mathbf{F}_{Sp}\frac{\partial \psi^i}{\partial \mathbf{E}_{Sp}}\,. \tag{8.27}$$

Thus, eqn $(4.17)_2$ is proved.

9. References

Aktin, R.J. and Craine, R.E.: Continuum theories of mixtures: Basic theory and historical development, Q. J. Mech. Appl. Math., 29 (1976a), 209–244.

Aktin, R.J. and Craine, R.E.: Continuum theories of mixtures: Applications, J. Inst. Maths. Applics., 17 (1976b), 153–207.

Bachmat, Y. and Bear, J.: Macroscopic modelling of transport phenomena in porous media. 1: The continuum approach, Transport in Porous Media, 1 (1986), 213–240.

Baer M.R. and Nunziato, J.W.: A two-phase mixture theory for the deflagration-to-detonation transition (DDT) in reactive granular materials, Int. J. Multiphase flow, 12 (1986), 861–889.

Bedford, A. and Drumheller, D.S.: Theories of immiscible and structured mixtures, Int. J. Engng. Sci., 21 (1983), 863–960.

de Boer, R.: On plastic deformation of soils, Int. J. Plasticity, 4 (1988), 371–391.

de Boer, R. and Dresenkamp, H.-T.: Contitutive equations for concrete in failure state, ASCE: J. Engrg. Mech., 115 (1989), 1591–1608.

de Boer, R. and Ehlers, W.: On the problem of fluid- and gas-filled elasto-plastic solids, Int. J. Solids Structures, 22 (1986a), 1231–1242.

de Boer, R. and Ehlers, W.: Theorie der Mehrkomponentenkontinua mit Anwendung auf bodenmechanische Probleme, Teil I, Forschungsberichte aus dem Fachbereich Bauwesen der Universität-GH-Essen, 40, Universität Essen 1986b.

de Boer, R. and Ehlers, W.: A historical review of the formulation of porous media theories, Acta Mech., 74 (1988), 1–8.

de Boer, R. and Ehlers, W.: Uplift, friction and capillarity – three fundamental effects for liquid-saturated porous solids, Int. J. Solids Structures, 26 (1990a), 43–57.

de Boer, R. and Ehlers, W.: The development of the concept of effective stresses, Acta Mech., 83 (1990b), 77–94.

de Boer, R., Ehlers, W., Kowalski, S. and Plischka, J.: Porous media – a survey of different approaches, Forschungsberichte aus dem Fachbereich Bauwesen der Universität-GH-Essen, 54, Universität Essen 1991.

de Boer, R. and Kowalski, S.J.: A plasticity theory for fluid-saturated porous solids, Int. J. Engng. Sci., 21 (1983), 1343–1357.

de Boer, R. and Lade, P.V.: Towards a general plasticity theory for empty and saturated porous solids, Forschungsberichte aus dem Fachbereich Bauwesen der Universität-GH-Essen, 55, Universität Essen 1991.

Bowen, R.M.: Theory of mixtures, in Eringen, C. (ed.), Continuum Physics, Vol. III, Academic Press, New York 1976, pp. 1–127.

Bowen, R.M.: Incompressible porous media models by use of the theory of mixtures, Int. J. Engng. Sci., 18 (1980), 1129–1148.

Bowen, R.M.: Compressible porous media models by use of the theory of mixtures, Int. J. Engng. Sci., 20 (1982), 697–735.

Bowen, R.M.: Diffusion models implied by the theory of mixtures, in Truesdell, C. (ed.), Rational Thermodynamics, 2nd Edn., Springer-Verlag, New York 1984, pp. 237–263.

Chaboche, J.L.: Sur les lois de comportement des matériaux sous sollicitations monotones ou cycliques, Rech. Aérosp., no. 1983-5 (1983), 363–375.

Cross, J.J.: Mixtures of fluids and isotropic solids, Arch. Mech., 25 (1973), 1025–1039.

Deasi, C.S.: Single surface yield and potential function plasticity models: A review, Computers and Geotechnics, 7 (1989), 319–335.

Desai, C.S.: Modelling and testing: Implementation of numerical models and their application in practice, in Desai, C.S. and Gioda, G. (eds.), Numerical Methods and Constitutive Modelling in Geomechanics, Springer-Verlag, Wien 1990, pp. 1–168.

Desai, C.S. and Salami, M.R.: Constitutive model for rocks, ASCE: J. Geotech. Engr. Div., 113 (1987), 407–423.

Dresenkamp, H.-T.: Stoffgleichungen für spröde und granulare Stoffe im Bruchzustand, Dissertation (Dr.-Ing.), Universität Essen 1987.

Drucker, D.C. and Prager, W.: Soil mechanics and plastic analysis of limit design, Quart. Appl. Math., 10 (1952), 157–165.

Drumheller, D.S.: The theoretical treatment of a porous solid using a mixture theory, Int. J. Solids Structures, 14 (1978), 441–456.

Ehlers, W.: On thermodynamics of elasto-plastic porous media, Arch. Mech., 41 (1989a), 73–93.

Ehlers, W.: Poröse Medien – ein kontinuumsmechanisches Modell auf Basis der Mischungstheorie, Forschungsberichte aus dem Fachbereich Bauwesen der Universität-GH-Essen, 47, Universität Essen 1989b.

Ehlers, W.: A general approach to porous media elasto-plasticity, in Khan, A.S. and Tokuda, M. (eds.), Advances in Plasticity, Proceedings of Plasticity '89, Pergamon Press, Oxford 1989c, pp. 19–22.

Ehlers, W.: Finite elasto-plasticity for liquid-saturated porous solids, in Fan, J. and Murakami, S. (eds.), Advances in Contitutive Laws for Engineering Materials, Proceedings of ICCLEM (Asia), International Academic Publishers, Beijing 1989d, pp. 439–444.

Ehlers, W.: Toward finite theories of liquid-saturated elasto-plastic porous media, Int. J. Plasticity, 7 (1991), 443–475.

Ehlers, W.: An elastoplasticity model in porous media theories, Transport in Porous Media, 9 (1992), 49–59.

Ehlers, W.: A new single surface plasticity model for brittle and granular solid materials, in preparation.

Flory, P.J.: Thermodynamic relations for high elastic materials, Trans. Faraday Soc., 57

(1961), 829–838.

Green, A.E. and Naghdi, P.M.: A general theory of an elastic-plastic continuum, Arch. Rational Mech. Anal., 18 (1965), 251–281; 19 (1965), 408.

Green, R.J.: A plasticity theory for porous solids, Int. J. Mech. Sci., 14 (1972), 215–224.

Hassanizadeh, M. and Gray, W.G.: General conservation equations for multi-phase systems: 1. Averaging procedure, Adv. Water Res., 2 (1979a), 131–144.

Hassanizadeh, M. and Gray, W.G.: General conservation equations for multi-phase systems: 2. Mass, momenta, energy and entropy equations, Adv. Water Res., 2 (1979a), 191-203.

Haupt, P.: On the concept of an intermediate configuration and its application to a representation of viscoelastic-plastic material behavior, Int. J. Plasticity, 1 (1985), 303–316.

Haupt, P. and Tsakmakis, Ch.: On the application of dual variables in continuum mechanics, Continuum Mech. Thermodyn., 1 (1989), 165–196.

Hutter, K.: The foundations of thermodynamics, its basic postulates and implications. A review of modern thermodynamics, Acta Mech., 27 (1977), 1–54.

Kelly, P.D.: A reacting continuum, Int. J. Engng. Sci., 2 (1964), 129–153.

Kim, M.K. and Lade, P.V.: Single hardening constitutive model for frictional materials, I. Plastic potential function, Computers and Geotechnics, 5 (1988), 307–324.

Kojić, M. and Cheatham, J.B.: Theory of plasticity of porous media with fluid flow, Soc. Petr. Eng. J., 14 (1974a), 263–270.

Kojić, M. and Cheatham, J.B.: Analysis of the influence of fluid flow on the plasticity of porous rock under an axially symmetric punch, Soc. Petr. Eng. J., 14 (1974b), 271–278.

Lade, P.V. and Duncan, J.M.: Cubical triaxial tests on cohesionless soil, ASCE: J. Soil Mech. Found. Div., 99 (1973), 793–812.

Lade, P.V.: Three-dimensional stress-strain behavior and modeling of soils, Schriftenreihe des Instituts für Grundbau, Wasserwesen und Verkehrswesen, Serie Grundbau, Heft 4, Ruhr-Universität Bochum 1979.

Macvean, D.B.: Die Elementararbeit in einem Kontinuum und Zuordnung von Spannungs- und Verzerrungstensoren, ZAMP, 19 (1968), 157–185.

von Mises, R.: Mechanik der plastischen Formänderung von Kristallen, ZAMM, 8 (1928), 161–185.

Morland, L.W.: A simple contitutive theory for a fluid-saturated porous solid, J. Geophys. Res., 77 (1972), 890–900.

Müller, I.: A thermodynamic theory of mixtures of fluids, Arch. Rational Mech. Anal., 28 (1968), 1–39.

Nigmatulin, R.I.: Spatial averaging in the mechanics of heterogeneous and dispersed systems, Int. J. Multiphase Flow, 5 (1979), 353–385.

Paulun, J.: Plastische Inkompressibilität bei großen Formänderungen, Acta Mech., 37 (1980), 43–51.

Phillips, A.: The foundations of thermoplasticity – experiments and theory, in Zeman, J. L. and Ziegler, F. (eds.), Topics in Applied Continuum Mechanics, Springer-Verlag, Wien 1974, pp. 1–21.

Phillips, A. and Eisenberg, M.: Observations on certain inequality conditions in plasticity , Int. J. Non-linear Mech., 1 (1966), 247–255.

Prévost, J.A.: Consolidation of anelastic porous media, ASCE: J. Engrg. Mech., 107 (1981), 169–186.

Rivlin, R.S.: Large elastic deformation of isotropic materials, I. Fundamental concepts, Phil. Trans. Roy. Soc. Lond. A, 240 (1948), 459–490.

Schad, H.: Nichtlineare Stoffgleichungen für Böden und ihre Verwendung bei der numerischen Analyse von Grundbauaufgaben, Mitteilung des Baugrundinstituts Stuttgart, Heft 10, Universität Stuttgart 1979.

Simo, J.C. and Pister, K.S.: Remarks on rate constitutive equations for finite deformation problems: Computational implications, Comput. Meths. Appl. Mech. Engrg., 46 (1984), 201–215.

Thamm, B.: Anfangssetzungen und Anfangsporenwasserüberdrücke eines normalverdichteten wassergesättigten Tones, Mitteilungen des Baugrundinstituts Stuttgart, Heft 1, Universität Stuttgart 1974.

Wang, C.-C.: Inhomogeneities in second-grade fluid bodies and isotropic solid bodies, Arch. Mech., 25 (1973), 765–780.

Wang, C.-C. and Truesdell, C.: Introduction to Rational Elasticity, Noordhoff International Publishing, Leyden 1973.

Yamada, Y. and Ishihara, K.: Anisotropic deformation characteristics of sand under three-dimensional stress conditions, JSSMFE: Soil and Foundations, 19 (1979), 79–94.

The Flow of Ice Sheets and Ice Shelves

L. W. Morland
School of Mathematics, University of East Anglia
Norwich NR4 7TJ, UK

A brief survey of the mathematical and numerical modelling of ice sheet and ice shelf flows is presented, emphasising the developments over the last decade. For the very slow gravity driven flow of natural ice masses over time scales of thousands of years, the ice is modelled by an incompressible non-linearly viscous fluid with a temperature dependent rate factor. The conservation laws of mass, linear momentum and energy are formulated, together with the free surface conditions and basal conditions for a grounded sheet and a floating shelf. A systematic reduction of the equations for both sheet and shelf is developed by the introduction of dimensionless variables and stretched co-ordinates, which are motivated by the forms of the sheet profiles sought, and the necessary assumptions are drawn out. Analytic progress is made for restricted flow configurations. The failure of depth-integrated equations to provide a valid approximation in more general configurations is noted. A binary mixture theory is developed for co-existing ice and water in extensive melt zones.

1. Introduction

The large-scale advance and retreat of the Earth's large ice sheets due to global climate changes take place on time-scales of thousands of years. This is the time-scale associated with significant movement of an ice sheet front and significant variation of its thickness under global changes which have occurred in the past. Surge conditions on the much shorter time scale of a year are strongly localized, and will not significantly influence the global picture, but will have effects on the local terrain. The overall problem is a complex thermo-mechanically coupled flow of the ice driven by gravity and subject to mechanical and thermal interactions with the atmosphere at the ice sheet surface and with the ground at the base, and with the ocean beneath a shelf. These notes discuss in detail the formulation of idealised descriptions, or conceptual models, of the ice flow and boundary conditions, to highlight the approximations and restrictions, and to outline the applications to ice sheet evolution.

An outline of ice sheet modelling over the past thirty years is presented in the next section to provide background for the course, though it may have more meaning after the course. The references provide a thread through the development of the subject, and though there will be many contributions not mentioned it is hoped all would appear as cross-references.

The lectures will focus on the applications of mathematical techniques to ice sheet modelling which have emerged over the last decade, and which are the basis of an increasing research endeavour to understand and predict the effects of ice sheet evolution on the world-wide environment. This view of glaciology from the standpoint of theoretical (thermo)mechanics is only one aspect of a wide subject, but is an essential cornerstone for the process of systematic analysis. The presentation will necessarily reflect a personal approach to the subject, with perhaps a different emphasis to that seen elsewhere, but it is hoped that there will be no inconsistencies in the arguments and conclusions, and that the theory will provide a sound framework for further development.

To describe the flow over the long time scales of ice sheet advance and retreat, or even over scales of years, the ice is modelled as an incompressible, non-linearly viscous heat conducting fluid, with a strongly temperature dependent rate factor. The velocity, stress and temperature fields within the ice sheet are governed by the conservation laws of mass, momentum and energy, subject to thermal and mechanical surface and base conditions. These are formulated first for a single phase cold ice sheet, with no interior melting or refreezing. The analysis of a grounded sheet in normalised dimensionless variables and stretched co-ordinates which reflect the dominant physical balances and long aspect ratio, determine a reduced model applicable to a wide class of problems. A detailed derivation demonstrates the mathematical approach and exposes the approximations and restrictions. The reduced model allows considerable analytic progress in the case of plane flow with a prescribed temperature field, which is illustrated, and is the basis for many numerical treatments. Depth integrated models which eliminate dependence on the vertical co-ordinate, using mean values over the thickness, have been widely used, but it is demonstrated that they require further, arbitrary, approximations, in addition to the reduced model restriction, and offer no additional simplification.

A floating ice shelf has basal conditions at the water interface distinct from the grounded sheet conditions, resulting in distinct velocity and stress patterns. The scaled variable and stretched co-ordinate technique again leads to a reduced model, with further simplification if the temperature field is prescribed. In general, the transition zone in which neither reduced model applies, may be extensive, not simply a few ice thicknesses, and as yet there are no valid simplified models to treat transition. Transition includes ice streams which are fast flowing grounded ice feeding into the shelves. An additional complication is the mobility of a grounding line as the sheet-shelf-crust configuration changes, with associated questions of stability or instability of the configuration. There are conflicting views on the present day stability of the West Antarctic Ice Sheet, because rigorous models to treat this wider problem of transition have still to be developed.

A final section discusses briefly a binary mixture theory to describe co-existing ice and water regions in a zone of melting and refreezing. Such regions can be mobile, and are features of the evolution to be determined (moving boundary problems). They will be significant during extended periods of climate warming. Theory and applications in this area are still in early stages.

It is clear that problem solving in the many aspects of ice sheet evolution, even where good reduced models exist, will rely on large scale computation. The development of accurate, efficient, numerical algorithms for three-dimensional space and time systems of partial differential equations, often with unknown moving boundaries, is itself a major discipline, which requires its own skills and experiences. This course makes no attempt to address this aspect, other than to reference various applications in the next section. Its purpose is to provide a sound theoretical mechanics formulation of the physical problems, and to show how mathematical techniques can provide valid simplifications in many situations, and also note where simplifications are not valid. Such theory underpins good numerical applications. Also, analytic progress in idealised problems can serve as test cases for numerical algorithms. When solutions are possible only by large scale computations, test cases are important, and it is also desirable that close numerical agreement between solutions to a problem by independent algorithms is achieved.

2. An Outline of Ice Sheet Modelling

The objective of an ice sheet model is to predict the evolution of the configuration of the ice sheet, and its stress, velocity and temperature fields, driven by gravity under prescribed mechanical and thermal conditions at the surface and base. In reality, the surface and basal conditions may involve interactions with evolving conditions in the atmosphere, bedrock, or ocean, but the complexity of such coupling can be set aside in the formulation and analysis of basic models. It is supposed, then, that the surface of the ice sheet is free of traction (a uniform isotropic atmospheric pressure is taken as zero stress) and subjected to accumulation or melting defined as an equivalent normal ice flux - the volume of ice crossing unit cross-section in unit time, and has a prescribed temperature or heat flux. These conditions depend, in general, on location and on elevation. At the base, a melting or freezing flux is prescribed and a temperature or heat flux. In addition, a tangential mechanical condition is required which prescribes the velocity or traction, or more generally a "sliding law" which relates velocity and traction for a grounded sheet. Note that the surface has one extra mechanical condition - the zero normal traction - which is required to determine the surface position (shape); this is a free surface or moving boundary problem in which part of the boundary of the ice sheet domain, on which there are prescribed conditions, must itself be determined as part of the solution. In the case of a floating ice shelf there is at the base both a zero tangential traction and a normal traction balancing the water pressure, but the base is also an unknown boundary. Whereas the shelf analysis is simpler than that of a grounded sheet, that for the transition zone between grounded sheet and floating shelf is more complex.

On the global length, time and velocity scales, ice particle accelerations are extremely small compared with gravitational acceleration and stress gradients, so that the stress field is essentially in equilibrium with gravity; that is, the linear momentum balance is reduced to an equilibrium equation. Except in recently deposited surface layers, the ice compression is negligible compared with the shears over time scales of interest, so an incompressibility approximation is made. On short time scales, minutes to days depending on stress level, the response of ice is non-linearly viscoelastic, either a solid behaviour remembering a reference configuration, or a fluid behaviour depending on history relative to the current configuration. On the longer time scales of ice sheet flow, beyond the time scales of laboratory testing, it is supposed that the ice flows as a non-linearly viscous fluid for which the shear stress is a function of the current shear strain-rate. There are ice field data, over years, which can be correlated with this assumed response, but verification rests on satisfactory conclusions drawn from model predictions. The viscosity not only depends on stress (or strain-rate), mositure content and fabric, but has a strong dependence on temperature, and this increases the complexity of the thermomechanical coupling between the energy and momentum balances.

The above model, with, or without, further simplifications, is a system of coupled non-linear partial differential equations for the stress and velocity components and temperature as functions of three space variables and time, acting over an unknown evolving ice sheet configuration. Analytical and numerical developments have proceeded by a series of special cases with varying degrees of ad hoc and rational approximations. Before moving to detailed formulations, a brief summary of the successive steps taken in the theoretical development is presented.

The simplest case for a grounded sheet is that of an isothermal symmetric sheet on a plane horizontal bed, in plane steady flow with a steady profile, which requires that the net accumulation and melting over the surface is zero. The flow is now described by two velocity components depending on the space variables, independent of time. NYE (1959) presented the first theoretical determination of the profile on the basis of a series of approximations, namely

(i) the horizontal velocity is independent of depth, depending only on horizontal distance, on the assertion that any significant shear motion takes place in a thin warmer basal layer,

(ii) the horizontal velocity (basal velocity) is a function of the basal shear stress,

(iii) the surface slope is small and stress gradients parallel to the surface are negligible, leading to a simple basal shear stress - free surface height and slope relation.

These lead to a first order differential equation for the surface height in terms of the surface accumulation function, which is simply integrated. The solution violates approximation (iii) near the margins. Note that the analysis is independent of the viscous property of the ice, eliminated by approximation (i). WEERTMAN (1961) modified Nye's solution by generalising the sliding law (ii). BODVARDSSON (1955) had earlier formulated an unsteady problem, obtained approximate solutions of the steady equations and constructed small time-dependent perturbations to model climate variation.

An analysis of the full equations for the same problem, and a solution, were given by MORLAND and JOHNSON (1980). The systematic approximation was based on a small parameter which defines the ratio of the vertical to horizontal length scales (or vertical to horizontal velocity magnitudes) in natural ice masses. This in turn governs the relative magnitudes of the vertical and horizontal gradients of the physical variables, and supports (iii) above, and leads to appropriate co-ordinate stretching to impose the necessary physical balances. The leading order equations (reduced model) can be integrated. The details of this approach in a more general context are presented later. The singular effects of a viscous power law (with infinite viscosity at zero stress) are seen, in contrast to a bounded viscosity law, and also the need for a particular form of sliding law behaviour at the margin (as the stress approaches zero).

The next step is to consider mean bed plane inclination, to contrast large ice sheets on horizontal mean bed planes with steep mountain glaciers on mean bed planes with finite (order unity) slopes, and then to allow bed topography (undulations) and investigate the influence of such factors on the surface shape. This extension of the model has led to a variety of treatments. The first, given by ROBIN (1967), assumes that

(iv) the bed slope and surface slope are everywhere small,

(v) The direct shear stress on planes parallel to the bed is negligible compared to longitudinal deviatoric stress.

Approximation (iv) is the extension of (iii) above, but (v) is entirely the reverse of the relative stress magnitudes imposed by NYE (1959) and supported by the MORLAND and JOHNSON (1980) analysis, and must be violated for a grounded sheet. It is a valid floating shelf result when the shear stress on both surface and base is zero. There have been three distinct small parameter expansion schemes to investigate the influence of bed topography. HUTTER (1981) assumes "slow variation" of surface (bed) undulations, characterised by small ice thickness compared with undulation wavelength and small undulation amplitude compared with ice thickness, which in practice is of very limited interest. HUTTER, LEGERER and SPRING (1981) assume that the amplitude is small compared with the ice thickness and that the wavelength is not less than the thickness, but theirs is a central zone analysis not extending to the margins. MORLAND and JOHNSON (1982) use their original small length ratio parameter and assume that the bed slope (relative to a mean bed plane) does not exceed the surface slope in magnitude. It therefore allows finite topographic amplitude (e.g. of order the ice thickness due to long-term crustal depression), but not a series of short wavelength undulations which would imply finite slope. The main result is that topography on a steep bed plane can penetrate to the surface, but small slope topography on a horizontal bed plane only has influence on the surface if its amplitude is large. None of the three reduced models allows finite slope undulations with wavelengths not greatly exceeding the thickness, which is a situation of practical importance.

Before turning to the development of a more general reduced model, it is important to recognise the rôle of the basal "sliding law". While the relative velocity between bed and surface is governed by the adopted viscous law for the ice shearing, the total surface velocity depends also on the sliding over the bed. For the global problem this must be interpreted as the tangential velocity at a conceptually smooth surface representing the basal boundary of the ice domain, ignoring the complicated small-scale bed structure and the associated complex flow in a thin basal layer. Even for a cold base with non-slip at the actual rock interface there can be a small mean velocity on the adopted basal contour. Also, there can be a contribution to this basal velocity from the movement of a deforming till layer. The sliding law, then, is a basal boundary condition which must reflect these processes, and hence depends on the nature of the bed. For the model, it is usually a prescription of tangential velocity in terms of basal stress, tangential and normal pressure, with non-slip or a limiting friction stress as special cases.

WEERTMAN (1957) proposed a regelation mechanism in which the melting and refreezing of the ice on upstream and downstream faces of undulations, due to pressure variation, produces a thin lubricating water layer. His treatment of the bed as a regular array of rectangular obstacles proceeded as a series of disconnected estimates of the various influences, offering no solution of the flow equations. It was later extended (WEERTMAN, 1964) to include cavitation on the downstream faces. A more elaborate treatment of a sinusoidal bed, including cavitation, was presented by LLIBOUTRY (1968), but in the same spirit. NYE (1969, 1970) and KAMB (1970) focus on the need for an exact flow and heat conduction solution, and construct half-space solutions on a small slope sinusoidal boundary for an incompressible linearly viscous (Newtonian) fluid, without cavitation. KAMB (1970) takes the linear solution as the starting point for an approximate treatment when the ice satisfies a non-linearly viscous power law. The main ingredients are a pressure-melting condition and zero tangential traction on the water covered interface, which has a source-sink distribution of latent heat to balance with the heat fluxes in the ice and ground, and the analysis proceeds by series expansions in the small slope parameter. MORLAND (1976a, 1976b) presented more rigorous complex variable analyses of the analogous problem incorporating a mean bed inclination and a free surface at a finite height (large compared with the undulation amplitude). The analysis determines the mean basal velocity (on the base plane) in terms of the mean drag on the bed, the pressure, and the profile and wavelength. It also predicts if, and where, cavitation occurs (negative pressure), which is strongly influenced by the form of an assumed "friction law" replacing the zero tangential traction assumption (MORLAND,1976b). More general predictions and interpretations for the non-linearly viscous case have been presented by FOWLER (1977, 1979a, 1981) using matched asymptotic expansions to construct a basal boundary layer solution.

A variety of reduced models to describe different situations in two and three-dimensional non-isothermal flows have been derived by appropriate expansion procedures in scaled variables; for example, FOWLER (1977, 1979b, 1991), FOWLER and LARSON (1978), HUTTER (1982a, 1982b, 1983), MORLAND (1984). Numerical applications first focused on plane flow, or axisymmetric flow, to eliminate the third space variable, in isothermal conditions, with prescribed temperature fields (not solving the energy balance), and in thermomechanically coupled conditions, for both steady conditions with no time dependence and the unsteady evolutionary problem. Steady uncoupled solutions have been calculated by MORLAND and SMITH (1984), MORLAND, SMITH and BOULTON (1984), BOULTON, SMITH and MORLAND (1984), unsteady uncoupled solutions by HINDMARSH, MORLAND, BOULTON and HUTTER (1987), steady coupled solutions by HUTTER, YAKOWITZ and SZIDAROVSKY (1986,

1987), and unsteady coupled solutions by HINDMARSH and HUTTER (1988) and HINDMARSH, BOULTON and HUTTER (1989). These latter applications are non-trivial computations, even without the dependence on the third spatial variable and without transverse flow, which are both features of realistic ice sheet evolution.

In the isothermal or prescribed temperature field reduced model, the equations can be integrated explicitly with respect to the depth co-ordinate (normal to the mean bed plane), reducing the independent variables by one. Thus steady plane flow reduces to an ordinary differential equation for the surface shape, while unsteady plane flow reduces to a parabolic partial differential equation for the evolving surface profile. However, in the thermomechanically coupled system, the equations cannot be explicitly integrated with respect to the depth co-ordinate. Integration through the thickness to obtain relations between mean (or total) values of the variables with dependence on one less space variable is an attractive proposition, but as will be shown in a later section this requires elimination or estimation of integrated quantities by arbitrary assumptions, which have not been assessed by comparisons with full solutions to typical problems. Nevertheless, the approach has been used to simplify applications, and in particular to investigate the effects of the longitudinal deviatoric stress neglected in the leading order global reduced model, with varying assumptions. Formal integrations are given by COLLINS (1968) and HUTTER (1983), and various applications by NYE (1969), BUDD (1969, 1970, 1971), OERLEMANS (1981), SHOEMAKER and MORLAND (1984), VAN DER VEEN (1987) and REEH (1988), among others. The full reduced model for three-dimensional unsteady flow has recently (HUYBRECHTS, 1992) been applied in a numerical study of the West Antarctic Ice Sheet.

The first model of a floating shelf is due to WEERTMAN (1957) who treats a floating slab of uniform thickness free of horizontal and vertical shear stress, to estimate the longitudinal deviatoric stress (stretching force) in balance with the integrated water pressure. Generalisations were made by THOMAS (1973a) and SANDERSON and DOAKE (1979), to incorporate lateral stretching and side drag, and by SANDERSON (1979) to include the thickness gradient in the longitudinal direction. A systematic analysis for the plane flow configuration (longitudinal flow only), exploiting the long aspect ratio, was presented by MORLAND and SHOEMAKER (1982), who derive reduced models and solutions for steady isothermal flow and for a particular non-isothermal case. They also demonstrate the inconsistencies of the earlier treatments incorporating lateral stretching. Shelf front conditions give only an integrated stress, due to the water pressure, but the zero tangential tractions on both surface and base (negligible viscous shear stress in the water) leads to a reduced model with longitudinal velocity independent of depth, and relations for the integrated stresses in the isothermal or prescribed temperature case. The thermomechanically coupled flow does not allow description in thickness integrated variables only. Axi-symmetric flow was analysed and illustrated by MORLAND and ZAINUDDIN (1987), and the two-dimensional spreading of an unconfined shelf was modelled by MORLAND (1987), both for prescribed temperature fields.

The importance of shelves is their interaction with the feeding sheets, complicated by the more complex flows across the transition zones and the grounding line mobility. The idealised reduced model for purely longitudinal flow or unconfined flow ignores the "back-forces" induced by side drag for shelves extending to bay shores, and by grounded zones in the bay. Such back forces have a strong influence on the sheet-shelf interaction. With no valid reduced models for transition and for grounded or side-attached shelves, the many treatments of these features and their consequences are open to question. They are preliminary attempts which may, or may not, prove to yield valid conclusions. An early issue was the stability of the West Antarctic Ice Sheet, with conflicting arguments and conclusions see, for example, HUGHES (1973, 1975), WEERTMAN (1976), MERCER (1978), THOMAS, SANDERSON and ROSE (1979), THOMAS and BENTLEY (1978), THOMAS (1979) and VAN DER VEEN (1987). Examples of substantial numerical attacks on shelf drag and transition problems are those of MACAYEAL (1987), MACAYEAL and BARCELON (1988), MACAYEAL and LANGE (1988), and HERTERICH (1987, 1988). I would also draw attention to the first major glaciology text (2nd edition) by PATERSON (1969), and the review articles by ROBIN and SWITHINBANK (1987), BENTLEY (1987), and HINDMARSH (1992) who kindly gave me his pre-publication manuscript.

Natural ice masses are not all cold throughout; some contain finite temperate regions in which ice and water phases co-exist at the melting point, and are termed polythermal glaciers. A glacier may be wholly temperate. The melting and refreezing process and maintenance of melting temperature in a temperate region influences the flow and temperature field in an adjacent cold region, and the location of the actual cold-temperate interface is itself part of the ice sheet evolution solution. The initial value-boundary value problem is now much more complex. The presence and movement of water through a temperate ice mass has always been recognised. Early treatments considered flow through discrete conduits with melting or refreezing at the channel walls, but the analysis of large polythermal ice masses requires a continuum model for the ice-water mixture which does not attempt to identify the actual constituent located at each point. This approach has been taken by FOWLER and LARSON (1978), HUTTER (1982b), FOWLER (1984), HUTTER, BLATTER and FUNK (1988), BLATTER and HUTTER (1991) and HUTTER (1992), for example. The final section describes biefly a binary mixture theory for co-existing ice-water.

3. Ice Properties

Even at the base of large ice sheets, at pressures up to $3 \times 10^7\, Nm^{-2}$, BUDD (1969) reports compressibility less than $4 \times 10^{-10}\, (Nm^{-2})^{-1}$, implying maximum compressions of order 10^{-2}, negligible in comparison with the large shear strains arising over flow times of hundreds of years and more. Incompressibility is thus an excellent approximation. Furthermore, the unchanging density of a given ice element is assumed to be uniform through the ice mass, ignoring variations in unconsolidated surface layers. The response of ice to uni-axial stressing at different temperatures is reviewed in detail by MELLOR (1980), which, due to incompressibility, reflects the shear (deviatoric) response, but on times scales of at most six months in laboratory testing. The rate of response increases with temperature, with the most significant changes near the melting temperature. The response is non-linearly viscoelastic (at each constant temperature). At constant stress there is an instantaneous (elastic) strain of similar magnitude to the compression, then a primary decelerating creep to a minimum strain-rate, referred to as secondary creep, followed by an accelerating tertiary creep. The time scale of each phase decreases as temperature increases, but even near to melting it is not evident that in the laboratory experiments the minimum strain-rate has been reached at the low shear stresses of order $10^5\ Nm^{-2}$ associated with ice sheets.

Viscoelastic models of differential and integral type, solid and fluid, have been formulated to describe the observed uni-axial stress responses (MORLAND, 1979; MORLAND and SPRING, 1981, 1982; SPRING and MORLAND, 1982, 1983; MORLAND, 1992a) on laboratory time scales, appropriate to many engineering applications. Furthermore, the influence of temperature on the rate of response has been modelled by a temperature-dependent rate-factor, equivalent to assuming a universal mechanical response on a pseudo-time scale dependent on temperature history (MORLAND, 1979).

For a description of the long-time response appropriate to ice sheet flow it has been convenient to assume a non-linearly viscous fluid response, with temperature dependent rate-factor, for the deviatoric behaviour, together with incompressibility. If σ is the Cauchy stress tensor with components measuring tractions per unit cross-section area in the current configuration, referred to rectangular Cartesian co-ordinates x_i $(i = 1, 2, 3)$, and p is the mean pressure, then the deviatoric stress, representing shear stress only, is

$$\sigma' = \sigma + p\mathbf{1} \tag{3.1}$$

where $\mathbf{1}$ is the unit tensor and

$$p = -tr(\sigma) = -\sigma_{kk}\,, \tag{3.2}$$

where the summation convention for a repeated suffix is used. The strain-rate tensor $\mathbf{D}$ with components measuring rates of stretch and angle change of, and between, line elements currently parallel to the co-ordinate axes, is defined by

$$D_{ij} = \frac{1}{2}\left[\frac{\partial v_i}{\partial x_j} + \frac{\partial v_j}{\partial x_i}\right], \tag{3.3}$$

where v_i $(i = 1, 2, 3)$ are the velocity components. Incompressibility implies equivalently that

$$\rho = \text{constant}\,, \qquad \text{div}\,\mathbf{v} = D_{kk} = 0, \tag{3.4}$$

where ρ is density, which together guarantee the conservation of mass; $\mathbf{D}$ is deviatoric. The pressure p is arbitrary, not determined by the deformation but by momentum and boundary conditions.

A general viscous deviatoric law, with temperature dependent rate factor $a(T)$, has equivalent forms (SMITH and MORLAND, 1981)

$$\begin{aligned} \hat{\mathbf{D}} &= \psi_1 \hat{\sigma'} + \psi_2 \left[\left(\hat{\sigma'}\right)^2 - \frac{2}{3} J_2 \mathbf{1}\right], \\ \hat{\sigma'} &= \phi_1 \hat{\mathbf{D}} + \phi_2 \left[\hat{\mathbf{D}}^2 - \frac{2}{3} I_2 \mathbf{1}\right], \end{aligned} \tag{3.5}$$

where

$$\sigma' = \sigma_0 \hat{\sigma}', \quad \mathbf{D} = a(T) D_0 \hat{\mathbf{D}}, \tag{3.6}$$

and the shear stress and strain units, σ_0 and D_0 respectively, are chosen to normalise the functions ψ_1, ψ_2, ϕ_1 and ϕ_2. Here ψ_1, ψ_2 are functions of the stress deviator invariants J_2, J_3, and ϕ_1, ϕ_2 are functions of the strain-rate invariants I_2, I_3, given by

$$\begin{aligned} J_2 &= \frac{1}{2} tr\left(\left[\hat{\sigma}'\right]^2\right), \quad J_3 = det\left(\hat{\sigma}'\right), \\ I_2 &= \frac{1}{2} tr\left(\left[\hat{\mathbf{D}}\right]^2\right), \quad I_3 = det\left(\hat{\mathbf{D}}\right). \end{aligned} \tag{3.7}$$

Other pairs of invariants are possible. Note that the forms (3.5) ensure that $tr(\mathbf{D})$ and $tr(\sigma')$ vanish.

A general form analogous to (3.5) was introduced by GLEN (1958), with private acknowledgement to F. Ursell, noting that combined shear and compression data (STEINEMANN, 1958) was not consistent with the simpler form proposed by NYE (1953):

$$\hat{\mathbf{D}} = \psi(J_2)\hat{\sigma}', \tag{3.8}$$

in which $\hat{\mathbf{D}}$ and $\hat{\sigma}'$ are parallel. However, the determination of two functions of two arguments is not possible from uni-axial data for minimum strain-rate as a function of stress, and few two-dimensional data exist, while adoption of the minimum strain-rate assumption ignores the possibility of significant tertiary creep effects. It is therefore customary to adopt the parallel form (3.8) with restricted dependence on one invariant, which has the equivalent form

$$\hat{\sigma}' = \phi(I_2)\hat{\mathbf{D}}, \tag{3.9}$$

where

$$\phi(I_2)\psi(J_2) \equiv 1 \tag{3.10}$$

and

$$I_2 = J_2\psi^2(J_2), \quad J_2 = I_2\phi^2(I_2). \tag{3.11}$$

In general the relations (3.10) and (3.11) do not allow explicit inversions between $\psi(J_2)$ and $\phi(I_2)$. For simple shear stress $\sigma_{12} = \tau$, other $\sigma_{ij} = 0$, with corresponding strain-rate $D_{12} = \dot{\gamma}$, other $D_{ij} = 0$, the invariants are $J_2 = (\tau/\sigma_0)^2$, $I_2 = (\dot{\gamma}/D_0 a)^2$, and the relations (3.8) and (3.9) become respectively

$$\dot{\gamma} = D_0 a \psi(J_2)\tau/\sigma_0, \quad \tau = \sigma_0\phi(I_2)\dot{\gamma}/D_0 a. \tag{3.12}$$

For uni-axial stress $\sigma_{11} = \sigma$, other $\sigma_{ij} = 0$, and axial strain-rate $D_{11} = \dot{\epsilon}$ with associated lateral strain-rates $D_{22} = D_{33} = -\frac{1}{2}\dot{\epsilon}$ by the incompressibility requirement (3.4), other $D_{ij} =$

0, then $p = -\frac{1}{3}\sigma$ so $\sigma'_{11} = \frac{2}{3}\sigma$, $\sigma'_{22} = \sigma'_{33} = -\frac{1}{3}\sigma$. Hence $J_2 = \frac{1}{3}(\sigma/\sigma_0)^2$, $I_2 = \frac{3}{4}(\dot{\epsilon}/D_0 a)^2$, and the relations (3.8) and (3.9) become, respectively

$$\dot{\epsilon} = \frac{2}{3} D_0 a \psi(J_2)\sigma/\sigma_0 \,, \qquad \sigma = \frac{2}{3}\sigma_0 \phi(I_2)\dot{\epsilon}/D_0 a \,. \tag{3.13}$$

If a stress dependent viscosity $\mu(J_2)$ is defined by the simple shear relation in incremental form:

$$\mathrm{d}\tau = 2\mu(\tau^2)\mathrm{d}\dot{\gamma} \,, \tag{3.14}$$

then by $(3.12)_1$,

$$\mu^{-1}(J_2) = \{2\psi(J_2) + 4J_2\psi'(J_2)\}\, aD_0/\sigma_0 \,. \tag{3.15}$$

where $\psi' = \mathrm{d}\psi/\mathrm{d}J_2$.

SMITH and MORLAND's (1981) comparison of a variety of uni-axial minimum strain-rate data at different temperatures and over different stress ranges shows that no single rate factor $a(T)$ can yield a reasonable universal mechanical law, but moreover that the data reveals wide discrepancies, in contrast to BUDD's (1969) conclusion that measured minimum strain-rates at given stress were within a factor 2. Laboratory results are sensitive to ice specimen preparation and to thermal and mechanical controls. Indirect deductions from ice shelf data, the natural laboratory, may well yield the best model expression (THOMAS, 1971). For illustration, a weighted least squares correlation of GLEN's (1955) uni-axial data at temperature $T_0 = 273.15\,K$ gave an excellent polynominal approximation

$$a(T_0)\psi(J_2) = 0.3336 + 0.32\,J_2 + 0.02963\,J_2^2 \tag{3.16}$$

for $0 \le J_2^{\frac{1}{2}} \le 5$ when

$$D_0 = 1\,yr^{-1} = 3.18 \times 10^{-8} s^{-1} \,, \quad \sigma_0 = 10^5\,Nm^{-2} \,. \tag{3.17}$$

Inverse sinh and tan expressions gave better correlations than polynomials for the inverse function $\phi(I_2)$. From the viscosity definition (3.15), at temperature T_0 the viscosities at zero stress and at shear stress σ_0 ($J_2 = 1$), a typical maximum in ice sheet flows, are approximately

$$\mu(0) \sim 5 \times 10^{12}\,Nm^{-2}s, \quad \mu(1) \sim 0.7 \times 10^{12}\,Nm^{-2}s \,, \tag{3.18}$$

showing a decrease by a factor 7.

GLEN's (1955) proposal for the axial relation was a power law, corresponding to

$$a(T_0)\psi(J_2) = 1.5k(3J_2)^{\frac{n-1}{2}} \tag{3.19}$$

with $k = 0.17$ and $n = 3.17$ to correlate data for the range $\sigma = 10^5 - 10^6\,Nm^{-2}$, but with poor agreement for the lower stresses relevant to ice sheet flow. In particular, at zero stress, the power law (3.19) with $n > 1$ gives vanishing $\psi(J_2)$ and $J_2\psi'(J_2)$ and hence unbounded viscosity and unbounded $\phi(I_2)$ as stress and strain-rate approach zero. While the scaling arguments leading to the reduced model are straightforward for bounded ϕ, they can be generalised to accommodate the power law behaviour (MORLAND and JOHNSON 1980, 1982). However, a weakly singular (passive to leading order) layer of high stress gradients arises near a free surface, which is physically unexpected, and the singular perturbation solution (JOHNSON and MCMEEKING, 1984) adds complexity and questionable reality. Nevertheless, Glen's algebraically simple power law continues to be used extensively despite these concerns for its low stress behaviour. It allows the explicit inversion

$$\phi(I_2) = \frac{2}{3}\left[\frac{a(T_0)}{k}\right]^{\frac{1}{n}}\left[\frac{4}{3}I_2\right]^{-\frac{n-1}{2n}} . \tag{3.20}$$

While the non-linearity in stress exhibits a decrease in viscosity by a factor 7 over the relevant shear stress range $0 - 10^5\ Nm^{-2}$ at the near melting temperature T_0, the influence of the temperature-dependent rate factor $a(T)$ is more dramatic. At constant stress, the viscosity μ is inversely proportional to $a(T)$, see (3.15), and $a(T)$ decreases by a factor of 55 as temperature decreases by 20 K below melting, and by a factor 1000 for a 40 K decrease, giving a corresponding increase in viscosity, with the most rapid variation near melting. Temperature changes from near melting to 20 K and more below are common in ice sheets. SMITH and MORLAND (1981) obtained a close correlation with MELLOR and TESTA's (1969) uni-axial data over a temperature range $212 - 273\ K$ with the form

$$a(T) = 0.7242\ \exp(11.9567\bar{T}) + 0.3438\ \exp(2.9494\bar{T})\,, \tag{3.21}$$

where

$$\bar{T} = (T - 273.15)/20 \tag{3.22}$$

when T has units of K, which is normalised at $T = 273.15\ K$ where $a = 1.068$. Interpretation of ice shelf data by THOMAS (1971, 1973b) with a solution based on Glen's law and isothermal conditions at the mean temperature suggested that natural ice needs a scale factor $a_0 \approx 0.2$ multiplying the laboratory function (3.21); that is, natural ice is more viscous (stiffer) by a factor 5. However, ice shelf solutions constructed by MORLAND and ZAINUDDIN (1987) incorporating a temperature profile through the thickness show that the longitudinal force balance is controlled by the colder ice near the surface, below the mean temperature, and that further scaling of $a(T)$ is not necessary. Model calculations can include an additional scale parameter a_0 to assess its influence, essentially scaling the product $a(T)\psi(J_2)$ which determines the strain-rate at given temperature and stress.

It has been common practice, which persists, to adopt the Arrhenius function

$$a(T) = A\ \exp(-Q/RT) \tag{3.23}$$

based on low (homologous) temperature physics not appropriate to natural ice masses within 40 K of melting (MELLOR, 1980), and which does not correlate with data over this range with single choices of the constant activation energy Q and gas constant R. It offers no advantage in the analysis nor in the computation.

Figure 2 illustrates the configuration and boundary conditions of an evolving ice shelf, showing a vertical cross-section $0xz$ with $0x$ horizontal ($\chi = 0$) and parallel to the mean flow direction from the grounded sheet. There is a transition region extending on both sides of the grounding line through which the flow and stress pattern changes. The surface conditions are a continuation of those for the grounded sheet, but conditions at the basal interface

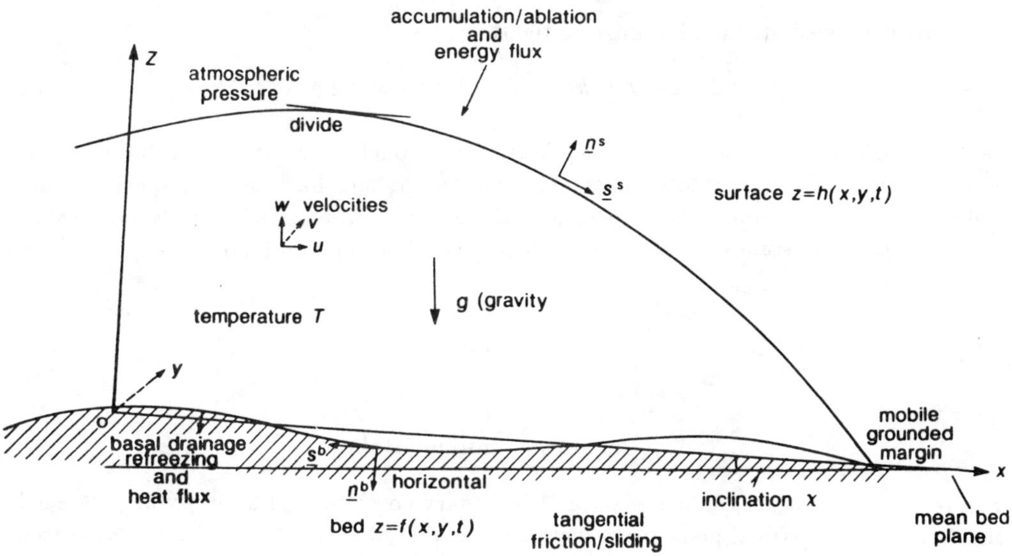

Figure 1: Vertical plane cross-section of a grounded ice sheet.

with the ocean become a melting point condition, with latent heat of melting or refreezing providing a heat flux balance, continuity of the normal traction with water pressure, and zero tangential tractions due to the negligible viscous shear stresses in the water. This section examines a single phase cold ice-sheet with no melting or refreezing (phase change) in the interior, but a binary mixture theory of ice-water in a temperate zone will be presented in the final section. It is also seen there that momentum and energy jump conditions at a slowly moving non-material surface with accumulation or ablation yield the proposed traction continuity and heat flux balance as excellent approximations. The present exposition draws on MORLAND (1984) and the allied Morland contributions discussed earlier.

Denote the surface by

$$S_h(x_1, x_2, x_3, t) = x_3 - h(x_1, x_2, t) = 0\,, \tag{4.1}$$

To complete the physical properties required in the thermomechanical balances, over the relatively small temperature range arising, the ice is assumed to have a constant specific heat C, and a constant thermal conductivity λ. Thus, the internal energy per unit mass is

$$E = CT, \tag{3.24}$$

and the heat flux $\hat{\mathbf{q}}$ is given by

$$\hat{\mathbf{q}} = -\lambda \,\mathrm{grad} T\,, \qquad \hat{q}_i = -\lambda \frac{\partial T}{\partial x_i} \quad (i = 1,2,3)\,. \tag{3.25}$$

The thermal constants are (PATERSON, 1969)

$$C = 2 \times 10^3\, Nm\, kg^{-1}\, K^{-1}\,, \qquad \lambda = 2.2\, N K^{-1} s^{-1}\,. \tag{3.26}$$

In summary, for global ice sheet analysis, the ice model is an incompressible non-linearly viscous fluid with temperature dependent rate factor, and has constant specific heat and constant thermal conductivity. It is clearly an idealisation, and lacks consistent correlation with shorter time- scale laboratory data. There is no better candidate.

4. Thermomechanical Balances

Figure 1 illustrates the configuration and boundary conditions of an evolving grounded ice sheet, showing a vertical plane cross-section Oxz where Ox is the line of steepest descent of the mean bed plane Oxy at angle χ to the horizontal, with Oy horizontal. For large sheets $\chi = 0$, and χ is generally small except on steep mountain glaciers. A subscript notation x_i $(i = 1,2,3)$ is also used for mathematical expediency, where $x_1 = x, x_2 = y, x_3 = z$. The surface $z = h(x,y,t)$ is unknown, and in particular the location of the margin, to be determined as part of the solution, so the problems are "unknown domain" or "moving boundary" which add considerable complexity to numerical solution. The bed $z = f(x,y,t)$ may be fixed, independent of t, but crustal deformation during the ice sheet evolution requires the time dependence. This implies a coupled problem between deforming ice and deforming crust, but the present description prescribes basal conditions for the ice. The flow is gravity driven, with the surface free of traction (atmospheric pressure is taken to be zero stress) but subjected to accumulation or ablation distribution, and heat flux or prescribed temperature (or combination). The base is subjected to a tangential stress-velocity condition (sliding law) as described earlier, and melting or refreezing, and heat flux. Strictly, the thermal and phase change conditions are coupled, but the present description supposes simple prescriptions. The problem is to determine the velocity, stress and temperature fields, and the ice sheet boundary, for given boundary conditions.

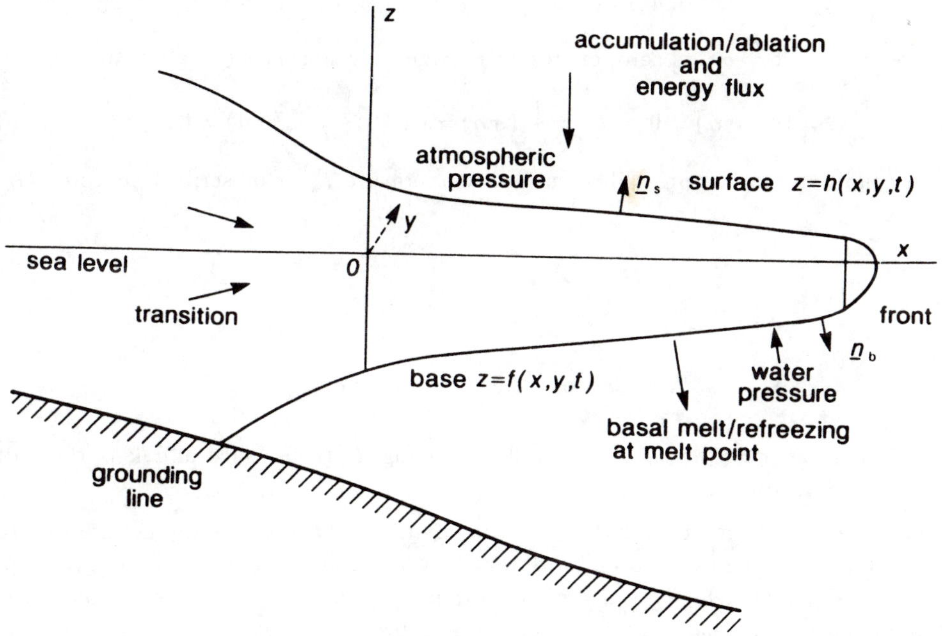

Figure 2: Vertical plane cross-section of a floating ice sheet.

then the unit outward normal **n** is given by

$$\mathbf{n} = \frac{\operatorname{grad} S_h}{\Delta_h}, \qquad (n_1, n_2, n_3) = \frac{1}{\Delta_h}\left[-\frac{\partial h}{\partial x_1}, -\frac{\partial h}{\partial x_2}, 1\right], \tag{4.2}$$

where

$$\Delta_h = \left|\operatorname{grad} S_h\right| = \left\{\left[\frac{\partial h}{\partial x_1}\right]^2 + \left[\frac{\partial h}{\partial x_2}\right]^2 + 1\right\}^{\frac{1}{2}}. \tag{4.3}$$

Let $\overset{1}{\mathbf{s}}, \overset{2}{\mathbf{s}}, \mathbf{n}$ be a right-hand system of orthogonal unit vectors so that $\overset{1}{\mathbf{s}}, \overset{2}{\mathbf{s}}$ form the tangent plane, and choose $\overset{1}{\mathbf{s}}$ to lie in the Ox_1x_3 plane with positive component in the x_1 direction.

Thus

$$\overset{1}{\mathrm{s}} = \frac{(n_3, 0, -n_1)}{(n_1^2 + n_3^2)^{\frac{1}{2}}}, \quad \overset{2}{\mathrm{s}} = \frac{(-n_1 n_2, n_1^2 + n_3^2, -n_2 n_3)}{(n_1^2 + n_3^2)^{\frac{1}{2}}}. \tag{4.4}$$

The surface traction $\hat{\mathbf{t}} = \boldsymbol{\sigma n}$ is zero, expressed by vanishing normal and tangential components

$$\hat{t}_n = \boldsymbol{n} \cdot (\boldsymbol{\sigma n}) = 0, \quad \hat{t}_{s1} = \overset{1}{\mathrm{s}} \cdot (\boldsymbol{\sigma n}) = 0, \quad \hat{t}_{s2} = \overset{2}{\mathrm{s}} \cdot (\boldsymbol{\sigma n}) = 0. \tag{4.5}$$

A thermal condition may be prescribed surface temperature T_h or prescribed heat flux G_h; thus

$$x_3 = h: \quad \begin{aligned} T &= T_h \\ \text{or} \quad \lambda \frac{\partial T}{\partial n} &= G_h \mathbf{k} \cdot \mathbf{n} \end{aligned} \tag{4.6}$$

where G_h is the vertically downward heat flux arriving at the surface and $\mathbf{k}$ is the unit vertically upward vector.

The kinematic accumulation or ablation condition at the surface relates the normal inward flux of ice q (the volume of ice frozen onto unit area of surface per unit time, negative for ablation) to the particle velocity and surface motion in space. Let u_n be the normal speed of a surface point (not ice particle) in space, in the $\mathbf{n}$ direction, then

$$\frac{\partial S_h}{\partial t} + u_n \mathbf{n} \cdot \operatorname{grad} S_h = 0, \tag{4.7}$$

which determines

$$u_n = -\frac{\partial S_h}{\partial t} \Delta_h. \tag{4.8}$$

The normal speed of the ice at the surface is $v_n = \mathbf{v} \cdot \mathbf{n}$, so the ice flux through the surface in the $\mathbf{n}$ direction is $v_n - u_n$, which is prescribed as $-q$. Thus, eliminating $S_h, \mathbf{n}$ and u_n by (4.1), (4.2) and (4.8) gives the surface accumulation condition

$$x_3 = h: \qquad \frac{\partial h}{\partial t} + v_1 \frac{\partial h}{\partial x_1} + v_2 \frac{\partial h}{\partial x_2} - v_3 = q \Delta_h. \tag{4.9}$$

A typical magnitude for q is $1\, myr^{-1} \approx 3 \times 10^{-8} ms^{-1}$.

For a floating shelf, or grounded sheet overlying a deforming crust, there is in general a time-dependent base

$$S_f(x_1, x_2, x_3, t) = f(x_1, x_2, x_3, t) - x_3 = 0, \tag{4.10}$$

with unit outward normal $\mathbf{n}$ defined by

$$\mathbf{n} = \frac{\operatorname{grad} S_f}{\Delta_f}, \quad (n_1, n_2, n_3) = \frac{1}{\Delta_f}\left[\frac{\partial f}{\partial x_1}, \frac{\partial f}{\partial x_2}, -1\right] \tag{4.11}$$

where $\Delta_f = |\operatorname{grad} S_f|$. Tangent vectors (4.4) again form a right-hand system $\overset{1}{\mathbf{s}}, \overset{2}{\mathbf{s}}, \mathbf{n}$ with $\overset{1}{\mathbf{s}}$ in the Ox_1x_2 plane but with positive component in the negative x_1 direction. If b is the normal outward flux of ice (melting if positive, refreezing if negative), then $v_n - u_n = b$ with the above terminology, where

$$u_n = -\frac{\partial S_f}{\partial t}\Delta_f, \tag{4.12}$$

and eliminating S_f, $\mathbf{n}$ and u_n by (4.10), (4.11) and (4.12) gives the basal melting condition

$$x_3 = f: \qquad \frac{\partial f}{\partial t} + v_1\frac{\partial f}{\partial x_1} + v_2\frac{\partial f}{\partial x_2} - v_3 = b\Delta_f\,. \tag{4.13}$$

A prescribed basal temperature T_f or vertically upward heat flux G_f arriving at the base (the geothermal heat flux) gives

$$x_3 = f: \qquad T = T_f \quad \text{or} \quad \lambda\frac{\partial T}{\partial n} = G_f\mathbf{k}\cdot\mathbf{n}\,. \tag{4.14}$$

For the floating shelf the condition is prescribed temperature at the melting point, while at a cold grounded base the heat flux condition would apply. When f is an unknown boundary to be determined by matching with a deforming crust or water traction, there are three traction conditions: the two tangential relations describing sliding or continuity with water traction, and continuity of normal traction. If there is a rigid bed, so f is prescribed, the bedrock stresses are arbitrary and normal traction continuity does not provide a boundary condition, compensating for the elimination of an unknown f.

As discussed earlier, the tangential sliding relations reflect a complex basal layer motion and interaction with small scale bed structure transposed to a smooth basal boundary contour for the large scale flow. Non-slip may be appropriate if ice velocities near cold rigid bedrock are negligible, but this gives rise to singular behaviour at the margins in the reduced model. More generally, a sliding law is prescribed which relates the tangential ice velocity (relative to the bedrock) to the tangential tractions and, on the basis of basal layer analyses (FOWLER 1979, 1981) and on the need to avoid margin singularities in the reduced model, to the normal pressure. In three-dimensional flow, the tangential traction and tangential velocity have directions, and MORLAND (1984) proposes a law in which they are parallel, so extending the common plane flow prescription. Furthermore, following MORLAND and JOHNSON (1980, 1982) and MORLAND and SMITH (1984), a relation separable in the tangential velocity and normal traction is suggested for the tangential traction. Thus, if $\mathbf{e}$ is the unit vector parallel to, and in the same sense as, the tangential relative (to bedrock) velocity

$$[\mathbf{v}_s] = [v_{s1}]\overset{1}{\mathbf{s}} + [v_{s2}]\overset{2}{\mathbf{s}} = [v_s]\mathbf{e}, \qquad (4.15)$$

with modulus $[v_s]$, then the tangential traction of the ice on the bed is $-\hat{t}_s\mathbf{e}$ where

$$\hat{\mathbf{t}}_s = \hat{t}_{s1}\overset{1}{\mathbf{s}} + \hat{t}_{s2}\overset{2}{\mathbf{s}} = \hat{t}_s\mathbf{e} \qquad (4.16)$$

It is convenient to adopt a separable law in the form

$$-\hat{t}_s = p_b\hat{\mu}(p_b)\eta([v_s]), \quad \geq 0, \qquad (4.17)$$

where

$$p_b = -\hat{t}_n, \quad \geq 0, \qquad (4.18)$$

is the normal pressure on the bed, which becomes zero at a margin. It is assumed that the ice moves in the direction of the tangential traction of the ice on the bed, $-\hat{t}_s\mathbf{e}$, so that $\hat{t}_s \leq 0$. It is found that the reduced model requires $\hat{t}_s \sim p_b$ as $p_b \to 0$ to avoid singular behaviour at the margin (MORLAND and JOHNSON, 1980, 1982), so $\hat{\mu}(p_b)$ in the form (4.17) is bounded and non-zero at $p_b = 0$. Following Nye's assumption that $v_s \propto \hat{t}_s^m$ for some constant $m \geq 1$, MORLAND and JOHNSON (1980, 1982) assumed $\eta(v_s) = v_s^{1/m}$ for a fixed bed, and MORLAND, SMITH and BOULTON (1984) adopted this form with $m = 1, 2, 3$, or 4 to construct $\hat{\mu}(p_b)$ from various ice sheet profiles, temperature fields and surface data, by inverting the reduced steady plane isothermal flow solution. This restricted form of solution is unlikely to be applicable to the unsteady three-dimensional flow over a real bed form with finite slopes. Furthermore, available data are limited in quantity, and were drawn from a variety of quite different ice sheets, so no universal sliding law emerged. In addition, a warm basal zone is expected to increase the velocity, so isothermal solutions are not appropriate. A modified law was deduced by MORLAND and SMITH (1984) with a prescribed temperature distribution solution, but there was little difference for small p_b near the margin. It has been concluded (BOULTON, SMITH and MORLAND, 1984) that height and span of the sheet are most sensitive to sliding changes near the margin, and much less sensitive to changes of a small sliding velocity predicted under the thicker sections of the sheet by an increasing $\hat{\mu}(p_b)$. In the absence of any established sliding law, a trial model would be

$$-\hat{t}_s = A(T)p_b[v_s] \qquad (4.19)$$

where $A(T)$ is a magnitude factor which could be expected to decrease as T increases, increasing $[v_s]$ for given $\hat{t}_s$ and p_b, while $[v_s]$ decreases as p_b increases at given $\hat{t}_s, T$. The actual magnitude of A dictates whether $[v_s]$ influences the solution; that is, whether $[v_s]$ is significant compared to the change of longitudinal velocity from base to surface due to the ice shearing, which requires that

$$A \approx \frac{\bar{\mu}}{dp_b} \qquad (4.20)$$

where $\bar{\mu}$ is the mean viscosity through the thickness d. The grounded sheet model assumes that basal sliding $[v_s]$ does not exceed the velocity change through the depth, which requires that A is not less than the magnitude (4.20). This will fail in a fast flowing ice stream feeding into a shelf. For a floating shelf, since the viscous shear stresses in the water are negligible compared to those in the ice, the tangential traction components are zero, replacing the relations (4.19) or (4.17).

Within the ice sheet, the stress, velocity and temperature are governed by the conservation laws. Mass conservation is simply

$$\rho = \text{constant} \qquad \text{or div } \mathbf{v} = 0 \,, \tag{4.21}$$

equivalent to the incompressibility condition. The mean pressure ρ is then arbitrary and only the deviatoric stress $\boldsymbol{\sigma}'$ is determined by the viscous law. For the very slow flows of natural ice masses, the acceleration terms are negligible in comparison with gravity and the stress gradients (the Reynolds number is extremely small), so linear momentum conservation becomes the equilibrium equations

$$\frac{\partial\sigma_{11}}{\partial x_1} + \frac{\partial\sigma_{12}}{\partial x_2} + \frac{\partial\sigma_{13}}{\partial x_3} + \rho g \sin\chi = 0 \,, \tag{4.22}$$

$$\frac{\partial\sigma_{12}}{\partial x_1} + \frac{\partial\sigma_{22}}{\partial x_2} + \frac{\partial\sigma_{23}}{\partial x_3} = 0 \,, \tag{4.23}$$

$$\frac{\partial\sigma_{13}}{\partial x_1} + \frac{\partial\sigma_{23}}{\partial x_2} + \frac{\partial\sigma_{33}}{\partial x_3} - \rho g \cos\chi = 0 \,. \tag{4.24}$$

Energy conservation requires

$$\frac{\partial T}{\partial t} + v_1\frac{\partial T}{\partial x_1} + v_2\frac{\partial T}{\partial x_2} + v_3\frac{\partial T}{\partial x_2} = \frac{\lambda}{\rho C}\nabla^2 T + \frac{1}{\rho C} tr(\boldsymbol{\sigma D}) \,, \tag{4.25}$$

which is a balance between local heating, advection of heat, diffusion of heat and stress working (dissipation). Temperate ice undergoing melting or refreezing is treated in the final section.

5. A Reduced Grounded Sheet Model

For large ice sheets with large aspect ratios, span to thickness, the system of balances, boundary conditions and properties simplify to a reduced model if gradients in the longitudinal directions are small compared to gradients through the thickness. The reduced model therefore is not valid for finite slope bed topography; that is, it cannot describe the flow over an undulation with slope much greater than the small surface slope. It is expected that isolated finite slope undulations do not significantly influence the global solution and that only the local flow details are lost, but this is not an established conclusion. Also, boundary conditions are necessarily restricted to variations over the span scale, and not the thickness scale, and again it is expected that an isolated rapid variation will only have a local effect.

First introduce dimensionless variables denoted by a superposed bar, with a length scale d_0 which is a thickness magnitude, a velocity scale q_0 which is a surface accumulation or basal melting magnitude, and the associated time scale d_0/q_0 and strain-rate scale q_0/d_0. Thus

$$(x_i, h, f) = d_0(\bar{x}_i, \bar{h}, \bar{f}), \tag{5.1}$$

$$(v_i, q, b) = q_0(\bar{v}_i, \bar{q}, \bar{b}), \tag{5.2}$$

$$t = d_0\bar{t}/q_0, \quad \mathbf{D} = q_0\bar{\mathbf{D}}/d_0. \tag{5.3}$$

The thickness co-ordinate $\bar{x}_3$ (or $\bar{z}$) and surface height $\bar{h}$ are then order unity, and the velocity component $\bar{v}_3$ (or $\bar{w}$) is supposed to be of order unity; the value of q_0 should be chosen accordingly. The time scale is then the time for a particle to descend from the surface at maximum thickness to the base. A maximum stress magnitude is the overburden pressure at the maximum thickness $\rho g d_0$, so choosing

$$(\sigma, \sigma', p) = \rho g d_0(\bar{\sigma}, \bar{\sigma}', \bar{p}) \tag{5.4}$$

ensures that the dimensionless stresses are order unity (or less). Since the temperature range in the ice sheet is typically melting to 40 K below (occasionally 60 K), the $\bar{T}$ definition (3.22) with scale 20 K is convenient. Note that the dimensionless magnitudes of the in (bed) plane co-ordinates and velocities, and the shear stresses and strain-rates, are not ascribed at this stage.

Now the relation (4.3) for Δ_h, with the analoguous relation for Δ_f, and relations (4.5) for surface tractions $\hat{\mathbf{t}}$, and (4.9), (4.13) for surface and base kinematic conditions, simply require the superposed bars on the respective variables. The thermal conditions (4.6) and (4.14) become

$$\bar{x}_3 = \bar{h}: \quad \bar{T} = \bar{T}_h \quad \text{or} \quad \frac{\partial \bar{T}}{\partial n} = \frac{G_h \mathbf{k}\cdot\mathbf{n}}{[20\,K]\lambda}, \tag{5.5}$$

$$\bar{x}_3 = \bar{f}: \quad \bar{T} = \bar{T}_f \quad \text{or} \quad \frac{\partial \bar{T}}{\partial n} = -\frac{G_f \mathbf{k}\cdot\mathbf{n}}{[20\,K]\lambda}, \tag{5.6}$$

and the model sliding law (4.19) becomes

$$-\bar{t}_s = \bar{A}(\bar{T})\bar{p}_b[\bar{v}_s], \quad \bar{A} = q_0 A. \tag{5.7}$$

The balance laws (4.21) - (4.25) become

$$\frac{\partial \bar{v}_1}{\partial \bar{x}_1} + \frac{\partial \bar{v}_2}{\partial \bar{x}_2} + \frac{\partial \bar{v}_3}{\partial \bar{x}_3} = 0, \tag{5.8}$$

$$\frac{\partial \bar{\sigma}_{11}}{\partial \bar{x}_1} + \frac{\partial \bar{\sigma}_{12}}{\partial \bar{x}_2} + \frac{\partial \bar{\sigma}_{13}}{\partial \bar{x}_3} + \sin\chi = 0, \tag{5.9}$$

$$\frac{\partial \bar{\sigma}_{12}}{\partial \bar{x}_1} + \frac{\partial \bar{\sigma}_{22}}{\partial \bar{x}_2} + \frac{\partial \bar{\sigma}_{23}}{\partial \bar{x}_3} = 0, \tag{5.10}$$

$$\frac{\partial \bar{\sigma}_{13}}{\partial \bar{x}_1} + \frac{\partial \bar{\sigma}_{23}}{\partial \bar{x}_2} + \frac{\partial \bar{\sigma}_{33}}{\partial \bar{x}_3} - \cos\chi = 0, \tag{5.11}$$

$$\frac{\partial \bar{T}}{\partial \bar{t}} + \bar{v}_1\frac{\partial \bar{T}}{\partial \bar{x}_1} + \bar{v}_2\frac{\partial \bar{T}}{\partial \bar{x}_2} + \bar{v}_3\frac{\partial \bar{T}}{\partial \bar{x}_3} = \beta\left[\frac{\partial^2 \bar{T}}{\partial \bar{x}_1^2} + \frac{\partial^2 \bar{T}}{\partial \bar{x}_2^2} + \frac{\partial^2 \bar{T}}{\partial \bar{x}_3^2}\right] + \alpha\, tr(\bar{\sigma}\bar{\mathbf{D}}), \tag{5.12}$$

where

$$\beta = \frac{\lambda}{\rho C d_0 q_0}, \quad \alpha = \frac{g d_0}{[20\,K]C}. \tag{5.13}$$

Finally, the viscous relations (3.8), (3.9) become

$$\bar{\mathbf{D}} = \nu^{-1}\bar{a}(\bar{T})\psi(J_2)\bar{\sigma}', \quad \bar{\sigma}' = \nu\phi(I_2)\bar{\mathbf{D}}/\bar{a}(\bar{T}), \tag{5.14}$$

where

$$a(T) = a_0\bar{a}(\bar{T}), \quad \nu = \frac{\sigma_0 q_0}{\rho g d_0^2 D_0 a_0}, \quad a_0 = a(273.15\,K). \tag{5.15}$$

Note that the polynomial approximation (3.16) with identity (3.10) and temperature factor (3.21) imply that $\psi(J_2)$ and $\phi(I_2)$ are non-zero and order unity over relevant shear stress ranges, which is used in the subsequent analysis of the balancing terms.

The dimensionless parameters β, α, ν depend on the ice properties λ and C given by (3.26), on the scale factors D_0, σ_0 given by (3.17) and on the gravity constant and ice density

$$g = 9.81\,ms^{-2}, \quad \rho = 918\,kg\,m^{-3}. \tag{5.16}$$

In addition they depend on a_0, which is unity in (3.21) but may be smaller for stiffer ice, and on the depth and velocity magnitudes d_0, q_0 which can vary between sheets and climatic conditions. Taking an average

$$q_0 = 1\,myr^{-1} = 3.18 \times 10^{-8}\,ms^{-1}, \tag{5.17}$$

and a thickness range $d_0 = 500m - 4000m$, gives the range of parameter values shown in *Table 1* for $a_0 = 1$. The immediate result is that

$$\nu \ll 1, \tag{5.18}$$

which is still valid if a smaller stiffening factor $a_0 \approx 0.2$ is adopted, and even for thin glaciers with $d_0 \approx 100m$ and for moderate increases of q_0. However, if large temperature drops occur in moderately thin sheets, with a/a_0 decreasing by a factor 10^{-3} in the cold surface ice, then ν/a may not be small, but

$$\frac{\nu}{a} \le 1, \quad \frac{\nu}{a} \ll 1 \quad \text{in warm basal layer.} \tag{5.19}$$

It is helpful to recapitulate the original MORLAND and JOHNSON (1980, 1982) approach for steady, plane, isothermal flow, which adopted $(\nu/a_m) = \hat{\nu} \ll 1$, where a_m represents a mean rate factor. Formal series expansions in $\hat{\nu}$ of all the mechanical equations, assuming that all the dimensionless (superbar) variables are order unity (or less), yields a leading order solution which describes an ice reservoir - a sheet with a horizontal free surface and no stress variation in the longitudinal direction, and never reaching a margin. It was evident that a bounded sheet with a margin must incorporate a horizontal pressure gradient, which in

$d_0(m)$	α	β	ν
500	0.125	0.08	4×10^{-5}
1000	0.25	0.04	10^{-5}
2000	0.5	0.02	2.5×10^{-6}
4000	1.0	0.01	6×10^{-7}

Table 1: Approximate values of α, β, ν

turn implies a leading order contribution from the shear stress gradient $\partial\bar{\sigma}_{13}/\partial\bar{x}_3$ in the longitudinal balance (5.9). This is achieved by a coordinate stretching

$$X_1 = \epsilon\bar{x}_1 \,, \quad X_3 = \bar{x}_3 \qquad (\epsilon \ll 1) \tag{5.20}$$

such that X_1 and X_3 derivatives have the same status, asserting that $\bar{x}_1$ gradients are an order ϵ smaller than $\bar{x}_3$ gradients. The consequent analysis showed that

$$\epsilon = \begin{cases} \nu^{\frac{1}{2}} & \text{if} \quad \chi \le \epsilon \\ \nu & \text{if} \quad \chi \sim 1 \,, \end{cases} \tag{5.21}$$

defined as small and finite bed slope cases. The latter reflects a steep mountain glacier, with distinct solution pattern from the large ice sheet with horizontal (or near horizontal) mean bed plane. MORLAND and JOHNSON (1980, 1982) generalised the scaling and perturbation analysis to the power law response with unbounded ϕ at zero stress, but it was noted that singularities arise at the surface - a passive boundary layer later analysed by JOHNSON and MCMEEKING (1984). This adds complications to both analysis and computation, and introduces an unnatural singular behaviour near the free surface, so there seems little merit in persisting with the power law response in the construction of tractable reduced models.

Returning to the thermomechanically coupled three-dimensional balances, including the significant non-uniformity of the rate factor $\bar{a}(\bar{T})$ in the presence of realistic temperature distributions with $\nu/\bar{a}$ not small in cold surface regions, MORLAND (1984) showed that an analogous reduced model can be derived on the basis of the valid weaker inequalities (5.19). The basic assumption is that gradients in $\bar{x}_1$ and $\bar{x}_2$ directions are small compared to gradients in the $\bar{x}_3$ direction - the small surface slope or large aspect ratio notion - quantified by introducing stretched co-ordinates

$$X_1 = \epsilon \bar{x}_1 \, , \quad X_2 = \epsilon \bar{x}_2 \, , \quad X_3 = \bar{x}_3 \, , \qquad (\epsilon \ll 1) \tag{5.22}$$

with X_1 and X_2 gradients having equal (or less) status than X_3 gradients. ϵ is therefore the greatest magnitude of the surface or basal slope (relative to the mean bed plane), which excludes a bed topography with slope greater than ϵ in magnitude. Mass conservation (5.8) immediately imposes the velocity scalings

$$V_1 = \epsilon \bar{v}_1 \, , \quad V_2 = \epsilon \bar{v}_2 \, , \quad V_3 = \bar{v}_3 \, , \tag{5.23}$$

where V_i are order unity, to retain the balance between all terms:

$$\frac{\partial V_1}{\partial X_1} + \frac{\partial V_2}{\partial X_2} + \frac{\partial V_3}{\partial X_3} = 0 \, . \tag{5.24}$$

Essentially, it is asserted that the stretching or compression normal to the bed plane makes a leading order contribution to the volume change, and must be balanced by compression or stretching in at least one in-plane direction; that is, "the vertical deformation cannot be negligible in a gravity driven flow". Cycling between the equilibrium balances and zero surface traction conditions, with the inequalities (5.19) and viscous relation (5.14), shows that

$$\bar{\sigma}_{13} \sim \max(\sin\chi, \epsilon) \, , \quad \bar{\sigma}_{23} \sim \epsilon \, , \tag{5.25}$$

$$\epsilon = \begin{cases} \nu^{\frac{1}{2}} & \text{if} \quad \sin\chi \leq \nu^{\frac{1}{2}} \\ \nu / \sin\chi & \text{if} \quad \sin\chi \geq \nu^{\frac{1}{2}} \, , \end{cases} \tag{5.26}$$

and that the in-plane deviatoric stresses are an order of magnitude smaller than the direct shear stresses:

$$\bar{\sigma}'_{12} \sim \bar{\sigma}'_{33} \sim \bar{\sigma}'_{11} \sim \epsilon \bar{\sigma}'_{13} \, , \quad \bar{\sigma}'_{22} \sim \epsilon \bar{\sigma}'_{23} \, . \tag{5.27}$$

Thus

$$\bar{\sigma}'_{11} \sim \bar{\sigma}'_{22} \sim \bar{\sigma}'_{33} \sim -\bar{p} \sim -1 \, . \tag{5.28}$$

By construction it is supposed that X_1 is the dominant in-plane direction, so that quantities associated with the X_2 direction may be smaller than the maximum magnitudes indicated above; for example, in plane flow in the vertical plane OX_1X_3 .

The leading order balances are therefore the complete incompressibility condition (5.24), the reduced equilibrium equations,

$$\epsilon \frac{\partial \bar{\sigma}_{11}}{\partial X_1} + \frac{\partial \bar{\sigma}_{13}}{\partial X_3} + \sin\chi = 0 \, , \tag{5.29}$$

$$\epsilon \frac{\partial \bar{\sigma}_{12}}{\partial X_2} + \frac{\partial \bar{\sigma}_{23}}{\partial X_3} = 0 \, , \tag{5.30}$$

$$\frac{\partial \bar{\sigma}_{33}}{\partial X_3} - \cos\chi = 0 \, , \tag{5.31}$$

and energy balance

$$\frac{\partial \bar{T}}{\partial \bar{t}} + V_1 \frac{\partial \bar{T}}{\partial X_1} + V_2 \frac{\partial \bar{T}}{\partial X_2} + V_3 \frac{\partial \bar{T}}{\partial X_3} = \beta \frac{\partial^2 \bar{T}}{\partial X_3^2} + \alpha \nu^{-1} \bar{a}(\bar{T}) \psi(J_2) \bar{J} \,, \tag{5.32}$$

where

$$J_2 = \left[\frac{\rho g d_0}{\sigma_0}\right]^2 \bar{J}, \quad \bar{J} = \bar{\sigma}_{13}^2 + \bar{\sigma}_{23}^2 \,. \tag{5.33}$$

Then leading order normal stresses are

$$\bar{\sigma}_{11} = \bar{\sigma}_{22} = \bar{\sigma}_{33} = -\bar{p} \sim 1 \,, \tag{5.34}$$

and the leading order viscous relations are

$$\frac{\partial V_1}{\partial X_3} = 2\epsilon \nu^{-1} \bar{a}(\bar{T}) \psi(J_2) \bar{\sigma}_{13} \sim 1 \,, \tag{5.35}$$

$$\frac{\partial V_2}{\partial X_3} = 2\epsilon \nu^{-1} \bar{a}(\bar{T}) \psi(J_2) \bar{\sigma}_{23} \sim 1 \,. \tag{5.36}$$

Note that $\bar{J} \sim \nu$ for a small bed slope, and $\bar{J} \sim \sin^2 \chi$ for a steep glacier, so that $J_2 \approx \rho g q_0/((\sigma_0 D_0 a_0))$ does not exceed unity in magnitude for large sheets as originally asserted. The energy dissipation term has magnitude $\alpha \bar{a}$ which is significant in a warm basal layer of a thick sheet, see *Table 1*, and contributes to the thermomechanical coupling.

On the surface $X_3 = H(X_1, X_2, \bar{t})$, to order ϵ:

$$\Delta_h = 1, \quad \mathrm{n} = \left[-\epsilon \frac{\partial H}{\partial X_1}, -\epsilon \frac{\partial H}{\partial X_2}, 1\right],$$
$$\overset{1}{\mathrm{s}} = \left[1, 0, \epsilon \frac{\partial H}{\partial X_1}\right], \quad \overset{2}{\mathrm{s}} = \left[0, 1, \epsilon \frac{\partial H}{\partial X_2}\right], \tag{5.37}$$

and the zero tractions (4.5) become

$$\bar{t}_n = \bar{\sigma}_{33} = 0 \,, \quad \bar{t}_{s1} = \bar{\sigma}_{13} = 0 \,, \quad \bar{t}_{s2} = \bar{\sigma}_{23} = 0 \,, \tag{5.38}$$

the thermal condition (5.5) becomes

$$\bar{T} = \bar{T}_h \quad \text{or} \quad \frac{\partial \bar{T}}{\partial X_3} = \frac{G_h}{[20\,K]\lambda} \,, \tag{5.39}$$

and the kinematic condition (4.9) becomes

$$\frac{\partial H}{\partial \bar{t}} + V_1 \frac{\partial H}{\partial X_1} + V_2 \frac{\partial H}{\partial X_2} - V_3 = \bar{q} \,. \tag{5.40}$$

On the base $X_3 = F(X_1, X_2, \bar{t})$, to order ϵ:

$$\Delta_f = 1, \quad \mathbf{n} = \left[\epsilon\frac{\partial F}{\partial X_1}, \epsilon\frac{\partial F}{\partial X_2}, -1\right],$$

$$\overset{1}{\mathbf{s}} = \left[1, 0, -\epsilon\frac{\partial F}{\partial X_1}\right], \quad \overset{2}{\mathbf{s}} = \left[0, 1, \epsilon\frac{\partial F}{\partial X_2}\right], \tag{5.41}$$

and the normal and tangential tractions become

$$\hat{\bar{t}}_n = -\bar{\sigma}_{33}, \quad \hat{\bar{\mathbf{t}}}_s = (\bar{\sigma}_{13}, -\bar{\sigma}_{23}, 0) = \hat{\bar{t}}_s \mathbf{e} \tag{5.42}$$

where by the sliding assumptions (4.15), (4.16),

$$\hat{\bar{t}}_s = \left[\bar{\sigma}_{13}^2 + \bar{\sigma}_{23}^2\right]^{\frac{1}{2}}, \quad \mathbf{e} = \left[\hat{V}_1^2 + \hat{V}_2^2\right]^{-\frac{1}{2}}\left[\hat{V}_1, -\hat{V}_2, 0\right], \tag{5.43}$$

where $\hat{V}_1, \hat{V}_2$ denote velocities relative to the bed. The model sliding law (5.7) is then

$$\bar{J} = \bar{A}^2(T)\left[\hat{V}_1^2 + \hat{V}_2^2\right]. \tag{5.44}$$

The thermal condition (5.6) becomes

$$\bar{T} = \bar{T}_f \quad \text{or} \quad \frac{\partial \bar{T}}{\partial X_3} = -\frac{G_f}{[20\,K]\lambda}, \tag{5.45}$$

and the kinematic condition (4.13) becomes

$$\frac{\partial F}{\partial \bar{t}} + V_1\frac{\partial F}{\partial X_1} + V_2\frac{\partial F}{\partial X_3} - V_3 = \bar{b}. \tag{5.46}$$

The third equilibrium equation (5.31) and first surface condition (5.38) allow the explicit integration

$$-\bar{\sigma}_{33} = \bar{p} = \cos\chi\left[H(X_1, X_2, \bar{t}) - X_3\right], \tag{5.47}$$

for both steady and unsteady conditions, showing that to leading order (to order ϵ in fact) the pressure and normal stress are simply in balance with gravity, independent of the viscous flow. Here the systematic approximations have yielded a result in accord with physical expectation.

The first equilibrium equation (5.29) now balances the shear stress gradient with the longitudinal pressure gradient and $\sin\chi$ if $\sin\chi \leq \nu^{\frac{1}{2}}$, but with $\sin\chi$ alone if $\sin\chi \gg \nu^{\frac{1}{2}}$, in view of the magnitudes of $\sigma_{13}, \sigma_{11}, \epsilon$ given by (5.25), (5.26) and (5.28), but explicit integration retaining all terms covers all cases. Thus, with the surface condition (5.38),

$$\bar{\sigma}_{13} = \left[\sin\chi - \epsilon\cos\chi\frac{\partial H}{\partial X_1}\right]\left[H(X_1, X_2, \bar{t}) - X_3\right].^{\dagger} \tag{5.48}$$

The second equilibrium equation (5.30) with surface condition (5.38) integrates to

$$\bar{\sigma}_{23} = -\epsilon \cos\chi \frac{\partial H}{\partial X_1}\Big[H(X_1, X_2, \bar{t}) - X_3\Big].^{\dagger} \tag{5.49}$$

The stress expressions (5.47), (5.48), (5.49) in terms of $H(X_1, X_2, \bar{t})$ replace the three equilibrium equations and three surface traction conditions, leaving two balance equations (5.24) and (5.32), and two viscous relations (5.35), (5.36) for three velocity components V_1, V_2, V_3, and temperature $\bar{T}$, subject to a thermal and kinematic condition on both surface and base and three traction conditions on an unknown base, two on a known base. This system is the reduced model. Before demonstrating the further analytic reduction possible in plane flow (or axi-symmetric flow), it is interesting to note some general features of the energy balance (5.32) deduced by MORLAND (1984).

First note that a mean geothermal heat flux (PATERSON, 1969) implies a basal temperature gradient $2.5 \times 10^{-2}\,Km^{-1}$ which corresponds to a range $0.6 - 5$ for $\partial\bar{T}/\partial X_3$ in (5.55) for d_0 in the range $500 - 4000\,m$, which is a leading order contribution. The time scale for $\bar{t}$ is d_0/q_0 which ranges from 500 to 4000 years with the adopted $q_0 = 1\,m\,yr^{-1}$, and this is the global response time for significant changes through the sheet. Seasonal forcing through thermal and kinematic changes at the surface over times of one year cannot induce global changes. If $t_0 = 1\,yr$, let

$$t = t_0 t^{\star}\,, \quad \bar{t} = \gamma t^{\star}\,, \quad \gamma = \frac{t_0 q_0}{d_0} \ll 1\,, \tag{5.50}$$

then an order unity $\partial\bar{T}/\partial t^{\star}$ in (5.42) requires a balance between local heating and diffusion, and possibly vertical advection, near the surface in a layer of thickness $2(\gamma\beta)^{\frac{1}{2}} d_0 = 2(\lambda t_0/\rho C)^{\frac{1}{2}} \approx 12\,m$, indpendent of d_0. Below this transient surface layer the temperature changes only on the longer time scale, so seasonal changes may be ignored in the global reduced model. Next, *Table 1* shows that β may be small for large sheets, not much larger than $\epsilon = \nu^{\frac{1}{2}}$, suggesting the presence of a basal thermal boundary layer in which diffusion balances advection. It is shown that the layer thickness is of order $\beta d_0, \beta^{\frac{1}{2}} d_0$ or $\beta^{\frac{1}{3}} d_0$ depending on basal velocity conditions, which indicates weak boundary layers with only moderate temperature gradients. These should be approximated satisfactorily by numerical methods.

6. Plane Flow

It is now clearer to use the notation (X, Z) and (U, W) in place of the subscripts 1 and 3, since there is no dependence on $X_2 = Y$ and $V_2 = V \equiv 0$. Note that the lateral strain rate $\bar{D}_{yy} = \partial V/\partial Y \equiv 0$, but this is not necessarily true on a vertical symmetry plane on which $V \equiv 0$ in a three-dimensional flow where V is varying with Y. The plane flow approximation is only valid when Y variation is negligible over a large lateral span. The reduced model becomes

$$\frac{\partial U}{\partial X} + \frac{\partial W}{\partial Z} = 0\,, \tag{6.1}$$

†The $\cos\chi$ factor, significant for steep beds, was omitted in MORLAND(1984).

$$\frac{\partial \bar{T}}{\partial \bar{t}} + U\frac{\partial \bar{T}}{\partial X} + W\frac{\partial \bar{T}}{\partial Z} = \beta\frac{\partial^2 \bar{T}}{\partial Z^2} + \alpha\nu^{-1}\bar{a}(\bar{T})\psi(J_2)\bar{\sigma}_{xz}^2 , \tag{6.2}$$

$$\frac{\partial U}{\partial Z} = 2\epsilon\nu^{-1}\bar{a}(\bar{T})\psi(J_2)\bar{\sigma}_{xz} , \tag{6.3}$$

for $U, W, \bar{T}$ as functions of $(X, Z, \bar{t})$, where

$$J_2 = \left[\frac{\rho g d_0}{\sigma_0}\right]^2 \bar{\sigma}_{xz}^2 , \tag{6.4}$$

$$\bar{\sigma}_{xz} = \left[\sin\chi - \epsilon\cos\chi\frac{\partial H}{\partial X}\right]\left[H(X,\bar{t}) - Z\right] , \tag{6.5}$$

$$-\bar{\sigma}_{zz} = -\bar{\sigma}_{yy} = -\bar{\sigma}_{xx} = \bar{p} = \cos\chi\left[H(X,\bar{t}) - Z\right] . \tag{6.6}$$

The surface conditions on $Z = H(X, \bar{t})$ are

$$\frac{\partial H}{\partial \bar{t}} + U\frac{\partial H}{\partial X} - W = \bar{q} , \tag{6.7}$$

$$\bar{T} = \bar{T}_h \quad \text{or} \quad \frac{\partial \bar{T}}{\partial Z} = \frac{G_h}{[20\, K]\lambda} , \tag{6.8}$$

and the base conditions on $Z = F(X, \bar{t})$ are

$$\frac{\partial F}{\partial \bar{t}} + U\frac{\partial F}{\partial X} - W = \bar{b} , \tag{6.9}$$

$$\bar{T} = \bar{T}_f \quad \text{or} \quad \frac{\partial \bar{T}}{\partial Z} = -\frac{G_f}{[20\, K]\lambda} , \tag{6.10}$$

with the model sliding law (5.54) giving

$$\sin\chi - \epsilon\cos\chi\frac{\partial H}{\partial X} = \cos\chi\bar{A}(\bar{T})\hat{U}_b , \tag{6.11}$$

where $\hat{U}_b$ is the ice velocity relative to the bed, and $\bar{p}_b = \cos\chi[H(X,\bar{t}) - F(X,\bar{t})]$ and $\bar{\sigma}_{xz|b} = [\sin\chi - \epsilon\cos\chi\partial H/\partial X][[H(X,\bar{t}) - F(X,\bar{t})]$ must match the pressure and shear traction in a deforming bed.

For the prescribed fixed bed surface $Z = F(X)$ the tractions are automatically continuous and (6.9) becomes a steady condition with $\partial F/\partial \bar{t} = 0$. The analysis is continued for this case. For a steep glacier (6.5) shows that

$$\bar{\sigma}_{xz} = \sin\chi\left[H(X,\bar{t}) - Z\right] , \tag{6.12}$$

which is independent of the slope $\partial H/\partial X$, so that both sliding law (6.11) and viscous relation (6.3) depend only on H and not on $\partial H/\partial X$, which simplify the system. The reduced model is a parabolic system of non-linear partial differential equations with the extra surface condition (6.7) necessary to determine an unknown surface including the locations of the margins, and of the divide where the surface is horizontal and $\sigma_{xz} \equiv 0, U \equiv 0$. Such problems are described as free surface, moving boundary, unknown domain, involving the prescription of boundary conditions on a surface which must be determined as part of the solution. Numerical treatments are therefore complicated and subject to accuracy difficulties. The numerical illustrations discussed earlier in section 2 are based on this reduced model.

In the steady state case the velocities and temperature are independent of time and the surface profile, $Z = H(X)$, is an equilibrium profile corresponding to balancing accumulation and melting zones on the surface with melting and refreezing zones at the base. Given the time scale d_0/q_0 of the global balances, it is not clear that any current ice sheet is steady (negligible $\partial/\partial\bar{t}$) on this scale, nor that sequences of steady states reflect a realistic evolution see, for example, constructions by BOULTON, SMITH and MORLAND (1984), and by HINDMARSH, MORLAND, BOULTON and HUTTER (1987). It has, however, allowed useful analytical and numerical progress.

Another simplifying feature is to uncouple the mechanics by prescribing a temperature field, steady or unsteady, instead of satisfying the energy balance, which allows explicit integrations to yield a differential equation for the profile $H(X,t)$. A prescribed temperature field, however reasonable in a large scale sense, will miss the more sensitive balances of the coupled system in thermally active zones, which in turn may have significant effects on the large scale solution. Nevertheless, analytic simplification and progress is useful, and at least generates solutions which can be used as test cases for more general numerical algorithms. MORLAND and SMITH (1984) developed the theory in the steady case, but it follows also for the unsteady case (MORLAND, 1984). The mass balance (6.1) is satisfied by introducing a stream function $\omega(X, Z, \bar{t})$ such that

$$U = \frac{\partial \omega}{\partial Z}, \qquad W = -\frac{\partial \omega}{\partial X}, \tag{6.13}$$

then the surface kinematic condition (6.7) becomes

$$\frac{\partial H}{\partial \bar{t}} + \frac{\partial}{\partial X}\left\{\omega\left[X, H(X,\bar{t}), \bar{t}\right]\right\} = \bar{q}, \tag{6.14}$$

and the steady basal kinematic condition (6.9) for $Z = F(X)$ becomes

$$\frac{\partial}{\partial X}\left\{\omega\left[X, F(X), \bar{t}\right]\right\} = \bar{b}. \tag{6.15}$$

Define

$$\Gamma = \frac{\partial H}{\partial X}, \qquad \zeta = \operatorname{sgn}(\tan\chi - \epsilon\Gamma), \tag{6.16}$$

then $\zeta = +1$ when the surface slope relative to the horizontal is negative and $\zeta = -1$ when it is positive; that is, $\zeta = +1$ and $\zeta = -1$ define the regions to the right and left of the divide respectively, in which $\bar{\sigma}_{xz}, U_b, U$ are respectively positive and negative. Introducing

$$\tau = \left\{ \begin{array}{lll} \zeta\nu^{-\frac{1}{2}}\bar{\sigma}_{xz}\,, & \theta = \rho g q_0/\sigma_0 D_0 a_0 & \text{for } \sin\chi \leq \nu^{-\frac{1}{2}} \\ \zeta\sin^{-1}\chi\bar{\sigma}_{xz}\,, & \theta = (\rho g d_0 \sin\chi/\sigma_0)^2 & \text{for } \sin\chi \geq \nu^{-\frac{1}{2}} \end{array} \right\}^{\dagger}, \tag{6.17}$$

the viscous relation (6.3) becomes

$$\frac{\partial^2 \omega}{\partial Z^2} = \frac{\partial U}{\partial Z} = 2\zeta\bar{a}(\bar{T})\tau\psi(\theta\tau^2)\,, \quad \tau \geq 0 \tag{6.18}$$

for both low and steep slope beds.

Now define

$$\begin{aligned} g(\tau) &= 2\tau\psi(\theta\tau^2)\,, \\ g_1(X,Z,\bar{t}) &= \int_F^Z \bar{a}(\bar{T}')g(\tau')\mathrm{d}Z'\,, \\ g_2(X,Z,\bar{t}) &= \int_F^Z g_1(X,Z',\bar{t})\mathrm{d}Z'\,, \end{aligned} \tag{6.19}$$

where $\bar{T}'$ and τ' denote values at $Z = Z'$, then integrating (6.18) twice with respect to Z from the fixed base $Z = F(X)$, where $U = U_b(X,\bar{t})$, gives

$$\left. \begin{aligned} U &= U_b(X,\bar{t}) + \zeta g_1(X,Z,\bar{t})\,, \\ \omega &= U_b(X,\bar{t})\big[Z - F(X)\big] + \zeta g_2(X,Z,\bar{t}) + \omega\big[X, F(X), \bar{t}\big] \end{aligned} \right\}. \tag{6.20}$$

The kinematic condition (6.14) now becomes

$$\frac{\partial H}{\partial \bar{t}} + \frac{\partial}{\partial X}\Big\{\big[H(X,\bar{t}) - F(X)\big]U_b(X,\bar{t}) + \zeta g_2\big[X, H(X,\bar{t}), \bar{t}\big]\Big\} = q - b\,, \tag{6.21}$$

when $\omega[x, F(x), t]$ is eliminated by (6.15). For a small slope bed τ depends on H and $\partial H/\partial X$ so that g_2 depends on $\partial H/\partial X$ and the spatial differential operator in (6.21) involves $H, \partial H/\partial X$ and $\partial^2 H/\partial X^2$. MORLAND (1984) shows that the coefficient of $\partial^2 H/\partial X^2$ is strictly negative, and hence (6.21) is parabolic with characteristics on which $\bar{t}$ = constant, and forward integration in time is stable. For a steep glacier and $\tau = \zeta(H - Z)$, the spatial operator is first order, involving only $\partial H/\partial X$.

---

† MORLAND (1984) used different factors in τ and θ for the steep bed $\sin\chi \geq \nu^{\frac{1}{2}}$.

The corresponding steady state equations are respectively second and first order ordinary differential equations for $H(X)$. Starting from one margin X_1 in the steep case determines the second margin X_2 when H returns to F, which guarantees the integral property of zero net mass flux since

$$\int_{X_1}^{X_2} (\bar{q} - \bar{b})\mathrm{d}X = \Big[(H - F)U_b\Big]_{X_1}^{X_2} + \zeta g_2\Big[X_2, F(X_2)\Big] - \zeta g_2\Big[X_1, F(X_1)\Big] = 0\,, \tag{6.22}$$

since g_2 vanishes at $Z = F$ by construction. This applies to both steep and small slope beds, but the second order equation for the small slope bed requires two boundary conditions. MORLAND and JOHNSON (1980, 1982) in the isothermal case, and MORLAND and SMITH (1984) in the prescribed temperature case, show how the behaviour of the differential equation near a margin imposes an additional restriction on margin values to recover a unique bounded surface slope solution. If Γ_m is the margin slope $\partial H/\partial X$ and β_m is the corresponding bed undulation slope $\partial F/\partial X$, and $\bar{q}, \bar{b}$ have margin values $\bar{q}_m, \bar{b}_m$ ($\bar{q}_m - \bar{b}_m < 0$), and $U_b = U_m$, then

$$(\Gamma_m - \beta_m)U_m = q_m - b_m\,. \tag{6.23}$$

Now the sliding law determines U_m in terms of the margin slope, so (6.23) is an algebraic equation for Γ_m, with both sides negative since $\Gamma_m - \beta_m \lessgtr 0$ and $U_m \gtrless 0$ at left and right-hand margins. It can be shown that there is a unique physically valid Γ_m satisfying these conditions for power law sliding. In particular, for the linear sliding law (6.11) with $\chi = \epsilon\chi_0$, $\chi_0 = \mathcal{O}(1)$,

$$U_m = \kappa(\chi_0 - \Gamma_m)\,, \quad \kappa > 0\,, \tag{6.24}$$

so (6.23) becomes a quadratic for Γ_m with the two roots defining the left and right margin slopes. Thus the steady case becomes an initial value problem for the second order differential equation starting at a margin with prescribed $H = F$ and $\partial H/\partial X = \Gamma_m$, which allows direct integration through the divide and on to the far margin. In the unsteady case there is no explicit expression for Γ_m since there is an additional term $\partial H/\partial \bar{t}$ to be included on the left-side of (6.23).

7. Depth Integrated Model

In view of the large longitudinal and lateral spans of ice sheets compared to thickness it is an attractive proposition to integrate the equations through the depth in an attempt to deduce relations between "mean" or "total" quantities which depend on one less spatial variable. A three-dimensional unsteady motion would then reduce to dependence on time and two space variables. The treatments described earlier focused on plane flow reducing to dependence on time and one longitudinal space variable, termed "flow line theories". Explicit integration of the system of partial differential equations with respect to the x_3 co-ordinate is, of course, not possible, and the simplified theories rest on arbitrary approximations which, even if satisfactory, have yet to be justified. A systematic integration of the balances (5.8) - (5.12) subject to the viscous relation (5.16) and boundary conditions (4.5), (4.9), (4.13), (5.5) - (5.7) in dimensionless (unscaled) variables displays the extent of the reductions and the associated extra terms which cannot be expressed in the mean variables.

Let greek subscripts α, β take values 1 and 2 so x_α are the in-bed plane co-ordinates, and define volume fluxes across the normal co-ordinate planes by

$$\bar{Q}_\alpha = \int_{\bar{f}}^{\bar{h}} \bar{v}_\alpha \mathrm{d}\bar{x}_3 \,, \tag{7.1}$$

and total stresses by

$$\bar{N}_{\alpha\beta} = \int_{\bar{f}}^{\bar{h}} \bar{\sigma}_{\alpha\beta} \mathrm{d}\bar{x}_3 \,, \tag{7.2}$$

then

$$\bar{T}_\alpha = \bar{N}_{\alpha\beta} n_\beta \tag{7.3}$$

define forces per unit width on the normal planes between base and surface. Analogous total stresses $N_{\alpha 3}$ are not used. The thickness is

$$\bar{d}(\bar{x}_\alpha, \bar{t}) = \bar{h}(\bar{x}_\alpha, \bar{t}) - \bar{f}(\bar{x}_\alpha, \bar{t}) \,, \tag{7.4}$$

and means through the thickness can be defined by dividing $\bar{Q}_\alpha$ and $\bar{N}_{\alpha\beta}$ by $\bar{d}$. Denote the surface and base gradients by

$$\gamma_\alpha = \frac{\partial \bar{h}}{\partial \bar{x}_\alpha} \,, \qquad \beta_\alpha = \frac{\partial \bar{f}}{\partial \bar{x}_\alpha} \,, \tag{7.5}$$

then for any $r(\bar{x}_\alpha, \bar{x}_3, \bar{t})$ with depth integral

$$R(\bar{x}_\alpha, \bar{t}) = \int_{\bar{f}}^{\bar{h}} r(\bar{x}_\alpha, \bar{x}_3, \bar{t}) \mathrm{d}\bar{x}_3 \,, \tag{7.6}$$

$$\frac{\partial R}{\partial \bar{x}_\alpha} = \int_{\bar{f}}^{\bar{h}} \frac{\partial r}{\partial \bar{x}_\alpha} \mathrm{d}\bar{x}_3 + \gamma_\alpha r|_{\bar{h}} - \beta_\alpha r|_{\bar{f}} \,. \tag{7.7}$$

On the surface $\bar{x}_3 = \bar{h}(\bar{x}_\alpha, \bar{t})$,

$$\mathrm{n} = \frac{1}{\Delta_h}(-\gamma_1, -\gamma_2, 1) \,, \tag{7.8}$$

and on the base $\bar{x}_3 = \bar{f}(\bar{x}_\alpha, \bar{t})$,

$$\mathrm{n} = \frac{1}{\Delta_f}(\beta_1, \beta_2, -1) \,, \tag{7.9}$$

where

$$\Delta_h = (1 + \gamma_\alpha \gamma_\alpha)^{\frac{1}{2}}\,, \qquad \Delta_f = (1 + \beta_\alpha \beta_\alpha)^{\frac{1}{2}}\,. \tag{7.10}$$

Thus, on the surface

$$\Delta_h \hat{t}_i = -\bar{\sigma}_{i\beta}\gamma_\beta + \bar{\sigma}_{i3} = 0\,, \tag{7.11}$$

and on the base

$$\Delta_f \hat{t}_i = \bar{\sigma}_{i\beta}\beta_\beta - \bar{\sigma}_{i3}\,. \tag{7.12}$$

Integrating the mass balance (5.8) subject to the surface and base kinematic conditions (4.9) and (4.13) shows that

$$\frac{\partial \bar{d}}{\partial \bar{t}} + \frac{\partial \bar{Q}_\alpha}{\partial \bar{x}_\alpha} = \bar{q}\Delta_h - \bar{b}\Delta_f\,, = q^\star\,, \tag{7.13}$$

where $q^\star$ is the net surface and base inward volume flux into a column on unit area of bed plane. Integrating the in-plane equilibrium equations (5.9), (5.10) subject to the surface and base traction condition (7.11), (7.12), shows that

$$\frac{\partial \bar{N}_{\alpha\beta}}{\partial \bar{x}_\beta} + \Delta_f \hat{t}_\alpha\big|_f + \bar{d}\sin\chi\,\delta_{1\alpha} = 0\,, \tag{7.14}$$

where δ_{ij} is the Kronecker delta, but the basal tractions $\hat{t}_\alpha\big|_f$ depend on the tangential tractions $\hat{\mathbf{t}}_s$ given by a sliding law, and on the normal traction $\hat{t}_n$ which involves solution of the equilibrium equation (5.11). The floating shelf situation is discussed in the next section. It is to be expected that even the in-plane stresses depend on all basal traction components when the bed undulation is not nearly parallel to the mean plane, and that progress is only made when small undulation slopes β_α are assumed. It is evident that the energy balance (5.12) allows no useful integration through the depth, and that an uncoupled prescribed temperature field is required. Usually a "mean" temperature uniform in $\bar{x}_3$ is adopted, and since realistic temperature variations make significant changes to the viscous response this has doubtful validity. Furthermore, even with a constant rate factor $\bar{a}$, the viscous relations (5.14) cannot be integrated to relate gradients of Q_α to $\bar{N}_{\alpha\beta}$ because of the stress and strain-rate dependent coefficients $\psi(J_2)$ and $\phi(I_2)$, so again "mean" coefficients uniform in $\bar{x}_3$ have been assumed, ignoring significant non-linearity. HUTTER (1983) has pursued the uncoupled plane flow analysis and commented critically on the various approximate treatments.

Given that the energy balance has no useful depth integration even in the reduced, stretched co-ordinates form (5.32), due to the significant heat advection terms, normal and in-plane, any integrated model must assume prescribed temperature and lose the effects of thermomechanical coupling. It has already been shown that the uncoupled reduced model, in which the direct shear stresses dominate the in-plane deviatoric stresses, then allows integration of the full system with respect to $\bar{x}_3 = X_3$ without further approximation. An uncoupled depth integrated model is only a useful modification if it can treat the system in which the full stretched co-ordinate reduction is not applicable, specifically, when the in-plane deviatoric stresses contribute to the balances. This would arise with finite slope bed topography, though that precludes any useful reduction, but it arises also in transition zones between reduced grounded sheet and floating shelf solutions. The reduced grounded sheet solution has negligible in-plane deviatoric stress compared with direct shear stress, while we see shortly that the reduced floating shelf solution has negligible direct shear stress (tangential tractions zero on surface and base) compared with the in-plane (horizontal) deviatoric stress which governs the horizontal extending deformation. A transition solution must allow equal magnitude deviatoric stresses, with the in-plane deviatoric stress at the shelf edge usually greater than the direct shear stress in the grounded sheet. Also the in-plane velocity components $\bar{v}_1, \bar{v}_2$ have gradients in $\bar{x}_1, \bar{x}_2$ larger than their gradients in $\bar{x}_3$ as the shelf is approached. A stretched co-ordinate scaling and perturbation analysis can be applied separately to sheet and shelf, but the corresponding velocity and stress components have very different relative magnitudes due to the distinct basal traction conditions. Existing integrated model treatments fail to describe such an exchange of pattern. More generally, such models make approximations which have not been justified, and which are inconsistent with the explicit depth integrations when applied to reduced model situations.

8. Ice Shelf Flow

Figures 2 and *3* show respectively vertical and horizontal sections of a floating ice shelf fed by ice flow from a grounded sheet. There may be an extensive transition zone between the flow governed by the reduced grounded sheet model and the reduced shelf flow dominated by the floating base condition which is supposed to start at $x = 0$. In *Figure 2* the vertical front is set back from the actual margin, representing the limit of smooth velocity and stress fields. The velocity and stress close to the actual front, which is typically non-smooth, will be complex, but such local patterns are assumed not to extend far back into the shelf. In the horizontal plan, *Figure 3*, this conceptual front is shown by the dashed line, and front conditions for the unconfined section are based on the transmission of force from the water pressure across the ignored zone. There will be side wall friction, possibly limiting non-slip, at the confined boundary sections. The reduced model does not account for drag forces imposed by grounded zones, which require a transition between floating and grounded base conditions. It is supposed that the ice flux from the grounded sheet - transition zone across $x = 0$ is prescribed, which is sensible for a completely unconfined shelf flow, but side wall or grounding resistance would influence the sheet flow. The solution of transition flow is a major problem.

The ice shelf domain is bounded by the unknown free surface $z = h(x, y, t)$, where (x, y, z) and (x_1, x_2, x_3) are equivalent, the unknown base $z = f(x, y, t)$, the rear edge AB $(x = 0)$, the side-wall sections AD and BC, and unconfined front CD. The balance laws (4.21), (4.22) with $\chi = 0$, and (4.25) hold through the domain. The zero traction condition (4.5), thermal condition (4.6) and accumulation condition (4.9) apply on the free surface $z = h$. On the base the temperature is at the melting point $T_M(p_w)$, where p_w is the water pressure (ignoring any influence of salinity), so (4.14) is replaced by the explicit thermal condition

$$x_3 = f: \qquad T = T_M(p_w). \tag{8.1}$$

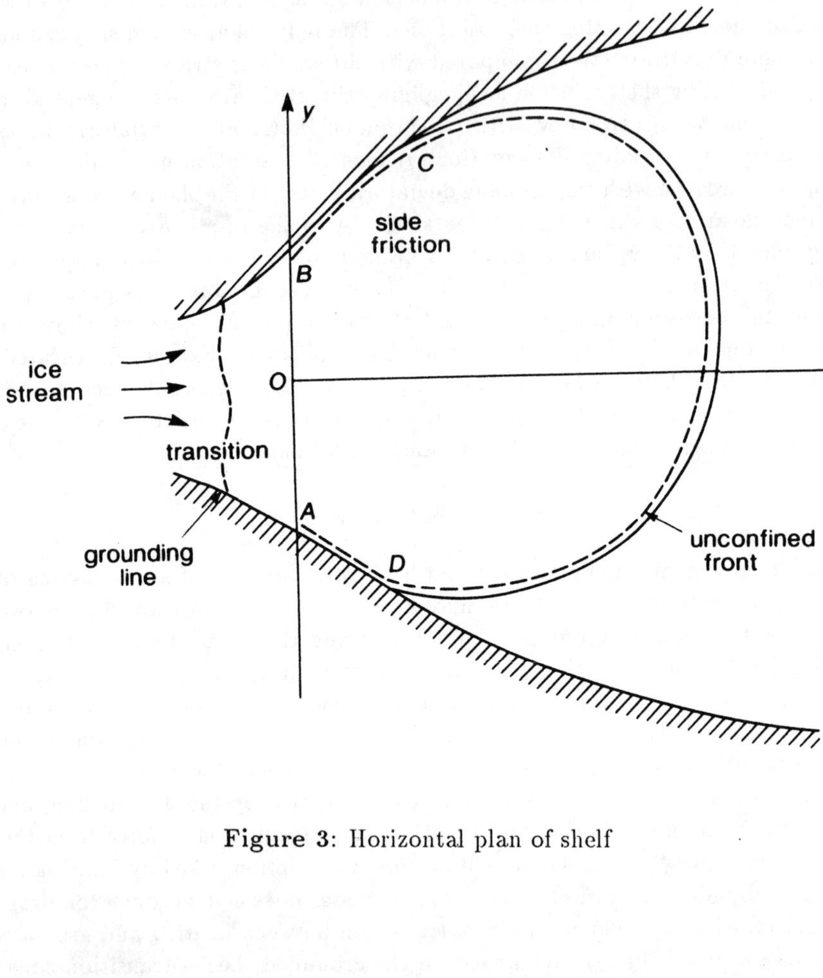

Figure 3: Horizontal plan of shelf

The basal melting and refreezing condition (4.13) again applies, though b is governed by the phase change occurring at the singular surface $z = f$ between ice and water. Noting that internal energy change at phase change is dominated by the latent heat L, and neglecting stress working and kinetic energy change with phase change, if follows that (see section 9) the melt flux

$$b = \left[\frac{\lambda}{L}\frac{\partial T}{\partial n}\right], \tag{8.2}$$

where $[-\lambda \partial T/\partial n]$ is the jump in heat flux from ice to water. Thus b involves the temperature solution in both ice and water.

By comparison with the stresses in the ice, the viscous shear stresses in the water are negligible, so continuity of traction with the water replaces a grounded sliding law (4.17) or (4.18) by zero tangential tractions

$$x_3 = f : \qquad \hat{t}_{s1} = \hat{t}_{s2} = 0 , \tag{8.3}$$

and normal traction continuity is given by

$$x_3 = f : \qquad \hat{t}_n = -p_n . \tag{8.4}$$

At the rear edge AB we should expect to prescribe the normal velocity $u = v_1$, but we find that the reduced model only requires a thickness integrated flux

$$x_1 = 0 : \qquad d_0(x_2)\bar{V}_1(x_2,t) = \int_f^h v_1 \mathrm{d}x_3 , \tag{8.5}$$

where

$$d = h - f \tag{8.6}$$

is the thickness with prescribed distribution $d_0(x_2)$ on AB. On the side sections AD, BC attached to bay sides there will be a friction law relating tangential traction to slip velocity, or limiting non-slip. The reduced model would only allow a thickness integrated prescription, but it is expected that depth variation would not persist far (on the horizontal length scales of the shelf) into the shelf, represented by the internal dashed boundary in *Figure 3*. An example condition is non-slip:

$$AD \text{ and } BC : \qquad v_1 = v_2 = 0 . \tag{8.7}$$

Finally, there is an unknown unconfined front CD, again represented by the internal dashed line in *Figure 3* to eliminate local non-smoothness. While calving (and ablation) will, at times, remove ice, following MORLAND (1987), it is reasonable to suppose that the conceptual front CD is a material surface, determined by the flow solution without any additional condition if it is prescribed at some initial time. Since the reduced model requires the horizontal span to be large in comparison with thickness and ignored side and front zones, it can only determine the evolution from a time at which a substantial shelf already exists, and not from its formation. It remains to determine the forces on the smooth vertical front represented by CD due to the water stress $-p_w\mathbf{n}$ acting over the actual front profile below sea level, with zero pressure above sea level. This is a major element of the reduced shelf model, and it is instructive to first display the arguments for the simpler plane flow case, to emphasize the necessary assumptions ignored in earlier derivations and motivate a corresponding derivation for a curved front with transverse flow. These analyses were introduced by MORLAND and ZAINUDDIN (1987) and MORLAND (1987).

Figure 4(a) illustrates the front zone ahead of the conceptual front $x = x_f$ when plane flow is assumed behind $x = x_f$, using co-ordinates (x, y, z) with unit axis vectors $\mathbf{i}, \mathbf{j}, \mathbf{k}$ respectively for clarity. While there is no dependence on y, and σ_{xy} and σ_{yz} are identically zero behind x_f, these properties do not generally hold in the frontal region ahead of x_f. Let l be the

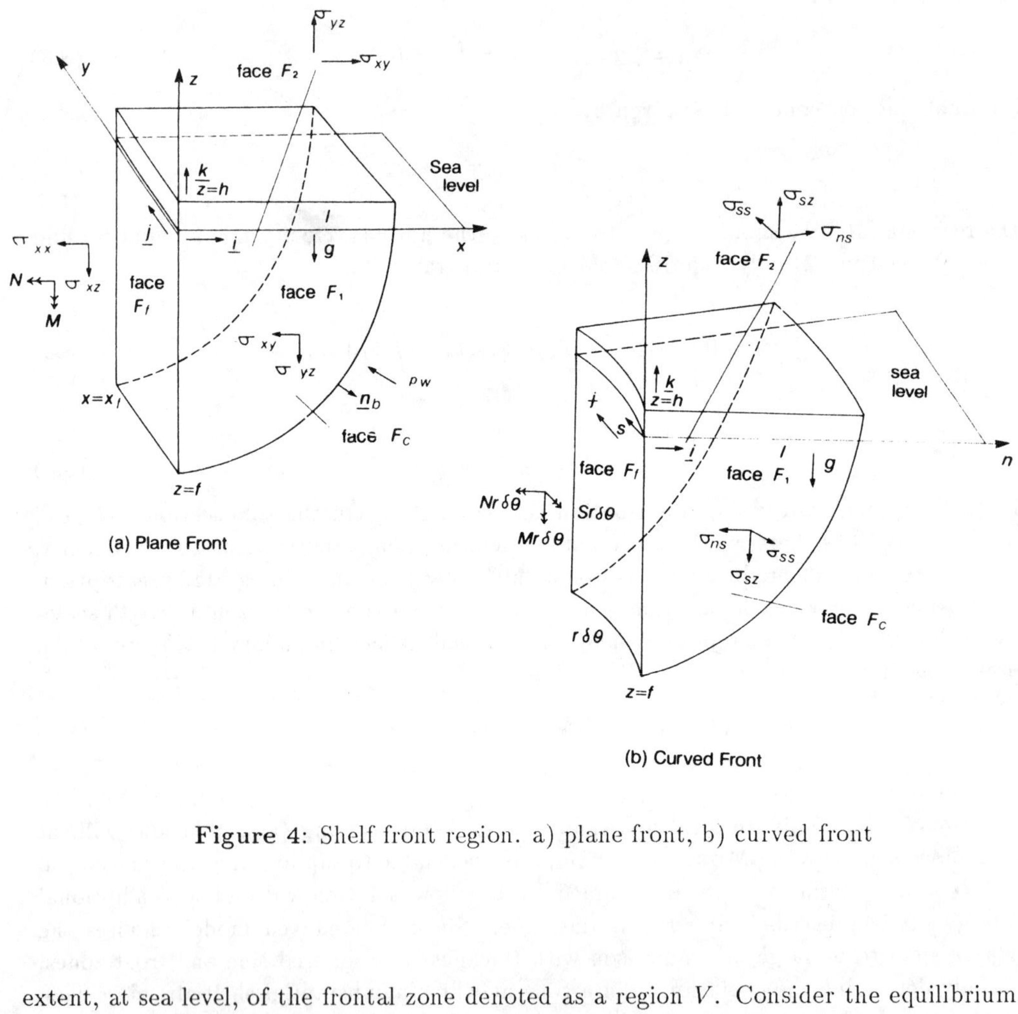

Figure 4: Shelf front region. a) plane front, b) curved front

extent, at sea level, of the frontal zone denoted as a region V. Consider the equilibrium of the ice in V. The total force per unit y span on the face F_f $(x = x_f)$ is

$$-N\mathbf{i} - M\mathbf{k} = -\int_f^h (\sigma_{xx}\mathbf{i} + \sigma_{xz}\mathbf{k})\,\mathrm{d}z\,. \tag{8.8}$$

The force per unit y span due to the water pressure over the actual front F_c^w below sea level $(z = 0)$ is

$$\mathbf{P}_w = -\int_{F_c^w} p_w \mathbf{n}_b \mathrm{d}S = \int_{F_c^w} \rho_w g z \mathbf{n}_b \mathrm{d}S \tag{8.9}$$

where $\mathbf{n}_b$ is the unit outward normal and ρ_w is the water density, and hydrostatic water pressure is assumed. Let the superscript w denote surfaces and regions of V restricted to parts below sea level $z = 0$, then V^w has a bounding surface $F_f^w \cup F_c^w \cup F_0 \cup F_1^w \cup F_2^w = S^w$, where F_0 is the face $z = 0$ within V. By the Divergence Theorem

$$\int_{V^w} \rho_w g \mathbf{k} \, dV = \int_{S^w} \rho_w g z \mathbf{n} \, dS \tag{8.10}$$

where $\mathbf{n}$ denotes the unit outward normal on each face of S^w. Now $z = 0$ on F_0 , and $\mathbf{n} = -\mathbf{i}$ on $F_f^w, \mathbf{n} = \pm\mathbf{j}$ on F_2^w, F_1^w respectively, $\mathbf{n} = \mathbf{n}_b$ on F_c^w. Thus

$$\begin{aligned} \mathbf{P}_w &= \rho_w g V^w \mathbf{k} + \int_{F_f^w} \rho_w g z \mathbf{i} \, dS + \int_{F_1^w - F_2^w} \rho_w g z \mathbf{j} \, dS \,, \\ &= \rho_w g V^w \mathbf{k} - \frac{1}{2} \rho_w g f^2(x_f, t) \mathbf{i} + \int_{F_1^w - F_2^w} \rho_w g z \mathbf{j} \, dS \,. \end{aligned} \tag{8.11}$$

Gravity contributes a force

$$\mathbf{G} = -\rho g V \mathbf{k} \,. \tag{8.12}$$

The force due to tractions on the faces F_1 and F_2 is

$$\mathbf{L} = \int_{F_2 - F_1} (\sigma_{xy}\mathbf{i} + \sigma_{yy}\mathbf{j} + \sigma_{yz}\mathbf{k}) \, dS \,. \tag{8.13}$$

Equilibrium requires that

$$\begin{aligned} N\mathbf{i} + M\mathbf{k} &= -\frac{1}{2}\rho_w g f^2 \mathbf{i} + g(\rho_w V^w - \rho^w V)\mathbf{k} + \mathbf{j} \int_{F_1 - F_2} \rho_w g z \, dS \\ &\quad + \int_{F_2 - F_1} (\sigma_{xy}\mathbf{i} + \sigma_{yy}\mathbf{j} + \sigma_{yz}\mathbf{k}) \, dS \,. \end{aligned} \tag{8.14}$$

Now the surfaces F_1, F_1^w, F_2, F_2^w, regions V and V^w, and tractions on those surfaces, are properties of the unknown complex frontal solution. Since there is no $\mathbf{j}$ component of traction on $x = x_f$ if the plane flow assumption behind x_f is justified, then the frontal solution must yield $F_1^w \approx F_2^w$ and $F_1 \approx F_2$ for most parts of the front, implying a profile varying little with y. In the reduced model solution for $x < x_f$ we will see that $\rho_w V^w = \rho V$ for sections between two planes $x =$ constant, but this does not follow for the frontal region in general. However, the reduced model integrated stress equations do not require a prescription for M, so we are concerned only with the $\mathbf{i}$ contribution of the final unknown integral. To neglect this requires $\sigma_{xy} \approx 0$ or $F_1 \approx F_2$ and $\partial\sigma_{xy}/\partial y \approx 0$, which are assertions that normal water pressure is acting over a front with little curvature in the y direction and so maintaining the plane flow condition to a good approximation. It is not possible to guarantee a negligible contribution from the unknown frontal terms in (8.14) by taking the extent l sufficiently large compared to the thickness d. We must therefore adopt this assumption and infer the conventional result

$$x = x_y : \qquad N = \int_f^h \sigma_{xx}\,\mathrm{d}Z = -\frac{1}{2}\rho_w g f^2 . \tag{8.15}$$

Turning to the case of two-dimensional horizontal flow, with a curved unconfined front CD in *Figure 3*, a frontal region is illustrated in *Figure 4(b)*. Here r is the local radius of curvature and the region has a span of arc length $r\delta\theta$, and extent l. Local parametric co-ordinates (n, s, z) define the normal and tangential (to CD) horizontal directions, and vertical direction with $\mathbf{i},\mathbf{j},\mathbf{k}$ again corresponding unit vectors based at the intersection of F_1, F_f and $z = 0$. Thus the normal and (horizontal) tangent directions on F_2 are rotated through an angle $\delta\theta$ from those on F_1, so that the traction on F_2 is $\mathbf{i}(\sigma_{ns}\cos\delta\theta - \sigma_{ss}\sin\delta\theta) + \mathbf{j}(\sigma_{ss}\cos\delta\theta + \sigma_{ns}\sin\delta\theta)$. Defining the force on F_f per unit arc length by

$$-N\mathbf{i} - S\mathbf{j} - M\mathbf{k} = -\int_f^h (\sigma_{nn}\mathbf{i} + \sigma_{ns}\mathbf{j} + \sigma_{nz}\mathbf{k})\,\mathrm{d}z\,, \tag{8.16}$$

leads to the equilibrium requirement

$$\begin{aligned}(N\mathbf{i} + S\mathbf{j} + M\mathbf{k})r\delta\theta \;=\;& -\frac{1}{2}\rho_w g f^2 \mathbf{i} r\delta\theta + g(\rho_w V^w - \rho V)\mathbf{k} \\ & +\mathbf{j}\int_{F_1^w - F_2^w} \rho_w g z\,\mathrm{d}S + \int_{F_2 - F_1} (\sigma_{ns}\mathbf{i} + \sigma_{ss}\mathbf{j} + \sigma_{sz}\mathbf{k})\,\mathrm{d}S \\ & +\mathbf{i}\int_{F_2^w} \rho_w g z\delta\theta\,\mathrm{d}S - \mathbf{i}\int_{F_2} \sigma_{ss}\delta\theta\,\mathrm{d}S + \mathbf{j}\int_{F_2} \sigma_{ns}\delta\theta\,\mathrm{d}S \end{aligned} \tag{8.17}$$

where $V^w, V, F_1^w - F_2^w$, and $F_2 - F_1$ have a factor $r\delta\theta$. Given that the stresses do not exceed the basal water pressure magnitude $\rho_w g d$, the last three additional terms do not exceed $\rho_w g d^2 l\delta\theta$ in magnitude, and can be neglected in comparison with the contribution $-\frac{1}{2}\rho_w g f^2 r\delta\theta$ provided that

$$l \ll r\,. \tag{8.18}$$

That is, the frontal extent l must be much smaller than the radius of curvature, so for l restricted to several thicknesses to define a sensible front contour CD we require a radius much greater than the thickness - a moderate curvature. Neglect of the unknown terms in $\mathbf{i}$ and $\mathbf{j}$, as before, requires slowly varying curvature so that $F_1^w - F_2^w$ and $F_2 - F_1$ and transverse gradients of σ_{ns}, σ_{ss} remain very small. With these assumptions:

$$CD: \qquad N = -\frac{1}{2}\rho_w g f^2\,, \qquad S = 0\,, \tag{8.19}$$

which prescribe the integrated horizontal tractions at the front. If a front does not have sufficient smoothness viewed on the large scale, then it is not clear how far the ignored frontal stresses penetrate into the shelf and violate the assumed front conditions.

In the dimensionless variables introduced in section 5 with length scale d_0 , velocity s-cale q_0, time scale d_0/q_0, the horizontal equations (7.14) for the depth integrated stresses $\bar{N}_{\alpha\beta}$ $(\alpha, \beta = 1, 2)$ become

$$\frac{\partial \bar{N}_{\alpha\beta}}{\partial \bar{x}_\beta} + \frac{\rho_w}{2\rho}\frac{\partial}{\partial \bar{x}_\alpha}(\bar{f}^2) = 0\,, \tag{8.20}$$

since $\chi = 0$, and on the base $\bar{z} = \bar{f}$:

$$\bar{z} = \bar{f}: \qquad \bar{t}_i = -\bar{p}_w n_i = \frac{\rho_w}{\rho}\bar{f} n_i\,, \tag{8.21}$$

where $\Delta_f n_\alpha = \partial \bar{f}/\partial \bar{x}_\alpha$ by (7.5), (7.9), (7.10). These are generalised plane stress equilibrium equations with body force $\frac{1}{2}\nabla(\bar{f}^2)$, unknown. As discussed in section 7, there are no explicit relations between $\bar{N}\alpha\beta$ and the mean horizontal velocities, and no explicit depth integration of the energy balance. On the unconfined front CD, the conditions (8.19) give boundary conditions

$$CD: \qquad \bar{N}_{\alpha\beta} n_\beta = -\frac{\rho_w}{2\rho} f^2 n_\alpha\,, \tag{8.22}$$

but side-wall drag and rear edge ice flux involve velocities.

Further simplification is possible only when horizontal derivatives are small compared to vertical derivatives, reflected by the co-ordinate and velocity scalings (5.22) and (5.23), which is supported by the very small thickness gradients of ice shelves away from grounded zones. However, the magnitude conclusions (5.25), (5.26) and (5.27) do not follow. Recall that the grounded sheet model assumed there was sufficient basal friction to balance a direct shear stress induced by the gravity driven viscous shearing. Here on the base there is zero tangential traction and normal water pressure given by (8.21), which to lead order become

$$X_3 = F: \qquad \bar{\sigma}_{13} = \bar{\sigma}_{23} = 0\,, \qquad \bar{\sigma}_{33} = \frac{\rho_w}{\rho}F\,, \tag{8.23}$$

analogous to the surface conditions (5.38) with $\bar{\sigma}_{13}$ and $\bar{\sigma}_{23}$ vanishing on both surface and base.

Now the viscous relations $(5.14)_2$ give

$$\begin{aligned} \bar{\sigma}_{13} &= \frac{\nu}{\bar{a}}\phi(I_2)\frac{1}{2\epsilon}\left[\frac{\partial V_1}{\partial X_3} + \epsilon^2\frac{\partial V_3}{\partial X_1}\right], \\ \bar{\sigma}_{23} &= \frac{\nu}{\bar{a}}\phi(I_2)\frac{1}{2\epsilon}\left[\frac{\partial V_2}{\partial X_3} + \epsilon^2\frac{\partial V_3}{\partial X_2}\right], \end{aligned} \tag{8.24}$$

where, by (3.6), (3.7), (5.3),

$$I_2 = \left[\frac{\delta}{\bar{a}}\right]^2 \bar{I}\,, \qquad \bar{I} = \frac{1}{2}tr([\bar{D}])\,, \qquad \delta = \frac{q_0}{a_0 D_0 d_0}\,, \tag{8.25}$$

while

$$\left.\begin{aligned} \bar{\sigma}_{12} &= \frac{1}{2}\frac{\nu}{\bar{a}}\phi(I_2)\left[\frac{\partial V_1}{\partial X_2} + \frac{\partial V_2}{\partial X_1}\right], & \bar{\sigma}'_{33} &= \frac{\nu}{\bar{a}}\phi(I_2)\frac{\partial V_3}{\partial X_3}, \\ \bar{\sigma}'_{11} &= \frac{\nu}{\bar{a}}\phi(I_2)\frac{\partial V_1}{\partial X_1}, & \bar{\sigma}'_{22} &= \frac{\nu}{\bar{a}}\phi(I_2)\frac{\partial V_2}{\partial X_2}, \end{aligned}\right\} \tag{8.26}$$

are order $\nu/\bar{a}$ everywhere. Expansion of the zero tangential traction conditions (4.5) on surface and base with normal and tangent vector components (5.37) and (5.41), and the magnitudes (8.26), show that $\bar{\sigma}_{13}$ and $\bar{\sigma}_{23}$ are order $\epsilon\nu/\bar{a}$ on surface and base, more specific than the lead order expressions (5.38) and (8.23). But the horizontal equilibrium equations show that $\partial\bar{\sigma}_{13}/\partial X_3$ and $\partial\bar{\sigma}_{23}/\partial X_3$ are order ϵ since $\bar{\sigma}_{11}, \bar{\sigma}_{22}, \bar{\sigma}_{33}$ are order unity, so, assuming no significant oscillation with X_3,

$$\bar{\sigma}_{13},\ \bar{\sigma}_{23} \sim \epsilon\nu/\bar{a} \tag{8.27}$$

through the shelf, from the warm base where $\nu\bar{a} \ll 1$ to order ϵ at the cold surface where $\nu/\bar{a} \sim 1$. Hence the relations (8.24) show that

$$\begin{aligned} V_1 &= V_1(X_1, X_2, \bar{t}) + \mathcal{O}(\epsilon^2), \\ V_2 &= V_2(X_1, X_2, \bar{t}) + \mathcal{O}(\epsilon^2), \end{aligned} \tag{8.28}$$

so the lead order horizontal velocities are independent of X_3 , in contrast to the grounded sheet results. Furthermore,

$$\bar{\sigma}_{13},\ \bar{\sigma}_{23} \sim \epsilon(\bar{\sigma}_{12}, \bar{\sigma}'_{11}, \bar{\sigma}'_{22}, \bar{\sigma}'_{33}), \tag{8.29}$$

reversing the relative magnitudes of the grounded sheet stresses.

Vertical equilibrium, the zero normal traction on the surface and normal water pressure on the base, give the lead order results

$$\bar{\sigma}_{33} = X_3 - H(X_1, X_2, \bar{t}), \tag{8.30}$$

$$H = \left[1 - \frac{\rho}{\rho_w}\right]\Delta, \qquad F = -\frac{\rho}{\rho_w}\Delta, \tag{8.31}$$

where

$$\bar{f} = F(X_1, X_2, \bar{t}), \quad \bar{h} = H(X_1, X_2, \bar{t}), \quad \bar{d} = \bar{h} - \bar{f} = \Delta(X_1, X_2, \bar{t}). \tag{8.32}$$

The vertical stress $\bar{\sigma}_{33}$ is simply the balance with gravity, as the grounded sheet result (5.47), but here the horizontal stresses $\bar{\sigma}_{11}$ and $\bar{\sigma}_{22}$ differ from $\bar{\sigma}_{33}$ when $\bar{\sigma}'_{11}, \bar{\sigma}'_{22}$ are not of lower magnitude, that is, when $\nu/\bar{a}$ is order unity near a cold surface, in contrast to the equivalence (5.34). With the simple dependences (8.28), the mass balance (5.24) can be explicitly integrated through the thickness to give

$$V_3 = (H - X_3)\left[\frac{\partial V_1}{\partial X_1} + \frac{\partial V_2}{\partial X_2}\right] + W_s(X_1, X_2, \bar{t}), \tag{8.33}$$

where W_s is the surface value of V_3 , and then

$$W_b(X_1,X_2,\bar{t}) = W_s(X_1,X_2,\bar{t}) + \Delta\left[\frac{\partial V_1}{\partial X_1} + \frac{\partial V_2}{\partial X_2}\right] \tag{8.34}$$

is the base value. Further, the lead order surface and base kinematic conditions (5.40) and (5.46) can be differenced to obtain

$$\frac{\partial \Delta}{\partial \bar{t}} + V_1\frac{\partial \Delta}{\partial X_1} + V_2\frac{\partial \Delta}{\partial X_2} - (W_s - W_b) = \bar{q} - \bar{b} = q^\star\,, \tag{8.35}$$

which, applying (8.34), becomes

$$\frac{\partial \Delta}{\partial \bar{t}} + \frac{\partial}{\partial X_1}(\Delta V_1) + \frac{\partial}{\partial X_2}(\Delta V_2) = q^\star\,. \tag{8.36}$$

Introducing the integrated stress combinations (MORLAND, 1987)

$$\hat{N}_{11} = \int_F^H (2\sigma'_{11} + \sigma'_{22})\mathrm{d}X_3\,, \qquad \hat{N}_{22} = \int_F^H (2\sigma'_{22} + \sigma'_{11})\mathrm{d}X_3\,, \tag{8.37}$$

and using the identities

$$\bar{\sigma}_{11} = 2\bar{\sigma}'_{11} + \bar{\sigma}'_{22} + \bar{\sigma}_{33}\,, \qquad \bar{\sigma}_{22} = 2\bar{\sigma}'_{22} + \bar{\sigma}'_{11} + \bar{\sigma}_{33}\,, \tag{8.38}$$

with the results (8.30), (8.31), to eliminate $\bar{\sigma}_{33}$, transforms the horizontal equilibrium equations (8.20) to

$$\left.\begin{aligned} \frac{\partial \hat{N}_{11}}{\partial X_1} + \frac{\partial \hat{N}_{12}}{\partial X_2} &= \tfrac{1}{2}\left[1 - \frac{\rho}{\rho_w}\right]\frac{\partial \Delta^2}{\partial X_1}\,, \\ \frac{\partial \hat{N}_{12}}{\partial X_1} + \frac{\partial \hat{N}_{22}}{\partial X_2} &= \tfrac{1}{2}\left[1 - \frac{\rho}{\rho_w}\right]\frac{\partial \Delta^2}{\partial X_2}\,, \end{aligned}\right\} \tag{8.39}$$

while the unconfined boundary conditions (8.22) become

$$CD: \quad \left.\begin{aligned} \hat{N}_{11}n_1 + \hat{N}_{12}n_2 &= \tfrac{1}{2}\left[1 - \frac{\rho}{\rho_w}\right]\Delta^2 n_1\,, \\ \hat{N}_{12}n_1 + \hat{N}_{22}n_2 &= \tfrac{1}{2}\left[1 - \frac{\rho}{\rho_w}\right]\Delta^2 n_2\,. \end{aligned}\right\} \tag{8.40}$$

The corresponding integrated stress-velocity gradient relations are

$$\left.\begin{aligned} \hat{N}_{11} &= \nu\left[2\frac{\partial V_1}{\partial X_1} + \frac{\partial V_2}{\partial X_2}\right]\int_F^H \frac{\phi(I_2)}{\bar{a}}\mathrm{d}X_3 \\ \hat{N}_{22} &= \nu\left[2\frac{\partial V_2}{\partial X_2} + \frac{\partial V_1}{\partial X_1}\right]\int_F^H \frac{\phi(I_2)}{\bar{a}}\mathrm{d}X_3 \\ 2\hat{N}_{12} &= \nu\left[\frac{\partial V_1}{\partial X_2} + \frac{\partial V_2}{\partial X_1}\right]\int_F^H \frac{\phi(I_2)}{\bar{a}}\mathrm{d}X_3 \end{aligned}\right\}, \tag{8.41}$$

where

$$I_2 = \left[\frac{\delta}{\bar{a}}\right]^2 \left[\left[\frac{\partial V_1}{\partial X_1}\right]^2 + \left[\frac{\partial V_2}{\partial X_2}\right]^2 + \frac{\partial V_1}{\partial X_1}\frac{\partial V_2}{\partial X_2} + \frac{1}{4}\left[\frac{\partial V_1}{\partial X_2} + \frac{\partial V_2}{\partial X_1}\right]^2\right], \tag{8.42}$$

and $\bar{a}$, depend on X_3 through the temperature field $\bar{T}(X_1, X_2, X_3, \bar{t})$, governed by the energy balance (5.32) with negligible stress working which becomes

$$\frac{\partial \bar{T}}{\partial \bar{t}} + V_1\frac{\partial \bar{T}}{\partial X_1} + V_2\frac{\partial \bar{T}}{\partial X_2} + \left\{[H - X_3]\left[\frac{\partial V_1}{\partial X_1} + \frac{\partial V_2}{\partial X_2}\right] + W_s\right\}\frac{\partial \bar{T}}{\partial X_3} = \beta\frac{\partial^2 \bar{T}}{\partial X_3^2}. \tag{8.43}$$

The thermomechanically coupled balances for $V_1, V_2, \hat{N}_{11}, \hat{N}_{22}, \bar{N}_{12}$ and $\bar{T}$ are (8.36), (8.39) and (8.43), with the stress relations (8.41) and boundary conditions. The X_3 dependence of $\bar{T}$ cannot be obtained without solution of the full coupled system, and further simplification of the mechanical balances is possible only if $\bar{T}$ is prescribed, not solving the energy balance (8.43), so that $\bar{a}$ has a known dependence in X_3 in the integrals (8.41). Simple prescriptions (MORLAND and SHOEMAKER, 1982) fail to yield satisfactory approximations of the balance (8.43), which has significant horizontal advection depending on the solutions V_1, V_2. However, effects of other features can be explored by such prescriptions. The greatest simplication is achieved when $\bar{T}$, and hence $\bar{a}$, depends only on the relative height above the base, $\eta = (X_3 - F)/\Delta$, and the function $\phi(I_2)$ is given by the power law (3.21), which includes the Newtonian fluid $\phi = 2\mu$ (constant) when $n = 1$. Then

$$\int_F^H \frac{\phi(I_2)}{\bar{a}}\mathrm{d}X_3 = \frac{2\Delta}{3}\left[\frac{a_0}{k}\right]^{\frac{1}{n}}\left[\frac{4}{3}\delta^2\bar{I}\right]^{-\frac{n-1}{2n}}\int_0^1 \bar{a}^{-\frac{1}{n}}\mathrm{d}\eta, \tag{8.44}$$

where the final integral is a constant.

I am not aware of any numerical solutions of this two-dimensional reduced model incorporating unconfined fronts. An early attempt to describe unconfined flow modified the vertical plane flow model with Δ and V_1 functions of $X_1, \bar{t}$ only, by assuming $\bar{D}_{22} \equiv \bar{D}_{11} = \partial V_1/\partial X_1$, so that $V_2 = X_2\bar{D}_{11}$ ($= 0$ on symmetry line $X_2 = 0$). To reconcile these restrictions with the above equations for all X_2 (not just an approximation within a few thicknesses of the symmetry line) requires $\partial V_1/\partial X_1 = D(t)$, independent of X_1, by (8.40), and hence $\bar{D}_{12}, \bar{N}_{12} \equiv 0$, and

$$\hat{N}_{11} = \hat{N}_{22} = 2\nu D(t)\int_F^H \frac{\phi(I_2)}{\bar{a}}\mathrm{d}X_3, \tag{8.45}$$

where $I_2 = 3[\delta D(T)/\bar{a}]^2$ depends on X_1 and X_2 only through $\bar{a}$. The non-trivial balance $(8.39)_1$ integrates explicitly to the boundary condition $(8.40)_1$ applied at all X_1 (with $\bar{N}_{12} \equiv 0$), which implies that

$$\int_F^H \frac{\phi(I_2)}{\bar{a}}\mathrm{d}X_3 \equiv \frac{(1 - \rho/\rho_w)}{6\nu D(\bar{t})}\Delta^2(X_1, \bar{t}). \tag{8.46}$$

This identity in X_1 appears possible only if Δ is independent of X_1, so H and F are independent of X_1, and $\bar{a}$ is independent of X_1, which is certainly required in the simplified case (8.44). Then the integral in (8.46) is simply $J(\bar{t})\Delta(\bar{t})$, and hence

$$6\nu D(\bar{t})J(\bar{t}) = (1 - \rho/\rho_w)\Delta(\bar{t}), \tag{8.47}$$

where by (8.36),

$$\Delta'(\bar{t}) + 2D(\bar{t})\Delta(\bar{t}) = q^\star, \tag{8.48}$$

with $q^\star = q^\star(t)$ necessarily, which determines $\Delta(t)$ given its initial value. Such a maintained uniform thickness shelf solution is of little general interest. The early treatment, in fact, simply applied the longitudinal balance (8.46) for non-uniform $\Delta(X_1), D(X_1)$ subject to (8.36) for steady isothermal flow, violating the transverse balance and front conditions (except near $X_2 = 0$).

The most common application of shelf theory has been for plane flow, in $0x_1x_3$, for which $v_2, \sigma_{12}, \sigma_{32}$ are identically zero, and all variables are independent of x_2. This describes flow in a bay with parallal frictionless sides, or possibly a central zone flow, away from the sides, emerging from a wide uniform stream. This simplified problem is described by the equations

$$\frac{\partial \Delta}{\partial \bar{t}} + \frac{\partial}{\partial X_1}(\Delta V_1) = q^\star, \tag{8.49}$$

$$\hat{N}_{11} = \frac{1}{2}(1 - \rho/\rho_w)\Delta^2 = 2\nu\frac{\partial V_1}{\partial X_1}\int_F^H \frac{\phi(I_2)}{\bar{a}}\mathrm{d}X_3, \tag{8.50}$$

$$\frac{\partial \bar{T}}{\partial \bar{t}} + V_1\frac{\partial \bar{T}}{\partial X_1}\left[[H - X_3]\frac{\partial V_1}{\partial X_1} + W_s\right]\frac{\partial \bar{T}}{\partial X_3} = \beta\frac{\partial^2 \bar{T}}{\partial X_3^2}, \tag{8.51}$$

where $F, H, \Delta, V_1, \hat{N}_{11}$ are functions of $(X_1, \bar{t}), T$ is a function of $(X_1, X_3, \bar{t})$, and

$$I_2 = \left[\frac{\delta}{\bar{a}}\right]^2 \bar{I}, \qquad \bar{I} = \left[\frac{\partial V_1}{\partial X_1}\right]^2, \tag{8.52}$$

subject to surface and base thermal conditions and to initial conditions $\bar{\Delta}(X_1, 0)$, $T(X_1, X_3, 0)$, and to rear edge conditions $\Delta(0, \bar{t})$, $V_1(0, \bar{t}) = U(\bar{t}), T(0, X_3, \bar{t})$. For the steady flow problem, $\Delta, V_1, \bar{T}$ independent of $\bar{t}$ (but the material front, and hence span of domain, changing), only the edge conditions $\Delta(0), V_1(0)$, and $\bar{T}(0, X_3)$ are prescribed. Note that the prescriptions $d(x_1, 0) = \Delta(\epsilon^{-1}x_1, 0)$ and $\bar{v}_1(0, t) = \epsilon^{-1}V_1(0, \bar{t})$ in unscaled variables determine an approximate ϵ in the unsteady problem; ϵ is not determined by the mechanical force balance (8.50), in contrast to the grounded sheet scaling.

The balance (8.50) implies a solution V_1 subject to the prescribed edge value $U = \epsilon\bar{V}_1(0, \bar{t})$ which maintains $\partial V_1/\partial X_1$ order unity. Given that longitudinal velocities can increase from rear edge to front by a factor of 10, we should choose $U \sim 0 \cdot 1$ and $\epsilon = 0 \cdot 1q_0/V_1(0, 0)$ say, implying $\epsilon = 10^{-3}$ for a typical $V_1(0, 0) \sim 100\,myr^{-1}$. With $\rho \sim 917\,kg\,m^{-3}, \rho_w \sim 1028\,kg\,m^{-3}$ (saline water), $\frac{1}{4}(1 - \rho/\rho_w) \sim 0.025 \gg \nu$, implying that the integral in (8.50) is much greater than unity, dominated by a region of cold ice where $\bar{a}$ is much less than unity. Note that the data correlations (3.16), (3.19) incorporate $a_0 = a(\bar{T}_0)$ in the definitions of ψ, and the balance (8.50) is independent of a_0 since $\nu\phi = \nu a_0(a_0\psi)^{-1}$ is independent of a_0. Increasing the viscosity from that given by the laboratory data can be achieved by using the same $a_0\psi$ with a decreased strain-rate unit D_0 , hence increased ν , as earlier suggested to match (8.50) with shelf data, but those estimates took no account of the strongly decreasing $\bar{a}$ from base to surface. For a given $\phi(I_2)$ and $\bar{a}(\bar{T})$, the solution of (8.49), (8.50), (8.51) determines $\partial V_1/\partial X_1$, but consistency with the reduced model assumptions requires that $\partial V_1/\partial X_1 = o(\epsilon^{-1})$. In practice, stresses N_{12} induced by lateral flow or side wall drag, and increased resistance due to grounding drag, yield a smaller longitudinal stretching force $\hat{N}_{11}$, and hence smaller $\partial V_1/\partial X_1$ for given ϕ and $\bar{a}$.

The most simple situation is steady flow with a prescription $q^\star = q^\star(X_1)$, so that

$$\Delta V_1 = \Delta_0 U + \int_0^{X_1} q^\star(X_1)\mathrm{d}X_1 = Q(X_1)\,, \tag{8.53}$$

where $\Delta_0 = \Delta(0)$, and hence

$$\frac{\mathrm{d}V_1}{\mathrm{d}X_1} = -\frac{Q}{\Delta^2}\frac{\mathrm{d}\Delta}{\mathrm{d}X} + \frac{q^\star}{\Delta}\,. \tag{8.54}$$

The steady coupled problem for Δ, V_1 and $\bar{T}(X_1, X_3)$ is still complex, and the uncoupled problem with prescribed $\bar{a}(X_1, X_3)$ still requires the solution of the simultaneous integro-differential equations (8.50), (8.54), with (8.52), for Δ, V_1. The commonly adopted case is constant $\bar{a} = \bar{a}_m$, which assumes that a mean rate factor $a_m = a_0\bar{a}_m$ through the depth can represent the strongly non-uniform factor for a realistic temperature distribution, equivalent to assuming isothermal condition at some mean temperature. The choice of a_m, or the mean temperature, is not clear, but we will show that it must correspond to the near surface cold ice. With constant $\bar{a} = \bar{a}_m$, (8.50) with (8.52) and (3.11) becomes

$$J_2^{\frac{1}{2}} = I_2^{\frac{1}{2}}\phi(I_2) = \frac{1}{4}\nu^{-1}\delta(1 - \rho/\rho_w)\Delta = R\Delta\,, \tag{8.55}$$

where $R = \frac{1}{4}\varphi^{-1}(1 - \rho/\rho_w) \sim 0\cdot 25 \rightarrow 2\cdot 5$ for $d_0 = 100 \rightarrow 1000\,m$, so R is order unity. Thus, again from (3.11),

$$I_2 = J_2\psi^2(J_2) = R^2\Delta^2\psi^2(R^2\Delta^2)\,, \tag{8.56}$$

and hence

$$\frac{\mathrm{d}V_1}{\mathrm{d}X_1} = \bar{a}_m\delta^{-1}R\Delta\psi(R^2\Delta^2)\,. \tag{8.57}$$

Now (8.54) yields the first order differential equation

$$Q\frac{\mathrm{d}\Delta}{\mathrm{d}X_1} - q^\star\Delta + \bar{a}_m\delta^{-1}R\Delta\psi(R^2\Delta^2)\Delta^3 = 0 \tag{8.58}$$

for $\Delta(X_1)$ subject to $\Delta(0) = \Delta_0$. Note that $\bar{a}_m\delta^{-1} = a_m D_0 d_0/q_0$ depends on the chosen rate factor a_m and the chosen strain-rate unit D_0 , and must not exceed unity in magnitude to achieve order unity $\mathrm{d}\Delta/\mathrm{d}X_1$ near the rear edge where $\Delta \sim \Delta_0 \sim 1$, implying sufficiently small $a_m D_0/q_0$ to counteract the factor d_0^2 in $\delta^{-1}R$. In fact, the balance (8.58) would predict large $\mathrm{d}\Delta/\mathrm{d}X_1$ near a rear edge with $d_0 \sim 1000\,m$, even for $a_m \sim 10^{-3}$, suggesting that this reduced model idealisation cannot extend to a grounding line with very thick ice. Recall that side drag and grounding resistance effectively decrease R, but the model then fails for other reasons. As illustration, apply Glen's power law (3.19), then (8.58) becomes

$$Q\frac{\mathrm{d}\Delta}{\mathrm{d}X_1} - q^\star\Delta + C\bar{a}_m\Delta^{n+2} = 0\,, \qquad C = \frac{1}{2}(3)^{\frac{n+1}{2}}kR^n(a_0\delta)^{-1}\,. \tag{8.59}$$

Alternatively, prescribing the $\bar{a}$ dependence on $\eta = (X_3 - F)/\Delta$ with the power law leads to the explicit integral (8.44), with the temperature influence governed by the constant

$$A = \int_0^1 \bar{a}^{-\frac{1}{n}}\mathrm{d}\eta\,. \tag{8.60}$$

For the case $\bar{a} \equiv \bar{a}_m$, $\quad A = \bar{a}_m^{-\frac{1}{n}}$. Now (8.50) gives

$$\frac{\mathrm{d}V_1}{\mathrm{d}X_1} = CA^{-n}\Delta^n\,, \tag{8.61}$$

and (8.54) gives

$$Q\frac{\mathrm{d}V_1}{\mathrm{d}X_1} - q^\star\Delta + CA^{-n}\Delta^{n+2} = 0\,, \tag{8.62}$$

which is simply (8.59) with $\bar{a}_m$ replaced by A^{-n}. With $n = 3$, $C \sim 1 \rightarrow 625 \rightarrow 10^4$ for $d_0 = 100 \rightarrow 500 \rightarrow 1000\,m$, so order unity $\mathrm{d}\Delta/\mathrm{d}X_1$ for $d_0 \sim 500\,m$ requires A^{-3} or $\bar{a}_m \sim 2\times10^{-3}$. The latter refers to very cold ice, while $A \sim 10$ implies small $\bar{a}$ over a large fraction of shelf depth. If $\bar{a} = e^{-r\eta}$ where $r = -ln[\bar{a}(1)] \sim 4\cdot6 \rightarrow 10$ for $\bar{a}(1) \sim 10^{-2} \rightarrow 10^{-3}$, then $A \sim 2\cdot4 \rightarrow 4$, $A^{-3} \sim 0\cdot07 \rightarrow 0\cdot016$, which would imply that $\mathrm{d}\Delta/\mathrm{d}X_1$ order unity requires $d_0 \lesssim 200 - 300\,m$. While a large thickness gradient arises at the rear edge starting point, calculations show that the differential equation solution rapidly progresses to a moderate gradient. Coupled solutions are required to assess the merits of such idealised temperature prescriptions, first for the plane flow restriction, and also unsteady solutions to assess the effects of different features of shelf evolution.

The major problem not covered by the reduced models is transition from grounded to floating basal conditions, which can extend for large distances. The reduced grounded sheet model is stress controlled, with simple linear variations with depth of $\bar{\sigma}_{33}, \bar{\sigma}_{13}, \bar{\sigma}_{12}$, and velocity variations with depth highly non-uniform to accommodate the significant variation of $\bar{a}$ with depth. The reduced shelf model is velocity controlled, with the horizontal velocities uniform with depth, and the deviatoric stresses changing in response to $\bar{a}$. This change of stress and flow pattern, abrupt or gradual, involves a transition region in which longitudinal and vertical gradients of velocity have equal importance, and while the co-ordinate stretching to reflect the small thickness gradient still applies over a gradual change, the structure of the partial differential balance equations is not simplified. Further understanding of the pattern change requires a numerical treatment.

9. Ice Sheets with Melt Zones

Natural ice masses may have extensive interior zones sufficiently close to the melting point at the local pressure for there to be a significant water presence (and water vapour), and this situation will be more common during periods of sustained climate warming. The overall flow properties of the local ice will be modified, but there is also a significant effect of latent heat on the energy balance as phase changes take place to maintain the melt temperature in zones of co-existing ice and water. This further influences the temperature distribution in adjacent cold ice regions. The location of an interface between cold ice and melt zone is itself a feature of the overall balance solution, and such unknown moving boundaries introduce significant complexity into numerical treatments. This final section will attempt only to formulate a model for describing the flow of co-existing ice and water in a melt zone - a two-constituent or binary mixture - and the interface conditions. Previous references cover the history of polythermal glacier modelling, and recent interest and activity is reflected by HUTTER's (1992) mixture theory treatment, presented at a Grenoble Conference in November 1990.

The theory of interacting continua - mixture theory - provides an elegant framework for describing the distinct flows and deformation of two or more constitutents when length scales of interest exceed the scale at which the individual constitutents are distinguished. A representative mixture element of mixture is therefore supposed to contain sufficient of each constituent that mean constituent velocities and densities sensibly describe their presence in the mixture. It is a reasonable assumption for an extensive melt zone with modest water presence, and offers the only tractable model on this large scale. The mixture is viewed as overlapping continuous distributions of each constituent, so that every point in the mixture is occupied by each constituent in the sense that there are continuously distributed particle velocity and partial density fields for each constituent. The conservation laws of mass, momentum and energy were given by TRUESDELL (1965 - a translation of his 1957 paper in Italian), and there followed a variety of abstract papers discussing thermodynamic restrictions on constitutive theories cast in mixture variables. The applications of the mixture concept to flow in porous media in a geophysical context is detailed in a substantial text by BEAR (1972). My interests have ranged from geotechnical applications to descriptions of snowpacks with ice grains, water and vapour, and have been concerned with relating mixture variables to intrinsic variables describing the state, and properties, of the actual constituent elements. These developments have been incorporated with a self-contained account of the concepts of mixture theory, and the derivation of conservation laws for a continuous region and across a singular surface, in MORLAND (1992b). The present account adopts the same notation and states the conservation laws without repeating the derivations. They are generalisations of those for a single continuum, but the following paragraph extracted from the Introduction emphasises a distinction which appears to be commonly ignored.

"The fundamental balance laws of mass, linear momentum, angular momentum and energy apply to any fixed body of each constituent, which, in a mixture, occupies an incoherent multiply-connected configuration. Mixture concepts allow a precise integral formulation and reduction to point differential form, in terms of the **partial variables**. Summing over constituents provides balances for the mixture of constituents, not for a **mixture body**. A coherent set of constituent elements forming a body of mixture at one time becomes incoherent as constituents diffuse, so we cannot define a continuous mixture body and apply mechanical and thermodynamical continuum principles to mixture elements. It is, however, helpful to introduce a **constituent matrix** as a continuous geometric configuration which envelops a given set of the constituent elements as the motion proceeds, but a changing population of the other constituents. This is a familiar concept for a connected porous solid deforming little from a reference configuration, but is also useful in describing the kinematics of mobile fluid constituents. All constituents have a symmetric entry in the mixture theory balances, distinctions appear only through constitutive relations for individual constituent properties and for interactions between constituents."

Let a superscript ν label variables associated with the ν-constitutent. Here $\nu = 1$ or i for ice, $= 2$ or w for water, but the general equations apply to an n-constitutent mixture ($\nu = 1, \ldots, n$). All variables are supposed to be continuous, and differentiable as necessary, functions of $(\mathbf{x}, t)$ throughout the mixture, except across singular surfaces which are discussed separately. Constituent ν has a volume fraction ϕ^ν, a partial density ρ^ν which is its mass per unit mixture volume, and a particle velocity $\mathbf{v}^\nu$ which represents a mean velocity over a mixture element. For consistency in the conservation laws $\mathbf{v}^\nu$ must be the intrinsic velocity to deliver the correct mass, momentum and energy fluxes; that is, a mean of the actual partial velocity (MORLAND, 1992b). Denote intrinsic variables, those associated with actual continuous elements of the constituent, by a further superscript $\star$ when necessary to distinguish them from mixture variables, conveniently termed partial variables. Then the intrinsic density $\rho^{\nu\star}$, masss per unit constituent volume, is related to ρ^ν by

$$\rho^\nu = \phi^\nu \rho^{\nu\star}, \tag{9.1}$$

and the mixture density is

$$\rho = \sum_{\nu=1}^{n} \rho^\nu . \tag{9.2}$$

The velocity $\mathbf{v}^\nu$ is defined precisely to be the vector such that $\rho^\nu \mathbf{v}^\nu$ is the mass flux of ν-constituent across a unit mixture cross-section (normal to the flow), then mass conservation requires

$$\frac{\partial \rho^\nu}{\partial t} + \operatorname{div}(\rho^\nu \mathbf{v}^\nu) = \frac{\mathrm{D}_\nu \rho^\nu}{\mathrm{D}t} + \rho^\nu \operatorname{div} \mathbf{v}^\nu = m^\nu , \tag{9.3}$$

where m^ν is the rate of mass transfer to ν from other constituents (phase change here) per unit mixture volume, and

$$\frac{\mathrm{D}_\nu}{\mathrm{D}t} = \frac{\partial}{\partial t} + v_j^\nu \frac{\partial}{\partial x_j} . \tag{9.4}$$

Denote the force supported by the ν-constituent per unit mixture area on a cross-section with unit normal $\mathbf{n}$ by the partial traction $\mathbf{t}^\nu(\mathbf{n})$. Let $\mathbf{b}^\nu$ be the external body force on unit mass of ν-constituent (commmonly gravity), and $\beta^\nu, \tilde{\beta}^\nu$ the interaction body forces per unit mixture mass due to other constituents, the former representing particle contact effects - drag - and the latter the thrust due to momentum transfer. Linear momentum conservation leads to

$$\frac{\partial \sigma_{ij}^\nu}{\partial x_j} + \rho^\nu b_i^\nu + \rho\beta_i^\nu + \rho\tilde{\beta}_i^\nu = \rho^\nu \frac{\mathrm{D}_\nu v_i^\nu}{\mathrm{D}t}, \qquad (9.5)$$

where $\boldsymbol{\sigma}^\nu$ is the partial stress tensor defined by

$$\boldsymbol{t}^\nu(n) = \boldsymbol{\sigma}^\nu \boldsymbol{n}. \qquad (9.6)$$

The total traction $\mathbf{t}(\mathbf{n})$ on unit mixture cross-section, and corresponding total stress $\boldsymbol{\sigma}$, are given by

$$\mathrm{t}(\mathrm{n}) = \sum_{\nu=1}^{\mathrm{n}} \mathrm{t}^\nu(\mathrm{n}) = \sigma\mathrm{n}, \qquad \sigma = \sum_{\nu=1}^{\mathrm{n}} \sigma^\nu. \qquad (9.7)$$

Since the contact forces between each pair of constituents are equal and opposite,

$$\sum_{\nu=1}^{n} = \beta^\nu = 0, \qquad (9.8)$$

while the linear momentum balances for individual constituents and the mixture of constituents shows that

$$\sum_{\nu=1}^{n} \tilde{\beta}^N = -\sum_{\nu=1}^{n} m^\nu \mathrm{v}^\nu. \qquad (9.9)$$

This does not determine each $\tilde{\beta}^\nu$, but a further postulate that the contribution to $\tilde{\beta}^\nu$ from mass transfer $m^{\nu\mu}$ from each other constitutent μ is proportional to that mass transfer and to their relative velocity, with a common relation for all pairs ν, μ gives the explicit result (MORLAND 1992b)

$$\rho\tilde{\beta}^\nu = -\frac{1}{2}\sum_{\mu=1}^{n} m^{\nu\mu}(v^\nu - v^\mu), \qquad (9.10)$$

where

$$m^\nu = \sum_{\mu=1}^{n} m^{\nu\mu}, \qquad m^{\nu\nu} = 0. \qquad (9.11)$$

and each $m^{\nu\mu}$ will be determined by the particular $\nu - \mu$ phase change (or chemical interaction). Angular momentum conservation leads to the symmetry of the total stress tensor, and to the symmetry of the partial stress tensor if there are no interaction couples, which I assume here, so that

$$\sigma_{ij} = \sigma_{ji}, \qquad \sigma_{ij}^\nu = \sigma_{ji}^\nu. \qquad (9.12)$$

Let U^ν be the internal energy (stored energy) per unit mass of ν-constituent, r^ν the rate of external energy supply (e.g. radiation) per unit mass of ν-constituent, $\mathbf{q}^\nu$ the energy flux per unit mixture cross-section into constituent ν (e.g. heat conduction), and let the rate of interaction energy transfer per unit mass of mixture to constituent ν from all other constituents be the sum of ψ^ν, independent of mass transfer, and $\tilde{\psi}^\nu$ due solely to mass transfer. Energy conservation leads to

$$\rho^\nu \frac{\mathrm{D}_\nu U^\nu}{\mathrm{D}t} = \rho^\nu r^\nu + \rho(\psi^\nu + \tilde{\psi}^\nu) - \mathrm{div}\mathbf{q}^\nu + tr(\boldsymbol{\sigma}^\nu \boldsymbol{D}^\nu). \tag{9.13}$$

Summing (9.13) over all constituents and comparing with energy conservation for the mixture shows that, in the absence of interaction couples, eliminating $\boldsymbol{\sigma}^\nu$ by (9.5),

$$\begin{aligned} \sum_{\nu=1}^{n} \rho\psi^\nu &= -\sum_{\nu=1}^{n} \rho\boldsymbol{\beta}^\nu \cdot \boldsymbol{v}^\nu, \\ \sum_{\nu=1}^{n} \rho\tilde{\psi}^\nu &= -\sum_{\nu=1}^{n} \left[m^\nu(U^\nu + \tfrac{1}{2}\mathbf{v}^\nu \cdot \mathbf{v}^\nu) + \rho\tilde{\boldsymbol{\beta}}^\nu \cdot \boldsymbol{v}^\nu\right]. \end{aligned} \tag{9.14}$$

With a decomposition of $\boldsymbol{\beta}^\nu$ into binary drags $\boldsymbol{\beta}^{\nu\mu} = -\boldsymbol{\beta}^{\mu\nu}$ between ν and μ-constituents, and a postulate that ψ^ν has a common linear dependence on the working of each $\boldsymbol{\beta}^{\nu\mu}$, there is an explicit result

$$\psi^\nu = -\frac{1}{2}\sum_{\mu=1}^{n} \boldsymbol{\beta}^{\nu\mu} \cdot (\mathbf{v}^\nu - \mathbf{v}^\mu). \tag{9.15}$$

A postulate that $\tilde{\psi}^\nu$ has a common linear dependence on $m^{\nu\mu}$ and on the internal energy difference $(U^\nu - U^\mu)$, gives the explicit result

$$\rho\tilde{\psi}^\nu = -\frac{1}{2}\sum_{\mu=1}^{n} m^{\nu\mu}(U^\nu - U^\mu). \tag{9.16}$$

Note that if there is phase change ($m^{\nu\mu} \neq 0$) between specific ν and μ-constituents, so that $T^\nu = T^\mu$ is a prescribed function of the other variables (e.g. melting point as a function of pressure) then the corresponding two energy equations determine a necessary contribution to $\mathrm{div}\mathbf{q}^\nu$ and $\mathrm{div}\mathbf{q}^\mu$ and determine $\tilde{\psi}^\nu + \tilde{\psi}^\mu$, and hence the rate of mass transfer $m^{\nu\mu}$ by (9.16), when all other variables are determined. In general, constituents are allowed distinct temperatures T^ν.

The conservation laws must be complemented by constitutive laws describing the constituent properties and interactions. Relations are needed for the stresses $\boldsymbol{\sigma}^\nu$, internal energies U^ν, volume fractions ϕ^ν, and interaction drags $\boldsymbol{\beta}^\nu$, given that mass transfers are determined by the phase change relations so that $\tilde{\psi}^\nu$ and $\tilde{\boldsymbol{\beta}}^\nu$ are determined. By definition U^ν is an intrinsic quantity for constituent ν, but $\boldsymbol{\sigma}^\nu$ is a partial variable related to mixture cross-sections, while $\boldsymbol{\beta}^\nu$ is a mixture property describing the interaction of constituents. If $\phi_A^\nu(\mathbf{n})$ is the area fraction of ν-constituent in a cross-section with unit normal $\mathbf{n}$, then the intrinsic traction $t^{\nu\star}(\mathbf{n})$ and stress $\boldsymbol{\sigma}^{\nu\star}$ are given by

$$\mathbf{t}^\nu(\mathbf{n}) = \phi_A^\nu(\mathbf{n})\mathbf{t}^{\nu\star}(\mathbf{n}), \qquad \sigma^\nu = \phi_A^\nu(\mathbf{n})\sigma^{\nu\star}, \tag{9.17}$$

introducing additional variables ϕ^ν_A. The common assumption has been

$$\phi^\nu_A = \phi^\nu \,, \tag{9.18}$$

and there have been a variety of procedures to derive this result, with different assumptions, but essentially it prohibits a dependence of area fraction on orientation $\mathbf{n}$ which implies an isotropic (non-structured) mixing of the constitutents on the length scale of interest. The result (9.18) is necessary for the identification of $\mathbf{v}^\nu$ with the intrinsic velocity, implicit in the momentum and energy equation, and (9.18) can also be derived by assuming this identification - see MORLAND (1992b) for a full discussion. Adopting (9.18),

$$\sigma^\nu = \phi^\nu \sigma^{\nu\star} \,. \tag{9.19}$$

Since $\mathbf{v}(\mathbf{x}, t)$ is a differentiable field, a partial strain-rate $\mathbf{D}^\nu$ and partial spin $\mathbf{W}^\nu$ defined by

$$D^\nu_{ij} = \frac{1}{2}\left[\frac{\partial v^\nu_i}{\partial x_j} + \frac{\partial v^\nu_j}{\partial x_i}\right] \,, \qquad W_{ij} = \frac{1}{2}\left[\frac{\partial v^\nu_i}{\partial x_j} - \frac{\partial v^\nu_j}{\partial x_i}\right] \,, \tag{9.20}$$

have the usual interpretations for line element motions, but an intrinsic strain-rate $\mathbf{D}^{\nu\star}$ and spin $\mathbf{W}^{\nu\star}$ must be defined by their physical interpretations since actual ν-elements are not continuously distributed. Now $tr\mathbf{D}^{\nu\star}$ is the rate of increase of volume per unit volume of a given ν-element, which is given by

$$tr\mathbf{D}^{\nu\star} = tr\mathbf{D}^\nu + \frac{1}{\phi^\nu}\frac{\mathrm{D}_\nu}{\mathrm{D}t}(\phi^\nu) - \frac{m^\nu}{\rho^\nu} \,, \tag{9.21}$$

and a simple assumption to complete the definitions is

$$\mathbf{D}^{\nu\star} - \frac{1}{3} tr\mathbf{D}^{\nu\star}\mathbf{1} = \mathbf{D}^\nu - \frac{1}{3} tr\mathbf{D}^\nu \mathbf{1} \,, \qquad \mathbf{W}^{\nu\star} = \mathbf{W}^\nu \,. \tag{9.22}$$

This shearing assumption has more appeal for fluids than for a structured solid matrix. The corresponding deformation gradient relation is

$$\mathbf{F}^{\nu\star} = \left[\frac{J^{\nu\star}}{J^\nu}\right]^{\frac{1}{3}} \mathbf{F}^\nu \,, \tag{9.23}$$

where

$$J^{\nu\star} = \det \mathbf{F}^{\nu\star} \,, \qquad J^\nu = \det \mathbf{F}^\nu \,, \tag{9.24}$$

and

$$J^{\nu\star} = \frac{\phi^\nu J^\nu}{\phi^\nu_0} \exp\left\{ - \int^t \frac{m^\nu}{\rho^\nu} \mathrm{d}t \right\} \tag{9.25}$$

with the integral following a ν-particle path. Infinitesimal strain relations are an appropriate linear approximation. A constitutive theory must relate each intrinsic stress $\sigma^{\nu\star}$ to appropriate intrinsic deformation variables and temperature to reflect the constituent response, perhaps modified by the presence of other constitutents. The partial stress laws require also a prescription of the volume fractions ϕ^ν, which are mixture properties describing how the constituents share the space as they are stressed and heated, subject to

$$\sum_{\nu=1}^{n} \phi^\nu = 1 . \tag{9.26}$$

In addition, each U^ν and β^ν must be prescribed.

Before presenting a model for an ice-water mixture, the jump conditions across a singular surface imposed by mass, momentum and energy conervation will be stated, required at internal interfaces and non-material external boundaries. Let the singular surface $S^\star$ across which variables or their derivatives have a possible discontinuity, have a normal velocity $u_n \mathbf{n}$, where $\mathbf{n}$ is a unit normal to the surface and u_n is its speed of propagation in the $\mathbf{n}$ direction. Mass conservation requires the normal mass flux $m_n^\star$ to be continuous across $S^\star$:

$$m_n^\star = \{\rho^\nu(\mathbf{v}^\nu \cdot \mathbf{n} - u_n)\}_1 = \{\rho^\nu(\mathbf{v}^\nu \cdot \mathbf{n} - u_n)\}_2 \tag{9.27}$$

where the notation $\{\,\}_1$, $\{\,\}_2$ denotes evaluation on sides 1 and 2 of $S^\star$ respectively where the direction $\mathbf{n}$ is defined to point from side 1 to side 2. That is, side 2 is ahead of a moving $S^\star$ if $u_n > 0$. Linear momentum conservation requires

$$[\mathbf{t}^\nu] = m_n^\star[\mathbf{v}^\nu] \tag{9.28}$$

where $[\,]$ denotes the jump across $S^\star$, say $\{\,\}_2 - \{\,\}_1$, and energy conservation requires

$$m_n^\star[U^\nu + \frac{1}{2}\mathbf{v}^\nu \cdot \mathbf{v}^\nu] = [\mathbf{t}^\nu \cdot \mathbf{v}^\nu - \mathbf{q}^\nu \cdot \mathbf{n}] . \tag{9.29}$$

At a material surface, $m_n^\star = 0$,

$$\{\mathbf{v}^\nu \cdot \mathbf{n}\}_1 = \{\mathbf{v}^\nu \cdot \mathbf{n}\}_2 = u_n . \tag{9.30}$$

which is continuity of normal particle velocity, but the jumps in tangential velocity and density are arbitrary. Also the traction $\mathbf{t}^\nu$, both normal and tangential components, is continuous, but continuity of the normal energy flux $\mathbf{q}^\nu \cdot \mathbf{n}$ requires continuity of $\mathbf{v}^\nu$, both normal and tangential components.

At a non-material surface, $m_n^\star \neq 0$, $S^\star$ moves relative to particles on both sides, and any traction component is continuous if, and only if, the corresponding particle velocity component is continuous. A discontinuity in $\mathbf{t}^\nu \cdot \mathbf{n}$ and $\mathbf{v}^\nu \cdot \mathbf{n}$ is called a shock wave, and a discontinuity in the tangential particle velocity is called a vortex sheet. The energy balance (9.29) then determines the normal flux jump $[\mathbf{q}^\nu \cdot \mathbf{n}]$ in terms of the internal energy jump $[U^\nu]$, which at a phase change front can be related to the latent heat L. The Stefan condition which neglects the density jump, and the kinetic energy and stress working jumps compared to the internal energy jump, is

$$[\mathbf{q}^\nu \cdot \mathbf{n}] = \pm\rho^\nu L[u_n - \mathbf{v}^\nu \cdot \mathbf{n}] \tag{9.31}$$

where the $\pm$ choice corresponds to $[U^\nu]\gtrless 0$, and a common further simplification is to neglect $\mathbf{v}^\nu$ compared to u_n. The latter is appropriate to a freezing or melting front advancing with significant speed, but not for a near-stationary interface.

We can now focus on a binary mixture of ice and water co-existing at a common melt temperature $T^i = T^w = T_M(p^{w\star})$, where $p^{w\star}$ is the intrinsic water pressure (pore pressure). Let n be the porosity, so

$$\phi^w = n\,, \qquad \phi^i = (1-n)\,, \tag{9.32}$$

and let

$$\beta^i = \beta\,, \quad \beta^w = -\beta\,, \quad m^w = m\,, \quad m^i = -m\,. \tag{9.33}$$

Assuming both ice and water are (intrinsically) incompressible with constant densities $\rho^{i\star} = \rho_i$, $\rho^{w\star} = \rho_w \quad (\rho_i < \rho_w)$, then

$$\rho^i = (1-n)\rho_i\,, \quad \rho^w = n\rho_w\,, \tag{9.34}$$

and $tr\mathbf{D}^{\mathrm{i}\star} = tr\mathbf{D}^{\mathrm{w}\star} = 0$, so by (9.21):

$$\begin{aligned} \mathrm{div}\mathbf{v}^{\mathrm{i}} &= tr\mathbf{D}^{\mathrm{i}} = \frac{1}{1-n}\frac{\mathrm{D}_i n}{\mathrm{D}t} - \frac{m}{(1-)\rho_i} \\ \mathrm{div}\mathbf{v}^{\mathrm{w}} &= tr\mathbf{D}^{\mathrm{w}} = -\frac{1}{n}\frac{\mathrm{D}_w n}{\mathrm{D}t} + \frac{m}{n\rho_w} \end{aligned} \tag{9.35}$$

which also follow from the mass conservation laws (9.3). We can suppose that the ice rate factor $a(T)$ remains at the constant value a_0 since T_M remains close to the melt point at zero pressure, but the ice viscosity may now have a moisture dependent rate factor $A(n)$, though there is no firm form of dependence (HUTTER, 1992). Note that HUTTER (1992) uses a moisture content

$$w = \frac{n\rho_w}{n\rho_w + (1-n)\rho_i} = \frac{\rho^w}{\rho}\,, \tag{9.36}$$

the water mass fraction, in preference to the volume fraction n for much of his model formulation. The partial stress law (3.8) for the ice becomes

$$(1-n)\left[\mathbf{D}^{\mathrm{i}} - \frac{1}{3}tr\mathbf{D}^{\mathrm{i}}\mathbf{1}\right] = A\psi(J_2^\star)\left[\sigma^i + p^i\mathbf{1}\right] \tag{9.37}$$

where $p^i = (1-n)p^{i\star}$ is an arbitrary partial ice pressure, $J_2^\star$ is the second invariant of the intrinsic deviatoric ice stress, and $tr\mathbf{D}^{\mathrm{i}}$ is given by (9.35). Similarly, assuming the water is an incompressible linearly viscous fluid with constant viscosity $\mu = 4 \times 10^{-4}\ Pa\,s$,

$$2\mu n[\mathbf{D}^{\mathrm{w}} - \frac{1}{3}tr\mathbf{D}^{\mathrm{w}}\mathbf{1}] = [\sigma^w + p^w\mathbf{1}] \tag{9.38}$$

where $p^w = np^{w\star}$ is an arbitrary partial water pressure and $tr\mathbf{D}^{\mathrm{w}}$ is given by (9.35). The stress laws are not complete until the variation of the mixture property n is related to other variables.

The conventional porous rock or soil theory for infinitesimal strain is based on a linear dependence of $n - n_0$ on the matrix and fluid pressures. Applying this to incremental changes of n in the absence of phase change, and taking the incompressibility limits for both constitutents, leads to

$$\frac{\mathrm{D}_i n}{\mathrm{D}t} = \kappa \frac{1-n}{n} \left\{ (1-n) \frac{\mathrm{D}_i p^w}{\mathrm{D}t} - n \frac{\mathrm{D}_i p^w}{\mathrm{D}t} \right\} \tag{9.39}$$

where changes are interpreted relative to the ice matrix, and where κ is a free draining ($p^w = 0$) compressibility. Alternatively, a viscous type response could be adopted:

$$\frac{\mathrm{D}_i n}{\mathrm{D}t} = \frac{1-n}{n\zeta} \left\{ (1-n) p^w - n p^i \right\} = -\frac{1-n}{\zeta} \left\{ p - p^{w\star} \right\}, \tag{9.40}$$

where $p = p^w + p^i$ is the total pressure and $p - p^{w\star} = \bar{p}$ is known as the effective pressure, so ζ can be interpreted as a bulk viscosity for the mixture. I am not aware of evidence to support either model for wet ice, nor what magnitudes would be appropriate for κ or ζ in trial models. κ or $1/\zeta = 0$, leaving n unchanged by stressing, is the simple case. Phase change does influence the volume fractions, and analysing the effects on an element of ice matrix, in the absence of any stress change induced variation, shows that

$$\frac{\mathrm{D}_i n}{\mathrm{D}t} = m \left\{ \frac{n}{\rho_i} + \frac{1-n}{\rho_w} \right\} = \frac{m\rho}{\rho_i \rho_w}. \tag{9.41}$$

In general, a stress induced contribution, such as (9.39) or (9.40), is added to (9.41), but (9.41) alone offers a law for n which supposes that phase change is the dominant effect. In the case of snowpacks there is significant compaction due to stressing. The mass conservation balances in the form (9.35), in conjunction with (9.41), give the explicit expressions

$$\operatorname{div} \mathbf{v}^{\mathrm{i}} = -m \frac{\rho_w - \rho_i}{\rho_w \rho_i}, \quad \operatorname{div} \mathbf{v}^{\mathrm{w}} = -m \frac{\rho_w - \rho_i}{\rho_w \rho_i} - \frac{\mathbf{v}^{\mathrm{w}} - \mathbf{v}^{\mathrm{i}}}{n} \cdot \operatorname{grad} n, \tag{9.42}$$

which are required in the stress laws (9.37), (9.38), while (9.42) replace the two mass conservation laws if (9.41) is the law for n. If there is a further contribution to (9.41), then there are further contributions to (9.42) following from (9.35), which could involve stress rates or stress, but always these are three equations relating $n, \mathbf{v}^{\mathrm{i}}, \mathbf{v}^{\mathrm{w}}$ and possibly stresses.

For the very low ice and water velocities the acceleration terms are negligible, and the external body force $\mathbf{b}$ is gravity $\mathbf{g}$, so (9.5) gives the equilibrium equations

$$\left. \begin{aligned} -\frac{\partial p^i}{\partial x_i} + \frac{\partial \sigma^{i'}_{ij}}{\partial x_j} + \rho^i g_i + \rho\beta_i + \rho\tilde{\beta}^i_i &= 0 \\ -\frac{\partial p^w}{\partial x_i} + \frac{\partial \sigma^{w'}_{ij}}{\partial x_j} + \rho^w g_i - \rho\beta_i + \rho\tilde{\beta}^w_i &= 0 \end{aligned} \right\} \tag{9.43}$$

where $\boldsymbol{\sigma}^{i'}$ and $\boldsymbol{\sigma}^{w'}$ are the deviatoric stresses given by (9.37), (9.38). The thrusts $\tilde{\boldsymbol{\beta}}^i, \tilde{\boldsymbol{\beta}}^w$ are given by (9.10), and it remains to prescribe the interaction drag $\boldsymbol{\beta}$ $(= \boldsymbol{\beta}^i = -\boldsymbol{\beta}^w)$.

Darcy's experiment for steady, slow axial flow of linearly viscous fluid through a porous solid of uniform porosity, driven by a constant intrinsic fluid pressure gradient, leads to an empirical linear relation between fluid flux and intrinsic pressure gradient. Comparison with the equilibrium equation (9.5) for the water implies that

$$\rho\beta = \frac{\mu n^2}{k}(\mathbf{v^w} - \mathbf{v^i}) - \frac{p^w}{n}\operatorname{grad} n \tag{9.44}$$

where μ is the water viscosity ($= 4 \times 10^{-4}\,Pas$) and where k is the permeability of the porous matrix (with unit: length squared). For ground materials k ranges from $10^{-7} \to 10^{-16}\,m^2$, but I am not aware of an established ice range, which must depend on the porosity. Note the dependence on $\operatorname{grad} n$ which would induce a positive seepage velocity $\mathbf{v^w} - \mathbf{v^i}$ in the direction of increasing porosity if the partial water pressure p^w is uniform; Darcy's result has been generalised to imply zero flux at zero intrinsic pressure gradient independent of porosity variation. The alternative interpretation would lead to non-zero intrinsic pressure gradient in a stationary fluid if porosity is non-uniform. While the porosity gradient term is usually negligible compared to pressure gradient in typical soil mechanics applications, it is possible here that phase change variation of n could be more significant. The distinct dispersion of two miscible fluids (common intrinsic pressure) within a porous matrix due to concentration gradient is postulated directly, necessarily ignoring momentum conservation (9.5), by BEAR (1972), but can be described by an appropriate additional contribution to β^ν (MORLAND, 1992b) to reflect interaction forces at the molecular scale. This is not relevant to water in ice pores. Alternative mechanisms to Darcy drag may exist, but we have no basis for a different model.

It is instructive to summarise the mechanical problem. There are two mass conservation equations (9.35), or (9.42) for the particular porosity law (9.41), and six equilibrium equations (9.43), with a porosity law such as (9.41), and ten deviatoric stress laws (9.37), (9.38), giving nineteen equations for the nineteen variables $n, \mathbf{v^i}, \mathbf{v^w}, \rho^i, \rho^w, \sigma^{i'}, \sigma^{w'}$. They involve the drag given by (9.44), and the thrusts

$$\rho\tilde{\beta}^i = \rho\tilde{\beta}^w = -\frac{1}{2}m(\mathbf{v^w} - \mathbf{v^i}), \tag{9.45}$$

given by (9.10), and the rate of mass transfer m which will be determined from energy conservation.

We now assume simple internal energy functions

$$U^w = C_w T, \qquad U^i = C_i T \tag{9.46}$$

where C_w and C_i are the constant specific heats at constant density, both approximately $2 \times 10^3\,J\,kg^{-1}K^{-1}$, and at the melt point $T = T_M$,

$$U^w - U^i = (C_w - C_i)T_M = L + \frac{p^{w*}}{\rho_w} - \frac{p^{i*}}{\rho_i} \tag{9.47}$$

where $L = 3{\cdot}3\times10^5\, J\, kg^{-1}$ is the latent heat. With the prescription of constant C_w, C_i, L, this determines $T_M(p^{w*}, p^{i*})$ when $p^{i*} \neq p^{w*}$ is allowed in the distinct ice and water motions, but it is common to prescribe an empirical $T_M(p^{w*})$ and ignore the central expression of (9.47). By (9.16)

$$\rho\tilde{\psi}^i = \rho\tilde{\psi}^w = -\frac{1}{2}m\left[L + \frac{p^{w*}}{\rho_w} - \frac{p^{i*}}{\rho_i}\right], \tag{9.48}$$

and by (9.15)

$$\psi^i = \psi^w = \frac{1}{2}\beta\cdot(\mathbf{v^w} - \mathbf{v^i}). \tag{9.49}$$

We will suppose there is no significant radiation deposit r^i, r^w (usually a near-surface effect), so the remaining contributions to the energy conservation (9.13) are the energy fluxes $\mathbf{q^i}, \mathbf{q^w}$. Recall that these incorporate flux within the constituent and across constituent interfaces, where the latter could be transferred to ψ^i, ψ^w, or vice versa. Since the earlier prescriptions limited ψ^ν to drag-working, we must allow a transfer term in addition to the internal heat conduction. Thus

$$\mathbf{q^i} = -(1-n)K_i \operatorname{grad} T + \tilde{\mathbf{q}}, \quad \mathbf{q^w} = -nK_w \operatorname{grad} T - \tilde{\mathbf{q}} \tag{9.50}$$

where K_i, K_w are the thermal conductivities of ice and water, both approximately $2\, J K^{-1}\, m^{-1}\, s^{-1}$.

Differencing and summing the two energy conservation equations (9.13) shows that

$$\rho^i C_i \frac{\mathrm{D}_i T}{\mathrm{D}t} \mp \rho^w C_w \frac{\mathrm{D}_w T}{\mathrm{D}t} - [(1-n)K_i \mp nK_w]\nabla^2 T - tr(\sigma^i \mathbf{D^i}) \mp tr(\sigma^w \mathbf{D^w})$$

$$= \begin{cases} -2\operatorname{div}\tilde{\mathbf{q}} \\ \rho\beta\cdot(\mathbf{v^w} - \mathbf{v^i}) - m\left[L + \dfrac{p^{w*}}{\rho_w} - \dfrac{p^{i*}}{\rho_i}\right], \end{cases} \tag{9.51}$$

which are two independent equations for T, the first involving the unknown $\tilde{\mathbf{q}}$, the second involving the unknown m. In the case of inert constitutents ($m = 0$) with common temperature T, a typical approximation for slow flows, the second is the mixture equation for T, and the first then implies non-zero $\operatorname{div}\tilde{\mathbf{q}}$ to maintain the common temperature. With phase change at prescribed

$$T = T_M(p^{w*}), \tag{9.52}$$

the first again determines the necessary $\operatorname{div}\tilde{\mathbf{q}}$, and the second determines the required mass transfer m to complete the thermomechanical system.

Since the water viscosity μ is negligible compared to ice viscosity, the deviatoric water stresses $\sigma^{w'}$ can be neglected compared to the deviatoric ice stress $\sigma^{i'}$, which is therefore the total stress, and the five relations (9.38) are redundant. The equilibrium equations (9.43) are now conveniently written in terms of total stress and water pressure, thus

$$\left.\begin{aligned} -\frac{\partial p}{\partial x_i} + \frac{\partial \sigma'_{ij}}{\partial x_j} + \rho g_i - m(v_i^w - v_i^i) = 0\,, \\ -\frac{\partial p^w}{\partial x_i} + \rho^w g_i - \frac{\mu n^2}{k}(v_i^w - v_i^i) + \frac{p^w}{n}\mathrm{grad}\, n - \tfrac{1}{2}m(v_i^w - v_i^i) = 0\,, \end{aligned}\right\} \tag{9.53}$$

where σ'_{ij} is given by the ice relation (9.37). The first equation is that of the cold ice therefore with the addition of the body thrust $m(\mathbf{v}^{\mathbf{w}} - \mathbf{v}^{\mathbf{i}})$, but recall that $tr\mathbf{D}^{\mathbf{i}}$ in relation (9.37) for $\sigma^{i'}$ also depends on m by (9.42). We need to estimate the possible magnitude of m from (9.51), and its contribution to the porosity variation (9.41), velocity divergences (9.42), and thrust (9.45), and the consequent coupling of the two equilibria (9.53).

Estimates of some terms in (9.51) require an estimate of $[\mathbf{v}] = [\mathbf{v}^{\mathbf{w}} - \mathbf{v}^{\mathbf{i}}]$, both vertical and horizontal components, which in turn needs a magnitude for k. Note that an impermeable limit $k \to 0$, describing for example non-connected pores, $\mathbf{v}^{\mathbf{w}} \to \mathbf{v}^{\mathbf{i}}$, requires a non-zero drag

$$\rho\boldsymbol{\beta} = \lim \frac{\mu n^2}{k}[\mathbf{v}^{\mathbf{w}} - \mathbf{v}^{\mathbf{i}}] \tag{9.54}$$

in (9.53) to maintain $\mathbf{v}^{\mathbf{w}} = \mathbf{v}^{\mathbf{i}}, \boldsymbol{\beta}$ replacing the known $\mathbf{v}^{\mathbf{w}}$ as the vector variable to be determined along with p^w. The coupling still exists since m, and hence n and div $\mathbf{v}^{\mathbf{i}}$, depend on p^w. For preliminary estimates suppose k has the earth material range $10^{-7} \to 10^{-16}\,m^2$, depending on n, then since vertical gradients of pressure and n will far exceed horizontal gradients (assuming the reduced model co-ordinate scaling applies), supposing that the drag is not dominated by the porosity gradient and that $\partial p^w/\partial x_3$ does not exceed $\rho^w g$, $(9.53)_2$ with (9.44) imply that

$$|\rho\beta_3| \lesssim 10^4\,Pa\,m^{-1}\,, \qquad |v_3^w - v_3^i| \lesssim 10 \to 10^{-8}\,ms^{-1} \tag{9.55}$$

where the minimum seepage of $1\,m\,yr^{-1}$ is similar to a vertical ice velocity. The high permeability result is not reasonable so consider a more likely upper limit $|\mathbf{v}^{\mathbf{w}}| \sim 1000\,|v^i| \sim 10^{-5}\,ms^{-1}$, corresponding to $k \sim 10^{-13}\,m^2$. Turning to the energy balance (9.51) with $\sigma^w = -p^w\mathbf{1}$, the drag working can therefore have a magnitude range

$$|\rho\boldsymbol{\beta}\cdot(\mathbf{v}^{\mathbf{w}} - \mathbf{v}^{\mathbf{i}})| \lesssim 10^5 \to 10^{-1} \to 10^{-4}\,Pa\,s^{-1}\,, \tag{9.56}$$

showing estimates for the three proposed k limits. Based on the cold ice results, with $||\sigma^{i'}|| \lesssim 10^5\,Pa$, the ice shear stress working has a limit

$$tr(\sigma^{i'}\,D^{i'}) \lesssim 10^{-2}\,Pa\,s^{-1}\,. \tag{9.57}$$

From (9.55) and (9.42) using a vertical length scale $d = 1000\,m$, and allowing a relative n change not greater than unity,

$$|\mathrm{div}\,\mathbf{v}^{\mathbf{w}}| \sim |\mathrm{div}\,\mathbf{v}^{\mathbf{i}}| \lesssim 10^{-2} \to 10^{-8} \to 10^{-11}\,s^{-1}\,, \tag{9.58}$$

so the pressure working range is

$$p^w \operatorname{div} \mathbf{v}^{\mathbf{w}} \sim p^i \operatorname{div} \mathbf{v}^{\mathbf{i}} \lesssim 10^5 \rightarrow 10^{-1} \rightarrow 10^{-4} \ Pa\,s^{-1} . \tag{9.59}$$

However, in the energy balance (9.51), the term $mL \sim 10^5 \tau \, kg^{-1} \times m$, while from (9.42)

$$p^i \operatorname{div} \mathbf{v}^{\mathbf{i}} \sim p^w \operatorname{div} \mathbf{v}^{\mathbf{w}} \sim 10^3 \tau \, kg^{-1} \, m \tag{9.60}$$

for $p^i \sim p^w \sim 10^7 \, Pa$, the pressure at depth, suggesting that the pressure working is not a dominant contribution in the balance.

Moving to the thermal terms, and adopting $\mathrm{d}T_M/\mathrm{d}p^{w\star} \sim 10^{-7} \, K \, Pa^{-1}$ (HUTTER, 1983), the vertical conduction is

$$K \frac{\partial^2 T}{\partial x_3^2} \sim 10^{-7} \, Pa\,s^{-1} , \tag{9.61}$$

while the advection in the water is

$$\rho C_w v_3^w \frac{\partial T}{\partial x_3} \sim 10^4 \rightarrow 10^{-2} \rightarrow 10^{-5} \, Pa\,s^{-1} , \tag{9.62}$$

and local heating in water and ice, $\rho C \, \partial T / \partial t$, should not exceed this. The dominant terms, then, are the drag working (9.56), the shear stress working in the ice (9.57), local heating, and advection in the water (9.62). The thermal contributions are negligible if $k \ll 10^{-13} \, m^2$. Since $p^{w\star}/\rho_w$ and $p^{i\star}/\rho_i \lesssim 10^4 \, Pa \, m^3 \, kg^{-1} = 10^4 \, J \, kg^{-1}$ and $L \sim 3 \times 10^5 \, J \, kg^{-1}$, the energy production magnitude (9.57) suggests that

$$m \lesssim 10^{-7} \, kg \, m^{-3} \, s^{-1} . \tag{9.63}$$

In turn, the porosity change due to mass transfer, (9.41), will not exceed 3×10^{-3} in a year, but for maintained warming over hundreds of years this change of porosity cannot be neglected. Within an annual cycle a constant porosity model may be satisfactory; that is, constant for each ice element, but possibly non-homogeneous through the ice. The contributions to $\operatorname{div} \mathbf{v}^{\mathbf{i}}, \operatorname{div} \mathbf{v}^{\mathbf{w}}$, (9.42), do not exceed $10^{-11} \, s^{-1}$, which are small compared to maximum shear strain rates in the ice of order $10^{-8} \, s^{-1}$, so approximations $\operatorname{div} \mathbf{v}^{\mathbf{i}} = \operatorname{div} \mathbf{v}^{\mathbf{w}} = 0$ are justified. Finally, comparing the thrust (9.45) with the drag (9.44), for each component

$$\frac{|\tilde{\beta}_i|}{|\beta_i|} \sim \frac{km}{2\mu n^2} , \lesssim \frac{10^{-4} k}{n^2} , \tag{9.64}$$

when the limit (9.63) is applied and k has unit m^2. Thus, for the entire range of possible k, even at very low porosity, the thrust can be neglected compared to the drag.

On the basis of these magnitude estimates, the model equations become

$$\operatorname{div} \mathbf{v}^{\mathbf{i}} = \operatorname{div} \mathbf{v}^{\mathbf{w}} = 0 , \tag{9.65}$$

$$-\frac{\partial p}{\partial x_i} + \frac{\partial \sigma'_{ij}}{\partial x_j} + \rho g_i = 0 , \tag{9.66}$$

$$-\frac{\partial p^w}{\partial x_i} + \rho^w g_i - \frac{\mu n^2}{k}(v_i^w - v_i^i) + \frac{\rho^w}{n}\frac{\partial n}{\partial x_i} = 0\,, \tag{9.67}$$

$$(1-n)D_{ij}^i = A\psi(J_2^\star)\sigma'_{ij} \tag{9.68}$$

$$\frac{\mathrm{D}_i n}{\mathrm{D}t} = \frac{m\rho}{\rho_i \rho_w}\,, \tag{9.69}$$

$$m\left[L + \frac{p^w}{n\rho_w} - \frac{p^i}{(1-n)\rho_i}\right] = \frac{\mu n^2}{k}(\mathbf{v}^\mathbf{w} - \mathbf{v}^\mathbf{i})^2 + tr(\sigma'\mathbf{D}^\mathbf{i}) - (1-n)\rho_i C_i \frac{\mathrm{D}_i T}{\mathrm{D}t} - n\rho_w C_w \frac{\mathrm{D}_w T}{\mathrm{D}t}\,, \tag{9.70}$$

where

$$D_{ij}^i = \frac{1}{2}\left[\frac{\partial v_i^i}{\partial x_j} + \frac{\partial v_j^i}{\partial x_i}\right], \qquad J_2^\star = \frac{1}{2}(1-n)^{-2}tr([\sigma']^2)\,, \tag{9.71}$$

$$\rho = (1-n)\rho_i + n\rho_w\,, \qquad \frac{\mathrm{D}_i}{\mathrm{D}t} = \frac{\partial}{\partial t} + v_j^i\frac{\partial}{\partial x_j}\,, \quad \frac{\mathrm{D}_w}{\mathrm{D}t} = \frac{\partial}{\partial t} + v_j^w\frac{\partial}{\partial x_j}\,, \tag{9.72}$$

$$T = T_M\left(\frac{p^w}{1-n}\right). \tag{9.73}$$

The flow variables are $\mathbf{v}^\mathbf{i}, \mathbf{v}^\mathbf{w}, \mathbf{D}^\mathbf{i}, p, \sigma', p^w, n, m, T$, twenty two in all, to be determined by the above twenty two equations. However, the equations (9.65) , (9.66), (9.68), (9.71) are analogous to the cold ice isothermal mechanical equations for $\mathbf{v}^\mathbf{i}, p, \sigma', \mathbf{D}^\mathbf{i}$, if n is known. Coupling with the water flow is only through n, and the ice pressure $p^{i^\star} = (p - p^w)/(1-n)$ is given only when n and p^w are determined. The reduced cold ice model based on the co-ordinate stretchings $\bar{x}_i = \epsilon^{-1}X_i \quad (i = 1,2)\ (\epsilon \ll 1)$ will apply, so the same explicit integrations for ρ, σ' follow, but not for $\mathbf{v}^\mathbf{i}$ because of $J_2^\star$, and boundary conditions concern matching with adjacent cold ice regions as well as external conditions. A further simplification

$$n \equiv n_0(\chi^i) \tag{9.74}$$

appears reasonable for annual cycles (not extended warming or cooling) where χ^i is a reference label for an ice particle; that is, n is prescribed. Then the ice equations including $\mathbf{v}^\mathbf{i}$ can be integrated independent of the water flow, but interface conditions still impose the coupling.

Melt zone problems are an area of theoretical glaciology research needing a substantial research effort.

References

BEAR, J. *Dynamics of Fluids in Porous Media.* Elsevier, New York (1972).

BENTLEY, C. R. *Antarctic ice streams: a review.* J. Geophys. Res. (1987) 92, 8843 - 8851.

BLATTER, H. and HUTTER, K. *Polythermal conditions in Arctic glaciers.* J. Glaciology (1991) 37, 261 - 269.

BODVARDSSON, G. *On the flow of ice sheets and glaciers.* Jökull (1955) 5, 1 - 8.

BOULTON, G. S., SMITH, G. D. and MORLAND, L. W. *The reconstruction of former ice sheets and their mass balance characteristics using a non-linearly viscous model.* J. Glaciology (1984) 30, 140 - 152.

BUDD, W. F. *The Dynamics of Ice Masses.* Australian Nat. Antarctic Res. Expeds Rep. A (IV) No 108 (1969).

BUDD, W. F. *Ice flow over bedrock perturbations.* J. Glaciology (1970) 9, 29 - 48.

BUDD, W. F. *Stress variations with ice flow over undulations.* J. Glaciology (1971) 10, 177 - 195.

COLLINS, I. F. *On the use of the equilibrium equation and flow law in relating the surface and bed topography of glaciers and ice sheets.* J. Glaciology (1968) 7, 199 - 204.

FOWLER, A. C. *Glacier Dynamics.* D. Phil. Thesis (1977) Oxford University.

FOWLER, A. C. *A mathematical approach to the theory of glacier sliding.* J. Glaciology (1979a) 23, 131 - 141.

FOWLER, A. C. *The use of a rational model in the mathematical analysis of a polythermal glacier.* J. Glaciology (1979b) 24, 443 - 456.

FOWLER, A. C. *A theoretical treatment of the sliding of glaciers in the absence of cavitation.* Phil. Trans. Roy. Soc. (1981) 298, 637 - 685.

FOWLER, A. C. *On the transport of moisture in polythermal glaciers.* Geophys. Astrophys. Fluid Dynamics (1984) 28, 99 - 140.

FOWLER, A. C. *Modelling ice sheet dynamics.* Geophys. Astrophys. Fluid Dynamics (1991) to appear.

FOWLER, A. C. and LARSON, D. A. *On the flow of polythermal glaciers. I.* Model and preliminary analysis. Proc. Roy. Soc. A. (1978) 363, 217 - 242.

GLEN, J. W. *The creep of polycrystalline ice.* Proc. Roy. Soc. A (1955) 228, 515 - 538.

GLEN, J. W. *The flow law of ice. A discussion of the assumptions made in glacier theory, their experimental foundations and consequences.* Proc. IASH Symp. 47, Chamonix (1958), 171 - 183.

HERTERICH, K. *On the flow within the transition zone between ice sheet and ice shelf.* Proc. Workshop Dynamics of The West Antarctic Ice Sheet, Utrecht 1985. Ed. Van der Veen, C. J. and Oerlemans, J. Reidel, Dordrecht (1987), 185 - 202.

HERTERICH, K. *A three-dimensional model of the Antarctic Ice Sheet.* Annals of Glaciology (1988) 11, 32 - 35.

HINDMARSH, R. C. A. *Modelling the dynamics of ice sheets.* Prog. Phys. Geog. (1992 submitted).

HINDMARSH, R. C. A., MORLAND, L. W., BOULTON, G. S. and HUTTER, K. *The unsteady plane flow of ice sheets: a parabolic problem with two moving boundaries.* Geophys. Astrophys. Fluid Dynamics (1987), 39, 183 - 225.

HINDMARSH, R. C. A., BOULTON, G. S. and HUTTER, K. *Modes of operation of thermo-mechanically coupled ice sheets.* Annals of Glaciology (1989) 12, 57 - 69.

HINDMARSH, R. C. A. and HUTTER, K. *Numerical fixed domain mapping solution of free-surface flows coupled with an evolving interior field.* Int. J. Num. Anal. Meth. Geomech. (1988) 12, 437 - 459.

HUGHES, T. J. *Is the West Antarctic Ice Sheet distintegrating?* J. Geophys. Res. (1973) 78, 7884 - 7910.

HUGHES, T. J. *The West Antarctic Ice Sheet: instability, disintegration and initiation of ice ages.* Rev. Geophys. Space Phys. (1975) 13, 502 - 526.

HUTTER, K. *The effect of longitudinal strain on the shear stress of an ice sheet: in defence of using stretched co-ordinates.* J. Glaciology (1981) 27, 39 - 56.

HUTTER, K. *Dynamics of glaciers and large ice masses.* Ann. Rev. Fluid Mech. (1982a) 14, 87 - 130.

HUTTER, K. *A mathematical model of polythermal glaciers and ice sheets.* Geophys. Astrophys. Fluid Dynamics (1982b) 21, 201 - 224.

HUTTER, K. *Theoretical Glaciology*, Reidel, Dordrecht (1983).

HUTTER, K. *Thermo-mechanically coupled ice sheet response.* Cold, polythermal, temperate. J. Glaciology in press (1993).

HUTTER, K., BLATTER, H. and FUNK, M. *A model computation of moisture content in polythermal glaciers.* J. Geophys. Res. (1988) 93, 12205 - 12214.

HUTTER, K., LEGERER, F. and SPRING, U. *First order stresses and deformation in glaciers and ice sheets.* J. Glaciology (1981) 27, 227 - 270.

HUTTER, K. , YAKOWITZ, S. and SZIDAROVSKY, F. *A numerical study of plane ice sheet flow.* J. Glaciology (1986) 32, 139 - 160.

HUTTER, K., YAKOWITZ, S. and SZIDAROVSKY, F. *Coupled thermomechanical response of an axisymmetric cold ice sheet.* Water Resources Res. (1987) 23, 1327 - 1339.

HUYBRECHTS, P. *The Antarctic Ice Sheet and environmental change: a three-dimensional modelling study.* Reports on Polar Research 99, Alfred Wegener Institute for Polar and Marine Research (1992).

JOHNSON, R. E. and MCMEEKING, R. M. *Near-surface flow in glaciers obeying Glen's law.* Quart. J. Mech. Appl. Math. (1984) 37, 273 - 291.

KAMB, W. B. *Sliding motion of glaciers: theory and observation.* Rev. Geophys. Space Phys. (1970) 8, 673 - 728.

LLIBOUTRY, L. A. *General theory of subglacial cavitation and sliding of temperate glaciers.* J. Glaciology (1968) 7, 21 - 58.

MACAYEAL, D. R. *Ice shelf back pressure: form drag versus dynamics drag.* Proc. Workshop Dynamics of The West Antarctic Ice Sheet, Utrecht 1985. Ed. Van der Veen, C. J. and Oerlemans, J. Reidel, Dordrecht (1987), 141 - 160.

MACAYEAL, D. R. and BARCILON, V. *Ice shelf response to ice stream discharge fluctuations, I Unconfined ice tongues.* J. Glaciology (1988) 34, 121 - 127.

MACAYEAL, D. R. and LANGE, M. A. *Ice shelf response to ice stream discharge fluctuations, II Ideal rectangular basin.* J. Glaciology (1988) 34, 128 - 135.

MELLOR, M. and TESTA, R. *Effect of temperature on the creep of ice.* J. Glaciology (1969) 8, 131 - 145.

MELLOR, M. *Mechanical properties of polycrystalline ice.* Proc. IUTAM Symp. Physics and Mechanics of Ice, Copenhagen 1979. Ed. Tryde, P. Springer-Verlag, Berlin (1980), 217 - 245.

MERCER, J. H. *West Antarctic Ice Sheet and CO_2 greenhouse effect: a threat of disaster.* Nature (1978) 271, 321 - 325.

MORLAND, L. W. *Glacier sliding down an inclined wavy bed.* J. Glaciology (1976a) 17, 447 - 462.

MORLAND, L. W. *Glacier sliding down an inclined wavy bed with friction.* J. Glaciology (1976b) 17, 463 - 477.

MORLAND, L. W. *Constitutive laws for ice.* Cold Reg. Sci. Tech. (1979) 1, 101 - 108.

MORLAND, L. W. *Thermomechanical balances of ice sheet flows.* Geophys. Astrophys. Fluid Dynamics (1984) 29, 237 - 266.

MORLAND, L. W. *Unconfined ice sheet flow.* Proc. Workshop Dynamics of The West Angarctic Ice Sheet, Utrecht 1985. Ed. Van der Veen, C. J. and Oerlemans, J. Reidel, Dordrecht (1987), 99 - 116.

MORLAND, L. W. *Non-linear viscoelastic response of ice.* Applied Oceans Research (1992a) 13, 31 - 38.

MORLAND, L. W. *Flow of viscous fluids through a porous deformable matrix.* Surveys in Geophys. (1992b) 13, 209-268.

MORLAND, L. W. and JOHNSON, I. R. *Steady motion of ice sheets.* J. Glaciology (1980) 25, 229 - 246.

MORLAND, L. W. and JOHNSON, I. R. *Effects of bed inclination and topography on steady isothermal ice sheets.* J. Glaciology (1982) 28, 71 - 90.

MORLAND, L. W. and SHOEMAKER, E. M. *Ice shelf balances.* Cold Reg. Sci. Tech. (1982) 5, 235 - 251.

MORLAND, L. W. and SMITH, G. D. *Influence of non-uniform temperature distribution on the steady motion of ice sheets.* J. Fluid Mech. (1984) 140, 113 - 133.

MORLAND, L. W., SMITH, G. D. and BOULTON, G. S . *Basal sliding relations deduced from ice sheet data.* J. Glaciology (1984) 30, 131 - 139.

MORLAND, L. W. and SPRING, U. *Viscoelastic fluid relation for the deformation of ice.* Cold Reg. Sci. Tech. (1981) 4, 225 - 268.

MORLAND, L. W. and SPRING, U. *Single integral representations for non-linear viscoelastic solids.* Mech. Mat. (1982) 1, 161 - 170.

MORLAND, L. W. and ZAINUDDIN, R. *Plane and radial ice-shelf flow.* Proc. Workshop Dynamics of The West Antarctic Ice Sheet, Utrecht 1985. Ed. Van der Veen, C. J. and Oerlemans, J. Reidel, Dordrecht (1987), 117 - 140.

NYE, J. F. *The flow law of ice from measurements in glacier tunnels, laboratory experiments and the Jungfrau firn borehole experiments.* Proc. Roy. Soc. A (1953) 219, 477 - 489.

NYE, J. F. *The motion of ice sheets and glaciers.* J. Glaciology (1959) 3, 493 - 507.

NYE, J. F. *A calculation on the sliding of ice over a wavy surface using a Newtonian viscous approximation.* Proc. Roy. Soc. A (1969a) 311, 445 - 467.

NYE, J. F. *The effect of longitudinal stress on the shear stress at the base of an ice sheet.* J. Glaciology (1969b) 8, 207 - 213.

NYE, J. F. *Glacier sliding without cavitation in a linear viscous approximation.* Proc. Roy. Soc. A (1970) 315, 381 - 403.

OERLEMANS, J. *Some basic experiments with a vertically integrated ice sheet model.* Tellus (1981) 33, 1 - 11.

PATERSON, W. S. B. *The Physics of Glaciers (2nd ed.).* Pergamon, Oxford (1969).

REEH, N. *A flow-line model for calculating the surface profile and velocity, strain-rate, and stress fields in an ice sheet.* J. Glaciology (1988) 34, 46 - 54.

ROBIN, G. DE Q. *Surface topography of ice sheets.* Nature (1967) 215, 1029 - 1032.

ROBIN, G. DE Q. *Formation, flow, and disintegrations of ice shelves.* J. Glaciology (1979) 24, 259 - 271.

ROBIN, G. DE Q. and SWITHINBANK, C. *Fifty years of progress in understanding ice sheets.* J. Glaciology (1987) Special Issue, 33 - 47.

SANDERSON, T. J. O. *Equilibrium profiles of ice shelves.* J. Glaciology (1979) 22, 435 - 459.

SANDERSON, T. J. O. and DOAKE, C. S. M. *Is vertical shear in an ice shelf negligible?* J. Glaciology (1979) 22, 285 - 292.

SHOEMAKER, E. M. and MORLAND, L. W. *A glacier flow model incorporating longitudinal deviatoric stresses.* J. Glaciology (1984) 30, 334 - 340.

SMITH, G. D. and MORLAND, L. W. *Viscous relations for the steady creep of polycrystalline ice.* Cold Reg. Sci. Tech. (1981) 5, 141 - 150.

SPRING, U. and MORLAND, L. W. *Viscoelastic solid relations for the deformation of ice.* Cold Reg. Sci. Tech. (1982) 5, 221 - 234.

SPRING, U. and MORLAND, L. W. *Integral representations for the viscoelastic deformation of ice.* Cold Reg. Sci. Tech. (1983) 6, 185 - 193.

STEINEMANN, S. *Flow and recrystallisation of ice.* Proc. IASH Symp. 39, Rome (1958), 449 - 462.

THOMAS, R. H. *Flow law of Antarctic ice shelves.* Nature, Physical Sciences (1971) 232, 85 - 87.

THOMAS, R. H. *The creep of ice shelves: theory.* J. Glaciology (1973a) 12, 45 - 53.

THOMAS, R. H. *The creep of ice shelves: interpretation of observed behaviour.* J. Glaciology (1973b) 12, 55 - 70.

THOMAS, R. H. *Dynamics of marine ice sheets.* J. Glaciology (1979) 24, 166 - 177.

THOMAS, R. H. *Ice shelves: a review.* J. Glaciology (1979) 24, 273 - 286.

THOMAS, R. H. and BENTLEY, C. R. *A model for the Holocene retreat of the West Antarctic Ice Sheet.* Quat. Res. (1978) 10, 150 - 170.

THOMAS, R. H., SANDERSON, T. J. O. and ROSE, K. E. *Effect of climate warming on the West Antarctic Ice Sheet.* Nature (1979) 277, 355 - 357.

TRUESDELL, C. *On the foundations of mechanics and energetics, in The Rational Mechanics of Materials.* Gordon and Breach, New York (1965), 293 - 304.

VAN DER VEEN, C. J. *Longitudinal stresses and basal sliding.* Proc. Workshop Dynamics of the West Antarctic Ice Sheet, Utrecht 1985. Ed. Van der Veen, C. J. and Oerlemans, J. Reidel, Dordrecht (1987), 223 - 248.

WEERTMAN, J. *Deformation of floating ice shelves.* J. Glaciology (1957) 3, 38 - 42.

WEERTMAN, J. *On the sliding of glaciers.* J. Glaciology (1957) 3, 33 - 38.

WEERTMAN, J. *Equilibrium profile of ice caps.* J. Glaciology (1961) 3, 953 - 964.

WEERTMAN, J. *The theory of glacier sliding.* J. Glaciology (1964) 5, 287 - 303.

WEERTMAN, J. *Stability of the junction of an ice sheet and an ice shelf.* J. Glaciology (1974) 13, 3 - 11.

WEERTMAN, J. *Glaciology's grand unsolved problem.* Nature (1976) 260, 284 - 286.

Mechanics of Granular Flows

Stuart B. Savage
Department of Civil Engineering and Applied Mechanics
McGill University, 817 Sherbrooke West, Montreal, Canada

The present paper deals with the flow of bulk solids comprised of discrete particles. Attention will be focused on understanding the interactions that take place between individual particles at the microstructural level as a means to establish the constitutive relationships (expressions for stresses, energy fluxes, energy dissipation, etc.) that govern the flow of granular materials. We begin with a review of the classical papers of Bagnold and proceed to discuss the various modes and regimes of granular flow. A dimensional analysis provides a physical background for the interpretation of subsequent presentations of theoretical and experimental work. By making analogies with gases and liquids at the molecular level it is possible to develop kinetic theories of granular flows. A simple analysis of this kind is briefly outlined. Recent attempts to consider the effects of boundary conditions, to include the effects of quasi-static stresses and interstitial fluids are also discussed. Molecular dynamics type computer simulations that are related to these kinetic theories are reviewed. As an application of granular flow theories to geophysical problems we discuss some analyses of avalanches of granular materials down inclines. Continuum approaches can be used to model rockfalls and ice avalanches in a very simple, but approximate way; they appear to yield the gross features of these flows. The time dependent evolution of avalanches and other granular flows can also be modelled by means of molecular dynamics type computer simulations. Examples of two and three–dimensional computations are described. Typically the avalanches are made up materials of different sizes. The shear that takes place over the depth of material results in a grading of particles in which larger particles move to the top and the front of the moving pile. A simple explanation of this phenomenon, based upon the segregation analysis of Savage & Lun (1988), is given.

1. Introduction

The terms *bulk solids* or *granular materials* refer to mixtures made up of discrete solid particles which are dispersed in a fluid phase. In situations involving the flow and yielding of such bulk materials, the typical solids concentrations are quite high and the solid constituents come in contact or near contact with their neighbors. Bulk solids occur in various forms: dry grains and powders, slurries, suspensions or pastes. Some examples are sand, ore, mineral concentrate, coal, crushed oil shale, grain, animal feed, foodstuffs, pharmaceuticals, granular snow, pack ice, and metal and ceramic powders. Although there are some instances and flow regimes in which the interstitial fluid can be important, the essential mechanical aspect of bulk solids is the interaction between the solid particles. As a result of studies in the area of soil mechanics, a great deal is known about granular materials in a static state or undergoing slow deformations. However, the present understanding of granular materials

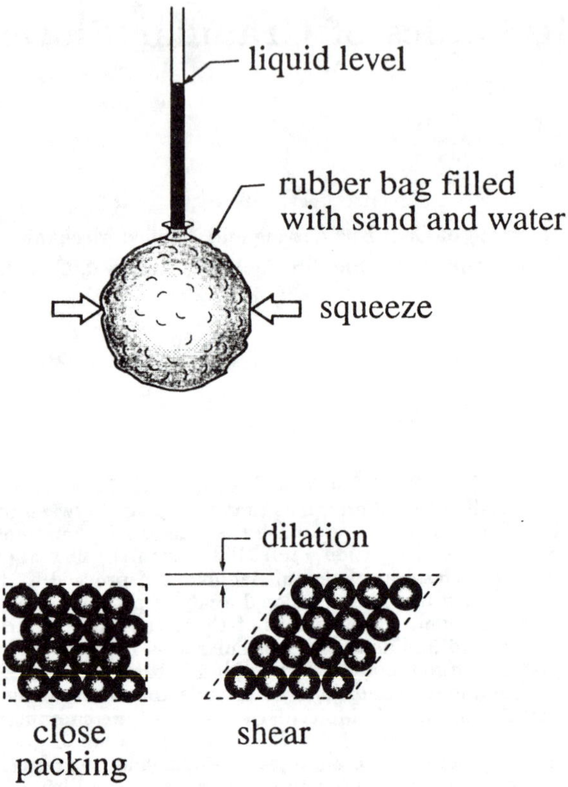

Fig. 1: Reynolds' apparatus used to demonstrate the phenomenon of dilatancy.

which are flowing rapidly is much more rudimentary. Because of its formidable nature, the scientific development of the subject has lagged behind that associated with simpler and more conventional, single phase, continuum fluids. There are many unsolved problems; in some cases even a qualitative understanding of the flow phenomenon is lacking.

There is a very wide range of technological and scientific problems that require an understanding of and the ability to predict the flow of granular materials. Examples involving the manufacture, processing, and handling of granular materials can be found almost everywhere, not only in traditional areas such as the pharmaceutical, foodstuff, detergent, chemical, and processing industries, but in new and emerging high tech fields involving techniques such as ultrastructural processing of ceramics, new methods of Xerography, and powder metal forming. The economic significance is huge, since the numbers and the sales volumes of the industries involved are so large.

Closely related to these industrial applications are problems involving the mechanics of granular materials in geophysical and environmental contexts. There are engineering problems associated, for example, with hydraulic dredging, disposal of solids waste in the form of underwater debris flows, handling of particulate materials (both dry and in a slurry form) in mineral processing, and problems involving resource development such as the determination of forces developed by fractured ice against oil exploration structures and ships in arctic waters. In addition, there are natural disasters and catastrophes involving debris flows and mud slides, rock avalanches, and pyroclastic flows. The latter have caused major global climate changes in addition to significant damage on a more local scale. One is interested not only in the likelihood of such events that can, in some cases, result from poor land management or overdevelopment, but in understanding the flow behaviour and the potential for property damage and personal injuries should such natural catastrophes occur.

Despite the enormous economic consequences, the analyses required for geophysical problems and engineering projects involving bulk solids are often based upon comparatively primitive *ad hoc* methods that lack a sound scientific basis. By comparison with more usual fluids, relatively little is known about the equations that govern bulk solids flow. At present, one is seldom in a position to perform rational predictions with the confidence that we have come to expect in other areas that have reached a more developed stage. For example, in the context of engineering, even the scale-up of small model tests to the full scale industrial prototype is fraught with uncertainty. The result is expensive trial and error development and the common failure to meet design performance and reliability standards.

Nevertheless, during the past 15 years significant steps have been made towards an improved understanding of granular flows (see reviews of Weighardt 1975, Savage 1984, 1989, Jaeger & Nagel 1992). Progress has been made through the development of kinetic theories that can be used to predict the characteristics of various flow fields (for example, Savage & Jeffrey 1981, Lun, *et al.* 1984, Jenkins & Richman 1985), through computer simulations of these flows (for example, Walton & Braun 1986, Walton, *et al.* 1987, Campbell 1990, Savage 1992), and by experimental investigations (for example, Savage 1979, Savage & Sayed 1984, Bridgwater, *et al.* 1985, Johnson & Jackson 1987, Ahn, *et al.* 1988, Sanders, *et al.* 1988). Advances based upon laboratory experiments have been hindered because of the difficulty of developing techniques to measure things like mean flow velocities, particle velocity fluctuations, and densities at 'points' in the interior of the flowing material without disturbing the flow field. The important role of collisional particle interactions has been recognized by workers in such diverse areas as materials handling (Sinclair & Jackson 1989), fluidized beds (Ding & Gidaspow 1990), sediment transport (Jenkins & McTigue 1990), geophysical flows of various kinds (Davies 1982, 1986, Iverson & Denlinger 1987, Denlinger 1987), and astrophysics (Araki 1991a, 1991b). This has led to new ways of looking at these problems and increased activity to incorporate these effects into theoretical models of the various flow phenomena.

There is also considerable new interest and active research on granular materials by the theoretical physics community. They have modelled sand avalanches to formulate a simple non-linear dynamical system that exhibits the phenomena of chaos and 'self organization' (Kadanoff, *et al.* 1989, Bak & Chen 1991). Other problems in the mechanics of granular materials also show a rich diversity of nonlinear behaviour deserving of attention (Baxter, *et al.* 1990, Bideau & Dodds 1992, Jaeger & Nagel 1992).

In the present paper attention will be focused on understanding the interactions that take place between particles at the microstructural level as a means to establish the constitutive relationships (expressions for stresses, energy fluxes, energy dissipation, etc.) that determine the overall large scale bulk flow fields. After a discussion of various granular flow regimes, granular flow kinetic theories, and numerical approaches of the molecular dynamics kind, these concepts will be applied to consider some problems of a geophysical nature.

2. Some physical concepts

2.1. A little history, sand clocks, and dilatancy

If we go back to the time of the Romans, we find that Lucretius put forth the idea that matter is made up of atoms in ceaseless motion, colliding and rebounding from one another. The atomistic view of nature had been proposed earlier by the Greek Philosophers (notably Democritus). This idea of looking at things at the level of individual grains is inherent in most of the current research dealing with flows of granular materials and is at the heart of the present series of lectures. The following is a quotation from the famous poem of Lucretius, 'De rerum natura' (which may be translated as "of the nature of things") that was written in 55 AD.

And you may scoop up poppy seed as easily
As water, which will also if you spill it,
Glide away with as ready a downward flow.

There described very clearly is the idea that particulate materials (bulk solids) can flow very much like a fluid.

Probably the earliest application of granular flows was the hour-glass or sand clock. These devices were in common use by the end of the 13th century for the measurement of the speed of ships and were used during the Middle Ages by scholars to regulate the routine of their studies and by the clergy to time their sermons (Balmer 1978). The fresco, titled "School Allegory of Good Government", which was painted in 1338-39 by Ambrogio Lorenzetti is the earliest known evidence of the existence of sand clocks. It shows the three allegorical figures of Magnanimity, Temperance and Justice; Temperance is holding the sand clock.

The earliest scientific studies of granular flows were performed by two fluid mechanicists, G. Hagen and Osborne Reynolds; but, little further work was done until more recent times. In 1852 Hagen performed experiments which showed that for granular flow, the flow rate through an orifice is independent of h, the head of material above the orifice, in contrast with the flow of a fluid where it depends upon $\sqrt{h}$.

In his memoir 'The Sub-mechanics of the Universe' (1903), Reynolds attempted to explain all known physical phenomena in terms of a mechanical system of uniform spherical grains in various kinds of packings. Reynolds had everything worked out unequivocally in great detail; for example, "the diameter of a grain in C.G.S. units, is $5.534 \times 10^{-18} = \sigma$, etc., etc." It was published (after some negotiations) by the Cambridge University Press as the third volume of his collected writings, but it had little impact because it was so obscure. Sir Horace Lamb wrote in his obituary notice of Reynolds, "Unfortunately, illness had already begun to impair his powers of expression, and the memoir as it stands is affected with omissions and discontinuities which render it unusually difficult to follow." Sir J.J. Thomson, one of his former students, was somewhat more frank, "...it was the most obscure of all his writings, as at the time his mind was beginning to fail ...".

Reynolds' main contribution to the mechanics of granular media was the idea of dilatancy which he defined as the property possessed by a mass of granular material to alter its volume in accordance with a change in arrangement of its grains. To illustrate the phenomenon of dilatancy, Reynolds put together an apparatus consisting of a rubber bag filled with a mixture of granular material and water which was purged of any air bubbles and then tied to a glass tube as shown in figure 1. When the bag is squeezed, one's naive intuition suggests that the water level in the tube would rise. But, instead, it goes down. What happens is that when squeezing the bag by hand, it is not possible to apply a uniform pressure all around the outside and shear deformations develop. From geometrical considerations the bulk has to expand and water rushes in to fill the void spaces created. This is shown schematically in figure 1b. In the initial state the spherical grains are closely packed, then when the bulk is sheared, particle overriding and a vertical expansion occurs. Reynolds used these ideas about dilatancy to explain a phenomenon which can be observed when walking along the seashore. When one's foot is planted on the moist sand, shear and dilation occur and the area around the foot immediately becomes dry. But, when the foot is raised, water is drawn in to occupy the additional voids created and a pool of water soon collects in the footprint.

2.2. Internal friction angle and quasi-static yielding

Dilatancy is one key concept for understanding the mechanics of granular materials. Another is the existence of an internal friction angle. This merely means that unlike a fluid, the granular material can be piled in a heap. The phenomenon is quite analogous to a block of weight W which can slide on a rough plane (figure 2). If the component of the weight tangent to the plane is greater than the frictional shear force S that can be developed, then sliding takes place. Expressing this condition in terms of a friction angle μ and a normal force N, we observe from figure 2 that sliding occurs when $(W \sin\theta = N \tan\theta) \geq (S = N \tan\mu)$, i.e. when $\theta \geq \mu$. A related kind of thing happens in the interior of the heap. Its behaviour can be approximated by the so-called *Mohr-Coulomb yield criterion* which states that yielding or failure or shear can occur on some plane, if the magnitude of the shear stress τ on that plane is equal to the normal stress σ times the tangent of the internal friction angle ϕ, i.e.

$$|\tau| = \sigma \tan\phi \,. \tag{2.1}$$

Typically, for spherical particles, such as glass beads, $\phi \simeq 24°$, and for more angular particles, such as crushed sand, ϕ is around 37°. Figure 3a shows the yield lines where the condition (2.1) is satisfied in $\tau - \sigma$ space for the simplest case of a cohesionless material with a constant internal friction angle. An alternative representation showing the yield lines in principal stress space is given in figure 3b. Note that the ratio of the major principal stress σ_1 to the minor principal stress σ_2 can be expressed in terms of the friction angle ϕ as

$$\frac{\sigma_1}{\sigma_2} = \frac{(1 + \sin\phi)}{(1 - \sin\phi)} \,. \tag{2.2}$$

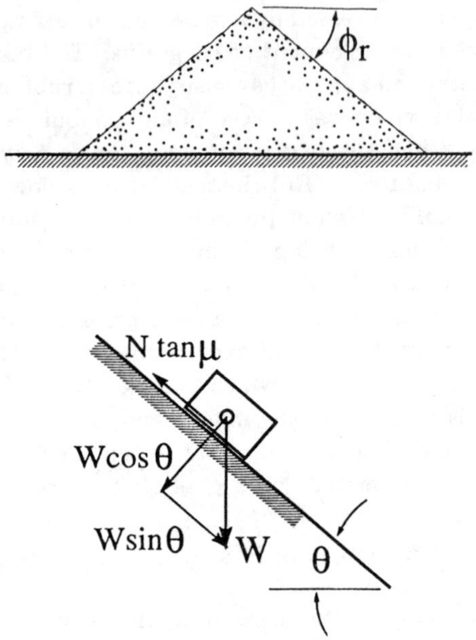

Fig. 2: Angle of repose and sliding of a block on a rough plane.

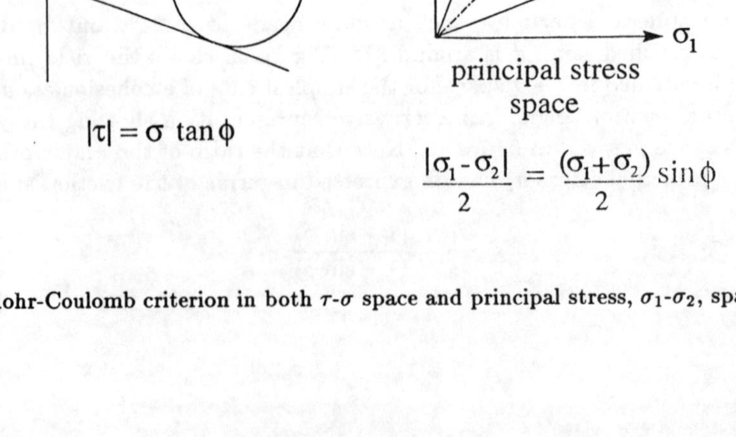

Fig. 3: Mohr-Coulomb criterion in both τ-σ space and principal stress, σ_1-σ_2, space.

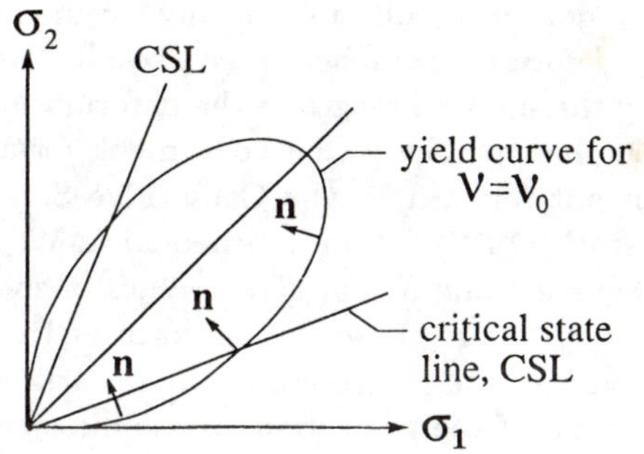

Fig. 4: Yield curve in principal stress σ_1-σ_2 space for constant value of solids fraction, $\nu = \nu_0$. Inward normals to yield curve and critical state lines (CSL) are shown.

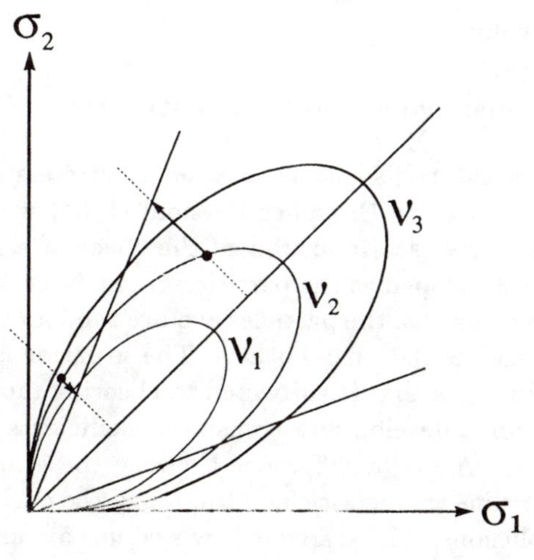

Fig. 5: Yield loci for three values of solids fraction. Two examples of deformations for constant values of mean pressure.

The Mohr-Coulomb yield criterion is often a reasonable approximation for continued deformations that take place at more or less uniform solids fractions ν (where ν is the volume of solids/total volume). However, as described above, when a highly compacted mass of granular material is initially deformed, dilation usually occurs. Alternatively, when a loosely packed mass is initially deformed, compaction can result. Models that can account for bulk density changes during these initial stages of the deformations involve further generalizations of the Mohr-Coulomb approach. One convenient formulation is the Critical State Soil Mechanics approach put forward by the Cambridge Soil Mechanics school, notably, Roscoe, Schofield and Wroth (1958) (also see Schofield & Wroth 1968). This can be described schematically as in figures 4 and 5. For given values of solids volume fractions ν, the yield loci in principal stress space are shown in the form of the solid teardrop shaped lines in figures 4 and 5. Now consider the projection of the inward normals to a yield locus on the principal diagonal (figure 4). If the projection on the diagonal is positive (outward) then dilation will occur during deformation, i.e. ν will decrease. If the projection is negative (inward), then compaction will occur during deformation and ν increases. At the so-called 'critical state', the normal is perpendicular to the principal diagonal and deformation occurs without change in volume. Figure 5 shows two cases of deformations at a constant mean stress ($\sigma_1 + \sigma_2 =$ const.), both starting from a value of solids fraction of ν_2. For the example corresponding to the larger value of the mean stress, the solids fraction is seen to increase as the stress difference $q = (\sigma_1 - \sigma_2)$ increases in magnitude, until the critical state line is reached. In the other example, corresponding to the smaller value of the mean stress, dilation occurs as the critical state line is approached. When the critical state is reached, deformations occur at constant volume.

2.3. Collisional generation of stresses

A third important concept is the collisional generation of stresses and the associated energy dissipation. This idea goes back to Brigadier Bagnold (1954) who was interested in problems of sediment transport in rivers. He examined the shear of a mass of grains and considered the stresses that were developed as the particles in one layer overtook those in an adjacent layer. As a result of the collisions, the particles acquire random velocity fluctuations and spins in addition to their mean translational motions. The situation is analogous to what happens at the molecular level in a gas that is subjected to shearing motion. As a result of the molecular interactions, pressures develop and transport coefficients such as viscosities, thermal conductivities, etc. exist. A major difference between the granular flows and the molecular flows is that the solid grains are inelastic and frictional; the particles' kinetic energy is dissipated into heat during collisions. The granular flow systems are extremely dissipative and in the absence of some driving mechanism the velocity fluctuations very quickly die away. To maintain the fluctuations we must supply energy. For example, this may come from shear work due to some mean shearing motion, or from the input of vibrational energy through the walls of a container.

In a paper that is now regarded as a classic in the field, Bagnold (1954) described experiments on solid-fluid mixtures that he performed in a concentric cylinder Couette flow device. He also developed simple analyses to predict the forms of the stresses that he measured in his experimental apparatus. Bagnold's experiments and flow classifications will be described in detail in the next section. This will be followed in §4 by a discussion of granular flow kinetic theories that can be regarded as sophisticated elaborations of Bagnold's ideas. In granular flows, dispersive normal stresses as well as shear stresses occur. It is found that for the case of simple shear flow the shear stresses can be related to the normal stresses through a dynamic friction angle ϕ_D, reminiscent of the Mohr-Coulomb yield criterion that has been described above for quasi-static deformations. Because of the generation of normal stresses there is a tendency for the material to dilate.

3. Granular flow regimes and constitutive behaviour

Granular flows can be classified in terms of different flow regimes, each having distinct characteristics. While the main focus of the present discussion is on the rapid flow regime in which the effects of particle grain–inertia are dominant, it is important for several reasons to point out the kinds of flow behaviour which exist in other flow regimes. For example, in some instances of particulate flow, portions of the granular material may be 'locked' together and act as a nearly rigid material; the mechanics of deformation of these portions is different in certain respects from those in which grain–inertia is important. In other instances, the interstitial fluid viscosity and density may be important. Some appreciation of these effects is necessary to better understand the experimental results and to recognize the limitations of the theoretical treatments. We shall first review Bagnold's pioneering work on granular flow and in particular examine his classifications for suspensions of solids in fluids[1]. Further classifications of the flow of *dry* granular materials are then discussed. Laboratory experiments performed to determine the constitutive behaviour are briefly described.

3.1. Flow regimes

3.1.1. Review of Bagnold's papers on fluid–solid mixtures

The present level of understanding of granular flows, for flow regimes in which grain inertia is important, owes much to the now classical papers of Bagnold (1954, 1956, 1966). His work was motivated by an interest in the mechanics of phenomena such as the bed load transport of sediment in rivers. Bagnold (1954) performed experiments on neutrally buoyant, uniform sized spherical wax beads suspended in Newtonian fluids (water and glycerine–water–alcohol mixtures) and sheared in a coaxial rotating cylinder apparatus. By using flexible rubber to form his inner cylinder wall he could measure both the torque and the normal stress in the radial direction when various concentrations of grains were sheared by rotating the rigid outer cylinder. Bagnold distinguished three different regimes of flow behaviour, which he termed *macro–viscous*, *transitional* and *grain–inertia*. Classification of a particular shear flow, $u(y)$ (cf. figure 6) depends upon the value of a dimensionless shear group (subsequently named by Hill (1966) as the *Bagnold number*)

$$N = \lambda^{1/2}\rho_f\sigma^2(du/dy)/\mu_f, \tag{3.3}$$

---

[1] Material in §3.1 is based upon work previously published in Savage (1984).

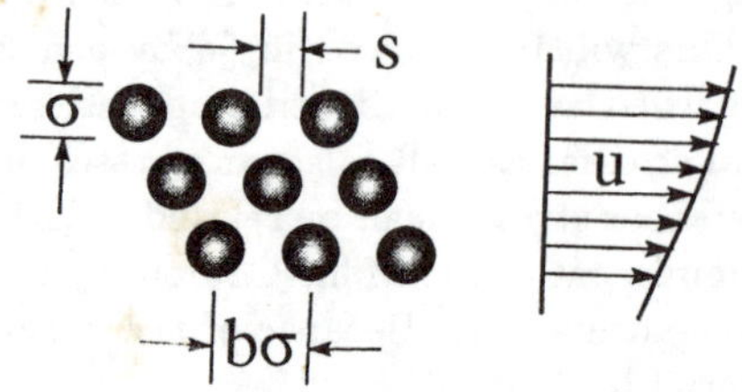

Fig. 6: Definition sketch for Bagnold's (1954) analysis of sheared granular material.

where ρ_f and μ_f are the mass density and viscosity of the interstitial fluid, σ is the particle diameter, du/dy is the velocity gradient and λ is the *linear concentration* of particles. If the mean distance between particle centers is defined as $b\sigma$ and the resulting free distance is s (figure 6) then

$$b = 1 + \frac{s}{\sigma} = 1 + \frac{1}{\lambda}, \tag{3.4}$$

and hence we can define $\lambda = \sigma/s$. Note that the solids fraction ν may be expressed in terms of λ as

$$\nu = \frac{\nu_\infty}{b^3} = \frac{\nu_\infty}{(1+1/\lambda)^3},$$

where ν_∞ is the maximum possible solids concentration when $\lambda = \infty$ (for uniform spheres $\nu_\infty = \pi/(3\sqrt{2}) = 0.7405$).

Bagnold derived simple analyses to explain the rheological behaviour in the two limiting regimes, the *macro-viscous* and *grain-inertia* regimes. In his *macro-viscous* regime, corresponding to small N, viscosity is dominant and the shear and normal stresses are linear functions of the velocity gradient, du/dy. Bagnold attributed the presence of the normal stress in the radial direction (which he called a 'dispersive' pressure) to a statistically preferred anisotropy in the spatial particle distributions as sketched in figure 6. More pertinent to the present discussion is his *grain-inertia* regime, corresponding to large values of N, in which the interstitial fluid plays a minor role and the major effects are due to particle–particle interactions. Bagnold argued that the main mechanism for momentum transfer is the succession of glancing collisions as the grains of one layer overtake those of the adjacent slower layer (cf. figure 6). Both the change in momentum during a single collision and the rate at which collisions occur are proportional to the relative velocity of the two layers, giving rise to stresses that depend upon the *square* of the shear rate. Sanders (1963) called this process the *Bagnold Effect.* Through an analysis that was roughly analogous to a simple kinetic gas theory he deduced that the normal stress in the y–direction was

$$p_{yy} = a\rho_p \lambda f(\lambda)\sigma^2 (du/dy)^2 \cos\phi_D, \tag{3.5}$$

and that the grain shear stress was

$$p_{xy} = p_{yy} \tan\phi_D, \tag{3.6}$$

where ρ_p is the mass density of the individual particles, f is an unknown function of λ, a is a constant and ϕ_D is an unknown dynamic friction angle dependent upon collision conditions. Connecting the two limiting flow regimes was Bagnold's *transitional* flow in which the dependence of the stresses upon shear-rate varied from a linear one corresponding to the *macro-viscous* regime to a square dependence predicted for the *grain-inertia* flow regime.

From his experiments, Bagnold was able to define the various flow regimes in terms of N; $N < 40$ corresponded to the *macro-viscous* regime, $N > 450$ was termed the *grain-inertia* regime, and the *transitional* region occupied the intermediate range of N. The limiting experimental flow behaviour was consistent with his predictions for the *macro-viscous* and *grain-inertia* regimes. In the *grain-inertia* regime, stresses varied with the square of the shear rate as in (3.5) and (3.6). For $\lambda < 14$, $f = \lambda$, and for $\lambda > 14$, f increased very rapidly with λ. Large λ corresponds to grains being close together; if λ is large enough the grains lock together and very large stresses are required to shear the bulk. The angle ϕ_D was found to vary only slightly with λ.

One interesting feature common to all flow regimes was the presence of a substantial radial normal 'grain' stress which was proportional to (and greater than) the shear stress. Such behaviour is reminiscent of the Coulomb yield or failure criterion first used to describe the stresses in certain soils under conditions of limiting equilibrium. Yielding is influenced by the hydrostatic pressure in the Coulomb failure criterion, which states that yielding will occur at a point on a plane element when

$$|\tau| = c + \sigma_n \tan\phi, \tag{3.7}$$

where τ and σ_n are respectively the shear stress and normal stress acting on element, c is the cohesion and ϕ is the internal angle of friction of the bulk material. For dry, coarse materials, the cohesion c is negligible and (3.7) takes on the same form as (3.6). Typical values for ϕ obtained during quasi-static yielding at low stress levels are close to the angles of repose, for example, about 24° for spherical glass beads and 37° for angular sand grains. It is interesting to compare the quasi-static ϕ for typical *spherical* particles with the *dynamic* friction angle ϕ_D obtained in Bagnold's experiments with spherical wax beads sheared in the *grain-inertia* regime. For $\lambda > 12$ ($\nu > 0.58$) he found that $\phi_D \simeq 22°$; and for $\lambda < 12$, ϕ_D tends to approach an angle of about 18° (see discussion in Bagnold 1973).

Bagnold (1954) applied his cylindrical shear cell results and his analyses for the stresses to study the problems of gravity flow of particulate matter down inclines of the type that occurs in rock falls and debris flows. His predictions for the surface velocity in such a flow in the *grain-inertia* regime were about 50% higher than measurements he made of the velocity of quartz sand flowing down a simple flume. He also investigated the flow of gravel, the interstices of which were filled with mud of slightly lower mass density; the predicted flow behaviour was compared with some rough field observations of Professor K. Terzaghi. The grain-inertia analysis was used to explain the sorting of grains of mixed sizes that may be observed in bed load sediment transport and submarine grain flows. This mechanism is also used in various devices for the mineral processing industry. In this type of sorting called *inverse grading*, the large particles drift towards the surface and the finer particles towards the bed. However, Bagnold's (1954) explanation for the grading process is not above criticism and Middleton (1970) (see Middleton & Hampton 1976) has suggested an alternative mechanism. (We shall examine these kinds of segregation problems in more detail in §5.3.) Bagnold further applied the results of his 1954 paper to study the bulldozing of a mass of dry sand and the 'singing' mechanism in desert sand dunes (Bagnold 1966) and to problems of bed-load sediment transport rivers (Bagnold 1956).

3.1.2. Modes of flow for dry cohesionless granular materials

The work of Bagnold discussed above considered the various regimes of flow that occur in a fluid-solid mixture when the Bagnold number, N, is varied over a wide range. It happens that there are regions in 'granular flow regime-space' in addition to those noted by Bagnold. For example, in *some* flows of dry (gaseous interstitial fluid), coarse particulate solids, the fluid phase can be neglected for essentially all values of the shear-rate. Here, the Bagnold number alone is insufficient to characterize the types of flow behaviour that are possible. In one limit, for large shear-rate, we do expect the *grain-inertia* type of flow corresponding to large N. But for very small shear-rates and high concentrations of coarse solids, one expects the kind of behaviour characteristic of the quasi-static deformation of a cohesionless soil and not the behaviour associated with Bagnold's *macro-viscous* flow. A *transitional* region joining the *quasi-static* type of flow to the *grain-inertia* flow must exist. In this subsection, we shall examine these flow regimes in more detail using dimensional analysis and shall discuss in physical terms the various modes of flow. For simplicity, electrostatic forces as well as the interstitial fluid phase will be neglected, although clearly there are situations where these effects are of considerable importance.

During the deformation of a dry granular material, mean stresses may be generated by a number of different mechanisms. In general, the instantaneous motions of particular grains, their translational velocities and spins, are different from the mean motion of the bulk. Individual particles may interact with one another in various ways, – in rigid clusters of particles which generate a network of contact forces through sustained rolling or sliding contacts, or by nearly instantaneous collisions during which linear and angular momentum are exchanged and energy is dissipated because of inelasticity and friction. The relative importance of these mechanisms may be used as the characteristics that define various flow regimes. Some insight into the functional dependence of the stresses upon the flow and material properties may be obtained through dimensional analysis. It is convenient to work in terms of the deviatoric stress

$$\widehat{\tau}_{ij} = \tau_{ij} - \frac{1}{3}\tau_{kk}\delta_{ij} = \tau_{ij} - p\delta_{ij}, \tag{3.8}$$

where τ_{ij} is the stress tensor (with compressive stresses being taken as positive) and p is the mean normal stress or pressure. We suppose that the particles have an instantaneous translational velocity $\mathbf{c}$, a mean velocity $\mathbf{u} = \langle \mathbf{c} \rangle$ and hence an instantaneous translational velocity fluctuation about the mean of $\mathbf{C} = \mathbf{c} - \mathbf{u}$. For simplicity let us consider a particular flow situation; for example, a simple shear flow in which the only non-zero component of the velocity gradient is $du/dy = u_{1,2}$. We anticipate that, in general, the deviatoric stress $\widehat{\tau}_{ij}$ and the pressure p may be expressed as

$$\hat{\tau}_{ij} = fn\left(\rho_p, u_{1,2}, \langle C^2\rangle^{1/2}, \langle \omega^2\rangle^{1/2}, p, \nu, \sigma, H, g, e, \mu, E, S\right), \tag{3.9}$$

and

$$p = fn\left(\rho_p, u_{1,2}, \langle C^2\rangle^{1/2}, \langle \omega^2\rangle^{1/2}, \nu, \sigma, H, P_B, g, e, \mu, E, S\right), \tag{3.10}$$

where ρ_p is the mass density of the individual solid particles, $\langle C^2\rangle^{1/2}$ and $\langle \omega^2\rangle^{1/2}$ are the root mean square of the translational and rotational velocity fluctuations arising from interparticle collisions, P_B is some reference or characteristic value of the normal stress ***applied*** at the boundary of the sheared region, ν is the solids fraction (volume of solids/total volume), σ the particle diameter, H the length scale for the width of shear flow, g the gravitational acceleration, e, μ and E are respectively the coefficients of restitution, the surface coefficient of friction and the modulus of elasticity of the solid particles, and S is the shape factor defining the angularity or sphericity of the solid particles. If we were to consider a Couette flow in which the material was sheared between two parallel plates a distance H apart, then we should also include in (3.9) and (3.10) information about the wall properties, for example, the coefficient of restitution e_w, the surface coefficient of friction μ_w, the modulus of elasticity E_w and some measure of the wall roughness or irregularity. Note that the pressure p has been included in the functional relationship for $\hat{\tau}_{ij}$ in order to explicitly recover the dependence of shear stress upon normal stress which is characteristic of frictional Coulomb–type materials familiar in soil mechanics. Dimensional homogeneity requires that

$$\frac{\hat{\tau}_{ij}}{\rho_p\sigma^2(u_{1,2})^2} = fn\left(\frac{p}{\rho_p\sigma^2(u_{1,2})^2}, \frac{\sigma u_{1,2}}{\langle C^2\rangle^{1/2}}, \frac{u_{1,2}}{\langle \omega^2\rangle^{1/2}}, \frac{g}{\sigma(u_{1,2})^2}, \nu, \frac{\sigma}{H}, \frac{p}{E}, e, \mu, S\right), \tag{3.11}$$

$$\frac{p}{\rho_p\sigma^2(u_{1,2})^2} = fn\left(\frac{\sigma u_{1,2}}{\langle C^2\rangle^{1/2}}, \frac{u_{1,2}}{\langle \omega^2\rangle^{1/2}}, \frac{g}{\sigma(u_{1,2})^2}, \frac{\rho_p\sigma^2(u_{1,2})^2}{P_B}, \frac{P_B}{E}, \nu, \frac{\sigma}{H}, e, \mu, S\right). \tag{3.12}$$

We shall first examine these relationships for two limiting cases and then attempt to describe the microscopic physical behaviour of the flow in the transition region between these two limiting flows.

1) *Fully dynamic, fluid-like grain–inertia regime* (high shear rates, moderate stress levels, low ν). Here the shear rates and stress levels are high enough such that the gravitational forces are negligible, i.e.

$$\frac{g}{\sigma(u_{1,2})^2} \rightarrow 0, \tag{3.13}$$

but low enough such that the particles can be regarded as rigid in connection with the determination of bulk density, thus

$$p/E \rightarrow 0. \tag{3.14}$$

The group $\rho_p\sigma^2(u_{1,2})^2/P_B$, which is proportional to the ratio of the dynamic or collisional stresses to the total boundary stresses, is the primary parameter which distinguishes the various flow regimes for dry granules. In the fully dynamic regime, $\rho_p\sigma^2(u_{1,2})^2/P_B$ is of order one. If the width of the shear layer H is large such that we can neglect the effects of finite particle size to shear layer thickness ratio, i.e. $\sigma/H \to 0$, then equations (3.11) and (3.12) reduce to

$$\frac{\widehat{\tau}_{ij}}{\rho_p\sigma^2(u_{1,2})^2} = fn\left(\frac{p}{\rho_p\sigma^2(u_{1,2})^2}, \frac{\sigma u_{1,2}}{\langle C^2\rangle^{1/2}}, \frac{u_{1,2}}{\langle\omega^2\rangle^{1/2}}, \nu, e, \mu, S\right), \tag{3.15}$$

and

$$\frac{p}{\rho_p\sigma^2(u_{1,2})^2} = fn\left(\frac{\sigma u_{1,2}}{\langle C^2\rangle^{1/2}}, \frac{u_{1,2}}{\langle\omega^2\rangle^{1/2}}, \nu, e, \mu, S\right). \tag{3.16}$$

In this case, the stresses, including those at the boundaries, are completely determined by the flow dynamics, and hence it is redundant to include $P_B/[\rho_p\sigma^2(u_{1,2})^2]$ in equation (3.16).

This flow regime corresponds to the 'grain–inertia' type of flow proposed by Bagnold (1954). Stresses are proportional to the square of both the particle diameter and shear rate. The shear stress depends upon the normal stress as found for dry frictional, *rate–independent*, Coulomb materials. At somewhat lower stress levels than those just mentioned, gravitational forces are important and we must include the parameters $g/[\sigma(u_{1,2})^2]$ and σ/H in (3.15) and (3.16).

Equations (3.15) and (3.16) exhibit behaviour characteristic of a non–linear viscous fluid, such as a Reiner–Rivlin fluid. In the present limiting case, bulk stresses are developed as a result of two mechanisms. At the very lowest concentrations, when particles are widely dispersed and large void spaces exist, momentum transport can occur through the translation of particles from one shear layer to another. The flow is analogous to a dilute gas described by the kinetic theories in which the mean free path is large compared to the particle diameter (Chapman & Cowling 1970). In the kinetic theories at higher concentrations, voids large enough to accept a particle occur less frequently and momentum occurs primarily by the continuous action of intermolecular forces (Temperley, *et al.* 1968, Faber 1972, Hansen & McDonald 1976). One might expect the granular flow mechanics in this region to be similar to the 'hard sphere models' used in statistical mechanical theories of the liquid state where the intermolecular forces are impulsive forces associated with particle collisions. We shall pursue this analogy for the development of kinetic theories of granular flow in §4.

2. *Quasi–static, rate–independent plastic regime* (vanishing shear rate, high ν, moderate to high stress levels). Let us first consider the stresses to be high enough such that the elasticity and deformation of the individual grains are important. Deformations are slow, the parameter $\rho_p\sigma^2(u_{1,2})^2/P_B \to 0$. Neglecting the effects of velocity fluctuations and eliminating the shear–rate dependence such that the stresses depend at most upon the sign of $u_{1,2}$ (or for more general deformations, upon $\mathbf{D}/(\mathrm{tr}\,\mathbf{D}\,\mathbf{D})^{1/2}$ where $\mathbf{D}$ is the rate deformation tensor) equations (3.11) and (3.12) may be reduced to

$$\frac{\widehat{\tau}_{ij}}{P_B} = fn\left(\frac{p}{P_B}, \frac{\rho g H}{P_B}, \frac{p}{E}, \frac{\sigma}{H}, \nu, \mu, S\right), \tag{3.17}$$

$$\frac{p}{P_B} = fn\left(\frac{\rho g H}{P_B}, \frac{P_B}{E}, \frac{\sigma}{H}, \nu, \mu, S\right). \tag{3.18}$$

The stresses here are governed not by the magnitude of the applied shear rate, but by the stresses applied to the boundaries, of which P_B is a characteristic value. This case corresponds to the plastic behaviour of a frictional Coulomb material of the kind that has been studied extensively in the context of soil mechanics (Sokolovski 1965, Schofield & Wroth 1968, Mandl & Fernández–Luque 1970, Salencon 1977, Spencer 1981) and discussed briefly in §2.2. Particles can stick together, roll or maintain sliding contact with one another for extended periods during the bulk deformation. Inertia forces are negligible and stresses are transmitted from one region to another through a network of contact forces. The yield stress depends upon the normal stress p, and p is related to ν as in the critical state theories of soil mechanics (Roscoe, *et al.* 1958). Although the velocity gradient does not appear in (3.17) and (3.18), it is implied that the bulk is undergoing large and continuous deformation. During the initial failure or yielding, soils can experience an increase or a decrease in volume depending upon the initial state of the material (Scott 1963). But with continued deformation or flow the material tends toward an asymptotic state in which no further volume change occurs. The voids ratio (volume of interstitial voids/volume of solids) associated with this state is called the *critical voids ratio*. For a particular soil, its value depends upon the mean stress level p. Note that a bulk material would not necessarily have a unique p–ν relationship if *not* continuously sheared. This can easily be understood by considering the example of a vertical cylinder containing particles in a static state of loose random packing. The overall solids concentration ν could be increased by vibrating the cylinder for a short period. Then the pressure p at a section near the base, for example, might be much the same as before although ν has been increased.

At stress levels low enough such that we can take $p/E \to 0$, the particles can be considered rigid and changes in ν come about solely because of particle rearrangements.

3. *Transitional regime.* In the two limiting flow regimes just discussed, the fully dynamic *grain–inertia* and the *quasi–static* plastic regimes, bulk stresses are generated by very different kinds of mechanisms. In the general case (described by (3.11) and (3.12)) the stresses may be developed by some combination of these rate–dependent and rate–independent mechanisms. If we started with a flow in the fully dynamic *grain–inertia* regime and gradually reduced the shear–rate while increasing the concentration, it is expected that the flow would pass through some *transitional* flow regime before reaching the *quasi–static* flow as $\mathbf{D} \to 0$. A proposal for the microscopic flow behaviour that might occur while passing through this transition regime is now presented. For discussion purposes, suppose that a Couette flow has been generated in the granular material contained between two rough parallel plates, the upper one moving relative to the lower one to generate a flow as shown in figure 7. Let us consider how the normal and shear stresses could be developed at the plate surfaces.

In the fully dynamic, grain–inertia regime (figure 7a) ν is low, the particles have a random fluctuating velocity component in addition to the mean shear $u_1(x_2)$ and momentum is transferred almost entirely by binary interparticle collisions. The mean (time–averaged) stresses are proportional to the square of the shear–rate as described by equations (3.15) and (3.16). Because of the random velocity fluctuations, there also exist fluctuations in the bulk density.

With increasing the mean bulk density (figure 7b) the bulk density fluctuations are such that clusters of particles (which are in rubbing contact) form and break up in random fashion. If the clusters remain rigid for a time they have the effect of increasing the effective shear–

rate locally in the adjacent sheared region and thus they increase dynamic stresses. The stresses associated with deformation of the clusters also have a dry friction, rate–independent component.

A further increase in mean concentration increases the frequency of formation as well as the size of the particle clusters (figure 7c). Some of the clusters are large enough to span the width of the shear region, forming columns between the two walls. Columns or chains of particles in contact, which form force networks that are lined up primarily along the major principal stress direction, have been observed in quasi–static photoelastic studies by Drescher & de Josselin de Jong (1972). It is proposed that similar chains form in the dynamic case. Deformation and collapse of these columns adds a Coulomb–type stress component to the dynamic part associated with the regions adjacent to the columns where there are strong particle fluctuations.

Increasing the mean bulk density still further (figure 7d) will force almost all the particles to be in continuous contact with their near neighbors (i.e., the whole region is occupied by the column–like structures described earlier). Reducing the shear–rate such that the accelerations generated during particle overriding are negligibly small leads to the quasi–static rate–independent behaviour. The stresses result from surface friction and interlocking between particles. The internal friction angle ϕ associated with this Coulomb–type behaviour depends upon the individual particle characteristics μ and S, and the solids concentration ν.

In our discussion of this *transitional regime*, so far we have neglected to include the effects of gravity. Gravity induces additional inhomogeneities in the vertical direction. Consider the case when the flow is driven by two rigid rough walls as described above and gravity acts in the vertical direction. When the shear–rates and concentrations are such that the collisional stresses are much larger than the stresses arising from the weight of the material, then the shear and normal stresses, the velocity fluctuations and the bulk density tend to be uniform across the flow width, or at least, symmetrical about the channel centreline. If the shear–rates are reduced, then the body forces are significant and particles tend to move to the lower wall, increasing the bulk density there. If the shear–rates are low enough or the height of material is large, the shearing is no longer uniform. Particles lock together to form a rigid, dead flow, lower layer and the shearing motion is confined to an upper layer. For *deep* free surface flows, as in an open channel, we can observe *at least* two different types of flows. For bed slopes very close to the angle of repose of the material, one finds that the motion consist of the avalanching of a thin layer of material of the order of 5 or 10 diameters thick. There is another mode for deep layers in which the upper layer is a plug which 'slides' on a thin 'fluidized' boundary shear layer next to the bed (Savage 1979, Hungr & Morgenstern 1984a).

3.1.3. Summary of limiting flow regimes

We have discussed *some* of the possible flow regimes for both fluid–solid mixtures and dry granular materials. Each of these flow regimes is characterized by the relative magnitudes of interstitial fluid viscosity μ_f, solids fraction ν, rate of deformation $\mathbf{D}$, and the associated parameters such as the Bagnold number $N = \lambda_{1,2}\rho_f\sigma^2(u_{1,2})/\mu$ and the group

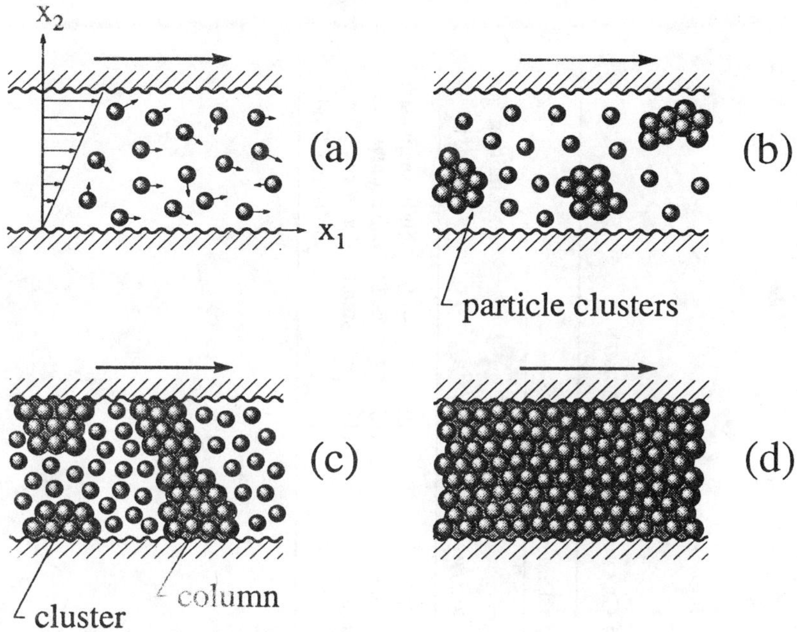

Fig. 7: Proposed microscopic description of transition as mean density and shear–rate are increased. (a) Fully dynamic grain–inertia regime with *viscous* behaviour, similar to molecules in a dense gas; (b) and (c) depict transitional regimes, and (d) depicts the quasi–static regime with a rate–independent behaviour, where most particles are in contact with near neighbors.

$\rho_p\sigma^2(u_{1,2})^2/P_B$. This information is summarized in Table 1. Also included are the values of the parameter τ_c/τ_t which is the fraction of some time interval τ_t that particles are in contact with one another. This ratio is obviously related to the group $\rho_p\sigma^2(u_{1,2})^2/P_B$. In the grain–inertia regime, collisions are almost instantaneous and $\tau_c/\tau_t \rightarrow 0$. In the quasi–static regime, particles experience enduring contacts and $\tau_c/\tau_t \rightarrow 1$.

It should be emphasized that Table 1 is incomplete. We have not included regimes where there is a large differential velocity between the solid and the fluid phases (the extreme example being fluidized bed flows), regimes where electrostatic forces are important, etc. Table 1 is also coarse in the sense that the flow regimes listed may be further subdivided. For example, there are sub–regimes within the *grain–inertia* regime, characterized by the dissipative properties of the materials. In a shear flow involving smooth, nearly elastic grains, $\sigma u_{1,2}/\langle C^2\rangle^{1/2}$ must be small to generate the dissipation required to maintain a steady flow; whereas for very rough inelastic particles, the parameters $\sigma u_{1,2}/\langle C^2\rangle^{1/2}$ and $u_{1,2}/\langle\omega^2\rangle^{1/2}$ are larger (Savage & Jeffrey 1981, Lun, *et al.* 1984).

3.2. Shear cell experiments

A number of experiments have been performed using annular shear cells to obtain information about the dependence of shear and normal stresses upon shear–rate and particle

CHARACTERISTICS OF SOME LIMITING GRANULAR FLOW REGIMES

	Physical quantities			Dimensionless groups			
Flow regimes	Viscosity μ_f	Solids fraction	Deformation rate $\mathbf{D}$	$N = \lambda^{1/2}\rho_f\sigma^2 u_{1,2}/\mu_f$	$\rho_p\sigma^2(u_{1,2})^2/P_B$	τ_c/τ_t	Limiting Prototype
Macroviscous	large	small	small	small	small	$\rightarrow 0$	dilute suspension
Quasi-static	small	large	small	usually small	small	$\simeq 1$	cohesionless soil
Grain-inertia	small	small to moderate	large	large	O(1)	small	dense gas or molecular fluid

Table 1: Characteristics of some limiting granular flow regimes.

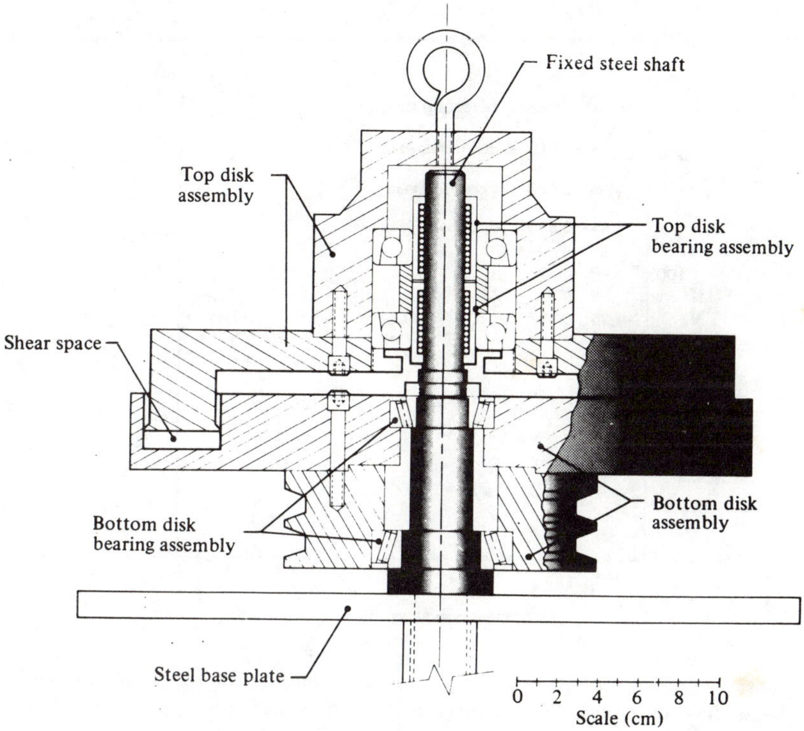

Fig. 8: Schematic diagram of annular shear cell of Savage & Sayed (1984).

concentration (Savage & Sayed 1984, Hanes & Inman 1985, Craig, *et al.* 1986, 1987, Stadler & Buggisch 1985, Stadler 1986). Most experimental arrangements are similar to that of Savage & Sayed 1984 (figure 8) which was the first shear cell designed to obtain granular flow rheological data at high shear-rates. Typically, granular material is contained in an annular trough in the lower half of the shear cell and capped by an annular ring attached to the upper half of the shear cell. The horizontal surfaces that come in contact with the granular material are roughened, so that when the two halves of the shear cell are rotated with respect to one another about a vertical axis, the granular material is sheared. Quite different kinds of results are obtained, depending upon how the tests are performed. The most common method (and the most convenient one for making comparisons with kinetic theories) is to perform the tests so as to maintain a constant volume (i.e., constant *mean* solids concentration). At high shear-rates it is found that stresses vary with the square of the shear-rate. Stresses also increase strongly with increase in solids concentration, but the *ratio* of shear to normal stress is *nearly* constant. These trends are the same as those predicted by the kinetic theories. At low shear-rates and high solids fraction the stresses depend upon the shear-rate raised to a power less than two. Some typical results (Savage & Sayed 1984) for tests at constant solids fraction are shown in figure 9. There $\overline{u}$ is the linear velocity of lower disk assembly at mid-annulus and H is the depth of the shear gap.

Another way to perform the experiments is measure the shear stresses while the normal

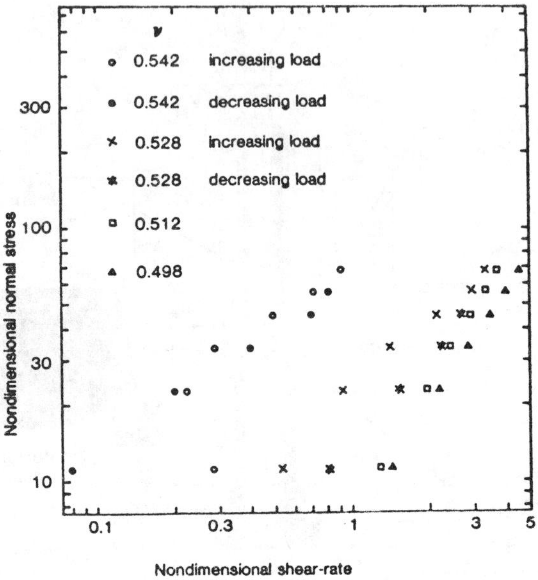

Fig. 9: Nondimensional normal stress $p_{yy}/(\rho_p g \sigma)$ versus nondimensional shear-rate $(\sigma/g)^{1/2}\overline{u}/H$ for polystyrene beads having a bimodal size distribution (30% 0.55 mm spheres and 70% 1.65 mm spheres).

load is kept constant (Hungr & Morgenstern 1984b, Stadler & Buggisch 1985, Stadler 1986). In such tests, when the shear-rate is increased, there is a tendency for the dispersive normal stress in the granular material to increase. But, since the normal load on the top plate is kept constant, the bed of particles expands vertically, and hence the mean solids fraction decreases. Under such conditions, it is found that the shear stress is nearly independent of shear-rate. This is quite consistent with the nearly constant dynamic friction angle (ratio of shear to normal stress). But, it is quite different from what would occur with a Newtonian fluid, and it led some of the early investigators to infer a complete lack of rate-dependence for sheared granular materials.

4. Granular flow kinetic theories and computer simulations

Bagnold's basic ideas about the importance of collisional interactions in the grain-inertia regime and the notion of expressing the particle motions in terms of a mean velocity and velocity fluctuations suggest an analogy with the kinetic theory of gases. The analogy has been exploited to develop much more elaborate kinetic theories of granular flows (Savage & Jeffrey 1981, Jenkins & Savage 1983, Lun, *et al.* 1984, Jenkins & Richman 1985a, 1985b, 1986, Farrell, *et al.* 1986, Lun & Savage 1986, 1987, Jenkins 1987a, 1987b, Jenkins & Mancini 1987, Lun & Savage 1987, Richman 1988). As concentrations in typical granular flows are often quite high, these kinetic theories have been patterned after statistical mechanical approaches previously developed for 'hard sphere' models for dense gases and liquids (Chapman

& Cowling 1970). The key difference in the case of granular flows is the need to include energy dissipation which occurs during collisions and sliding contacts between rough, inelastic, granules. Some important results that emerge from these granular kinetic theories are the evolution equations for the particle velocity fluctuations and spins.

The kinetic energy associated with the translational velocity fluctuations has an obvious analogy with the definition of the temperature in a gas at the molecular level. The term 'granular temperature' has been associated with the mean square particle velocity fluctuations. As a consequence of the particles' collisions, there is a continuous dissipation of the kinetic energy associated with the velocity fluctuations. This collisional energy dissipation results in an increase of the more familiar *thermodynamic temperature* of the individual grains. Because of the significant collisional energy dissipation, shear work or some other means of energy input (for example, vibration) is essential to provide the energy necessary to maintain the velocity fluctuations; otherwise the granular temperature quickly decays. This is quite unlike what happens, for example, in a gas at the molecular level, where the velocity fluctuations of the molecules (and hence, the *kinetic temperature*) are maintained even in the absence of shearing.

4.1. Kinetic theory concepts; negligible interstitial fluid effects

Let us first describe an approach for the simple case of uniform, nearly elastic, smooth, spherical particles based upon the treatment of Lun, *et al.* (1984). We take the particles to be hard, such that the collisions can be regarded as instantaneous, and hence we need consider only binary collisions.

4.1.1. Conservation equations and integral forms for the constitutive equations for stress, flux of fluctuation energy and dissipation

The development of the so-called *hydrodynamic* or *fluid–mechanical* equations and the integral forms for stresses, energy fluxes and the collisional rate of energy dissipation per unit volume will be patterned after the approach that has been used in the kinetic theory of dense fluids (Chapman & Cowling 1970, Ferziger & Kaper 1972).

Consider a fixed volume element $d\mathbf{r}$ centred at $\mathbf{r}$ that contains $n(\mathbf{r},t)d\mathbf{r}$ particles, where n is the number density of particles. The ensemble average of the single-particle quantity ψ is defined as

$$\langle\psi\rangle = \frac{1}{n}\int \psi f^{(1)}(\mathbf{r},\mathbf{c};t)\,d\mathbf{c}, \tag{4.19}$$

where $f^{(1)}(\mathbf{r},\mathbf{c};t)$ is the usual single-particle velocity distribution function defined such that $f^{(1)}\delta\mathbf{c}$ is the differential number of particles per unit volume with velocities within the range $\mathbf{c}$ and $\mathbf{c}+\delta\mathbf{c}$. Similarly, a complete pair-distribution function $f^{(2)}(\mathbf{r}_1,\mathbf{c}_1;\mathbf{r}_2,\mathbf{c}_2;t)$ may be defined such that $f^{(2)}(\mathbf{r}_1,\mathbf{c}_1;\mathbf{r}_2,\mathbf{c}_2;t)$ may be defined such that $f^{(2)}(\mathbf{r}_1,\mathbf{c}_1;\mathbf{r}_2,\mathbf{c}_2;t)\delta\mathbf{r}_1\delta\mathbf{r}_2\delta\mathbf{c}_1\delta\mathbf{c}_2$ is the probability of finding a pair of particles in the volume elements $\delta\mathbf{r}_1,\delta\mathbf{r}_2$ centred on the points $\mathbf{r}_1,\mathbf{r}_2$ and having velocities within the ranges $\mathbf{c}_1$ and $\mathbf{c}_1+\delta\mathbf{c}_1$, and $\mathbf{c}_2$ and $\mathbf{c}_2+\delta\mathbf{c}_2$.

The rate of change of $\langle n\psi\rangle$ may be expressed as

$$\frac{\partial}{\partial t}\langle n\psi\rangle = n\langle D\psi\rangle - \nabla\cdot\langle n\mathbf{c}\psi\rangle + \phi_c, \tag{4.20}$$

where

$$D\psi = \frac{d\mathbf{c}}{dt}\cdot\frac{\partial\psi}{\partial\mathbf{c}} = \frac{\mathbf{F}}{m}\cdot\frac{\partial\psi}{\partial\mathbf{c}}, \tag{4.21}$$

$\mathbf{F}$ is the external force on a particle of mass m, and ϕ_c is the collisional rate of increase of the mean of ψ per unit volume (Reif 1965).

To obtain more explicit expressions for the collisional term ϕ_c, we consider binary collisions between hard, smooth but inelastic particles of uniform diameter σ. At the instant of a collision between between particles labelled 1 and 2 we take the centre O_2 of the second particle to be located at position $\mathbf{r}$ and the centre O_1 of particle 1 to be at $\mathbf{r}-\sigma\mathbf{k}$, where $\mathbf{k}$ is the unit vector along the line of centres from O_1 to O_2. In time δt prior to the collision, particle 1 moved through a distance $\mathbf{c}_{12}\delta t$ relative to particle 2, where $\mathbf{c}_{12}=\mathbf{c}_1-\mathbf{c}_2$. For a collision to occur within δt, then, O_1 must lie within the volume $\sigma^2\delta\mathbf{k}(\mathbf{c}_{12}\cdot\mathbf{k})\delta t$.

The probable number of collisions such that O_2 lies within the volume $\delta\mathbf{r}$ and in which $\mathbf{c}_1, \mathbf{c}_2$ and $\mathbf{k}$ lie within the ranges $\delta\mathbf{c}_1, \delta\mathbf{c}_2$ and $\delta\mathbf{k}$ is

$$\sigma^2(\mathbf{c}_{12}\cdot\mathbf{k})f^{(2)}(\mathbf{r}-\sigma\mathbf{k},\mathbf{c}_1;\mathbf{r},\mathbf{c}_2;t)\,\delta\mathbf{k}\,\delta\mathbf{c}_1\,\delta\mathbf{c}_2\,\delta\mathbf{r}\,\delta t.$$

During a collision, particle 2 gains a quantity $\psi_2'-\psi_2$ of ψ, where primed and unprimed quantities refer to values after and before the collision. Considering only particles that are about to collide (i.e. taking $\mathbf{c}_{12}\cdot\mathbf{k}>0$), we find that the collisional rate of increase for the mean of ψ per unit volume

$$\phi_c = \sigma^2\int_{\mathbf{c}_{12}\cdot\mathbf{k}>0}(\psi_2'-\psi_2)(\mathbf{c}_{12}\cdot\mathbf{k})f^{(2)}(\mathbf{r}-\sigma\mathbf{k},\mathbf{c}_1;\mathbf{r},\mathbf{c}_2;t)\,d\mathbf{k}\,d\mathbf{c}_1\,d\mathbf{c}_2. \tag{4.22}$$

This may be put in a more convenient and physically significant form. Interchanging the roles of the colliding particles by interchanging the subscripts 1 and 2 and replacing $\mathbf{k}$ by $-\mathbf{k}$, (4.22) may be written as

$$\phi_c = \sigma^2\int_{\mathbf{c}_{12}\cdot\mathbf{k}>0}(\psi_1'-\psi_1)(\mathbf{c}_{12}\cdot\mathbf{k})f^{(2)}(\mathbf{r},\mathbf{c}_1;\mathbf{r}+\sigma\mathbf{k},\mathbf{c}_2;t)\,d\mathbf{k}\,d\mathbf{c}_1\,d\mathbf{c}_2. \tag{4.23}$$

After expanding the pair-distribution function $f^{(2)}(\mathbf{r}-\sigma\mathbf{k},\mathbf{c}_1,\mathbf{r},\mathbf{c}_2;t)$ in a Taylor series and rearranging, we may write

$$\begin{aligned} f^{(2)}(\mathbf{r},\mathbf{c}_1;\mathbf{r}+\sigma\mathbf{k},\mathbf{c}_2;t) &= f^{(2)}(\mathbf{r}-\sigma\mathbf{k},\mathbf{c}_1;\mathbf{r},\mathbf{c}_2;t)\\ &+[\sigma\mathbf{k}\cdot\nabla-\frac{1}{2!}(\sigma\mathbf{k}\cdot\nabla)^2+\frac{1}{3!}(\sigma\mathbf{k}\cdot\nabla)^3+\dots\,]\,f^{(2)}(\mathbf{r},\mathbf{c}_1;\mathbf{r}+\sigma\mathbf{k},\mathbf{c}_2;t). \end{aligned} \tag{4.24}$$

Substituting (4.24) in (4.23) and adding this to (4.22), we obtain

$$\phi_c = -\nabla\cdot\boldsymbol{\theta}+\chi, \tag{4.25}$$

where the collisional transfer contribution

$$\boldsymbol{\theta} = -\frac{\sigma^3}{2}\int_{\mathbf{c}_{12}\cdot\mathbf{k}>0}(\psi_1' - \psi_1)(\mathbf{c}_{12}\cdot\mathbf{k})\mathbf{k}\left[1 - \frac{1}{2!}\sigma\mathbf{k}\cdot\nabla + \frac{1}{3!}(\sigma\mathbf{k}\cdot\nabla)^3 + ...\right]$$
$$\times f^{(2)}(\mathbf{r},\mathbf{c}_1;\mathbf{r}+\sigma\mathbf{k},\mathbf{c}_2;t)\, d\mathbf{k}\, d\mathbf{c}_1\, d\mathbf{c}_2\,, \tag{4.26}$$

and the 'source-like' contribution

$$\chi = \frac{\sigma^2}{2}\int_{\mathbf{c}_{12}\cdot\mathbf{k}>0}(\psi_2' + \psi_1' - \psi_2 - \psi_1)(\mathbf{c}_{12}\cdot\mathbf{k})f^{(2)}(\mathbf{r}-\sigma\mathbf{k},\mathbf{c}_1;\mathbf{r},\mathbf{c}_2;t)\, d\mathbf{k}\, d\mathbf{c}_1\, d\mathbf{c}_2\,. \tag{4.27}$$

Note that, when ψ is a summational invariant, χ is 0. When ψ is taken to be the translational kinetic energy $\frac{1}{2}mc^2$, then $\chi = -\gamma$, where γ is the rate of energy dissipation per unit volume due to inelastic collisions of the smooth particles. Also note that to first order

$$\boldsymbol{\theta} \approx -\frac{\sigma^3}{2}\int_{\mathbf{c}_{12}\cdot\mathbf{k}>0}(\psi_1' - \psi_1)(\mathbf{c}_{12}\cdot\mathbf{k})\mathbf{k}\, f^{(2)}(\mathbf{r}-\frac{1}{2}\sigma\mathbf{k},\mathbf{c}_1;\mathbf{r}+\frac{1}{2}\sigma\mathbf{k},\mathbf{c}_2\; t)\, d\mathbf{k}\, d\mathbf{c}_1\, d\mathbf{c}_2\,, \tag{4.28}$$

which is the same expression as given by Chapman & Cowling (1970, § 16.4) for the collisional transfer of 'molecular' properties. The expression for $\boldsymbol{\theta}$ given by (4.26) differs from the corresponding expression of Jenkins & Savage (1983); the difference is due to an incorrect spatial shifting of $f^{(2)}$ and the neglect of some higher-order terms in Jenkins & Savage (1983). However, because of the simple form used by Jenkins & Savage for $f^{(2)}$, their final integrated expressions for the constitutive equations are no different from those that are obtained using (4.26) and (4.27).

Considering the particles to be smooth but inelastic with a coefficient of restitution e (Jenkins & Savage 1983), one finds that the component of relative velocity $\mathbf{c}_{12}$ in the direction of $\mathbf{k}$ is changed during a collision such that

$$\mathbf{k}\cdot\mathbf{c}_{12}' = -e(\mathbf{k}\cdot\mathbf{c}_{12})\,. \tag{4.29}$$

The translational kinetic–energy change during a collision is

$$\begin{aligned}\Delta E &= \frac{1}{2}m[(c_1'^2 + c_2'^2) - (c_1^2 + c_2^2)]\,,\\ &= -\frac{1}{4}m(1-e^2)(\mathbf{k}\cdot\mathbf{c}_{12})^2\,.\end{aligned} \tag{4.30}$$

By taking ψ to be $m, m\mathbf{c}$ and $\frac{1}{2}mc^2$ in (4.20) and (4.25)–(4.27) we obtain the usual hydrodynamic equations

$$\frac{d\rho}{dt} = -\rho\nabla\cdot\mathbf{u}\,, \tag{4.31}$$

$$\rho\frac{d\mathbf{u}}{dt} = \rho\mathbf{b} - \nabla\cdot\mathbf{p}\,, \tag{4.32}$$

$$\frac{3}{2}\rho\frac{dT}{dt} = -\mathbf{p} : \nabla\mathbf{u} - \nabla\cdot\mathbf{q} - \gamma\,. \tag{4.33}$$

In these equations $\rho = mn = \nu\rho_p$ is the bulk mass density, ν is the bulk solids fraction, ρ_p is the mass density of an individual particle, $\mathbf{u} = \langle\mathbf{c}\rangle$ is the bulk velocity, $\mathbf{p}$ is the pressure tensor, $\gamma = -\chi(\frac{1}{2}mc^2)$ is the collisional rate of dissipation per unit volume and $\mathbf{b}$ is the body force per unit mass. Finally, $\frac{3}{2}T = \frac{1}{2}\langle C^2\rangle$ is the specific kinetic energy of the velocity fluctuations (or for short the translational fluctuation energy) where $\mathbf{C} = \mathbf{c} - \mathbf{u}$, $C = |\mathbf{C}|$ and $\mathbf{q}$ is the flux of fluctuation energy. The quantity T has become known as the granular temperature.

The total pressure tensor $\mathbf{p}$ is the sum of a kinetic part

$$\mathbf{p}_k = \rho\langle\mathbf{CC}\rangle\,, \tag{4.34}$$

and a collisional part

$$\mathbf{p}_c = \boldsymbol{\theta}(m\mathbf{C})\,. \tag{4.35}$$

Similarly, the flux of fluctuation energy $\mathbf{q}$ is the sum of a kinetic part

$$\mathbf{q}_k = \frac{1}{2}\rho\langle C^2\mathbf{C}\rangle\,, \tag{4.36}$$

and a collisional part

$$\mathbf{q}_c = \boldsymbol{\theta}(\frac{1}{2}mC^2)\,. \tag{4.37}$$

4.1.2. Constitutive equations for smooth, nearly elastic spherical particles

It is possible to determine detailed expressions for the constitutive equations for the case of dissipative granular particles by following the approach of either the Chapman-Enskog or the moment methods. These methods are based on the assumption (explicit or implicit) that gradients of the mean-flow properties such as velocity, temperature and bulk density are in some sense small. The small-gradient assumption in the present context implies that the energy dissipated during a collision is small. Hence, in the approach that follows we assume that the particles are nearly elastic, i.e. the coefficient of restitution e is nearly unity. This analysis for inelastic particles should reduce to the standard Enskog results for dense hard-sphere fluids (Chapman & Cowling 1970, Ferziger & Kaper 1972, Hirschfelder, Curtis & Bird 1954, Davis 1973) in the limit as $e \to 1$.

Instead of using the procedure described by Chapman & Cowling (1970), which relies upon the Boltzmann equation, we use a simple moment method based on the Maxwell transport equation (4.20) and (4.25)–(4.27). This is done in the spirit of the approach used in the early work of Chapman (1916), which has been described in elementary terms for the dilute-gas case in the books of Present (1958), Guggenheim (1960) and Reif (1965).

Following the usual practice we assume that the singlet distribution function $f^{(1)}$ can be written in the form

$$f^{(1)} = f^{(0)}(1 + \phi)\,, \tag{4.38}$$

where the perturbation $\phi \ll 1$, and

$$f^{(0)}(\mathbf{r},\mathbf{c},t) = \frac{n}{(2\pi T)^{\frac{3}{2}}} \exp\left(-\frac{(\mathbf{c}-\mathbf{u})^2}{2T}\right) \tag{4.39}$$

is the *local* equilibrium Maxwellian single-particle velocity distribution function. From the standard kinetic theories we anticipate that a good approximation to the perturbation ϕ is the trial function

$$\phi = a_1(\mathbf{CC} - \frac{1}{3}C^2\mathbf{I}):\nabla\mathbf{u} + a_2\left(\frac{5}{2} - \frac{C^2}{2T}\right)\mathbf{C}\cdot\nabla\ln T + a_3\left(\frac{5}{2} - \frac{C^2}{2T}\right)\mathbf{C}\cdot\nabla\ln n\,, \tag{4.40}$$

where

$$\mathbf{C} = \mathbf{c} - \mathbf{u}\,. \tag{4.41}$$

The quantities a_1, a_2 and a_3 appearing in (4.40) can be determined by satisfying various moment equations generated by (4.20) and (4.25)–(4.27). The first two terms in (4.40) are of the standard form, but the last term is not present in the usual theories. The need for the third term in (4.3) is suggested when one attempts to satisfy the moment equations. The functional dependence of this third term upon C is evident from the scalar nature of ϕ and the necessary compatibility with definition for the mean velocity $\mathbf{u} = \langle\mathbf{c}\rangle$. However, subsequent examination of the order of the various terms in (4.40) after the expressions for a_1, a_2 and a_3 have been determined will reveal that the term involving a_3 is of higher order than the other two terms and that $a_3 \to 0$ as $e \to 1$.

We make the Enskog assumption that the pair-distribution function

$$f^{(2)}(\mathbf{r}_1,\mathbf{c}_1;\mathbf{r}_2,\mathbf{c}_2;t) = g_0(\sigma)f^{(1)}(\mathbf{r}-\sigma\mathbf{k},\mathbf{c}_1;t)f^{(1)}(\mathbf{r},\mathbf{c}_2;t) \tag{4.42}$$

just prior to collision, and take the equilibrium radial distribution function $g_0(\sigma)$ to be that due to Carnaham & Starling (1969) given by

$$g_0(\nu) = \frac{(2-\nu)}{2(1-\nu)^3}\,. \tag{4.43}$$

The coefficient a_1, appearing in (4.40) may be determined by considering the case of $\nabla\mathbf{u} = \mathbf{e}_y\mathbf{e}_x du/dy$, $du/dy = \text{const}$, $T = \text{const}$ and $n = \text{const}$, (where $\mathbf{u} = (u,v,w)$) and taking $\psi = c_x c_y$ in (4.20) and (4.25)–(4.27). Similarly a_2 and a_3 may be determined by considering $\mathbf{u} = 0$, and T and n to have gradients in the x-direction and taking $\psi = c^2 c_x$. Thus we find

$$a_1 = -\frac{\mu_i}{\rho T^2 g_0}\left[1 + \frac{8}{5}\eta(3\eta-2)\nu g_0\right], \tag{4.44}$$

$$a_2 = \frac{2\lambda_i}{5\rho T g_0}\left[1 + \frac{12\eta^2}{5}(4\eta-3)\nu g_0\right], \tag{4.45}$$

$$a_3 = \frac{24\lambda_i \eta}{25\rho T g_0}(2\eta - 1)(\eta - 1)\frac{d}{d\nu}(\nu^2 g_0)\,, \tag{4.46}$$

where

$$\eta = \frac{1}{2}(1+e)\,, \tag{4.47}$$

$$\mu_i = \frac{\mu}{\eta(2-\eta)}, \quad \lambda_i = \frac{8\lambda}{\eta(41-33\eta)}\,, \tag{4.48}$$

$$\mu = \frac{5m(T/\pi)^{\frac{1}{2}}}{16\sigma^2}, \quad \lambda = \frac{75m(T/\pi)^{\frac{1}{2}}}{64\sigma^2}\,. \tag{4.49}$$

The expressions for μ and λ may be identified as the shear viscosity and thermal conductivity for perfectly elastic particles ($e = 1, \eta = 1$) at *dilute concentrations.* They check with the usual first approximation results from the hard-sphere kinetic theory (see Davis 1973). The symbols μ_i and λ_i denote the corresponding shear viscosity and granular thermal conductivity for the case of *inelastic* particles at dilute concentrations. However, the analysis has been based upon the assumption of small dissipation, such that $(1-e) \ll 1$, as well as the assumption that the gradients of mean velocity, granular temperature and solids fraction are all small. The expressions for a_1, a_2, a_3, μ_i and λ_i can be written in terms of the small parameter $(1-e)$. Jenkins (1987b) has pointed out that if we consider the perturbation ϕ in (4.40) to involve only first order gradients, then to be consistent we must use the zeroth order approximations for a_1, a_2 and a_3, and not the full expressions (4.43)–(4.45). Then, to zeroth order $a_3 = 0$, $\mu_i = \mu$, $\lambda_i = \lambda$, and the expressions for a_1 and a_2 reduce to

$$a_1 = -\frac{\mu}{\rho T^2 g_0}\left[1 + \frac{8}{5}\nu g_0\right]\,, \tag{4.50}$$

$$a_2 = \frac{2\lambda}{5\rho T g_0}\left[1 + \frac{12}{5}\nu g_0\right]\,. \tag{4.51}$$

These results are exactly the same as those for perfectly elastic particles; the effects of the inelastic particles will appear only in the next approximation to ϕ involving higher order gradients of mean velocity, temperature, density, etc.

Using (4.40), (4.50) and (4.51) in (4.26), (4.27) and (4.34)-(4.37) we can calculate the constitutive equations for stress, translational fluctuation-energy flux, and the collisional rate of dissipation per unit volume. The kinetic and collisional contributions to the total stress tensor $\mathbf{p}$ and fluctuation energy fluxes $\mathbf{q}$ are the same as the classical results for dense hard-sphere fluids corresponding to the first terms in the Sonine polynomial expansions (Chapman & Cowling 1970, Ferziger & Kaper 1972, Hirschfelder, *et al.* 1954, Davis 1973), i.e.

$$\mathbf{p}_k = \rho T \mathbf{I} - \frac{2\mu}{g_0}\left[1 + \frac{8}{5}\nu g_0\right]\mathbf{S}, \tag{4.52}$$

$$\mathbf{p}_c = 4\rho T \nu g_0 \mathbf{I} - \frac{16\mu\nu}{5}\left[1 + \frac{8}{5}\nu g_0\right]\mathbf{S} - \frac{256}{5\pi}\mu\nu^2 g_0\left[\frac{6}{5}\mathbf{S} + (\nabla\cdot\mathbf{u})\,\mathbf{I}\right]\,, \tag{4.53}$$

where

$$\mathbf{S} = \frac{1}{2}(u_{i,j} + u_{j,i}) - \frac{1}{3}u_{k,k}\delta_{ij}. \tag{4.54}$$

The total stress tensor may be written as

$$\mathbf{p} = \mathbf{p}_k + \mathbf{p}_c = [\rho T(1+4\nu g_0) - \mu_b \nabla\cdot \mathbf{u}]\mathbf{I} - \left[\frac{2\mu}{g_0}(1+\frac{8}{5}\nu g_0)^2 + \frac{6}{5}\mu_b\right]\mathbf{S}\,, \tag{4.55}$$

where

$$\mu_b = \frac{256\mu\nu^2 g_0}{5\pi}\,, \tag{4.56}$$

is the bulk viscosity.

The kinetic and collisional contributions to the vector flux of translational fluctuation kinetic energy are

$$\mathbf{q}_k = -\frac{\lambda}{g_0}\left[1+\frac{12}{5}\nu g_0\right]\nabla T\,, \tag{4.57}$$

$$\mathbf{q}_c = -\frac{12\lambda\nu}{5}\left[1+\frac{12}{5}\nu g_0 + \frac{128}{15\pi}\nu g_0\right]\nabla T\,, \tag{4.58}$$

and the total 'thermal' flux vector is

$$\mathbf{q} = \mathbf{q}_k + \mathbf{q}_c = -\frac{\lambda}{g_0}\left[(1+\frac{12}{5}\nu g_0)^2 + \frac{512}{25\pi}(\nu g_0)^2\right]\nabla T\,. \tag{4.59}$$

The expression for the collisional rate of energy dissipation per unit volume carried out to the same order of approximation as the expressions for stress and energy flux is

$$\gamma = \frac{24}{\pi^{\frac{1}{2}}}(1-e)\frac{\rho_p\nu^2}{\sigma}g_0 T^{\frac{3}{2}}\,, \tag{4.60}$$

which vanishes as $e \to 1$. To *first order*, this is the only expression which differs from the classical results for elastic particles.

4.1.3. Simple shear flow

Let us now use these results to consider the case of simple uniform shear flow ($du/dy =$ const.), having uniform translational granular temperature T and bulk density ρ. From the expression for the stress tensor (4.55) it may be seen that the normal stresses are isotropic and depend linearly upon the granular temperature T, i.e.

$$p_{yy} = p_{xx} = p_{zz} = \rho T(1+4\nu g_0)\,, \tag{4.61}$$

whereas the shear stress p_{xy} depends upon $T^{1/2}$,

$$p_{xy} = p_{yx} = -F(\nu)\frac{5}{96}\rho_p(\pi T)^{\frac{1}{2}}\sigma\frac{du}{dy}\,, \tag{4.62}$$

where

$$F(\nu) = \frac{1}{g_0}\left[1+\frac{8}{5}\nu g_0\right]^2 + \frac{768}{25\pi}\nu^2 g_0\,. \tag{4.63}$$

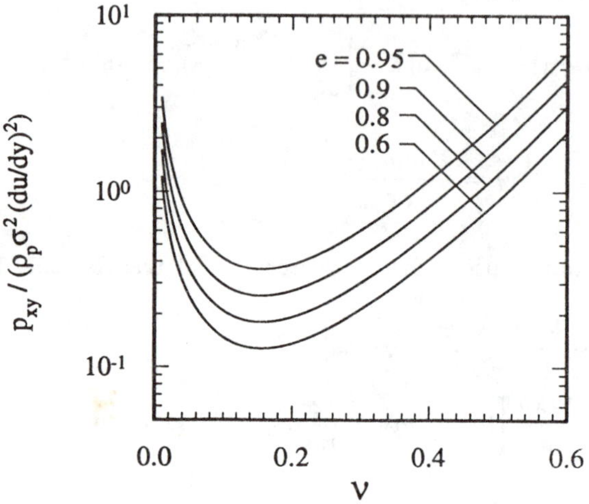

Fig. 10: Kinetic theory predictions of nondimensional shear stress versus solids fraction for different values of coefficient of restitution. Smooth, slightly inelastic, spherical particles in a uniform simple shear flow (first order results deduced from Lun, *et al.* 1984).

When the energy equation (4.33) (which reduces to $p_{xy} du/dy + \gamma = 0$ for this uniform homogeneous shear flow) is considered, we can determine the magnitude of granular temperature for a particular shear–rate and particle coefficient of restitution e. It is convenient to express this relationship in terms of a nondimensional parameter R which was first defined by Savage & Jeffrey (1981)

$$R = \frac{\sigma |du/dy|}{(3T)^{\frac{1}{2}}} = \left[\frac{768}{5\pi} \frac{\nu^2 g_0}{F(\nu)} \right]^{\frac{1}{2}} . \tag{4.64}$$

We can define a dynamic friction angle ϕ_D whose tangent is the ratio of the shear to normal stress

$$\frac{|p_{xy}|}{p_{yy}} = \tan \phi_D = \frac{5(1-e) g_0 F(\nu)}{4(1 + 4\nu g_0)} . \tag{4.65}$$

From the above it can be seen that the granular temperature T is proportional to the *square* of the shear–rate du/dy. Hence, we recover Bagnold's result for the *grain–inertia* regime that both the shear and normal stresses depend upon the square of the shear–rate and that the ratio of the shear to normal stress is nearly constant for fixed concentration ν. Some typical results for shear stresses from the above expressions are shown in figure 10. In general, stresses decrease with decreases in coefficient of restitution e. At high concentrations, it is found, as expected, that the dominant contribution to the stresses comes from the collisional interactions. But, at low concentrations, the collisional stresses are negligible compared to the *kinetic* or *transport terms* resulting from the transport of momentum by particle translation across shear layers. The predicted stress levels (at high concentrations) are in the same range as the stresses measured in annular shear cell experiments (Savage & Sayed 1984, Hanes & Inman 1985, Craig, *et al.* 1986).

Jenkins & Savage (1983) performed a simple analysis that made use of the *ad hoc* assumption of Savage & Jeffrey (1981) for the pair distribution function. There $f^{(2)}$ was assumed to be the product of two (unperturbed) Maxwellian single particle velocity distribution functions and a particle pair spatial distribution function. This assumption makes the evaluation of the constitutive relations quite straightforward, but it can only recover the collisional contributions. In that simple approach, the stresses increase in monotonic fashion with concentration. When the kinetic–stress contributions are included as in the more complete analysis just described, the stresses have *high* values at low concentration. The stresses *decrease* with increasing concentration to a minimum, and thereafter increase with increasing concentration and follow the same trends as the collisional stresses alone. This kind of behaviour at low ν may seem implausible, but it can be explained simply in physical terms as follows. As ν decreases, the contributions of the collisional stresses becomes a smaller fraction of the total stresses and the kinetic contribution becomes dominant. Consider the shear rate du/dy to be fixed as ν decreases. From the usual arguments for Maxwell's law for a dilute gas (Jeans 1967, p.164), the shear viscosity and thus the shear stress will be independent of bulk density. However, the dissipation γ has a stronger dependence on granular temperature T than does the shear stress. Therefore, when ν is decreased, T must increase to provide the dissipation required for the energy balance. Thus the increase in T causes the shear stress to increase as ν is decreased while keeping shear–rate fixed. Note, for example, that in figure 10 the stresses appear to increase without limit as the concentrations $\nu \rightarrow 0$. In real flows this would not occur for two reasons. The analysis is based upon the assumption of a constant coefficient of restitution, whereas in reality e decreases with increased impact velocity. A real flow would take place between walls spaced a finite distance apart. As bulk density is decreased, the mean free path could not increase without limit since the particles would collide with the confining walls.

4.1.4. More refined models of particle interactions

In the above example we considered the simplest model possible to represent the particles. While it captures many of the essential features of the granular flows, more complex kinetic theories have been developed in attempts to refine the modelling of the particle interactions. These have included studies of (a) the effects of particle surface roughness (Jenkins & Richman 1985b, Lun & Savage 1987); (b) the effects of coefficients of restitution that depended upon particle impact velocity (Lun & Savage 1986); and (c) highly inelastic particles (Richman 1988) that give rise to granular temperatures which are anisotropic (recall that the Maxwellian velocity distribution function (4.39) that was used in the above analysis for nearly elastic particles involves an isotropic granular temperature). Granular flows often involve particles which are not uniform and kinetic theories have been developed to consider binary mixtures of particles of different sizes (cf. Farrell, *et al.* 1986, Jenkins & Mancini 1987).

4.1.5. Quasi-static stresses

The kinetic theories just discussed have been based upon the assumption of instantaneous binary collisions and are appropriate for the grain inertia regime at moderate concentrations. There are many engineering applications and natural flow situations in which the concentrations are fairly high and the particle interactions are not instantaneous impulsive forces. At high concentrations, particles can come into contact for extended periods; this results in the generation of sustained normal and Coulomb frictional forces. Such conditions often arise when gravitational forces are significant. Although such mixed type flow regimes can be treated quite easily by the 'soft' particle computer simulations, it is not obvious how to extend the analytical kinetic theory models in a rigorous way to handle these kinds of particle interactions. The only analytical attempts to do so have involved a rather crude patching together of results from analyses for the quasi–static and grain–inertia flow regimes mentioned above. Simple examples of this may be found in Savage (1983) and in the more detailed analyses of Johnson & Jackson (1987) and Johnson, Nott & Jackson 1990. Savage (1983) suggested that, for the case of a shear flow in a gravitational field, one might represent the total stresses as the linear sum of a rate–independent, dry friction part plus a rate–dependent 'viscous' part obtained from the high shear rate granular flow kinetic theories. Because of its practical importance, this flow regime is obviously one that deserves increased attention.

The inclusion of the effects of long time frictional rubbing contacts as well as the proper treatment of the boundary conditions are essential if one is to solve boundary value problems in a physically realistic way. Their importance is illustrated by comparing the success achieved by Johnson & Jackson (1987) in their shear cell simulations and later chute flow simulations of Johnson, *et al.* (1990) with the problems experienced by Szidarovszky, *et al.* (1987) in analogous chute flow computations. The very limited range of conditions for which the latter workers (Szidarovszky, *et al.* 1987) were able to obtain solutions, and the extreme difficulties that were encountered in obtaining them, resulted from their failure to include the above effects.

4.1.6. Boundary conditions

The previous sections have not touched upon the question of boundary conditions; the above uniform, simple shear flow example corresponded to an 'infinite' flow field. All real granular flows take place adjacent to solid boundaries and/or 'dead flow' zones, and can sometimes entail free surfaces. Velocity 'slip' and granular temperature 'jumps' at such boundary surfaces can occur and particle fluctuation kinetic energy can be supplied or drained away at the boundary 'surfaces'. The application of boundary conditions such as 'no–slip' and specified values for the granular temperature or its gradient, as might be done for a continuum fluid, are often only crude first approximations to what actually occurs, and in some cases they can be seriously in error. The appropriate treatment of boundary conditions has been the subject of recent studies (Hui, *et al.* 1984, Jenkins & Richman 1986, Hanes, *et al.* 1988, Richman 1988, Richman & Chou 1988, Johnson & Jackson 1987, Johnson, *et al.* 1990).

Some investigators have considered flows next to 'rough' impenetrable walls made up of hemispherical or disk–like particles that were assumed to be randomly arranged with constant *mean* spacing along the wall. Boundary conditions were determined by consideration of the particle collisional interactions using kinetic theory analyses analogous to those carried out for the particles in the interior. Here we will sketch the essence of the approach described by Johnson & Jackson (1987). It includes the frictional stress contributions $\mathbf{p}_{fr}$ as well as the collisional stress contributions $\mathbf{p}_c$; thus, while it is similar to the other treatments, it is somewhat more general in conception (if not the detailed analysis). The basic idea is to consider a rectangular control volume of unit area parallel to the wall which encompasses both the wall 'particles' and the interior flow particles. Two boundary conditions are obtained by considering the momentum and energy balances over the control volume in the limit as the depth (perpendicular to the wall) of the control volume tends to zero. A slip velocity $\mathbf{u}_{sl}$ is defined as the difference between the velocity of the interior particles at the 'wall' and the velocity of the wall itself, i.e. $\mathbf{u}_{sl} = \mathbf{u} - \mathbf{u}_w$.

Consideration of momentum fluxes and stresses yields the momentum condition acting in the direction of the slip velocity

$$\frac{\mathbf{u}_{sl} \cdot (\mathbf{p}_c + \mathbf{p}_{fr}) \cdot \mathbf{n}}{|\mathbf{u}_{sl}|} + M + N_{fr} \tan\delta = 0 , \tag{4.66}$$

where $\mathbf{n}$ is the inward unit normal vector at the wall, N_{fr} is the normal frictional component of stress at the wall, and δ is the angle of friction between the 'wall' and the interior particulate material. The first term in (4.66) corresponds to the total (collisional and frictional) stress in the interior, the second term M is the collisional rate of momentum transfer per unit wall area, and the last term is the magnitude of the tangential frictional stress component at the wall.

The overall energy balance leads to the collisional energy condition

$$-\mathbf{n} \cdot \mathbf{q} = \Gamma + \mathbf{u}_{sl} \cdot \mathbf{S}_c , \tag{4.67}$$

where the left hand side of (4.67) corresponds to the normal component of the particle fluctuation (pseudo–thermal) energy flux at the wall, Γ is the rate of energy dissipation per unit wall area due to collisions with the boundary, and the last term is the (collisional) slip work term which is the product of the slip velocity $\mathbf{u}_{sl}$ and $\mathbf{S}_c$, the force per unit area on the boundary due to particle collisions with the boundary.

Problems of Couette flow of granular material between rough walls and gravity driven granular flows down inclined chutes have been treated (cf. Jenkins & Richman 1986, Hanes, *et al.* 1988, Richman 1988, Richman & Chou 1988, Johnson & Jackson 1987, Johnson, *et al.* 1990) by making use of these kinds of boundary conditions in conjunction with the conservation and constitutive equations (such as (4.31)–(4.33), (4.55), (4.59), (4.60), etc.).

4.2. Attempts to include interstitial fluid effects

4.2.1. Slow flows of concentrated suspensions

Jenkins & McTigue (1990) have given heuristic arguments (following Haff's (1983) treatment for the grain–inertia regime) to establish the form of the constitutive relations for concentrated suspensions at low shear rates corresponding to Bagnold's macro–viscous regime.

One of the important features of this flow regime was the existence of a dispersive normal stress which depended linearly upon shear–rate and was proportional to the shear stress. As we have seen for the grain–inertia regime, the particle velocity fluctuations (and hence, the granular temperature) play an essential role in establishing the overall dynamics and in determining the form and magnitude of transport coefficients. To solve boundary value problems, it is necessary to solve the particle fluctuation energy equation in conjunction with the conservation of mass and linear momentum equations. Jenkins & McTigue assume that the particle phase of the concentrated suspension can be treated as a continuum fluid for which the usual conservation of mass and momentum equations (4.31) and (4.32) hold. They argue, in a somewhat more heuristic way, that the velocity fluctuations are governed by a fluctuation energy equation of the form of equation (4.33) given previously for the grain–inertia regime, where the fluctuation energy flux there was given by (4.59).

By dimensional arguments, they establish the form of the transport coefficients for low particle Reynolds numbers, $R_p = \rho_f T^{1/2}\sigma/\mu_f$, where ρ_f and μ_f are the interstitial fluid density and viscosity. For small R_p, inertial forces generated during collisional interactions are of less significance and viscous interaction forces dominate. Hence, the density ρ_p should not appear in the constitutive expressions and we replace ρ_p by the dimensionally equivalent group $\mu_f/(\sigma T^{1/2})$. Because of the finite size effects at high concentrations the interstitial fluid experiences a shear–rate greater than the apparent shear–rate by roughly a factor of (σ/s), where s is the free space between particles. In order to generate the dispersive normal stresses, the Reynolds number must be large enough or other effects must present such that the flow is not reversible; i.e. the viscous interparticle forces developed when particles approach must be different from that when they separate. By considering the form of the constitutive relations in the grain–inertia regime and eliminating ρ as just noted, they obtain results as follows. Examining (4.55) that was derived for the grain–inertia regime, it was argued that, for the macro–viscous regime, the dispersive normal stress should be proportional to $\mu_f(\sigma/s)T^{1/2}/\sigma$ and that the effective viscosity should be proportional to $\mu_f\sigma/s$. Eliminating ρ_p from (4.59) and (4.60), one finds, for the macro–viscous regime, that the conductivity for fluctuation energy is proportional to $\mu_f\sigma/s$ and the rate of dissipation of fluctuation energy per unit volume is proportional to $\mu_f(\sigma/s)T/\sigma^2$. By considering the simple case of a steady, uniform, simple shear flow du/dy, it is found that the pseudo–temperature $T \sim \sigma^2(du/dy)^2$. As a result, both the normal stress and shear stress depend linearly upon the shear–rate du/dy and the ratio of shear to normal stress is constant, consistent with the experimental observations in Bagnold's macro–viscous regime.

Plausible and consistent functional forms for the transport coefficients have been obtained by these dimensional and heuristic arguments. However, the dependencies on solids fraction ν and the numerical constants of proportionality are lacking. Further, more detailed computations, analogous to those performed in the grain–inertia kinetic theories, may provide this information.

4.2.2. Kinetic theories for gas–solid mixtures

We now consider fluid effects at the other extreme, away from the macro–viscous regime. In the applications to gas–solid systems, the particles can be considered as relatively massive.

By making modifications to the granular flow kinetic theories developed for the grain–inertia regime to account for the presence of a low density interstitial fluid, the effective solids pressure and the effective solids viscosity can be calculated in a systematic way.

The approach of Sinclair & Jackson (1989), for gas–particle flow in a vertical pipe, makes use of the axial momentum equation for the gas phase, in conjunction with the axial and radial momentum equations and fluctuation energy equation for the solids phase. The interaction between the gas and particle phase is restricted to a mutual drag force term in the axial momentum equations which depends upon the local solids concentration and the relative velocity between the gas and the particles. The constitutive relations for the particle phase stress, pseudo–thermal energy flux and dissipation rate were based upon the granular flow kinetic theory of Lun, *et al.* (1984). The theory of Sinclair & Jackson (1989) was able to generate a surprisingly complex variety of behaviours, some of which are observed in practice but not produced by simpler analyses. Because of the lack of detailed experimental measurements, it was not possible to make quantitative comparisons with the theory. It was suggested (Sinclair & Jackson 1989) that such a comparison might have been premature because of the neglect in the analysis of important factors such fluid turbulence.

A related two–phase flow analysis has been developed by Ding & Gidaspow (1990) to treat bubbling fluidized beds. Again the solids viscosities and stresses were obtained from the granular flow kinetic theory of Lun, *et al.* (1984). Particle drag effects were included in momentum equations for both phases and an additional term was added to the particle pseudo–thermal energy equation to account for dissipation due to particle fluctuations in the viscous interstitial fluid. The resulting set of governing equations for the gas and solids phases were integrated numerically and the predicted formation of bubbles and solids flow patterns were in agreement with experimental observations.

4.3. Computer simulations

In parallel with the above mentioned work on kinetic theory analyses, there have been computer simulations (Campbell 1990) of granular flows that are analogous to molecular dynamics calculations used in the studies of liquids and dense fluids at the molecular level. In some cases, they have been based upon the same kinds of modelling of the particle interactions as in the kinetic theories, but in other cases these interactions have been treated in more complex and more realistic ways. However, the important attribute of the simulations is that they are completely deterministic calculations based upon the solution of Newton's laws of motion. They don't rely upon any assumptions such as 'molecular chaos', small perturbations of equilibrium Maxwellian velocity distributions functions, binary collisions, product forms for the pair velocity distribution functions, etc., that are used in the kinetic theories. In fact, they can be used to check the validity of such assumptions. In general, it has been found that there is quite close agreement between the computer simulations and kinetic theories, except under extreme conditions where the assumptions in the kinetic theories are obviously inappropriate. Examples of such differences are very high concentration flows in thin shear gaps where particles tend to flow in distinct layers (Savage & Dai 1992), or attempts to predict flows of very dissipative particle with kinetic theories intended for nearly elastic particles.

Some simulations (Campbell & Brennen 1985, Campbell 1989, 1990, Campbell & Gong 1986, Savage 1992, Walton, *et al.* 1991) have modelled the grains as 'hard' particles, such that the collisions can be regarded as nearly instantaneous and binary. This usually results in fairly efficient computational schemes, since the time step can be chosen as the (shortest) time between collisions. While there are some indications that the 'hard' particle models that are based on instantaneous contacts can predict some of the general features of flows at quite high concentrations, they obviously cannot be used to treat flows which are truly in the continuous, multiple contact, quasi-static regime. Such flows can be handled by computer simulations that consider 'soft' particles; these models involve more complex kinds of interparticle interactions of finite durations (Walton & Braun 1986, Walton, *et al.* 1987, Walton, *et al.* 1991). Multiple contacts, inelasticity and surface friction can be handled, but short time steps are required to accurately calculate the dynamics of the impact and rebound process, and hence, the computations are usually more time consuming than the 'hard' particle approaches. These types of computations are important because many real physical problems involve long term rubbing contacts and no analytical treatments of this flow regime have yet been developed.

Some simulations have involved flows of finite thickness down inclined chutes and Couette flows generated by the motion of two rough bounding walls. Calculations are typically performed within a finite computational 'box' and 'periodic' boundary conditions are used on the 'box' surfaces perpendicular to the solid bounding walls to effectively extend the region of computation. Other simulations have involved shear flows over effectively infinite flow fields by use of the Lees–Edwards (1972) type periodic boundary conditions on the 'box' surfaces parallel to the shear planes.

While these kinds of deterministic simulations are the most direct and exact, they are also expensive in terms of computer time and memory. Related, but much simpler and faster, calculations can be carried out by using Monte-Carlo methods (Hopkins & Shen 1988). Typically, in this approach it is not possible to calculate all the details of the structure of the flow field and certain assumptions such as the mean shear rate, etc., are required.

Simulations are very useful for checking the results of analytical (kinetic theory) approaches as well as for suggesting ways for them to be modified and improved. While direct simulations, involving even only a few thousand particles (most previous studies have considered only a few *hundred* particles), are computationally very intensive and presently require supercomputer power, one can expect that, at the current rate of improvements in computer performance, in a few years they will not seem so daunting.

5. Granular avalanches; a geophysical example of granular flows

In the Alpine regions, rockfalls pose a threat to life and property. When rock material is dislodged on steep slopes, it accelerates down the slope, reaches a maximum velocity, and then decelerates as the the angle of inclination of the bed decreases; eventually bed friction brings the material to rest. The moving streams of debris, in the form of numerous discrete blocks and fragments that are generated during the rockfall, have been termed "strutzstroms". Erisman (1979) has made various comparisons to give one an appreciation for the enormous energies involved in these events. For example, the prehistoric Flims landslide in Switzerland had a volume of about $1.2 \times 10^{10}\,\mathrm{m}^3$ and probably took place over a period of a couple of minutes. Its motion resulted in the dissipation of an amount of energy equal to the *total present day energy consumption of the world for a period of 10 hours.*

Sturtzstroms often spread out into very thin layers and flow on surfaces that are much less inclined than the angle of repose of the debris material. For over 100 years, since Albert Heim (1882, 1932) observed and described the Elm rockfall in Switzerland, attempts have been made to explain the apparent fluid-like behaviour and high mobility of these slides. Various hypotheses have included fluidization by an upward current of air (Kent 1965), a kind of hovercraft action (Shreve 1966, 1968a, 1968b), fluidization by high pressure steam (Goguel 1978), mechanical fluidization aided by the presence of interstitial dust (Hsü 1975), vibrational fluidization by large earthquakes (McSaveney 1978), lubrication by a layer of molten rock (Erisman 1979, 1986) acoustic fluidization (Melosh 1986), and the presence of a thin layer of vigorously fluctuating particles beneath a densely packed overburden (Dent 1986). All of these hypotheses are at best controversial and none have been universally accepted. Few have been accompanied by detailed computations.

It has been observed that the effective bed friction coefficient of rockfalls is size dependent (Heim 1932, Scheidegger 1975, Hsü 1978, Lucchitta 1978, Davies 1982, and Li Tianchi 1983). The ratio of the total vertical fall height, H, to the total horizontal travel, L, (both measured from top of scar to tip of rockslide) decreases with increasing total volume of the rock material for volumes greater about $10^6\,\mathrm{m}^3$. The explanation for this increased mobility with increased volume is still unknown.

5.1. Continuum treatment of granular avalanches

It is possible to describe the apparent fluid–like behaviour if we analyze the rockfall as a granular flow. Let us consider the simple example of the two–dimensional motion of a finite mass of granular material down a rough curved bed and outline the treatment of Savage & Hutter (1989, 1991). In those analyses, the granular material was treated as an incompressible, cohesionless continuum obeying a Mohr-Coulomb yield criterion. The pile of material was assumed to be long and thin, the streamwise velocity profile nearly uniform, and the curvature of the bed was taken to be moderate (figure 11). By integrating over the thickness of the pile, the following *nondimensional* depth-averaged equations of motion were obtained

$$\frac{\partial h}{\partial t} + \frac{\partial (h\overline{u})}{\partial x} = 0\,, \tag{5.68}$$

and

$$\frac{\partial \overline{u}}{\partial t} + \overline{u}\frac{\partial \overline{u}}{\partial x} = \sin\zeta - \tan\delta\,\mathrm{sgn}(\overline{u})\left(\cos\zeta + \lambda\kappa\overline{u}^2\right) - \epsilon\, k_{actpass}\cos\zeta\,\frac{\partial h}{\partial x}\,, \tag{5.69}$$

where $\overline{u}$ is the depth average streamwise velocity defined by

$$\overline{u} = \int_0^h u\,dy\,, \tag{5.70}$$

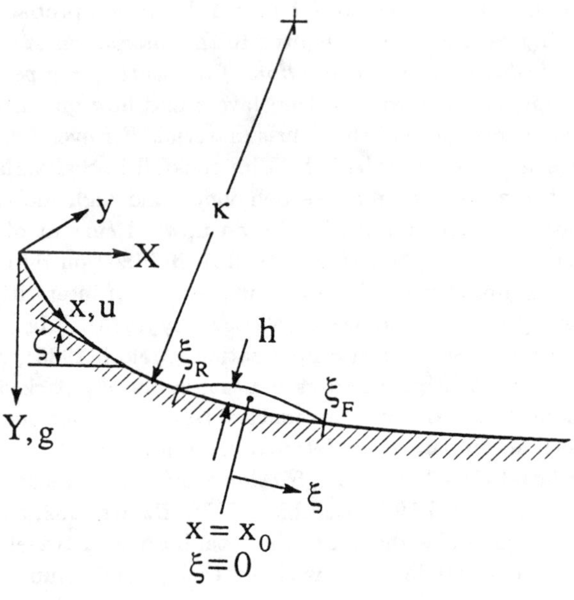

Fig. 11: **Definition sketch for finite mass of granular material moving down a rough curved bed.**

h is the local depth, ζ is the local bed inclination angle, κ is the local curvature of the bed profile, t and x are the nondimensional time and streamwise distance, $\epsilon = H/L$ is the ratio of the characteristic depth H to length L of the pile, and $\lambda = L/R$ is the ratio of the characteristic length L to the bed radius of curvature R. In equations (5.68) and (5.69) the depth was made nondimensional by dividing by the maximum value of the initial physical depth H, the streamwise distance was nondimensionalized by the initial physical pile length L, the streamwise velocity was divided by $\sqrt{gL}$ and time was divided by $\sqrt{L/g}$, where g is the gravitational acceleration.

Following the assumption of Coulomb-like behaviour, the shear stress at the bed was related to the the normal stress there by means of a bed friction angle δ. In accordance with the behaviour discussed in earlier sections, whereby the dynamic friction angle ϕ_D for *dry* granular materials is found to be only weakly dependent on concentration and shear–rate, the bed friction was approximated as a constant, independent of the velocity (or in other cases very nearly so). Following a common soil mechanics practice, the normal stresses parallel and perpendicular to the bed were related to one another through an *earth pressure coefficient*, $k_{actpass}$, which may take on the active or passive values,

$$\left.\begin{matrix} k_{act} \\ k_{pass} \end{matrix}\right\} = 2\left[1 \mp \sqrt{1-(1+\tan^2\delta)\cos^2\phi}\right]/\cos^2\phi - 1\,, \qquad \text{for } \partial\overline{u}/\partial x = \begin{matrix} >0 \\ <0 \end{matrix} \quad (5.71)$$

It was found quite difficult to obtain accurate and reliable numerical solutions to this set of equations; however, analytical similarity solutions (Savage & Noguchi 1988, Savage & Hutter

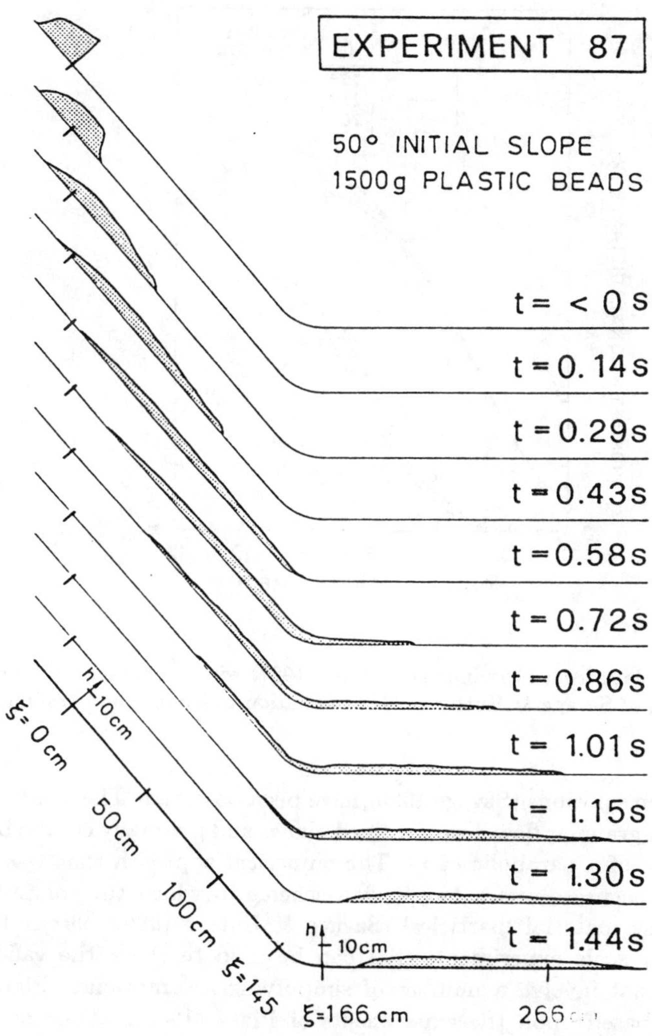

Fig. 12: Evolution of a pile of plastic particles released from rest on a 50° incline. Data obtained from photographs of laboratory experiments of C. Pluss (1987).

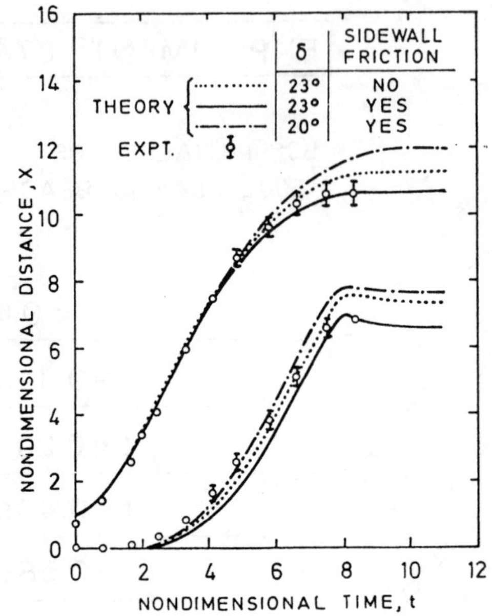

Fig. 13: Comparison of data from experiments of Pluss (1987) with predictions of Lagrangian finite difference computations of Savage & Hutter (1991). Variation of leading and trailing edges of pile with time.

1989) and perturbation solutions (Savage 1990) have been obtained. The similarity solutions were found for cases of granular flow down inclined planes and gradually curved beds; in them the pile has the shape of a parabolic cap . The numerical approach that was found to be the most simple and accurate was a Lagrangian scheme in which the computational grid was advected with the material 'particles' (Savage & Hutter 1989, 1991). It is possible to perform laboratory scale experiments that can be used to check the validity of these theoretical analyses that involve a number of simplifying assumptions. Figure 12 shows some typical results based upon the experiments of Pluss (1987). They are taken from snapshots at various times of the profiles of an evolving pile of plastic beads flowing down a rough inclined bed. Figure 13 compares the measured temporal evolution of the front and rear edges of the avalanche with those predicted by the Lagrangian finite difference scheme.

Most notable is the rapid evolution into a very thin layer of material. The predicted positions of the front and rear of the pile, as well as its depth profile, are in good agreement with the measurements. The analysis has made use of no unusual fluidization mechanism. We have merely assumed a very simple and ordinary Coulomb-type behaviour; the spreading and fluid-like behaviour result directly from the dynamical equations of motion.

It is possible to extend these studies to consider physically more realistic cases of three-dimensional motions in which no side walls are present and lateral spreading is permitted.

Fig. 14: Avalanche of plastic beads released from rest on an inclined plane. Pile has spread out both laterally and longitudinally into a thin layer.

Figures 14 and 15 show results of experiments performed by the present author, C. Bucher and K. Hutter at E.T.H., Zürich in which a mass of granular material having the initial shape of a hemispherical cap was released on rough, plane inclines and the subsequent evolution of the pile was studied. Figure 14 shows a three–quarter view of the evolving pile some time after release. The pile has spread out into a thin layer, with sharp bounding edges. None of the experiments showed any evidence of leading edge steepening of the kind that can sometimes occur in nonlinear shallow water waves. Figure 15 shows a sequence of photos taken by a 35 mm camera with a high speed motor drive that was placed with its axis perpendicular to the plane bed. The very regular and systematic spreading from the initial hemispherical cap shape and the tendency to evolve into a teardrop shaped planform are evident. The above simple two–dimensional granular avalanche analyses can be extended in a straightforward way to handle three–dimensional spreading. When the computational scheme is carefully set up, taking proper account of the initial development phase, one finds good agreement between the experimental measurements and predicted timewise development of the planform shapes. Lang (1992) has performed both experiments and analyses along the same lines to consider the case of three–dimensional spreading of granular materials that are released on steep slopes and eventually come to rest on the horizontal segment of the curved bed.

5.2. Discrete particle modelling of granular avalanches

Although the above simple depth–averaged models of the granular avalanches appear to predict the gross features, such as the overall depth and spreading, with some success, the approach is not easily generalized to permit more detailed examination of the avalanche development. These models were based upon the assumptions of uniform density, of the use of *quasi-static* soil mechanics type *active* and *passive* stress states for the determination of the lateral normal stresses, constant *quasi-static* internal friction angle, the use of either a constant or an empirically based velocity dependence for the bed friction angle, the assumption of a continuum and the applicability of the above quasi-static continuum concepts even when the concentrations are quite dilute, etc. In view of this, the degree of success that these simple continuum models have achieved is noteworthy. It is possible to develop discrete particle simulations of granular avalanches which model the interactions between the individual particles. We can study idealized two-dimensional problems using disks (or cylinders) or perform physically more realistic computations involving three-dimensional particles.

5.2.1. Computational models

For simplicity, let us consider random-sized, circular disks or spherical particles rather than irregular polygonal shapes. The particle interactions can be modelled using the ratchet spring model (Walton & Braun 1986) for the normal forces and a simple Coulomb friction model for the tangential forces.The ratchet spring models use different spring constants for the loading or compaction phase and for the unloading or recovery phase of the collision. Figure 16 shows two ratchet contact models to incorporate normal contact dissipation. When using Model 1, for low coefficients of restitution, the spring constant during the recovery or expansion phase of the contact, will be very large. This means that very small time steps must be used to compute the particle interactions and the program is very inefficient. An alternative approach is Model 2 (figure 16). Again during the compaction phase the spring constant is k_1, but when the recovery phase begins and the particle centres begin to move apart, the force immediately drops to a lower value. During the recovery phase the normal contact forces correspond to those associated with a linear spring of constant $k_2 < k_1$. The area on the force-displacement diagram that is enclosed during a collision is associated with the normal contact energy dissipation. For Model 1, we find that the normal coefficient of restitution $e = \sqrt{k_1/k_2}$ and for Model 2, $e = \sqrt{k_2/k_1}$. The particles are again assumed to be rotationally rigid and the tangential shear force F_s is related to the normal contact force F_n through a Coulomb-type friction coefficient μ, i.e. $F_s = \mu F_n$. When slip occurs between two particles during a contact, the frictional shear forces generated act in directions that tend to oppose the relative slip between the particles. It is straightforward to write down the equations for the individual particle's linear and angular accelerations. For the present case we neglect the interstitial fluid. Hence, the only forces acting on a particle are the interparticle contact forces and gravity. We use a Verlet algorithm to compute velocities and spatial displacements, and speed up the computations by making use of a 'nearest neighbor' list.

Figure 17 illustrates the application of the boundary conditions for the two and three-dimensional computations. For the two-dimensional case of the flow of disk-like particles, the 'rough' bed was made up of random size circular disks. Three-dimensional computations involving spherical particles were also performed for the case of flows which were

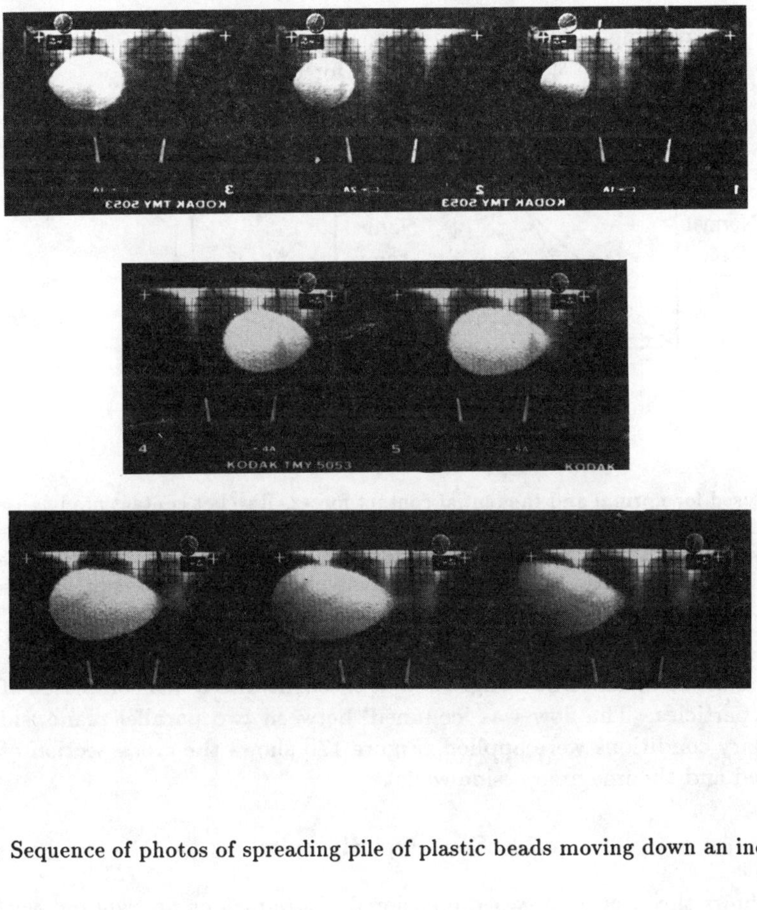

Fig. 15: Sequence of photos of spreading pile of plastic beads moving down an inclined plane.

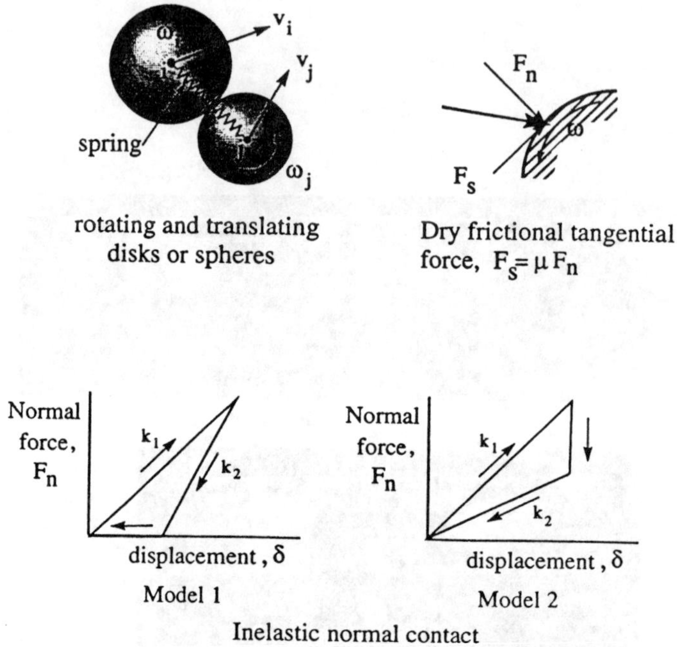

Fig. 16: Models used for normal and tangential contact forces. Ratchet contact models 1 and 2 yield normal contact forces F_n for inelastic collisions. Model 2 is computationally more efficient for low coefficients of restitution. The magnitude of the tangential contact (shear) force F_s is determined by a simple Coulomb friction model ($F_s = \mu F_n$) and the direction is such as to oppose the relative slip between the sliding particles.

two–dimensional in the mean. The 'rough' bed in these instances was comprised of random–sized spherical particles. The flow was 'confined' between two parallel plane 'sides' where periodic boundary conditions were applied. Figure 17b shows the cross–section of flow, the fixed 'rough' bed and the imaginary 'side walls'.

5.2.2. Some typical results

Figure 18 shows the results of two–dimensional computations for random sized circular disks that were initially placed in a wedge shape sitting on top of a 45° inclined part of the bed. The particles were released from rest; they then flowed down the incline as a result of gravitational forces, and eventually came to rest on the horizontal flat segment of the bed. The particle positions are shown at regular time intervals to give some sense of the motion. The general flow patterns are quite similar to the laboratory experiments of Pluss (1987) shown in figure 12. Note that the final rest position is one in which the pile is long and thin and the leading edge is sharper than the tail. It can also be seen that some of the particles become quite agitated, bounce away from the main mass and a few are able to travel quite far

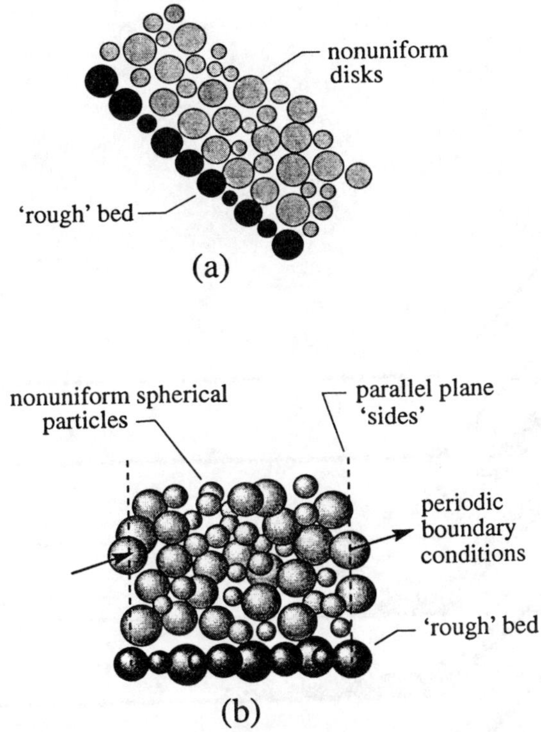

Fig. 17: Boundary conditions for avalanche computations. (a) Two–dimensional flows of disks down 'rough' bed made up of disks. (b) Three–dimensional computations down 'rough' bed of random–sized spherical particles for the case of mean two–dimensional flow. Cross–section of flow between plane 'sides' where periodic boundary conditions are applied.

upstream. Similar behaviour is often observed in laboratory scale experiments with gravel and stones.

Figure 19 shows the results of similar computations performed for the release of a trapezoidal shaped pile of random sized circular disks down the same bed. The temporal development of the pile is quite similar to the previous example and the depth profile in final rest position resembles that of figure 18.

Three–dimensional computations using random sized spherical particles give similar results. Figure 20 shows some particles in their final rest position for the case of 1600 particles which flowed down a a horizontal flat. Notice that to some degree the particles have been graded with the smaller particles residing at the bottom and the larger particles on top.

5.3. Segregation phenomena

5.3.1. Observations of particle size segregation

In systems of granular materials, the individual grains usually differ from each other in

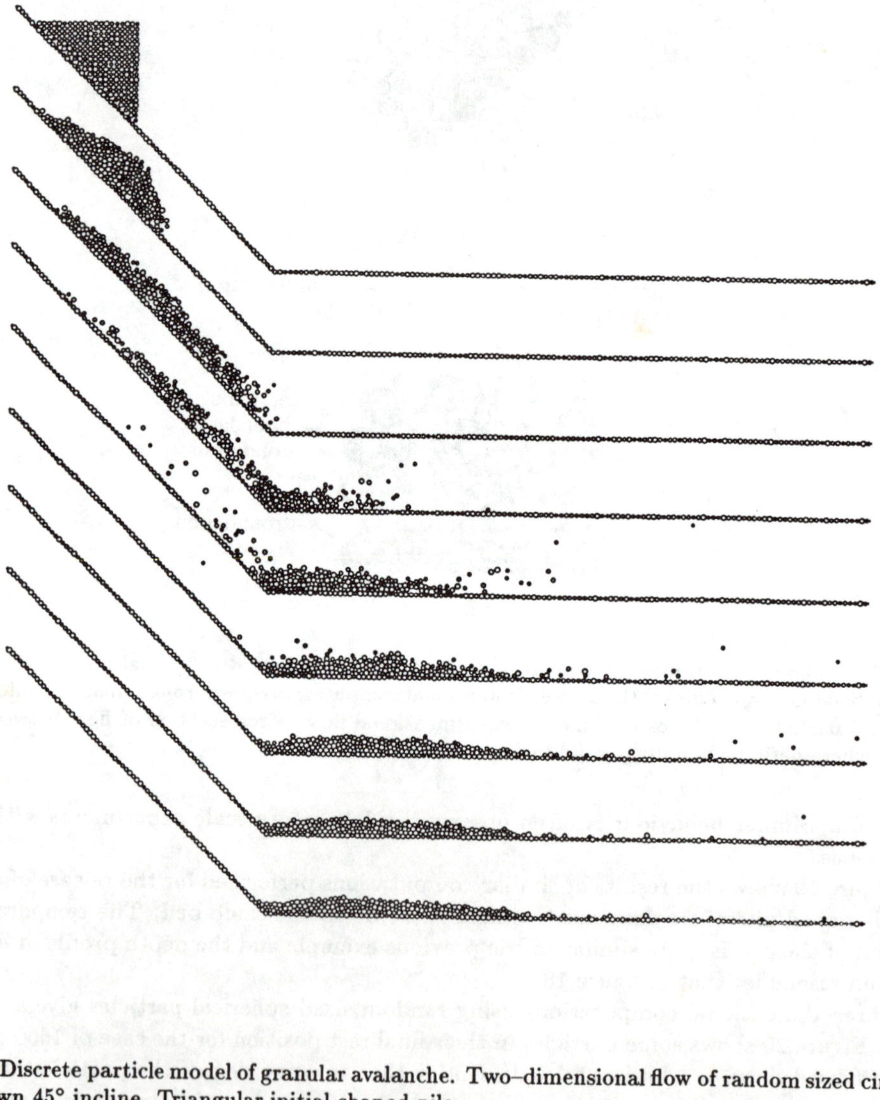

Fig. 18: Discrete particle model of granular avalanche. Two-dimensional flow of random sized circular disks down 45° incline. Triangular initial shaped pile.

terms of size, density, surface characteristics, shape, etc. When such a mass of material is agitated or deformed in the presence of a gravitational field, segregation or grading of the particles can occur and particles having the same or similar properties tend to collect together in one part of the system (Savage 1987). Devices such as pinched sluices, Humphreys spirals and Reichert cones are widely used in the mineral processing industry for sorting materials of different sizes, materials of different densities, concentrates from tailings, etc. These devices take advantage of the mechanism of gravity separation when granular materials, in the form of slurries or in the dry state, are sheared as they flow down inclined surfaces. In the geological context there is a closely related phenomenon associated with the deposition of sediments that is termed 'reverse' or 'inverse' grading (Middleton 1970, Middleton & Hampton 1976, Sallenger 1979, Naylor 1980). The result is the *opposite* of 'normal grading' in which the finer particles are found in the upper layers of a lake or river bed and the coarse ones are lower down. Normal grading is due to the differences in particle sedimentation velocities, i.e. when a mass of particles is discharged into the water, the faster falling (large) particles reach the bed first, and the slowest falling (fine) particles are deposited last and form the top of the bed.

The size segregation that occurred in the simulations shown in figure 20 are commonly seen in field events. In mudflows which exhibit surging behaviour, large stones and boulders collect at the front of the surge wave (Davies 1986, Li Jian & Luo Defu 1981). One can also observe size segregation in laboratory granular avalanche experiments. Figure 21 shows a typical final rest state for laboratory experiments of granular avalanches between parallel walls performed by James Vallance at McGill. This test involved the release of an initially homogeneous mixture of dry gravel and sand on a rough 50° incline. During the travel down the rough curved bed, shear developed over the depth of flow, generating a recirculating flow within the pile and segregation occurred. Very obvious is the movement of the larger particles to the front and top of the pile.

Davies (1988) has performed some very interesting laboratory experiments with a moving-bed channel to investigate steady-state surge waves in both dry granular materials, and mixtures of solid particles and water or other more viscous liquids. At particular inclinations and upward velocities of the inclined movable bed, a finite amount of the granular solid-fluid mixture, when placed on the upstream moving bed, would evolve into a steady, roughly 'tear-drop' shaped mass. The recirculating motions could be observed. When a small amount of larger particles were poured on top of the pile, after a short time these particle moved to the top front of the pile.

5.3.2. Random fluctuating sieve model of segregation in shear flows

Bagnold (1954) has attempted to give a physical explanation for inverse grading based upon his 'grain–inertia' theory which predicted that stresses were proportional to $\sigma^2(du/dy)^2$. His concept was that the *larger* particles migrate to regions where the shear–rate, du/dy, is small (i.e. to the free surface). Bagnold's suggestion has been treated in more detail by Takahashi (1980); but, as Iverson & Denlinger (1979) have noted, the explanation is based upon circular arguments and is not entirely satisfactory. Furthermore, as recent measurements of velocity profiles in granular chute flows have shown, the shear–rates are often smaller near

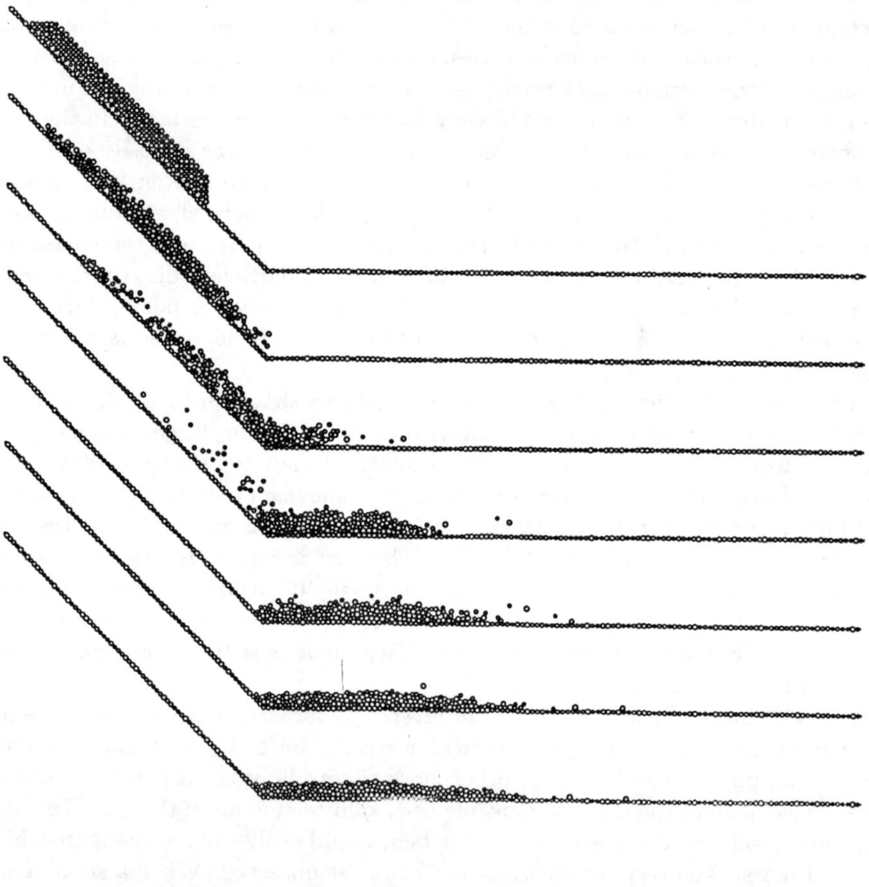

Fig. 19: Same as figure 18, but for trapezoidal initial pile shape.

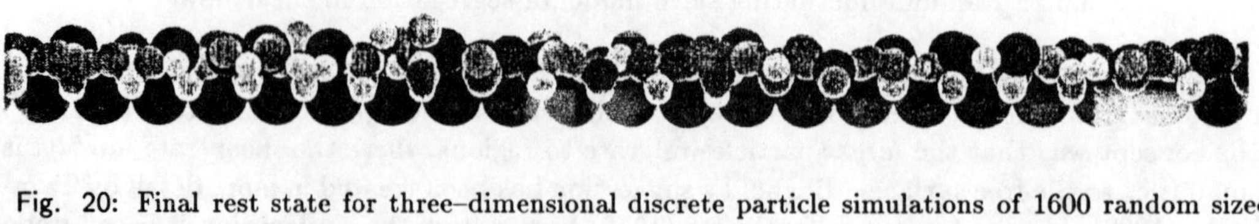

Fig. 20: Final rest state for three-dimensional discrete particle simulations of 1600 random sized spherical particles. Size segregation is evident.

Fig. 21: Strong segregation evident in final rest position of granular avalanche comprised of gravel and sand (courtesy of J. Vallance).

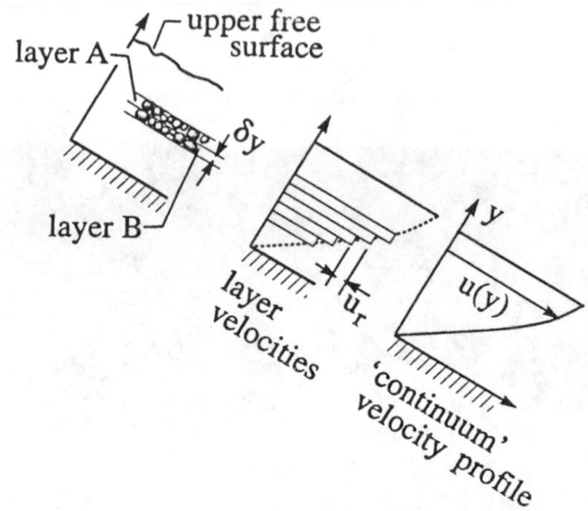

Fig. 22: Flow of particles in layers down a rough inclined plane.

the bed than near the top free surface, in contradiction to Bagnold's assumption.

A more plausible, alternative mechanism of 'kinetic sieving' was proposed by Middleton (1970). This idea was developed in quantitative terms by Savage & Lun (1988), who presented an analysis for flow of a binary mixture of small and large spherical particles of equal mass density down a inclined chute. A brief presentation of that analysis, which is an elaboration of the statistical mechanical interpretation of percolation proposed by Cooke & Bridgwater (1979) is now given.

The model is restricted to relatively high particle concentrations, and slow shearing flows where particles are in nearly continuous contact with one another, overriding each other during the shear process. Thus, we neglect diffusive processes which are of greater importance in the low density, rapid granular flow regime in which particles interact vigorously and primarily through nearly instantaneous collisions. Assume that the flow takes place in layers which are in motion relative to one another as a result of the mean shear rate Γ developed by the rough lower boundary (see figure 22). As a result of overriding of adjacent layers and the continual rearrangement of particles within a layer, the contact force network and the void spaces are undergoing continual random changes. At any instant, there will be a distribution of void spaces in a given layer (figure 23). If a void space is large enough, then a particle from the layer above can fall into it as the layers move relative to one another. For a given overall solids concentration, the probability of finding a hole that a small particle can fall into is larger than the probability of finding a hole that a large particle can fall into. Hence, there is a tendency for particles to segregate out, with fines at the bottom and coarse ones at the top. Savage & Lun (1988) termed this the 'random fluctuating sieve' mechanism. Through the use of information entropy concepts, they obtained the net percolation velocities, q_S and q_L, of the small and large particles. Comparisons of their predicted percolation velocity with the experiments of Bridgwater, *et al.* (1978) showed reasonably good agreement for the variation of the nondimensional percolation velocity, $q_S/(\Gamma\sigma_L)$, with the diameter ratio,

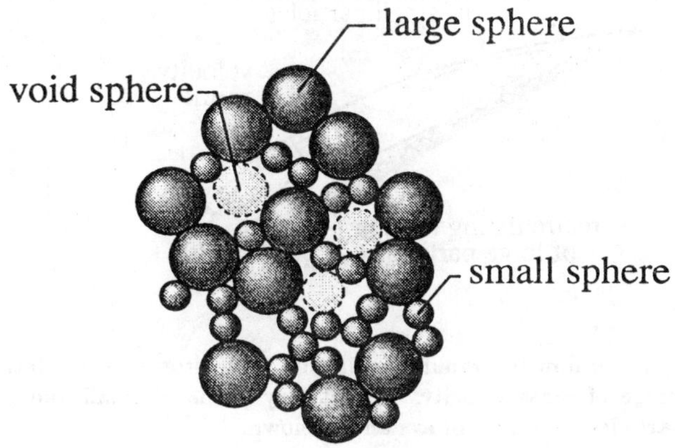

Fig. 23: Random continuous network for binary mixture of small and large spherical particles in a layer (after Savage & Lun 1988).

σ_S/σ_L, of small to large particles. The model was applied to analyze the size segregation process during the flow down a roughened inclined chute. Uniformly mixed coarse and fine particles of equal mass density entered the upstream end of the chute. Besides the percolation flux of particles, an opposing flux was imposed to satisfy the assumed condition that there is no overall mass flux in the direction normal to the plane of the inclined chute. The mass conservation equation for the small particles was solved by the method of characteristics to determine concentration profiles and distances for complete separation of the fine from the coarse particles. Savage & Lun (1988) also performed experiments with a binary sized mixture of spherical polystyrene beads, and concentration profiles were determined at various downstream distances. Reasonable agreement was found between the measured and predicted concentration profiles, and the fine from coarse separation distances.

5.3.3. Particle segregation in granular avalanches

Given the above physical description of the 'inverse grading' process, we can explain the accumulation of the larger size particulate material at the front of the granular avalanche that can be observed laboratory experiments as shown in figure 21, as well as in field events. Figure 24 shows a schematic diagram of both the large and small particle recirculating motions in the granular avalanche. The recirculating loop traced out by the large particles is smaller than that taken by the smaller particles. Consider the motion of a large particle starting at the nose near the bed. Because of the rough bed, the particle is moved rearward toward the tail. However, due to the kinetic sieving mechanism it quickly rises through the surrounding smaller particles to the surface layer, where the mean motion is toward the nose. Because of the upward rise, it never travels longitudinally very far, and so remains confined to the nose region. In his movable bed laboratory experiments, Davies (1988) found that when the maximum height of the wave was deep, the grains were dispersed more uniformly over the length of the pile. For these thicker surges, the upper layer of material is nearly a

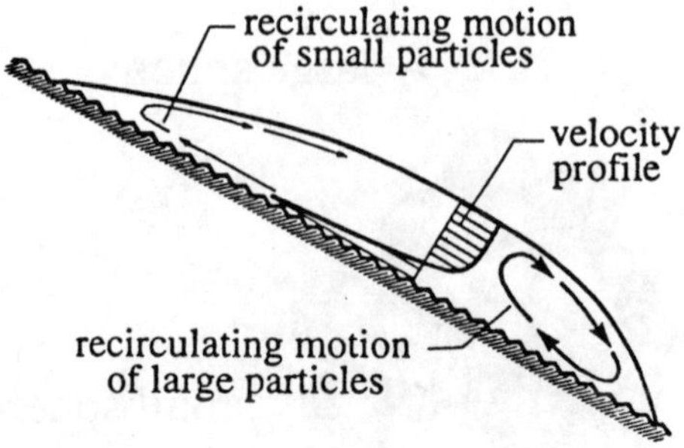

Fig. 24: Schematic diagram of flow in granular avalanche. Velocities shown relative to a coordinate system moving with centre of mass velocity. Circulatory paths of small and large particles and accumulation of large particles near nose of avalanche shown.

non-shearing plug region. Shear is essential for the operation of the kinetic sieve mechanism; thus, particles will not rise through such a plug region and are more likely to be carried further upstream from the nose region.

Acknowledgements: The author is grateful to the Natural Sciences and Engineering Research Council (NSERC) for financial support of this work through an NSERC Operating Grant. I am also grateful to the Academic Press for granting permission to reprint material in §3.1 that was previously published as part of a chapter in Vol. 24 of *Advances in Applied Mechanics*, 1984.

References

AHN, H., BRENNEN, C.E. & SABERSKY, R.H. 1988 Experiments on chute flows of granular materials. *Micromechanics of Granular Materials*, (eds. M. Satake & J.T. Jenkins), Elsevier, 339–348.

ARAKI, S. 1991a The dynamics of particle disks, III Dense and spinning particle disks. *Icarus*, **90**, 139–171.

ARAKI, S. 1991b Dynamics of planetary rings. *American Scientist*, **79**, 44–59.

BAGNOLD, R.A. 1954 Experiments on a gravity free dispersion of large solid spheres in a Newtonian fluid under shear. *Proc. R. Soc. London, Ser. A* **225**, 49–63.

BAGNOLD, R.A. 1956 The flow of cohesionless grains in fluids. *Phil. Trans. R. Soc. London, Ser. A* **249**, 235–97.

BAGNOLD, R.A. 1966 The shearing and dilation of dry sand and the 'Singing' mechanism. *Proc. R. Soc. London, Ser. A* **295**, 219–32.

BAGNOLD, R.A. 1973 The nature of saltation and of "Bed-Load" transport in water. *Proc. R. Soc., London, Ser. A* **332**, 473–504.

BAK, P. & CHEN, K. 1991 Self-organized criticality. *Scientific American* **264**, 46–54.

BALMER, R.T. 1978 The operation of sand clocks and their medieval development. *Technol. Culture* **19**, 615–632.

BAXTER, G.W., BEHRINGER, R.P., FAGERT, T. & JOHNSON, G.A. 1990 Pattern formation and time-dependence in flowing sand. *Two Phase Flows and Waves* (eds. D.D. Joseph & D.G. Schaeffer), Springer, New York, 1–29.

BIDEAU, D. & DODDS, J.A. (Eds.) 1992 *Physics of Granular Media.* Nova Science Publishers, New York.

BRIDGWATER, J., COOKE, M.H. & DRAHUN, J.A. 1985 Strain induced percolation. *Instn. Chem. Engrs. Symposium Series No. 69*, 171–191.

BRIDGWATER, J., COOKE, M.H., & SCOTT, A.M. 1978 Inter-particle percolation: Equipment development and mean percolation velocities. *Trans. Instn Chem. Engrs.* **56**, 157-167.

CAMPBELL, C.S. 1989 The stress tensor for simple shear flows of a granular material. *J. Fluid Mech.* **203**, 449–473.

CAMPBELL, C.S. 1990 Rapid granular flows. *Ann. Rev. Fluid Mech.* **22**, 57–92.

CAMPBELL, C.S. & BRENNEN, C.E. 1985 Computer simulation of granular shear flows. *J. Fluid Mech.* **151**, 167–188.

CAMPBELL, C.S. & GONG, A. 1986 The stress tensor in a two-dimensional granular shear flow. *J. Fluid Mech.* **164**, 107–125.

CARNAHAN, N.F. & STARLING, K.E. 1969 Equations of state for non-attracting rigid spheres. *J. Chem. Physics* **42**, 635–636.

CHAPMAN, S. 1916 *On the kinetic theory of a gas. Phil. Trans. R. Soc. Lond.* A**217**, 115–197.

CHAPMAN, S. & COWLING, T.G. 1970 *The Mathematical Theory of Non-Uniform Gases.* 3rd edn. Cambridge University Press.

COOKE, M.H., & BRIDGWATER, J. 1979 Interparticle percolation: A statistical mechanical interpretation. *Ind. Engng. Chem. Fundam.* **18**, 25-27.

CRAIG, K., BUCKHOLTZ, R.H. & DOMATO, G. 1986 An experimental study of the rapid flow of dry cohesionless metal powders. *J. Appl. Mech.* **53**, 935–942.

CRAIG, K., BUCKHOLTZ, R.H. & DOMATO, G. 1987 Rapid shear flow of a dry magnetic powder in an externally applied magnetic field: An experimental study. *Phys. Fluids* **30**, 1993–1999.

DAVIES, T.R.H. 1982 Spreading of rock avalanche debris by mechanical fluidization. *Rock Mech.*, **15**, 9–29.

DAVIES, T.R.H. 1986 Large debris flows: A macro–viscous phenomenon. *Acta Mechanica*, **63**, 161–178.

DAVIES, T.R.H. 1988 Debris flow surges - A laboratory investigation. Mitteilung No. 96 der Versuchsanstalt für Wasserbau, Hydrologie und Glaziologie an der ETH, 122 pp.

DAVIS, H.T. 1973 Kinetic theory of dense fluids and liquids revisited. *Adv Chem. Phys.* **24**, 257–343.

DENT, J.D. 1986 Flow properties of granular materials large overburden loads. *Acta Mechanica* **64**, 111-122.

DING, J. & GIDASPOW, D. 1990 A bubbling fluidization model using kinetic theory of granular flow. *AIChE Journ.* **36**, 523–538.

DRESCHER, A. & DE JOSSELIN DE JONG, G. 1972 Photoelastic verification of a mechanical model for the flow of a granular material. *J. Mech. Phys. Solids* **20**, 337–51.

ERISMANN, T.H. 1979 Mechanisms of large landslides. *Rock Mechanics* **12**, 15-46.

ERISMANN, T. 1986 Flowing, rolling, bouncing, sliding: Synopsis of basic mechanisms. *Acta Mechanica* **64**, 101-110.

FABER, T.E. 1972 *An Introduction to the Theory of Liquid Metals.* Cambridge University Press, London and New York.

FARRELL, M., LUN, C.K.K. & SAVAGE, S.B. 1986 A simple kinetic theory for granular flow of binary mixtures of smooth, inelastic, spherical particles. *Acta Mech.* **63**, 45–60.

FERZIGER, J.H. & KAPER, H.G. 1972 *Mathematical Theory of Transport Processes in Gases.* Elsevier, Amsterdam.

GOGUEL, J. 1978 Scale dependent rockslide mechanisms. In *Rockslides and Avalanches, Vol. 1* (ed. B. Voight), Elsevier, 167-180.

GUGGENHEIM, E.A. 1960 *Elements of the Kinetic Theory of Gases.* Pergamon, Oxford.

HAFF, P.K. 1983 Grain flow as a fluid-mechanical phenomenon. *J. Fluid Mech.* **134**, 401–430.

HAGEN, G. 1852 Druck und Bewegung des Trockenen Sandes. *Berl Monatsb. Akad. d. Wiss.*, S35–S42.

HANES, D.M. & INMAN, D.L. 1985 Observations of rapidly flowing granular-fluid mixtures. *J. Fluid Mech.* **150**, 357–380.

HANES, D.M., JENKINS, J.T., & RICHMAN, M.W. 1988 The thickness of steady plane shear flows of circular disks driven by identical boundaries. *J. Appl. Mech.* **110**, 969–974.

HANSEN, J.P., & MCDONALD, I.R. 1976 *Theory of Simple Liquids.* Academic Press, New York.

HEIM, A. 1882 Der Bergsturz von Elm. *Deutsch Geol. Gesell. Zeitschrift* **34**, 74-115.

HEIM, A. 1932 Bergsturz und Menscheleben. *Beiblatt zur Vierteljahresschrift der Natf. Ges. Zurich* **20**, 1-218.

HILL, H.M. 1966 Bed forms due to a fluid stream. *J. Hydr. Div. Proc. ASCE* **92**, 127–43.

HIRSCHFELDER, J.O., CURTIS, C.R. & BIRD, R.B. 1954 *The Molecular Theory of Gases and Liquids.* Wiley, New York.

HOPKINS, M.A. & SHEN, H. 1988 A Monte–Carlo simulation of a rapid simple shear flow of granular materials. *Micromechanics of Granular Materials* (eds. M. Satake & J.T. Jenkins), Elsevier, 349–358.

HSÜ, K. 1975 On sturzstroms - catastrophic debris streams generated by rockfalls. *Geol. Soc. Am. Bull.* **86**, 129-140.

HSÜ, K. 1978 Albert Heim: Observations on landslides and relevance to modern interpretations. In *Rockslides and Avalanches, Vol. 1* (ed. B. Voight), Elsevier, 69-93.

HUI, K., HAFF, P.K., UNGAR, J.E. & JACKSON, R. 1984 Boundary conditions for high shear grain flows. *J. Fluid Mech.* **145**, 223–233.

HUNGR, O. & MORGENSTERN, N.R. 1984a Experiments on the flow behaviour of granular materials at high velocity in an open channel flow. *Geotechnique* **34**, 405–413.

HUNGR, O. & MORGENSTERN, N.R. 1984b High velocity ring shear tests on sand. *Geotechnique* **34**, 415–421.

IVERSON, R.M. & DENLINGER, R.P. 1987 The physics of debris flows – a conceptual assessment. *Erosion and Sedimentation in the Pacific Rim (Proceedings of the Corvallis Symposium)* IAHS Publ. No. 165.

JAEGER, H.M. & NAGEL, S.R. 1992 Physics of the granular state. *Science* **255**, 1523–1531.

JEANS, J. 1967 *An Introduction to the Kinetic Theory of Gases.* Cambridge University Press.

JENKINS, J.T. 1987a Balance laws and constitutive relations for rapid flows of granular materials. *Constitutive Models of Deformations* (eds. J. Chandra & R. Srivastav), SIAM, Philadelphia, 109.

JENKINS, J.T. 1987b Rapid flows of granular materials. *Non-classical Continuum Mechanics: Abstract Techniques and Applications* (eds. R.J. Kops & A.A. Lacey), Cambridge Univ. Press, 213–224.

JENKINS, J.T. & MANCINI, F. 1987 Balance laws and constitutive relations for plane flows of a dense, binary mixture of smooth, nearly elastic circular disks. *J. Appl. Mech.* **54**, 27–34.

JENKINS, J.T. & MCTIGUE, D.F. 1990 Transport processes in concentrated suspensions: The role of particle fluctuations. *Two Phase Flows and Waves* (eds. D.D. Joseph & D.G. Schaeffer), Springer, New York, 70–79.

JENKINS, J.T. & RICHMAN, M.W. 1985a Grad's 13–moment system for a dense gas of inelastic spheres. *Arch. Rat. Mech. Anal.* **87**, 355–377.

JENKINS, J.T. & RICHMAN, M.W. 1985b Kinetic theory for plane flows of a dense gas of identical, rough, inelastic, circular disks. *Phys. Fluids* **28**, 3485–3494.

JENKINS, J.T. & RICHMAN, M.W. 1986 Boundary conditions for plane flows of smooth nearly elastic, circular disks. *J. Fluid Mech.* **171**, 53-69.

JENKINS, J.T. & SAVAGE, S.B. 1983 A theory for the rapid flow of identical, smooth, nearly elastic particles. *J. Fluid Mech.* **130**, 186–202.

JOHNSON, P.C. & JACKSON, R. 1987 Frictional–collisional constitutive relations for granular materials, with application to plane shearing. *J. Fluid Mech.* **176**, 67–93.

JOHNSON, P.C., NOTT, P. & JACKSON, R. 1990 Frictional-collisional equations equations of motion for particulate flows and their application to chutes. *J. Fluid Mech.* **210**, 501–535.

KADANOFF, L.P., NAGEL, S.R., WU, L. & ZHOU, S. 1989 Scaling and universality in avalanches. *Phys. Rev. A* **39**, 6524–6537.

KENT, P.E. 1965 The transport mechanism in catastrophic rockfalls. *J. Geol.* **74**, 79-83.

LANG, R.M. 1992 An experimental and analytical study on gravity driven free surface flows of cohesionless granular material. Ph.D. Dissertation, Technische Hochschule Darmstadt, Germany.

LEES, A.W. & EDWARDS, S.F. 1972 The computer study of transport processes under extreme conditions. *J. Phys.* **C5**, 1921–1929.

LI JIAN & LUO DEFU 1981 The formation and characteristics of mudflow and flood in the mountain area of the Dachao River and its prevention. *Zeit. Geomorph.* **25**, 470-484.

LI TIANCHI 1983 A mathematical model for predicting the extent of a major rockfall. *Z. Geomorph.* **27**, 473-482.

LUN, C.K.K. & SAVAGE, S.B. 1986 The effects of an impact velocity dependent coefficient of restitution on stresses developed by sheared granular materials. *Acta Mech.* **63**, 15–44.

LUCCHITTA, B.K. 1978 A large landslide on Mars. *Geol. Soc. Amer. Bull.* **89**, 1601-1609.

LUN, C.K.K. & SAVAGE, S.B. 1987 A simple kinetic theory for granular flow of rough, inelastic, spherical particles. *J. Appl. Mech.* **54**, 47–53.

LUN, C.K.K., SAVAGE, S.B., JEFFREY, D.J. & CHEPURNIY, N. 1984 Kinetic theories for granular flow: inelastic particles in Couette flow and slightly inelastic particles in a general flowfield. *J. Fluid Mech.* **140**, 223–256.

MANDL, G. & FERNANDEZ-LUQUE, R. 1970 Fully developed plastic shear flow of granular materials. *Geotech.* **20**, 277–307.

MCSAVENEY, M.J. 1978 Sherman Glacier rock avalanche, Alaska, U.S.A., In *Rockslides and Avalanches, Vol. 1* (ed. B. Voight), Elsevier, 197-258.

MELOSH, J. 1986 The physics of very large landslides. *Acta Mechanica* **64**, 89-99.

MIDDLETON, G.V. 1970 Experimental studies related to problems of Flysch sedimentation. In *Flysch Sedimentology in North America* (ed. J. Lajoie), 253–72. Geol. Assoc. Can. Spec. Rep. 7.

MIDDLETON, G.V. & HAMPTON, M.A. 1976 Subaqueous sediment transport and deposition by sediment gravity flows. In *Marine Sediment Transport and Environmental Management*, (eds. D.J. Stanley & D.J.P. Swift), 197–218, Wiley, N.Y.

NAYLOR, M.A. 1980 The origin of inverse grading in muddy debris flow deposits - A review. *J. Sedimentary Petrology* **50**, 1111-1116.

PLÜSS, CH. 1987 Experiments on granular avalanches. Diplomarbeit, Abt X, Eidg. Techn. Hochschule, Zürich, 113 pp.

PRESENT, R.D. 1958 *Kinetic Theory of Gases.* McGraw-Hill, New York.

REIF, F. 1965 *Fundamentals of Statistical and Thermal Physics.* McGraw-Hill, New York.

RICHMAN, M.W. 1988 Boundary conditions based upon a modified Maxwellian velocity distribution for flows of identical, smooth nearly elastic, spheres. *Acta Mechanica* **75**, 227–240.

RICHMAN, M.W. 1989 The source of second moment in dilute granular flows of highly inelastic spheres. *J. Rheology* **33**, 1293–1306.

RICHMAN, M.W. & CHOU, C.S. 1988 Boundary effects on granular shear flows of smooth disks. *Journ. Appl. Math. Physics (ZAMP)* **39**, 885-901.

REYNOLDS, O. 1885 On the dilatancy of media composed of rigid particles in contact. *Phil. Mag. Ser. 5*, **20**, 469–481.

REYNOLDS, O. 1903 *The Sub-mechanics of the Universe.* Cambridge University Press.

ROSCOE, K.H., SCHOFIELD, A.N. & WROTH, C.P. 1958 On the yielding of soils. *Geotech.* **8**, 22–53.

SALENCON, J. 1977 *Applications of the Theory of Plasticity in Soil Mechanics.* Wiley, N.Y.

SALLENGER, A.H. 1979 Inverse grading and hydraulic equivalence in grain-flow deposits. *J. Sedimentary Petrology* **49**, 553-562.

SANDERS, J.E. 1963 Concepts of fluid mechanics provided by primary sedimentary sedimentary structures. *J. Sedim. Petrology* **33**, 173–9.

SANDERS, B.E., HOPKINS, M.A. & ACKERMANN, N.A. 1988 Physical experiments and numerical simulation of two dimensional chute flow. *Micromechanics of Granular Materials*, (eds. M. Satake & J.T. Jenkins), Elsevier, 359–366.

SAVAGE, S. B. 1979 Gravity flow of cohesionless granular materials in chutes and channels. *J. Fluid Mech.* **92**, 53–96.

SAVAGE, S.B. 1983 Granular flows down rough inclines - review and extension. *Mech. of Granular Materials: New Models and Constitutive Relations* (eds. J.T. Jenkins & M. Satake), Elsevier, 261–82.

SAVAGE, S.B. 1984 The mechanics of rapid granular flows. *Advances in Applied Mechanics* **24** (eds. T.Y Wu & J. Hutchinson), Academic, 289–366.

SAVAGE, S.B. 1987 Interparticle percolation and segregation in granular materials: A review. In *Developments in Engineering Mechanics* (ed. A.P.S. Selvadurai, A.P.S.),Elsevier, Amsterdam 347-363.

SAVAGE, S.B. 1989 Flow of granular materials. *Theoretical and Applied Mechanics* (eds. P. Germain, M. Piau & D. Caillerie), Elsevier, 241–266.

SAVAGE, S.B., 1990 Symbolic computation of the flow of granular avalanches. *Journal of Symbolic Computation* **9**, 515-530.

SAVAGE, S.B. 1992 Numerical simulations of Couette flow of granular materials; Spatio–temporal coherence and $1/f$ noise. *Physics of Granular Media*, (eds. D. Bideau & J.A. Dodds), Nova Science Publishers, New York, 343–362.

SAVAGE, S.B. & DAI, R. 1992 Some aspects of bounded and unbounded shear flows of granular materials. *U.S.-Japan meeting on Micromechanics of Granular Materials, Clarkson University, August 5-9, 1991* (eds. M. Satake and H. Shen), Elsevier, Amsterdam, (in press).

SAVAGE, S.B. & HUTTER, K. 1989 The motion of a finite mass of granular material down a rough incline. *J. Fluid Mech.* **199**, 177-215.

SAVAGE, S.B. & HUTTER, K. 1991 The dynamics of avalanches of granular materials from initiation to runout. Part I: Analysis. *Acta Mechanica* **86**, 201-223.

SAVAGE, S.B. & JEFFREY, D.J. 1981 The stress tensor in a granular flow at high shear rates. *J. Fluid Mech.* **110**, 255–272.

SAVAGE, S.B. & NOHGUCHI, Y. 1988 Similarity solutions for avalanches of granular materials down curved beds. *Acta Mechanica* **75**, 153-174.

SAVAGE, S. B. & SAYED, M. 1984 Stresses developed by dry cohesionless granular materials sheared in an annular shear cell. *J. Fluid Mech.* **142**, 391–430.

SCHEIDEGGER, A.E. 1975 *Physical Aspects of Natural Catastrophies.* Elsevier.

SHREVE, R.L. 1966 Sherman landslide, Alaska. *Science* **154**, 1639-1643.

SHREVE, R.L. 1968a The Blackhawk Landslide. *Geol. Soc. Am., Spec. Paper* **108**, 47 pp.

SHREVE, R.L. 1968b Leakage and fluidization in air-lubricated avalanches. *Geol. Soc. Am. Bull.* **79**, 653-658.

SINCLAIR, J.L. & JACKSON, R. 1989 Gas-particle flow in a vertical pipe with particle–particle interactions. *AIChE Journ.* **35**, 1473–1486.

SCHOFIELD, A.N. & WROTH, C.P. 1968 *Critical State Soil Mechanics.* McGraw–Hill, New York.

SCOTT, R.F. 1963 *Principles of Soil Mechanics.* Addison-Wesley, Reading.

SINCLAIR, J.L. & JACKSON, R. 1989 Gas–particle flow in a vertical pipe with particle–particle interactions. *AIChE Journ.* **35**, 1473–1486.

SOKOLOVSKI, V.V. 1965 *Statics of Granular Media.* Pergamon, Oxford.

SPENCER, A.J.M. 1981 Deformation of an ideal granular material. In *Mechanics of Solids, Rodney Hill 60th Anniv. Volume.* (eds. H.G. Hopkins & J.J. Sewell). Pergamon, Oxford.

STADLER, R. 1986 Stationäres, schnelles Fliessen von dicht gepackten trockenen und feuchten Schüttgütern. Dr.-Ing. Dissertation, Univ. Karlsruhe, Karlsruhe, West Germany.

STADLER, R. & BUGGISCH, H. 1985 Influence of the deformation rate on shear stress in bulk solids, theoretical aspects and experimental results. *Reliable Flow of Particulate Solids*, (EFCE Publication Series No. 49, Bergen, Norway), pp. 15.

SZIDAROVSZKY, F., HUTTER, K. & YAKOWITZ, S. 1987 A Numerical Study of Steady Plane Granular Chute Flows Using the Jenkins–Savage Model and Its Extension. *Int. J. Num. Meth. Engrg.* **24**, 1993–2015.

TAKAHASHI, T. 1980 Debris flow on prismatic channel. *Journ. Hydr. Div., ASCE* **106**, 381-395.

TEMPERLEY, H.N.V., ROWLINSON, J.S., & RUSHBROOKE, G.S. 1968 *Physics of Simple Liquids.* North Holland.

WALTON, O.R. & BRAUN, R.L. 1986 Viscosity and temperature calculations for assemblies of inelastic frictional disks. *J. Rheology* **30**, 949–980.

WALTON, O.R., BRAUN, R.L., MALLON, R.G., & CERVELLI, D.M. 1987 Particle–dynamics calculations of gravity flow of inelastic, frictional spheres. *Micromechanics of Granular Materials* (eds. M. Satake & J.T. Jenkins), Elsevier, 153–162.

WALTON, O.R., KIM, H., & ROSATO, A. 1991 Micro-structure and stress difference in shearing flows of granular materials. Proc. ASCE Eng. Mech. Div. Conf., Columbus, Ohio, May 19-21, 1991.

WEIGHARDT, K. 1975 Experiments in granular flow. *Ann. Rev. Fluid Mech.* **7**, 89–114.